HANDBOOK OF

Microlithography, Micromachining, and Microfabrication

Volume 1: MICROLITHOGRAPHY

EDITORIAL BOARD

SPIE PRESS MONOGRAPH PM39

IEE MATERIALS AND DEVICES SERIES 12
Series Editors: Professor A. J. Moses
Dr. John Wood

HANDBOOK OF

Microlithography, Micromachining, and Microfabrication

Volume 1: MICROLITHOGRAPHY

P. Rai-Choudhury, *Editor*

SPIE OPTICAL ENGINEERING PRESS

A Publication of SPIE—The International Society for Optical Engineering
Bellingham, Washington USA

THE INSTITUTION OF ELECTRICAL ENGINEERS
London, UK

Library of Congress Cataloging-in-Publication Data

Handbook of microlithography, micromachining, & microfabrication / P. Rai
    -Choudhury, editor.
        p.   cm.
    Includes bibliographical references and index.
    ISBN 0-8194-2378-5 (v. I)
    I. Microlithography.   2. Micromachining.   3. Microfabrication.
 I. Rai-Choudhury, P.
    TK7836.H3423   1997
    670—dc21   96-40237
                                                                                 CIP

Copublished by

SPIE—The International Society for Optical Engineering
P.O. Box 10
Bellingham, Washington 98227-0010
Phone: 360/676-3290
Fax: 360/647-1445
Email: spie@spie.org
WWW: http://www.spie.org/
SPIE Press Monograph PM39: ISBN 0-8194-2378-5

The Institution of Electrical Engineers
Michael Faraday House
Six Hills Way, Stevenage, Herts.
SG1 2AY United Kingdom
Phone: +44 (0)1438 313311
Fax: +44 (0)1438 360079
Email: books@iee.org.uk
WWW: http://www.iee.org.uk
IEE Materials and Devices Series 12: ISBN 0-85296-906-6

CONTENTS

PREFACE

Microlithography and microfabrication are rapidly finding applications in many areas, from sensors and actuators to biomedical devices, in addition to their uses in microelectronics device manufacturing. Lithography is the key technology that has driven the dynamic growth of the IC industry over the past two decades. To date, optical lithography continues to be the mainstream technology for the IC industry, and is being used in production by leading-edge high-volume manufacturers to support 0.25-μm minimum feature size. Although the exposure system using 193-nm optical lithography is expected to extend to 0.13 μm, the industry remains undecided as to the choice of an exposure system beyond 0.13 μm. The options include extreme ultraviolet (EUV or projection x-ray), e-beam projection, massive parallel direct write, and 1X proximity x-ray. The field of lithography will continue to be very dynamic, and demands an authoritative handbook for process development and production to aid in the training of scientists and engineers.

Microlithography and micromachining are also driving microelectromechanical systems (MEMS) technology, which is rapidly developing. Within the next decade the cost of micromachined devices will drop to the point where there will be an explosive demand for these devices for use in such industries as automotive, chemical, aircraft, and disposable medical products. MEMS will also find applications for in-situ process monitoring, environmental health and safety monitoring, and numerous other sensor and actuator systems. Use of lithography for fabrication of many microelectromechanical devices frequently requires processing procedures that range from the fabrication of high-aspect-ratio structures down to ultrafine structures.

Although there are a number of books on lithography, a need exists to compile all the diverse information into an easily accessible handbook-type format. SPIE Press is publishing the handbook of Microlithography, Micromachining, and Microfabrication in two volumes. Volume 1 addresses microlithography, and Volume 2 covers micro-machining and microfabrication. Volume 1 focuses on the application of microlithography techniques in microelectronics manufacturing. We hope it will be a useful tutorial introduction to the key microlithography technologies for researchers and engineers who are not necessarily experts in the field, as well as a good sourcebook for those who are.

Acknowledgments

I would like to thank all the participating authors, editorial board members, and reviewers for contributing to this handbook. I would also like to thank Mary Horan, Susan Price, and Dixie Cheek of the SPIE Press for editing and organizing the book, Eric Pepper for encouragement to compile this handbook, and Stephanie Harmon for administrative support. Special thanks must go to my wife, Margaret, for all her expertise with the English language, as well as for her support while I was working on this handbook.

P. Rai-Choudhury
January 1997

HANDBOOK OF

Microlithography, Micromachining, and Microfabrication

Volume 1: MICROLITHOGRAPHY

Introduction

Burn J. Lin
Linnovation

P. Rai-Choudhury
SPIE—The International Society for Optical Engineering

Microlithography is a term developed to denote a particular branch of lithography that is specifically applied to integrated circuit fabrication. The term came into use shortly after the invention of the integrated circuit (IC) in 1958. Starting from fabricating ICs with dimensions in the hundreds of micrometers, industry is now poised to make circuits whose critical dimension is orders of magnitude smaller, with dimension control in hundredths of a micrometer. The number of transistors has grown by eight orders of magnitude, and is now approaching nine. The fast pace of progress makes it difficult to capture the most advanced achievements into a comprehensive reference book, but there is a continuous need for such documentation. A significant effort has been made by the authors of this handbook to review the state of the art, so relevant materials are assembled consistently within a single volume. Expert microlithographers can marvel at the accomplishments made in the field of microlithography, and continue to build upon their ever-expanding technology, while those entering the IC fabrication industry can have a tutorial overview of all the major microlithography technologies.

The rapid development in microlithography started in the early 1970s and has spanned over two decades. Microlithography began as optical lithography. The resist on the wafer was delineated by replicating the pattern on the mask. This pattern on the mask was made by exposing with a focused spot of light from a microscopic objective brought to the locations requiring exposure. This spot-scanning technique was a substantial improvement over the primitive technique of cutting and pasting a mask pattern before it is reduced to a usable mask. Even with the spot-scanning mask technology, reduction imaging continued to be used in the mask-making process, leaving the wafer-exposing equipment simple by remaining in the 1X domain. It was soon realized that generating an electron beam, which is much smaller than an optical spot, and moving it rapidly, is inherently easier. As a result, e-beam lithography quickly replaced optical lithography in mask making, except for the recent development in masking using multiple scanning laser spots. A review of optical projection imaging is presented by Levinson and Arnold in Chapter 1. E-beam systems are covered by McCord and Rooks in Chapter 2, and Skinner et al. treat mask fabrication issues in Chapter 5.

Many attempts have been made to replace optical lithography because of the concern that optical lithography will no longer be capable of achieving the smaller geometries that modern ICs demand. E-beam, x-ray, ion-beam, and extreme-ultraviolet (EUV) have been proposed, funded, and developed as possible technologies to meet these requirements. Progress and achievements are reported in chapters 2, 3 (Cerrina), and 8 (Peckerar et al.).

In the early 1970s, the critical dimension of ICs was in the regime of 2–5 µm. Wafer imaging was performed by replicating a mask, by placing the photoresist-coated wafer in contact with the mask and exposing with broadband and near-UV light in the spectrum between 300 and 450 nm. To reduce wear and tear of the mask that is subject to repeated contact with the wafer, the emulsion mask has been replaced with the chromium mask, which also helps to improve the edge definition. With these advantages, the chromium mask has become the workhorse of the IC industry. To further reduce wear and tear to the mask under repeated contact, the wafer was placed in close proximity (at a distance of 10–25 µm) to the mask instead of in hard contact with it, and optical proximity printing was born. The mask-to-wafer gap has become indispensable to maintain a profitable wafer yield. To maintain this gap and still continue to reduce the minimum feature size (MFS), the exposure wavelength was reduced to deep-UV centering at 250 nm, and further to soft x-ray on the order of 1 nm in wavelength.

The wavelength reduction and e-beam mask-making activities necessitated resist development for wafer-masking materials. Resists are required that react to the higher energy beams, maintain a usable throughput, and possess good processing characteristics, such as adhesion, uniformity, etch resistance, thermal stability, stripability, and long shelf life. Even within the main energy spectrum for microlithography, the photoresist continues to be improved by providing a choice of polarity, better developing characteristics, higher sensitivity, lower defects, and higher consistency. In Chapter 4 on deep-UV resists, Allen et al. describe the vast amount of work required to develop the resist for microlithography. Chapters 1 and 2 contain information on resists in the optical and e-beam disciplines, respectively. Chapter 8 discusses the status and requirements of resists for manufacturing devices smaller than 0.1 µm.

Optical lithography did not remain long in the proximity/contact printing phase. In 1974, 1X full-wafer projection printing using an all-reflective system was introduced. Because of the total separation between the mask and the wafer and better alignment, 1X projection all-reflective printing superseded proximity printing for minimum feature size, down to the vicinity of 1.5 µm. Attempts were made to extend 1X full-wafer projection printing to 1 µm in the 1980s, using deep-UV exposure. Even though optical imaging may be made to work in the 1X full-wafer projection printing regime, the need for a stringent tolerance 1X mask has switched the bulk of sub-1.5-µm imaging to reduction step-and-repeat projection printing, which was introduced in 1978 and has remained viable after six generations of MFS reduction. For economic reasons, 1X step-and-repeat projection printing, introduced in 1980, is still used to pattern all noncritical levels. Recently, the reduction step-and-repeat method is experiencing difficulties in further reducing the MFS, but still increasing the field size to accommodate the larger chips that usually follow higher densities. In the areas of large field sizes and high resolution, the reduction step-and-scan system, introduced in 1989, is starting to replace step-and-repeat printing. Because of the move from reflective systems with large bandwidth to narrow-bandwidth refractive systems used for step-and-repeat, multiple reflections within the resist layer becomes an important issue to understand and to overcome. Chapters 1 and 7 discuss this issue.

Even in the era of proximity printing, it was recognized that experimental work is expensive. The experiments also tend to be empirical, unless simulation of the physical phenomena of imaging is performed to assist in understanding the process and to guide the experiments. Leading the charge to turn microlithography from black magic to science is the work in theorizing and understanding proximity printing, even in the theoretically difficult region of exact contact, the resist exposure and development mechanism, two-dimensional partially coherent imaging for optical projection printing, x-ray partial coherent imaging, and e-beam proximity effects. Most of the simulation activities are reported by Neureuther and Mack in Chapter 7 on optical lithography modeling. E-beam and x-ray modeling are covered in Chapters 2 and 3, respectively.

To further move microlithography into the realm of well-understood science, the placement and size of the images produced have to be measured to a high degree of precision and accuracy regardless of the type of microlithography imaging system producing them. Metrology itself has become a science, an indispensable part of microlithography, and even a bottleneck at times. Metrology standards and the measurement of overlay and critical linewidths are covered by Lauchlan et al. in Chapter 6.

Looking further into the future, Chapter 8 considers the limits of each discipline for making devices with 0.1 µm minimum feature size and smaller, including quantum effects. These disciplines include optical, e-beam, and x-ray lithography, resists, metrology, yield assessment, ionizing radiation effects, and dry-etch damages. An interesting proposal is made to use massively parallel arrays of atomic force microscopes (AFMs) for quantum device manufacturing.

Instead of viewing them as photons, electrons, or ions, microlithographic systems can be separated into pattern-generating systems and replicating systems. The pattern generating systems take a mask design in software form and expose a physical pattern on the mask blank. Flexibility of the system makes it suitable to generate the mask for mass replication. This feature also encourages many lithographers to speed up the system to use as a maskless direct-write tool for wafer exposure. However, the sequential nature of pattern generation makes it difficult to compete with the throughput of a replication system, regardless of how much faster the individual beam can be accelerated. Attempts have been made to use many sequential pattern generations in parallel. A noted accomplishment is in pattern generation using multiple laser beams. However, the throughput is still not competitive to that of a replicating system. Attempts in multiple e-beams have had limited success. Therefore, it appears that pattern-generating systems are best suited for mask making, and replication systems are best suited for reproducing the mask patterns. To bridge the advantages of the two techniques one possibility is to use cell projection, as reported in Chapter 2.

The replication systems can be separated into two main categories, 1X and reduction systems. The 1X systems usually require simple imaging optics. For example, the 1X projection printing system enjoys the symmetry of optics and short focal length in both the mask and wafer sides, resulting in a small, low-aberration imaging lens. The proximity printing system, the ultimate example, uses no imaging system. The basic problem of a 1X replication system is that the 1X mask has to be built with stringent specifications to achieve an identical linewidth tolerance and overlay performance on the wafer that a reduction system can produce. This is illustrated in Table 1, which shows the linewidth tolerance in percentage.

TABLE 1 Linewidth tolerance components in 1X and 4X replication systems.

	1X	4X	1X perfect wafer lithography
Resist image on mask	8%	1.6 %	8%
Etched image on mask	8%	1.6%	8%
Resist image on wafer	10%	10%	0
Etched image on wafer	10%	10%	10%
RSS total	18.1%	14.3%	15%

Using the same 8% linewidth tolerance on the 1X and 4X mask results in a much larger total linewidth tolerance on the wafer. Even a perfect wafer imaging system producing zero linewidth tolerance cannot produce the result better than a 4X system. To be competitive, the 1X imaging system has to be specified for linewidth tolerance identical to the 4X system. This results in the need to specify the linewidth tolerance 4X better on the mask. At the forefront of MFS, everything is pushed to its limit, including the mask-making capability. There is no margin in mask making to accommodate the requirement of a 1X system.

Similarly, the feature placement error requirement on a 1X mask is too stringent, as seen in Table 2. The placement error is the displacement of the feature from its ideal location that is caused by uncertainty of the beam position during mask making. An extremely good e-beam system has a placement error on the order of 50 nm. Considering that the overlay is between two levels, the overlay error contribution from the mask placement error is the root-mean-square of the errors induced in two masking levels to be aligned to each other, reflecting the statistical nature of the error.

TABLE 2 Overlay error components in 1X and 4X replication systems.

	1X	4X
Mask writer placement	72 nm (RSS between 2 levels)	18 nm (RSS between 2 levels)
Wafer alignment error	50 nm	50 nm
Stepper table error	30 nm (RSS between 2 levels)	30 nm (RSS between 2 levels)
Lens distortion	15 nm	30 nm
RSS total	**94 nm**	**68 nm**

This root-mean-square procedure applies as well to the stepper table error. During step-and-repeat imaging of the wafer, a table stepping error causes the mask image to be misplaced on the wafer, leading to misalignment to the image on the previous or subsequent level that is also subject to a table stepping error. The price to pay for allowing a larger overlay budget for a 1X system is too high.

Even though it requires a 1X mask, the proximity soft x-ray imaging system has been given attention and funding since the first system was used for making experimental magnetic bubble circuits in the mid-1970s and the first exposure with a storage ring in the late 1970s. In the two decades that followed, tremendous progress was made in mask making, mask repair techniques, the storage ring source, the beam line, the step-and-repeat alignment station, and resists. Chapter 3 provides excellent treatment of this subject. Other than proximity soft x-ray, there is no other 1X system being developed for submicrometer microlithography. The field of replication microlithography has narrowed down to candidates capable of reduction imaging. There are four systems, shown in Table 3 together with their advantages and concerns.

TABLE 3 Reduction replication systems and their advantages and concerns.

System	Advantages	Concerns
UV projection	Available light source Available mask technology—mask making, defect detection, and repair Available lens material, design, and fabrication capabilities Available resists	Depth of focus Alignment
EUV projection	Large depth of focus Potential for high resolution	Light source Lens Multilayer reflective mask Resists Window material Alignment
E-beam projection	Available light source Large depth of focus Potential for high resolution	Mask E-beam optics Proximity effects Radiation damage Resists Charging effects
Ion-beam projection	Large depth of focus Potential for high resolution	Ion-beam optics Mask Radiation damage Charging effects

A natural extension of UV projection lithography is to continue to reduce the wavelength from 365 nm, the Hg i-line, to 248 nm, the KrF excimer laser line, then to 193 nm, the ArF excimer laser line. Below 193 nm, several difficulties arise: (1) The atmosphere absorbs too much light; the imaging system has to be in vacuum. (2) Transmissive material becomes rare; reflective optics is required. (3) An entirely new set of resist requirements is inevitable. To date, the EUV wavelengths between 10 to 70 nm have been proposed; some have been demonstrated. To develop one of these systems for production there has to be (1) a light source brighter than a storage ring, (2) coating and window materials capable of withstanding such strong radiation, (3) high-reflectivity, low-defect multilayer reflective coatings, and (4) reflective lenses capable of reduction and field sizes competitive to UV lenses.

Another reduction possibility is to use e-beam for replication using a special mask such as the SCALPEL (scattering with angular limitation for projection electron lithography) system covered in Chapter 2. Unlike EUV systems, there apparently is a good e-beam source and the mask problem is not as insurmountable as that of EUV. The e-beam projection lens is easier than EUV, but is still a long way from matching the field size of a high-performance UV lens. It also suffers from traditional e-beam problems such as proximity effects, radiation damage, charging, and lack of a good resist.

Reduction ion-beam projection also has been demonstrated. It has negligible proximity effects, and a resist of required sensitivity is not difficult to develop. However, ion-beam optics are much harder and bulkier than existing systems. Also, continuous ion bombardment to the mask can be detrimental.

The next reduction system that may succeed a 193 nm step-and-scan optical reduction system for making 0.13 μm MFS is likely to be SCALPEL, if all its concerns are addressed. However, a reduction system may no longer be needed, because e-beam cell projection in combination with a shaped beam (as discussed in Chapter 2) is probably closer to successfully producing 0.13 μm MFS.

Historically, microlithography has been an innovative and rapidly advancing technology. To date, any effort to project its future often inspired activities to break through the limits identified by the projection. With the great changes in IC technology of the past two decades, it's difficult to project, or even imagine, the possibilities and great strides that will be made over the next two.

CHAPTER 1
Optical Lithography

Harry J. Levinson
Advanced Micro Devices

William H. Arnold
Advanced Micro Devices

CONTENTS

1.1 INTRODUCTION

In 1979, *Electronics* magazine reported that optical lithography would be a passing fancy superseded by direct write electron beam lithography by the year 1985.[1] It was admitted in a follow-up article, written for that same magazine in 1985, that the demise of optical lithography had been predicted prematurely and that it would take until 1994 for shipments of optical wafer steppers to be of lower volume than those of x-ray step-and-repeat systems.[2] It was expected that optical lithography, having reached a resolution limit of 0.5 μm, would need to be replaced. It is now 1996, and optical lithography is still going strong.

Much of optical lithography's longevity and productivity may be attributed to the extent to which fundamental problems have been identified, understood, and addressed. It is the authors' intention that this chapter should contribute toward an understanding by providing a clear explication of engineering problems inherent to optical lithography, along with a review of many of the contributions towards achieving this understanding and some of the ingenious solutions that have been devised along the way.

For the manufacture of integrated circuits, the lithographer has responsibility for creating the patterns of the desired sizes and shapes, overlaying these to prior patterns, and ensuring that the patterns are defect free. These three topics are the subjects of this chapter.

1.2 IMAGING

The basic problem of imaging is shown in Fig. 1.1. Light from an illumination source passes through a photomask, which defines the images. The simple photomask illustrated here consists of areas that are completely opaque and complementary areas that are transparent. In this example, the clear area is a long space of uniform width, and the optical and resist profiles shown in Fig. 1.1 are resulting cross sections. Some of the light that passes through the mask continues through a lens, which projects an image of the mask pattern onto a wafer. The wafer is coated with a photosensitive film, photoresist, that undergoes a chemical reaction upon exposure to light. After exposure, the wafer is baked and developed, leaving regions covered by photoresist and complementary regions that are not covered. The imaging objective of microlithography is for the resist features to be well-defined and sized within specifications. The challenge of imaging results from the shape of the light intensity distribution at the wafer plane, which lacks a clearly defined edge (Fig. 1.1) and is therefore not well defined. If the light intensity distribution had the shape shown in Fig. 1.2, there would be no problem, because a clear delineation would exist between areas of the resist exposed to light and unexposed areas.

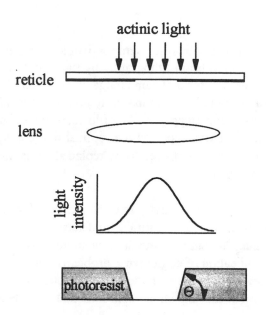

FIG. 1.1 An illustration of the imaging process. Light passes through a reticle. The resulting pattern is imaged onto a photoresist covered wafer by a lens. The finite resolution of the lens results in a light intensity distribution which does not have clearly defined edges.

Creating resist images, within manufacturing tolerances and on a repeatable basis, from less well defined optical profiles is the challenge confronting microlithographers. Optics and photoresist are the two ingredients of image formation. A lithographic exposure system is essentially a camera in which the photoresist plays the role of the film. Following the develop process, a pattern is formed in the photoresist. The quality of the image is determined by the resolution power of the optics, the focusing accuracy, the contrast of the resist process, and an assortment of other variables that will cause the final image to be a less than perfect reproduction of the features on the mask. Just as common film for cameras will produce pictures of varying qualities, depending upon the film's photospeed, contrast and graininess, as well as the focus setting of the camera, the final image produced photolithographically will be affected by the resist process and the optics used.

An understanding of the independent contributions from each and their interactions is essential for the photolithographer. For example, consider the situation in which a production process has started to produce poorly defined patterns. Is the problem due to the resist process or the exposure system? An understanding of the various contributions to the final image would enable the lithography engineer to resolve the problem expeditiously. For a new process, advanced optics would provide high resolution, at the expense of capital, while sophisticated resist processes might extend the capabilities of any optics, but with a possible increase in process complexity and other issues which will be discussed later. By appreciating the roles that the resist and optics play in image formation, the photolithography engineer can design the most cost effective manufacturing line. An overview of the lithographic process will be presented in this section, whereby the separate contributions from the optics and the resist process can be seen. Each subject will then be discussed in more detail in subsequent sections.

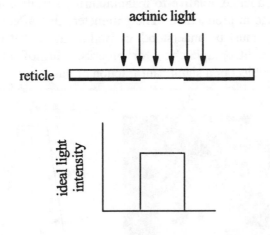

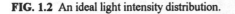

FIG. 1.2 An ideal light intensity distribution.

Two characteristics of the final resist pattern are of primary interest, the size and the shape. For reasons of yield, device performance and functionality, a process that is capable of producing the smallest possible linewidths is desired. The requirements on the shape of the resist are usually imposed by the postlithographic processing. Consider, for example, the situation in which the photoresist is to serve as a mask for ion implantation. If the edges of the resist are nearly vertical, the implantation will be clearly delineated by the resist edge, resulting in a sharp doping profile. On the other hand, if the resist has considerable slope, the high energy ions will penetrate the partial thickness of resist at the edges of the pattern and the resulting doping profile will be graded. The relevance of the resist edge slope and its effect on the doping profile depends upon the overall process requirements. Another common example in which the slope of the resist is important occurs when the resist is to be used as a mask for plasma or reactive ion etching, and the etching process erodes the photoresist. The slope of the photoresist can be transferred to the etched layer, which may or may not be desirable. There is no purely lithographic consideration that determines what the optimum process is. However, once the requirements of the final resist pattern are determined from considerations of the postlithographic processing, the masking process can be specified. The most preferred resist profiles are ones that are nearly vertical. Since these are generally the most difficult to produce, as well as the most desired, our discussion will be oriented towards such profiles.

Regardless of the specific requirements of the size and shape of the resist pattern, which may vary from one technology to another, all lithographic processes must be consistent and reproducible, relative to manufacturing specifications, in order to be appropriate for use in production. The parameters that affect process uniformity and consistency must be understood, as well as those that limit the ultimate performance of a lithographic tool or process. Identification of those parameters will be the subject of a significant fraction of this chapter.

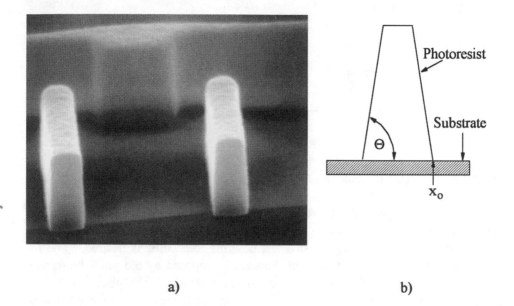

a)   b)

FIG. 1.3 a) Cross sections of Apex E, printed on a Micrascan II. b) Idealized trapezoid cross section.

A typical resist profile is shown in Fig. 1.3. The shape of the cross section is often idealized as a trapezoid. Three dimensions are of most interest, the width of the resist line at the resist-substrate interface, the slope of the sidewall, and the maximum thickness of the resist film after development.

Of course, actual resist profiles often depart significantly from the idealized trapezoid. The Semiconductor Equipment and Materials International (SEMI) standard that defines linewidth[3] accounts for such departures, and linewidth is defined to be a function of the height from the resist substrate-interface (Fig. 1.4). Throughout the next section, if the word "linewidth" is used with no further clarification, it is understood to be the dimension L measured at the resist-substrate interface. This definition has been chosen for three reasons. First, it is the width of the resist line that is of greatest relevance to the final result achieved after the post-lithographic processing. For example, in a highly selective etch, where there is little resist erosion, all other dimensions of the resist line have negligible influence on the resultant etch. Moreover, in the presence of a sloped resist profile, the definition adopted here for linewidth is unambiguous. Finally, the value of the linewidth is decoupled from the slope of the resist line.

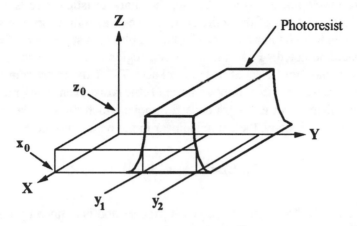

FIG. 1.4 The SEMI standard definition for linewidth. Linewidths are given by the quantities, $y_2 - y_1$, defined at the distances x_0 along the resist line and at height z_0.

1.2.1 The contributions of physics and chemistry

With these basic concepts in hand, we can now start to explore the photolithographic process through some simple models[4,5]. The first step is to assign some numerical values to the light intensity profile depicted in Fig. 1.1. The light intensity is modulated by the mask and further modified by the optics of the lithography equipment. Because we are considering only a long space of glass in the photomask, the light intensity of I at the plane of the wafer is only a function $I(x)$ of one variable, where x is the distance along the direction perpendicular to the space. Also, we will take the substrate which is coated by the resist film to be non-reflecting, to simplify the discussion. Effects due to substrate reflectivity will be addressed in a later section. Our basic assumption will be that the thickness $T(x)$ of photoresist that remains after development is determined by the exposure energy dose $E(x) = I(x) * t$, where t is the exposure time:

$$T(x) = T_E(E) = T_E(E(x)) \tag{1.1}$$

This is a reasonable assumption for thin photoresist. For positive photoresist, the greater the exposure the thinner the resist film that remains, while for negative resist, thicker resist remains with higher exposures. The current discussion will be restricted to positive photoresists.

A typical plot of $T_E(E)$ is shown in Fig. 1.5 and is called the "characteristic curve" of the resist, following similarly named curves used for characterizing photographic films[6]. Such a curve may be obtained by exposing large areas on a wafer with different exposure energies and then measuring the residual thicknesses in those areas after development. To be precise, the characteristic curve is a property not only of the resist but of the entire resist process, as will be seen in subsequent discussions in this chapter and the chapter on photoresist. Several features of the curve should be noted. First, there is a flat region at low exposure, where the final thickness may not be equal to the thickness T_0 of the resist film before exposure, since the developer may remove some of the resist even though it is unexposed. Second, there is an exposure dose E_0 above which the photoresist film is completely removed. Finally, the curve is linear in the region around E_0:

$$T_E(E) = T_0 \, \gamma \, \ln\left(\frac{E_0}{E}\right) \qquad (1.2)$$

The slope γ is called the "contrast" of the resist process and is defined by Eq. 1.2. Typical values for γ are shown in Table 1.1, though it should be noted that the resist contrast is a function of the entire resist process, bake temperatures and time, developer normality and time, etc., not just the resist itself. Contrast will be discussed in more detail in a later section.

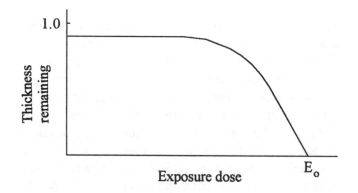

Fig. 1.5 A typical characteristic curve for photoresist, showing photoresist thickness remaining after develop as a function of exposure dose. The exposure dose is plotted on a logarithmic scale to produce a curve that is approximately linear in the vicinity of E_0.

The dependence of the final resist pattern on the resist process can be seen in the following analysis. The slope of the resist sidewall is simply

$$\frac{dT}{dx} = \tan\theta \qquad (1.3)$$

where the derivative is evaluated at the point x_o where the resist profile and the substrate intersect (Fig. 1.3) Note also that

$$E_o = E(x_o). \tag{1.4}$$

Eq. 1.3 can be expressed as[7]

$$\frac{dT}{dx} = \frac{dT}{dE}\frac{dE}{dx}. \tag{1.5}$$

Eq. 1.5 neatly divides the factors that determine the resist profile. The first factor, dT_E/dE, is a characteristic of the photoresist and development process, independent of the exposure tool, while the second factor, dE/dx, is completely determined by the optics of the exposure system. One obtains dT_E/dE by differentiating a curve such as the one shown in Fig. 1.5, but on a linear scale. The function $E(x)$ will be discussed shortly.

TABLE 1.1: Resist contrast for selected photoresists.

Resist	Exposure wavelength	Contrast (γ)	Reference
AZ 1470	g-line	5.8	8
Shipley 511	i-line	6.9	9
Apex E	DUV	4.7	10
TOK IP3000	i-line	6.8	11

In the vicinity of x_o, $T_E(E)$ is described by Eq. 1.2. This results in the following expression for the slope of the resist profile

$$\tan\theta = \frac{dT}{dx} = -T_o\,\gamma\,\frac{1}{E(x)}\,\frac{dE(x)}{dx}. \tag{1.6}$$

Our simple model is based upon the assumption that resist development behavior measured in large exposed areas can be applied directly to situations where the light intensity is modulated over small dimensions. Within the limits to which our assumption (that $T(x)$ is solely a function of $E(x)$) is valid, the dependence of the profile slope is cleanly separated into the contributions from the optics (the factor $1/E\,dE/dx$) and from the resist process (represented by γ), and each can be studied independently of the other.

Theoretical models of image formation in projection optics have been developed, usually starting from the Hopkins formulation for partially coherent imaging[12], which is based on scalar wave diffraction theory from physical optics.   These models provide the engineer with the capability for calculating $E(x)$ for various configurations of optics. Let us first consider the optics term $1/E \ dE/dx$. $E(x)$ has been calculated for a "perfect" lens using a masking simulation program, PROLITH[13,14], and the results are shown in Figs. 1.6 and 1.7. All profiles are shown in normalized units so that for each dimension the mask edges occur at +/-0.5. The normalized derivation is shown only around the region of interest, close to the edge of the mask feature. As one can see, for fixed optics the "sharpness" of the optical intensity profile degrades and the value of $1/E \ dE/dx$ at the edge of the mask feature is clearly reduced as the feature size is decreased. Lens aberrations or defocus would also reduce the magnitude of $1/E \ dE/dx$, while higher resolution optics would increase its value. Since Eq. 1.6 is non-linear, substantial degradation of the optical contribution to the profile slope can be tolerated if the contrast γ of the resist process is large enough.

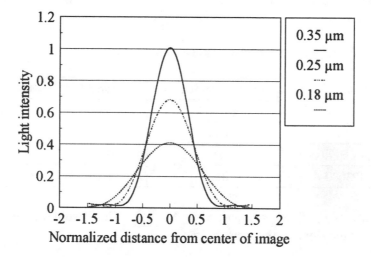

FIG. 1.6 Calculated light intensity distributions for isolated spaces of varying sizes, using parameters for an aberration-free 0.5 NA lens, with a partial coherence of $\sigma = 0.6$, at a wavelength of 248 nm.

Another situation in which a high contrast resist process is advantageous arises when one is trying to print over a surface that is not flat. Since resist, when spun on the wafer, tends to planarize[15,16,17], masking over an underlying feature is equivalent to patterning photoresist of varying thickness. This can be seen in Fig. 1.8.

If the thickness of the photoresist is changed from T_o to $T_o + \Delta T$, with all other aspects of the lithographic process remaining equal, then the edge of the photoresist line moves from x_o to $x_o + \Delta x$, and the minimum exposure energy required to clear the resist to the substrate changes from E_o to $E_o + \Delta E$. After exposing the thicker resist with energy E_o the amount of resist remaining is ΔT.

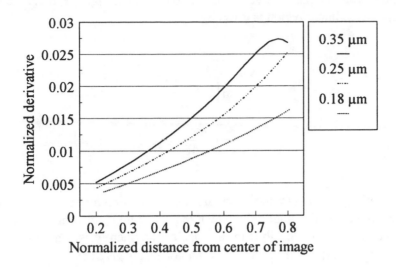

FIG 1.7 Calculated normalized derivatives $\frac{1}{E_o}\frac{dE}{dx}$ of the intensity distributions shown in Fig. 1.6.

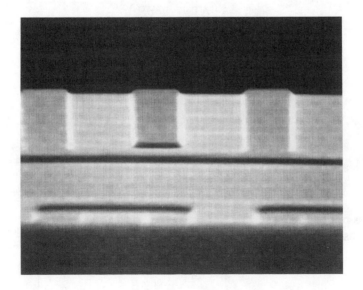

FIG. 1.8 Resist (dark material at the top of the figure) spun over topography has a relatively planar top surface.

Drawing a new characteristic curve for thicker photoresist, Fig. 1.9, it can be seen that the slope in the linear region is given by

$$\gamma = \frac{\dfrac{\Delta T}{T_0 + \Delta T}}{\ln\left(1 + \dfrac{\Delta E}{E_0}\right)} \qquad (1.7)$$

$$\approx \frac{\Delta T}{T_0 \dfrac{\Delta E}{E_0}} \qquad (1.8)$$

for small variations. (It is assumed here that the resist is on a non-reflecting substrate. The more complex situation in which the substrate is reflecting will be discussed in Section 1.2.4.) The shift in the edge of the photoresist line that occurs due to the change in the thickness ΔT of the photoresist is given by

$$\Delta x = \frac{\Delta E}{dE/dx}, \qquad (1.9)$$

where the derivative is evaluated at the point x_0. From Eq. 1.8, this becomes

$$\Delta x = \left(\frac{\Delta T}{\gamma T}\right)\left(\frac{1}{E_0}\frac{dE}{dx}\right)^{-1}. \qquad (1.10)$$

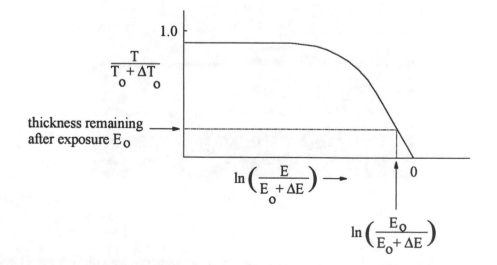

FIG. 1.9 Characteristic curve for thicker photoresist.

Again, there is a separation between the optical term, which reappears in the form of the normalized derivative $1/E\ dE/dx$, and the photoresist, whose contribution is expressed through the quantity γ. From Eq. 1.10 it can be seen that a high contrast resist process provides the benefit of linewidth control over topography, in addition to steep resist profiles.

One might think that the optimum resist process is one in which the contrast is maximized. Unfortunately, there are some circumstances in which the contrast of the resist process has little effect in improving performance or latitude, and there are some situations in which an excessively large γ will lead to a process that is difficult to control. This latter situation will arise in our discussion of thin film optical effects. A parameter for which the contrast of the resist process affects process control only in limited circumstances is exposure latitude, i.e., the sensitivity of the linewidth to the exposure dose. This can be seen through our simple model. If the exposure dose is changed fractionally from E to $(1 + f)E$, then the edge of the resist line is moved from x_o to $x_o + \Delta x$. The edge of the resist line is still determined by Eq. 1.4:

$$(1 + f)\, E(x_o + \Delta x) = E_o = E(x_o) . \tag{1.11}$$

Accordingly,

$$\Delta x = -f\, E(x_0) \left(\frac{dE}{dx} \right)^{-1} . \tag{1.12}$$

Letting

$$f = \frac{\Delta E}{E} \tag{1.13}$$

we obtain an expression for the change in linewidth:

$$\Delta L = 2 \frac{\Delta E(x)}{E(x)} \left(\frac{1}{E(x)} \frac{d}{dx} E(x) \right)^{-1} , \tag{1.14}$$

where all expressions are evaluated for x at the line edge. Our expression for exposure latitude Eq. 1.14 contains only factors involving the exposure optics and is independent of the photoresist process and its characteristics. This result is valid over a large range of processing parameters, particularly in the limits of thin or high contrast photoresist. Note that Eq. 1.14 could be rewritten with $E(x)$ replaced by the intensity function $I(x)$. Finally, the process engineer is usually interested in knowing what fractional change in exposure will keep linewidths within specifications, which are usually stated as a fraction of nominal linewidth. From the preceding, an expression for this exposure latitude is given by:

$$\frac{\Delta E}{E} = \frac{\Delta L}{L}\left[2 L \left(\frac{1}{E(x)}\frac{d}{dx}E(x) \right) \right].$$ (1.15)

If one disregards technical rigor (a logarithm of a parameter with dimension, such as length, mass, etc., is not a well defined quantity.) one can rewrite this equation as

$$\frac{\Delta E}{E} = \frac{\Delta L}{L} 2 \left[L \frac{d\ln E}{dx} \right].$$ (1.16)

The quantity in brackets is referred to as the normalized log-slope. It is a good metric of image quality since optics with higher normalized log-slopes will have greater exposure latitude, stepper resist profiles, etc.

The above results show how the optics and resist process play separate and independent roles in image formation. The study of optics and its role in photolithography reduces to the analysis of $E(x)$ and the parameters that determine its values. Similarly, the part that photoresist plays is related to its development rate as a function of exposure. In all circumstances, the performance of the lithographic process will be determined by practical issues as well, such as resist adhesion. The analysis of the lithographer's "tools of the trade," resist and optics, will be the subject of the remaining sections on imaging.

1.2.2 Aerial image considerations

The origin of the image intensity profile of Fig. 1.1 is the physical phenomenon of diffraction, a subject that was studied extensively in the 19th century and is well-understood[18]. Diffraction effects are those that result from the wave nature of light. The resolution of optical tools is limited fundamentally by the physical phenomenon of diffraction, and as device geometries shrink, the lithography engineer must ultimately contend with this barrier imposed by the laws of physics. This is illustrated in Fig. 1.6. For a fixed set of optics, the aerial image in the plane of best focus is shown for features of varying sizes. As the feature size shrinks, the edge acuity of the light intensity distribution degrades. At some point, one must be able to say that features are no longer resolved, but it is clear from Fig. 1.6 that there is a gradual transition from "resolved" to "unresolved."

A definition of resolution is not obvious, because of this lack of a clear delineation between "resolved" and "unresolved." Simple diffraction analyses lead to the most frequently cited quantitative definition of resolution, the Rayleigh criterion, which will be introduced shortly. While sufficiently instructive to justify its consideration, it does not provide a criterion directly applicable to the situation encountered in photolithography and should be used with care. Our discussion of the Rayleigh criterion emphasizes its assumptions and therefore its applicability.

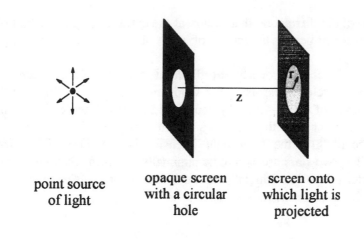

point source
of light

opaque screen
with a circular
hole

screen onto
which light is
projected

FIG. 1.10 A circular aperture illuminated by a point source of light.

The phenomenon can be appreciated by considering the following situation, in which uniform illumination is normally incident on a masking screen in which there is a circular hole (Fig. 1.10). Such a situation would arise, for example, if the light were produced by a point source located far from the masking screen. The light that passes through the hole illuminates another screen. In the absence of diffraction, the incident light would simply pass through the aperture and produce a circular illuminated spot on the second screen. Because of diffraction, in addition to the normally incident rays of light, some light propagates at divergent angles.

The problem of an illuminated circular aperture has been solved analytically taking into account diffraction effects[18], and the solution for the light intensity is plotted in Fig. 1.11. Rather than a circular beam propagating in a single direction, the light is distributed over angles that are functions of the wavelength of the light and the radius of the circular aperture. For light of wavelength λ the light intensity distribution at radius r on the imaging screen is given by

$$I(x) = I_0 \left(2 \frac{J_1(x)}{x} \right)^2 , \tag{1.17}$$

where $x = kdr/2z$, $k = 2\pi/\lambda$, d is the diameter of the aperture, z is the distance between the two screens, and J_1 is the first order Bessel function. I_0 is the intensity at the peak of the distribution. It should be noted that shorter wavelengths and larger apertures lead to less angular divergence. The light intensity distribution given by Eq. 1.17 is called the Airy pattern, after G. B. Airy, who first derived it[19]. Because of the diffraction that occurs at the entrance pupil or edges of a lens, producing divergent beams, the images of point objects are focused Airy patterns.

Rayleigh used the above property of diffraction to establish a criterion for the resolving power of telescopes, using the following argument[20]. Suppose there are two point sources of light — stars — separated by a small angle. Being independent sources, the light from the stars does not interfere, i.e., they are mutually incoherent. Accordingly, the total light intensity is the sum of the individual intensities. The two stars are said to be minimally resolved when the maximum of the Airy pattern from one star falls on the first zero of the Airy pattern from the other star, which occurs when

$$\frac{x}{2\pi} = 0.61. \tag{1.18}$$

The resulting total light intensity due to the two stars with an angular separation given by $x = 2kdr/2z = 0.61 \cdot 2\pi$ is plotted in Fig. 1.12, where two peaks can be distinguished clearly.

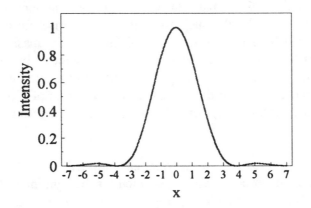

FIG. 1.11 The light intensity distribution from a point source projected through a circular aperture in an opaque screen. The parameter x is defined on the next page.

For focusing optics, the minimum resolved distance d between the peak of the Airy distribution and its first zero, can be related using criteria that must be satisfied for focusing optics, resulting in

$$d = 0.61 \frac{\lambda}{n \sin \theta}, \tag{1.19}$$

where n is the index of refraction of the medium surrounding the lens ($n \approx 1$ for air), λ is the wavelength of the light, and 2θ is the angle subtended by the lens (Fig. 1.13). The quantity $n \sin \theta$ is called the numerical aperture (NA) of the lens.

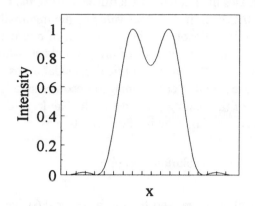

FIG. 1.12 Light intensity distribution of two light sources projected through a circular aperture.

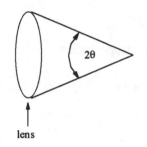

FIG. 1.13 The angle subtended by a lens determines its numerical aperture.

The minimum separation given by Eq. 1.19 is referred to as the Rayleigh resolution. Examples for wavelengths typically used in semiconductor processing are given in Table 1.2. As can be seen from the table, a high resolution (small feature size) can be achieved with a large numerical aperture or a short wavelength. As we will discuss shortly, when focus is taken into consideration, these two paths to a given resolution are not equivalent.

It is worth restating that the Rayleigh criteria were obtained by considering the imaging of point sources of light, which is not the problem encountered by photolithographers. The geometries on photomasks that must be imaged onto the front surface of wafers are extended in size, generally approximating the size and shape of a circuit element. The resolution problem is the one shown in Fig. 1.6. As the features become smaller the optical profile becomes flatter and it becomes increasingly difficult to locate the "edge" of a feature. As will be discussed later, linewidth control is also directly related to the slope of the optical intensity distribution. Although they were derived for a situation different from that encountered by lithographers, the Rayleigh criteria are nevertheless useful as scaling equations.

For example, if one increases the numerical aperture by 20%, the resolution is also expected to change by 20%, even if the exact value of the minimally resolved feature is not given by Eq. 1.19. Resolution in the context of microelectronics processing is also dependent upon the properties of the photoresist, while the Rayleigh criterion was derived solely on the basis of optical considerations. As resists improve, the minimum feature size that can be imaged becomes smaller, for fixed optics. To take advantage of the scaling property of the Rayleigh criterion, and the dependence on resist processes, Eq. 1.19 is often rewritten as

$$\text{resolution} = k_1 \frac{\lambda}{\text{NA}} \qquad (1.20)$$

where NA is the numerical aperture, and the prefactor of 0.61 has been replaced. k_1 is usually referred to as a "k-factor" for the process. For a given resist system, Eq. 1.20 provides scaling, and the prefactor would change if a different photoresist process were used.

TABLE 1.2 Typical parameters for stepper optics.

Wavelength	Numerical aperture	Rayleigh Resolution	Light source
436 nm	0.30	0.89 μm	Hg arc lamp (g-line)
436 nm	0.54	0.49 μm	Hg arc lamp (g-line)
365 nm	0.45	0.49 μm	Hg arc lamp (i-line)
365 nm	0.60	0.37 μm	Hg arc lamp (i-line)
248 nm	0.35	0.43 μm	Hg arc lamp or KrF
248 nm	0.50	0.30 μm	Hg arc lamp or KrF
248 nm	0.60	0.25 μm	Hg arc lamp or KrF
248 nm	0.70	0.42 μm	Hg arc lamp or KrF
193 nm	0.50	0.24 μm	ArF excimer laser
193 nm	0.60	0.20 μm	ArF excimer laser

The Rayleigh criterion, Eq. 1.19, assumes that the degradation of resolution results entirely from diffraction, i.e., the lenses are free from aberrations and imperfections. Such optics are called "diffraction limited" because the design and manufacture of the lenses are so good that the dominant limit to resolution is diffraction. The extent to which stepper lenses approach the diffraction limit is determined by comparing calculated and measured optical profiles. There are a

number of commercially available software packages for calculating image profiles. Methods for measuring optical images are given in Refs. 21, 22 and 23.
Modern lenses are typically found to be very near the diffraction limit, but deviations from the limit are observable.

Optical systems, designed and built by mere mortals, are never perfect. Departures of the optics from diffraction limited imaging are caused by "aberrations,"
and can be considered from two perspectives[12,18]. The first involves those factors
that affect the shape of optical profiles. These will affect the resolving capabilities
of lenses and will be discussed in this section. The second class of aberrant behavior, termed "distortion," affects two dimensional patterns and will reappear in
the discussions on overlay and alignment. It should be noted that the primary aberrations refer to deficiencies in the optics that could be corrected by design, not
necessarily manufacture.

The primary aberrations affect resolution in three important ways. First, by their
very nature, many aberrations degrade resolution most for points far from the optical axis. The most commonly encountered of such aberrations are field curvature and distortion. As a result, lenses will tend to have higher resolution in the
center of their fields. It is difficult to design a high resolution lens for which the
resolution is maintained over a large field. Another type of aberration that degrades imaging is astigmatism, which causes points to appear as lines when imaged. Alternatively, astigmatism causes the focus of lines to depend upon their
orientation relative to the optical axis, so they will be considered again in the section on focus. An example of astigmatism is shown in Fig. 1.14, where one can
see that the vertical spaces are well resolved but the horizontal spaces are bridged.

It is an extremely difficult task to design a lens free from the above aberrations
even at a single wavelength. The inability of a lens to focus light over a range of
wavelengths, which results from the variation in the refractive indices of the
glasses used to make the lens, is called chromatic aberration[12]. An advantage of
reflective optics is their freedom from chromatic aberration, since their focusing
properties are independent of the wavelength. For systems that use refractive optics, the illuminator must be designed consistently with the bandwidth requirements of the projection optics, keeping in mind that most sources of light, such as
mercury arc lamps, produce a broad spectrum of light. Early lithography systems,
such as the Perkin-Elmer Micralign, used substantially reflective optics that could
image over a broad range of wavelengths, including multiple peaks of the mercury
arc lamp emission spectrum. However, it proved to be very difficult to build reflective optics with high numerical apertures, and a transition was made on lithography equipment to refractive optics. Lenses using refractive optics have been
designed which are corrected at two wavelengths corresponding to emission peaks
of mercury arc lamps. Unfortunately, the first of these to be characterized was so

poorly corrected for the continuum between the peaks that no advantage was realized.[24] Subsequent lenses were made that were superior in the continuum region,[8] but two wavelength imaging with refractive optics had acquired such a bad reputation by that time that the approach was abandoned.

FIG. 1.14 An example of astigmatism. The vertical spaces in the resist are well resolved, but the horizontal lines are so defocused that they are bridged.

In the deep UV, the small number of materials that can be used for making lenses has made it difficult to correct for chromatic aberrations, and extremely narrow bandwidths are required for refractive DUV lenses[25]. The problem of chromatic aberration will reappear in the section on alignment, where it presents a problem for through-the-lens imaging for the purposes of alignment. Another difficulty with exposure systems that use monochromatic illumination is their susceptibility to substrate thin film optical effects, a topic that will be discussed later.

In the analysis leading to Eq. 1.19, the intensities, rather than the amplitudes, of the two sources of light (stars) could be added because they were mutually incoherent. Suppose instead that the two light sources were completely coherent and in phase. This situation was first considered by Abbe[26], who determined that the minimum resolution is given by

$$d = 0.77 \frac{\lambda}{n \sin \theta} \tag{1.21}$$

when the light is coherent. Except for the value of the prefactor, the two conditions for resolution, Eqs. 1.19 and 1.21, are the same. In both criteria, the resolution is directly proportional to the wavelength and inversely proportional to the numerical aperture. Higher resolution through increased numerical aperture requires optics that are physically larger, as can be seen in Fig. 1.13. This leads to practical design and manufacturing problems. Thus, there is a compromise between high resolution and field size, since both require larger optics.

It should be pointed out that resolution is independent of magnification, a factor that does not appear in any of the above equations. In properly designed microscopes, there is a natural relationship between resolution and magnification, so the two tend to be associated. The historical transition from full field scanning systems to higher resolution steppers involved an increase in the reduction ratio of the optics, but this increase was not required to achieve the higher resolution. In lithographic equipment, resolution and magnification are independent quantities. Very high resolution 1× magnification systems have been built.[27]

There are some advantages to reduction optics. The reticle will have manufacturing variations, and these will be replicated onto the wafer. For optical systems with $1/N$ reduction (meaning that features on the wafer are $1/N$ times the size of the corresponding reticle geometries), these variations will be reduced by the factor of $1/N$. Reduction optics will also have a similar benefit with respect to reducing sensitivities to defects on the reticles and variations in overlay.[28,29,30]

The application of these criteria to the situation encountered in photolithography is complicated by the fact that the illumination used with production tools is often neither completely incoherent nor coherent. Light that has coherence properties between incoherent and coherent is called partially coherent. Analyses of optical systems with partially coherent illumination are considerably more complicated than the situations discussed above, and simple criteria for resolution such as Eqs. 1.19 - 1.21 do not result. Nevertheless, the similarity of the above expressions leads to the semi-quantitative conclusion that the resolution should vary directly with wavelength and inversely with numerical aperture. Assigning a more definite number for resolution must result from another approach.

The resolution limits of optical systems can be appreciated from another approach by considering the system shown in Fig. 1.15. A reticle with a diffraction grating with a periodicity $2d$ (equal lines and spaces) is illuminated by a source of coherent light of wavelength λ (in vacuum). For an angle of incidence of θ_i, the grating diffracts the light into various beams whose directions are given by

$$\sin(\theta) - \sin(\theta_i) = \frac{m\lambda}{2nd}, \quad m = 0, \pm1, \pm2, \cdots \qquad (1.22)$$

Consider a lens with a collection angle of θ_o used to image the grating and on-axis illumination ($\theta_i = 0$). Only a finite number of diffraction beams will be collected by the lens because of Eq. 1.22. Coherent light diffracts a grating into beams that correspond to the object's Fourier components, and a diffraction limited lens will recombine those beams that pass through the entrance pupil of the lens. As light from large angles is collected, the image will consist of more terms in the Fourier series expansion of the light intensity distribution of the grating (see Fig. 1.16):

$$I(x) = \frac{I_o}{2} - \frac{I_o}{\pi} \sum_{m=-\infty}^{\infty} \frac{(-1)^{|m|}}{2|m|-1} \cos \frac{(2|m|-1)\pi x}{d} . \qquad (1.23)$$

The zeroth term of the series expansion involves only the intensity of the illumination and contains no spatial information. The first-order terms are required for the dimensions of the grating to be reproduced in the image. The more terms retained in the expansion Eq. 1.23, the more the image will resemble a rectangular grating (Fig. 1.16). Since the first two terms in the expansion of $I(x)$ must be retained to exhibit any spatial variation, θ in Eq. 1.22 must be large enough to capture diffracted beams for $m \geq 1$. Eq. 1.22 then leads to the following expression for a minimally resolved feature:

$$d = 0.5 \frac{\lambda}{n \sin \theta_o} . \qquad (1.24)$$

Again, the resolution is found to be directly proportional to the wavelength of the light and inversely proportional to the numerical aperture. Only the prefactor is different among Eqs. 1.19, 1.20, 1.21 and 1.24.

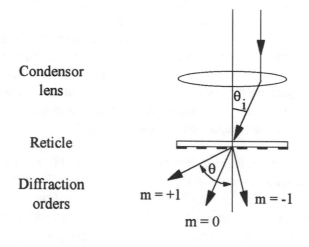

Condensor lens

Reticle

Diffraction orders

$m = +1$

$m = -1$

θ

$m = 0$

θ_i

FIG. 1.15 Situation with coherent light and a diffraction grating on the reticle.

The above analysis contains a curious feature associated with coherent illumination. For a particular set of optics, grating dimensions are resolved to a certain point and then not resolved at all for smaller features. This behavior of optics is familiar to most photolithographers, where the performance of the optics falls off very rapidly as the rated resolution limits of the lenses are approached. The extremely sharp cut-off in resolving power of the last example is an artifact of the complete coherence of the light, which led to a situation in which terms of the expansion Eq. 1.23 were retained with their complete magnitudes or not retained at all. For incoherent light, the magnitudes of the various terms in the image decrease more gradually as the resolution limit is approached. The amount by which sinusoidally varying intensities of incoherent light diminish as they pass through an optical system is called the modulation transfer function (MTF)[31] and provides a useful way to characterize the resolving capabilities of optics.

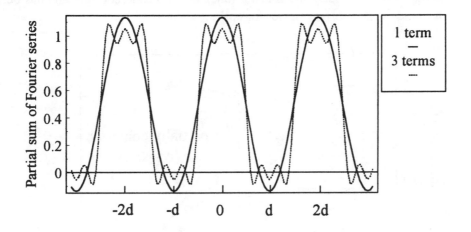

FIG 1.16 Partial sums of Eq. 1.24. The more terms added, the closer a rectangular grating is approximated.

The transfer characteristics — the proportion of transmitted light — of a general lens for coherent and incoherent light are shown in Fig. 1.17. In the case of diffraction limited optics, the transfer characteristics can be plotted in terms of normalized spatial frequencies, v/v_o, where $v_o = NA/\lambda$. For coherent imaging, up to the resolution limit of the lens, Fourier components are completely transferred by the lens, with no transfer above that limit. For incoherent light, Fourier components are transferred partially and up to a larger value than for coherent light.

Unfortunately, we must again contend with the fact that the illumination used in microlithography is neither completely coherent nor incoherent, and the curves shown in Fig. 1.17 for coherent or incoherent light do not properly characterize the transfer properties of mask features onto wafers. The transfer function concept has been adapted for characterizing systems with partially coherent illumination in the following way.

Suppose the mask pattern consists of a grating of a particular periodicity $2d$ (Fig. 1.15). After passing through the optical system, the light intensity profile will appear as shown in Fig. 1.18. Near the resolution limit, the image consists of only two terms in the series expansion (Eq. 1.23): a constant plus a single cosine that has the same periodicity as the mask pattern. The minimum light intensity, I_{min}, will be greater than zero, and the peak light intensity, I_{max}, will be less than the illumination intensity. The optical contrast or modulation is defined as[32,33]

$$C = \frac{I_{max} - I_{min}}{I_{max} + I_{min}} \qquad (1.25)$$

by analogy with the modulation transfer function for incoherent light and has been used as a metric of resolving power for optics.

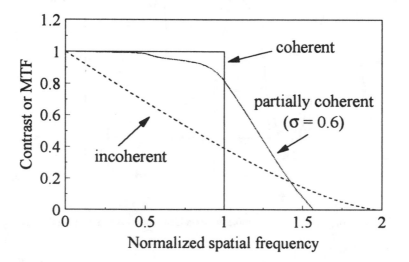

FIG 1.17 Transfer characteristics of a diffraction limited lens. Coherent light is given by $\sigma = 0$, while incoherent light is given by $\sigma = \infty$. Contrast is defined in Eq. 1.25.

The optical contrast is plotted in Fig. 1.17 for varying degrees of partial coherence. The spatial frequencies are related to feature sizes by the relationship

$$\upsilon = \frac{1}{2d} . \qquad (1.26)$$

As shown in Fig. 1.17, there is a sharp cut-off for image contrast with completely coherent illumination, with no image contrast resulting for features smaller than 0.5 NA/λ. For incoherent and partially coherent light, the image contrast degrades and there is not a similar distinction between features which can be imaged with good contrast and those which will not.

When operating in a region where the optical contrast curve is relatively flat, the imaging variations among features of different sizes is small. As the resolution limit of the lens is approached, the imaging capabilities of the lens fall off quickly.

The greater the coherence of the light, the greater the optical contrast will be up to the resolution limit, but the cut-off is for smaller features. Patterns in photoresist can be generated for optical contrasts less than 1.0 but significantly greater than zero, the exact number depending upon the resist system. The partial coherence of exposure tools is chosen to provide the optical contrast necessary for patterning minimum sized features in single layers of photoresist. Other metrics for optical image quality are used more extensively today, and these will be discussed shortly.

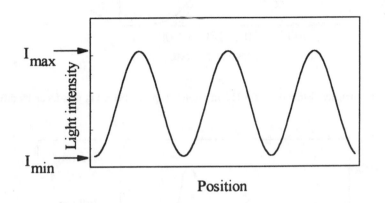

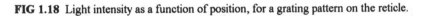

FIG 1.18 Light intensity as a function of position, for a grating pattern on the reticle.

The preceding discussion has demonstrated several general properties of the resolving capabilities of optical systems. First, and foremost, the resolution of an optical system is limited by the physical phenomenon of diffraction. This limit may not be realized in actual optical systems because of aberrations. The sharpness of the resolution "edge" of a lens is dependent upon the degree of coherence of the illumination, and for exposure systems that have been used in the past, the fall-off in optical contrast is steep as the resolution limit is approached, but not perfectly sharp.

Another feature optical systems that makes the preceding discussion useful is linearity. Patterns which do not consist of gratings can still be analyzed in terms of the behavior of individual Fourier components[31]. The imaging of patterns of arbitrary shapes can be calculated from linear combinations of grating patterns.

Having considered the general properties of resolution, it is time to return to the model presented earlier, in which the effects of the optics arise through the light energy distribution curve at the wafer, $E(x)$, and the normalized derivative $1/E_o \, (dE/dx)$. The effects of resolution on these functions can be easily calculated and are presented in Figs. 1.19 and 1.20. Not only does increased resolution permit the printing of smaller feature sizes, but it affects the printing of the larger features that can be resolved with optics of lower resolution.

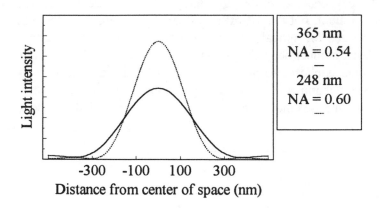

FIG. 1.19 Light intensity distribution of a 350 nm isolated space, imaged at best focus by diffraction limited optics.

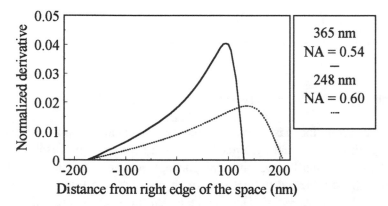

FIG. 1.20 The normalized derivative of the light intensity distributions shown in Fig. 1.19. The normalized derivative is given in units of μm^{-1}.

Consider a 350 nm feature. The normalized derivative at the edge of the space increases from 0.009 μm^{-1} for the 0.54 numerical aperture lens with 365 nm wavelength illumination to 0.018 μm^{-1} for the 0.60 numerical aperture lens at 248 nm wavelength illumination, an increase by a factor of 2 (Fig. 1.20). According to the results of the previous section, this translates into direct increases in exposure latitude and linewidth control. Such behavior has led to the proposal that resolution be defined by exposure latitude.

Of course, life is never fair, and higher resolution optics are never acquired simply to make the photolithographer's job easier. The challenge is simply shifted to smaller features, where the normalized derivative remains small. As a consequence, a related metric, with an additional normalization related to the nominal linewidth L is often used, namely the normalized log-slope (NLS) introduced in Eq. 1.16:

$$NLS = L \frac{1}{E_o} \frac{dE}{dx}. \tag{1.27}$$

From Eq. 1.16 one can see that the fractional change in linewidth is related to a fractional change in exposure dose by:

$$\frac{\Delta L}{L} = 2 \, (NLS)^{-1} \frac{\Delta E}{E}. \tag{1.28}$$

As the normalized log-slope decreases, so does the exposure latitude, i.e., the change in linewidth due to a change in exposure energy. From Eqs. 1.6 and 1.10, one can see similar improvements in resist profile slope and linewidth control subject to variations in resist thickness. The normalized log-slope is therefore a very convenient measure of the optical contribution to process latitude. However, as discussed in the next paragraph, the normalized log-slope is not determined uniquely by the parameters of the optics, such as exposure wavelength, numerical aperture, degree of partial coherence, aberrations, etc.

The value of the normalized derivative, $1/E_o$ (dE/dx), which affects the exposure latitude, resist wall profile, and linewidth control over steps, depends not only on the resolution of the optics but on the print bias as well, that is, the point at which the derivative is evaluated. The value of $1/E_o$ (dE/dx) at the mask edge is different from that at other positions. The normalized derivative is evaluated at the position x_o, which is the edge of the resist line, which may not correspond to the edge of the feature on the mask. The resist patterns may have dimensions different from the ones on the mask. If the (positive) resist lines are printed smaller than the ones on the mask, then they are termed "overexposed," and if they are larger they are called "underexposed." Resist patterns that have the same dimensions as on the mask are termed "nominal." As a result of either over- or underexposure, the quantities that depend upon the normalized derivative, such as exposure latitude, will be functions of the print bias. This is seen quite clearly in Fig. 1.20, where the best exposure latitude occurs for overexposure. As will be seen in a later section, this result is true only so long as focus is well maintained, and the situation in which the optics are significantly defocused is more complex.

At this point the astute reader will have noted that the term "resolution" has not been given a definition. Since the ability of lithography to resolve small features is a driving force in the microelectronics industry, a definition of this key quantity might appear to be in order. The use of theoretically based criteria, such as those discussed in this chapter, are useful for gaining insight into the lithography process and estimating capability, but they are not completely adequate because they fail to account for imperfect lens design and manufacture. They are also insufficient for other reasons. For example, a lithography process might be capable of exposing a feature which is 0.35 μm on the reticle down to 0.25 μm on the wafer but might be unable to image a feature that is 0.25 μm on the reticle. Depending upon one's perspective, the resolution of the lens could be considered to be either 0.35 or 0.25 μm. This results from inclusion of an additional variable, print bias. Similarly, optics have different capabilities for imaging isolated lines, isolated spaces, square contacts, or isolated islands. Should resolution be considered for one type of feature or for all features, with or without print bias? The answer to these questions is user dependent, making it impossible to assign the parameter "resolution" to particular optics, or even a particular lithography process, in a way that meets everyone's needs.

One can speak as to whether particular features are resolved, within context. For a particular circumstance, limits can be placed on linewidth, slope, resist thickness, etc. Once these limits are specified, a feature can be said to be resolved so long as it fits within the limits. However, the size and shape of resist patterns that can be considered acceptable — in the most important circumstance, integrated circuit manufacturing lines — depend upon the requirements of post-lithographic processing.

For evaluating and comparing optical systems this problem may be circumvented by choosing a particular resist process as a reference process, with preestablished requirements for linewidth variations, slope, etc. Other resist processes can be characterized as well by comparison with the reference process. Elegant definitions of resolution may be appropriate in a laboratory environment, while the masking engineer needs quantities that can be determined with the tools at hand. The resolving capabilities of a lithography exposure system are usually determined by inspecting photoresist patterns. By using such a method, the contributions from the resist process influence the patterning. It is appropriate at this point to discuss photoresist.

1.2.3 Photoresists — operational considerations

Before we can continue to describe the interplay between optics and resist, some characteristics of resist must be discussed. Resist chemistry is covered in extensive detail in Chapter 4 of this handbook. In this chapter, resist will be considered from an operational perspective. We will begin by discussing the work of F. H. Dill and coworkers,[34] which firmly established lithography as an engineering science.

Photoresist is a multicomponent material. The active ingredient in the resist, the photoactive compound, undergoes a chemical reaction upon exposure to light. Other components may absorb light but do not undergo photochemical reactions.

The intensity of light, I, passing in the z-direction through a material varies according to Lambert's law:

$$\frac{dI}{dz} = -\alpha\, I(z), \tag{1.29}$$

where α is the absorption coefficient of the material. Integrating this equation for a homogeneous medium leads to

$$I(z) = I_0\, e^{-\alpha z}, \tag{1.30}$$

where I_0 is the intensity of the light at $z = 0$. If absorption is caused by the presence of an absorbing molecule, α depends linearly on the concentration c of that molecular constituent over a wide range of circumstances:

$$\alpha = a\, c, \tag{1.31}$$

where a is the molar absorption coefficient. When a layer of photoresist is exposed to light, several things happen. First, because of the photochemical reaction in the resist, the absorption coefficient will change:

$$\alpha = a_{PAC}\, c + a_P(c_0 - c) + \Sigma a_R c_R, \tag{1.32}$$

where a_{PAC} is the molar absorption coefficient of the photoactive compound, a_P is the molar absorption coefficient for the material that results from the photochemical reaction of the photoactive compound, and the a_R are the molar absorption coefficients for all of the other materials in the photoresist, which are assumed to be unchanged upon exposure to light. The initial concentration of the photoactive compound is represented by c_0, while c_R represents the concentrations of all other constituents. In the paper of Dill et al., Eq. 1.32 was written in the form of:

$$\alpha = AM + B, \tag{1.33}$$

where
$$A = (a_{PAC} - a_P)\, c_0 \tag{1.34}$$

$$B = \Sigma a_R c_R + a_P c_0 \tag{1.35}$$

$$M = c/c_0. \tag{1.36}$$

For unexposed resist, $M = 1$, while $M = 0$ for completely bleached resist. A and B are easily measurable for a given photoresist. A represents the optical absorption that changes with exposure, while B represents the absorption of the components of the resist that do not change their absorption upon exposure to light.

The model is complete with the addition of one final equation. Not all photons absorbed by the photoactive compound will result in a chemical reaction. This quantum efficiency is taken into account by the coefficient C:

$$\frac{\partial M}{\partial t} = -CIM, \tag{1.37}$$

where t represents time. The parameters A, B, and C have become known as the Dill parameters for the resist. Typical values are given in Table 1.3.

Once a film of resist is exposed to light, the absorption is no longer uniform throughout the film, because light near the top of the resist will be more bleached, having received more light than near the bottom. Although the determination of resist exposure throughout the depth of the resist film is not amenable to a closed form solution, the above equations can be used to calculate $M(x,y,z,t)$ using a computer. Resist profiles can be calculated if the development rate R is known as a function of M, and the rate is usually measured directly for particular resist and developer systems.

Table 1.3 Dill parameters for some commercially available photoresists[35,36,37].

Resist	A (μm^{-1})	B (μm^{-1})	C (cm^2/mJ)	Exposure wavelength	Reference
Shipley 511-A	0.85	0.04	0.016	i-line	35
AZ 1470	0.56	0.03	0.010	g-line	36
AZ 1470	0.88	0.22	0.010	i-line	36
TOK IP-3400	0.75	0.11	0.016	i-line	37
Apex-E	-0.01	0.36	0.012	248 nm	35

The first widely available program for calculating resist profiles was SAMPLE[38], developed by various researchers at the University of California, Berkeley. Since the introduction of SAMPLE, numerous software packages have become available. With tools for calculating and predicting behavior, the basic criteria of a science are met, and lithography has progressed rapidly since the advent of predictive computer modeling. Additional details on modeling are contained in Chapter 7 of this handbook.

Chris Mack[39] made a very interesting observation by integrating Eq. 1.12 with respect to time, giving

$$M = e^{-CIt} \tag{1.38}$$

and then differentiating this with respect to x. This resulted in the following:

$$\frac{\partial M}{\partial x} = M \ln(M) \, \frac{1}{I} \frac{\partial I}{\partial x}. \tag{1.39}$$

This expression represents the variation in the photoactive compound following exposure, in the limit of thin resist. The modulation in x results from the pattern on the photomask. What is interesting about this equation is that the factor $M\ln(M)$ has a maximum, when $M = 0.37$. A process that is designed to have line edges coincident with this maximum has an enhanced opportunity for improved performance.

In the previous section, resist properties were incorporated into a simple model through a single parameter, γ, the resist contrast. It is useful to revisit contrast and explore some of its properties. Consider once again the situation where a resist film is exposed by uniform illumination and then developed. The development rate, R, is directly related to the time rate of change of the resist thickness T:

$$R = -\frac{dT}{dt} \tag{1.40}$$

With T as the resist thickness remaining after development,

$$T_0 - T = T_0 - \int_0^t R(M)dt' \tag{1.41}$$

where T_0 is the initial resist thickness. By definition, γ is given by:

$$\gamma = \frac{1}{T_0}\frac{\partial T}{\partial(\ln \varepsilon)} \ @ \ \varepsilon = 1 \tag{1.42}$$

where $\varepsilon = \frac{E}{E_0}$, E is the exposure dose, and E_0 is the minimum dose to clear the resist. This is the same parameter discussed in section 1.2.1. From Eqs. 1.41 and 1.42,

$$\gamma = -\frac{\partial}{T_0\partial(\ln \varepsilon)}\int_0^t R(M)dt' \ @ \ \varepsilon = 1 \tag{1.43}$$

$$= -\frac{1}{T_0}\int_0^t \frac{\partial R}{\partial M}\frac{\partial M}{\partial \varepsilon}dt' \ @ \ \varepsilon = 1 \ . \tag{1.44}$$

This is an interesting expression. The integrand is divided into two factors. The first contains the development properties of the resist and the second factor is determined by the exposure. For example, a dyed resist would be more absorbing, reducing the change in M near the resist-substrate interface during exposure, thereby reducing γ. This also explains the observation that measured contrast shows variations with resist thickness[40] rather than being an intrinsic property of the resist chemistry.

Alternative definitions for contrast have been proposed. An expression similar to Eq. 1.44 has been derived:[41,42,43]

$$\gamma = \frac{R(T_o)}{T_o} \int_0^{T_o} \left(\frac{\partial \ln R}{\partial \ln E} \right) \frac{dz}{R(z)}, \tag{1.45}$$

where z is the depth in the resist film. (Again, technical rigor has been less than complete, with the taking of logarithms of non-dimensionless quantities.) The expression

$$\frac{\partial \ln R}{\partial \ln E} = \gamma_{th} \tag{1.46}$$

for the "theoretical contrast" has gained acceptance as a measure of resist performance[44,45]. Although there are difficulties associated with using this metric, such as its dependence on exposure dose, it has been found to be a good indicator of how well resists will perform[46]. Eq. 1.46 has several advantages over the earlier expression for contrast, introduced in Section 1.2.6. It is dependent upon the dissolution properties of the photoresist and does not have the dependencies on resist film thickness found in the earlier definitions.

1.2.4 Thin film optics

If the substrate were to have optical properties identical to photoresist, the light intensity would decrease gradually from the top of the resist to the bottom as a consequence of optical absorption. This decrease follows a simple exponential, at least at the beginning of exposure when the resist had homogeneous optical properties:

$$I(z) = I_o \exp(-\alpha z), \tag{1.47}$$

where I_o is the intensity of the incident light, α is the optical absorption coefficient, and z is the depth within the resist film, with $z = 0$ being the top surface. Typical values for α can be determined from the A and B parameters given in Table 1.3, along with the relationship of Eq. 1.33. From these values one can see that the spatial scale for variation of the light intensity, due to conventional absorption of light propagating through an absorbing medium, is tenths of a micron.

The phenomenon of resist bleaching during exposure does not change this behavior significantly. In the limit where $B \rightarrow 0$, Eqs. 1.29, 1.33, and 1.37 can be solved exactly to give the light intensity and amount of remaining photoactive compound at depth z in the resist, after t seconds of exposure:

$$I(z, t) = \frac{I_0}{\left[1 - e^{-CI_0 t}(1 - e^{Az}) \right]} \tag{1.48}$$

$$M(z, t) = \frac{1}{\left[1 - e^{-Az}(1 - e^{CI_0 t}) \right]}, \tag{1.49}$$

where I_o is the intensity of the incident light. At $t = 0$, Eq. 1.48 reduces to Eq. 1.30. For the case of $B \neq 0$, $\alpha(z)$ must be determined iteratively on a computer[47].

The presence of the substrate underneath the photoresist has a significant effect on the light intensity distribution within the photoresist film, compared to the picture just presented. This can be appreciated by considering the situation depicted in Fig. 1.21. Light uniformly illuminates a substrate covered by photoresist. Typical substrates are silicon covered with various films: silicon dioxide, silicon nitride, aluminum, various silicides, titanium nitride, etc. In this example the films are uniform and have large spatial extent. At the interface between each pair of films some of the incident light is reflected, while the remainder is transmitted.

Most important, within the photoresist film the light consists of an incident component and a reflected one. There are two significant consequences of this geometry:

1) The light intensity varies rapidly, in the vertical direction, within the photoresist film;

2) The amount of light energy coupled into the photoresist film has a strong dependence on the thickness of the various films in the stack.

The first property, the rapid variation of the light intensity within the photoresist film, results from the interference between the incident and reflected light within the resist. The variation in light intensity is sinusoidal, where the spacing between adjacent maxima and minima are separated by very nearly $\lambda/4n$, where n is the real part of the refractive index of the photoresist. For typical values of λ and n, this quarter-wave separation is on the order of hundredths of a micron, a full order of magnitude smaller than one predicts from simple optical absorption. The effect of this is shown in Fig. 1.22, where the light intensity is plotted as a function of depth in the photoresist film for two situations, one for a silicon substrate and the other where the substrate is matched optically to the photoresist. The rapidly varying light distribution within the depth of the photoresist is referred to as a "standing wave."

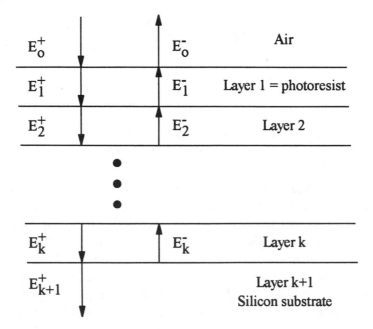

FIG. 1.21 Configuration of the thin film optical problem.

The consequence of standing waves of light intensity throughout the depth of the photoresist film is having alternating levels of resist with high and low exposure. For positive resist, the high exposure regions will develop quickly, while the low exposure regions will develop more gradually. Standing waves are visible in micrographs of resist features, where the resist sidewall has a ridged appearance because of the alternating layers of resist that have developed at different rates. (Fig. 1.23).

Standing waves constrain the process, because there must be sufficient exposure for the resist in the least-exposed standing wave to still develop out. For highly reflective films, such as aluminum, this dose may be several times greater than the resist might require in the absence of standing wave effects. This exaggerates the problems of substrates with topography, where regions requiring low doses will be over-exposed in the attempt to clear out the last standing waves in other areas on the wafers.

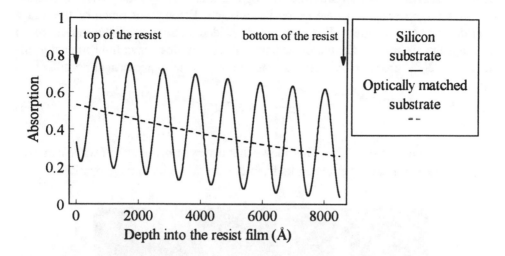

FIG. 1.22 Light intensity (λ = 365 nm) throughout the depth of an 8500 Å thick film of photoresist. The calculation of absorption was performed for the resist at the initiation of exposure, i.e., the resist was unbleached and had uniform optical properties.

The most common method to reduce the effect of standing waves is the post-exposure bake[8,48,49]. Following exposure, the photoactive compound varies in concentration in direct proportion to the light intensity. In photoresists where the photoactive compound is not bonded to the resin, diffusion of the photoactive compound may be induced by baking the resist. Diffusion will take place in the direction of the largest concentration gradient[50,51] which is typically between standing wave maxima and minimum, because of the short distances ($\lambda/4n$) over which the standing waves occur. Because standing waves can severely degrade lithographic performance, most resists today are designed to be processed with a post-exposure bake.

One might think that this diffusion will also degrade the resist profile, because photoactive compound will also diffuse laterally between exposed and unexposed regions. However, the lateral gradient is usually much smaller than the one across standing waves. One can estimate the point at which image degradation occurs due to lateral diffusion as follows. For image degradation the horizontal gradients of the photo-active compound must be less than the gradients across standing waves. As a first approximation, these gradients can be estimated from the gradients in the light intensity. At the edge of a feature, such as a line or space, the gradient of the light is

$$\frac{1}{I_0}\frac{dI}{dx} \approx 1 \text{ to } 2 \ \mu m^{-1}. \tag{1.50}$$

Here the derivative is normalized to the light intensity that would exist in the middle of a large transparent feature on the reticle. Throughout much of the image this derivative is smaller. For example, it is identically zero in the middle of an isolated line or space. On the other hand, if the standing wave light intensity differs by 20% from maximum to minimum, then the vertical gradient is approximately

$$\frac{0.2}{(\lambda/4n)} = 3.7 \ \mu m^{-1} \qquad (1.51)$$

for $\lambda = 365$ nm and $n = 1.7$. This shows that a very small standing wave can give a larger vertical gradient than results from the lateral optical image profile.

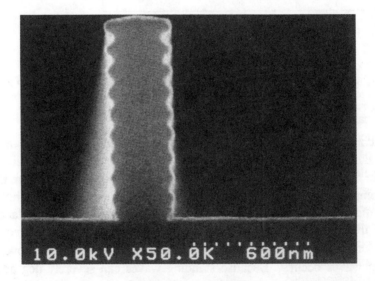

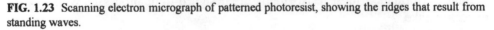

FIG. 1.23 Scanning electron micrograph of patterned photoresist, showing the ridges that result from standing waves.

The efficiency of the post-exposure bake depends upon the diffusion coefficient of the photoactive compound, which depends upon many variables: the size of the photoactive compound, the resin, the amount of residual solvent, etc.[52] This is particularly apparent for post-exposure bake temperatures in the neighborhood of the resist's glass transition temperature[53]. Because the density of the resin and the amount of residual solvent are affected by the softbake of the resist, there is often a significant interaction between the softbake and the post-exposure bake processes.[52,54]

In addition to the standing wave throughout the depth of the photoresist film, there is another consequence of thin film interference effects: the amount of light absorbed in the resist has a strong functional dependence on the thickness of the substrate and resist films. Consider the total light energy in the photoresist, integrated through the depth of the photoresist film:

$$E_{total} = \int_0^{T_o} E(z)dz \qquad (1.52)$$

This quantity is plotted in Fig. 1.24 as a function of resist thickness for resist on an oxide layer on silicon. For 1.0 μm thick resist, a change in the oxide thickness from 2500 Å to 2800 Å resulted in a 14% decrease in the integrated light absorption. This situation has been simulated, and the corresponding linewidths are also shown. Again there are oscillations, with the $\lambda/4n$ minimum-to-maximum spacing characteristic of standing wave phenomena, with linewidth minima corresponding to light absorption maxima.

While post-exposure bakes can reduce the variations of $M(z)$ within the resist film, they have no effect on the integrated light energy (Eq. 1.52). The curves shown in Fig. 1.24 are known as "swing curves." The phenomena exemplified in Fig. 1.24 have significant consequences for microlithography. Because linewidths vary with resist thickness variations on the order of a quarter wave, resist thickness must be controlled to levels much less than $\lambda/4n$, which is less than 100 nm for typical exposure wavelengths and photoresist. This can be an impossible task for process architectures that can have virtually any topography. The standing wave phenomenon places severe requirements on control of thin film thickness, as seen from Fig. 1.24, where a small change in the thickness of the oxide layer changes the amount of integrated light energy absorbed and the linewidth as a consequence. This constraint on substrate films is particularly severe for materials that have a large index of refraction, such as polysilicon. For the Hg lamp i-line, the index of refraction is 4.9, so the quarter wave maximum-to-minimum spacing is only 15 nm for a polysilicon film. It may not be practical to control films to tolerances that are a fraction of that.

There are a number of solutions that are used to address the problem of varying integrated light intensity caused by the standing wave effect:

1) Centering the resist thickness at a standing wave extremum.
2) Dyed resists.
3) Bottom antireflection coatings.
4) Top antireflection coatings.
5) Use of multiple wavelength light sources.
6) Multilayer resist processes.
7) Surface imaging resists.

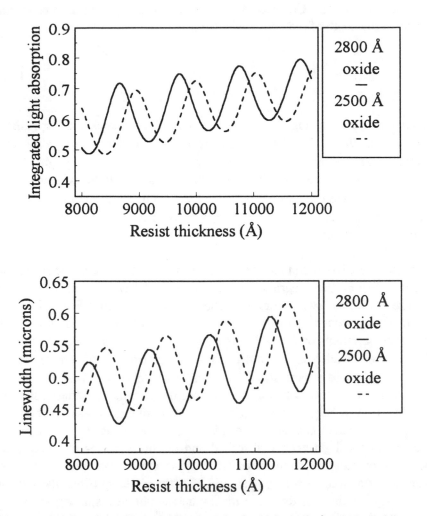

FIG. 1.24 Standing wave effects as a function of resist thickness for λ = 365 nm (i-line) at the initiation of exposure. The top curves give the integrated light absorption (Eq. 1.52), while the lower curves are linewidths for 0.5 μm nominal features exposed at a fixed exposure dose, calculated using PROLITH.

One clearly has the minimal sensitivity to resist thickness variations when the process is centered at an extremum of the curves shown in Fig. 1.24. On flat surfaces, choosing the resist thickness to correspond to a swing curve maximum or minimum results in the smallest linewidth variations due to variations in resist thickness. When the topography heights are a significant fraction of the quarter wavelength, or greater, operating at a swing curve extremum is not an option for immunizing the process against standing wave effects.

Operating at a minimum of a swing curve (Fig. 1.24) has certain advantages. Swing curves such as Fig. 1.24 are usually dominated by light rays at near normal incidence. By choosing a minimum, oblique rays are more strongly coupled into the resist, since they travel further through the resist film and represent higher points on the swing curve. These oblique rays contain the higher spatial frequencies of the image. Other differences between swing curve maxima and minima can be appreciated by examining Figure 1.25, where the light intensity is plotted as a function of depth in resist films corresponding to a swing curve absorption maximum (resist thickness = 8900 Å for the parameters used to calculate the results shown in Figs. 1.22 and 1.24) and a swing curve minimum (resist thickness = 8365 Å). At a swing curve maximum, the light intensity coupled into the resist is greatest, thereby minimizing exposure time. For a swing curve minimum on a silicon substrate, the average dose is similar to that obtained on an optically matched substrate. The amplitudes of the intensity variations through the depth of the resist film, relative to the average intensity, are fairly similar. In both cases, there are absorption minima at the resist-substrate interface.

The calculations of light intensity shown in Figures 1.22 and 1.24 were for resist that was just beginning to be exposed. The standing wave variations are less for the unexposed resist, as compared to the completely bleached situation. This effect motivated people to intentionally add non-bleaching dyes into the resist as a means of reducing standing wave effects[55]. There are other reasons to use dyed resists, in addition to reducing standing wave effects, as will be discussed shortly. However, adding a dye to the resist reduces its performance in terms of its ability to produce vertical sidewalls and form small features, which compromises the utility of this approach. Numerous references of the early work on dyed resist can be found in Reference 56.

The best way to avoid standing wave effects is to have a non-reflecting substrate. If the process architecture requires the use of reflective materials, such as metals for electrical interconnections, the substrate can be coated with an anti-reflection coating (ARC) to suppress reflections. Commonly used anti-reflection coatings are amorphous silicon on aluminum[57, 58] for g-line lithography, titanium nitride for i-line[59, 60], silicon nitride and silicon oxynitride for i-line and DUV[61], and various spin-coated organic materials[62-64].

Consider again the situation where a resist film is coated over a silicon dioxide film on a silicon surface and is exposed to uniform illumination (Fig. 1.24). As one can see in Fig. 1.24, the amplitude of the intensity variations changes little with oxide thickness; only the phase is shifted. In this case, silicon dioxide is unsuitable as an anti-reflection coating. This is a general property when substrate films have large indices of refraction, particularly with large imaginary components: Anti-reflection coatings must absorb light.

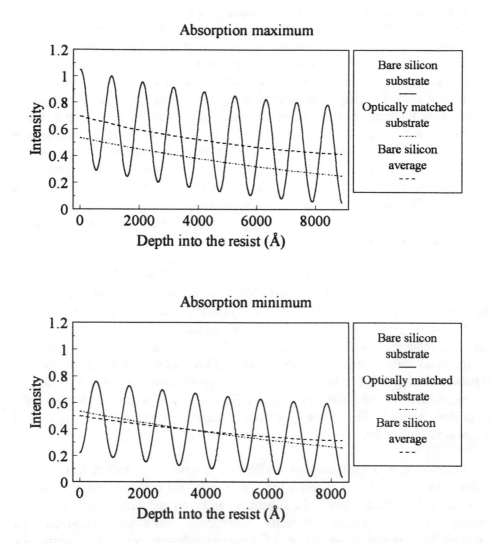

FIG. 1.25. Light intensity through the depth of resist films representing swing curve absorption minima and maxima.

The requirement that anti-reflection coatings must absorb light results from two factors:

1) The anti-reflection coating must function over a range of resist thicknesses, since variations in resist thickness result from circuit topography as well as variations in the process.

2) The refractive indices of metallic substrates have large imaginary components, which introduces another condition in the equation which must be satisfied by anti-reflection coatings.

Consider the situation in Fig. 1.21, where layer 2 is an anti-reflection coating and layer 3 is a metal. Then the reflected light is given by:

$$\frac{\left|E_1^-\right|}{\left|E_1^+\right|} = \frac{\left|\rho_{1,2} + \rho_{2,3}\exp\left(-i4\pi n_2 d_2/\lambda\right)\right|}{\left|1 + \rho_{1,2}\rho_{2,3}\exp\left(-i4\pi n_2 d_2/\lambda\right)\right|}, \tag{1.53}$$

where

$$\rho_{i,j} = \frac{n_i - n_j}{n_i + n_j}. \tag{1.54}$$

The condition for an anti-reflection coating is therefore

$$\rho_{1,2} + \rho_{2,3}\exp\left(-i4\pi n_2 d_2/\lambda\right) = 0. \tag{1.55}$$

When all refractive indices are real, the above equation leads to the well-known conditions for an anti-reflection coating:

$$n_2 = \sqrt{n_1 n_3} \tag{1.56}$$

$$d_2 = \frac{m\lambda}{4n_2}, \tag{1.57}$$

where m is any positive integer. These equations are used for anti-reflection coatings on refractive optical elements. When the refractive index of layer 3 has a large imaginary component, Eq. 1.55 becomes two equations, one for the real part and one for the imaginary part. This necessitates another factor that can be varied, such as the imaginary component in the refractive index of layer 2, the anti-reflection coating.

If a material is too optically absorbing, it will become a reflective surface. Consequently, there is an optimal level of absorption for anti-reflection coatings. What this optimum is can be estimated from the following considerations. Thin films (200 - 1200 Å) are desirable in order to maintain process simplicity during the steps that remove the anti-reflection coating. For a thin film to attenuate normally incident light on two passes (incident and upon reflection) through the film (Fig. 1.21) to 10% or less, the following condition must be met:

$$\rho\, e^{-2t\frac{4\pi\kappa}{\lambda}} \le 0.1, \tag{1.58}$$

where t is the thickness of the film, κ is the imaginary part of the refractive index, ρ is the reflectance between the underlying substrate and the anti-reflection coating, and λ is the wavelength of the light. For $\rho = 0.7$, $t = 700$ Å, and $\lambda = 365$ nm (i-line) this implies that $\kappa > 0.4$. On the other hand, if κ becomes too large, then the thin film itself will become too reflective. The reflectance from a semi-infinite thick layer of material is plotted in Fig. 1.26. From Fig. 1.26 and Eq. 1.58 one obtains

$$0.4 \leq \kappa \leq 0.7 \text{ to } 1.2 \tag{1.59}$$

in order to maintain reflectance back into the photoresist below 10%. The upper bound on κ depends upon the degree to which the real parts of the resist and anti-reflection coating are matched. Also note that the lower bound scales with the wavelength, according to Eq. 1.58. Not surprisingly, most materials in use as anti-reflection coatings have values of κ that fall in this range. For TiN, a good inorganic anti-reflection coating for i-line lithography, $n = 2.01 - 1.11i$, while $n = 1.90 - 0.41i$ for BARLi, an organic i-line ARC.

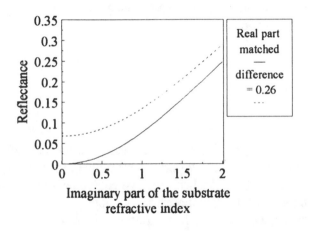

FIG. 1.26 Reflectance from a semi-infinite absorbing substrate, calculated as the square of Eq. 1.54. The real part of the resist's refractive index was $n_{real} = 1.74$.

Bottom anti-reflection coatings address all of the problems associated with reflective substrates: linewidth variations over topography (assuming the process results in adequate coverage of the anti-reflection coating over they topography), standing waves within the resist film, and the problem of notching, which will be discussed shortly. However, there are disadvantages as well. First, the anti-reflection coating must be deposited or coated; it is an additional processing step that adds costs and potential for defects. (It may be argued that the deposition does not represent an additional step, because an adhesion promotion step is eliminated by the use of a spin-coated organic ARC. However, the ARC materials and equipment are usually much more expensive than those required for conventional adhesion promotion.) Second, it must be etched. Some spin-coated ARCs will develop out[65], but the develop step depends critically on bake temperatures and can be difficult when there is topography and accompanying variations in ARC thickness. Developing is not anisotropic, and the resulting undercut limits the use of ARCs for extremely small geometries. In general, the ARC must be anisotropically etched out of the exposed areas. Finally, the ARC usually must be removed from the unetched areas, since it is rarely a film that is part of the process architecture, other than to assist the lithography. (TiN on aluminum is a notable exception to this, where TiN is used for hillock suppression[66] and as a barrier metal.) Organic ARCs can usually be removed in the same process used to remove remaining photoresist, but inorganic ARCs require a special etch, which must be compatible with the other exposed films. In spite of the expense of additional processing, the advantages for lithography are so great that anti-reflection coatings can often be justified.

There is at least one disadvantage to a non-reflective substrate. This can be appreciated from Fig. 1.25. Calculated in this figure are the light intensity averages connecting the midpoints between swing curve minima and maxima. These average curves represent, to a degree, the effective exposure following a post-exposure bake. Comparing the average curves to the light absorption for resist on a non-reflecting substrate, one sees that reflected light does give some advantage in offsetting the bulk light absorption effect. This effect is strongest for absorbing resists, where the standing wave effect is already somewhat suppressed.

Finally, a top anti-reflection coating[67,68] can be used by making use of an approach that is very common in optics. Consider the situation where resist directly covers an aluminum substrate. In such a circumstance very little light is transmitted into the substrate, and the reflectance is nearly the complement of the absorption (Fig. 1.27), which is the quantity relevant to lithography. The reflectance is given by the expression[69]

$$R = \frac{\rho_1^2 + 2\rho_1\rho_2\cos 2\delta + \rho_2^2}{1 + 2\rho_1\rho_2\cos 2\delta + \rho_1^2\rho_2^2} , \qquad (1.60)$$

where ρ_1 is the reflectivity between the air and the resist, ρ_2 is the reflectivity between the resist and aluminum and

$$\delta = \frac{2\pi}{\lambda} n\, d \qquad (1.61)$$

for normal incidence. The thickness and refractive index of the resist are d and n, respectively.

The oscillatory behavior of the reflectance (and absorption) results from the factor $2\rho_1\rho_2\cos 2\delta$. This factor can be suppressed by a bottom anti-reflection coating

$(\rho_2 \to 0)$ or a top anti-reflection coating $(\rho_1 \to 0)$. A top anti-reflection coating is a film that has the following properties[18]:

$$\text{refractive index} = \sqrt{n} \qquad (1.62)$$

$$d = \frac{(k+1)\lambda}{4n}, \qquad (1.63)$$

where k is a non-negative integer. Top anti-reflection coatings address situations where variations in the absorption of light in the resist is modulated by changes in the thickness of non-absorbing or moderately absorbing films. The resist itself can be one of these films. Because top anti-reflection coatings do not eliminate reflection from the substrate, they are less effective for highly reflective substrates and situations where the reflectivity of the substrate is highly variable, as occurs when metal interconnects have already been formed, surrounded by insulating oxides.

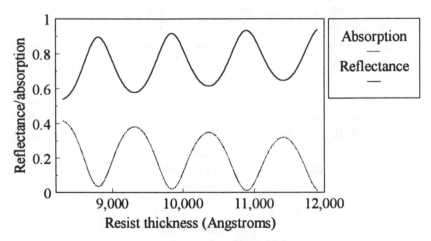

FIG. 1.27 Absorption and reflectance for i-line resist on aluminum (at the beginning of exposure). Note that absorption and reflectance are complementary.

One way to minimize the impact of the standing wave effect is to use a source with multiple wavelengths. Minima from one wavelength can be canceled out by maxima from another. This works well in terms of reducing variations in the integrated intensity, but a standing wave still exists at the interface between resist and a metallic substrate, as shall now be explained. Reflected light from a highly reflective surface maintains most of the magnitude of the incident light but undergoes a phase reversal. (In Eq. 1.54, $\rho_{i,j} \to -1$, as $n_j \gg n_i$, where n_j is the refractive index of the reflective layer and n_i is the index of refraction of the photoresist.) For a perfectly reflective surface 100% of the light is reflected and the phase reversal is 180°, while actual surfaces, which are less than perfect reflectors, will have magnitudes and phases according to the degree to which the substrate approaches a perfect reflector. All wavelengths of interest to optical lithography are reflected from metals, silicon, and many silicides, so all wavelengths undergo a phase reversal at the resist-reflector interface. For distances on the order of several quarter waves, most wavelengths are still relatively in phase, so the standing waves interfere constructively near the resist-reflector interface. Further away from the interface, the standing waves for different wavelengths will go out of phase.

One of the most egregious problems that results from highly reflective substrates is that of notching. The problem occurs when there is topography, as depicted in Fig. 1.28. Light is reflected from edges and slopes into regions that are intended to be unexposed. This results in notching of resist lines, as shown in Fig. 1.29. It is useful to calculate the distance over which scattered light can be effective in producing notches. Consider the characteristic curve shown in Fig. 1.5. Note that there is negligible resist loss for exposure doses less than some fraction f of E_o. Values for f are larger for higher contrast resist, with typical values of 0.7. Wafers typically receive doses E_e that are about twice E_o. Notching does not occur as long as the scattered dose E_s meets the following condition:

$$E_s \leq f\,E_o \leq 0.7\,E_o = 0.35\,E_e. \tag{1.64}$$

If the reflectance from the substrate is ρ, then the scattered dose at a distance d from a step is given by:

$$E_s = E_e\,\rho\exp(-\alpha(d+t)), \tag{1.65}$$

where α is the absorption coefficient for the resist and t is the resist thickness. From the above two equations, there will be no notching so long as

$$\rho\exp(-\alpha(d+t)) \leq 0.35. \tag{1.66}$$

One obvious way to reduce the distance d over which notching may occur is to increase the absorption α. This can be done by adding a dye into the resist that does not bleach with exposure, and such an approach has been used extensively[70]. The benefit of having dye in the resist was already discussed as a method for reducing the effects of standing waves.

Values for ρ can be as high as 0.9 for specular aluminum, and a typical value for α is 0.6 μm^{-1}. For these parameters, no notching occurs for $d+t \geq 1.5\,\mu m$. From these considerations, one can see that the utility of dyed resists is diminishing for deep sub-micron lithography. Moreover, as discussed in the section on photoresists, optical absorption reduces the performance of the resist.

Side view   Top view

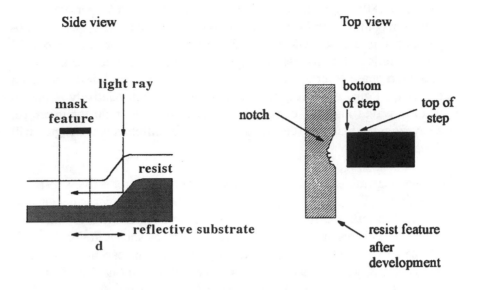

FIG. 1.28 Situation for reflective notching. Light reflects from substrate topography into areas in which exposure is not desired.

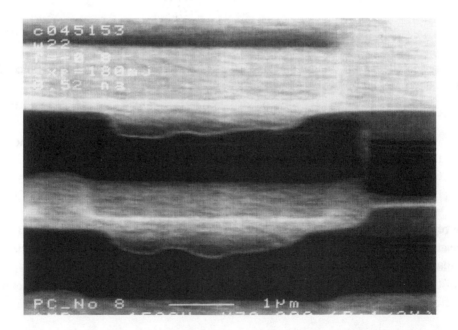

FIG. 1.29 Example of reflective notching, for i-line resist on polysilicon. In the area of the topography the resist is notched. Standing waves resulting from substrate reflectivity are also evident.

1.2.5 Focus

In earlier sections, the light intensity distribution was considered only in the plane of best focus. From common experience, one knows that defocus reduces the clarity of optical images. The problem is particularly acute in optical lithography, where the depth of focus — the range over which there are clear optical images — is very small, to the extent that it becomes of concern as to whether optical wafer steppers are capable of maintaining the image in focus. The calculation of the intensity distribution for the image of a point source of light, discussed earlier (Fig. 1.11), has been extended to the situation in which a lens is used to image the point source of light. In this case, the circular aperture shown in Fig. 1.10 holds a diffraction-limited lens. The resulting calculated light intensity profiles in different focal planes are shown in Fig. 1.30. With defocus, the peak intensity diminishes and more light is diffracted away from the center spot.

Using the criteria that the peak intensity should not decrease by more than 20%, the following expression results for a depth of focus (DOF):

$$DOF = \pm 0.5 \frac{\lambda}{(NA)^2}.$$ (1.67)

Over this range of focus, the peak intensity remains within 20% of the peak value for best focus. This expression is usually referred to as the Rayleigh depth of focus; one Rayleigh unit of defocus is $0.5\lambda/NA^2$. Just as the Rayleigh criterion for resolution was generalized, Eq. 1.67 is often written in the form

$$DOF = k_2 \frac{\lambda}{(NA)^2}.$$ (1.68)

Eq. 1.68 is expressed as a total range, instead of distances from a mid-point, because the symmetry of the aerial image is broken by the photoresist or optical aberrations in actual application, and the depth of focus is typically asymmetric in high resolution optical lithography. This asymmetry will be discussed later in this section. The Rayleigh expressions for resolution and focus both depend upon only two parameters, the wavelength and the numerical aperture, and a pair of constants. One can then relate resolution R and depth of focus:

$$DOF = (constant) \frac{R^2}{\lambda}.$$ (1.69)

Within the framework of this analysis one can seen that the depth of focus will diminish as one attempts to print smaller features (smaller R). At one time, improper interpretations of the Rayleigh criteria led to the conclusion that optical lithography would not be able to break the 1.0 μm resolution barrier, when in fact photolithography has produced working large scale devices with 250 nm minimum features and is expected to continue to be the dominant method of patterning feature design rules down to 130 nm.   Just as the Rayleigh criterion was insufficient for defining the resolution of optics, in the context of photolithography, there are additional considerations to be examined for understanding focus and the depth of focus.

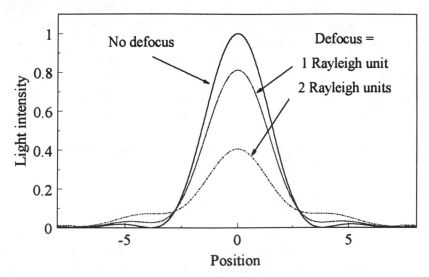

FIG. 1.30. Light intensity profile of a point source of light imaged by a circular diffraction-limited lens, at different planes of focus. The horizontal axis is given in units of $2\pi NA/\lambda$. One Rayleigh unit of defocus is $0.5\lambda/NA^2$.

The Rayleigh criteria might lead one to a false conclusion that the depth of focus will diminish whenever one attempts to improve the resolution of the optics. That this is not true can be appreciated from the following argument. Suppose one has a lens that has a resolution of 1.0 μm, i.e., it will not image features smaller than this. One could say that such a lens has zero depth of focus for submicron features. A higher resolution lens capable of submicron resolution will therefore have greater depth of focus for such smaller features. This line of reasoning led to the observation that there is an optimum numerical aperture that is a function of feature size[5,71 - 75].

The DOF of the lens depends on the feature size and type being imaged. As the NA is increased, the limiting resolution becomes finer but the depth of focus for larger features becomes smaller. The optimum NA and partial coherence values depend on the size and type of geometry being imaged, as well as the type of illumination and degree of coherence. For example, the optimum NA to image a contact hole is higher than that needed for a grating of equal lines and spaces of the same size on the reticle, and greater depth of focus is achieved with higher values of σ. As will be discussed, the dependence of imaging on feature size, shape and surroundings have led to wafer steppers with variable parameters, such as numerical aperture. Because resolution is not useful without adequate depth of focus, the concept of practical resolution[76,77] - the resolution achieved with some specified depth of focus — is useful for defining the optimum NA, σ, and other process parameters. For the situation encountered in lithography, the effects of defocus can be observed readily. Shown in Fig 1.31 are cross sections of lines and spaces of resist as focus is varied. As can be seen, the sidewall slope increases for large defocus. As shown in Fig. 1.32, linewidth also varies with focus, and the sensitivity of linewidth to defocus is a function of the exposure dose and print bias[78].

Typical specified depths of focus for various wafer steppers are given in Table 1.4. The amount of defocus that results in significant changes in either slope or linewidth should be noted: it is ranges between 1.5 and 0.7 µm, and is decreasing over time. This should be compared to the amount of focus variation that can be induced within a process (Table 1.5). All parameters in Table 1.5, except for the metrology figure[79], refer to variations across an exposure field. In the following discussion, it will be shown that the relevant value for circuit topography is the actual height of circuit features divided by the index of refraction of the photoresist.

Table 1.4 Typical depths of focus for various wafer steppers.

Stepper	Resolution	Depth of focus	NA	Wavelength	Year
GCA 4800	1.25 µm	1.5 µm	0.28	g-line	1980
Nikon	1.0 µm	1.5 µm	0.35	g-line	1984
Canon	0.8 µm	1.2 µm	0.43	g-line	1987
ASML 5000/50	0.5 µm	1.0 µm	0.48	i-line	1990
Micrascan 2+	0.30 µm	0.8 µm	0.50	DUV	1995
ASML 5500/300	0.25 µm	0.7 µm	0.57	DUV	1996

The actual focal range over which imaging must be good is typically somewhere between the sum total of the parameters in Table 1.5 and the root-sum-of-squares (RSS) total, since some of the components tend to act randomly (wafer non-flatness, stepper focus control), while others are systematic components (circuit topography). As one can see, the values for these parameters are comparable to the depth of focus for high resolution optics.

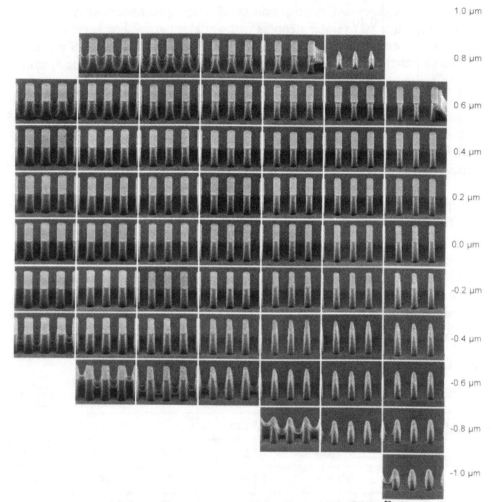

FIG. 1.31 Resist profiles for imaging at best focus and outside the depth of focus[80].

One common misconception is that optical lithography will be limited by diffraction when the depth of focus becomes less than the thickness of the photoresist. Such a misunderstanding results from not using a rigorous definition of the depth of focus. Just as for resolution, the depth of focus will depend upon the resist system as well as the optics, and non-lithography considerations will also determine what types of resist profiles are acceptable. The definition of the depth of focus begins with the specification of the system: optics and the resist process. The latter will include the thickness of the resist. The depth of focus is the range over which the focal plane can be moved in a direction parallel to the optical axis and have the resist profiles and linewidths remain within specifications.

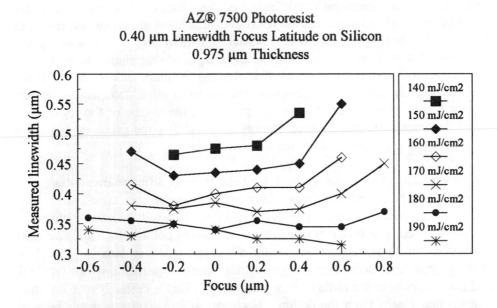

AZ® 7500 Photoresist
0.40 μm Linewidth Focus Latitude on Silicon
0.975 μm Thickness

FIG. 1.32 Linewidth as a function of focus for dense lines[80]. The process conditions were: softbake and post-exposure bake at 110° C for 60 seconds, puddle development for 52 seconds in AZ® 300 MIF developer at 22° C, exposed on a Nikon 0.54 NA i-line wafer stepper. Curves such as these are referred to as Bossung curves[78].

Table 1.5 Parameters in a typical focus budget.

Parameter	Typical values (range, in μm)
Chuck non-flatness	0.10
Wafer non-flatness	0.20
Stepper focus control	0.20
Circuit topography	0.20
Metrology for determining focus	0.10
Total (sum)	0.80
Total (RSS)	0.37

Parameters that are typically considered for characterizing the resist images are linewidth, slope of the resist profile and resist loss from "unexposed" areas. From such a definition it is clear that it is quite possible to speak of having a process with a 1.0 μm thick resist layer and a 0.5 μm depth of focus. The 0.5 μm is the range over which the 1.0 μm layer of resist is moved, and images printed in that resist film remain within specifications.

There is not a consensus on the definition of "best focus." For optics where the Rayleigh depth of focus exceeds the resist thickness, most effects of defocus, such as linewidth and slope variation, behave symmetrically about some plane. Moreover, the smallest features are typically imaged at that plane. In such a circumstance, it is reasonable to define that plane as the plane of best focus. However, for high numerical aperture optics, asymmetry is introduced, and the choice of best focus becomes less obvious. In particular, the middle of the region over which linewidths are within specifications is no longer the place at which the smallest features are resolved nor the slope maximized.

At a given point in an exposure field, long objects with different orientations may have different best depths of focus. This results from the aberration called astigmatism. Astigmatism is quantified as the difference between the best depths of focus for geometries of different orientations, at a given point in the exposure field. For a cylindrically symmetric lens used in a step-and-repeat system, one would expect astigmatism to occur between geometries parallel to a radius vector (sagittal) and perpendicular to the radius vector (tangential). This is certainly true for astigmatism that results from the design. However, astigmatism can occur between geometries of any orientation in any part of the field, because of lens manufacturing imperfections. Astigmatism has been observed in the center of the field, where none would be possible for a truly cylindrically symmetric system. An example of astigmatism was shown in Fig. 1.14.

Another aberration that can result in different best depths of focus for different geometries is field curvature. As a result of this aberration, the surface of best focus is a section of a sphere rather than a plane, and focus is not constant throughout an exposure field. The range of best focus positions for all geometries over the entire exposure field is called the field flatness and includes both astigmatism and field curvature.

The depth of focus of a given feature, in a particular orientation, can be determined for a given set of optics and resist process. However, this does not represent the depth of focus that can be used for making integrated circuits, because integrated devices typically have geometries of many different orientations distributed over the area of the chip. It has become accepted practice to distinguish between the usable depth of focus (UDOF) and the individual depth of focus (IDOF)[81]. The individual depth of focus is the depth of focus at any one point in the image field, for one particular orientation. The usable depth of focus is the common range of all IDOF's over the printable area and all orientations. The concept of usable depth of focus follows from the understanding that, in the practical world of integrated circuit manufacturing, imaging needs to be good for all points within a field to ensure complete circuit functionality. This philosophy, where all points within a field need to be good simultaneously, is discussed in more detail in

Ref. 82 in the context of overlay, but the basic concepts are applicable to all lithographic parameters, including focus. Aberrations such as astigmatism and field curvature reduce the usable depth of focus. As the diffraction limit is approached it is clearly important that these aberrations be minimized since there is not much depth of focus to work with.

The models used to illustrate the concepts of resolution can be extended to considerations of focus. The light intensity distribution can be calculated for planes other than the plane of best focus; examples are shown in Fig. 1.33. As can be seen, the optical intensity profiles degrade with defocus. As seen in Fig. 1.33, there are positions, labeled x_c, where the exposure dose changes little, or not at all, with defocus. Processes that are set up to have the edges of developed resist features correspond to these positions will generally have small changes in linewidth with defocus. The positions x_c are known as the conjugate points[83], and they are usually close to the nominal line edges, i.e., the resist linewidths have approximately the same dimensions as those on the reticle (adjusted for the reduction ratio of the projection optics.). Earlier, it was seen that the normalized derivative of the optical intensity profile at best focus was increased by having an exposure bias. As derived, parameters such as exposure latitude, linewidth control over topography and resist sidewall slope are improved with a large normalized derivative. In the absence of defocus, one would conclude that a print bias would lead to an optimized process.

In the presence of defocus, the normalized derivative behaves differently, as shown in Fig. 1.34. At best focus, the normalized derivative increases for a significant amount of print bias. The magnitude of the normalized slope diminishes at all print biases for defocus, but the degradation is greater away from the conjugate point. For sufficiently large defocus, the normalized derivative actually decreases with print bias. Thus, one is led to a significantly different conclusion concerning the advantage of large print bias when defocus is a significant concern.

The thin resist model of Section 1.2.1 can be extended to include defocus. In this case the light intensity profile becomes a function of two variables:

$$E(x) \rightarrow E(x, \zeta) \tag{1.70}$$

where x refers to the distance perpendicular to the (long) resist lines and spaces, and ζ is the amount of defocus. The condition that determines the position of the edge of the photoresist is

$$E_0 = E(x_0, \zeta) = E(x_0 + \Delta x, \zeta + \Delta \zeta) \tag{1.71}$$

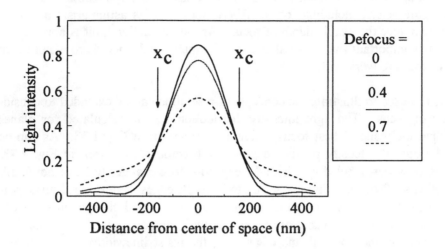

FIG. 1.33 Calculated light intensity distributions at different focus settings for a 300 nm space exposed at λ = 248 nm, NA = 0.5 and σ = 0.6. Defocus is measured in units of microns.

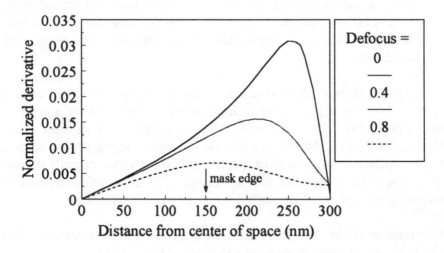

FIG. 1.34 Normalized derivative of the light intensity profiles shown in Fig. 1.33. The normalized derivative is in units of μm^{-1} and defocus is measured in units of microns.

Expanding the right side of Eq. 1.71 in a Taylor series about the plane of best focus,

$$\Delta x = -\tfrac{1}{2}(\Delta\zeta)^2 \left| \frac{1}{E_o}\frac{\partial^2 E}{\partial\zeta^2} \right| \left[\frac{1}{E_o}\frac{\partial E}{\partial x} \right]^{-1}. \qquad (1.72)$$

Derivatives in x are evaluated at the point x_o and derivatives in ζ are evaluated at best focus, $\zeta = 0$. The linear term terms in ζ are zero because

$$\frac{\partial E}{\partial \zeta} = 0 \qquad (1.73)$$

in the plane of best focus, and higher order terms in Δx were dropped from Eq. 1.72. The symmetry of the aerial image about the plane of best focus is expressed by Eq. 173. In the presence of photoresist or certain optical aberrations this symmetry is broken. This is not particularly problematic where the thin resist model is a valid approximation, but it can be relevant for situations involving high numerical apertures and thick (relative to the Rayleigh depth of focus) photoresist.

In prior discussions, process latitude was improved as the normalized derivative

$$\frac{1}{E_o}\frac{\partial E}{\partial x} \qquad (1.74)$$

was increased. From Eq. 1.72, it can be seen that operating points with large normalized derivative may occur at points where the defocus term

$$\frac{1}{E_o}\frac{\partial^2 E}{\partial \zeta^2} \qquad (1.75)$$

may decrease an offset benefit from enhancement of the factor (1.74).

Typical resist thicknesses (≈ 1.0 μm) are on the order of the Rayleigh depths of focus of contemporary high resolution lenses. Consequently, one might expect that the variation in the light intensity distribution in the direction parallel to the optical axis should be relevant when calculating resist profiles. In this case the thin resist model will not predict accurately. This can be appreciated from simple geometrical optics. Consider a point source of light focused by a lens, as shown in Fig. 1.35. In the limit of geometrical optics, in which diffraction effects are ignored, the light is focused back to a point by the lens, but in any plane other than the plane of best focus, the image is a circle. The distance along the optical axis in which such circles are an acceptably small diameter is the depth of focus. If diameters of the focused image vary appreciably throughout the thickness of the resist film, then a more sophisticated model than has been considered for the optical image needs to be considered.

For thick resist, let us first refer to Fig. 1.36. Suppose that a maximum image radius r is acceptable. For an optical image in air, the depth of focus would then be

$$DOF = \frac{2r}{\tan \theta_o},$$ (1.76)

where θ_o is the largest angle for incident rays of light and is related to the numerical aperture by

$$\sin \theta_o = NA.$$ (1.77)

When a thick layer of resist is placed in the vicinity of best focus, refraction becomes a factor. In Fig. 1.36, two focused light rays converge to the point F_1 instead of the focal point F_o because of refraction at the air-resist interface.

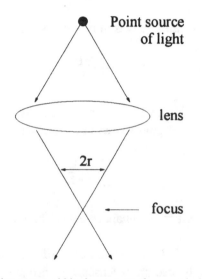

Point source of light

lens

2r

focus

FIG. 1.35 Imaging of a point source within the context of geometrical optics. Away from the plane of best focus the image is a circle of radius r.

The angle θ_1 between the normal and the refracted ray is related to θ_o by Snell's law:

$$\sin(\theta_1) = \frac{\sin(\theta_o)}{n},$$ (1.78)

where n is the index of refraction of the photoresist. The depth into the photoresist through which these light rays can travel before they diverge a lateral distance greater than $2r$ is increased from Eq. 1.76 to

$$DOF' = \frac{2r}{\tan \theta_1}.$$ (1.79)

This leads to an effective increase in the depth of focus, over the aerial image, by a factor of

$$\frac{\tan(\theta_o)}{\tan(\theta_1)}.$$

(1.80)

For small angles, this is approximately equal to the photoresists's index of refraction, n. Typical values for n are in the neighborhood of 1.7.

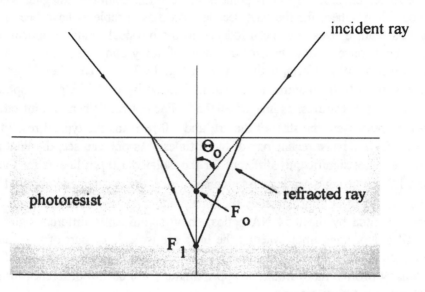

FIG. 1.36. Refraction of light at the air-resist interface and its effect on the position of focus and the spreading of the image due to defocus.

This simple analysis leads to the prediction that imaging will be asymmetric about best focus. As light is focused away from the resist, the defocus is reduced by a factor of $1/n$, but not when the light is focused into the resist. This prediction, based upon simple geometrical arguments, has been verified by full vectorial diffraction calculations[84].

A method for increasing the depth of focus (FLEX: focus-latitude enhancement exposure) has been proposed that involves partial exposures at different focus positions.[85,86] With FLEX, sharply focused images are superimposed with defocused ones throughout the depth of the resist film. The depth of focus for imaging small features such as contact holes is extended by doing partial exposures at multiple locations along the optical axis. The appropriate choice of exposure and focus steps can lead to a 3 or 4x increase in contact hole depth of focus, but the depth of focus for line features is typically not improved. FLEX has been demonstrated to improve the depth of focus, but with a loss of ultimate resolution[87] by the multiply defocused exposures. As we shall see in a later section, this is a general property of most attempts to improve the depth of focus that do not address the fundamental issue of diffraction at the reticle.

1.2.6 Optical proximity effects

The topics covered up to this point have been the dominant imaging concerns of photolithographers for the past decade. As these problems have been mastered, and as feature sizes have been reduced to the physical limits of optical technologies, new concerns have become the focus of lithography engineers. One of these, optical proximity effects, is illustrated in Fig. 1.37. The calculated light intensity profile for 1.0 μm features is shown, as imaged by 0.35 NA g-line optics, along with 0.35 μm features, as imaged by 0.54 i-line optics. When first introduced, the g-line optics were the state of the art, and 1.0 μm was the typical resolution supported by the resist technology of those times. As one can see, the light intensity profile is not significantly different for the isolated 1.0 μm line as for the 1.0 μm line/1.0 μm spacing grating. As a consequence, the isolated and grouped lines are expected to print approximately the same width. On the other hand, the 0.35 μm lines, printed by the 0.54 NA optics, show significantly different light intensity profiles depending upon whether the lines are isolated or part of a grating structure. In a situation where the linewidth is determined by the threshold of the light intensity profile, there would be approximately a 0.09 μm difference in linewidth between the two situations.

This is an example of the *proximity* effect, where the size of a line is dependent upon its proximity to other geometries. Actual situations are more complicated, and the magnitude of the proximity effect depends upon the resist process as well as the optics. An example of the optical proximity effect occurs in memories, where the interconnects and gates are very dense in the memory array but are relatively isolated in the periphery of the device. The isolated lines of photoresist are sometimes found to be much smaller than their counterparts in the memory array, and other times larger, even though their mask dimensions are the same. An example of the process dependence of proximity effects was illustrated by the formation of patterns, consisting of isolated lines and spaces, and gratings, using single and a multilayer resist systems[88]. Significantly different proximity-dependent print biases were seen between the single layer resist and the multilayer system, which provided the photoresist with a "substrate" that was non-reflective.

Proximity effects are only one type of situation in which features print differently from mask dimensions. Print bias, the difference between the mask dimension and the printed dimension, may be different for features of different sizes. This effect is referred to as "linearity." Both linearity and proximity effects need to be considered when producing integrated circuits, since geometries of many different shapes and sizes will exist on a typical masking level.

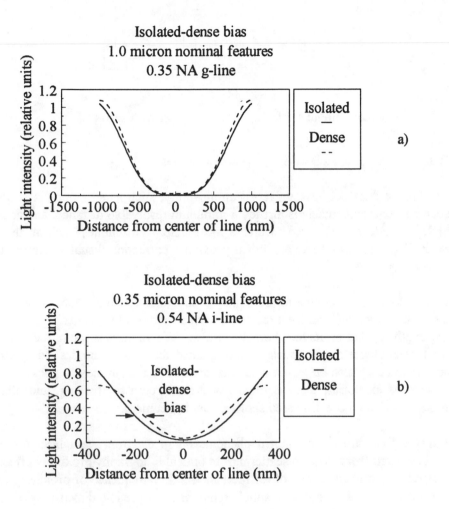

FIG. 1.37 Light intensity profiles for isolated and dense spaces. There is little proximity effect when the optics are far from the diffraction limit (a), but there is a considerable proximity effect when the optics are pushed to the diffraction limit (b).

A technique related to proximity effects involves the use of modified shapes or adjacent subresolution geometries to improve imaging. An example of this is the use of serifs on the corners of contacts[89]. For contacts with dimensions near the resolution limit of the optics, a square pattern on the reticle will print more nearly as a circle. Additional geometries on the corners (Fig. 1.38) will help to square the corners of the contacts.

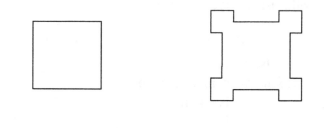

Contact without serifs   Contact with serifs

FIG. 1.38 Contacts with and without serifs, as drawn on the reticle.

This technique does not have universal applicability, such as situations in which the spacing between contacts is already a minimum resolvable distance, and additional layout effort is required to create the reticle. Techniques such as those shown in Fig. 1.38 are known as optical proximity correction, usually referred to by the acronym OPC.

In addition to corner rounding for nominally rectangular features, there is also the issue of line shortening[90], where a rectangle whose width (W) is printed to size will have a length (L) that prints too short (Fig. 1.39). Where there is room, rectangles can be biased longer on the reticle to give printed features of the desired length. Compensation may also be required for corner rounding[91]. In closely packed circuits, such as memories, there is often insufficient room for biasing, and other techniques, such as those discussed shortly, may be required.

Proximity effects are a well-known phenomenon in electron beam lithography, where they result from electron scattering. In optical lithography proximity effects are caused by the phenomenon of diffraction. As a consequence of proximity effects, printed features do not have simple relationships to reticle dimensions. This creates a situation in which it is difficult to fabricate a photomask where the designer gets what he wants on the wafer. This increased complexity is related directly to the fact that optical lithography has been pushed to diffraction limits. The situation is further complicated by the fact that steppers have more degrees of freedom, such as variable numerical aperture and partial coherence, than they did in the past, and print biases and proximity effects will depend upon which parameter values are chosen by the process engineer. As will be seen in the following discussions, techniques and methods that have been developed, or are the subjects of current research and development, to extend the capabilities of optical lithography all involve new complexities related to printing features of different sizes and configurations. These new techniques are all directed towards addressing the effects of diffraction, which limit the depth of focus as resolution is moved to smaller feature sizes.

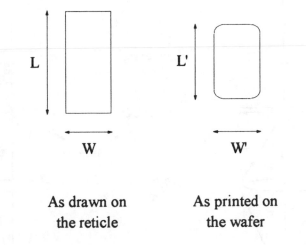

As drawn on
the reticle

As printed on
the wafer

FIG. 1.39 Line shortening, where W = W', but L' < L.

1.2.7 Off-axis illumination

One method, or group of methods, for improving lithography capabilities involves modification of the illumination of the reticle. As discussed earlier, light (coherent) that illuminates a grating is diffracted in very specific directions according to Eq. 1.23. For normally incident light, sufficiently small dimensions result in a situation where all beams except the zeroth order are diffracted outside the entrance pupil of the imaging optics (Fig. 1.40).

A single beam is a plane wave, containing no spatial information; at least two interfering beams are required for pattern imaging. Alternatively, consider the situation in which the illumination is not normally incident. For such off-axis illumination it is possible for the zeroth order light and one first order beam to enter the entrance pupil of the imaging optics. From this simple analysis, one might expect to enhance image contrast by eliminating the illumination with small angles of incidence, since those light rays contribute only to the background light intensity without providing spatial modulation. From Fig. 1.40, one might also expect that off-axis illumination should improve depth of focus even when, for conventional illumination, the first order diffracted beams travel through the entrance pupil, because the angular spread of the light rays in the off-axis situation is less than that for conventional illumination. The benefits of off-axis illumination for enhancing image contrast have been known among optical microscopists. The concept was introduced to lithographers by Fehrs et. al.[92] and Mack[93] and has since been explored extensively[94-97].

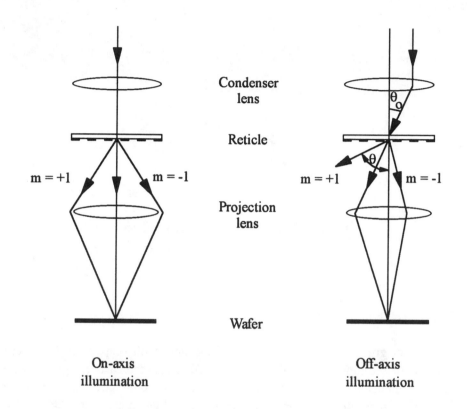

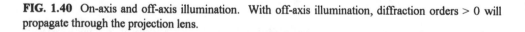

FIG. 1.40 On-axis and off-axis illumination. With off-axis illumination, diffraction orders > 0 will propagate through the projection lens.

If we consider the zero and one first order diffracted rays, the light intensity at the wafer plane is given by

$$I(x, z) = 2a_0^2 E_0^2 + 2a_1^2 E_0^2 \cos\left(\left(\sqrt{1 - (\sin\theta_i - \lambda/p)^2} - \cos\theta_i\right)\frac{2\pi z}{p}\right)\cos\left(\frac{2\pi x}{p}\right),$$

$$(1.81)$$

where a_0 is the fraction of the E-field in the zero order beam, a_1 is the fraction of the E-field in the first order beam, E_0 is the amplitude of the incident E-field, p is the pitch of the grating, x is the lateral distance along the wafer, and z is the amount of defocus. This equation is independent of z, i.e., independent of focus, when

$$\cos \theta_i = \sqrt{1 - (\sin \theta_i - \lambda/p)^2} \; , \qquad (1.82)$$

which occurs when

$$\sin (\theta_i) = \frac{\lambda}{2p} \; . \qquad (1.83)$$

This indicates that optimum parameters, such as the angle of incidence for the illumination, are feature size dependent. The optimum angles for off-axis illumination are discussed in Ref. 98.

Results for off-axis illumination from more detailed analyses and measurements have found the following:

1) Enhancement is seen for features smaller than the coherent cut-off limit. Above this limit there is no difference between conventional and off-axis illumination.

2) The depth of focus improvement for grouped features is much greater than for isolated features.

3) Grouped/isolated feature print bias is affected by off-axis illumination.

As one can see from the preceding analysis of the imaging of a diffraction grating, off-axis illumination enhances imaging for gratings that have lines and spaces running perpendicular to the plane containing the illuminating rays of light. An aperture such as the one shown in Fig. 1.41(a) will improve imaging for gratings oriented in one direction but degrade them for gratings oriented in the perpendicular direction. Quadrapole illumination [Fig. 1.41(b)] works well when all features are oriented left-right or up-down but will distort features of other orientations[99]. The annular aperture shown in Fig. 1.41(c) will improve features of all orientations. Particularly good results have been obtained with illumination sources that include some on-axis illumination, though at lower intensity than for the off-axis light[99,100].

1.2.8 Pupil plane filtering and Super-FLEX

In the Super-FLEX method, originating with Fukuda et al.[101,102], the amplitudes for two images with different focal planes are superposed, and the phase difference between the two images is controlled in such a way that the DOF can be increased. Fukuda has proposed two methods to practically attain such a composite image. One involves placing a special amplitude and phase filter in the pupil plane of the stepper[101]. The second method involves modulating the pattern on the mask to produce a composite image[102].

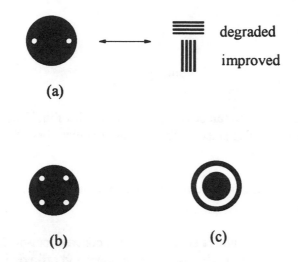

(a)

degraded

improved

(b)   (c)

FIG. 1.41 Different types of off-axis illumination apertures. The aperture with holes at the edges or corners — (a) and (b) — will improve lines oriented in one direction but degrade imaging of the lines in the orthogonal direction

Von Bunau and co-workers at Stanford University have explored the limits to extending the depth of focus using apodization, or pupil plane filtering[103]. They showed that the DOF for contact holes can be increased greatly, but at the expense of reduced peak intensity and increased energy in the sidelobes of the aerial image. Resolution enhancement by combining oblique illumination and a transmittance-adjusted pupil filter that has a conjugate shape to the secondary light source has been reported by workers at NTT[104].

Pupil plane filtering is of only theoretical interest today since the pupil plane in microlithographic lenses is usually somewhere inaccessible in the optical path and can't be reached unless the lens is disassembled — which of course inactivates the lens. Since different mask types require different filters for optimum performance, one can't just build a filter into the lens at the pupil plane like another fixed optical element.

1.2.9 Phase-shifting masks

Another method for overcoming the restrictions imposed by diffraction is phase shifting. While there are now many types of phase-shifting masks, they all employ the same basic concept, which is well illustrated by the original version introduced by Levenson et al.[105] , and it is used here to introduce the subject. Consider a grating that is imaged by optics with a particular numerical aperture and partial coherence. Imaging is degraded because light from clear areas on the mask is diffracted into regions that ideally would be completely dark. The nominally dark region has light diffracted into it from the space on both the left and right.

The idea behind the alternating phase-shifting mask is to modify the reticle so that alternating clear regions also cause the light to be *phase-shifted* 180° (Fig. 1.42). As a consequence, the light diffracted into the nominally dark area from the clear area to the left will interfere destructively with the light diffracted from the right clear area. This improves image contrast, as shown in Fig. 1.43. Interference is more effective with a high degree of coherence[106]. All phase-shifting masks employ this same characteristic, where the destructive interference of light of opposite phases is used to improve image contrast.

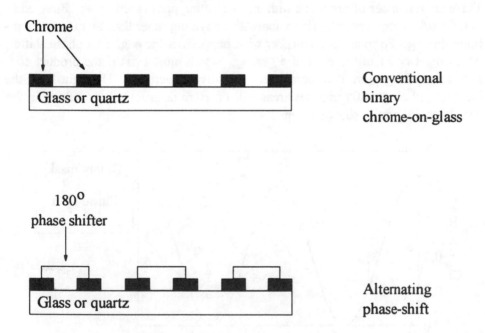

FIG. 1.42 Conventional binary chrome-on-glass reticle and alternating, or Levenson, phase-shifting mask.

Light that travels a distance a in air will change phase by

$$\phi_a = \frac{2\pi}{\lambda}a ,$$

(1.84)

where λ is the wavelength of the light in vacuum. Through the phase-shifting material the phase will change by an amount

$$\phi_n = \frac{2\pi}{\lambda}na\,, \qquad (1.85)$$

where n is the index of refraction of the phase shifter. The difference in phase shift caused by the phase-shifter is then

$$\Delta\phi = \frac{2\pi}{\lambda}a(n-1). \qquad (1.86)$$

To achieve the condition $\Delta\phi = 180° = \pi$ rad, we have the following relationship for the thickness a of the phase-shifting layer:

$$a = \frac{\lambda}{2(n-1)}. \qquad (1.87)$$

There are a number of problems with the alternating phase shift mask. First, additional work is required in order to apply it to anything other than an effectively infinite grating. For example, consider what happens at the edge of a phase shifter, which must occur at the edge of a grating, which must exist if the product consists of anything more than completely repetitive structures. The light from the clear 0° and clear 180° adjacent areas will interfere destructively, resulting in the light intensity profile shown in Fig. 1.44.

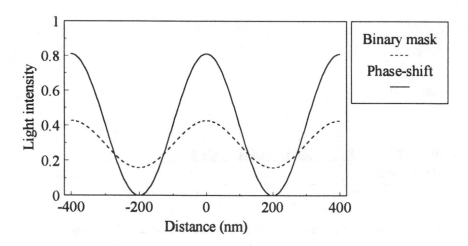

FIG. 1.43 Light intensity distribution of a 400 nm pitch grating with equal lines and spaces, imaged with 0.5 NA optics at a wavelength of 248 nm. For the binary mask image, $\sigma = 0.6$, while $\sigma = 0.3$ for the alternating phase-shifting image.

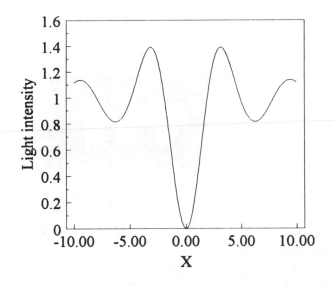

FIG. 1.44 Light intensity distribution at a phase edge. $X = 2\pi NAx/\lambda$, where x is the physical distance from the phase edge measured in the same units as λ. The phase edge occurs at $X = 0$.

Such an optical image will print into photoresist (see Fig. 1.45). Implementation of alternating phase-shifting masks must deal with these edge shifters. Three approaches are taken. In the first, additional features are introduced between the 0° and 180° phase areas to produce phase edges of <<180°, which do not print. Alternatively, a second mask is used to expose the phase edges away. Neither approach is completely satisfactory. The first method requires additional steps in the reticle fabrication process, while the second reduces stepper productivity. However, the improved capabilities that result from the alternating phase shift type of mask may justify these additional processing steps.

A third approach involves the use of negative photoresist. As one can see from Fig. 1.45, creating islands of resist using negative resist and phase-shifting masks avoids the phase edge problem. This is one of the reasons that negative resists are again being used for leading edge lithography.

The light intensity distribution shown in Fig. 1.44 can be understood as follows[107]. As discussed in Section 1.2.2, the image of a diffraction limited lens for coherent illumination can be obtained from the Fourier transform of the mask pattern. In the case of the phase edge, the mask pattern has unit intensity but half of the pattern has a phase 180° (π rad) different from the other half:

$$T(k) = \frac{i}{\pi k}.$$

$$(1.88)$$

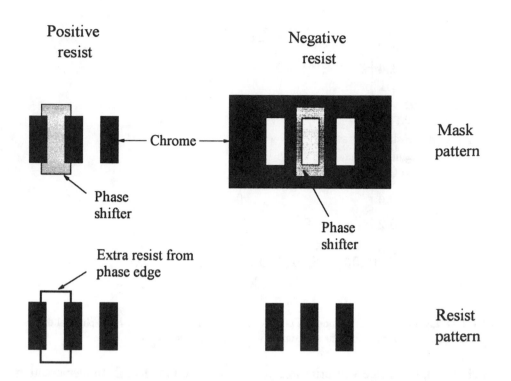

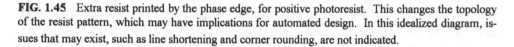

FIG. 1.45 Extra resist printed by the phase edge, for positive photoresist. This changes the topology of the resist pattern, which may have implications for automated design. In this idealized diagram, issues that may exist, such as line shortening and corner rounding, are not indicated.

The intensity for this pattern, for coherent light, is given by

$$I(x) = \frac{4}{\pi^2} Si^2\left(\frac{2\pi NA x}{\lambda}\right), \tag{1.89}$$

where

$$Si(\theta) = \int_0^\theta \frac{\sin(z)}{z} dz. \tag{1.90}$$

From Eq. 1.89, one can estimate that the width of the aerial image at 0.25 intensity is approximately

$$Width = 0.25 \frac{\lambda}{NA}. \tag{1.91}$$

The prefactor of 0.25 in Eq. 1.91 should be compared with the prefactors in Equations 1.19, 1.21, and 1.24. With the phase edge, one can obtain over a 50% reduction in the size of a printable feature, compared to non-phase shifted imaging.

Phase-edge photomasks show considerable promise for delineating extremely narrow lines[108], which would find use in patterning gates and interconnects.

One of the problems of a phase-shifting mask with the construction shown in Fig. 1.42 is the potential for the phase shifter to affect light amplitude as well as phase. This will occur when the phase shifter has an index of refraction different from that of quartz. The most common method of circumventing this problem is to replace the phase-shifting material with a recess into the quartz. This latter approach requires well controlled etches so that the appropriate depth may be achieved uniformly with only a timed etch.

In addition to the phase-edge and alternating phase-shifting masks, several other versions of phase-shifting masks have been introduced, and the most common ones are listed in Table 1.6. One of these, the edge shifter, was just discussed, where it was seen as an undesirable complication for applying the alternating phase-shifting mask and as a powerful method for printing isolated lines.

The *rim shift* phase-shifting mask[109] was developed to address the problem of overlay in the mask fabrication process, through the creation of a self-aligned phase-shifting structure. This is illustrated in Fig. 1.46a. In the fabrication of the rim-shifter, the reticle is patterned with resist, and the phase-shifting material is etched anisotropically. Then the chrome is etched so it undercuts the phase- shifting material. Among the issues are rim-size optimization and sensitivities for print biases, slope in the light intensity profile, and side-lobe intensities[110], leading back to reticle fabrication concerns.

Outrigger phase-shifting masks[111] (Fig. 1.46b) involve an approach that requires some degree of overlay between the step that patterns the chrome and the step that patterns the phase shifter, but in a mode where the edge of the phase shifter is placed on chrome, as in the case of the alternating phase-shifting mask. These types of masks can be considered as hybrids between phase-shifting masks and optical proximity corrections of the type discussed in the preceding section.

An extremely attractive type of phase-shifting mask, from the perspective of mask fabrication, is the *attenuated phase-shifting mask*[112,113]. In this of type mask, the non-clear areas are partially transmitting. This can be achieved, for example, with thin layers of chrome. Other materials, such as CrO, $CrON$, $MoSiO$, and $MoSiON$[114,115], have also been used. With suitable processing, one can achieve 180° phase difference between the clear and non-clear areas, often with an etch of the quartz to an appropriate depth. The partial transmission of the non-clear areas is a problem with this type of mask. For example, the threshold exposure dose for significant resist loss must be less than the amount of light that "leaks" through. An optimization of the normalized slope of the aerial image may have unacceptable levels of light in nominally "dark" areas of the mask (Fig. 1.47). This places requirements on the resist and on the transmittance of the partially transmitting areas on the reticle. The attenuated phase-shifting mask is attractive because it can be created in a single exposure step and is therefore relatively easy to fabricate.

Table 1.6 Types of phase-shifting masks

Type of phase shift mask	Alternate names	Category	Applicable situations
Alternating	Levenson	strong	grouped lines and spaces
Rim shift		weak	contacts, isolated features
Attenuated	leaky-chrome, half-tone, embedded	weak	contacts, isolated features
Phase-edge	chrome-less, unattenuated	strong	narrow lines
Outrigger	subresolution, additional aperture	weak	isolated features

As phase-shifting technology becomes more commonplace, a name is needed for the conventional types of masks, which consists only of opaque areas (usually chrome) and quartz or glass. Because conventional masks have only completely clear or completely opaque areas, they are usually referred to as binary intensity masks or chrome-on-glass.

Phase-shifting masks are also classified as "strong" or "weak," according to their ability to suppress the zero-order diffraction component. This is seen in Fig. 1.43, where, without a phase-shifting mask, zero intensity cannot be achieved at any point because of a zero-order component background. Phase-shifting masks other than the alternating or chrome-less are less capable at eliminating the zero-order diffraction component. The combination of rim-shifted and attenuated phase-shifting masks and off-axis illumination has been found to be a powerful synergy[116], since each method is only partially capable of reducing the zero-order light component by itself.

From the preceding discussions, it is clear that a particular type of phase-shifting mask or alternative illumination is applicable in specific situations but not in others. For example, alternating phase-shifting masks are useful for dense lines and spaces but not for isolated contacts. Table 1.6 lists different types of phase-shifting methods and the situations in which they result in enhanced lithographic performance. Similarly, off-axis illumination was found to improve imaging of dense lines and spaces but not particularly useful for patterning isolated features. FLEX improves the depth of focus for contacts but not lines and spaces and was found to reduce ultimate contrast. This appears to be a general feature of many newly developed methods for enhancing optical lithography: the improvement is restricted to particular types of features; there is no "magic bullet."

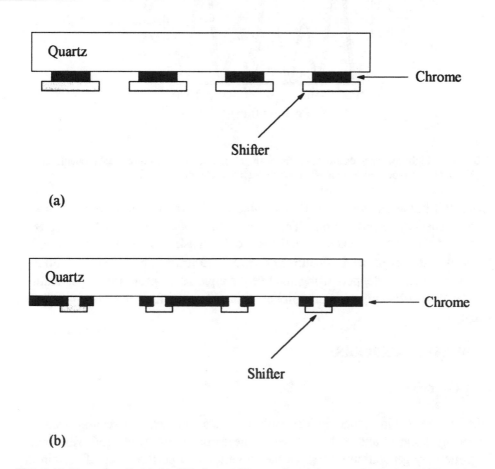

FIG. 1.46 Examples of a (a) rim shift phase-shifting mask and (b) an outrigger phase-shifting mask.

The fabrication of phase-shifting masks requires additional processing compared to that necessary for making binary photomasks. In particular, the alternating phase-shifting mask requires at least two patterning steps. This introduces the need for overlay capability in the mask making process. The introduction of additional types of features on the reticles imposes new demands on inspection capabilities. Such new requirements have significant effects with respect to the capitalization required for making advanced photomasks. Additional information on the issues associated with the fabrication of phase shift masks is discussed at length in the chapter on photomasks.

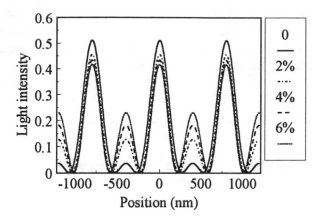

FIG. 1.47 Light intensity distributions from an attenuated phase-shifting mask, calculated with PROLITH/2 for various levels of transmission through the "chrome."

Phase-shifting masks are presently the subjects of development efforts and are not used extensively in production. This is a function of the immaturity of the technology, its inherent complexity, and the lack of mask making infrastructure, but primarily the lack of need. Features down to 0.25 μm can be imaged with binary masks on existing steppers using relatively simple resist processes. For features smaller than 0.25 μm, it is quite probable that phase-shifting technology will find a role.

1.3 WAFER STEPPERS

1.3.1 Overview

This section will introduce optical wafer steppers and step-and-scan systems that are being designed and built to serve the needs of the 64 Mbit, 256 Mbit, and 1 Gbit memory generations. It traces the evolution and performance of certain key subsystems of the modern reduction stepper: the wafer stage, the alignment system, the illuminator, and the reduction lens, from the viewpoint of the user. Stepper productivity and cost of ownership is discussed. Finally, future directions in which the technology of wafer steppers might evolve are outlined.

The wafer stepper appeared as a tool for silicon IC fabrication at several companies, including Philips, Thomson CSF, and IBM [117,118,119] in the late 1970s. These machines evolved from photomask step-and-repeat systems[120], which were optical pattern generators used to step masks for contact, proximity, and 1:1 scanning projection aligners. The GCA Corporation of North Andover, Mass., a pattern generator manufacturer, introduced the stepper commercially in 1977 with the 4800DSW™ (Direct Step on Wafer)[121]. This stepper, which handled 3, 4, or 5 inch wafers, was equipped with a 10× reduction, 0.28 NA, g-line lens, with a maximum square field of 10 mm, supplied by Carl Zeiss of Oberkochen, Germany. The stepper could achieve mixed overlay of ±0.7 μm and resolution of 1.25 μm lines and spaces over 3 μm depth of focus. This stepper was used by many companies to develop the 64K DRAM device. Its list price was about $300,000.

Other suppliers such as Nikon, Canon, SVG Lithography (SVGL) and ASML have displaced GCA in the market today. Production wafer steppers in 1996 can handle 5, 6, or 8 inch wafers, have i-line lenses with half the reduction ratio (5×), twice the NA (0.6), and twice the field size (≥ 22 mm square). These lenses can achieve 0.5 μm production resolution over > 1.5 μm DOF and 0.4 μm with > 1.0 μm DOF[122,123,124]. Mixed machine overlay is better than 120 nm. The list price of these steppers is $3 million. See Table 1.7 for a comparison of the 4800 DSW with its descendants today, and target performance goals for a 1997 deep UV step-and-scan system.

The key requirements for steppers and step-scan systems for the 64 Mbit/256 Mbit/1 Gbit generations are outlined in Table 1.8, taken from the NTRS Workshop Lithography Working Group Report[125]. The timing for the introduction of new generations of device density is dictated by Moore's law, which states that the number of transistors per chip doubles every 18 months. The semiconductor industry's adherence to Moore's law is the driving force that propels lithography forward, and to sustain the roadmap it requires lithography to continue to meet ever greater challenges.

1.3.2 National Technology Roadmap — Stepper Requirements

Table 1.8 gives the requirements for key lithography capabilities such as resolution, total overlay, and field size. The stepper must be considered as the key component of a system that also includes the resist process, the reticle technology, and the metrology technique, all of which are integrated to meet the requirements of the device technology.

1.3.3 Overview of the stepper as a system

A stepper has a high speed stage that "steps" the wafer precisely with respect to the imaging optics and the IC reticle, moving the distances necessary to exactly repeat the image field in a Cartesian grid and thus fill the wafer surface. In the typical reduction stepper the stage travels in the horizontal plane underneath a fixed, vertically mounted lens. Once the wafer is placed on the exposure chuck and stepped under the lens, it is aligned by automatic systems that detect wafer targets optically and move the stage in small increments to correct the wafer position with respect to the ideal image field. The wafer is also positioned in the vertical axis by autofocus systems, which in modern steppers also include the capability to pivot the vacuum chuck that holds the wafer during exposure in order to reduce any net tilt in the wafer surface due to chuck or wafer flatness errors[123, 126].

TABLE 1.7 - Evolution of the Wafer Stepper

	1977 GCA 4800 DSW	1995 i-line	1997 DUV: Step Scan
M:1	10×	5×	4×
Wavelength	g-line: 436 nm	i-line: 365 nm	DUV: 248 nm
Lens	0.28 NA	0.60 NA	0.60 NA
Resolution	1.25 μm	0.40 μm	0.25 μm
Field size	10 mm sq.	22 mm sq.	26 mm × 32 mm
Depth of focus	4.0 μm	1.0 μm	0.70 μm
Alignment	± 0.50 μm	± 0.06 μm	± 0.03 μm
Stage accuracy	100 nm	30 nm	15 nm
Lens distortion	250 nm	50 nm	30 nm
Wafer size	3, 4, 5 inch	5, 6, 8 inch	6, 8, 12 inch
Throughput	20 wph (4")	60 wph (6")	60 wph (8")
Cost	$300,000	$4,000,000	$8,000,000

N-times demagnified images of the IC pattern are projected onto the wafer surface by the microlithographic lens, mounted only a few millimeters above the wafer surface. N can range from 1 to 5, with 5× reduction the most common but 4× being adopted for the 0.25 μm generation. In the most common type, the lens is a long (up to 1000 mm) cylindrical metal jacket that contains a series of glass optical elements and is held rigidly in place by the supporting body. An illumination system exposes light through a chrome-on-glass reticle that has the IC pattern etched into the chrome.

The illuminator for an i-line stepper consists of a mercury arc lamp surrounded by an ellipsoidal mirror which concentrates light into an condensing optical system that collimates and filters the light for uniform illumination of the reticle. To improve resolution, KrF excimer lasers operating at 248 nm or ArF excimer lasers operating at 193 nm can be used for illumination.

Table 1.8 NTRS Roadmap: Requirements of Lithography.

Year of First Shipment (on production tooling)	1998	2001	2004	2007	2010
Technology generation (μm)	0.25	0.18	0.13	0.10	0.07
Product application					
DRAM (bits)	256M	1G	4G	16G	64G
Microprocessor (logic transistors/cm^2)	7M	13M	25M	50M	90M
Applications-specific integrated circuits (transistors/cm^2 auto layout)	4M	7M	12M	25M	40M
Minimum feature size (μm)					
Isolated line (microprocessor gates)	0.24	0.16	0.11	0.08	0.05
Dense Lines (DRAM half pitch)	0.25	0.18	0.13	0.10	0.07
Contact/vias	0.28	0.20	0.14	0.11	0.08
Development capability (minimum feature size, μm)	0.16	0.11	0.08	0.05	0.03
Gate CD control at post etch (nm, 3σ)*	24	16	11	8	5
Overlay (nm, mean + 3σ)*	85	60	45	35	25
DRAM chip size (mm^2, 2:1 aspect ratio), 1st shipment	280	420	640	960	1400
1st shrink (1 year later)	230	340	520	780	1130
2nd shrink (2 years later)	190	270	420	630	920
Microprocessor chip size (mm^2, 1:1 aspect ratio), 1st shipment	300	360	430	520	620
1st shrink	240	290	350	420	500
2nd shrink	200	240	280	340	410
Field size (mm)	22×22	25×50	25×50	25×50	25×50
Field size (mm^2)	484	1250	1250	1250	1250
Depth of focus (usable) (μm) (full field, +/- 10% exposure)	0.80	0.70	0.60	0.50	0.50
Defect density, lithography only (per layer/m^2)	320	135	60	30	15
@ defect size (μm)	0.08	0.06	0.04	0.03	0.02
Mask size (inches) (quartz)	6×6	9×9	9x9	9x9	9x9

* Requirements scale with resolution for shrinks

The illuminator for an i-line stepper consists of a mercury arc lamp surrounded by an ellipsoidal mirror which concentrates light into an condensing optical system that collimates and filters the light for uniform illumination of the reticle. To improve resolution, KrF excimer lasers operating at 248 nm or ArF excimer lasers operating at 193 nm can be used for illumination.

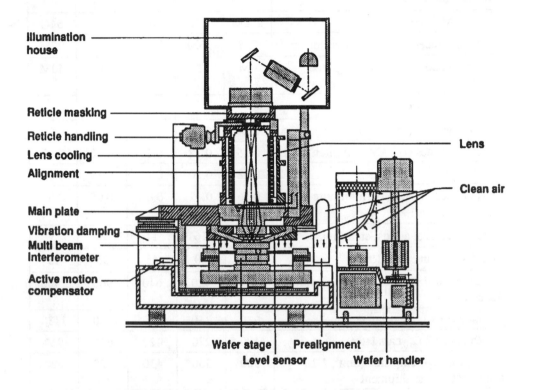

FIG. 1.48 ASM-L PAS5500 wafer stepper (from Ref. 123, Fig. 4).

1.3.4 Reduction lenses

The modern i-line lens has 25-35 glass elements held firmly in a steel cylindrical jacket. The lens may be a meter in length and weigh 300 kg or more. Reduction microlithographic lens design has generally followed the "double Gaussian" form first described by Glatzel of Carl Zeiss. Lens design is often a tradeoff between complexity and manufacturing tolerances on one hand and aberration correction on the other. Important issues in lens fabrication are the purity, homogeneity, and spectral transmission of glass materials, the precision to which spherical and aspheric surfaces can be ground and polished, and the centration and spacing of elements, among many others.

According to Eqs. 1.19 and 1.21, smaller features may be resolved by decreasing the wavelength or increasing the numerical aperture of the lenses. The numerical aperture may be increased by increasing the diameter of the lens or decreasing the focal length (Fig. 1.13). There are practical limits to these factors. The transmission properties of various types of optical glass are shown in Fig. 1.49[129,130].

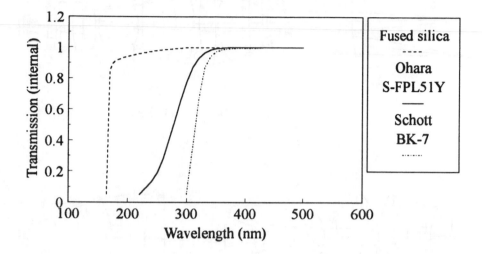

FIG. 1.49 Transmission for several types of optical glasses.

Only quartz (fused silica) and calcium fluoride are suitable for wavelengths below 350 nm, and these too have their limits. Moreover, in order to compensate for the various aberrations, lens elements with different indices of refraction need to be used. An example of this is shown in Fig. 1.50, where several different types of glass were used for making the lens. At shorter wavelengths, ingredients that change the glass's index of refraction also absorb the light. Lenses containing absorbing materials may heat up and expand during exposure, causing defocus[131], changes in magnification and other aberrations. The focal length of the lens is limited by the practical necessity of moving wafers underneath it without hitting it. Lenses of large diameters make design and manufacture more difficult, again practical issues. As was discussed in the section on focus, although it may be possible to make a lens of arbitrarily high resolution, it may not be usable for production equipment.

By design, stepper lenses need to operate with near diffraction limited performance and with near zero distortion. In balancing aberrations it was found that neglecting correction for chromatic aberration led to simpler designs. Thus one of the characteristics of the stepper lens is its operating wavelength and bandwidth. The bandwidth for i-line, the 365 nm line of the mercury arc spectrum, is about 4 to 6 nm in modern stepper designs. A KrF excimer laser stepper with a lens made completely from fused silica, the only glass with high enough transmission to be used, has a bandwidth of only 1 pm or less, due to the large dispersion of fused silica in the deep UV.

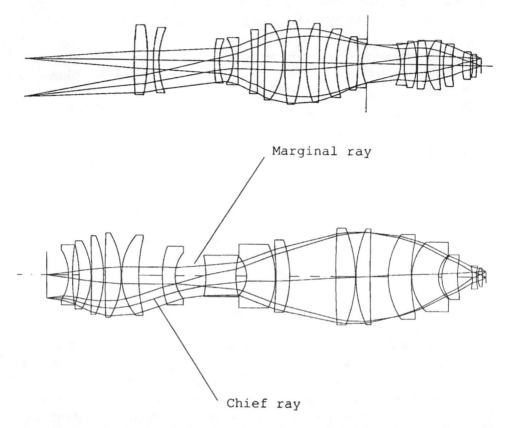

FIG. 1.50 Examples of fully refractive microlithographic lens design. a) NA = 0.38, λ = 405 nm, 5× reduction, from Braat (Ref. 127), b) NA = 0.42, λ = 365 nm, 5× reduction, from Williamson (Ref. 201).

Many lens manufacturers use phase measuring interferometry (PMI) to determine the form of the emerging wavefront from the assembled lens, information that can then be compared to the diffraction limited wavefront in order to measure the residual aberrations. The PMI was developed by Bruning and coworkers at AT&T Bell Laboratories[132].

For several device generations, lens manufacturers have been able to improve both NA and field size while maintaining diffraction limited performance, as is illustrated in Table 1.9. Fig. 1.51 shows the steady progression of lens pixel count (the circular area divided by a square one minimum feature wide) for 248 nm lenses from five manufacturers (Zeiss, Nikon, Canon, Tropel, and SVGL).

The imaging performance of a lens is limited by diffraction effects ultimately, but aberrations are also present to some degree in all lenses due to limitations in the lens fabrication process. Gortych and Williamson[133] investigated the effects of higher-order aberrations on imaging by computer simulation for several examples of aberrated wavefronts. They showed how lower- and higher-order spherical aberrations, coma, and astigmatism can affect the shape and position of exposure-defocus diagrams and concluded that "complete characterization of a lithographic lens requires aerial image simulations for a variety of mask features based on wavefronts calculated from the design, as well as measured on a PMI after the lens is assembled".

Table 1.9 Microlithographic Lenses for IC Production.

Minimum Features (μm)	NA	Wavelength	k_1	Resolution (μm)	Depth of Focus* (μm)	Pixel Count
1.00	0.38	g-line	0.80	0.92	3.02	2.4×10^8
0.70	0.40	i-line	0.75	0.70	2.28	3.8×10^8
0.50	0.48	i-line	0.60	0.46	1.58	6.1×10^8
0.35	0.60	i-line	0.50	0.35	1.01	2.0×10^9
0.25	0.60	KrF	0.60	0.25	0.69	1.3×10^{10}
0.18	0.60	ArF	0.50	0.17	0.54	2.5×10^{10}

* $k_2 = 1.0$

The higher the NA, the more specialized the lens becomes, with superior resolution and DOF over a small range of feature sizes. In order to reduce the overall cost of lithography there is a lot of pressure to use steppers over more than one generation of manufacturing. This economic incentive has led to the development of steppers with variable NA lenses as well as advanced illuminators with adjustable partial coherence and off-axis illumination[72,124,134]. These lenses allow the user to modify the aerial image by changing the NA and partial coherence to achieve the best depth of focus and/or exposure latitude for a given masking level.

To follow the pace required by the IC industry, lens manufacturers have had to use more glass elements and elements with ever greater diameters (~250 mm maximum). Wide field i-line lenses are extremely expensive (~$500,000). KrF lenses are even more expensive (≥$1 million) due to the cost of the optical quality fused silica glass. Microlithographic lenses will continue to increase sharply in cost as the field size requirements increase. Also, as the size of the lens increases, the stepper body must add more dynamic compensation in order to hold the lens in a vibration free environment. Thus further increases in pixel count by increasing field size will be ultimately limited by economics.

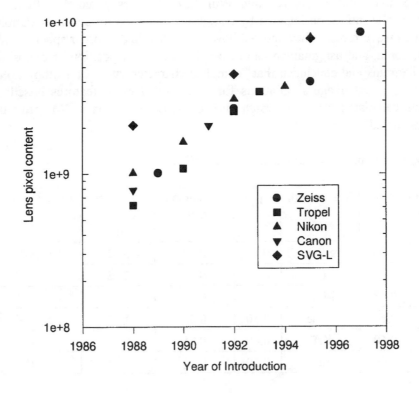

FIG. 1.51 Progression of lens pixel counts versus time for 248nm lenses.

As a result of the sharp increase in lens size and cost, steppers will gradually be replaced by step-and-scan tools for very large chip sizes. Markle has written on the concept of the scanning sub-field projector, better known as the step-and-scan system[135,136]. Step-and-scan tools can extend the image field of a lens by scanning the wafer and reticle through a small illuminated field of view corresponding to the field of the projection system. The velocities of the reticle and wafer stages, which are scanned synchronously, are chosen such that the ratio of the velocities is equal to the reduction ratio. In principle, the length of the die in the scan direction is limited only by the reduction ratio and the reticle writer's maximum writeble area. The image field of the lens can thus be restricted to sizes that can be achieved economically.

This use of a restricted field has important implications. Since only a rectangular strip of the lens will be used, the elements can be rotated with respect to each other, or "clocked", or the whole lens itself can be clocked, to find the best image performance. This tends to make final assembly of the lens go faster and achieve better final performance. In addition, many aberrations are proportional to powers of the field diameter and grow steeply near the field edges, which can be avoided when using the restricted field. Overall, the lens performance can be improved for both CD control and image placement when a restricted field can be scanned versus using the full image field.

1.3.5 Illumination Systems

Wafer steppers use partially coherent illumination, supplied by a Kohler type illuminator using either a mercury arc lamp or a laser as the light source[137]. The degree of partial coherence is a measure of how much of the entrance pupil of the optical system is filled. The geometry for defining the partial coherence is shown in Fig. 1.52. Simulation and experimental work[72,124,138] suggest that i-line steppers operate best in the range $\sigma = 0.6$ to 0.8 when imaging 0.35 μm lines and spaces and lower and $\sigma = 0.3$ to 0.5 for contact holes. Modern i-line steppers allow the user to vary σ in order to achieve best results.

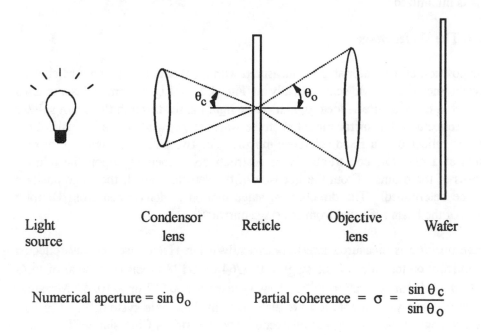

| Light source | Condensor lens | Reticle | Objective lens | Wafer |

Numerical aperture $= \sin \theta_O$ Partial coherence $= \sigma = \dfrac{\sin \theta_C}{\sin \theta_O}$

FIG. 1.52 Geometry for establishing the partial coherence σ.

Applying the results of annular illumination in microscopy, the concept of oblique or off-axis illumination in 1989[92,93] was introduced to resolution and DOF improvement in wafer steppers. As introduced in Section 1.2.7, the reticle is illuminated with a ring or annulus of light brought in at an oblique angle. This emphasizes passage of high spatial frequencies through the lens at the expense of light loss (thus a throughput hit) and increased sidelobes on images. Annular imaging tends to improve the depth of focus for most types of features at all but the lowest spatial frequencies[97,139].

In 1992 workers from both Nikon and Canon reported on quadrupole illumination schemes to impart some phase-shifting qualities to stepper imaging without the need for phase-shifting masks [95,139]. Quadrupole illumination is so named for the four beams that illuminate the mask, typically formed by placing a screen with four circular apertures in front of the fly's eye element in the illuminator path.

Modern optical steppers offer several different illumination conditions in order to achieve the best resolution and DOF for a reticle with given feature types and sizes. One stepper[135] employs an automatic carousel with different aperture types (annular, quadrupole, conventional) that can be selected by the user from the keyboard. A challenge for stepper suppliers is to design these systems so that light loss is minimized.

1.3.6 The Wafer Stage

The position of the wafer stage is measured with great accuracy using a Michelson interferometer[18], shown schematically in Fig. 1.53. Light from a laser passes through a beam splitter. Ideally, half of the light passes through the beam splitter to be reflected off a mirror mounted on the wafer stage, while the other half of the light is reflected to a fixed reference mirror. The two beams of light are recombined at a detector, constructively or destructively, depending upon the relative phases of the beams. From the intensity of the detector signal, the stage position can be determined. The direction of stage motion is determined from Doppler shifts of the beam reflected from the stage mirror[140].

Stage position is measured interferometrically using HeNe laser beams reflected off mirrors on the sides of the stage with $\lambda/64$ or $\lambda/128$ precision typical of tools up to the 0.5 μm generation[141,142]. New systems use $\lambda/512$ or $\lambda/1024$. Measurements of stage yaw or rotation are also made in three axis systems[143]. The best current stages have six degrees of freedom and can step a Cartesian grid accurate to about 25 nm and repeatable to about 15 nm, 3σ[123].

The effective wavelength of the light in air can vary because of environmental changes (Table 1.10). Interferometer errors due to temperature- or pressure-induced fluctuations in the optical path length of the air surrounding the wafer stage can be reduced by shielding the path from air currents and thermal gradients. However, since the stage must be able to move rapidly and freely within a large area, it is impossible to shield the entire path length. Lis has described an air-turbulence compensated interferometer[144] that reduces the error due to environmental fluctuations dramatically. Systematic position errors due to deviations from ideal flatness of the stage mirrors can be characterized and compensated for in software.

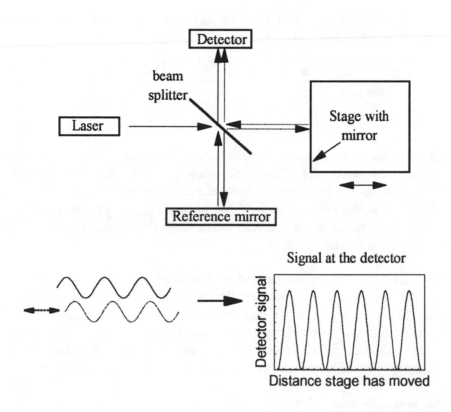

FIG. 1.53 The Michelson laser interferometer used to measure the position of the stage.

In order to minimize the dead time between successive exposures, one must improve the stage jerk, acceleration, maximum velocity, and settling time (the time required to stop and settle to a tolerable level of vibration). A simple analysis of stage motion is done through computing the average time per step of a given distance, e.g., one full field width: 20 mm. This time is typically in the range 250 to 500 ms.

Tradeoffs in the design of stages involve balancing the needs for stage speed and positional accuracy. High stage acceleration and velocity are most easily accomplished with a low mass stage, but wafer size, thermal stability, and dimensional rigidity dictate a lower limit to the size of the stage and the type of materials that can be used in its construction. A high performance stage design is found on the ASM-L 5500 series stepper, which employs a Zerodur stage that rides on a pneumatic airfoot over granite[123]. The stage is driven by an H-frame linear electric motor and is capable of high acceleration and peak velocity.

Table 1.10 Changes in environmental parameters that will result in 1 part-per-million changes in the wavelength of light in air[143].

Air temperature	1° C
Pressure	2.5 mm Hg
Relative humidity	80%

Since the imaging performance of the stepper can be severely affected by vibrations[145], the rapid movements of the wafer stage must be effectively decoupled from the optical column through pneumatic isolation and active damping mechanisms. Vibrational analysis is done by CAD in the design of new wafer stepper bodies to find and minimize natural resonances at the low frequencies (0 to 200 Hz) that can blur projected images.

With the advent of scanning reticle and wafer stages, control of vibrations assumes an even larger role. The mechanical problem of scanning a reticle stage at four times the velocity of a wafer stage moving in the opposite direction while keeping the positions of the two stages synchronized to an error of less than 10 nm requires the highest degree of active vibration control available. Stepper throughput is ultimately limited by the stepper's overlay error at the highest scan rates[146].

1.3.7 Alignment Systems

The original GCA 4800DSW had an off-axis, white-light microscope that could inspect the wafer at high magnification with a binocular viewing head or the image could be displayed on a black and white TV monitor. A pair of alignment marks, consisting of a segmented cross plus a second cross rotated 45° (an X), separated by three inches on the wafer could be viewed separately and moved by joystick control of the stage to fall within fiducial marks etched into the microscope objectives. The reticle was aligned separately to the optical column via a separate alignment microscope and manipulator system. Once the wafer was aligned to the fiducial marks, by human operators and joystick in the first systems and later by automatic alignment systems, it would be moved a known distance of a few inches to correspond to the designed location for exposure. If the reticle and wafer were perfectly aligned and if all the distances were as designed, and the wafer stage blindstepped the right distances, then the wafer would be well aligned. However, the baseline (the distance between the align position and the expose position) in this stepper was unstable with respect to temperature and varied widely in time, requiring periodic calibration by running test wafers for registration checks[147]. The baseline error severely limited both the overlay performance and the productivity of this stepper.

Commercial attempts to eliminate baseline type errors and directly align reticle to wafer images were made in the following generation of steppers[148,149]. But there were new problems, which included:

1) Relatively low NA of the stepper lens compared to a dedicated alignment microscope, yielding a lower contrast image of the wafer alignment target.

2) Monochromatic illumination used for imaging leads to severe thin film interference effects for wafer reflectance signals when faced with the normal range of thickness variation in films like resist and nitride or oxide[150].

3) Asymmetrical resist coating profiles severely distort line edge profiles when viewed in monochromatic light, leading to alignment offsets that depend on wafer position[151].

4) Antireflection coatings used to improve linewidth control make targets invisible at actinic wavelengths. Alignment at non-actinic wavelengths requires correction of chromatic aberration of the projection lens.

5) The advent of chemical mechanical polishing for wafer planarization has brought new challenges for alignment systems. Wafer alignment targets can be given asymmetrical profiles by the polishing process, or destroyed altogether.

The IBM stepper had a broadband, dark field alignment system that worked off-axis but had a small baseline distance, having its optics built into the stepper imaging optics[119]. This alignment system was later adopted by the GCA Corp. to replace the TV bright field system[152]. These systems work well on most process levels but still have problems. For example, when the surface texture of the wafer becomes rough (e.g., metal layers like Al or W), a lot of light is scattered into the dark field cone, reducing SNR. A good reference detailing optimization of dark field alignment on a stepper is given by Smith and Helbert[153].

The Philips SIRE stepper used a laser interferometric phase grating alignment system[117, 154] to directly align reticle and wafer images through the main stepper lens. Auxiliary optics were used to compensate for optical path differences between 633 nm light as used by the alignment system and 405 nm light for which the imaging lens was designed. A unique feature of this alignment system is that it is not dependent on the line edge profile, but rather on the phase depth of the wafer alignment mark, designed to give a reflected wave at 633 nm a phase shift of π radians. This has proved to be very robust compared to line edge sensing alignment systems in the face of the normal process variation that led to asymmetrical edge profiles.

The Philips (later, ASM-L) stepper uses the so-called zero level alignment strategy, in which the wafer has alignment marks etched directly into bare silicon before any other processing. All subsequent masks are aligned to the zero level marks. This strategy has subtle advantages (tertiary or higher alignment errors of one layer to another, simple strategy to implement) and disadvantages (extra processing step, does not allow ultimate in possible alignment accuracy of one layer to another since all alignments are secondary). With a few improvements, the ASM-L alignment system has been utilized on successive stepper models up to the current 5500 model, using the same global alignment marks throughout[123].

Early commercial steppers tried to improve overlay by aligning each stepper field separately, so-called die-by-die (D×D) alignment[149,150]. This was a misguided effort since it was found that using D×D alignment seriously decreased stepper throughput without improving alignment accuracy. Since D×D alignment marks are typically smaller than global marks, the time spent to acquire the marks has to be much lower in order to keep throughput up. Also, there was wide variation across the wafer in the alignment signals for reasons explained in points (2) and (3) above, which resulted in larger alignment errors on average than did two point global alignment.

Nikon introduced the concept of enhanced global alignment (EGA), which used Nikon's standard laser scanning alignment (LSA) system to acquire the position of 5 to 10 alignment marks across the wafer and then computed a least squares fit to the data[155,156]. The global alignment of the wafer is then corrected on the basis of the errors computed. This method has proved to be extraordinarily effective and has been adopted on commercial steppers today as a standard alignment scheme. Enhanced global alignment will give better overlay than D×D alignment whenever the precision of the laser stage is greater than the precision of the stepper's ability to capture alignment signals. In particular, it will have reduced sensitivity to degraded individual alignment targets.

Canon has perfected a broadband TV alignment system that uses a high resolution CCD camera system to align the wafer mark to the reticle mark[157]. The broadband illumination helps to reduce thin film interference effects, while the CCD camera has extremely small effective pixel sizes. However, the broadband illumination dictates a separate optical system from the projection lens, so this is an off-axis alignment system and must rely on baseline stability and autocalibration to achieve tight registration. An excellent article that reviews several types of global alignment methods has been written by van den Brink, Linders, and Wittekoek[158].

1.3.8 Stepper Total Overlay

The total overlay error expected when random steppers are mixed on the fab line is a key performance parameter. After resolution, total overlay error is the dominant term in determining lithography-limited die size[159]. The NTRS roadmap specifies the overlay requirements for 350 and 250 nm devices as 100 and 75 nm, respectively[133].

Models for stepper overlay error started with Perloff's work[156], which described interfield registration errors as the linear sum of errors in translation, rotation, and wafer expansion, as well as residuals, some part of which are due to stage stepping errors and some due to metrology. This model was expanded by McMillen and Ryden[160] for step-and-repeat systems to include intrafield errors, adding terms for image translation, rotation, magnification, trapezoid, and third order distortion, as given Eqs. 1.92 and 1.93:

$$\delta x = T_x + \Theta y + Mx + t_1 x^2 + t_2 xy - Dx(x^2 + y^2) \tag{1.92}$$

$$\delta y = T_y - \Theta x + My + t_1 xy + t_2 y^2 - Dy(x^2 + y^2), \tag{1.93}$$

where δx and δy are the overlay errors along the x and y axes, T_x and T_y are translation errors, Θ is the coefficient for field rotation, M accounts for reduction errors or wafer expansion and contraction, t_1 and t_2 are trapezoid coefficients, and D accounts for third order distortion. These overlay errors in Eqs. 1.92 and 1.93 occur within each exposure field at position (x,y) where the field center is the origin.

There is a similar model for step-and-scan systems:

$$\delta x = T_x + \Theta_y y + M_x x \tag{1.94}$$

$$\delta y = T_y - \Theta_x x + M_y y. \tag{1.95}$$

For step-and-scan systems, there are two independent field rotations and magnifications. The independent rotations lead to an error referred to as skew. Distortion is usually low in modern step-and-scan systems.

The characteristic overlay errors of wafer steppers are illustrated in Fig. 1.54, which shows that the analysis can be separated into intrafield and interfield, or grid, errors. Fig. 1.55 shows characteristic grid errors, including those made by the stepping stage and wafer alignment system, while Fig. 1.56 illustrates common intrafield errors, chiefly caused by the stepper optical system and reticle writing errors.

A comprehensive attempt to quantify stepper total overlay has been made by van den Brink and colleagues at ASM-L in a series of papers[161,162,163]. Zavecz has recently traced the development of the stepper overlay model[164]. Arnold has attempted to show how overlay error distributions can be estimated from knowledge of the component errors and their variation[165]. The idea of quantifying overlay in terms of the percentage of stepper fields that contain no error greater than the specification, the good fields rule, was advanced [82,165, 166].

Lens placement errors are the dominant term in the total overlay error today. Distortion has been nearly eliminated in the design ($\leq$ 10 nm) at each of the leading suppliers, but manufacturing errors in lens assembly and glass inhomogeneities are responsible for random distortions on the order of 30-50 nm over the full field in each lens. When two lens images are overlaid, the worst case error is thus expected to be about 50 - 70 nm. With a total overlay budget of 100 nm at 0.35 µm and 75 nm at 0.25 µm, there is increasing pressure on the suppliers to reduce this number significantly. Lens distortion is an example of a systematic error, which changes little over time and appears with the same magnitude at the same location in the stepping field or on the wafer each time.

In step-and-scan lithography some systematic overlay errors, notably lens distortion, are averaged in the direction of the scan motion. Other systematic lens errors including field curvature are also improved in the scan direction[167]. Pattern registration in reticle writing is another significant term in stepper total overlay.

Almost all overlay errors have a systematic component as well as a random component. Consider the following typical example. A wafer is selected from the production line on which several steppers are used interchangeably, i.e., any stepper can be used to image a given masking layer, to produce ICs. One takes measurements of total overlay error seen on a given wafer, from field to field. Each field will have a common vector map, with an average translation error, rotation error, degree of reduction error, and so forth. These averages correspond to a systematic component of the error. However, there is also a variation of the component errors from field to field, and this variation is considered to be a random component of the error.

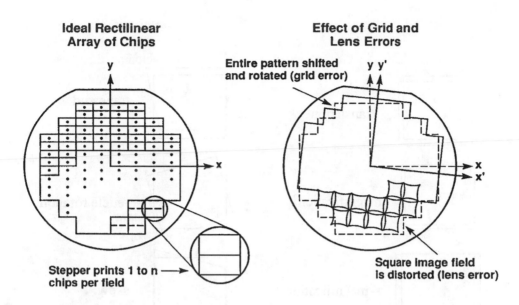

Ideal Rectilinear Array of Chips

Effect of Grid and Lens Errors

Entire pattern shifted and rotated (grid error)

Stepper prints 1 to n → chips per field

Square image field is distorted (lens error)

FIG. 1.54 Characteristic stepper overlay errors.

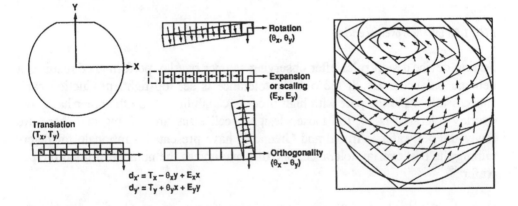

Rotation (θ_x, θ_y)

Expansion or scaling (E_x, E_y)

Translation (T_x, T_y)

Orthogonality ($\theta_x - \theta_y$)

$$d_{x'} = T_x - \theta_x y + E_x x$$
$$d_{y'} = T_y + \theta_y x + E_y y$$

FIG. 1.55 Interfield (grid) overlay errors.

Considered from this viewpoint, many of the systematic errors can be largely zeroed out through adjustments to the stepper. For example, the relative magnification difference between two steppers can be measured and compensated for. However, some systematic errors defy compensation, the most notable example being distortion, which results in an inherently random error across the lens field. In addition, it is impossible to totally eliminate systematic components due to metrology errors and to limits in mechanical and/or electrical adjustments to the stepper. The magnitude of this uncompensated systematic error defines the limit of overlay.

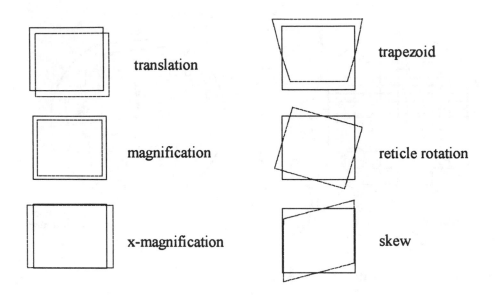

translation

trapezoid

magnification

reticle rotation

x-magnification

skew

FIG. 1.56 Intrafield overlay errors.

Setting device design rules after observing stepper overlay performance requires a delicate balancing act. If the overlay tolerance is set too tight, production can't successfully make the part with high process capability. If, on the other hand, the overlay tolerance is set too loose, designed cell sizes are too big and there are fewer die per wafer. Arnold and Greeneich have presented a rationale for determining the best overlay tolerance, one that maximizes the number of good die per wafer[159].

Consider the standard problem of determining the tolerance when covering a contact hole or via with a metal line. One of the geometrical design rules provides that the contact hole must be completely covered by the metal line or else the device will fail. Relative pattern overlay error from the metal to contact layer is typically the biggest contributor to the tolerance, but one must also consider the random linewidth control at both contact and metal masking and etching steps, as well as the biases or other systematic errors at each step.

If the overlay error can be reduced, one can use a tighter tolerance and thus achieve a smaller metal pitch, which in many cases is an important determinant of device die size. Die size in turn is a major factor in the overall cost of producing a given device since die size determines how many parts will be made on each wafer and also the relative susceptibility to defects.

On the other hand, if the tolerance does not make sufficient allowance for the over-lay error, then a high percentage of the devices may fail, as might occur when the contact hole plug conductor is exposed to metal etch chemistries, which could re-sult in contact "opens". Since some overlay errors, such as distortion, are system-atic and can show up in every stepping field at a given magnitude, the yield can go to zero if every die contains a systematic error greater than the tolerance.

Given a particular device and set of geometrical design rules, as well as the proc-ess defect density, the systematic and random components of the overlay, and those of the feature size control, analysis can be used to estimate the number of good die per wafer (GDW) as a function of the tolerance[159]. GDW is zero up to a level about equal to the systematic error, climbs to a peak at a tolerance equal to the overall systematic error plus Z times the overall random component ($Z \leq 3$), and then falls off gradually as the tolerance is further increased. In order to maxi-mize GDW, one tries to work near the peak of the curve. In practical cases, the yields fall off more rapidly in the direction of smaller tolerances.

1.3.9 Stepper Productivity and Cost of Ownership

Since the stepper has always been one of the most expensive pieces of capital equipment in the IC fab, and in most areas the number of steppers is the measure of the overall capacity of the fab, stepper productivity has long been scrutinized with extra interest. The rated throughput as quoted by the manufacturer is a maxi-mum number for a given simple pattern with the stepper fully utilized. In the real world, the stepper steps many different patterns, many unoptimized for maximum field utilization, is sometimes not being used due to lack of work on hand or to lack of an operator to run the machine, is sometimes down for maintenance, is sit-ting idle while the results of test wafers are analyzed, is limited by the rate at which the track it's interfaced with can process wafers, and so forth. If an IC manufacturer achieves 70% of the raw throughput as net, this is held to be pretty good.

The throughput R of the stepper (sometimes referred to as raw throughput) can be computed from Eqs. 1.96 and 1.97:

$$R \text{ (wafers per hour)} = \frac{3600}{\text{time to process one wafer}} \tag{1.96}$$

$$= \frac{3600}{t_{oh} + N_e(t_e + t_s)}, \tag{1.97}$$

where the other parameters are defined below. Net stepper throughput can be esti-mated by calculating the raw throughput[168] multiplied by the percentage that the stepper is utilized. Important areas to focus on when trying to improve the raw throughput rate of a stepper are

1) the overhead time, t_{oh} (secs): reduce the time necessary to remove the exposed wafer, to prealign the next wafer, to transfer the wafer to the exposure chuck, and to fine align the wafer;

2) the exposure time, t_e (secs) = D/I_0, D is the required dose in mJ/cm², and I is the illumination intensity in mW/cm²: must improve illumination power and/or resist sensitivity;

3) the stepping time, t_s (secs): must improve stage acceleration and velocity, stage settling time; and

4) reduce the number of required steps to cover the wafer, N_e, by increasing the field size.

Representative throughput and parameter values for i-line steppers, which are taken from a comparative study of i-line, deep UV, and soft x-ray lithography cost of ownership[169.,170.] are listed in Table 1.11.

Table 1.11 Representative i-line throughput parameters.

Parameter	Value
Power to the wafer plane	2.0 - 2.3 W
Intensity to the wafer	500 - 570 mW/cm²
Field size	4.0 - 5.1 cm²
Die size	0.8 × 0.9 cm²
Dose requirement	200 mJ/cm²
Exposure time (t_e)	0.35 - 0.40 s
Step time (t_s)	0.30 - 0.48 s
Steps (full field)	65 - 71 steps
Steps (die per field)	66 - 99 steps
Overhead time (t_{oh})	24 - 28 s
Throughput (full field)	41 - 52 wph
Throughput (2/3 field)	33 - 51 wph

One of the most important items to consider after raw stepper throughput is its utilization, i.e., the percentage of time it is stepping product at the raw rate. Fig. 1.57 shows lithography costs for three different steppers as a function of tool utilization, with and without reticles and resist costs added. A common detractor from stepper productivity is the inefficient use of the stepping field, as occurs when the die size does not evenly divide the image field. Fig. 1.58 shows graphically how the costs of lithography are increased on the three i-line steppers when the field is only 60% utilized.

A key to improving stepper utilization is to eliminate the need for test wafers to calibrate alignment, required exposure, or focus setting before printing a wafer lot. Early steppers were notoriously unstable in time so that baseline or focus setting had to be reacquired periodically, wasting production time. Stability of steppers has improved a lot over the 1980s and 1990s as a result of customer demand. In addition, steppers now have in-situ calibration devices,[135,171,172,173] some modeled after the original Censor Auto-Cal system,[150] which could be used to calibrate both stepper overlay and focus setting.

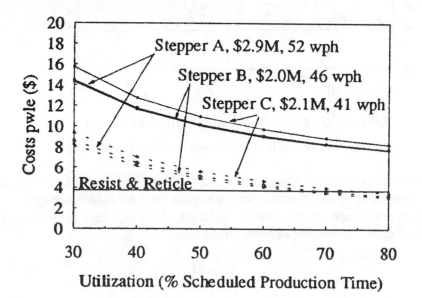

FIG. 1.57 Cost per wafer layer for three i-line steppers, with varying raw throughput[169].

The calculation for step-and-scan throughput starts with Eq. 1.97 with modified terms for the exposure time t_e and for stepping time t_s. Consider an illuminated rectangular aperture of height H and width W being scanned at uniform velocity v in the +x direction. A reticle field of height H and width L will be exposed in a time

$$t_e = \frac{L+W}{v}, \tag{1.98}$$

but we know that $t_e = D/I_0$ in order to properly expose the resist. So we can relate the scan speed of the wafer stage to the intensity, the required dose, and the field length and scan slit width by

$$v = \frac{(L+W)I_0}{D}.$$ (1.99)

If there are edge fields that only partially cover the wafer, the scan length for those fields will be less than L. This is different than in the step-and-repeat case, where all fields have the same exposure time, regardless of size.

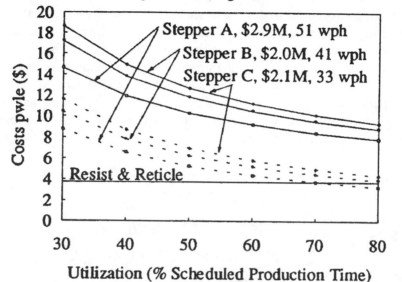

FIG. 1.58 Cost per wafer layer for three i-line steppers with inefficient field utilization[169].

On the other hand, there is reduced advantage for large fields in the step-and-scan configuration, relative to step-and-repeat, assuming constant light intensity regardless of field size. Suppose one doubles the length of the exposure field, which results in having 1/2 the number of fields on the wafer:

$$L \rightarrow 2L$$ (1.100)

$$N \rightarrow \frac{N}{2}.$$ (1.101)

Then, the time to process one wafer is given by

$$t_{oh} + \frac{N}{2}\left(t_s + \frac{W}{v}\right) + NI,$$ (1.102)

where t_s is a modified step-and-settle time, accounting for the longer field size. There is an improvement in throughput related to less stepping, but there is a factor independent of scan length, because longer fields need more time for scanning.

This is different than in the step-and-repeat case, where larger, and therefore fewer, fields directly reduce the exposure time, so long as the light intensity is held constant.

The average intensity in the wafer plane is related to the laser power P, the transmission of the optical system T, and the area of the illuminated slit:

$$I_0 = \frac{PT}{HW}, \tag{1.103}$$

so the scan speed can be written as

$$v = PT \frac{L+W}{DHW}. \tag{1.104}$$

The number of laser pulses that will expose the resist with scan speed v, slit width W, and laser repetition rate f is

$$N_p = \frac{fW}{v}. \tag{1.105}$$

The laser power P is related to the energy per pulse E_p and the repetition rate f by

$$P = E_p f, \tag{1.106}$$

so the intensity I_0 in the wafer plane is

$$I_0 = \frac{E_p f T}{HW}, \tag{1.107}$$

and the dose per pulse is

$$D_p = \frac{E_p T}{HW} = \frac{I_0}{f}, \tag{1.108}$$

while the total dose delivered to the wafer is

$$D = N_p D_p. \tag{1.109}$$

From the statistical nature of KrF and ArF sources, the energy per pulse is a random variable with an average E_p and a standard deviation σ_E. The error in the total dose is estimated by the standard error of the mean σ_E / N_p. One can calculate the number of pulses needed to achieve a given level of dose control given the pulse-to-pulse energy variation exhibited by the laser.

While stepper throughput is usually the most important contributor to the overall cost of ownership of a lithography process, other important factors are the cost of materials (wafers, reticles, resist, developer), the cost of other support equipment (track, metrology, etc.), labor, utilities and other overhead expenses. The utilization and throughput dependent lithography cost per wafer level exposure (pwle) in dollars[170] is calculated from

$$C_{UTD} = \frac{C_{LPH}}{T_p U} , \tag{1.110}$$

where U is the fraction of scheduled production time during which the system is actually in use, and C_{LPH} is the lithography cost per hour, which is given by

$$C_{LPH} = C_{ED} + C_L + C_{FP} + C_M. \tag{1.111}$$

In Eq. 1.111, C_{ED} is the capital equipment depreciation cost per hour; C_L is the per hour labor cost of running the lithography system (stepper plus track, typically); C_{FP} is the cleanroom cost per hour associated with the system footprint; and C_M is the per hour cost of maintaining the system. The utilization and throughput independent cost pwle is given by

$$C_{UTI} = C_C + C_R , \tag{1.112}$$

where C_C is the pwle cost of consumables, such as resist, and C_R is the pwle reticle cost. The total pwle cost is then given by

$$C_{TOT} = C_{UTD} + C_{UTI}. \tag{1.113}$$

The reticle, its cost, and its usage are important issues for overall lithography cost of ownership[169, 170]. Fig. 1.59 shows the cost per wafer level for 248 nm lithography and for 1× proximity x-ray lithography as a function of the usage of the reticle in terms of wafers exposed, and for two different reticle costs each[170].

Fig. 1.60 shows a study done in three AMD IC production fabs of the number of wafers out per reticle for all the devices processed over the span of a year[170]. The average number was only about 1800 wafers exposed per reticle, in the transition zone between reticle-dominated cost and reticle-free cost. Assuming an average of 2000 wafers exposed per reticle, the cost of one lithography layer is compared for i-line, 248 nm deep UV, and 1× x-ray lithography as a function of net stepper throughput in Fig. 1.61[169]. The two different x-ray curves correspond to cases in which the synchrotron storage ring is depreciated over five and ten years.

Impact of Reticle Cost and Usage

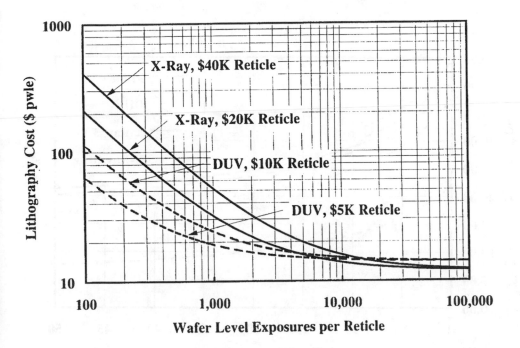

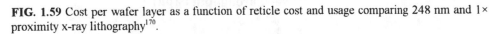

FIG. 1.59 Cost per wafer layer as a function of reticle cost and usage comparing 248 nm and 1×
proximity x-ray lithography[170].

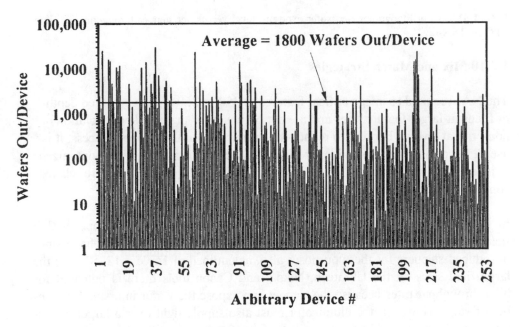

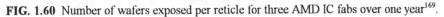

FIG. 1.60 Number of wafers exposed per reticle for three AMD IC fabs over one year[169].

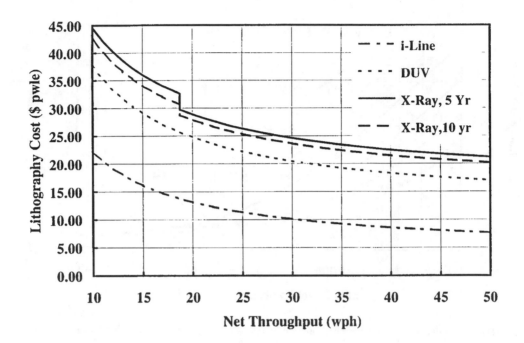

FIG. 1.61 Cost of lithography per wafer exposure level versus net stepper throughput — i-line, 248 nm, 1× x-ray[169].

1.3.10 Mix and Match Strategies

The NAs required to image critical features at 350 or 250 nm have limited depth of focus at larger feature sizes and are not optimum for printing low resolution or non-critical layers. In order to reduce the overall cost of the stepper process, it is useful to consider non-critical layer steppers. These machines would be optimized to image implant masks, pads, and other masks where the design rules are relaxed compared to the critical layers such as poly gate, contacts, and metal.

In 1992 Ultratech Stepper introduced the 2244i, a 1× catadioptric stepper with a field size of 22 × 44 mm, to meet these requirements. Other suppliers have since introduced steppers for these purposes also, as outlined in Table 1.12. Due to the large field sizes on the Ultratech, Nikon, and Canon tools there is potential for high throughput rates because the stepper can expose the wafer in a smaller number of steps. However, the illuminator must also supply light over a larger exposing area. When the field area expands by four, for the same illuminator design the light intensity will drop by about the same amount. In practice, the Nikon 4425i is reported to have 1/3 to 1/2 the light intensity in the wafer plane as the corresponding 22 mm field i-line stepper. Total stepping time is reduced, due to the decreased number of steps and settles, even though each step is twice as long. Since overlay requirements are relaxed, settling time can be shortened.

In order to fully realize the potential of these tools the throughput rates must be very high and the cost of the reticles must be lower than those for non-critical layers at 4 or 5×. In practice, it is in many cases difficult to implement mix-and-match strategies using steppers with different magnification ratios due to reticle-making problems, database errors, machine-matching errors, varying sign conventions, and other common errors.

Table 1.12 Non-critical layer steppers.

	Ultratech Saturn	Nikon 4425i	Canon 3000iW	ASML 5500/22
Reduction	1:1	2.5:1	2:1	5:1
NA	0.35	0.30	0.24	0.40
Wavelength (nm)	365	365	365	365
Resolution ($k_1 = 0.7$)	0.73 μm	0.85 μm	1.06 μm	0.70 μm
Field size	22 mm × 44 mm	44 mm × 44 mm	50 mm × 50 mm	22 mm × 22 mm

1.4 STEPPERS FOR THE 0.35 μm TO 0.18 μm GENERATIONS

1.4.1   0.35 μm /64 Mbit Generation: High NA i-Line and DUV Introduction

At most companies, the initial production of 64Mbit DRAM and similar density ICs is being done by extending the current high NA i-line technology used for 0.5 μm/16 Mbit manufacturing. Variable NA lenses coupled with thin i-line resists were used to push the familiar technology to the limits. The use of optical proximity correction on reticles, phase-shifting mask technology, and off-axis illumination to expand the exposure/defocus process window were all explored seriously for the first time in addressing the new issues for the 0.35 μm generation. Terasawa and coworkers at Hitachi gave the first convincing demonstration that i-line lithography could be driven to the 0.35 μm level[111], by demonstrating the feasibility of phase-shifting technology.

The first use of deep UV reduction stepper technology using step-and-scan technology was also adopted for the 0.35 µm generation of microprocessors. This was done primarily for the improved gate CD control that could be achieved in the higher k_1, scan-averaged process. However, the majority of the work at 0.35 µm is done by i-line steppers. As early as the 1990 IEDM, both Fujitsu and Matsushita reported[174,175] attempts to make the 64 Mbit DRAM with i-line phase shift lithography. Hitachi reported[176] the use of i-line phase-shifting lithography to make a 16Mbit SRAM, also at the 1990 IEDM.

The production i-line stepper featured a variable numerical aperture lens, with a maximum of 0.60 or greater, with a correspondingly wide image field, 22 mm on a side or greater[124,135]. Nakagawa et al.[187] gave the area of the Fujitsu 64 Mbit DRAM at 0.3 µm design rules as 11.2×19.94 mm^2. Thus two chips could be accommodated by a circular field with 31 mm diameter. The illuminator allowed the partial coherence of the imaging system to be adjustable. It also allowed quadrupole or annular illumination conditions. Refinement of the current art accompanied by the promise of imaging at greater than the Rayleigh depth of focus with advanced illuminators and/or phase-shifting mask technology brings the 0.35 µm generation within the capability of i-line. even though the process k_1 was less than 0.6.

Less critical mask layers with looser resolution and overlay requirements can be printed on lower NA i-line steppers (NA ~ 0.3-0.4) in order to achieve better depth of focus. A relaxed NA allows larger imaging fields, so these tools can also have throughput advantages, potentially reducing cost of ownership.

While showing much promise, deep UV lithography is only now being introduced to production in the 350 nm generation, and that only in the leading edge high speed CMOS logic industry, i.e., the U.S. microprocessor manufacturers. Progress in deep UV has been delayed by two major problems:

1) the lack of a DUV positive resist with acceptable stability, good performance on product substrates, and reasonable cost, and

2) the added cost of purchasing and operating the deep UV stepper, compared to i-line.

The most common form of deep UV lithography features excimer laser illumination and all fused silica projection optics, which were first developed at AT&T Bell Laboratories by Pol, and co-workers[177]. In a competing technology, the innovative Micrascan step-scan system[178] uses a mixture of glass and mirror optics and high power Hg arc lamp illumination to achieve imaging in the deep UV.

Deep UV lithography requires new resist chemistries since the familiar diazo napthaquinone resist used for a decade in g- and i-line steppers is too absorbing to produce images with vertical profiles at 248 nm. IBM has led progress in research and prototype applications with chemically amplified resists[179,180,181], but the materials are not yet mature. They are easily contaminated by environmental traces of base materials [e.g., HMDS and N-methyl pyrilidone (NMP)][182]. In addition, they are very transparent, which leads to serious reflective notching on poly and metal layers[183]. Most substrates are more reflective at 248 nm than at i-line or g-line. As a result, the use of bottom anti-reflection coatings for deep UV lithography began to become fairly common.

Deep UV lithography started to find its way onto the fab floor in the 0.35 μm generation for limited application at critical mask layers once resist performance on real substrates improved and once people understood that the extra costs can be justified by increased functional die yields or bin distributions due to, for example, improved critical dimension control.

1.4.2   0.25 μm/256 Mbit Generation: KrF Step-and-Scan

As minimum feature sizes shrink to less than 0.35 μm, i-line steppers are likely to give way to KrF excimer laser steppers and step-scan systems. At 248 nm wavelength the resolution of the stepper is improved by almost 50% over i-line for the same NA. The third generation of these steppers has recently become commercially available for initial 0.25 μm IC development and prototyping.

Die size 256 Mbit DRAM started at about 25 mm × 12 mm. The NTRS roadmap specifies a field size of 26 mm square. Since the maximum writing area on reticle writing systems is currently limited to about 130 mm square on a six inch quartz blank, it is difficult to continue to use the 5× reduction ratio commonly in use through the 16 Mbit DRAM generation. The 0.25 μm generation of exposure tools use 4× reduction. The mask contribution to CD and overlay error is thus larger by 20% compared to 5×, but this contribution is still less than that contributed by either the exposure tool or attributable to the process. However, the shortage of reticle writing equipment capable for the 0.25 μm generation of IC technology threatened to slow the development of 4× systems, particularly in Japan.

A major difficulty in extending the full field excimer laser stepper technology to the 0.25 μm level is the field size requirement of 26 mm square. An all-refractive, full field lens with NA 0.6 and 26 mm square image field with diffraction limited performance at 248 nm is estimated to weigh over 1800 lbs and cost over a million dollars by itself (estimate from SVG Lithography Systems, Inc.). Both the extreme mass and cost of the lens are forbidding, so much so that stepper suppliers have been designing alternatives to full-field stepping with conventional lens designs. Innovative full field designs have been proposed for the 0.25 μm generation. Markle has devised the 1:1 Half-Dyson deep UV lens, which is designed to achieve 0.25 μm resolution over a 25 mm square field, at 0.7 NA and 250 nm wavelength[184]. 1× optical systems are not popular at leading edge resolutions, however, due to the difficulty in making masks for such steppers. Other alternatives include field stitching and lens scanning.

The die-stitching method was reported by workers at Hitachi to make a 17.3 mm by 8.2 mm 16 Mbit DRAM with a first generation i-line stepper with a 10 mm square lens field[185], before the wide field lenses now available appeared. Stitching allows the use of well corrected small field lenses to image large die sizes. Stitching, however, is not an attractive option because of the need for two masks at every layer, overlay errors that occur at the stitching boundary, and sharply reduced throughput due to reticle exchanges needed for every wafer, among other reasons.

In 1989 Buckley and Karatzis[178] introduced the Micrascan, essentially a scanning subfield projector operating at 250 nm. Using a 4× reduction catadioptric system (one that uses both refractive and reflective optical elements), the Micrascan I scanned a slit 2 mm wide and 22 mm long over a 32.5 mm wide area (at the wafer plane), using vertical reticle and wafer stages. Once a full field is scanned, the Micrascan steps the wafer to the next field, which is scanned, and so forth. The system offered impressive resolution, field size, and focus-on-the-fly capabilities. In practice, the asymmetrical optical system was difficult to assemble and align to the tolerances needed for diffraction limited imaging.

A second generation Micrascan system, Micrascan II, was introduced with higher NA (0.50 vs. 0.357 on MS I) for use in 0.35 μm production[186]. The Micrascan II optical design (shown in Fig. 1.62) considerably simpler and less difficult to build than its predecessor, features a large primary mirror and a large beam splitter cube. The NA of the system is limited by the size of the beam splitter cube. The use of the aspheric mirror element allows a designed bandwidth of 4 nm, as compared to the 2 pm bandwidth for an all fused silica reduction stepper lens at 248 nm. Thus the Micrascan II can use a Hg arc lamp for the illumination source, which would be far too weak a source to use with an all-refractive lens design.

A high NA (0.63) version of this optical system has recently been designed for 0.25 μm prototyping. The bandwidth of 300 pm is wide enough to allow the use of an unnarrowed KrF laser, potentially boosting the light intensity in the wafer plane and thus improving scan speeds and throughput. In addition to being able to use unnarrowed lasers, the catadioptric system enjoys several other advantages compared to all-refractive fused silica lenses. It contains only about a dozen elements compared to 20 or more for the excimer laser stepper lens, which makes it both less massive and less costly to build. Consider the comparison of the NA 0.63 catadioptric lens cross-section with that of the equivalent all-refractive lens, shown in Fig. 1.50. It also achieves much broader stability with respect to temperature or pressure changes than is possible with all-refractive lens systems. This is illustrated in Table 1.13[187].

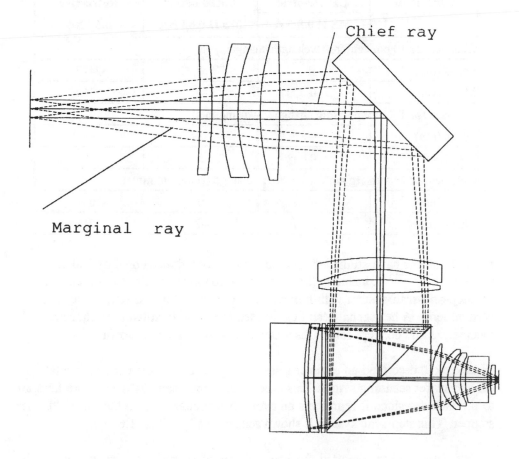

FIG. 1.62 4×, NA = 0.50, λ = 248 nm, catadioptric lens used in the Micrascan II, from Williamson [188].

Another difficulty with exposure systems that use monochromatic illumination is their susceptibility to substrate thin film optical effects, a topic that was discussed in Section 1.2.4. Reflective optics are advantageous because broadband illumination can be used with them. In practice, high numerical aperture fully reflective optics are difficult to design and build, and microlithographic lenses that contain reflective elements also contain some refractive lenses as well. Such mixed reflective and refractive lenses are called "catadioptric." An example was shown in Fig. 1.50.

Table 1.13 Comparison of sensitivities of catadioptric and refractive lenses to changes in wavelength, temperature, and pressure.

Parameter	Catadioptric MS II 0.5 NA	Catadioptric MS II 0.63 NA	Refractive 0.63 NA
Changes for 10 pm center-wavelength change:			
Focus (nm)	−1	0	2427
Mag (nm)	0	−1	−105
Changes for 1° C temperature change (Invar mounts):			
Focus (nm)	−40	−39	−5008
Mag (nm)	0	5	215
Changes for 20 mm Hg ambient pressure change (vented mounts):			
Focus (nm)	−3	0	5062
Mag (nm)	0	−2	−218

Several authors have studied the possibility of a two-dimensional optical scanning system[189,190,191,192,193] including Jain[192,193], who reported on a reduction excimer laser step-and-scan system with horizontal scanning wafer and reticle stages and a vertical lens. A hexagonal scan field is used to smoothly stitch two adjacent fields together. Jain's concept for the laser step-and-scan system is shown in Fig. 1.63.

Wittekoek[194] described an excimer laser step-and-scan system with horizontal reticle and wafer scanning stages. This type of step-and-scan system is more familiar to the industry since it represents an evolutionary change from the full field laser stepper. This step-scan system is shown schematically in Fig. 1.64 [194].

In late 1993, Nikon reported the construction of a lens scanning system[195], the NSR S201A. It is likely that difficulties in making higher resolution, wider field reduction optics as well as larger area reduction reticles will lead to more common use of step-and-scan techniques in the future. A schematic rendition of the Nikon step-and-scan tool that appeared in an industry trade paper is shown in Fig. 1.65.[196]

The use of pulsed excimer lasers with scanning exposure requires careful attention to dose uniformity. Since the pulse-to-pulse repeatability of the laser is only about 5%, a single pulse cannot be used to expose the resist and achieve good CD control, so averaging over several pulses or an active dose control method is required.

This means the scanning stage cannot move a significant fraction of the slit width in time $1/f$, or it will cause the resist to be exposed by too few pulses for good CD control. Synchronization of the movement of the scanning stage with the firing of the laser pulses is important in order to allow each field to receive the same number of laser pulses.

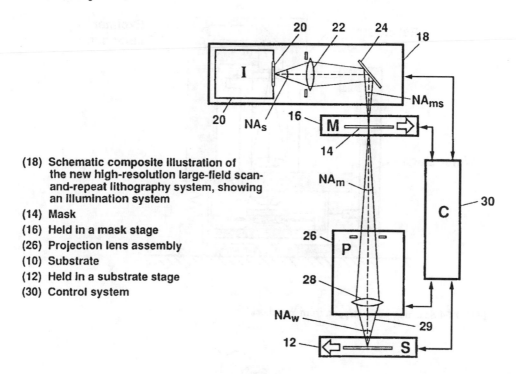

(18) **Schematic composite illustration of the new high-resolution large-field scan-and-repeat lithography system, showing an illumination system**

(14) **Mask**

(16) **Held in a mask stage**

(26) **Projection lens assembly**

(10) **Substrate**

(12) **Held in a substrate stage**

(30) **Control system**

FIG. 1.63 Reduction excimer laser step-and-scan system[192, 193].

Overlay control becomes increasingly complex in step-and-scan tools compared with conventional steppers, while at the same time the requirements tighten. The problem of measuring the positions of two stages moving in opposite directions at speeds on the order 100 mm/s and synchronizing their positions to within 10 nm so that overlay rules can be met is a significant challenge to modern interferometry, digital computing, and mechanical control systems.

These improvements to the basic lithographic tool, along with the improved optical projection system, cost a lot of money. The cost of a leading edge stepper has doubled every three to four years since the introduction of the wafer stepper around 1979.[169] Cost of ownership is a major issue for 0.25 µm lithography; reductions in the cost of lithography will be very difficult to achieve unless several improvements are made to the existing deep UV lithography cost structure.

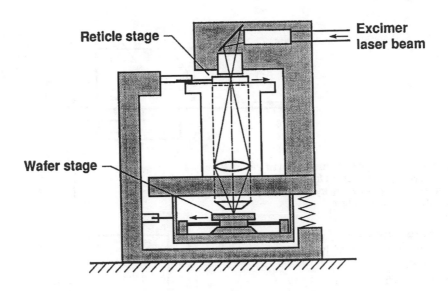

FIG. 1.64 Scanning wafer stepper, from Wittekoek[194].

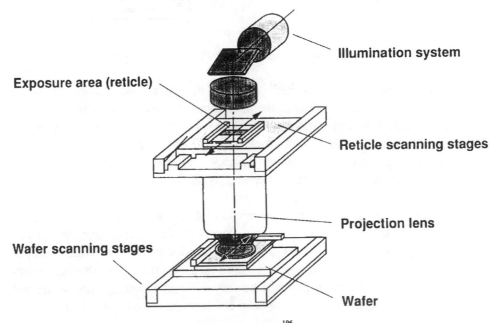

FIG. 1.65 Nikon NSR S201A step-and-repeat system[196]

1.4.3 0.18 µm Optical Lithography

IC manufacturers, having just introduced 248 nm processes into the masking area to make 0.25 µm devices, will be loath to switch again with the 0.18 µm generation, for 1 Gbit DRAM production. Unfortunately, at 0.18 µm feature size the usable depth of focus for high NA 248 nm imaging is simulated to be less than 0.5 µm. It is not known whether the process window for 248 nm will be sufficient to allow extension to two generations of use. However, every attempt to extend 248 nm will surely be made by IC manufacturers.

Hutchinson, Oldham and co-workers at the University of California at Berkeley[197] have studied the potential use of 213 nm, as generated by the fifth harmonic of Nd:YAG. The optical properties of fused silica are much the same as they are at 248 nm, so standard reduction lenses can still be fabricated. However, the percentage reduction in wavelength from 248 nm is very modest.

An attractive optical alternative to 248 nm uses the ArF excimer laser at 193 nm as the exposure source[198]. Workers in Japan reported first attempts to make ArF reduction steppers[199]. While the use of 193 nm has the theoretical potential to give IC makers both greater depth of focus at 0.18 µm and greater limiting resolution, perhaps even to 0.13 µm, there are serious technical issues. An ARPA-funded project at MIT's Lincoln Laboratories has shown encouraging initial results in adapting the SVG-L Micrascan II catadioptric lens (Fig. 1.62) and scanning mechanism to 193 nm exposure[200]. The tradeoffs in using a catadioptric system over a dioptric system at 193 nm are discussed later.

The major technical issues for the development of 193 nm lithography have been:

1) The transmission of fused silica is reduced at 193 nm compared to 248 nm (~1% absorption per cm at 193 nm), leading to formidable lens design and fabrication issues. Less path length through glass is required to keep transmission losses reasonable and aspheric elements may have to be used.

2) Laser damage to fused silica at 193 nm may limit the use of all-refractive (dioptric) or catadioptric lenses.

3) Development of resist materials for 193 nm exposure which have good lithographic and etch resistance in common plasma chemistries (CF_4, HBr, CCl_4, etc.)

Two main types of damage to fused silica have been identified: color center formation (which leads to increased absorption and fluorescence in the glass), and compaction (which leads to changes in local refractive index). Glass damage studies by MIT Lincoln Labs and UC Berkeley show the degree of color center formation in fused silica can be controlled through glass fabrication. Glass compaction is proportional to $(NI^2)^{0.7}$, where N is the number of laser pulses, and I is the intensity. Compaction of fused silica lenses will limit the laser fluence which may establish a limit for throughput. Indeed, 193 nm lithography will likely be much slower than 248 nm due to the relatively low power available in the wafer plane for resist exposure from the first generation of systems.

Achromatization of all-refractive designs using CaF_2 in a small number of the elements which must withstand the highest laser power densities is a promising path provided CaF_2 can be procured with suitable refractive index homogeneity and can be polished to the required finish.

Lens heating effects, so important in i-line lithography and largely absent in 248 nm lithography, will likely be quite severe in the extension of 193 nm lithography to 0.13 μm. CaF_2 has poor thermal stability (18 ppm expansion per degree C as compared with 0.5 ppm for fused silica). However, the change of refractive index with temperature is opposite in sign from its thermal expansion, offering potential compensation for thermal changes.

It is not yet apparent whether fully-refractive (dioptric) lens designs will work for high throughput lithography using 193 nm. Dioptric achromats must be studied carefully for shifts in focus and magnification with lens heating. In addition to absorption effects, there is likely to be lot of light lost in the optical system due to imperfect transmission coatings at each glass interface.

The wide color correction of the MSIII catadioptric lens design makes it uniquely suited to the adaptation of 193 nm as the exposure wavelength, since it can use a partially narrowed laser (fwhm ~ 50 pm). This allows a less expensive and less complex laser.

The catadioptric lens design also has fused silica components, notably the large beamsplitter cube, which leads to lens heating problems due to light absorption at 193 nm[201]. Glass damage in the form of compaction can occur over long periods of exposure at high intensity[202]. Thus intensity in the wafer plane is limited to the threshold at which glass damage begins to occur, which in turns limits the throughput of the stepper[201]. It does not seem to be possible to build a large beamsplitting cube from CaF_2.

The current state of resists sensitive to 193 nm exposure is immature compared to either 248 nm or i-line materials. Workers from Fujitsu[203] proposed the use of resist materials based on alicyclic polymers that were reasonably transparent at 193 nm. The etch resistance of the resist could approach that of novolac resist with the incorporation of 50 mol% of adamantyl methacrylate in the base polymer, but as the percentage is increased the resolution of the resist is degraded. Progress has been made elsewhere in developing resists with good resolution and high sensitivity (10-20 mJ/cm2) based on polymethracrylates. Photoacid generators are mixed with the methacrylate polymer to form a resist with chemical amplification properties and which can be developed in alkaline solutions common to the photo room today. Workers at IBM and MIT Lincoln Labs have developed methacrylate terpolymer materials that can show high resolution and vertical profiles, but have substantially reduced etch resistance compared to novolak or PVP based resists[204].

Excellent image formation in silylated resist systems at 193 nm have also been demonstrated. However, top surface imaging resists thus far have required long exposures (~100 mJ/cm2), which limit their interest to applications where high cost-of-ownership can be justified. In addition, line edge roughness remains a key problem with top surface imaging processes.

Brown[205] has addressed the problems of finding the huge amounts of funding that are required to bring 193 nm lithography to the point of successful commercial insertion in IC manufacturing lines for 0.18 μm, estimating the total bill as about $350 million.

1.5 YIELD

In addition to creating features of appropriate shape and size, and overlaying these to a previous layer, the microlithographer must create patterns with low levels of defects. As technology advances, the minimum tolerable defect size must decrease, scaling with the minimum feature sizes of the circuits. In addition, the defect density must also decrease, scaling the (increasing) size of integrated circuits. Fortunately, as requirements have become formidable, a powerful tool set has been developed. These tools include hardware as well as operating methodologies.

The dominant sources of defects in resist patterns originate in the materials and processes associated with photoresist application and development. This situation has resulted from significant improvements in cleanroom air cleanliness and operator garmenting. Defects can be controlled by the use of photochemicals manufactured with low particulate levels, point-of-use filtration, and careful control of the equipment used to apply the resist and developer. The key element of any defect control program in a lithography operation must involve the measurement of defects on wafers that have resist films on them. This leads to an issue which must be addressed at the beginning of all defect control programs — the assurance that defect measurements on wafers coated with resist films are reliable. The primary problem with such measurements has its origin in thin film optical effects, identical physics to that which was found earlier to affect imaging.

The nature of the thin film optical problem associated with particle detection can be appreciated once one understands how particles are detected. Consider, for example, the detection of defects based upon the scattering of laser light. For particles larger than the wavelength of the light, Mie scattering[18] predominates, which is characterized by scattering primarily in the forward direction. For particles on top of surfaces, detection is accomplished by the detection of forward scattered light, after it has been reflected back from the substrate (Fig. 1.66). Certain combinations of wavelengths and thin films can lead to "invisible" defects.

There are also process issues related to yield that must be considered. For example, development of isolated contact holes may be incomplete because of inadequate wetting or agitation. Improvement is possible by adjustments to the process or use of developers with better surfactants.

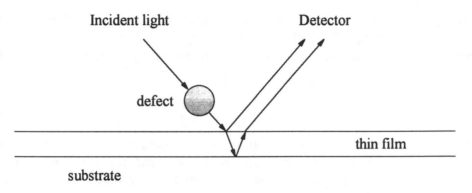

FIG. 1.66 Scattering of light from a defect (particle) on the surface of a wafer.

Another problem found with coatings is associated with a mismatch between photoresist and organic anti-reflection coatings. Again, this type of problem can be ameliorated, or avoided altogether, by modifications of the spin-on materials or adjustments to the coating processes.

1.6 THE LIMITS OF OPTICAL LITHOGRAPHY

As mentioned at the beginning of this chapter, the demise of optical lithography has been predicted because of fundamental physics — inadequate depth of focus as resolution is increased, caused by diffraction. These early predictions were wrong, because they assumed that optical lithography was already nearly diffraction limited, and extrapolations were based upon that assumption. The extent to which these assumptions were off the mark could be demonstrated by the application of today's superior lens design and manufacturing capability. Until recently, these capabilities have been applied only to leading edge, high numerical aperture lenses, but with the introduction of very large field systems for mix-and-match applications, actual diffraction limits for smaller numerical apertures can be observed. For example, Nikon has recently introduced an 0.3 NA i-line lens with its 4425i stepper. This lens has a resolution of 0.7 μm and over 5 μm depth of focus, while ten years ago, similar lenses had 1.0 μm resolution and 1.5 μm depth of focus. Extrapolating from these two different sets of capabilities, one arrives at significantly different conclusions. While these early predictions were incorrect in detail, the fact remains that diffraction imposes limits on the potential of optical lithography. At what feature size does diffraction finally overcome ingenuity and invention?

One can improve depth of focus at a given resolution, by decreasing wavelength, as seen by Eq. 1.69. Early indications are that 193 nm excimer laser lithography is achievable. Early resist formulations show that single layer resist systems can be synthesized, and imaging systems have been built. While the resists need to be improved, and there are a number of practical issues regarding 193 nm optics to be addressed, particularly damage, there do not appear to be any fundamental reasons that 193 nm cannot be made into a practical reality. For the remainder of this discussion, let us assume that this will be the case.

Recently, 250 nm features have been imaged in 0.8 μm thick films of photoresist using phase-shifting masks and off-axis i-line illumination, with reasonable depths of focus. Assuming that one obtains the same proportional improvement using phase-shifting and alternative illumination methodologies at 193 nm wavelengths, we can expect to be able to process wafers with 130 nm features using optical lithography.

For features smaller than 130 nm there are no obvious solutions, but there are a number of possibilities, particularly if resist systems with thin imaging layers can be developed. One of the problems with using optical lithography at wavelengths between 20 and 193 nm is that virtually all materials are highly absorbing, making thick resist layers impossible. With thin imaging layers, the issue becomes one only of making the optics. It should be noted that the mechanical stability of photoresist patterns becomes an issue when the height of features becomes greater than four times the features widths, particularly with wet development, though some progress has been made in this area[206, 207]. "Thick" resist systems are assumed to have a 4:1 aspect ratio. For 130 nm features, the resist can be no more than 520 nm thick. The nature of optical lithography is highly dependent upon etch capability and selectivities to photoresist. With thin imaging layers there is expected to be some improvement even for wavelengths shorter than 193 nm.

The emerging field of "wavefront engineering", an umbrella term coined to include all the techniques that seek to modify the effects of diffraction in order to improve resolution and depth of focus[208], will face significant challenges at the 0.18 μm level and below to forestall the long-forecasted "death of optics". Wavefront engineering has certain benefits for specific feature types, dependent on the optical configuration of stepper and mask. The challenge will be for the commercial stepper manufacturers to build systems that allow quick and reliable changes between configurations without throughput or performance degradation. A challenge for the maskmakers will be to build defect free and low cost reticles.

Optical lithographers are optimistic that high NA, 193 nm projection coupled with resolution enhancements such as phase-shifting reticles and off-axis illumination will allow extension to 0.13 μm[209,210], with volume production predicted to begin by 2004. How little depth of focus and exposure latitude is ultimately required for high yield IC manufacturing is still an open question. Lithographers have recently started looking at the potential of ultraviolet lasers operating at 157 nm and 126 nm to provide an optical path to 0.1 micron. Calcium fluoride still transmits light at 157 nm, so dioptric and catadioptric lens designs are still potentially possible. If the glass problems turn out to be intractable, all-reflective optics designs may become attractive. High NA reflective designs with reasonable field size have only recently become available, however. At 126 nm, reflecting elements coated with aluminum are still usable. Extreme ultraviolet lithography has spurred interest in new reflecting designs, but these are typically low NA (~0.1) due to the short wavelengths involved (13 nm).

Overlay is another critical parameter whose limit needs to be considered. Unlike resolution, which is limited fundamentally by the phenomenon of diffraction, overlay is limited by practical considerations and capability. The limits of overlay can be appreciated by considering the factors that affect overlay. For a stepper-to-itself, these factors are listed in Table 1.14.

Of these parameters, stage precision is the most fundamental, as it has a direct impact on overlay and it also contributes significantly to other factors, such as alignment target acquisition and control of grid errors. The significance of stage precision was discussed in Section 1.3, where stages were seen to be fundamental to the workings of wafer steppers. For a stepper-to-itself, the contribution of stage precision to overlay has been estimated to be as high as 80% of the total overlay capability of the stepper.[211] Improvements to wafer stage technology translate directly to improved overlay capability. From the recent work by Lis on stages that can directly measure and compensate for air turbulence[144], significant improvements in stepper overlay can be expected in the near future.

Table 1.14 Factors which affect stepper overlay capability.

Stage precision
Alignment target acquisition
Control of correctable errors
Non-correctable errors which drift over time
Reticle registration errors

After stage precision, alignment target acquisition is the parameter which has the most significant impact on total stepper overlay capability. One of the challenges for lithographers and manufacturers of wafer steppers is achieving alignment target acquisition on processed wafers which is as good as that realized on ideal substrates. Metal and grainy polysilicon surfaces have long presented problems for stepper alignment systems, and more recently problems have been encountered with wafers which have been through chemical-mechanical polish. One solution to this problem has been to align all layers to a "zero-layer" which has very consistent alignment targets. While such alignment schemes result in all critical alignments being indirect, superior performance can often result if alignment target acquisition is too variable for direct alignments. The limits for overlay will be reached when alignment systems are completely insensitive to the substrate.

Control of correctable errors depends upon the ability of the stepper to make accurate mechanical adjustments. For example, after measuring the positions of alignment targets and computing what the grid and field magnification adjustments should be, the stepper must be adjusted physically. Grid adjustments depend entirely upon the stage precision and are included in the direct overlay contributions by the stage. Intrafield adjustments require the movement of reticles or lens elements, and the actuators involved will have some limiting accuracy.

An example of non-correctable overlay errors which drift over time occurs on steppers that do not adjust trapezoid errors through software. For a machine-to-itself, non-correctable sources of overlay error are anachronisms. For example, trapezoid error is negligible on steppers with doubly-telecentric lenses, which are found on most modern steppers.

An estimate for a practical limit to overlay can be made by creating a hypothetical hybrid machine, taking the best attributes from all makes of steppers, and assuming certain improvements to stage precision and other mechanical adjustments based upon current research and development. These estimates are presented in

Table 1.15. In estimating the ultimate future performance it is also assumed that optimal performance is obtained on process layers. The total overlay result is based upon simulations of overlay using the stage precision, alignment target acquisition and control of correctables as inputs to the model. Enhanced global alignment is assumed with direct measurements of intrafield errors on both the wafer and reticle.

When multiple steppers are used there are systematic lens and stage matching errors which are introduced, as discussed in Section 1.3. The impact of these factors can be estimated through overlay simulation. For example, suppose that lens placement errors lead to a - 20 nm X registration error in the upper left part of the exposure field and +20 nm in the upper right part of the field. For a stepper that otherwise has modeled 75 nm overlay capability, to itself, the total overlay distribution grows to 85 nm 3σ. Thus, the overall effect, on total overlay, of a maximum lens placement error Δ between two steppers is approximately $\Delta/2$. Improvements in lens manufacturing technology, registration metrology and the use of step-and-scan configurations are all contributing to reductions in the contributions of lens placement errors to total overlay. Lenses with less than 20 nm maximum lens placement errors have already been made, and it is not unreasonable to expect further improvements to less than 10 nm. Grid errors between matched machines have already been reduced to less than 5 nm through software. The biggest limitation to further reduction has been the noise in the data from stage precision. Assuming significant improvements in stage precision in the near future, grid matching can be taken to effectively zero. Taking the estimated future overlay capability for a stepper-to-itself from Table 1.15, the total overlay capability that should be expected in the near future for matched steppers is 30 nm. This level of performance is sufficient to meet the SIA roadmap requirements for 100 nm technology, but not 70 nm technology. The overlay requirements for 70 nm technology may require the dedication of wafer lots to individual steppers, which will reduce overall stepper productivity and introduce another source of costs.

There seems little chance for the introduction of non-optical lithographies at the 0.18 µm level due to unproven feasibility for production and enormous initial investments required to shift to new, unfamiliar technologies. The industry is likely to forestall the introduction of post-optical technologies such as proximity x-ray or electron cell projection lithography until the 0.13 µm or 0.1 µm level, unless niche applications are found where the payback is worth the effort.

Assuming that non-optical lithography technologies will first find their way into the fab to image critical mask levels at the 0.13 µm level, the industry has less than five years to choose one or two likely alternatives and develop them to the point of being ready for insertion. In order to succeed optical lithography for volume device manufacturing, several important technical and business requirements will have to be met, which include:

- high resolution over large areas,

- tight level-to-level overlay,

- no device damage,

- cost of ownership not to exceed ~ 35% of total IC manufacturing cost, and

- well-developed infrastructure for resists, masks, support equipment, and the exposure tools.

Most importantly, all these requirements must be met simultaneously. Only optical lithography to date has been able to achieve this high level of integration. It is unclear if any other technology will be able to really replace it for mass production of devices with features too small for optics.

Table 1.15. Overlay capability for an ideal hybrid stepper, to itself, compared to current technology.

	Current	Estimated future
Stage precision	17 nm, 3σ	4 nm, 3σ
Alignment target acquisition	25 nm, 3σ	8 nm, 3σ
Control of correctibles	0.15 ppm for grid parameters 1.5 ppm for intrafield parameters	0.04 ppm for grid parameters 0.4 ppm for intrafield parameters
Total stepper contribution	45 nm	15 nm
Reticle registration errors (4X reduction stepper)	15 nm	5 nm
Total overlay performance	60 nm	20 nm

1.7 SUMMARY

In this chapter we have reviewed the state of the lithography art. Issues inherent to optical lithography have been discussed, particularly those that represent intrinsic limitations to optical micropatterning technologies. A question often asked is: What are the limits to optical lithography? Ten years ago the authors of this chapter predicted that optical lithography would be capable of supporting at least 0.5 μm semiconductor technology[4] because it was clear that such capability was possible by straightforward (but by no means trivial) extension of the existing technology. It was a matter of implementing controls to address the process issues that were known. During the intervening years since that prediction 0.5 μm technology, and smaller, has gone into production.

Extrapolating from today's technology, it appears that 130 nm lithography will be achieved using optical means. This will result from continuing improvements in exposure equipment, photoresists, process control, and the implementation of emerging techniques, such as off-axis illumination and phase shifting. As we have seen, our ultimate capabilities appear to be limited only by the ingenuity of the rational human mind, which always seems to take us farther than anyone ever expects.

1.8 ACKNOWLEDGMENTS

The authors would like to thank Chris Lyons for Figs. 1.23 and 1.29, AZ Photoresist Products unit of Hoechst Celanese Corporation for Fig. 1.31 and the data for Fig. 1.32, ASM-Lithography for contributing Figs. 1.48 and 1.63, SVG Lithography for contributing Table 1.13 and Fig. 1.50 and Fig. 1.62, and Dr. Kathy Early for allowing us to use Table 1.11 and Figures 1.57-1.60.

1. A. C. Tobey, "Wafer stepper steps up yield and resolution in IC lithography," Electronics, pp. 109 - 112, Aug. 16, 1979

2. J. Lyman, "Optical lithography refuses to die," Electronics, pp. 36 - 42, Oct. 7, 1985.

3. "Specification for metrology pattern cells for integrated circuit manufacture," SEMI Book of Standards, Micropatterning Volume, pp. 91 - 109, 1992.

4. H. J. Levinson and W. H. Arnold, "Focus: The critical parameter for submicron lithography," J. Vac. Sci. Technol. B5, pp. 293 - 298, 1987.

5. W. H. Arnold and H. J. Levinson, "Focus: The critical parameter for submicron lithography, Part 2," SPIE Vol. 772, pp. 21 - 34, 1987.

6. L. Larmore, *Introduction to Photographic Principles*, Second edition, Dover Publications, New York, 1965.

7. P. D. Blais, ""Edge Acuity and Resolution in Positive Type Photoresist Systems," Solid State Technol. Vol. 20, 76 - 85, August, 1977.

8. W. H. Arnold and H. J. Levinson, "High resolution optical lithography using an optimized single layer photoresist process," Kodak Microelectronics Seminar, pp. 80 - 92, 1983.

9. Shipley Co., Newton, Mass., private communication.

10. Current authors, unpublished.

11. Nick Eib, private communication.

12. H. H. Hopkins, *Wave Theory of Aberrations*, Clarendon Press, 1950.

13. C. A. Mack, "PROLITH: A comprehensive optical lithography model," SPIE Vol. 538, Optical Microlithography IV, pp. 207 - 220, 1985.

14. C. A. Mack, "Comparison of Scalar and Vector Modeling of Image Formation in Photoresist," SPIE Vol. 2440, pp. 381 - 394, 1995.

15. L. E. Stillwagon, R. G. Larson, and G. N. Taylor, "Planarization of Substrate Topography by Spin Coating," J. Electrochem. Soc., 134, 2030 - 2037, 1991.

16. D. B. LaVergne and D. C. Hofer, "Modeling Planarization with Polymers," SPIE Vol. 539, pp. 115 - 122, 1985.

17. L. K. White, "Approximating Spun-On, Thin Film Planarization Properties on Complex Topography," J. Electrochem. Soc., 132, pp. 168 - 172, 1985.

18. M. Born, and E. Wolf, *Principles of Optics*, 6th edition, Pergamon Press, New York, 1980.

19. G. B. Airy, Trans. Camb. Phil. Soc., Vol. 5, p. 283, 1835.

20. Lord Rayleigh, Phil. Mag. Vol. 8(5), p. 261, 1879.

21. T. Brunner, and R. R. Allen, "*In situ* measurement of an image during lithographic exposure," IEEE Electr. Dev. Lett., EDL-6, No. 7, pp. 329 - 331, 1985.

22. T. Brunner and R. R. Allen, "*In situ* resolution and overlay measurement on a stepper," SPIE Vol. 565, pp. 6 - 13, 1985.

23. A. K. Pfau, R. Hsu and W. G. Oldham, "A Two-Dimensional High-Resolution Stepper Image Monitor," SPIE Vol. 1674, pp. 182 - 192, 1992.

24. A. R. Neureuther, P. K. Jain, and W. G. Oldham, "Factors affecting linewidth control including multiple wavelength exposure and chromatic aberrations," SPIE Vol. 275, pp. 48 - 53, 1981.

25. McCleary, R. W., P. J. Tompkins, M. D. Dunn, K. F. Walsh, J. F. Conway, and R. P. Mueller, "Performance of a KrF excimer laser stepper," pp. 396 - 399, SPIE Vol. 922, 1988.

26. E. Abbe, Archiv. f. Mikraskopische Anat., Vol 9, p. 413, 1873.

27. G. Owen, R. F. W. Pease, D. A. Markle, and A. Grenville, J. Vac. Sci. Technol. B10(6), 1993.

28. H. J. Levinson, as discussed in Semiconductor International, Vol. 6 (12), pp. 22 -24, 1983.

29. J. N. Wiley, "Effect of stepper resolution on the printability of sub-micron 5× reticle defects," SPIE Vol. 1088, pp. 58 - 73, 1989.

30. G. Arthur, B. Martin, F. Goodall, and I. Loader, "Printability of sub-micron 5× reticle defects at g-line, i-line and duv exposure wavelengths," SPIE Vol. 2196, pp. 241 - 252, 1994.

31. J. W. Goodman, *Introduction to Fourier Optics*, McGraw Hill, New York, 1968.

32. W. J. Smith, *Modern Optical Engineering*, McGraw Hill, New York, 1966.

33. W. Oldham, P. Jain, A. Neureuther, C. Ting, H. Binder, "Contrast in high-performance projection optics," Kodak Microelectronics Seminar, pp. 75 - 80, 1981.

34. F. H. Dill, W. P. Hornberger, P. S. Hauge and J. M. Shaw, "Characterization of Positive Photoresist," IEEE Trans. Electr. Dev., ED-22, No. 7, pp. 445 - 452, 1975.

35. Shipley Microelectronics Product Catalog, Shipley Company, Newton, Mass.

36. M. Exterkamp, W. Wong, H. Damar, A. R. Neureuther, C. H. Ting, W. G. Oldham, "Resist characterization: procedures, parameters, and profiles," SPIE Vol. 334, pp. 182 - 187, 1982.

37. Tokyo Ohka, private communication.

38. W. Oldham, S.Nandgaonkar, A. Neureuther, and M. O'Toole, "A General Simulator for VLSI Lithography and Etching Processes: Part I - Application to Projection Lithography," IEEE Trans. Electr. Dev. ED-26, pp. 717 - 722 (1970).

39. C. A. Mack, "Photoresist process optimization," KTI Microelectronics Seminar, pp. 153 - 167, 1987.

40. P. Luehrmann and G. Goodwin, "Photoresist process optimization and control using image contrast," KTI Microelectronics Seminar, pp. 279 - 292, 1989.

41. C. A. Mack, "Photolithographic optimization using photoresist contrast," KTI Microelectronics Seminar, pp. 1 - 12, 1990.

42. S. V. Babu and S. Srinivasan, "Optical density and contrast of positive resists," SPIE Vol. 539, pp. 36 - 43, 1985.

43. V. Srinivasan and S. V. Babu, "Effect of Process Parameters on Contrast of Positive Photoresists: Calculation," J. Electrochem. Soc., Vol. 133, pp. 1686 - 1690, 1986.
44. M. P. C.Watts, "Optical positive resist processing. II. Experimental and analytical model evaluation of process control," SPIE Vol. 539, pp. 21 - 28, 1985.
45. P. Spragg, R. Hurditch, M. Toukhy, J. Helbert, and S. Malhotra, "The reliability of contrast and dissolution rate-derived parameters as predictors of photoresist performance," SPIE Vol. 1466, pp. 283- 296, 1991.
46. R. Hurditch and J. Ferri, "Investigation of Positive Resist Performance Based on the Model of a Dissolution Switch," OCG Microelectronics Seminar, pp. 71 - 90, 1993.
47. C. A. Mack, "Analytical expression for the standing wave intensity in photoresist," Applied Optics, Vol. 25, pp 1958 - 1961, 1986.
48. E. J. Walker, "Reduction of photoresist standing-wave effects by post- exposure bake," IEEE Trans. Electr. Dev. ED-22, pp. 464 - 466, 1975.
49. T. Batchelder and J. Piatt, "Bake effects in positive photoresist," Solid State Technol., pp. 211 - 217, August, 1983.
50. J. Crank, *The Mathematics of Diffusion*, 2nd ed., Oxford University Press, 1975.
51. D. A. Bernard, "Simulation of post-exposure bake effects on photolithographic performance of a resist film," Philips J. Res. 42, pp. 566-582, 1987.
52. P. Trefonas, B. K. Daniels, M. J. Eller and A. Zampini, "Examination of the mechanism of the post exposure bake effect," SPIE Vol. 920, pp. 203 - 211, 1988.
53. J. Crank and G. S. Park, *Diffusion in Polymers*, Academic Press, London, 1968.
54. K. Phan, M. Templeton and E. Sum, "A Systematic Approach for I-line Manufacturing Resist Optimization," SPIE Vol. 1087, pp. 279 - 289, 1989.
55. I. I. Bol, "High-resolution optical lithography using dyed single-layer resist," Kodak Microelectronics Seminar, Interface '84, pp. 19 - 22, 1984.
56. C. A. Mack, "Dispelling the myths about dyed photoresist," Solid State Tech., pp. 125- 130, January, 1988.
57. K. Harrison and C. Takemoto, "The use of Anti-reflection Coatings for Photoresist Linewidth Control," Kodak Microelectronics Seminar, Interface '83, pp. 107 - 111 (1983).
58. H. Van den Berg and J. van Staden, "Antireflection coatings on metal layers for photolithographic purposes," J. Appl. Phys. 50(3), pp. 1212 - 1214, 1979.
59. W. H. Arnold, M. Farnaam, and J. Sliwa, US Patent Number 4,820,611, "Titanium Nitride as an Antireflection Coating on Highly Reflective Layers for Photolithography."
60. C. Nölscher, L. Mader, and M. Schneegans, "High Contrast Single Layer resists and Anti-reflection Layers: an Alternative to Multilayer Resist Techniques," SPIE Vol. 1086, pp. 242 - 250 (1989).

61. T. Ogawa, H. Nakano, T. Gocho, T. Tsumori, "SiOxNy:H, high performance anti-reflection layer for the current and future optical lithography," SPIE Vol. 2197, pp. 722 - 732, 1994.

62. T. Brewer, R. Carlson, and J. Arnold, "The reduction of the standing wave effects in positive photoresists," Jour. Appl. Photogr. Eng., Vol 7, no. 6., pp. 184 - 186, 1981.

63. J. Lamb and M. G. Moss, "Expanding photolithography process latitude with organic AR coatings," Sol. State. Technol., pp. 79 - 83, September, 1993.

64. S. Kaplan, "Linewidth control over topography using a spin-on AR coating," KTI Microelectronics Seminar, pp. 307 - 314, 1990.

65. B. Martin, A. N. Odell, and J. E. Lamb III, "Improved bake latitude organic anti-reflection coatings for high resolution metallisation lithography," SPIE Vol. 1086, pp. 543 - 554, 1989.

66. M. Rocke and M. Schneegans, "Titanium nitride for anti-reflection control and hillock suppression on aluminum silicon metalization," J. Vac. Sci. Technol. B6(4), pp. 1113 - 1115, 1988.

67. T. Brunner, "Optimization of optical properties of resist processes," SPIE Proceedings Vol 1466, pp. 297 - 308, 1991.

68. T. Tanaka, N. Hasegawa, H. Shiraishi and S. Okazaki, "A new photolithography technique with anti-relfection coating on resist: ARCOR," J. Electrochem. Soc., Vol. 137, pp. 3900 - 3905, 1990.

69. O. M. Heavens, *Optical Properties of Thin Solid Films*, Dover, New York, 1955.

70. A. V. Brown and W. H. Arnold, "Optimization of resist optical density for high resolution lithography on reflective surfaces," SPIE Vol. 539, Advances in Resist and Processing II, pp. 259- 266, 1985.

71. C. Nölscher, L. Mader, S. Guttenberger, and W. Arden, "Search for the optimum numerical aperture," Microel. Eng. Vol. 11, pp. 161 - 166, 1990.

72. K. Yamanaka, H. Iwasaki, H. Nozue, and K. Kasama, "NA and σ optimization for high-NA i-line lithography," SPIE Vol. 1927, pp. 310 - 319, 1993.

73. W. N. Partlo, S. G. Olson, C. Sparkes, and J. E. Connors, "Optimizing NA and sigma for subhalf-micrometer lithography," SPIE Vol. 1927, pp. 320 - 331, 1993.

74. B. Lin, "The optimum numerical aperture for optical projection micro- lithography", SPIE Vol. 1463, pp. 42 - 53, 1991.

75. A. Suzuki et al. "Intelligent optical system of a new stepper", SPIE, Vol. 772, pp. 58-65, 1987.

76. H. Ohtsuka, K. Abe, Y. Itok and T. Taguchi, "Quantitative Evaluation Method of Conjugate Point for Practical Evaluation of Wafer Stepper," SPIE Vol. 1088, pp. 124 - 133, 1989.

77. H. Fukuda, A. Imai, T. Terasawa, S. Okazaki, "New approach to resolution limit and advanced image formation techniques in optical lithography", IEEE Trans. El. Dev., Vol. 38, No. 1, 67-75, 1991.

78. J. W. Bossung, "Projection printing characterization," SPIE Vol. 100, pp. 80 - 84, 1977.

79. Brunner, T. A., A. L. Martin, R. M. Martino, C. P. Ausschnitt, T. H. Newman, and M. S. Hibbs, "Quantitative stepper metrology using the focus monitor test mask," SPIE Vol. 2197, pp. 541 - 549, 1994.

80. AZ Photoresist Products unit of Hoechst Celanese Corporation.

81. B. Katz, J. Greeneich, M. Bigelow, A. Katz, F. van Hout, and J. Coolsen, "High Numerical Aperture I-line Stepper," SPIE Vol. 1264, pp. 94 - 126, 1990.

82. H. J. Levinson and R. Rice, "Overlay Tolerances for VLSI Using Wafer Steppers," SPIE Vol. 922, pp. 82 - 93, 1988.

83. S. A. Lis, "Processing Issues and Solutions to Conjugate Lithography," Kodak Microelectronics Seminar, pp. 117 - 136, 1986.

84. M. S. Yeung, "Modeling high numerical aperture optical lithography," SPIE Vol. 922, pp. 149 - 167, 1988.

85. T. Hayashida, H. Fukuda, N. Hasegawa, "A novel method for improving the defocus tolerance in step-and-repeat photolithography", Proc. SPIE Vol. 772, 66-71, 1987.

86. H. Fukuda, N. Hasegawa, T. Tanaka, and T. Hayashida, "A new method for enhancing focus latitude in optical lithography: FLEX," IEEE EDL-8(4), pp. 179 - 180, 1987.

87. C. A. Spence, D. C. Cole and B. B. Peck, "Using Multiple Focal Planes to Enhance Depth of Focus," SPIE Vol. 1674, pp. 285 - 295, 1992.

88. E. Ong, B. Singh, A. Neureuther, and R. Ferguson, "Comparison of Proximity Effects in Contrast Enhancement Layer and Bilayer Resist Processes," J. Vac. Sci. Technol., 1886.

89. A. Starikov, "Use of a single size square serif for variable print bias compensation in microlithography: method, design and practice," SPIE Vol. 1088, Optical/Laser Microlithography II, pp. 34 - 46, 1989.

90. P. Chien and M. Chen, "Proximity Effects in Submicron Optical Lithography," SPIE Vol. 772, pp. 35 - 40, 1987.

91. P. Pforr, A. K. Wong, K. Ronse, L. Van den Hove, A. Yen, S. R. Palmer, G. E. Fuller and O. W. Otto, "Feature biasing versus feature-assisted lithography: a comparison of proximity correction methods for 0.5*(λ/NA) lithography," SPIE Vol. 2440, pp. 150 - 170, 1995.

92. D. L. Fehrs, H. B. Lovering, R. T. Scruton, "Illuminator modification of an optical aligner," KTI Microelectronics Seminar, pp. 217- 230, 1989.

93. C. A. Mack, "Optimum stepper performance through image manipulation," KTI Microelectronics Seminar, pp. 209 - 216, 1989.

94. K. Kamon, T. Miyamoto, M. Yasuhito, H. Nagata, M. Tanaka, and K. Horie, "Photolithography system using annular illumination," Jap. Journ. Appl. Phys., Vol. 30, No. 11B, pp. 3021 - 3029, 1991.

95. N. Shiraishi, S. Hirukawa, V. Takeuchi, N. Magome, "New imaging technique for 64M-DRAM," SPIE Vol. 1674, pp. 741 - 752, 1992.

94. K. Kamon, T. Miyamoto, M. Yasuhito, H. Nagata, M. Tanaka, and K. Horie, "Photolithography system using annular illumination," Jap. Journ. Appl. Phys., Vol. 30, No. 11B, pp. 3021 - 3029, 1991.

95. N. Shiraishi, S. Hirukawa, V. Takeuchi, N. Magome, "New imaging technique for 64M-DRAM," SPIE Vol. 1674, pp. 741 - 752, 1992.

96. W. N. Partlo, P. J. Thompkins, P. G. Dewa, P. F. Michaloski, "Depth of focus and resolution enhancement for i-line and deep-UV lithography using annular illumination," SPIE Vol 1927, pp. 137 - 157, 1993.

97. K. Tounai, H. Tanabe, H. Nozue, and K. Kasama, "Resolution improvement with annular illumination," SPIE Vol. 1674, pp. 753 - 764, 1992.

98. K. Tounai, S. Hashimoto, S. Shiraki, and K. Kasama, "Optimization of Modified Illumination for 0.25-µm Resist Patterning," SPIE Vol. 2197, pp. 31 - 41, 1994.

99. T. Ogawa, M. Uematsu, T. Ishimaru, M. Kimura, T. Tsumori, "Effective Light Source Optimization with the Modified Beam for Depth-of-focus, Enhancements," SPIE Vol. 2197, pp. 19 - 30, 1994.

100. T. Ogawa, M. Uematsu, F. Uesawa, M. Kimura, H. Shimizu, and T. Oda, "Sub-quarter-micrometer Optical Lithography with Practical Superresolution Technique," SPIE Vol. 2440, pp. 772 - 783, 1994.

101. H. Fukuda, T. Terasawa, S. Okazaki, "Spatial filtering for depth of focus and resolution enhancement in optical lithography", J. Vac. Sci. Technol. B9 (6), pp. 3113 - 3116, 1991.

102. H. Fukuda, "Axial image superposing (Super-FLEX) effect using the mask modulation method for optical lithography", JJAP, Vol. 30, 3037-3042, 1991; Proceedings of the 4th Micro Process Conference (Kanazawa, Japan), paper B-9-1, 190-91, 1991.

103. R. von Bunau, G. Owen, R.F. Pease, "Depth of focus enhancement in optical lithography", J. Vac. Sci. Technol. B 10(6), pp. 3047 - 3054, Nov/Dec 1992.

104. T. Horiuchi, K. Harada, S. Matsuo, Y. Takeuchi, E. Tamechika, Y. Mimura, "Resolution enhancement by oblique illumination optical lithography using a transmittance-adjusted pupil filter", Jpn. J. Appl. Phys. Vol. 34, Part. 1, No. 3, pp. 1698 - 1708, 1995.

105. M. D. Levenson, N. S. Viswanathan, R. A. Simpson, "Improving resolution in photolithography with a phase-shifting mask," IEEE Trans. Electr. Dev., ED-29(12), pp. 1828 - 1836, 1982.

106. K. Hashimoto, K. Kawano, S. Inoue, S. Itoh, and M. Nakase, "Effect of Coherence Factor σ and Shifter Arrangement for the Levenson-Type Phase-Shifting Mask," Jpn. J. Appl. Phys., Vol. 31, pp. 4150 - 4154, 1992.

107. C. A. Mack, "Fundamental Issues in Phase-Shifting Mask Technology," OCG Microelectronics Conference, pp. 23 - 35, 1993.

108. T. Tanaka, S, Uchino, N. Hasegawa, T. Yamanaka, T. Terasawa, and S. Okazaki, "A Novel Optical Lithography Technique Using the Phase-Shifter Fringe," Vol. 30, Jpn. J. of App. Phys., pp. 1131- 1136, 1991.

109. A. Nitayama, T. Sato, K. Hashimoto, F. Shigemitsu, and M. Nakase, "New Phase Shifting Mask with Self-aligned Phase Shifters for a Quarter Micron Photolithography," IEDM Tech. Dig., pp. 57 - 60, 1989.

110. Z. Cui, P. D. Prewett, and B. Martin, "Optimization of Rim Phase Shift Masks for Contact Holes," Microelctr. Engr. Vol. 23, pp. 147 - 150, 1994.

111. T. Terasawa, N. Hasegawa, T. Kurosaki, and T. Tanaka, "0.3-μm Optical Lithography Using a Phase-shifting Mask," SPIE Vol. 1088, pp. 25 - 33, 1989.

112. T. Terasawa, N. Hasegawa, H. Fukuda, and S. Katagiri, "Imaging Characteristics of Multi-Phase-Shifting and Halftone Phase-Shifting Masks," Jpn. J. App. Phys. Vol. 30, pp. 2991 - 2997, 1991.

113. B. J. Lin, "The Attenuated Phase-Shifting Mask," Solid State Technol., pp. 43 - 47, January, 1992.

114. M. Nakajima, N. Yoshioka, J. Miyazaki, H. Kusunose, K. Hosono, H. Morimoto, W. Wakamiya, M. Murayama, Y. Watakabe, and K. Tsukamoto, "Attenuated Phase-shifting Mask with a Single-layer Absorptive Shifter of CrO, CrON, MoSiO, and MoSiON Film," SPIE Vol. 2197, pp. 111 - 121, 1994.

115. G. Dao, G. Liu, R. Hainsey, J. Farnsworth, Y. Tokoro, S. Kawada, T. Yamamoto, N. Yoshioka, A. Chiba and H. Morimoto, "248 nm DUV MoSiON Embedded Phase-Shifting Mask for 0.25 Micrometer Lithography," BACUS Photomask News, Vol. 11, Issue 8, pp. 1 - 9 , 1995.

116. T. A. Brunner, "Rim Phase-shift Mask Combined with Off-axis Illumination: A path for 0.5λ/NA Geometries," SPIE Vol. 1927, pp. 54 - 62, 1993.

117. A. Bouwer, G. Bouwhuis, H. van Heek, S. Wittekoek, "The silicon repeater", Philips Tech. Rev., Vol. 37, No. 11, 349, 1977; see also S. Wittekoek, "Optical aspects of the silicon repeater", Philips Technical Review, Vol. 41, p 268, 1983/84.

118. M. Lacombat, A. Gerard, G. Dubroeucq, M. Chartier, "Photorepetition et projection directe", Rev. Tech. Thomson-CSF, Vol. 9, No. 2, 337, 1977; see also Hans Binder and Michel Lacombat, "Step-and-repeat projection printing for VLSI circuit fabrication", IEEE Trans. El. Dev., Vol. ED-26, No. 4, pp. 698-704, April 1979.

119. J. Wilcyncski, "Optical step-and-repeat camera with dark field automatic alignment," J. Vac. Sci. Technol., Vol. 16, pp. 1929-1933, 1979.

120. F. Klosterman, "A step-and-repeat camera for photomasks," Philips Tech. Rev., Vol. 30, No. 3, 57, 1969.

121. J. Roussel, "Step-and-repeat wafer imaging", Solid State Tech., 67, May 1978.

122. K. Takahashi, M. Ohta, T. Kojima, M. Noguchi, "New i-line lens for half-micron lithography", proc. SPIE, Vol. 1463, 1991.

123. M. van den Brink, B. Katz, S. Wittekoek, "New 0.54 aperture i-line wafer stepper with field by field leveling combined with global alignment", Proc. SPIE, Vol. 1463, pp. 709 - 724, 1991.

124. S. Stalnaker, P. van Oorschot, "Characterization of an advanced i-line lens with variable NA and coherence for 0.35 μm lithography", OCG Microlithography Seminar, pp. 157 - 173, 1993.

125. "Lithography", in The National Technology Roadmap for Semiconductors, Semiconductor Industry Association, 1994.

126. K. Suwa, K. Ushida, "Optical stepper with a high numerical aperture i-line lens and a field by field leveling system", Proc. SPIE, Vol. 922, pp. 270 - 276, 1988.

127. J. Braat, "Quality of microlithographic projection lenses", Proc. SPIE, Vol. 811, 22-30, 1987.

128. E. Glatzel, "New lenses for microlithography", Proc. SPIE, Vol. 237, pp. 310 - 320, 1980.

129. Ohara i-line Glasses, Ohara Corporation, Kanagawa, Japan, June, 1993.

130. Schott Catalog of Optical Glass No. 10000 on Floppy Disk, Edition 10/92, Schott Glass Technologies, Duryea, PA.

131. T. A. Brunner, S. Cheng, A. E. Norton, "A stepper image monitor for precise setup and characterization," SPIE Vol. 922. pp. 377 - 327, 1988.

132. J.H. Bruning, D.R. Herriott, J.E. Gallagher, D.P. Rosenfeld, A.D. White, and D.J. Brangaccio, "Digital wavefront measuring interferometer for testing optical surfaces and lenses," Applied Optics, Vol. 13, No. 11, pp. 2693 - 2703, November 1974.

133. J. E. Gortych and D. Williamson, "Effects of Higher-order Aberrations on the Process Window," SPIE Vol. 1463, pp. 368 - 381, 1991.

134. M. Nei, H. Kawai, "Reduction Projection Type Exposure Device: NSR2005i9C", Denshi Zairyo, pp80-84, March 1993.

135. D. Markle, "The future and potential of optical scanning systems", Solid State Technology, pp.159 - 166, Sept. 1984.

136. D. Markle, "Submicron 1:1 Optical Lithography", Semiconductor International, 137-142, May 1986.

137. H. Lovering, "Optics for microprojection", U.C. Berkeley Short Course, Feb. 11, 1980.

138. M.K. Templeton, E. Barouch, U. Hollerbach, S. Orszag, "Suitability of high NA i-line steppers with oblique illumination for linewidth control in 0.35 μm complex circuit patterns", Proc. SPIE, Vol. 1927, pp. 427, 1993.

139. M. Noguchi, M. Muraki, Y. Iwasaki, A. Suzuki, "Subhalf micron lithography system with phase-shifting effect", Proc. SPIE, Vol. 1674, pp.92 - 104, 1992.

140. Hewlett-Packard Co., "High performance motion control for precision equipment," 1990.

141. H. de Lang, G. Bouwhuis, "Displacement measurement with a laser interferometer", Philips Tech. Rev., Vol. 30, 160, 1969

142. A. Rude, M. Ward, "Laser transducer system for high accuracy machine positioning", Hewlett Packard Journal, 2, Feb. 1976

143. S. Wittekoek, H. Linders, H. Stover, et al., "Precision wafer stepper alignment and metrology using diffraction gratings and laser interferometry", Proc. SPIE, Vol. 565, pp. 22 - 31, 1985

144. Steven A. Lis, "An air-turbulence compensated interferometer", Proc. SPIE, Vol. 2440, 891-901, 1995.

145. B.J. Lin, "Vibration tolerance in optical imaging", Proc. SPIE, Vol. 1088, 106-114, 1989.

146. J. Bischoff, W. Henke, J. van der Werf, P. Dirksen, "Simulations on Step&Scan Optical Lithography", SPIE vol. 2197, pp. 953 - 964, 1994.

147. W.C. Schneider, "Testing the Mann type 4800DSW wafer stepper", SPIE Vol. 174, pp. 6-14, 1979.

148. A. Stephanakis, H. Coleman, "Mix and match - 10× reduction wafer steppers", Proc. SPIE, Vol. 334, 132-138, 1982.

149. Herbert E. Mayer, Ernst W. Loebach, "Improvement of overlay and focusing accuracy of wafer step-and-repeat aligners by automatic calibration", SPIE Vol. 470, pp.178 - 184, 1984.

150. S. Cosentino, J. Schaper, J. Peavey, "Application of thin film reflectance calculations to linewidth measurements for HMOS circuit fabrication", Kodak Microelectronics Seminar, pp. 34 - 39, 1983.

151. K.A. Chivers, "A modified photoresist spin process for a field-by-field alignment system", Kodak Microelectronics Seminar, pp. 44 - 51, 1984.

152. D.R. Beaulieu, P.Hellebrekers, "Dark field technology - a practical approach to local alignment", Proc. SPIE, Vol. 772, pp. 142 - 149, 1987.

153. C. Smith, J. Helbert, "Improving dark field alignment through target and mapping software optimization", Microlithography World, pp. 5 - 9, Apr-June 1993.

154. G. Bouwhuis, S. Wittekoek, "Automatic alignment system for optical projection printing", IEEE Trans. El. Dev., Vol. ED-26, No.4, 723-728, 1979.

155. S. Murakami, T. Matsura, M. Ogawa, M. Uehara, "Laser step alignment for a wafer stepper", SPIE Vol. 538, pp. 9 - 16, 1985.

156. D. S. Perloff, "A four point electrical measurement technique for characterizing mask superposition errors on semiconductor wafers", IEEE Sol. St. Circ., Vol. SC-13, No. 4, pp.436 - 444, 1978.

157. S. Uzawa, A. Suzuki, N. Ayata, "A new alignment system for submicron stepper", SPIE, Vol. 1261, pp. 325 - 331 1990.

158. M. van den Brink, H. Linders, S. Wittekoek, "Direct-referencing automatic two-points reticle-to-wafer alignment using a projection column servo system", SPIE, Vol. 633, 60-71, 1986.

159. W.H. Arnold, J. Greeneich, "The impact of stepper overlay on advanced IC design rules", OCG Microlithography Seminar, pp. 87 - 100, 1993.

160. D. McMillen, W. D. Ryden, "Analysis of image field placement deviations of a 5× microlithographic reduction lens", SPIE, Vol. 334, pp.78 - 89, 1982.

161. M. van den Brink et al., "Matching performance for multiple wafer steppers using an advanced metrology procedure", SPIE, Vol. 923, pp. 180-197, 1988.

162. M. van den Brink et al., "Matching management of multiple wafer steppers using a stable standard and a matching simulator", SPIE, Vol. 1087, pp. 218 - 232, 1989.

163. M. van den Brink et al, "Matching of multiple wafer steppers for 0.35 μm lithography using advanced optimization schemes", SPIE, Vol. 1926, pp.188 - 207, 1993.

164. T. E. Zavecz, "Machine models and registration", SPIE CR52, pp. 134 - 159, 1993.

165. W. H. Arnold, "Overlay simulator for wafer steppers", SPIE, Vol. 922, pp. 94 - 105, 1988

166. W. H. Arnold, "Steppers: their envelopes of performance", Semiconductor International, 61, Feb. 1991

167. H. Sewell, "Step and Scan: the maturing technology," SPIE Vol. 2440, pp. 49 - 60 1995.

168. H. Stover, "Near-term case for 5× versus 10× wafer steppers", Proc. SPIE, Vol. 334, 60-69, 1982

169. K. Early, W.H. Arnold, "Cost of ownership for soft x-ray lithography", OSA Meeting, Monterey, CA, May 1993.

170. K. Early, W.H. Arnold, "Cost of ownership for 1× proximity x-ray lithography", Proc. SPIE, Vol. 2087, pp. 340 - 349, 1993.

171. T. Brunner, S. Cheng, A.E. Norton, "A stepper image monitor for precise setup and characterization", Proc. SPIE, Vol. 922, p. 366, 1988.

172. "Optical reduction stepper with built-in automated metrology", Solid State Technology, pp. 467 - 468, October 1989.

173. M. van den Brink, H. Franken, S. Wittekoek, T. Fahner, "Automatic on-line wafer stepper calibration system", Proc. SPIE, Vol. 1261, pp 298-314, 1990.

174. K. Nakagawa, M. Taguchi, T. Ema, "Fabrication of 64M DRAM with i-line phase-shift lithography", paper 33-1, Proc. of the International Electron Devices Meeting (IEDM), San Francisco, 1990.

175. H. Watanabe, Y. Todokoro, M. Inoue, "Transparent phase shift mask", paper 33-2, IEDM 1990.

176. T. Yamanaka, N. Hasegawa, T. Tanaka, K. Ishibashi, T. Hashimoto, A. Shimizu, N. Hashimoto, K. Sasaki, T. Nishida, E. Takeda, "A 5.9 μm² super low-power SRAM cell using a new phase-shift lithography", paper 18-3, IEDM 1990.

177. V.Pol, J.H. Bennewitz, G.C. Escher, M. Feldman, V.A.Firtion, T.E. Jewell, B.E. Wilcomb, J.T. Clemens,"Excimer laser based lithography: a deep-ultraviolet wafer stepper", Proc. SPIE, Vol. 633, 6-16, 1986.

178. J. Buckley, C. Karatzis, "Step and scan: a systems overview of a new lithography tool", Proc. SPIE, Vol. 1088, 424, 1989.

179. Hiroshi Ito, C. Grant Willson,"Chemical amplification in the design of dry developing resist materials", Polymer Engineering and Science, Vol. 23, 18, 1012-1018, Dec. 1983.

180. R.L. Woods, C.F. Lyons, R. Mueller, J. Conway, "Practical half-micron lithography with a 10× KrF excimer laser stepper", KTI Microelectronics Seminar, pp. 341, 1988

181. S. Holmes, R. Levy, A. Bergendahl, K. Holland, J. Maltabes, S. Knight, K.C. Norris, D. Poley, "Deep ultraviolet lithography for 500 nm devices", SPIE Vol. 1264, pp. 61-70, 1990

182. S. MacDonald, N. Clecak, H. Wendt, C.G. Willson, C. Snyder, C.J. Knors, N. B. Deyoe, J. Maltabes, J. Morrow, A.E. McGuire, S. Holmes, "Airborne chemical contamination of a chemically amplified resist", Proc. SPIE, Vol. 1466, pp. 2 - 12, 1991

183. D.D. Dunn, J.A. Bruce, M.S. Hibbs, "DUV photolithography linewidth variation from reflective substrates", SPIE, Vol. 1463, pp. 8 - 15, 1991

184. H. Jeong, D. Markle, G. Owen, R. F. W. Pease, and A. Grenville, "An Optical Projection System for Gigabit DRAMs," EIPB93.

185. Y. Kawamoto, S. Kimura, N. Hasegawa, A. Hiraiwa, T. Kure, T. Nishida, M. Aoki, H. Sunami, K. Itoh, "A half-micron technology for an experimental 16 Mbit DRAM using i-line stepper", VLSI Symposium, Paper III-1, pp. 17-18, 1988.

186. M. Barrick, et al., "Performance of a 0.5 NA broadband DUV step-and-scan system", Proc. SPIE, Vol. 1927, pp. 595-607, 1993

187. Fig. 1.49 and Table 1.13 courtesy of SVG Lithography Systems, Inc.

188. US Patent # 4,953,960, "Optical reduction system", David Williamson, inventor, filed July 1988 and granted Sept. 1990.

189. J.H. Bruning and A.D. White, "Compact Image Projection Apparatus", US Patent No. 4,171,870, Oct. 23, 1979.

190. A.D. White, "Photolithographic Projection Apparatus Using Light in the Far Ultraviolet", US Patent No. 4,302,079, Nov. 24, 1981.

191. B.J. Lin, "The paths to subhalf-micrometer optical lithography", Proc. SPIE, Vol. 922, 256-269, 1988.

192. K. Jain, "Scan and repeat high resolution projection lithography system", US Patent No. 4,924,257, issued May 8, 1990.

193. K. Jain, "A novel high-resolution large-field scan-and-repeat projection lithography system", SPIE Vol. 1463, pp. 666 - 677, 1991.

194. S. Wittekoek, "Optical lithography: present status and continuation below 0.25 μm", Microcircuit Engineering, 1993.

195. K. Nozaki, "Scan and EB Lithography Rapidly Emerging", Nikkei Microdevices, 54-59, September 1993.

196. Electronics News, July 10, 1995, p. 8.

197. J. M. Hutchinson, W.N. Partlo, R. Hsu, W. G. Oldham,"213 nm lithography", Microelectronic Engineering, Vol. 21, pp. 15-18, 1993.

198. R. Sandstrom, "Argon fluoride excimer laser source for sub-0.25 μm optical lithography", SPIE, Vol. 1463, pp. 610-616, 1991

199. N. Nomura et al., Microelectronic Engineering, Vol. 11, 183, 1990; see also N. Nomura, K. Yamashita, M. Endo, and M. Sasago, "Lithography beyond 64Mb", Microelectronic Engineering, Vol. 21, p. 3, 1993.

200. M. Rothschild, R. Goodman, M. Hartney, M. Horn, R. Kunz, J. Sedlacek, D. Shaver, "Photolithography at 193 nm", J. Vac. Sci. Technol. B 10(6), pp. 2989-2996, Nov/Dec 1992.

201. N. Harned, J. McClay, J. Shamaly, "Laser damage impact on lithography system throughput", J.Quantum Elec., Vol 1, pp. 837 - 840, 1995.

202. R.E. Schenker, L. Eichner, H. Vaidya, S. Vaidya, W.G. Oldham, "Degradation of fused silica at 193 nm and 213 nm", SPIE, Vol. 2440, pp. 118-125, 1995.

203. Y. Kaimoto, K. Nozaki, S. Takechi, N. Abe, SPIE, Vol. 1672, p. 66, 1992.

204. R.R. Kunz, R.D. Allen, W.D. Hinsberg, G.M. Wallraff, "Acid-catalyzed single-layer resists for ArF lithography", Proc. SPIE, Vol. 1925, 167-175, 1993.

205. K. Brown, "SEMATECH and the national technology roadmap: needs and challenges", Proc. SPIE, Vol. 2440, 33-39, 1995.

206. T. Tanaka, M. Morigami, H. Oizumi, T. Soga, T. Ogawa, F. Murai, "Prevention of resist pattern collapse by resist heating during rinsing," J. Electrochem. Soc., Vol. 141, No. 12, Dec., 1994.

207. T. Tanaka, M. Morigami, H. Oizumi, T. Ogawa, S. Uchino, "Prevention of resist pattern collapse by flood exposure during rinse process," Jpn. J. Appl. Phys., Vol. 33, Part 2, No. 12B, pp. L1803 - L1805, 15 Dec. 1994.

208. M.D. Levenson, "Wavefront engineering for photolithography", Physics Today, pp. 28 - 36, July 1993.

209. Luc van den Hove, paper E2/1, presented at Micro and NanoEngineering 1994, Davos, Switzerland, Sept. 1994.

210. M. David Levenson, "Extending Optical Lithography to the Gigabit era", Microlithography World, pp5-13, Autumn 1994.

211. S. Stalnaker, L. Straaijer, J. Stoeldraijer, and A. Katz, "System and Metrology Improvements to Achieve 85 nm Overlay," presented a Semicon/Korea 91 Technical Symposium, 1991.

CHAPTER 2
Electron Beam Lithography

Mark A. McCord
Stanford University

Michael J. Rooks
Cornell University

CONTENTS

2.1 INTRODUCTION

2.1.1 Definition and Historical Perspective

Electron beam lithography (EBL) is a specialized technique for creating the extremely fine patterns (much smaller than can be seen by the naked eye) required by the modern electronics industry for integrated circuits. Derived from the early scanning electron microscopes, the technique in brief consists of scanning a beam of electrons across a surface covered with a resist film sensitive to those electrons, thus depositing energy in the desired pattern in the resist film. The process of forming the beam of electrons and scanning it across a surface is very similar to what happens inside the everyday television or CRT display, but EBL typically has three orders of magnitude better resolution. The main attributes of the technology are 1) it is capable of very high resolution, almost to the atomic level; 2) it is a flexible technique that can work with a variety of materials and an almost infinite number of patterns; 3) it is slow, being one or more orders of magnitude slower than optical lithography; and 4) it is expensive and complicated – electron beam lithography tools can cost many millions of dollars and require frequent service to stay properly maintained.

The first electron beam lithography machines, based on the scanning electron microscope (SEM), were developed in the late 1960s. Shortly thereafter came the discovery that the common polymer PMMA (polymethyl methacrylate) made an excellent electron beam resist.[1] It is remarkable that even today, despite sweeping technological advances, extensive development of commercial EBL, and a myriad of positive and negative tone resists, much work continues to be done with PMMA resist on converted SEMs.

Fig. 2.1 shows a block diagram of a typical electron beam lithography tool. The column is responsible for forming and controlling the electron beam.

Underneath the column is a chamber containing a stage for moving the sample around and facilities for loading and unloading it. Associated with the chamber is a vacuum system needed to maintain an appropriate vacuum level throughout the machine and also during the load and unload cycles. A set of control electronics supplies power and signals to the various parts of the machine. Finally, the system is controlled by a computer, which may be anything from a personal computer to a mainframe. The computer handles such diverse functions as setting up an exposure job, loading and unloading the sample, aligning and focusing the electron beam, and sending pattern data to the pattern generator. The part of the computer and electronics used to handle pattern data is sometimes referred to as the datapath. Fig. 2.2 shows a picture of a typical commercial EBL system including the column, chamber, and control electronics.

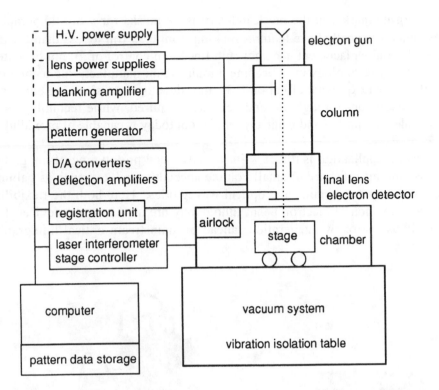

FIG. 2.1. Block diagram showing the major components of a typical electron beam lithography system.

2.1.2 Applications

Currently, electron beam lithography is used principally in support of the integrated circuit industry, where it has three niche markets. The first is in mask-making, typically the chrome-on-glass masks used by optical lithography tools. It is the preferred technique for masks because of its flexibility in providing rapid turnaround of a finished part described only by a computer CAD file. The ability to meet stringent linewidth control and pattern placement specifications, on the order of 50 nm each, is a remarkable achievement.

Because optical steppers usually reduce the mask dimensions by 4× or 5×, resolution is not critical, with minimum mask dimensions currently in the one to two μm range. The masks that are produced are used mainly for the fabrication of integrated circuits, although other applications such as disk drive heads and flat panel displays also make use of such masks.

An emerging market in the mask industry is 1× masks for x-ray lithography. These masks typically have features ranging from 0.25 μm to less than 0.1 μm and will require placement accuracy and linewidth control of 20 nm or better. Should x-ray technology ever become a mainstream manufacturing technique, it will have an explosive effect on EBL tool development since the combination of resolution, throughput, and accuracy required, while technologically achievable, are far beyond what any single tool today is capable of providing.

The second application is direct write for advanced prototyping of integrated circuits[2] and manufacture of small volume specialty products, such as gallium arsenide integrated circuits and optical waveguides. Here both the flexibility and the resolution of electron beam lithography are used to make devices that are perhaps one or two generations ahead of mainstream optical lithography techniques.

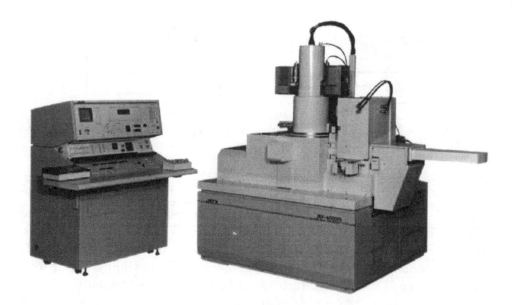

FIG. 2.2. Photograph of a commercial electron beam lithography tool. (courtesy of JEOL Ltd.)

Finally, EBL is used for research into the scaling limits of integrated circuits (Fig. 2.3)[3] and studies of quantum effects and other novel physics phenomena at very small dimensions. Here the resolution of EBL makes it the tool of choice. A typical application is the study of the Aharanov-Bohm effect,[4-6] where electrons traveling along two different paths about a micrometer in length can interfere constructively or destructively, depending on the strength of an applied magnetic field. Other applications include devices to study ballistic electron effects, quantization of electron energy levels in very small

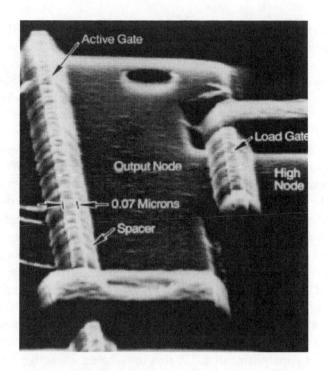

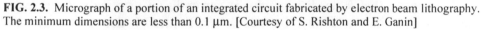

FIG. 2.3. Micrograph of a portion of an integrated circuit fabricated by electron beam lithography. The minimum dimensions are less than 0.1 μm. [Courtesy of S. Rishton and E. Ganin]

structures,[7,8] and single electron transistors. To see these effects typically requires minimum feature sizes of 100 nm or less as well as operation at cryogenic temperatures.

2.1.3 Alternative Techniques

It is prudent to consider possible alternatives before committing to EBL technology. For chrome-on-glass optical mask fabrication, there are optical mask writers available that are based either on optical reduction of rectangular shapes formed by framing blades or by multiple individually controlled round laser beams. Although at present EBL is technologically ahead of optical mask writers, this may not continue in the future. However, EBL will continue to provide a resolution advantage over the optical mask writers which may be important for advanced masks using phase shift or optical proximity correction. For 1× mask fabrication (i.e. x-ray), EBL will continue to be the most attractive option.

Optical lithography using lenses that reduce a mask image onto a target (much like an enlarger in photography) is the technique used almost exclusively for all semiconductor integrated circuit manufacturing. Currently, the minimum feature sizes that are printed in production are a few tenths of a micrometer. For volume production, optical lithography is much cheaper than EBL, primarily because of the high throughput of the optical tools. However, if just a few samples are being made, the mask cost (a few thousand dollars) becomes excessive, and the use of EBL is justified. Today optical tools can print 0.25 μm features in development laboratories, and 0.18 μm should be possible within a few years.

By using tricks, optical lithography can be extended to 0.1 μm or even smaller. Some possible tricks include overexposing/overdeveloping, phase shift and phase edge masks, and edge shadowing.[9] The problem with these tricks is that they may not be capable of exposing arbitrary patterns, although they may be useful for making isolated transistor gates or other simple sparse patterns. Another specialized optical technique can be used to fabricate gratings with periods as small as 0.2 μm by interfering two laser beams at the surface of the sample.[10] Again, the pattern choice is very restricted, although imaginative use of blockout and trim masks may allow for the fabrication of simple devices.

X-ray proximity printing may be a useful lithographic technique for sub-0.25 μm features.[11] Again, it requires a mask made by EBL, and since the mask is 1× this can be a formidable challenge. However, if the throughput required exceeds the limited capabilities of EBL, this may be an attractive option. The disadvantage is that x-ray lithography is currently an extremely expensive proposition and the availability of good masks is limited. It also requires either a custom built x-ray source and stepper or access to a synchrotron storage ring to do the exposures. With care, x-ray lithography can also be extended to the sub-0.1 μm regime.[12]

The final technique to be discussed is ion beam lithography. The resolution, throughput, cost, and complexity of ion beam systems is on par with EBL. There are a couple of disadvantages, namely, limits on the thickness of resist that can be exposed and possible damage to the sample from ion bombardment. One advantage of ion beam lithography is the lack of a proximity effect, which causes problems with linewidth control in EBL. Another advantage is the possibility of in situ doping if the proper ion species are available and in situ material removal by ion beam assisted etching. The main reason that ion beam lithography is not currently widely practiced is simply that the tools have not reached the same advanced stage of development as those of EBL.

Finally, it should also be noted that modern computer simulation tools, together with a detailed understanding of the underlying physics, in many cases allows one to accurately predict exploratory device characteristics without ever having to build actual hardware. This is especially true for silicon transistors.

2.2 ELEMENTS OF ELECTRON OPTICS

2.2.1 Introduction

The part of the EBL system that forms the electron beam is normally referred to as the column. An EBL column (Fig. 2.4) typically consists of an electron source, two or more lenses, a mechanism for deflecting the beam, a blanker for turning the beam on and off, a stigmator for correcting any astigmatism in the beam, apertures for helping to define the beam, alignment systems for centering the beam in the column, and finally, an electron detector for assisting with focusing and locating marks on the sample. The optical axis (Z) is parallel to the electron beam, while X and Y are parallel to the plane of the sample.

Electron optics are a very close analog of light optics, and most of the principles of an electron beam column (except for the rotation of the image) can be understood by thinking of the electrons as rays of light and the electron optical components as simply their optical counterparts. In order to operate an EBL machine, generally it is not necessary to understand the underlying math and physics, so they will not be discussed here although several excellent texts are available should the reader desire more information.[13,14] In addition, computer programs are available that allow easy and accurate design and simulation of optical components and columns.[15]

2.2.2 Electron Sources

Electrons may be emitted from a conducting material either by heating it to the point where the electrons have sufficient energy to overcome the work function barrier of the conductor (thermionic sources) or by applying an electric field sufficiently strong that they tunnel through the barrier (field emission sources). Three key parameters of the source are the virtual source size, its brightness (expressed in amperes per square centimeter per steradian), and the energy spread of the emitted electrons (measured in electron volts).

The size of the source is important since this determines the amount of demagnification the lenses must provide in order to form a small spot at the target. Brightness can be compared to intensity in light optics, so the brighter the electron source, the higher the current in the electron beam. A beam with a wide energy spread (which is undesirable, as will be shown in the section on lenses) is similar to white light, while a beam with a narrow energy spread would be comparable to monochromatic light. Although the energy spread of the source is important, space charge interactions between electrons further increase the energy spread of the beam as it moves down the column (Boersch effect).[16] An electron source is usually combined with two or more electrodes to control the emission properties, as shown in Fig. 2.5.[17]

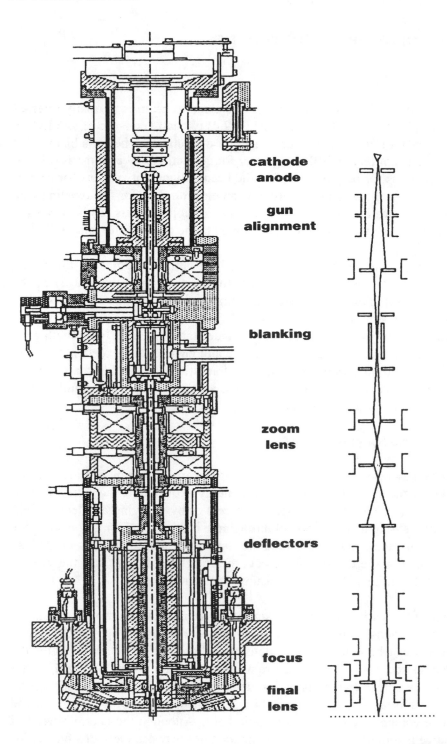

cathode
anode

gun
alignment

blanking

zoom
lens

deflectors

focus

final
lens

FIG. 2.4. Cross-section drawing of a typical electron beam column along with a raytrace of the electrons as they pass through the various electron optical components. (Courtesy of Leica Lithography Systems Ltd.)

Table 2.1 summarizes the properties of common sources. For many years the standard thermionic electron source for lithography optics was a loop of tungsten wire heated white hot by passing a current it. Tungsten was chosen for its ability to withstand high temperatures without melting or

source type	brightness (A/cm²/sr)	source size	energy spread (eV)	vacuum requirements (Torr)
tungsten thermionic	~10^5	~25 μm	2-3	10^{-6}
LaB$_6$	~10^6	~10 μm	2-3	10^{-8}
thermal FE	~10^8	~20 nm	0.9	10^{-9}
W & cold FE	~10^9	5 nm	0.22	10^{-10}

TABLE 2.1 Properties of the electron sources commonly used in electron beam lithography tools.

evaporating. Unfortunately, this source was not very bright and also had a large energy spread caused by the very high operating temperature (2700 K). More recently, lanthanum hexaboride has become the cathode of choice; due to a very low work function, a high brightness is obtained at an operating temperature of around 1800 K. The beam current delivered by thermionic sources depends on the temperature of the cathode. Higher temperatures can deliver greater beam current, but the tradeoff is an exponentially decreasing lifetime due to thermal evaporation of the cathode material.

Field emission sources typically consist of a tungsten needle sharpened to a point, with a radius less than 1 μm. The sharp tip helps provide the extremely high electric fields needed to pull electrons out of the metal. Although cold field emission sources have become common in electron microscopes, they have seen little use in EBL due to their instability with regard to short term noise as well as long term drift, which is a much more serious problem for lithography than microscopy. The noise is caused by atoms that adsorb onto the surface of the tip, affecting its work function and thus causing large changes in the emission current. Heating the tip momentarily (flashing) can clean it, but new atoms and molecules quickly readsorb even in the best of vacuums. In addition, atoms may be ionized by the electron beam and subsequently accelerated back into the tip, causing physical sputtering of the tip itself. To minimize the current fluctuations, the electron source must be operated in an extreme ultra high vacuum environment, 10^{-10} Torr or better.

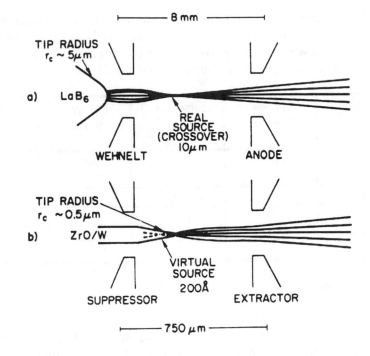

FIG. 2.5. Electrode structure and relevant dimensions for a) LaB$_6$ gun and b) thermal field emission gun. The electrodes are circularly symmetric about the optical axis. The Wehnelt and suppressor are biased negative with respect to the cathode, while the anode and extractor are positively biased. [From Gesley,[17] 1989]

A technology that is now available to EBL (as well as in many electron microscopes) is the thermal field emission source. It combines the sharp tungsten needle of the field emission source and the heating of the thermal source. Because the tip operates at a temperature of about 1800 K, it is less sensitive to gases in the environment and can achieve stable operation for months at a time. Although thermal field emitter is the common name, it is more properly called a Schottky emitter since the electrons escape over the work function barrier by thermal excitation. It features a brightness almost as high as the cold field emission sources, a very small virtual source size, and a moderate energy spread. The tungsten is usually coated with a layer of zirconium oxide to reduce the work function barrier. A heated reservoir of zirconium oxide in the electron gun continuously replenishes material evaporated from the tip. It requires a vacuum in the range of 10^{-9} Torr, which, although much better than required for the thermionic sources, is readily achievable with modern vacuum technology. (A light bakeout might be required to remove water vapor after the system has been vented.) LaB$_6$ sources are still preferred for shaped beam systems since the total current provided by the thermal field emission source is inadequate for this application.

2.2.3 Electron Lenses

Electrons can be focused either by electrostatic forces or magnetic forces. Although electron lenses in principle behave the same as optical lenses, there are differences. Except in some special cases, electron lenses can be made only to converge, not diverge. Also, the quality of electron lenses is not nearly as good as optical lenses in terms of aberrations. The relatively poor quality of electron lenses restricts the field size and convergence angle (or numerical aperture) that can be used. The two types of aberrations critical to EBL are spherical aberrations, where the outer zones of the lens focus more strongly than the inner zones, and chromatic aberrations, where electrons of slightly different energies get focused at different image planes. Both types of aberrations can be minimized by reducing the convergence angle of the system so that electrons are confined to the center of the lenses, at the cost of greatly reduced beam current. A magnetic lens is formed from two circularly symmetric iron (or some other high permeability material) polepieces with a copper winding in-between. Fig. 2.6 shows a cross-section through a typical magnetic lens, along with some magnetic flux lines. The divergence of the magnetic flux along the optical axis

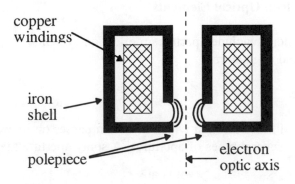

FIG. 2.6. Cross-section through a magnetic lens with lines showing the magnetic field distribution.

imparts a force on electrons back towards the optical (Z) axis, resulting in focusing action. The magnetic field also causes a rotation of the electrons (and the image) about the Z axis in a corkscrew fashion. Although this does not affect the performance of the lens, it does impact the design, alignment, and operation of the system. For instance, the deflection system must be rotated physically with respect to the stage coordinates. Also, when aligning a column, X and Y displacement in the upper regions of the column will not correspond to the same X and Y displacement at the target. Finally, changes in focus or changes in the height of the sample can cause a slight rotation in the deflection coordinates. This must be properly corrected or stitching and overlay errors will result. Magnetic lenses, particularly the final lens, may be liquid-cooled to maintain a controlled temperature, which is critical for stable operation of a system.

Electrostatic lenses have worse aberrations than magnetic lenses, so they are not as commonly used. They are most often found in the gun region as a condenser lens since they can be combined with the extractor or anode used to pull electrons out of the cathode, and they are easily made for ultrahigh vacuum use and are bakeout compatible. Also, aberrations in the condenser lens tend to be less important; system performance is usually dominated by the aberrations of the final lens. A simple electrostatic lens, as shown in Fig, 2.7, consists of three consecutive elements like apertures, the outer two being at ground potential and the inner at some other (variable) potential that controls the lens strength. The electric potentials set up by such a lens tend to pull an electron that is traveling away from the optical axis back towards the axis, resulting in the focusing action.

2.2.4 Other Electron Optical Elements

Other optical elements include apertures, deflection systems, alignment coils, blanking plates, and stigmators.

2.2.4.1 Apertures

Apertures are small holes through which the beam passes on its way down the column. There are several types of apertures. A spray aperture may be used to

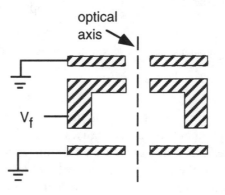

FIG. 2.7. Cross-section through an electrostatic Einzel lens. The focus of the lens is controlled by the voltage applied to the center electrode.

stop any stray electrons without materially affecting the beam itself. A blanking aperture is used to turn the beam on and off; by deflecting the beam away from the aperture hole, the aperture intercepts the beam when not writing. A beam limiting aperture has two effects: it sets the beam convergence angle α (measured as the half-angle of the beam at the target) through which electrons can pass through the system, controlling the effect of lens aberrations and thus resolution, and also sets the beam current. A beam limiting aperture is normally set in an X-Y stage to allow it to be centered, or aligned, with respect to the optical axis. It is best to have a beam limiting aperture as close to the gun as possible to limit the effects of space charge caused by electron - electron repulsion.

Apertures may be heated to help prevent the formation of contamination deposits, which can degrade the resolution of the system. If not heated, the apertures typically need to be cleaned or replaced every few months. With platinum apertures, cleaning is easily accomplished by heating the aperture orange hot in a clean-burning flame. Shaped beam systems also have one or more shaping apertures, which can be square or have more complicated shapes to allow the formation of a variety of beam shapes, such as triangles, etc.

2.2.4.2 Electron beam deflection

Deflection of the electron beam is used to scan the beam across the surface of the sample. As with lenses, it can be done either magnetically or electrostatically. The coils or plates are arranged so that the fields are perpendicular to the optical axis, as shown in Fig. 2.8(a). Deflecting the beam off axis introduces additional aberrations that cause the beam diameter to deteriorate, and deviations from linearity in X and Y increase as the amount of deflection increases. These effects limit the maximum field or deflection size that can be used. As with lenses, magnetic deflection introduces fewer distortions than electrostatic deflection. Double magnetic deflection using a pair of matched coils is sometimes used to further reduce deflection aberrations. However, electrostatic deflection can achieve much higher speeds since the inductance of the magnetic deflection coils limits their frequency response, and eddy currents introduced by the magnetic fields may further limit the speed of magnetic deflection. Since deflection systems are frequently placed inside the final lens, care must be taken to prevent the fields from interacting with conducting metal parts. Usually the final lens will be shielded with ferrite to minimize eddy currents. Some tools use multiple deflection systems, where high speed, short range deflection is done electrostatically while long range deflection is magnetic. In either case, the field size of the tool is limited by aberrations of the deflection system; some tools introduce dynamic corrections to the deflection, focus, and stigmators in order to increase the maximum field size, at the cost of additional complexity.

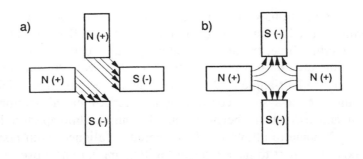

FIG. 2.8. Schematic showing the magnetic (electrostatic) field distribution for a) a simple beam deflector or alignment device energized for diagonal deflection and b) a stigmator. The optical axis is perpendicular to the plane of the page.

2.2.4.3 Beam blanking

Blanking, or turning the beam on and off, is usually accomplished with a pair of plates set up as a simple electrostatic deflector. One or both of the plates are connected to a blanking amplifier with a fast response time. To turn the beam off, a voltage is applied across the plates which sweeps the beam off axis until it is intercepted by a downstream aperture. If possible, the blanking is arranged to be conjugate so that, to first order, the beam at the target does not move while the blanking plates are activated. Otherwise, the beam would leave streaks in the resist as it was blanked. The simplest way to ensure conjugate blanking is to arrange the column so that the blanking plates are centered at an intermediate focal point, or crossover. In very high speed systems, more elaborate blanking systems involving multiple sets of plates and delay lines may be required to prevent beam motion during the blanking and unblanking processes.[14]

2.2.4.4 Stigmators

A stigmator is a special type of lens used to compensate for imperfections in the construction and alignment of the EBL column. These imperfections can result in astigmatism, where the beam focuses in different directions at different lens settings; the shape of a nominally round beam becomes oblong, with the direction of the principal axis dependent on the focus setting, resulting in smeared images in the resist. The stigmator cancels out the effect of astigmatism, forcing the beam back into its optimum shape. Stigmators may be either electrostatic or magnetic and consist of four or more poles (eight is typical) arranged around the optical axis. They can be made by changing the connections to a deflector, as shown in Fig. 2.8(b). With proper mixing of the electrical signals, a single deflector may sometimes perform multiple functions, including beam deflection, stigmation, alignment, and blanking.

2.2.5 Other column components

A number of other components may be found in the column, which although not important to the electron optics are nonetheless critical to the operation of the system. A Faraday cage located below the final beam limiting aperture is used to measure the beam current in order to ensure the correct dose for resist exposure. It can be either incorporated directly on the stage or a separate movable assembly in the column. The column will also typically have an isolation valve that allows the chamber to be vented for maintenance while the gun is still under vacuum and operational. All parts of an electron beam column exposed to the beam must be conductive or charging will cause unwanted displacements of the beam. Often a conductive liner tube will be placed in parts of the column to shield the beam from insulating components.

Finally, the system needs a method of detecting the electrons for focusing, deflection calibration, and alignment mark detection. Usually this is a silicon solid state detector similar to a solar cell, mounted on the end of the objective lens just above the sample. Channel plate detectors and scintillators with photomultiplier tubes may also be used. Unlike scanning electron microscopes, which image with low voltage secondary electrons, EBL systems normally detect high energy backscattered electrons since these electrons can more easily penetrate the resist film. The signal from low energy secondary electrons may be obscured by the resist.

2.2.6 Resolution

There are several factors that determine the resolution of an electron beam system. First is the virtual source size d_v divided by the demagnification of the column, M^1, resulting in a beam diameter of $d_g = d_v/M^1$. In systems with a zoom condenser lens arrangement, the demagnification of the source can be varied, but increasing the demagnification also reduces the available beam current.

If the optics of the column were otherwise ideal, this simple geometry would determine the beam diameter. Unfortunately, lenses are far from perfect. Spherical aberrations result from the tendency of the outer zones of the lenses to focus more strongly than the center of the lens. The resultant diameter is $d_s = 1/2C_s\alpha^3$, where C_s is the spherical aberration coefficient of the final lens and α is the convergence half-angle of the beam at the target. Using an aperture to limit the convergence angle thus reduces this effect, at the expense of reduced beam current. Chromatic aberrations result from lower energy electrons being focused more strongly than higher energy electrons. For a chromatically limited beam, the diameter is $d_c = C_c \alpha \Delta V / V_b$, where C_c is the chromatic aberration coefficient, ΔV is the energy spread of the electrons, and V_b is the beam voltage.

Finally, quantum mechanics gives the electron a wavelength $\lambda = 1.2/(V_b)^{1/2}$ nm; although much smaller than the wavelength of light (0.008 nm at 25 kV), this wavelength can still limit the beam diameter by classical diffraction effects in very high resolution systems. For a diffraction limited beam, the diameter is given by $d_d = 0.6\lambda/\alpha$. To determine the theoretical beam size of a system, the contributions from various sources can be added in quadrature: $d = (d_g^2 + d_s^2 + d_c^2 + d_d^2)^{1/2}$.

The diagram in Fig. 2.9 shows how these sources contribute in a typical column. In systems with thermionic sources, spherical aberrations tend to be the limiting factor for beam diameter, while chromatic aberrations dominate in field emission systems. For a given beam current, there will be an optimum

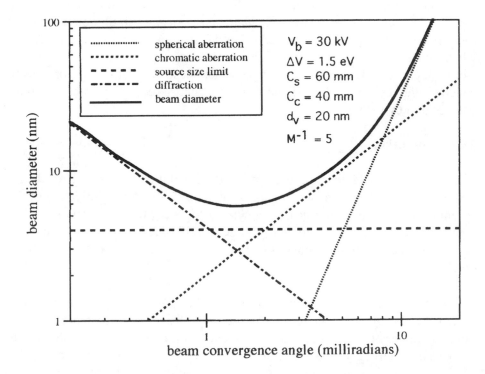

FIG. 2.9. A plot showing resolution as a function of beam convergence angle for an electron beam column at 30 kV. The plot assumes an energy spread of 1.5 eV, a source diameter of 20 nm, and a fixed demagnification of 5.

combination of convergence angle and system demagnification. Resolution can generally be improved in most systems by using a smaller beam limiting aperture, at the expense of reduced beam current and throughput. In systems where the demagnification can be varied, increasing the demagnification will also improve resolution, at the expense of reduced beam current.

2.3 ELECTRON-SOLID INTERACTIONS

Although electron beam lithography tools are capable of forming extremely fine probes, things become more complex when the electrons hit the workpiece. As the electrons penetrate the resist, they experience many small angle scattering events (forward scattering), which tend to broaden the initial beam diameter. As the electrons penetrate through the resist into the substrate, they occasionally undergo large angle scattering events (backscattering). The backscattered electrons cause the proximity effect,[18] where the dose that a pattern feature receives is affected by electrons scattering from other features nearby. During this process the electrons are continuously slowing down, producing a cascade of low voltage electrons called secondary electrons.

Fig. 2.10 shows some computer simulations of electron scattering in typical samples.[19] The combination of forward and backscattered electrons results in an energy deposition profile in the resist that is typically modeled as a sum of two Gaussian distributions, where α is the width of the forward scattering distribution, β is the width of the backscattering distribution, and η_e is the intensity of the backscattered energy relative to the forward scattered energy. Fig. 2.11 shows an example of a simulated energy profile.

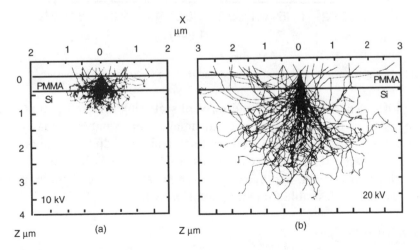

FIG. 2.10. Monte Carlo simulation of electron scattering in resist on a silicon substrate at a) 10 kV and b) 20 kV. [From Kyser and Viswanathan[19] 1975]

2.3.1 Forward Scattering

As the electrons penetrate the resist, some fraction of them will undergo small angle scattering events, which can result in a significantly broader beam profile at the bottom of the resist than at the top. The increase in effective beam diameter in nanometers due to forward scattering is given empirically by the formula $d_f = 0.9(R_t/V_b)^{1.5}$, where R_t is the resist thickness in nanometers and V_b is the beam voltage in kilovolts. Forward scattering is minimized by using the thinnest possible resist and the highest available accelerating voltage.

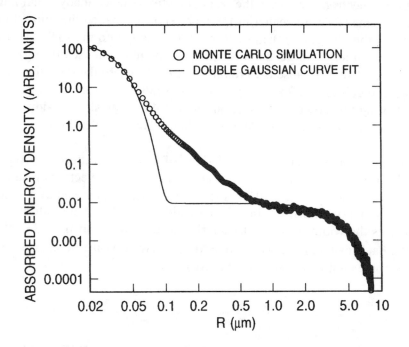

FIG. 2.11. Simulated profile of the energy absorbed from an electron beam exposure.

Although it is generally best to avoid forward scattering effects when possible, in some instances they may be used to advantage. For example, it may be possible to tailor the resist sidewall angle in thick resist by adjusting the development time.[20] As the time increases, the resist sidewall profile will go from a positive slope, to vertical, and eventually to a negative, or retrograde, profile, which is especially desirable for pattern transfer by liftoff.

2.3.2 Backscattering

As the electrons continue to penetrate through the resist into the substrate, many of them will experience large angle scattering events. These electrons may return back through the resist at a significant distance from the incident beam, causing additional resist exposure. This is called the electron beam proximity effect. The range of the electrons (defined here as the distance a typical electron travels in the bulk material before losing all its energy) depends on both the energy of the primary electrons and the type of substrate. Fig. 2.12 shows a plot of electron range as a function of energy for three common materials.[21]

The fraction of electrons that are backscattered, η, is roughly independent of beam energy, although it does depend on the substrate material, with low atomic number materials giving less backscatter. Typical values of η range from 0.17 for silicon to 0.50 for tungsten and gold. Experimentally, η is only loosely related to η_e, the backscatter energy deposited in the resist as modeled by a double Gaussian. Values for η_e tend to be about twice η.

2.3.3 Secondary Electrons

As the primary electrons slow down, much of their energy is dissipated in the form of secondary electrons with energies from 2 to 50 eV. They are responsible for the bulk of the actual resist exposure process. Since their range in resist is only a few nanometers, they contribute little to the proximity effect. Instead, the net result can be considered to be an effective widening of the beam diameter by roughly 10 nm. This largely accounts for the minimum practical resolution of 20 nm observed in the highest resolution electron beam systems and contributes (along with forward scattering) to the bias that is seen in positive resist systems, where the exposed features develop larger than the size they were nominally written.

A small fraction of secondary electrons may have significant energies, on the order of 1 keV. These so-called fast secondaries can contribute to the proximity effect in the range of a few tenths of a micron. Experimentally and theoretically, the distribution of these electrons can be fit well by a third Gaussian with a range intermediate between the forward scattering distribution and the backscattering distribution.

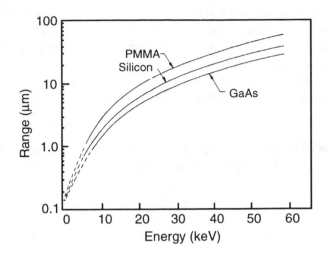

FIG. 2.12. Electron range as a function of beam energy for PMMA resist, silicon, and gallium arsenide. [From Brewer, 1980]

2.3.4 Modeling

Electron scattering in resists and substrates can be modeled with reasonable accuracy by assuming that the electrons continuously slow, down as described by the Bethe equation,[22] while undergoing elastic scattering, as described by the screened Rutherford formula.[23] Since the different materials and geometries make analytic solutions difficult, Monte Carlo techniques, where a large number of random electrons are simulated, are commonly used. The input to the program contains such parameters as the electron energy, beam diameter, and film thicknesses and densities, while the output is a plot of energy deposited in the resist as a function of the distance from the center of the beam.

Curve fitting with Gaussians and other functions to the simulated energy distribution may also be employed. In order to get good statistics, the energy deposition for a large number (10,000 to 100,000) of electrons must be simulated, which can take a few minutes to an hour or so on a personal computer. Software for Monte Carlo simulation of electron irradiation is available from several sources.[24,25,26,27] Such simulations are often used to generate input parameters for proximity effect correction programs (see next section). Alternatively, experimental data can be obtained by measuring the diameter of exposed resist from a point exposure of the beam at various doses[28] or by measuring the linewidths of various types of test patterns such as the "tower" pattern.[29]

2.4 PROXIMITY EFFECT

2.4.1 Introduction

The net result of the electron scattering discussed in the previous section is that the dose delivered by the electron beam tool is not confined to the shapes that the tool writes, resulting in pattern specific linewidth variations known as the proximity effect. For example, a narrow line between two large exposed areas may receive so many scattered electrons that it can actually develop away (in positive resist) while a small isolated feature may lose so much of its dose due to scattering that it develops incompletely. Fig. 2.13 shows an example of what happens to a test pattern when proximity effects are not corrected.[30]

2.4.2 Proximity Effect Avoidance

Many different schemes have been devised to minimize the proximity effect. If a pattern has fairly uniform density and linewidth, all that may be required is to adjust the overall dose until the patterns come out the proper size. This method typically works well for isolated transistor gate structures. Using higher contrast resists can help minimize the linewidth variations. Multilevel resists, in which a thin top layer is sensitive to electrons and the pattern developed in it is transferred by dry etching into a thicker underlying layer, reduce the forward scattering effect, at the cost of an increase in process complexity.

Higher beam voltages, from 50 kV to 100 kV or more, also minimize forward scattering, although in some cases this can increase the backscattering. When writing on very thin membranes such as used for x-ray masks, higher voltages reduce the backscatter contribution as well since the majority of electrons pass completely through the membrane.[31]

Conversely, by going to very low beam energies, where the electron range is smaller than the minimum feature size, the proximity effect can be eliminated.[32] The penalty is that the thickness of a single layer resist must also be less than the minimum feature size so that the electrons can expose the entire film thickness. The electron-optical design is much harder for low voltage systems since the electrons are more difficult to focus into a small spot and are more sensitive to stray electrostatic and magnetic fields. However, this is the current approach in optical maskmaking, where a 10 kV beam is used to expose 0.3 µm thick resist with 1 µm minimum features on a 5× mask. In more advanced studies, a 1.5 kV beam has been used to expose 70 nm thick resist with 0.15 µm minimum features.[33] A technique that can be used in conjunction with this approach in order to increase the usable range of electron energy is to place a layer with a high atomic number, such as tungsten, underneath the re-

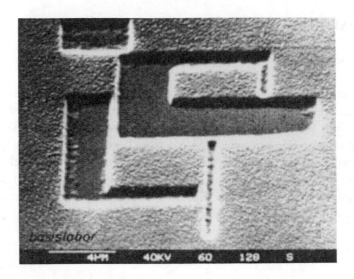

FIG. 2.13. SEM micrograph of a positive resist pattern on silicon exposed with a 20 kV electron beam demonstrates the proximity effect, where small isolated exposed areas receive less dose relative to larger or more densely exposed areas. [From Kratschmer,[30] 1981]

sist. This has the effect of further limiting the range of the backscattered electrons.

2.4.3 Proximity Effect Correction

2.4.3.1 Dose modulation

The most common technique of proximity correction is dose modulation, where each individual shape in the pattern is assigned a dose such that (in theory) the shape prints at its correct size. The calculations needed to solve the shape-to-shape interactions are computationally very time consuming. Although the actual effect of electron scattering is to increase the dose received by large areas, for practical reasons proximity correction is normally thought of in terms of the large areas receiving a base dose of unity, with the smaller and/or isolated features receiving a larger dose to compensate.

Several different algorithms have been used. In the self-consistent technique, the effect of each shape on all other shapes within the scattering range of the electrons is calculated. The solution can be found by solving a large number of simultaneous equations;[34] unfortunately, this approach becomes unwieldy as the number of shapes increases and their size decreases. An alternative is to define a grid and compute the interaction of the pattern shapes with the grid and vice versa;[35] however, the accuracy and flexibility of this technique may be limited. An optimal solution may also be arrived at by an iterative approach.[36] Finally, neural network techniques have been applied to the problem of proximity correction;[37] while not an attractive technique when implemented on a digital computer, it might be advantageous if specialized neural network processors become a commercial reality. Many of the algorithms in use assume that the energy distribution has a double Gaussian distribution as discussed in Sec. 2.3.

2.4.3.2 Pattern biasing

A computationally similar approach to dose modulation is pattern biasing.[38,39] In this approach, the extra dose that dense patterns receive is compensated for by slightly reducing their size. This technique has the advantage that it can be implemented on EBL systems that are not capable of dose modulation. However, the technique does not have the dynamic range that dose modulation has; patterns that contain both very isolated features and very dense features will have reduced process latitude compared to when dose modulation is used, since the isolated features will be under-dosed while the dense features will be over-dosed. Pattern biasing cannot be applied to features with dimensions close to the scale of the pixel spacing of the e-beam system.

2.4.3.3 GHOST

A third technique for proximity correction, GHOST,[40] has the advantage of not requiring any computation at all. The inverse tone of the pattern is written with a defocused beam designed to mimic the shape of the backscatter distribution (Fig. 2.14). The dose of the GHOST pattern, $\eta_e/(1+\eta_e)$, is also set to match the large area backscatter dose. After the defocussed inverse image is written, the pattern will have a roughly uniform background dose. GHOST is perhaps an underutilized technique; under ideal conditions it can give superb linewidth control.[41] Its disadvantages are the extra data preparation and writing time, a slight to moderate loss of contrast in the resist image, and a slight loss in minimum resolution compared to dose modulation due to the fact that GHOST does not properly correct for forward scattering.

2.4.3.4 Software

A number of companies for some time have had proprietary software for proximity correction.[25,42,43] Just recently, commercial proximity packages have become available, or are about to become available.[44,45] At present, these are limited in their accuracy, speed, and data volume capability; while excellent for correcting small research patterns, they may have difficulties with complex

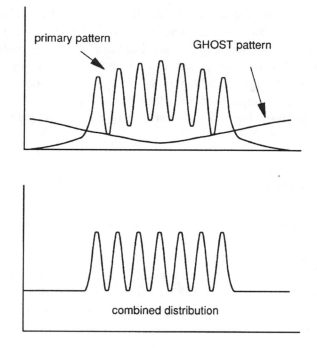

FIG. 2.14. Schematic showing how the GHOST technique can be used to correct for the proximity effect. The top curves show the energy distribution in the resist for a group of seven lines from the primary exposure and from the GHOST exposure. The bottom curve is the resulting final energy distribution, showing the dose equalization for all the lines.

chips. Finally, several packages have been developed at university and government laboratories, some of which might be available to an adventurous user with excessive amounts of free time.[38,46]

2.5 SYSTEMS

2.5.1 Environment

For best results, systems should be installed in a clean, quiet environment. 60 Hz noise is pervasive in most systems. To minimize this, careful consideration must be paid to the grounding of the system components to prevent ground loops. Also, analog and digital grounds should be kept separate as much as possible to minimize high frequency noise components. One useful method for tracking noise problems is to place the beam on the edge of a mark and monitor the electron detector output with a spectrum analyzer while disconnecting various suspect noise sources.

Acoustical noise can be a significant problem, especially in systems with field-emission electron sources. In such systems the demagnification of the field emission source, and thus the demagnification of vibrations, is much less than that of LaB_6 systems. Stray magnetic fields are also a common problem. Mechanical pumps, transformers, and fluorescent lights should be moved at least 10 ft from the column if possible. The system should be well isolated from mechanical vibrations with a pneumatic table; ideally, it should also be located on the ground floor. Finally, the temperature should be well controlled, ideally to within a tenth of a degree. This is particularly important if good placement accuracy is required.

This section begins with a description of the smallest e-beam systems -- namely, SEM conversions -- and proceeds to the largest commercial mask production tools. We conclude the section with a listing of e-beam fabrication services.

2.5.2 SEM and STEM Conversions

Any tool for microscopy -- optical, electron, or scanning probe -- may be adapted to work in reverse; that is, for writing instead of reading. Converted electron microscopes suffer the same limitations as light microscopes used for photolithography, namely, a small field of view and low throughput. Nevertheless, for a subset of research and R&D applications, converted SEMs offer a relatively inexpensive solution.

Of the many custom designed SEM conversions, most use a single set of digital-to-analog converters (DACs), from 12 to 16 bits wide, to drive the scan coils of the microscope. The beam is modulated with an electrostatic or magnetic beam blanker, which is usually located near a crossover of the beam. Alternatively, the beam can be blanked magnetically by biasing the gun alignment coils or not blanked at all. In the later case, the beam must be "dumped" to unused sections of the pattern. Figure 2.15 illustrates the "vector scan" method, in which shapes are filled with a raster pattern and the beam jumps from one shape to the next via a direct vector. By taking over the scan coils and beam blanking, a SEM can be used as a simple but high resolution lithography tool.

SEM conversions have evolved greatly in the past twenty years, primarily due to improvements in small computers and commercially available DAC boards. Early designs used relatively slow computers that sent primitive shapes (rectangles, trapezoids, and lines) to custom hardware. The custom pattern generator filled in the shapes by calculating coordinates inside the shapes and feeding these numbers to the DACs. While this approach is still the best way to avoid data transmission bottlenecks (and is used in commercial systems), inexpensive SEM conversions can now rely on the CPU to generate the shape filling data. A typical configuration uses an Intel CPU based PC, with a DAC card plugged into an ISA bus. In this case, the CPU can generate data much faster than it can be transmitted over an ISA bus.

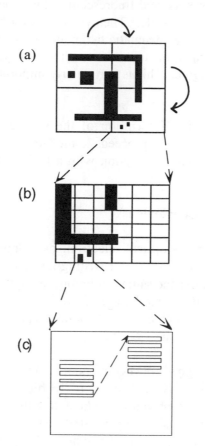

(a)

(b)

(c)

FIG 2.15 The vector-scan writing strategy. (a) Patterns are split into "fields". The stage moves from field to field, as shown by the arrows. Full patterns are stitched together from multiple fields. (b) In many vector-scan systems the fields are further tiled into subfields. A major DAC (16 bits) deflects the beam (a small "Gaussian" spot) to a subfield boundary, and a faster DAC (12 bits) deflects the beam within a subfield. SEM conversion kits typically do not include the faster 12-bit DAC. (c) The primitive shape is filled in by rastering the spot. Between shapes the beam is turned off ("blanked") and is deflected in a direct vector to the next shape. An alternative deflection strategy (not shown) is to use the major DAC to deflect the beam to the origin of each primitive shape

The bus limits the deflection speed to around 100 kHz, that is, to a dwell time per point of 10 μs.

What dwell time is required? With a 16-bit DAC and a SEM viewing field of 100 μm, the size of a pixel (the smallest logically addressable element of an exposure field) is 100 μm/2^{16}=1.5 nm, and its area A is the square of this. The charge delivered to this pixel in a time t is It, where I is the beam current. This must equal the dose times the pixel area. Given a beam current I on the order of 50 pA and a required dose D around 200 $\mu C/cm^2$ (typical for PMMA), we have a pixel dwell time

$$t = DA / I = 9 \times 10^{-8} \text{ s}, \tag{2.1}$$

or a deflection speed of 11 MHz. This being impossible with an ISA bus, we must either space out the exposure points, apply a short strobe to the beam blanker, or use a combination of the two. When the exposure points are spaced every n pixels (that is, when the 2^{16} available exposure points are reduced by a factor of n) then the "pixel area" and thus the dwell time is increased by a factor of n^2. Note that the *placement* of features can still be specified to a precision of 2^{16} within the writing field, while the shapes are filled in with a more coarse grid.

In the above example, we can set n to 11 so that the dwell time is increased to 1.1×10^{-5} s (91 kHz), increasing the pitch of exposure points to 16.5 nm. This spacing is a good match to the resolution of PMMA, and allows fine lines to be defined without any bumps due to pixelization. However, when we require 100 times the current (5000 pA in this example), the exposure point spacing must be increased by a factor of 10, possibly leading to rough edges. Some pattern generators (see Sect. 2.5.3.1) avoid this problem by allowing different exposure point spacings in the X and Y (or in the r and θ) directions, thereby allowing a larger exposure point spacing in the less critical dimension.

To use a SEM *without* a beam blanker, one must consider the large exposure point spacing required for common resists. Lack of a beam blanker leads to the additional problem of artifacts from the settling of scan coils and exposure at beam dump sites. Many SEM manufacturers offer factory-installed beam blankers. Retrofitted blankers are also sold by Raith GmbH[47]

The scan coils of a SEM are designed for imaging in a raster pattern and so are not commonly optimized for the random placements of a vector scan pattern generator. Settling times are typically around 10 μs for a JEOL 840 to as long as 1 ms for the Hitachi S800, where the bandwidth of the scan coils has been purposely limited to reduce noise in the imaging system. Thus, it is important to consider the bandwidth of the deflection system when purchasing a SEM for beamwriting.

The other major limitation of a SEM is its stage. Being designed for flexible imaging applications, SEM stages are not flat, and even when equipped with stepper motor control are no more accurate than ~1 to 5 μm. Periodic alignment marks can be used to stitch fields accurately, but this requires extra processing as well as the use of photolithography for printing alignment marks. The mark mask would presumably be fabricated on a commercial system with a laser-controlled stage. Fortunately, alignment with a converted SEM can be quite accurate, especially when using Moiré patterns for manual alignment. Automated alignment in the center of a SEM writing field is at least as good as in large commercial systems. Alignment at the edges of a SEM field will be compromised by distortions, which are typically much larger than in dedicated e-beam systems. Laser-controlled stages can be purchased for SEMs, but these are usually beyond the budgets of small research groups.

Electron beam lithography requires a flat sample close to the objective lens, making secondary electron imaging difficult with an ordinary Everhart-Thornley detector (a scintillator-photomultiplier in the chamber). A few high end SEMs are equipped with a detector above the objective lens or can be equipped with a microchannel plate on the pole-piece. These types of detectors are a great advantage for lithography since they allow the operator to decrease the working distance, and thus the spot size, while keeping the sample flat and in focus.

With patterning speed limited by beam settling and bus speed, it is clear that inexpensive SEM conversions cannot match the high speed writing of dedicated e-beam systems. However, a SEM based lithography system can provide adequate results for a wide variety of applications, at a small fraction of the cost of a dedicated system. The number of applications is limited by stitching, alignment, and automation. Practical applications include small numbers of quantum devices (metal lines, junctions, SQUIDs, split gates), small numbers of transistors, small area gratings, small masks, tests of resists, and direct deposition. The main limitations with SEM lithography are observed with writing over large areas, or when deflection speed and throughput are critical. Specifically, difficulties with stitching and/or distortions due to the electron optics of the microscope can become significant. SEMs are not practical for most mask making, integration of many devices over many fields, large area gratings, multifield optical devices, or any application requiring a large substrate.

2.5.3 Commercial SEM Conversion Systems

2.5.3.1 Nanometer Pattern Generation System (NPGS)

The SEM conversion kit sold by J.C. Nabity Lithography Systems[48] is built around a Windows-based PC-compatible with an ISA bus. A 16 bit multifunction board from Data Translation[49] is used to generate the X and Y beam deflections and to program a second board which provides the signals for blanking control. The beam is deflected from shape to shape in a writing field ("vector scan" mode), with the unique feature that the raster for filling arbitrary polygons can be defined by the user. Arbitrary polygons can be designed with up to 200 vertices and the user can specify the raster to be parallel to any side of the polygon. A unique feature of the NPGS is that the user has control over the exposure spot spacing in X and Y, allowing the critical dimension (e.g. perpendicular to grating lines) to be filled with greater accuracy (see Sect. 2.5.2). Circles and circular arcs are swept using a "polar coordinate" approach, with user control of the exposure spot spacing in r and θ. As with any ISA system, the data throughput is limited to around 100 kHz; and like most pattern generators, exposure points filling the features can be spaced by multiples of the DAC resolution (2^{16}) while still allowing full resolution for feature placement.

To provide for lower doses at reasonable currents, the Nabity system strobes the blanker at each exposure point.[50] For systems without a beam blanker, the Nabity Pattern Generation System (NPGS) can be programmed to "dump" the beam at user-defined locations within the writing field; however, this imposes significant limitations on the exposure spot spacing or on the lowest deliverable dose for a given beam current (refer to discussion above).

Mark alignment on the NPGS is performed by calculating the correlation between the measured mark image and the user-defined mark pattern. Signal processing such as averaging and edge enhancement can be executed before the alignment correlation, allowing the use of low contrast or rough marks. If the user supplies precisely defined marks (usually printed with a mask made on a commercial maskmaking tool) then NPGS can be used to correct for global rotation, scaling, and nonorthogonality. NPGS can control motorized stages, providing fully automated sample movement and pattern alignment. However, SEM stages are typically orders of magnitude slower than those of dedicated e-beam tools, and do not provide feedback to the deflection system (see Sect. 2.5.4).

Angled lines, polygons, and arbitrarily shaped features are all supported, and data can be imported in common e-beam formats: GDSII (Stream), CIF, and a subset of DXF (AutoCAD.)

2.5.3.2 Raith pattern generators

The Proxy-Writer SEM conversion kit is Raith's low end PC-based pattern generator. Like the Nabity system, the Proxy-Writer is a vector-scan system. Unlike the Nabity NPGS, the Proxy-Writer has only manual alignment, and patterns are limited to single writing fields. Corrections for rotation, shift, and orthogonality are applied to single fields (with single patterns); these corrections are not applied globally to correct the workpiece rotation and stage nonorthogonality. The unusual feature of this simple system is its support for exposure simulation and semiautomatic proximity effect correction. Pattern data can be generated with the simple CAD program included or imported from a DXF (AutoCAD) file.

The higher end Raith system, known as Elphy-Plus, supports the full range of e-beam operations, including control of a laser-controlled stage and corrections for workpiece rotation, gain, and orthogonality. The laser stage, also manufactured by Raith, allows field stitching to better than 0.1 μm. While the primary control is still a PC-compatible computer, the limitations of the ISA bus are circumvented by using a separate computer and integrated DAC as the pattern generator. In this way, the PC transmits only the coordinates of the corners of a shape, and the patterning hardware generates all of the internal points for exposure. Data throughput is thereby increased to 2.6 MHz (0.4 μs/point minimum); however, many SEM deflection systems will be limited to less than 1 MHz due to the inductance of the coils and low pass filters in the imaging system. The Elphy-Plus system supports fully automated mark detection and field stitching. All standard e-beam data formats are supported.

Useful features of the Raith Elphy-Plus system include support of data representation in polar coordinates (greatly reducing the data required to represent circles), bit-mapped pattern exposure, and a "path writing" mode. In the path writing mode, the beam is steered in a circular pattern (defining the width of a line) while the *stage* is moved over the length of the line or curve. This is a relatively slow way of writing a long line but avoids spatially localized stitching errors. Instead, the placement and drift errors are averaged over the length of the feature. The Raith Elphy-Plus is not only available for SEM conversions but is also used as the pattern generator for Leica's LION-LV1 e-beam system (see below.)

Even the most expensive SEM conversion kit will be limited by the SEM's slow magnetic deflection, large distortion, and small stage. Next, we look at fully integrated commercial systems.

2.5.3.3 Leica EBL Nanowriter

Somewhere between a converted SEM and a full featured e-beam system is the Leica EBL Nanowriter (Fig. 2.16). This system takes its electron gun and upper column from the Leica EBPG e-beam system, its deflection and imaging systems from the Leica 400 SEM series,[51] and adds custom pattern generation hardware. The pattern generator uses 16-bit DACs and has a deflection rate up to 1 MHz for vector scan operation. With an optional laser stage (5.3 nm resolution) this system costs substantially less than large e-beam systems and competes more directly with the high end Raith Elphy-Plus. Without a laser stage, the EBL will suffer from the same limitations as SEM conversions, namely, lack of stage flatness and the need for alignment marks for calibration. The system is available with a LaB_6 or Schottky thermal field emitter (TFE), and acceleration up to 100 kV. The system is unusual in offering such high voltage and a TFE emitter in a low cost system.

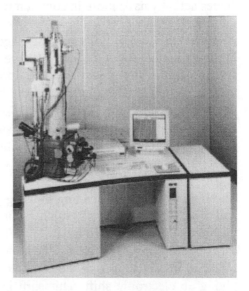

FIG. 2.16 Leica EBL-100, shown here with a 100 kV LaB6 electron source and a conventional SEM stage. The system is also available with a TFE source and laser-controlled stage. (Courtesy of Leica Lithography Systems Ltd.)

2.5.4 Gaussian vector scan systems

Like the converted SEMs, Gaussian vector scan systems use the writing strategy of stopping in each field, deflecting the beam from shape to shape, and filling in the shapes with a raster pattern. Large commercial systems, however, break the deflection into two (or more) sections, usually making use of a 16-bit DAC for "subfield" placement, and a faster 12-bit DAC for deflection inside the subfield (see Fig. 2.15). This is the scheme used in systems from JEOL, and some of the systems from Leica. Leica's EBPG series, and the Vector Scan (VS) tools built by IBM use an alternative technique: the slower DACs are used for placing the origin of each primitive shape and the faster DACs are used for filling in the shape. In addition to deflecting the beam with separate DACs, systems from Hitachi and Leica use these separate DACs to drive physically separate deflectors (magnetic or electrostatic). JEOL systems, in contrast, use a single stage electrostatic deflector.

Single stage deflectors have fewer problems with matching deflections of the "fast" and "slow" electronics, but sacrifice some speed.

The largest distinction of these commercial Gaussian spot systems (and in fact all commercial e-beam systems) is the use of high precision laser-controlled stages. Stage controllers from Hewlett-Packard or Zygo use the Zeeman effect to split the line of a He-Ne laser. The split-frequency laser beam is reflected off a mirror attached to the stage, and the beat frequency from the two lines is measured by high speed electronics. When the stage moves, the beat frequency shifts according to the Doppler effect, and the stage position is calculated by integrating the beat counts. While often referred to as "interferometers," these stages actually have more in common with radar speed guns.

Analysis of multiple points on the stage mirror allows the measurement of X, Y, and rotation about Z (yaw). Stage precision is often given in terms of a fraction of the laser's wavelength; a precision of $\lambda/128 = 5$ nm is commonly used in commercial systems, and the best stages now use $\lambda/1024 = 0.6$ nm. Even though the controller reports the stage location to this precision, the accuracy of the stage is limited by unmeasured rotations about the X and Y axes, and by bow in the mirrors. These nonlinearities, called "runout", limit the absolute placement accuracy to the order of 0.1 μm over 5 cm of stage travel.

The high precision in reading the stage position means that the stage motors and drive do *not* have to be highly refined. In fact, simple capstan motors and push rods have been used at IBM.[52,53] The stage controller receives a target location from a computer, drives the motors to a point close to this location, then sends an interrupt back to the computer and corrects the field position by applying an electronic shift. This shift is applied continuously, in real time, to compensate also for stage drift and low frequency vibration. In comparison, the laser stage built by Raith for SEM conversions applies corrections to relatively slow piezoelectric translators on the stage itself. By moving and measuring an alignment mark at various locations in the writing field, laser stages are used to calibrate the deflection gain, deflection linearity, and field distortion; that is, the stage is used as an absolute reference, and the deflection amplifiers are calibrated using the stage controller.

Other common features of commercial systems include a flat stage, a fixed working distance (contrasting with a SEM), and automated substrate handling. A flat stage keeps the sample in focus but requires the use of a detector either on or above the objective pole-piece. Most commonly, a microchannel plate or a set of silicon diodes is mounted on the pole-piece.

	JC Nabity Lithography Systems	*Raith GmbH*	*Leica Lithography Systems Ltd.*
Model	NPGS	Elphy-Plus	EBL Nanowriter
Alignment	Automated or manual	Automated or manual	Automated
Stitching	Automated, accuracy limited by stage	Automated, 0.1 μm accuracy with laser stage	Automated, with laser stage
Field	Variable, but < 100 μm typically, for 0.1 μm and low distortion, < 500 μm for 0.25 μm features	Variable, but < 100 μm typically, for 0.1 μm and low distortion, < 500 μm for 0.25 μm features	Variable, up to 2 mm at 50 kV
Energy	0-40 kV for typical SEM, but depends on target instrument	0-40 kV for typical SEM, but depends on target instrument	10 to 100 kV
DAC Speed	Low, > 10 μs per exposure point (100 kHz)	Mid-range, > 0.4 μs per exposure point (2.6 Mhz) but may be limited by the SEM deflectors	Mid-range, > 1 μs per exposure point (1 MHz)
Throughput limited by	Settling time of scan coils, transmission rate of ISA bus	Settling time of scan coils	Settling time of scan coils
Stage	Support for any automated stage	Optional laser controlled	Optional laser controlled
Control computer	PC compatible, ISA bus, DOS/Windows	PC compatible, DOS/Windows	PC compatible
Cost	Low, < $50 K, < $30K to universities - for pattern generator only (SEM purchased separately)	Mid-range, > $100 K for pattern generator only (SEM purchased separately)	Mid- to high-range, > $1000 K for a complete lithography system
Contact	406-587-0848, 406-586-9514 fax. JCNABITY@AOL.COM	Germany: (49) 0231-97-50-000, -005 fax. USA: 516-293-0870, -0187 fax	USA: 708-405-0213, -0147 fax. UK: 44-223-411-411, -211-310 fax

Table 2.1. Characteristics of SEM-based lithography systems. In all cases the resolution is high, depending (for Nabity and Raith) on the chosen SEM. All of these systems have relatively small stage motion, ~ 2 in. The Nabity and Raith devices are add-on products, while the Leica Nanowriter is an integrated system.

The market niche for commercial Gaussian spot high resolution e-beam tools has been primarily in research, and to a lesser extent for small-scale production of MMICs, high-speed T-gate transistors, and integrated optics.

2.5.4.1 JEOL systems

JEOL's popular JBX-5DII Gaussian vector scan system uses a LaB_6 emitter running at either 25 or 50 kV. Figure 2.17 shows the 5DII with two condenser lenses and two objective lenses. Only one of the objectives is used at a time; the operator has the choice of using the long working distance lens for a field size of 800 μm, or the short working distance lens, for an 80 μm field at 50 kV. (The fields are twice as large at 25 kV.) The pattern generator runs at 6 MHz (> 0.167 μs per exposure point) and the stage has a precision of $\lambda/1024 = 0.6$nm. As with all commercial systems, alignment, field stitching, and sample handling are fully automated. In fact, one drawback for research purposes is that there is no manual mode of operation. The system is capable of aligning to within 40 nm (2σ) and writing 30 nm wide features over an entire 5 in. wafer or mask plate. JEOL systems are known for their simple, high quality sample holders. The 5DII is one of the highest resolution (though not one of the fastest) e-beam tools in the LaB_6 class.

JEOL's JBX-6000 implements a number of improvements on the 5DII. The LaB_6 emitter is replaced with a thermal field emitter, eliminating the need for one of the condenser lenses. The pattern generator speed is increased to 12 Mhz, and the PDP-11 controller is replaced with a VAX. The system uses the same set of two objective lenses, and for a given objective lens the magnification is fixed (that is, the DAC's deflection is not scaled with the field size). As can be seen in the graph of figure 2.18, the ultimate spot size is somewhat improved over that of the LaB_6 machine, but more importantly, the current density at smaller spot sizes is greatly improved. The JBX-6000 runs at 25 kV or 50 kV.

With higher current density comes the property that the probe size is sometimes *smaller* than a pixel. For example, consider a pixel grid of spacing 0.0025 μm. If the rastering beam skips every n grid points, then the pixel area is $(n \times 0.0025 \text{ μm})^2$. With a current of 10 nA and a dose of 200 μC/cm^2, we must have $(n \times 0.0025 \text{ μm})^2 \times 200 \text{ μC/cm}^2 = 10$ nA × (exposure time for one pixel), and since the minimum exposure time is $1/(12 \text{ MHz}) = 0.08$ μs, the smallest value of n is 9. In this case the pixel spacing is 22.5 nm and the spot size, according to Fig. 2.18 is 12 nm. In this example the pixel spacing is *larger* than the spot size, and the exposed features may develop as a lumpy set of connected dots. The problem will be even more pronounced when using high speed resists, large field sizes, and larger currents. One solution would be to implement a faster pattern generator; however, JEOL's approach is to retain the superior noise immunity of the 12 MHz deflector and instead to use less current when necessary, or to increase the spot size by using a larger aperture. Alternatively, one can purposely *defocus* the beam. The NPGS system (see Sect. 2.5.3.1) attacks the problem by allowing different pixel spacings in X and Y (or in r and θ).

It is interesting to note that future high resolution systems under development at Hitachi[54] are likely to resemble the JEOL Gaussian-spot tools, with field sizes ≤ 500 μm and a single stage electrostatic deflector. Small fields avoid the complexities of dynamic focus and astigmatism corrections, and allow the short working distance needed to reduce the spot size. Single stage deflectors limit the bandwidth (speed) of the system, but improve intrafield stitching between deflections of coarse and fine DACs. The design tradeoff is clearly between high speed and high accuracy.

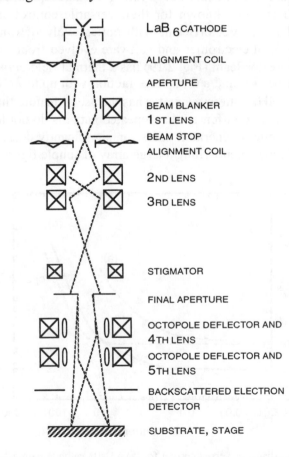

FIG. 2.17 Schematic of the JEOL JBX-5DII system with LaB$_6$ emitter. The system features two objective lenses for two different working distances (courtesy of JEOL Ltd.).

2.5.4.2 Leica Lithography Systems

Electron beam systems from Leica Lithography Systems Ltd. (LLS) are a combination of products previously manufactured by Cambridge Instruments, the electron beam lithography division of Philips, and most recently products from the former Jenoptik Microlit Division. Leica sells eight different models of Gaussian spot vector scan machines (the EBL Nanowriter has been described above). Systems in the mid-range of resolution include the EBML-300, a LaB_6 tool directly evolved from the Cambridge line, and the EBPG-5, a LaB_6 machine evolved from the Philips line. The EBPG-5 is comparable to the JEOL JBX-5DII in resolution but has accelerating voltage up to 100 kV. The EBMLand EBPG are both known for their versatile control software. On Leica's high end is the VectorBeam, with optics evolved from the Philips EBPG line and control electronics and software evolved from the Cambridge EBML line. The VectorBeam (Fig. 2.19) has a thermal field emission electron source running at 100 kV and a 6 in. stage motion with up to $\lambda/1024 = 0.6$ nm precision. The 25 MHz pattern generator has the useful feature that it is able to hold a small pattern in a buffer, so that repeated patterns do not have to be retransmitted to the pattern generator. This can significantly decrease the transmission overhead time when writing a large array of simple figures.

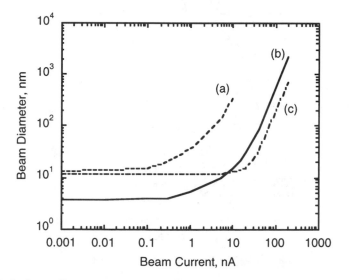

FIG 2.18 Probe beam diameter versus current for (a) a LaB_6 cathode with a 120 μm objective aperture, (b) a thermal field-emission (TFE) cathode with a 40 μm objective aperture, and (c) a thermal field-emission cathode with a 100 μm objective aperture. Data is from JEOL Gaussian-spot e-beam systems using 50 kV acceleration and a short working distance objective ("5th lens") (courtesy of JEOL Ltd.).

Leica e-beam tools are also distinguished from those of JEOL by their use of a single objective lens (one working distance), and scaleable writing fields with $2^{15}=32768$ or $2^{16}=65536$ pixels across the field. In the case of the EBML-300, field sizes up to 3.2 mm may be used, although the benefit of using such a large field is debatable.

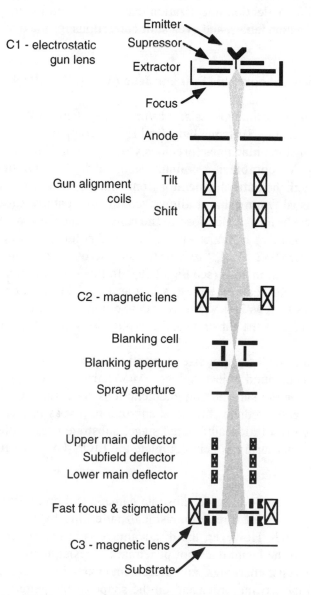

FIG. 2.19 Schematic of the Leica VectorBeam 100 kV column with a thermally assisted field emission electron source (courtesy of Leica Lithography Systems Ltd.)

The largest systems from Leica are also equipped with 100 kV TFE emitters, and have stages with up to 8 in. travel. Additional features include a glancing-angle laser height sensor for dynamic field size corrections, and dynamic focus/astigmatism corrections -- features more commonly found on high speed maskmaking tools. Systems using large writing fields, with deflection angles exceeding 5 to 10 milliradians, make use of a number of higher order corrections including deflection linearization maps, field rotation maps, dynamic focus and stigmation tables, and even shift corrections for the dynamic focus coil.

2.5.4.3 Leica Lithographie Systeme Jena (Jenoptik) LION

One of the most unique Gaussian vector scan systems is the LION-LV1 from Leica Lithographie Systeme Jena GmbH[55], a company better known for its large mask making machines (previously sold only in Eastern Bloc countries). The LION-LV1 combines a column designed by ICT GmbH (Heimstetten, Germany) with the pattern generator from Raith GmbH. This pattern generator has the unusual feature that it allows "continuous path control" of curves. In this mode the beam is held close to the center of the field while stage motion defines the shape of a Bezier curve. The ICT column is very similar to that used in the Leo 982 SEM,[51] except for the use of a beam blanker and higher bandwidth deflection coils (see Fig. 2.20). In this system, proximity effects are avoided by using beam energies as low as 1 to 2 keV. Although the voltage may be set as high as 20 kV, the system's selling point is low voltage -- avoiding both damage to the substrate and complications due to the proximity effect.

The column provides a spot size as small as 5 nm at 1 kV, through the use of an unusual compound objective lens. An electrostatic lens produces a diverging field, while the surrounding magnetic lens converges the beam. The complementary lenses reduce chromatic aberration, just as in a compound optical lens. A high resolution automated stage, substrate cassette loader, and substrate height measuring system complete the LION-LV1 as a full-featured system.

Low voltage operation avoids substrate damage and proximity effects, and offers the capability of three dimensional patterning by tailoring the electron penetration depth. However, the disadvantage is in greatly complicated resist processing. If the beam does not penetrate the resist, there will be significant effects from resist charging,[56] and placement errors due to charging may be dependent on the writing order and on the shape of the pattern itself. Charging may be avoided by using a resist trilayer with a conducting center (e.g., PMMA on Ti on polyimide), or by using a conducting overlayer (see sect. 2.7.1). Increased processing is required also for removing the resist layer over alignment marks. In a production environment this complexity adds significantly to the cost of ownership.

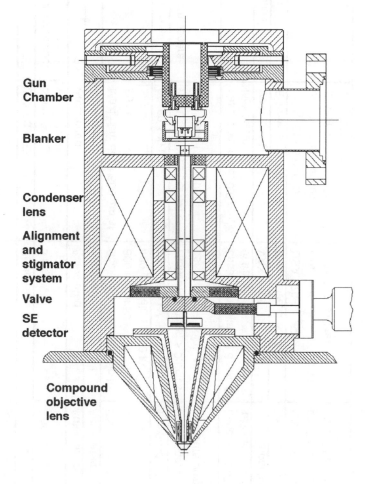

Gun
Chamber

Blanker

Condenser
lens

Alignment
and
stigmator
system

Valve

SE
detector

Compound
objective
lens

FIG. 2.20 Low-voltage column developed by ICT GmbH, used in the LION e-beam system from Leica. The beam blanker is directly above the anode. The objective lens combines electrostatic and magnetic elements to reduce the net chromatic aberration. Beam diameter at 1kV is approximately 5nm. (Courtesy of LLS Jena GmbH.)

	JEOL Inc.	Leica Ltd.	Leica Lithographie Systeme Jena GmbH
Model	JBX-6000FS	VectorBeam	LION-LV1
Resolution (minimum spot size)	5 nm	8 nm	5 nm
Alignment	Automated	Automated	Automated
Stitching	Automated	Automated	Automated
Field	Maximum 80 or 800 μm at 50 kV	Scaleable, 16 bits in up to 800 μm at 50 kV or 400 μm field at 100 kV	Scaleable, 16 bits
Energy	25, 50, 100 kV	10 to 100 kV	1 to 20 kV
Speed of pattern generation	High, > 0.08 μs per exposure point (12 MHz)	Highest of class, > 0.04 μs per exposure point (25MHz)	Mid-range, > 0.4 μs per exposure point (2.6 MHz)
Stage	Laser controlled, 0.6 nm, 6 inch travel	Laser controlled, 0.6 nm, 6 in travel	Laser controlled, 2.5 nm, 162 mm travel
Control computer	VAX VMS	VAX VMS	PC compatible
Cost	Expensive, > $3M	Expensive, > $3M	Expensive, > $1M
Contact	USA: 518-535-5900, Japan:1-2 Musashino 3-chome, Akishima Tokyo 196, 0425-42-2187	USA: 708-405-0213, -0147 fax. UK: 44-223-411-411, -211-310 fax	USA: 708-405-0213, -0147 fax. UK: 44-223-411-123, -211-310 fax

Table 2.2 Comparison of Gaussian-spot, vector-scan systems. All of these systems are equipped with thermally-assisted (Schottky) field emission electron sources.

without sacrificing speed is to implement a "graybeam" strategy, where the pixels on edges of features have dwell times and placements modulated on a per-pixel basis. This allows the bulk of a pattern to be written on a fast, coarse grid while edges are written with a finer resolution.[60]

2.5.5.2 Lepton EBES4

The EBES4 mask writer from Lepton Inc.[61] also uses a Gaussian spot, with a patterning strategy similar to that of the high resolution machines. In this system the coarse/fine DAC beam placement is augmented with an extra (third) deflection stage, and the mask plate is moved continuously, using the laser stage controller to provide continuous correction to the stage position. Unlike the high resolution JEOL machines, each stage of deflection has a separate telecentric deflector (instead of simply a separate set of DACs) for high speed operation. Patterns are separated into stripes (similar to writing fields) 256 μm wide (see Fig. 2.23). These stripes are separated into 32 μm subfields ("cells") which are further subdivided into 2 μm sub-subfields ("microfigures"). A spot of 0.125 μm diameter fills in the microfigure with a raster pattern.

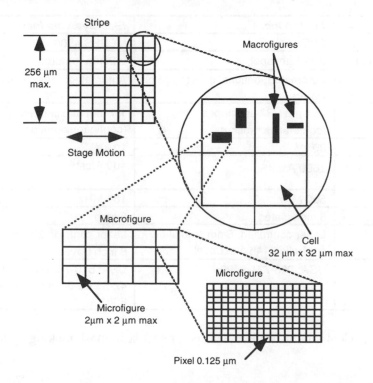

FIG. 2.23 Writing strategy of the Lepton EBES4 mask writing tool: pattern data is cut into stripes 256 μm wide. The stripes are fractured into smaller cells containing macrofigures. The macrofigures are split into even smaller microfigures which are finally written as a set of pixels.[62] (Courtesy of Lepton Inc.)

The entire EBES tool has been designed for high speed, with a current of 250 nA delivered in a 0.125 μm spot for a current density at the sample of 1600 A/cm^2. The EBES4 column uses a TFE electron gun operating at 20 kV and a single beam crossover at the center of a high-speed beam blanker.[63] The pattern generator operates at up to 500 MHz, and the high overall throughput allows production of a 16 Mbit DRAM mask in 30 min.[64]

A robot arm is used to load mask plates from a magazine module to the alignment and temperature equilibration chambers, and later to the exposure chamber. The internal mask carrier is made from the glass ceramic Zerodur™, which minimizes substrate temperature variations during exposure. The EBES4 automatically loads each mask plate into the carrier, establishes electrical contact to the substrate, and verifies the contact resistance.

The EBES4 mask writer has a spot size of 0.12 μm, uniformity to 50 nm (3σ), stitching error of 40 nm, and repeatability (overlay accuracy) of 30 nm over a 6 in. reticle.

	Lepton Inc.	Etec Systems Inc.
Model	EBES4	MEBES 4500
Resolution	0.125 μm spot	0.25μm features
Alignment	Automated, optional direct write on wafers	Automated, mask writing only
Field	256 μm × 32 μm stripes (continuous motion)	1.1 mm maximum stripe length (continuous motion)
Energy	20 kV	10 kV
Current density	1600 A/cm^2	400 A/cm^2
Speed	500 MHz	160 MHz
Samples	6 inch plates	8 inch
Stage	Laser controlled, 5 nm resolution controller, 146 mm travel	Laser controlled, 6.6nm (λ/96) 6 inch travel
Contact	USA: 908-771-9490	USA: 510-783-9210 France: 33-42-58-68-94 Japan: 81-425-27-8381

Table 2.3 Comparison of Gaussian spot, raster scan mask making systems.

2.5.6 Shaped Spot and Cell Projection Systems

All of the e-beam tools described above focus the beam into a small spot, and shapes are formed by rastering the beam. This spot is the reduced image of the source, often referred to as the "gun crossover," which has a current intensity profile resembling a Gaussian distribution. The time needed to paint a shape can be eliminated by forming the electron beam into primitive shapes (rectangles and triangles) and then exposing large areas with single "shots" of the beam. The optics of these shaped spot systems is shown schematically in Fig. 2.24. The upper aperture typically uses a square to form two sides of a rectangle, and the overlap of the lower aperture defines its length and width. More complex shapes are fractured into rectangles and triangles before exposure.

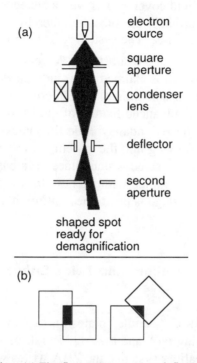

FIG. 2.24 Shaped spot optics. (a) The first square aperture defines two sides of a rectangle, and the second square aperture defines the other two sides. Deflectors determine the overlap and thus the length and width of the rectangle, as shown in (b). By deflecting over a corner, triangles can be formed. The shaped spot image would be further demagnified by magnetic lenses lower in the column.

Because of their increased parallelism over Gaussian raster-scan tools, shaped spot systems are much faster. However, throughput is still limited by the remaining serialism, by stage movements, and in a few cases by data transfer times. Shaped spot systems can readily be extended to 0.15 µm resolution (compared to the 0.25 µm resolution of the Gaussian raster beam systems). While there is no well defined standard for the comparison of throughput, we can say that the throughput of shaped spot machines remains under 10 wafers/hour -- making them superior to Gaussian systems but not competitive with optical steppers which produce, typically, 40 to 80 wafers/hour. The market for high speed shaped spot systems remains in maskmaking, direct-write prototyping, and low volume production of 0.15 μm scale features.

2.5.6.1 IBM EL-4

Shaped spot systems have been pioneered, but never sold, by IBM. The latest version, EL-4, combines an extraordinarily large number of lenses[65-67] (Fig. 2.25) with a unique three-stage deflection for optimum speed. The final lens, termed a variable axis immersion lens (VAIL) provides minimized off-axis aberrations (or maximum field coverage) as well as telecentric beam positioning, with the beam landing normal to the substrate, thereby reducing stitching errors due to substrate height variance. The system runs at 75 kV with a LaB_6 emitter, providing up to 50 A/cm^2 at the substrate. Wafers are held on the stage by electrostatic clamping, which is claimed to provide improved flatness, superior thermal stability, and lower contamination than conventional front-surface reference wafer chucks. An advanced feature of the EL-4 is its use of redundant data registers and a cyclic redundancy code for checking the validity of the many gigabytes of data flowing into the system. Another unique feature is the use of a servo guided planar stage which slides on a base plate without guide rails, moved by push rods coupled with friction drives to servo motors outside the vacuum chamber. The stage is positioned entirely through feedback from a multi-axis laser controller.[68,69]

2.5.6.2 Etec Systems Excaliber and Leica Lithographie Systeme Jena ZBA 31/32

The Leica Jena ZBA 31/32[70] handles plates up to 7 in and wafers up to 8 in. The "31" is a maskmaking tool, and the "32" is a direct-write instrument. Like Etec's AEBLE and Excaliber systems, the ZBA writes while the stage is moving. The ZBA delivers 20 A/cm^2. Its continuous stage motion and cassette-to-cassette wafer loader give it relatively high throughput when using high speed resist.

The latest generation of commercial shaped spot systems will offer resolution to 0.1 μm. Under development at Etec is the "Excaliber," with a field emission source, larger stage, and higher resolution than its predecessor, the AEBLE-150. The Excaliber system incorporates a number of features from IBM's EL-4, such as telecentric deflection and the sliding chuck ("wayless") stage with yaw compensation. Unlike EL-4, the Excaliber will keep the field size below 1 mm, thereby decreasing beam settling times while the stage moves continuously.

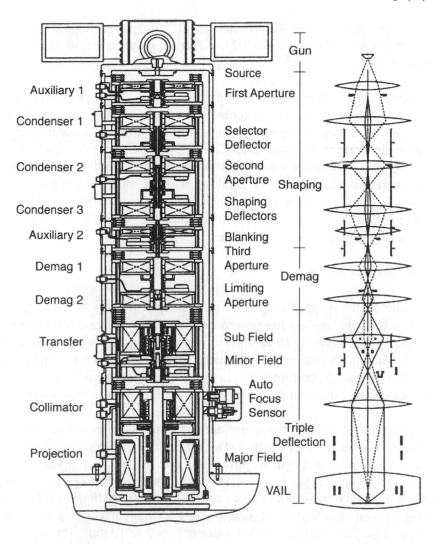

FIG. 2.25 Schematic of the IBM EL-4 column for shaped-beam lithography. On the right, the dashed ray trace corresponds to the source, and the solid trace to the shaped spot.[66] (Courtesy of IBM Corp.)

2.5.6.3 JEOL shaped spot systems

JEOL's JBX-8600DV[71] provides 0.1 µm resolution at 30 A/cm^2 for direct-write applications. The system uses two stage electrostatic deflection, and handles 6 in. wafers. The JBX-7000MVII[72] has been developed as a 4× reticle making system for 256 Mbyte DRAM class devices. As with most shaped spot systems, the JEOL machines can create a map of distortion values for the deflection so that patterns can be mapped more precisely onto optically-generated features. The JBX 7000MVII handles up to 7 in. plates with a laser stage measurement unit of 0.6 nm ($\lambda/1024$). Overlay accuracy is 30 nm (3σ) and placement accuracy is 40 nm.

The attention to absolute pattern placement accuracy is always much more extensive in dedicated maskmaking tools than in direct-write machines. To control thermal expansion of the plates, temperature monitoring and stabilization is far more elaborate. Like other manufacturers, JEOL creates a map of stage nonlinearity by measuring a set of marks, turning the plate in 90° increments and measuring the set again. The resulting stage distortion map is used to reduce the runout due to imperfections in the stage mirror. In fact, each individual plate holder has its own specific distortion table, which is identified automatically by reading a bar code on the cassette.

2.5.6.4 Cell Projection

The throughput of shaped beam tools is primarily limited by the average beam current in the spot, and by the pattern density. The average beam current for cell projection is modestly larger than for variable shaped beams. Both are limited by Coulomb interaction to a few microamperes. However, by replacing the simple beam shaping aperture with a more complex pattern, a "cell projection" system can greatly increase the pattern density without sacrificing throughput.

In cell projection systems the upper deflector steers the beam into one of a number of hole patterns, or "cells." The shaped beam is deflected back to the center of the column and is demagnified by another lens, forming an image on the substrate. The shaping aperture is made of a silicon membrane, around 20 µm thick, patterned with holes and coated with gold or platinum. To maintain small aberrations and high resolution, the cell is demagnified by a factor of 20 to 100, and the final cell size on the wafer is only 2 to 10 µm. The wafer containing these patterns also contains a simple rectangular aperture for general purpose pattern generation in a standard shaped spot mode. While a number of cell patterns may be placed on the beam shaping wafer, it is clear that the cell projection technique is advantageous and economical only for highly repetitive designs with small unit cells, namely, memory chips. Patterns for cell projection will require proximity correction by shape modification[38,39] or through a variation of the GHOST technique[40] (see Sect. 2.4.3).

To achieve throughput comparable to that of optical steppers, cell projection tools must reduce the shot count by a factor of around 100. Current machines have achieved shot reductions on the order of a factor of 10 and have throughputs of less than 10 wafer levels/hour (for a 6 in. wafer populated with 256 Mbyte DRAMs, $\sim 10^9$ shots/chip).

IBM,[73] Hitachi,[74,75] Toshiba,[76] Fujitsu,[77,78] and Leica have developed cell projection tools targeted for 256 Mbyte DRAM manufacture. Leica's "WePrint 200" instrument is a modified version of the ZBA-32. Hitachi also offers a cell projection/shaped spot system for sale: the HL-800D. Common features of cell projection systems include continuous stage motion[79] and resolution around 0.2 μm. Hitachi's HL-800D reduces the cell reticle by a factor of 25, while Fujitsu uses a factor of 100 and Toshiba uses a factor of 40. The final demagnified cell size is kept below ~10 μm to reduce aberrations.[76] Space charge effects also reduce the feature edge sharpness, but these can be compensated by using a current-dependent dynamic refocusing of the image.[75,80-82] Cell projection has not yet achieved the throughput of optical steppers but as a transitional technology may provide the resolution needed for near-term 256 Mbyte DRAM production.

2.5.7 SCALPEL

Cell projection uses small reticle areas to avoid spherical aberration and to minimize space charge effects. A natural extension of the idea would be to separate a large pattern into many small sections, etch each section into its own area of the aperture wafer, and then select and stitch the patterns together using a set of two deflectors. There are a number of limitations to this extension of cell projection: (1) 20 μm of silicon is needed to stop 50 kV electrons,[83] so the pattern must include deep holes. Because the aspect ratio of these holes is limited, lines can be no wider than ~2 μm; therefore, the electron optics must demagnify the pattern by a factor of at least 20 to produce linewidths of 0.1 μm. This limits the area available for cell patterns. (2) Multiply connected (e.g., doughnut shaped) patterns require complementary stencil masks, so the throughput and available pattern area is further reduced. (3) Residual stress in the stencil mask will distort the mask in a pattern-dependent way, and since stencil masks absorb most of the electron energy, the changing temperature will also cause similar pattern-dependent distortions.[84]

Instead of using an absorbing mask, Koops and Grob[85] proposed and researchers at AT&T Bell Laboratories[86-88] (now known as Lucent Techologies) later implemented the idea of using a *scattering* mask to produce a high contrast image with a technique commonly used in transmission electron microscopy. Figure 2.26 illustrates the technique "<u>s</u>cattering with <u>a</u>ngular <u>l</u>imitation in <u>p</u>rojection <u>e</u>lectron beam <u>l</u>ithography," or SCALPEL. Electrons traveling through a thin (typically 150 nm) silicon nitride membrane are focused by a lens and pass through an aperture (the "back focal plane filter"). Electrons scattered by

	IBM Corp.	Etec Systems Inc.	JEOL Inc.	Leica Lithographie Systeme GmbH	Hitachi
Model	EL-4	Excaliber *under development*	JBX-7000MVII	ZBA 31/32 *WePrint-200*	HL-800D *Cell Projection*
Resolution	0.15 µm features, 50 nm CD control	0.12 µm	0.2 to 0.5 µm	0.2 µm, 30 nm CD uniformity	0.25 µm, 50 nm CD uniformity
Alignment	Automated	Automated	Automated	Automated	Automated
Field	10 mm maximum	1 mm	1.5 mm	1.3 mm	
Energy	75 kV	100 kV	20 kV	20 kV	50 kV
Speed	~2-3 wafers/hour			2-5 wafers/hour	<10 wafers/hour in cell projection mode
Samples	8 in.	8 in.	up to 7 in. plate	8 in.	8 in.
Stage	"wayless" stage: electrostatic clamping, sliding chuck, servo powered, laser control with yaw compensation	"wayless" stage with sliding chuck, yaw compensation	laser controlled conventional stage	laser controlled conventional stage, cassette-to-cassette automated loading	laser controlled conventional stage
Cost	not for sale	*system under development*	high, > $3M	high, > $3M	high, > $3M
Contact	n/a	USA: 510-783-9210 France: 33-42-58-68-94 Japan: 81-425-27-8381	USA: 518-535-5900 Japan: 81-425-43-1111	USA: 708-405-0213, -0147 fax. UK: 44-223-411-123, -211-310 fax	USA: 415-244-7594, 415-244-7612 fax, or in Japan, 81-3-5294-2061

Table 2.4 Comparison of shaped spot systems.

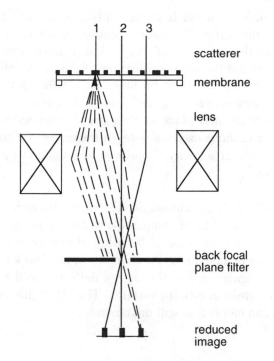

scatterer

membrane

lens

back focal
plane filter

reduced
image

FIG. 2.26 Schematic of the SCALPEL technique.[87] Electrons (1) that hit the scatterer (the patterns on the mask) are scattered, and most are filtered out by the aperture. Electrons traveling through the membrane (2,3) are demagnified through the aperture and form a high contrast image on the substrate. The mask is a pattern of tungsten supported on a low stress silicon nitride membrane. The membrane is supported on a silicon wafer, with periodic silicon support struts (not shown.) (Courtesy of Lucent Technologies Inc..)

the adsorber (typically 50 nm of Au or W) are most likely not to pass through the aperture. By choosing an optimal accelerating voltage (95 kV) for the membrane thickness (100 nm of low-stress silicon nitride) and adsorber (50 nm W), the contrast at the substrate can be as high as 95%, with a transmission of 55%.[89]

If the focal plane aperture includes an annular ring, then some of the "dark field" electrons pass through to expose the resist. The unfocused dark field image of the mask can thereby be used to provide a background dose correction to compensate for proximity effect, using a technique similar to GHOST[40] (see Sec. 2.4.3.3). Although this compensation scheme is still in the design stage, it holds the promise of proximity effect correction without any loss of throughput.[90]

As in cell projection, the mask is sequentially scanned and the image shifted and reduced onto the wafer. However, because the scattering features can be much thinner than the holes of cell projection, patterns can be fabricated at smaller dimensions and the demagnification of the mask can be decreased to 5×. A much larger chip can then be fabricated, with up to 2×10^{10} pixels.[91] Massive support struts between the "cells" are not imaged onto the wafer since the patterns are shifted into place as they are illuminated. While the mask structure is similar to those used for x-ray lithography, the support struts provide greater dimensional stability,[84] and use of reduction optics makes mask fabrication simpler.

The throughput of a fully-developed SCALPEL tools (which to date has only been modeled) is expected to be comparable to that of an optical stepper, while delivering resolution on the scale of 0.1 μm. However, several questions remain concerning its practical use: At energies in the 100 kV range resists are proportionally less sensitive, and the energy delivered to the substrate will be larger than in conventional e-beam systems. The effect this may have on transistor thresholds and mobility is still unknown.

2.5.8 Other E-Beam System Research

2.5.8.1 STM writing

The scanning tunneling microscope (STM) has been used to write nanometer-sized patterns in research experiments. It simply consists of a sharp tip used as a field emission cathode that is scanned a few nanometers above the surface of the sample. Resolution is obtained not by lenses but rather by keeping the tip so close to the surface that the electrons do not have a chance to diverge.
However, the technique is severely limited in writing speed and the resist thickness it can expose, and has seen only a few very limited applications. STM lithography is discussed in Sect. 8.8.3, and in the review article by Shedd and Russel.[92]

2.5.8.2 Parallel beam architectures - microcolumns

In addition to the projection systems described above, several other new architectures have been proposed for increasing the parallelism of e-beam lithography. One proposal is to build an integrated matrix of electron sources, producing an array of parallel beams within one column.[93,94] In contrast, researchers at NTT have proposed the use of an array of micromachined beam blankers and objective lenses, illuminated by a single high-current electron gun.[95] Other researchers are developing discrete components for miniaturized single-beam electron sources and columns.[96-98]

In an ongoing effort at IBM, researchers are seeking to shrink the lenses and other optical components to micrometer sizes using micromachining techniques, thereby building a high-performance, low voltage electron beam column.[96,99,100] Low-voltage has both advantages and disadvantages over high-voltage lithography (see Sect. 2.5.4.3) but is required here simply because of the small size of the components. In this design an entire e-beam column is only several millimeters high, assembled from micromachined silicon membranes supported on anodically bonded silicon and pyrex wafers. This concept is still in the early development stages.

Microcolumn research seeks to provide exposure parallelism by building an array of small columns. If they can be produced cheaply enough, maintenance would be simplified by the use of disposable electron optics. Although the optics may be inexpensive, the control system for a large array of columns may be very expensive. While many technical hurdles have already been overcome, the ultimate success of beam arrays may be decided solely by economics.

2.5.9 Electron Beam Fabrication Services

In addition to commercial mask vendors, many institutions offer services on large, high resolution e-beam tools. Payment for services varies widely, from purely collaborative work to hourly fees or contracts. Public access to many fabrication services is provided in the U.S. by the National Nanofabrication Users Network (NNUN), based primarily at Cornell and Stanford universities. Services provided through this network and the list of other sites changes so often that it is more appropriate to refer the reader to the World Wide Web page, http://www.cnf.cornell.edu/, which provides information about the services of the NNUN and other nonaffiliated U.S. fabrication centers. A list of mask vendors can be found in the *Semiconductor International Buyer's Guide*.

2.6 DATA PREPARATION

2.6.1 Pattern Structure

Preparation of pattern data for electron beam lithography may begin with a high level symbolic or mathematical description of a circuit, with the algorithmic description of a pattern (e.g. a Fresnel lens), or with a simple geometric layout. A computer aided design (CAD) program is usually used to lay out or at least inspect the pattern and to generate output in a standard exchange format. A separate program is then used to convert the intermediate format to machine-specific form. This last step can be quite involved since in most cases all hierarchy must by removed ("flattened"), polygons must be reduced to primitive shapes (e.g., trapezoids or triangles and rectangles), and the pattern must be fractured into fields, subfields, and even sub-subfields.

For shaped beam machines, or if the data is to be proximity corrected, medium and large sized shapes should be "sleeved", so that the edges of shapes are exposed separately from the interiors. For shaped beam machines this allows the edges to be exposed with a small shaped size that has better resolution; for proximity corrected patterns, this allows finer control over the dose delivered to the shapes. Frequently, a bias (also known as sizing) may be applied to the pattern shapes to account for resist characteristics or process steps that affect the final device linewidth.

For Gaussian beam machines, a reasonable pixel size must be selected. A good compromise is usually to use a pixel size of about half the beam diameter. Larger pixel sizes may speed up throughput, while smaller pixel sizes will reduce line edge roughness and improve feature size control. The machine field size is usually a fixed multiple of the pixel size. Field sizes may range from less than 100 μm for high resolution, high accuracy work to more than 1 mm for high speed, low resolution lithography.

When designing a device such as a transistor, you would organize the fabrication in a set of steps; e.g., mesa, ohmics, gate, etc. Each step is assigned to a "layer" in the CAD tool, and multiple layers are displayed as overlapping patterns (usually in different colors). Much later on, the layers will be split apart into separate pattern files. Some of these layers may be patterned with photolithography, some with e-beam. For example, you may design the geometry of each layer and place all of this information in the transistor "cell". Now you can put this cell at a number of other locations to create, say, a NAND logic gate. If you have not simply *copied* the transistor but rather have created *instances* of the cell (somewhat like a function called in a program) then any modifications in the transistor cell will be instantiated all over the NAND gate. The NAND gate is now a higher level cell, which can be used as part of, say, a half-adder. The hierarchy of an entire circuit is continued in this way. Of course, when building circuits from a standard technology such as CMOS, all of the basic component cells are usually purchased as part of the CAD program (a library of cells), and may even be placed and connected automatically as part of a symbolic CAD package.

2.6.2 Avoiding Trouble Spots

An e-beam lithographer would be unlikely to use any high level design tools. Rather, the lithographer must deal with data at the lower, geometrical level. If the scale of critical dimensions is far larger than the e-beam tool's placement errors, then the designer is free to place features anywhere. For instance, a set of 5× reticles with 5 μm design rules and 0.5 μm overlay error budget will demand little (except stability) of a commercial e-beam system. However, when the design requires a direct-write e-beam layer with 0.05 μm alignment, the placement of alignment marks becomes critical, and e-beam stitching errors can significantly affect device performance and yield. It is important for the designer to consider the limitations of the e-beam system before laying out any pattern.

Consider the case of a pattern targeted for a high resolution Gaussian beam system, such as the Leica EBPG or the JEOL-JBX series. For high resolution work the writing field may be as small as 80 μm. Larger patterns are formed by moving the sample and stitching fields together. Field stitching errors will be around 20 nm, so any fine lines in the pattern (e.g., a narrow gate) should not be placed at a field boundary.

2.6.3 Alignment Marks

Electron-beam lithography may be used to pattern optical masks and their corresponding alignment marks; steppers and contact aligners have specific design requirements for these marks. However, we will discuss here only the marks used for direct-write e-beam layers. There are two phases of alignment: (1) correction for the placement and rotation of the wafer (or piece) and (2) correction for the placement of individual chips on the wafer. The e-beam tool aligns each pattern file (in its final fractured form) to a mark before writing the pattern.

If your alignment tolerance is greater than ~0.5 μm, then the individual chip alignment will not be necessary. Global alignment -- that is, correction for the placement and rotation of the workpiece -- can use marks which are separate and larger than those used for chip alignment. Large global alignment marks are useful for the exposure of full wafers since the machine can be programmed to search for the first mark. Typical marks used for global alignment are large crosses of width 2 to 6 μm and length ~100 to 200 μm, placed at the top, bottom, left, and right sides of the wafer, as illustrated in Fig. 2.27. Alternatively, a few of the marks used for chip alignment could also be used for global alignment; this would allow global alignment on small pieces of a wafer. Alignment to chip marks is especially useful as a diagnostic of the maskmaking tool, allowing the measurement of displacements as a function of chip location.

For large patterns that take a long time to write, it may improve registration and placement accuracy if the machine stops periodically (every 5 to 10 minutes is typical) to reregister to the alignment marks. This corrects for thermal or other drifts that can occur during the writing process. For single level processes or maskmaking, reregistering to a single mark is sufficient to correct for drift.

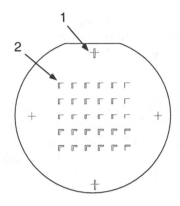

FIG 2.27 Alignment marks used for electron-beam lithography. Marks are typically etched pits in the wafer, or high-Z metal such as gold, platinum or tungsten. Global marks (1) are used to correct for the overall shift, rotation, and gain of the wafer, and chip marks (2) are used to correct for the placement of individual patterns. Chip marks can be used to correct for individual chip rotation and gain, to better match a badly adjusted optical stepper. The marks are not shown to scale. Typical wafer marks (1) are 200 μm long, and typical chip marks (2) are 10 μm long.

The size of a chip may be on the order of centimeters, and in photolithography the chips or entire wafers are aligned at once. While e-beam systems can align to global marks alone, the best tolerance (<0.1 μm) will be achieved when the alignment marks are within several hundred micrometers of the critical region. The designer may therefore wish to split the e-beam layer into smaller sections so that critical regions can be aligned individually. If these critical regions (e.g., gates) are arranged in a regular pattern, then arranging the sequence of e-beam writing will be simple. If the critical regions are placed randomly in the chip, the designer will have a time-consuming job of arranging the e-beam sequence and avoiding field boundaries

Alignment marks must be patterned in previous steps of the device fabrication. A "zero level" is sometimes used for the sole purpose of placing robust alignment marks on the sample before any actual device data are written. Typically the designer includes a photolithography step simply for patterning alignment marks as trenches to be etched into the substrate. The best alignment of layer 2 to layer 1 will be achieved when layer 1 contains the marks used for aligning layer 2 and when the marks are as close as possible to critical areas. If the material of layer 1 is unacceptable for alignment (e.g., a 20 nm thick metal layer) then both layers will have to be aligned to a third reference pattern (the "zero level"). Alignment to a third layer adds a factor of ~1.4 to the overlay error.

Well designed marks are commonly destroyed by processing. For example, ohmic metalizations become very rough when annealed. The rough marks are fine for optical alignment, but the lumps may cause the e-beam alignment hardware to trigger at the wrong locations. A good solution to this problem is to fabricate alignment marks as deep etched trenches (deeper than 1 μm). Plasma-etched or wet-etched trenches may be used. Such pits will not change after high temperature processing (unless material is deposited in them), and (unlike Au) are compatible with MOS processing. Other examples of effective alignment marks are W on Ti, Pt on Ti, and Au on Cr. Au is compatible with GaAs processing, but to maintain a smooth film, the alignment marks must be patterned after the annealing steps. In each of these cases the Ti or Cr provides improved adhesion to the substrate. A 200 nm thick layer of Pt or Au provides a good alignment signal, and 10 to 20 nm of Ti or Cr under the high-Z material provides improved substrate adhesion. Metal films can be patterned with very smooth edges by a liftoff process using a bilayer of PMMA and P(MMA/MAA) (see Sect. 2.7.4.2). In all cases, the designer must consider the thickness, roughness, and process compatibility of the material used for e-beam alignment marks, as well as the mark shape required for specific e-beam tools.

2.6.4 CAD Programs

CAD programs range from the very expensive schematic capture tools for VLSI to simple and inexpensive polygon editors. At the high end are widely used circuit capture, simulation, and layout tool sets from Cadence[101] (see http://www.cadence.com) and Mentor Graphics.[102] Other high-end packages are sold by Silvar Lisco,[103] Integrated Silicon Systems,[104] and a number of other vendors.[105] These tools run almost exclusively on UNIX workstations, and generate the standard intermediate format GDSII (also known as "Calma Stream" format) as well as the machine-specific MEBES format. Software tools in these sets include analog and digital simulators, silicon compilers, schematic capture, wire routers, design-rule checkers, and extensive cell libraries for CMOS, BiCMOS, and bipolar technologies.

In the mid-range of expense are the programs from Design Workshop[106] (DW2000) and Tanner Research[107] (L-Edit). Design Workshop implements a fully-functional graphical editor with the unusual feature of providing not only GDSII format, but also output in machine-specific formats for MEBES, JEOL, and Leica systems. DW2000 includes an integrated command language for algorithmic pattern definition. Design Workshop runs under the Macintosh OS, UNIX, and Windows NT. The Tanner Research tools run on PC compatibles, Macintoshes, and several UNIX workstations; output is in CIF or GDSII. Both Design Workshop and Tanner Research have implemented a less extensive set of companion tools (rule checkers, routers, simulators, etc.) and concentrate on the core graphical editors.

Inexpensive graphical editors include AutoCAD and other general-purpose CAD tools for PC compatibles and the Macintosh. AutoCAD and other similar programs generate DXF format, which must be converted to GDSII with a separate program.[108] AutoCAD has the disadvantage that it was not designed for lithography and so can generate patterns (such as 3D structures) that cannot be rendered by e-beam systems. Also, DXF format does not support "datatype" tags, which are used to specify individual dose values for geometrical shapes. Datatype tags are important when compensating (manually or automatically) for the proximity effect (see Sect. 2.4).

At the very low end are the free programs from UC Berkeley: Magic and OCT/VEM, which run on UNIX workstations. Magic is a widely used program geared for MOSIS-compatible CMOS processing. Magic is restricted to rectangles at right angles ("Manhattan geometry") and has no support for polygons. The VEM polygon editor in conjunction with the OCT database manager provides support for polygons. A number of companion simulation and routing tools also work with the OCT database but are distributed "as is," and without support. While these programs are distributed for only a shipping fee,[109] the real cost is the time and expertise required for installation and for working around bugs. Magic and VEM generate patterns in CIF format, which is supported by some mask vendors or may be translated to GDSII.

2.6.5 Intermediate Formats

2.6.5.1 GDSII

GDSII, also known as "Calma Stream", was originally developed by the Calma division of General Electric. Rights to the Calma products have changed hands several times, and are now owned by Cadence Design Systems. GDSII is by far the most stable, comprehensive, and widely used format for lithography. GDSII is a binary format that supports a hierarchical library of structures (called "cells"). Cells may contain a number of objects, including:

- Boundary, which may be used to represent polygons or rectangles,
- Box, which may be used to represent rotated rectangles,
- Path, which may be used to represent wires,
- Text, for annotation either on the CAD screen or the device,
- Sref, to include an instance of one structure (cell) inside another, and
- Aref, similar to Sref but providing an array instance of a cell.

There are 64 available Layers, numbered 0 to 63. Each primitive object (Boundary, etc.) lies on one of these layers. Each layer number typically represents one mask or electron-beam exposure step in a process.

A specification of GDSII format appears in the appendix to this chapter, portions of which are reprinted by permission of Cadence Design Systems.

2.6.5.2 CIF

The Caltech Intermediate Format, or CIF 2.0, is specified officially in *A Guide to LSI Implementation,* Second Edition, by R. W. Hon and C. H. Sequin,[110] and a nearly identical description appears in *Introduction to VLSI Systems*, by C. Mead and L. Conway.[111] This format is far simpler than GDSII and has the advantage that it is readable, using only ASCII characters. While providing nearly all of the functionality of GDSII, there are a few differences:

- Names of cells are not supported. Instead, cells are numbered.
- Datatypes are not supported. These are commonly used to assign different doses within a pattern. Therefore, proximity effect correction requires patterns to be split into multiple layers.
- There is no limit on the number of vertices in a polygon; therefore, CIF interpreters either set arbitrary values or simply run out of memory.
- The array structure (a square array of cells of n × m elements) is not supported and so the users of CIF have invented extensions to the format. These extensions have not been added to the CIF standard.

CIF is widely used by universities using the Berkeley CAD tools to design circuits for the MOSIS integrated circuit foundry service.[112] MOSIS requires a number of sensible restrictions on CIF data:[113]

- Polygons (P) must have at least three points; other than this, arbitrary polygons are accepted.
- Wires (W) must have at least one point.
- Round Flashes (R) must have a non-zero diameter.
- The "delete definition" (DD) command is not allowed.
- Symbols (cells) may not be redefined.
- Lines are limited to 509 characters of text.
- The following ASCII characters should not be used as "blanks": square brackets ([]), single quotes ('), and periods (.).
- User extensions are allowed but ignored. Wires are extended beyond the two extreme endpoints by half the wire width.
- The comment layer has a name ending with the letter "X". All geometry on this layer is read by MOSIS but is totally ignored; however, any syntax error in this layer may cause the CIF file to be rejected.
- The bonding pad layer is named "XP" in all technologies.

2.6.5.3 DXF

DXF format is produced by the program AutoCAD as well as by a number of other inexpensive CAD programs for Windows/DOS and the Macintosh. These programs were not designed for lithography and so contain structures (e.g. three-dimensional figures) that have no meaning in this area. Also, the common jargon (e.g., "cell") has been replaced with less familiar terminology (e.g. "block"). Like CIF, this format does not support datatype numbers. DXF is useful only after it has been translated into GDSII by a program such as that sold by Artwork Conversion Software[108] or those of various mask vendors.

In DXF there can be considerable confusion over such issues as whether an enclosed line represents a polygon or an actual line. Translation programs support different subsets of DXF and translate the structures into GDSII using various sets of rules. Users of DXF are advised to submit sample patterns for conversion before investing a lot of time in CAD work, and to bear in mind that the DXF file used for one vendor may not work at all for a different vendor. Therefore, the cost of data conversion should be considered when choosing an *apparently* inexpensive CAD tool.

2.6.5.4 PG3600

PG3600 and its predecessor PG3000 are used primarily by optical pattern generators built by GCA. These reticle printers use a high brightness lamp and a variable rectangular shutter to print patterns onto mask plates. The rectangle can be rotated to create angled features, and rectangular "flashes" are often overlapped to create curves, circles, and other shapes. Because of its popularity in reticle generation, many e-beam systems support the use of PG3600, even though the format would normally be considered low-level and machine specific. There are a number of disadvantages over GDSII:

- Overlaps must be removed by the conversion software. This can be very time consuming.
- The format is formally a specification for 9-track tape, using the EBCDIC character set. Some conversion programs require the disk format to use EBCDIC, and some allow a mapping into ASCII.
- Polygons in the CAD program are translated into overlapping rotated rectangles. This process is prone to error.
- Like CIF, datatypes are not supported, and so features with different doses must appear in different layers.
- A hierarchy of cells is not supported. The pattern must be "flat" and so may use a great deal of disk space.

2.6.6 Low-Level Formats

Conversion from one of the above formats to a machine-specific format usually involves flattening the hierarchy of cells, fracturing polygons into primitive shapes, and splitting the pattern into fields and subfields. The resulting machine-specific formats (e.g., MEBES, JEOL51, and BPD) usually use far more disk space than the hierarchical forms. These files must be carefully checked for software errors and may require manipulation for sizing, tone-reversal, mirroring, and so on. One way of verifying a conversion is simply to convert the low level format *back* to GDSII so that it can be displayed with the original CAD tool. Unfortunately, the pattern would have lost its cell structure, so the data set may be too large for the graphical editor. A special class of display and manipulation software is required that can handle very large, flat data sets.

The CATS program from Transcription Enterprises[114] and CAPROX from Sigma-C[115] offer not only viewing and manipulation of machine formats, but also will fracture GDSII directly into these formats. These conversion programs support machine formats from Etec Systems (MEBES, AEBLE), Hitachi, JEOL, Leica, GCA, and others. Operations include Boolean functions, tone reversal, rotation, sizing, and overlap removal. Sigma-C also offers a hierarchical proximity effect correction program. This software is an important alternative to the converters sold by e-beam manufacturers.

JEBCAD[116] is a less extensive, and less expensive, tool for viewing and manipulating JEOL and Leica formats. JEBCAD will read in GDSII, J01, SPD, and several low-level fractured formats; it will output machine formats for JEOL and Leica systems. Operations in JEBCAD include adding and deleting polygons, moving, copying, and adding arrays of objects.

Design Workshop[117] provides one of the most economical ways of producing machine specific formats for JEOL, Leica, and MEBES tools. DW2000's low-level fracturing modules are quite slow compared to alternative software, but are available at a small fraction of the cost.

2.7 RESISTS

Electron beam resists are the recording and transfer media for e-beam lithography. This section is not intended as a review of research in resists or as a guide to resist chemistry; for this, the reader is referred to Chap. 4 and to several review papers.[118-122] Instead, we present here a few standard resist systems and some useful recipes for processing and pattern transfer. The commercially available resists described here are summarized in Table 2.5.

The usual resists are polymers dissolved in a liquid solvent. Liquid resist is dropped onto the substrate, which is then spun at 1000 to 6000 rpm to form a coating.[123] Further details on resist application can be found in Chap. 4. After baking out the casting solvent, electron exposure modifies the resist, leaving it either more soluble (positive) or less soluble (negative) in developer. This pattern is transferred to the substrate either through an etching process (plasma or wet chemical) or by "liftoff" of material. In the liftoff process a material is

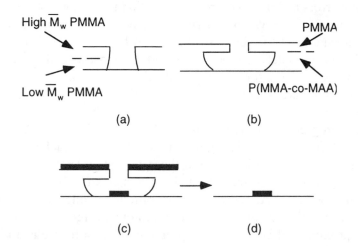

FIG 2.28 Two bilayer e-beam resist structures. (a) A high molecular weight PMMA is spun on top of a slightly more sensitive bottom layer of low molecular weight PMMA. The resist is developed in methyl isobutyl ketone:isopropanol (MIBK:IPA), typically 1:3, giving a slight undercut. (b) PMMA is spun on top of the copolymer P(MMA-co-MAA). The structure is typically developed in MIBK:IPA 1:1, giving a large undercut. In this case, MIBK develops PMMA and IPA develops the P(MMA-co-MAA). In the liftoff process metal is evaporated as shown in (c). The resist is then removed in a liquid solvent, leaving the pattern (d). Solvents such as acetone and methylene chloride are used to dissolve the resist.

evaporated from a small source onto the substrate and resist, as shown in Fig. 2.28. The resist is washed away in a solvent such as acetone or NMP (photoresist stripper). An undercut resist profile (as shown) aids in the liftoff process by providing a clean separation of the material.

If we expose a positive resist to a range of doses and then develop the pattern and plot the average film thickness versus dose, we have a graph as shown in Fig. 2.29. The sensitivity of the resist is defined as the point at which all of the film is removed. Ideally, the film thickness would drop abruptly to zero at the critical dose. In practice, the thickness line drops with a finite slope. If D_1 is the largest dose at which no film is lost [actually, the extrapolation of the linear portion of Fig. 2.29(a) to 100%] and if D_2 is the dose at which all of the film is lost [again, actually the extrapolation seen in Fig. 2.29(a)], then we define the contrast γ of the resist by

$$\gamma = \left| \log_{10}(D_2/D_1) \right|^{-1}. \tag{2.2}$$

The same expression defines the contrast of a negative resist (the film is retained where irradiated), when D_1 and D_2 are the points shown in Fig. 2.29(b).

A higher contrast resist will usually have a wider process latitude as well as more vertical sidewall profiles. In order to help minimize bias and proximity effects, positive resists should usually be exposed and/or developed as lightly as possible while still adequately clearing the resist down to the substrate for all features. In electron beam lithography, especially at beam voltages of 50 kV or more, it is possible to make resist structures with very high aspect ratios. Unfortunately, when the aspect ratio exceeds roughly 5:1, most resists undergo mechanical failure (features will fall over) during development, due primarily

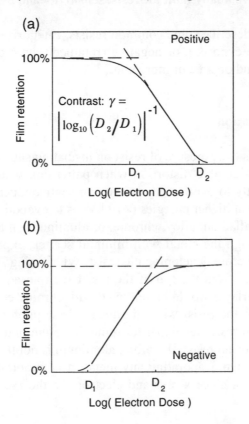

FIG. 2.29 Film thickness versus exposure dose for (a) positive and (b) negative resist. Contrast (γ) is defined as the slope of the linear portion of the falling (or rising) section of the curve.

to surface tension in the rinse portion of the development sequence.[124] Recently, commercial software for simulating electron-beam exposure of polymer resists has become available.[125]

The primary goals of e-beam lithography are high resolution and high speed (high sensitivity). Unfortunately, the highest resolution resists are usually the least sensitive. We can see a reason for this trend when we consider the limit of resist sensitivity. If a very sensitive resist has a critical dose of 0.1 $\mu C/cm^2$, and a pixel is 0.1 μm on a side, then only 62 electrons are needed to expose the pixel.[126] At this sensitivity, even small changes in the number of electrons will cause variations in the dose delivered to each pixel. If the sensitivity is increased further, then the number of electrons in each pixel becomes too small to allow an even exposure of the pattern. To look at it another way, if we wish to *decrease* the pixel size, then the resist will have to be made *less* sensitive to avoid statistical variations in the exposure. Although there is room for improving the sensitivity of both high and low resolution resists, the statistics of resist exposure will eventually limit the resist sensitivity and exposure rate.

In the following we describe some common resists, categorized as either positive (removed where exposed), or negative (retained where exposed), single layer or multilayer, and organic or inorganic.

2.7.1 Charge Dissipation

A common problem is the exposure of resist on insulating substrates. Substrate charging causes considerable distortion when patterning insulators and may contribute significantly to overlay errors even on semiconductors.[56] A simple solution for exposure at higher energies (>10 kV) is to evaporate a thin (10nm) layer of gold, gold-palladium alloy, chrome, or aluminum on top of the resist. Electrons travel through the metal with minimal scatter, exposing the resist. The film is removed before developing the resist. When using Au or Au/Pd, the metal film is removed from the top of the resist with an aqueous KI/I solution.[127] A chrome overlayer would be removed with chrome etch.[128] Aluminum can be removed from the resist with an aqueous base photoresist developer. Acid mixtures or photoresist developer for removing aluminum will sometimes react with exposed e-beam resist; therefore, aluminum is not the best choice for charge dissipation. When evaporating any metal, it is important *not* to use an electron gun evaporator since x-rays and electrons in the evaporator will expose the resist.

Another approach to charge dissipation is the use of a conducting polymer, either as a planarizing layer under the resist or as a coating over the resist. The commercial polymers TQV (Nitto Chemical Industry) and ESPACER100 (Showa Denko) have been used for this purpose.[129,130] Both are coated at a thickness of about 55 nm and have a sheet resistance around 20 MΩ/$\square$. TQV uses cyclohexanone as the casting solvent, which swells and dissolves novolac resins (present in most photoresists and SAL), and so a water-soluble PVA (polyvinyl alcohol) layer is needed to separate the resist from the TQV. ESPACER100 has the advantage that it is soluble in water and so can be coated directly onto many resists. TQV is removed with methyl isobutyl ketone/isopropanol (MIBK/IPA), the developer used for PMMA. ESPACER is removed in water. Other water soluble conducting polymers can be prepared from polyaniline doped with onium or triflate salts.[131,132]

2.7.2 Positive Resists

In the simplest positive resists, electron irradiation breaks polymer backbone bonds, leaving fragments of lower molecular weight. A solvent developer selectively washes away the lower molecular weight fragments, thus forming a positive tone pattern in the resist film.

2.7.2.1 PMMA

Polymethyl methacrylate (PMMA) was one of the first materials developed for e-beam lithography.[133,134] It is the standard positive e-beam resist and remains one of the highest resolution resists available. PMMA is usually purchased[135] in two high molecular weight forms (496 K or 950 K) in a casting solvent such as chlorobenzene or anisole. PMMA is spun onto the substrate and baked at 170 to 200 °C for 1 to 2 hours. Electron beam exposure breaks the polymer into fragments that are dissolved preferentially by a developer such as MIBK. MIBK alone is too strong a developer and removes some of the unexposed resist. Therefore, the developer is usually diluted by mixing in a weaker developer such as IPA. A mixture of 1 part MIBK to 3 parts IPA produces very high contrast[136] but low sensitivity. By making the developer stronger, say, 1:1 MIBK:IPA, the sensitivity is improved significantly with only a small loss of contrast.

The sensitivity of PMMA also scales roughly with electron acceleration voltage, with the critical dose at 50 kV being roughly twice that of exposures at 25 kV. Fortunately, electron guns are proportionally brighter at higher energies, providing twice the current in the same spot size at 50 kV. When using 50 kV electrons and 1:3 MIBK:IPA developer, the critical dose is around 350 μC/cm^2. Most positive resists will show a bias of 20 to 150 nm (i.e. a hole in the resist will be larger than the electron beam size), depending on the resist type, thickness, and contrast and development conditions and beam voltage.

When exposed to more than 10 times the optimal positive dose, PMMA will crosslink, forming a negative resist. It is simple to see this effect after having exposed one spot for an extended time (for instance, when focusing on a mark).

The center of the spot will be crosslinked, leaving resist on the substrate, while the surrounding area is exposed positively and is washed away. In its positive mode, PMMA has an intrinsic resolution of less than 10 nm.[137] In negative mode, the resolution is at least 50 nm. By exposing PMMA (or any resist) on a thin membrane, the exposure due to secondary electrons can be greatly reduced and the process latitude thereby increased. PMMA has poor resistance to plasma etching, compared to novolac-based photoresists. Nevertheless, it has been used successfully as a mask for the etching of silicon nitride[138] and silicon dioxide,[139] with 1:1 etch selectivity. PMMA also makes a very effective mask for chemically assisted ion beam etching of GaAs and AlGaAs.[140]

EXAMPLE PROCESS: PMMA POSITIVE EXPOSURE AND LIFTOFF

1. Start with 496K PMMA, 4% solids in chlorobenzene. Pour resist onto a Si wafer and spin at 2500 rpm for 40 to 60 seconds.
2. Bake in an oven or on a hotplate at 180 °C for 1 h.
    Thickness after baking: 300 nm.
3. Expose in e-beam system at 50 kV, with doses between 300 and 500 $\mu C/cm^2$. (Other accelerating voltages may be used. The dose scales roughly with the voltage.)
4. Develop for 1 min in 1:3 MIBK:IPA. Rinse in IPA. Blow dry with nitrogen.
5. Optional descum in a barrel etcher: 150W, 0.6 Torr O_2.
6. Mount in evaporator and pump down to 2×10^{-6} Torr.
7. Evaporate 10 nm Cr, then 100 nm Au.
8. Remove from evaporator, soak sample in methelyne chloride for ~10 min.
9. Agitate substrate and methylene chloride with an ultrasonic cleaner for ~1 min to complete the liftoff. Rinse in IPA. Blow dry.[141]

2.7.2.2 EBR-9

EBR-9 is an acrylate-based resist, poly(2,2,2-trifluoroethyl-α-chloroacrylate),[142] sold by Toray Inc.[143] This resist is 10 times faster than PMMA, ~10 $\mu C/cm^2$ at 20 kV. Its resolution is unfortunately more than 10 times worse than that of PMMA, ~0.2 μm. EBR-9 excels for mask writing applications, not because of its speed (PBS is faster) but because of its long shelf life, lack of swelling in developer, and large process latitude.

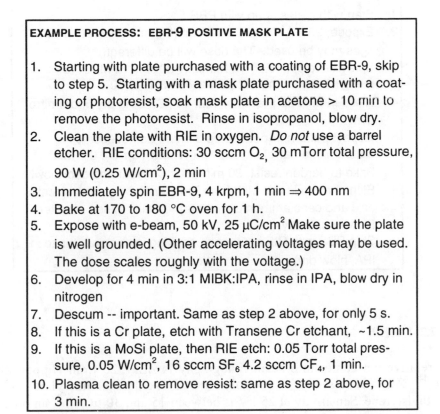

EXAMPLE PROCESS: EBR-9 POSITIVE MASK PLATE

1. Starting with plate purchased with a coating of EBR-9, skip to step 5. Starting with a mask plate purchased with a coating of photoresist, soak mask plate in acetone > 10 min to remove the photoresist. Rinse in isopropanol, blow dry.
2. Clean the plate with RIE in oxygen. *Do not* use a barrel etcher. RIE conditions: 30 sccm O_2, 30 mTorr total pressure, 90 W (0.25 W/cm^2), 2 min
3. Immediately spin EBR-9, 4 krpm, 1 min $\Rightarrow$ 400 nm
4. Bake at 170 to 180 °C oven for 1 h.
5. Expose with e-beam, 50 kV, 25 μC/cm^2 Make sure the plate is well grounded. (Other accelerating voltages may be used. The dose scales roughly with the voltage.)
6. Develop for 4 min in 3:1 MIBK:IPA, rinse in IPA, blow dry in nitrogen
7. Descum -- important. Same as step 2 above, for only 5 s.
8. If this is a Cr plate, etch with Transene Cr etchant, ~1.5 min.
9. If this is a MoSi plate, then RIE etch: 0.05 Torr total pressure, 0.05 W/cm^2, 16 sccm SF_6 4.2 sccm CF_4, 1 min.
10. Plasma clean to remove resist: same as step 2 above, for 3 min.

2.7.2.3 PBS

Poly(butene-1-sulfone) is a common high-speed positive resist used widely for mask plate patterning. For high-volume mask plate production, the sensitivity of 1 to 2 μC/cm^2 is a significant advantage over other positive resists. However, the processing of PBS is difficult and the only advantage is the speed of exposure. Plates must be spray developed at a tightly controlled temperature and humidity.[144] Contrast is poor, with $\gamma \sim 2$. For small to medium scale mask production, the time required for plate processing can make PBS slower than some photoresists.[145] (See Sect. 2.7.2.5.)

EXAMPLE PROCESS: PBS POSITIVE MASK PLATE

1. Start with plates spun with PBS.[146]
2. Expose, 25 kV, 1.0 to 1.6 μC/cm^2 (Other accelerating voltages may be used. The dose will be different.)
3. Spray develop, 10±1 °C, humidity 30±1%, in MIAK (5-methyl-2-hexanone) : 2-*pentanone* 3:1[147] ~30 s.
4. Rinse in MIAK:2-propanol 3:2, 10 °C. Spin dry under nitrogen.
5. Inspect pattern, repeat steps 3 and 4 as necessary.
6. Descum in a barrel etcher, 150 W, 0.6 Torr O$_2$, 0.5 min.
7. Bake to harden resist, 30 min 120 °C. Heat and cool slowly.
8. Etch chrome in wet etch from Transene or Cyantek (acetic acid and ceric ammonium nitrate) ~1 min. Rinse in water. Blow or spin dry.
9. Strip PBS with RIE in O$_2$ or by soaking in acetone. (rinse in IPA, blow dry).

2.7.2.4 ZEP

A relative newcomer to e-beam lithography is ZEP-520 from Nippon Zeon Co.[148] ZEP consists of a copolymer of α-chloromethacrylate and α-methylstyrene. Sensitivity at 25 kV is between 15 and 30 μC/cm^2, an order of magnitude faster than PMMA and comparable to the speed of EBR-9. Unlike EBR-9, the resolution of ZEP is very high -- close to that of PMMA. ZEP has about the same contrast as PMMA. Lines of width 10 nm with pitch 50 nm have been fabricated with this resist.[149,150] The etch resistance of ZEP in CF$_4$ RIE is around 2.5 times better than that of PMMA but is still less than that of novolac-based photoresists. ZEP is reported to have a long shelf life.[150] One disadvantage in using this resist is that (like PMMA) its sensitivity to electrons makes it difficult to inspect with a SEM. Resist lines shift and swell under high magnification SEM viewing, so it is necessary to judge the resolution of the resist by inspecting the etched patterns.

EXAMPLE PROCESS: ZEP PATTERNING OF SiO$_2$ HOLES

1. Prepare oxidized Si wafer. Spin ZEP-520 at 5 krpm for thickness 300 nm.
2. Bake at 170 °C, 2 min.

(continued)

---

3. Expose at 25 kV, 15 to 30 $\mu C/cm^2$ (Other accelerating voltages may be used. The dose will be different.)
4. Develop in xylene:p-dioxane (20:1) for 2 min. Blow dry.
5. Descum in barrel etcher, 0.6 Torr of oxygen, 150W, 1 min.
6. Etch oxide in 4 min intervals (to avoid resist flow) 15 mTorr total pressure, 42 sccm CF_4, 5 sccm H_2, 0.03 W/cm^2; oxide etches at ~15 nm/min.
7. Remove residual resist with oxygen RIE: 30 sccm O_2, 30 mTorr total pressure, 0.25 W/cm^2, 5 min.

2.7.2.5 Photoresists as e-beam resists

Most photoresists can be exposed by e-beam, although the chemistry is quite different from that of UV exposure.[151] Because electrons cause both positive exposure and cross-linking at the same time, a photoresist film exposed with electrons must be developed with a strong developer for "positive" behavior, or, the same film can be blanket-exposed with UV light and then developed in a weak developer for "negative" behavior. One of the best photoresists for positive e-beam exposure is AZ5206.[152,145] This resist has sensitivity around 6 $\mu C/cm^2$, contrast $\gamma=4$, and good etch resistance. With resolution around 0.25 μm and very simple processing, AZ5206 is one of the best alternatives for high-speed mask production.

EXAMPLE PROCESS: AZ5206 POSITIVE MASK PLATE

1. Soak mask plate in acetone > 10 min to remove the original photoresist. Rinse in isopropanol, blow dry.
2. Clean the plate with RIE in oxygen. *Do not* use a barrel etcher. RIE conditions: 30 sccm O_2, 30 mTorr total pressure, 90 W (0.25 W/cm^2), 5 min.
3. Immediately spin AZ5206, 3 krpm.
4. Bake at 80 °C for 30 min.
5. Expose with e-beam, 10 kV, 6 $\mu C/cm^2$, Make sure the plate is well grounded. (Other accelerating voltages may be used, but the dose will be different.)
6. Develop for 60 s in KLK PPD 401 developer. Rinse in water.
7. Descum - important Same as step 2 above, for only 5 seconds, Or use a barrel etcher, 0.6 Torr oxygen, 150W, 1 min.

(continued)

> 8. If this is a Cr plate, etch with Transene Cr etchant, ~1.5 min. If this is a MoSi plate, then RIE etch: 0.05 Torr total pressure, 0.05 W/cm^2, 16 sccm SF_6, 4.2 sccm CF_4, 1 min.
> 9. Plasma clean to remove resist: same as step 2 above, for 3 min.

Other UV sensitive resists used for e-beam include EBR900[153] from Toray,[143] (8 $\mu C/cm^2$ at 20 kV), the chemically amplified resist ARCH[154] from OCG,[155] (8-16 $\mu C/cm^2$ at 50 kV), and the deep-UV resists UVIII and UVN from Shipley.[156,157] The latest offerings from Shipley have been optimized for DUV (248 nm) exposure, and have higher resolution than that of AZ5206. The use of DUV resists allows exposure by both photons and electrons in the same film, thereby reducing e-beam exposure time.

2.7.3 Negative Resists

Negative resists work by cross-linking the polymer chains together, rendering them less soluble in the developer. Negative resists tend to have less bias (often zero) than positive resists. However, they tend to have problems with scum (insoluble residue in exposed areas), swelling during development, and bridging between features.

A reasonable starting point for developing a negative resist process is to choose a development time twice as long as the time needed to clear the unexposed resist and an exposure dose just sufficient to ensure acceptable resist thickness loss on all features (e.g., no more than 10%). From there, fine tuning of development time, dose, and postexposure bake conditions may be needed to optimize feature sizes, improve critical dimension control, and minimize resist scum.

2.7.3.1 COP

COP is an epoxy copolymer of glycidyl methacrylate and ethyl acrylate, P(GMA-co-EA), commonly used for negative exposure of mask plates.[147,122] This is a very high speed resist, 0.3 $\mu C/cm^2$ at 10 kV, with relatively poor resolution (1 μm).[158] COP also has relatively poor plasma etch resistance and requires spray development to avoid swelling. Because cross-linking occurs by cationic initiation and chain reaction, the cross-linking continues after exposure. Therefore, the size of features depends on the time between exposure and development. Unless speed is very critical, COP is probably not a good choice for a negative resist.

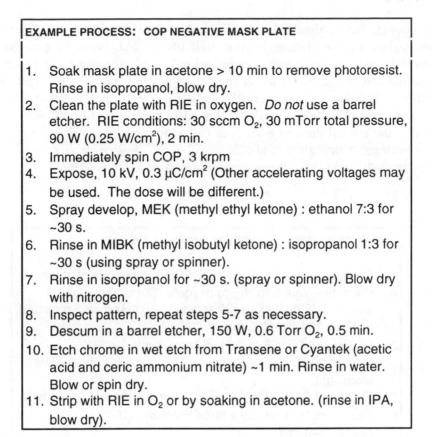

EXAMPLE PROCESS: COP NEGATIVE MASK PLATE

1. Soak mask plate in acetone > 10 min to remove photoresist. Rinse in isopropanol, blow dry.
2. Clean the plate with RIE in oxygen. *Do not* use a barrel etcher. RIE conditions: 30 sccm O_2, 30 mTorr total pressure, 90 W (0.25 W/cm^2), 2 min.
3. Immediately spin COP, 3 krpm
4. Expose, 10 kV, 0.3 µC/cm^2 (Other accelerating voltages may be used. The dose will be different.)
5. Spray develop, MEK (methyl ethyl ketone) : ethanol 7:3 for ~30 s.
6. Rinse in MIBK (methyl isobutyl ketone) : isopropanol 1:3 for ~30 s (using spray or spinner).
7. Rinse in isopropanol for ~30 s. (spray or spinner). Blow dry with nitrogen.
8. Inspect pattern, repeat steps 5-7 as necessary.
9. Descum in a barrel etcher, 150 W, 0.6 Torr O_2, 0.5 min.
10. Etch chrome in wet etch from Transene or Cyantek (acetic acid and ceric ammonium nitrate) ~1 min. Rinse in water. Blow or spin dry.
11. Strip with RIE in O_2 or by soaking in acetone. (rinse in IPA, blow dry).

2.7.3.2 Shipley SAL

Shipley Inc.[156] produces the popular SAL resist, which comes in a variety of versions and viscocities. SAL has three components: a base polymer, an acid generator, and a crosslinking agent. After exposure, a baking cycle enhances reaction and diffusion of the acid catalyst, leading to resist hardening by cross-linking. Common alkaline photoresist developers will dissolve the unexposed regions. The acid reaction and diffusion processes are important factors in determining the resolution,[159] and a tightly controlled postexposure baking process is required. The postexposure bake is usually on a feedback-controlled hotplate with a suction holder to ensure good thermal contact. The extent of the cross-linking reaction is therefore affected by the thermal conductivity of the sample and by the cooling rate after the bake. Resolution of 30 nm has been demonstrated at very low voltage,[160] and 50 nm wide lines have been fabricated using high voltage.[161] SAL-606 has 0.1 µm resolution in 0.4 µm thick films, exposed with 40 keV electrons at 8.4 µC/cm^2.

The novolac base polymer has etching properties similar to those of positive photoresists. Unlike photoresist, the shelf life of SAL is on the order of six months at room temperature. Refrigeration extends the shelf life to several years, but care is required to avoid condensation when the resist is dispensed to smaller containers. SAL is a sensitive resist, 7 to 9 $\mu C/cm^2$ at either 20 or 40 kV, and so is suitable for mask writing. It is interesting to note that, unlike PMMA, the critical dose of SAL does not scale proportionately with accelerating voltage. Although it is not as sensitive as other negative resists (COP, CMS, or GMC) SAL has far better process latitude and resolution.

EXAMPLE PROCESS: SAL NEGATIVE MASK PLATE

1. Soak mask plate in acetone > 10 min to remove photoresist.
2. Clean the plate with RIE in oxygen. *Do not* use a barrel etcher. RIE conditions: 30 sccm O_2, 30 mTorr total pressure, 90 W (0.25 W/cm^2), 5 min.
3. Immediately spin SAL-601, 4 krpm, 1 min.
4. Bake in 90 °C oven for 10 min. This resist is not sensitive to room light.
5. Expose at 50 kV, 11 $\mu C/cm^2$. Be sure the plate is grounded.
6. Post-bake for 1 min on a large hotplate, 115 °C.
7. Cool for > 6 min.
8. Develop for 6 min in Shipley MF312:water (1:1) Be sure to check for underdevelopment.
9. Descum 30 s with oxygen RIE: same as step 2, 10 s.
10. Etch with Transene or Cyantek Cr etchant, ~1.5 min.
11. Plasma clean to remove resist: Same as step 2, 5 min.

2.7.3.3 Noncommercial negative resists: P(SI-CMS) and EPTR

Although not yet commercialized, a very promising negative resist is P(SI-CMS), which combines the high speed of CMS (chloromethylstyrene) with the etch resistance of SI (trimethylsilylmethyl methacrylate). This resist offers at least 10 times the plasma etch resistance of SAL.[162-164] Its silicon component gives excellent resistance to etching in an oxygen plasma by forming a surface layer of silicon oxide. The sensitivity is similar to that of SAL (~10 $\mu C/cm^2$ at 40 kV) but the resolution is around 0.2 μm. P(SI-CMS) will be a good choice when etch resistance is more important than resolution.

The epoxy type resist EPTR[165-167] developed at IBM is a combination of a no-volac epoxy resin (*o*-cresol novolac glycidyl ether) and an onium salt (triphenylsulfonium hexafluoroantimonate) photoinitiator. EPTR is a high-speed resist (6 µC/cm^2 at 50 kV) with relatively high contrast (γ = 6.4) and high resolution (50 nm). While the resolution of EPTR is comparable to that of Shipley SAL, the epoxy formulation allows EPTR to be extended to layer thicknesses exceeding 200 µm.[168] The high aspect ratio and thicknesses accessible with EPTR make it uniquely suited for micromechanical applications.

	Tone	Resolution nm	Sensitivity µC/cm^2	Developer	Contact Reference
PMMA	Positive	10	100.0	MIBK:IPA	135
EBR-9	Positive	200	10.0	MIBK:IPA	143
PBS	Positive	250	1.0	MIAK : 2-pentanone 3:1	147
ZEP	Positive	10	30.0	xylene:p-dioxane	148
AZ5206	Positive	250	6.0	KLK PPD 401	152
COP	Negative	1000	0.3	MEK : ethanol 7:3	147
SAL-606	Negative	100	8.4	MF312:water	156

Table 2.5. Comparison of commercially available electron beam resists.

2.7.4 Multilayer Systems

2.7.4.1 Low/high molecular weight PMMA

Multilayer resist systems are useful for several purposes: when an enhanced undercut is needed for lifting off metal, when rough surface structure requires planarization, and when a thin imaging (top) layer is needed for high resolution. Figure 2.28 showed the simplest bilayer technique, where a high molecular weight PMMA is spun on top of a low molecular weight PMMA. The low weight PMMA is more sensitive than the top layer, so the resist develops with an enhanced undercut. At high energies (>20 kV), thin PMMA (<0.5 μm) will not normally develop an undercut profile; the best resist profile will be perpendicular to the substrate. The moderate undercut from this technique is useful when liftoff is required from densely packed features.

The two-layer PMMA technique was patented in 1976 by Moreau and Ting[169] and was later improved by Mackie and Beaumont[170] by the use of a weak solvent (xylene) for the top layer of PMMA. Use of a weak solvent prevents intermixing of the two layers. A further refinement of the technique[171] substituted MIBK, a solvent of intermediate strength, for the xylene. PMMA of various molecular weights dissolved in MIBK can now be purchased commercially.[172]

EXAMPLE PROCESS: LIFTOFF OF THIN METAL WITH PMMA BILAYER

1. Clean wafer, on the spinner, by spraying with acetone, then isopropanol. Spin dry.
2. Spin 495 K MW PMMA, 2% (in any solvent) 4 krpm for 30 s., for a thickness ~50 nm.
3. Bake at 170-180C for 1 h.
4. Spin 950 K MW PMMA, 2% in MIBK 4krpm for 30 s., for a thickness ~50 nm.
5. Bake at 170-180C for 1 h.
6. Expose at 50 kV, 350 to 450 $\mu C/cm^2$.
7. Develop in MIBK:IPA, 1:3 for 1 min. Rinse in IPA, blow dry.
8. Optionally, remove surface oxide of GaAs with 10 s dip in $NH_4OH : H_2O$ (1:15). Blow dry.
9. Evaporate 15 nm of Au:Pd (3:2) alloy, 2×10^{-6} Torr, base pressure, 0.5 nm/s.
10. Lift off by soaking in methylene chloride. Optionally, finish with mild ultrasonic agitation.

2.7.4.2 PMMA/copolymer

A larger undercut resist profile is often needed for lifting off thicker metal layers. One of the first bilayer systems was developed by Hatzakis.[173] In this technique a high sensitivity copolymer of methyl methacrylate and methacrylic acid [P(MMA-MAA)][174] is spun on top of PMMA. The exposed copolymer is soluble in polar solvents such as alcohols and ethers but insoluble in nonpolar solvents such as chlorobenzene. A developer such as ethoxyethanol/isopropanol is used on the top (imaging) layer, stopping at the PMMA. Next, a strong solvent such as chlorobenzene or toluene is used on the bottom layer.

This technique has been used to fabricate 1 μm memory arrays with thick gate metalizations.

A more common use of P(MMA-MAA) is as the *bottom* layer, with PMMA on top. In this case the higher speed of the copolymer is traded for the higher resolution of PMMA.[175] For simplicity a single developer is used -- the nonpolar solvent working on the PMMA and the polar solvent developing the copolymer. Effective developer combinations include ethylene glycol monoethyl ether : methanol (3:7) and MIBK:IPA (1:1). The undercut of this process is so large that it can be used to form free-standing bridges of PMMA, a technique developed by Dolan[176] and used extensively for the fabrication of very small superconducting tunnel junctions. Other shadowing and "step edge" techniques for fabricating small lines and junctions are covered in the chapter by Howard and Prober.[177] The polymer PMGI (polydimethylglutarimide) is used for the same purpose as P(MAA-MAA).[178,179]

2.7.4.3 Trilayer systems

Bilayer techniques using P(MMA-MAA) or PMGI work well because the polar/nonpolar combination avoids intermixing of the layers. Almost any two polymers can be combined in a multilayer if they are separated by a barrier such as Ti, SiO_2, aluminum, or germanium,[175,177] forming a so-called trilayer resist. After the top layer is exposed and developed, the pattern is transferred to the interlayer by RIE in CF_4 (or by Cl_2 in the case of aluminum). The interlayer serves as an excellent mask for RIE in oxygen. The straight etch profile available from oxygen RIE allows the fabrication of densely packed, high aspect ratio resist profiles. Such resist profiles can then be used for liftoff or for further etching into the substrate.

If we start with Hatzakis's bilayer scheme (PMMA on the bottom and copolymer on the top) and then add another top layer of PMMA, we have a structure that can be developed into a mushroom shape,[180] as shown in Fig. 2.30. In this technique a heavy dose is given to the central line and a lighter dose to the sides. Mutually exclusive developers are used to form the "T-gate" shape, and a thick layer of metal is lifted off. This technique is widely used to form MESFET gates with low capacitance and low leakage (from the small contact area) and low resistance (from the large metal cross-section).

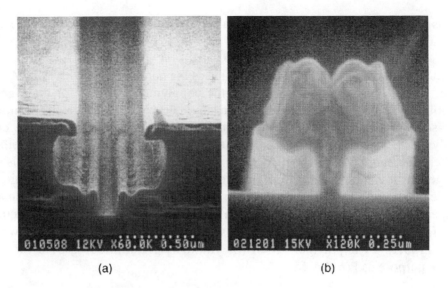

(a)   (b)

FIG. 2.30 (a) Resist cross-section (PMMA on P(MMA-MAA) on PMMA) for the lift-off of a "T" shaped gate. (b) Metal gate lifted off on GaAs. (Courtesy of R. C. Tiberio et al.[180])

2.7.5 Inorganic and Contamination Resists

Some of the first high-resolution e-beam exposures were made with "contamination lithography" -- by simply using the electron beam to crack contaminants sorbed onto the substrate. These carbonaceous and silicaceous contaminants are produced from oil in the vacuum pumps or from organic residue on the sample surface. By using the contamination as a mask for ion milling, wires as narrow as 50 nm were made in the 1960s.[181] Later, the technique was used for fabricating nanometer-scale superconducting devices[182] and metal lines for the study of electron transport in mesoscopic devices.[183]

The dose required for the deposition of contamination depends on how much oil and other contaminants are in the vacuum system (an untrapped diffusion pump provides an ample supply), but the dose is very high, typically in the range of 0.1 to 1 C/cm^2. The high dose limits its application to very sparse patterns. Cracked hydrocarbons provide poor selectivity for etching or milling, so the choice of metals is also limited (for instance, it is not practical to pattern aluminum this way). The contamination can be easily cleaned by heating the substrate to ~100 °C.

Another technique for producing nanometer-scale patterns -- again using doses on the order of 1 C/cm^2 -- is the use of metal fluorides. A high current density of electrons causes the dissociation of materials such as AlF_3, MgF_2, NaCl, LiF, KCl, and CaF_2[184] at doses around 10 to 20 C/cm^2. At lower doses (1 to 3 C/cm^2) AlF_3 acts as a negative resist, developed in water.[185]

One reason for the very high resolution is that these materials are modified by the primary beam of electrons and are insensitive to the much larger spread of secondary electrons. The highest resolution patterns were formed in NaCl crystals, where 50 keV electrons were used to drill holes of ~1.5 nm diameter,[186] but the patterns could not be transferred to any useful material. While negatively exposed AlF_3 makes an excellent etch mask[185] for fluorine-based RIE, the process has not been applied to any useful devices. Recent research in metallic compound resists[187,188] has concentrated on mixing AlF_3 and LiF to reduce the dose needed for dissociation, to provide more uniform films, and to expose these films with the lower current density and lower voltage (20 to 50 kV) available in common e-beam exposure tools. Slots in these films of width 5 nm have been made with 30 keV electrons.[188] At doses similar to those of the metal fluorides, silicon dioxide[189] has also been used for nanometer-scale patterning.

2.7.6 Other Research: Scanning Probes and Thin Imaging Layers

A great deal of research in electron-beam exposure of nanometer-scale patterns is in the field of scanning probe microscopy (SPM), which is covered in Sect. 8.3.3. For an excellent review of SPM lithography, see also the review article by Shedd and Russel.[92]

At low voltage (1 kV) and at higher energies, self-assembled monolayer[190,191] films have demonstrated high resolution but suffer from a very high defect density and difficulty in pattern transfer. Very thin films with lower defect density have been fabricated with Langmuir-Blodgett techniques.[192] Such thin imaging layers are important for low voltage[193] exposures and in-situ processing. However, the imaging layer must be transferred into an intermediate film which is subsequently used as the etch or liftoff mask. This process adds substantially to the cost and complexity of processing. An alternative approach to generating a thin imaging layer on top of a thick resist is the use of surface silylation. In the PRIME silylation process[194-196,193] electron beam exposure *prevents* the subsequent silylation of (attachment of silicon containing molecules to) the resist surface. The silylated regions act as a mask for oxygen plasma etching of the resist film.

2.8 ACKNOWLEDGMENTS

The authors would like to thank the many people who helped in the editing, proofreading, and checking of this chapter; primarily the SPIE reviewer and copy editor, and including also Sylvia Chanak (Cadence), Dennis Costello (Cornell), Mark Gesley (Etec), George Lanzarotta (Raith), Alex Liddle (Lucent), François Marquis (Design Workshop), Beth Moseley (Hitachi), Joseph Nabity (Nabity Lithography Systems), Yasutoshi Nakagawa (JEOL), Hans Pfeiffer (IBM), Rainer Plontke (Leica Jena), John Poreda (Lepton), and Bernard Wallman (Leica).

2.9 Appendix: GDSII Stream Format

The following is a description of the GDSII Stream data format (Release 6.0), the most commonly used format for electron beam lithography and photomask production.[197] This appendix omits the description of tape formatting, since disk files and disk file images on tape and other media are now the norm.[198]

The pattern data is considered to be contained in a "library" of "cells". Cells may contain geometrical objects such as polygons (boundaries), paths, and other cells. Objects in the cell are assigned to "layers" of the design. Different layers typically represent different processing steps for exposure on separate mask plates. Geometrical objects may also be tagged with "datatypes", which can be used for any purpose, but are most commonly used to group together similarly sized objects for compensation of the proximity effect.

There is no explicitly stated limit to the level of hierarchy (the degree of cell nesting); however, most CAD programs impose a limit of around 32 levels. GDSII interpreters will either impose such a limit explicitly, or will impose an implicit limit by running out of memory during recursive operations.

2.9.1 Order of records:

A GDSII Stream file has a great deal of flexibility, but must contain at least the following:

1. A header record
2. One or more Stream records
3. Library name record
4. End of library token

An example of a common record order (see below for record descriptions) follows:

HEADER	version number
BGNLIB	last modification date
LIBNAME	library name
GENERATIONS	see below
UNITS	data units
BGNSTR	begin structure
STRNAME	structure name
BOUNDARY	begin boundary (polygon)
LAYER	layer number
DATATYPE	a label associated with this item
XY	coordinates
ENDEL	end of element
(etc.)	
ENDSTR	end of structure (cell)
ENDLIB	end of library

2.9.2 Record description

The GDSII Stream file format is composed of variable length records. The minimum record length is four bytes. Records can be infinitely long. The first four bytes of a record are the header. The first two bytes of the header contain a count (in eight-bit bytes) of the total record length. The count tells you where one record ends and another begins. The next record begins immediately after the last byte included in the count. The third byte of the header is the record type (also known as a "token"). The fourth byte of the header describes the type of data contained within the record (see table below). The fifth through last bytes of a record are data.

2.9.3 Data type description

The data type value is found in the fourth byte of the record. Possible types and values are:

Data Type	Value
No data present	0
Bit array	1
Two-byte signed integer	2
four-byte signed integer	3
Four-byte real (not used)	4
Eight-byte real	5
ASCII string	6

Two- and four-byte signed integers use the usual twos complement format for negative values. The more significant bytes appear first in the file, so that by default no byte swapping is required when reading the integers with a big-endian CPU (e.g., Intel processors). Byte swapping is required when reading or writing integers with a little-endian machine, such as a VAX.

Real numbers are *not* represented in IEEE format. A floating point number is made up of three parts: the sign, the exponent, and the mantissa. The value of the number is defined to be (mantissa) $\times$ (16)$^{(exponent)}$. If "S" is the sign bit, "E" are exponent bits, and "M" are mantissa bits then an 8-byte real number has the format

$$\text{SEEEEEEE MMMMMMMM MMMMMMMM MMMMMMMM}$$
$$\text{MMMMMMMM MMMMMMMM MMMMMMMM MMMMMMMM}$$

The exponent is in "excess 64" notation; that is, the 7-bit field shows a number that is 64 greater than the actual exponent. The mantissa is always a positive fraction greater than or equal to 1/16 and less than 1. For an 8-byte real, the mantissa is in bits 8 to 63. The decimal point of the binary mantissa is just to the left of bit 8. Bit 8 represents the value 1/2, bit 9 represents 1/4, and so on.

In order to keep the mantissa in the range of 1/16 to 1, the results of floating point arithmetic are *normalized*. Normalization is a process whereby the mantissa is shifted left one hex digit at a time until its left *four* bits represent a non-zero quantity. For every hex digit shifted, the exponent is decreased by one. Since the mantssa is shifted four bits at a time, it is possible for the left three bits of a normalized mantissa to be zero. A zero value is represented by a number with all bits zero. The representation of negative numbers is the same as that of positive numbers, except that the highest order bit is 1, not 0.

2.9.4 Record types

Records are always an even number of bytes long. If a character string is an odd number of bytes long it is padded with a null character. The following is a list of record types. The first two numbers in brackets are the record type and the last two numbers in brackets are the data type (see the table above). *Note that the data type (e.g. "two-byte signed integer") refers to the type of data to follow in the record, not to the number of bytes in the record.* The first two bytes of the record *header* contain a count (in eight-bit bytes) of the total record length. The third byte of the header is the record type (also known as a "token") shown below, and the fourth byte is the data type. All record numbers are shown in hexidecimal. For example, in the HEADER record, "00" is the token, and "02" is the data type.

HEADER
[0002]

Two-byte signed integer: contains data representing the GDSII version number. Values are 0, 3, 4, 5, and 600. With release 6.0 the bersion number changes to three digits.

BGNLIB
[0102]

Two-byte signed integer: contains last modification time of library (two bytes each for the year, month, day, hour, minute, and second) as well as time of last access (same format) and marks beginning of library.

word 1	0x1C (hex) # bytes in record
word 2	0x0102 (the token for BGNLIB)
word 3	year of last modification
words 4,5,6,7,8	month, day, hour, minute, second
word 9	year of last access time
word 10,11,12,13,14	month, day, hour, minute, second

LIBNAME
[0206]

ASCII string: contains a string which is the library name. The string must adhere to CDOS file name conventions for length and valid characters, and may contain file extensions such as ".db".

UNITS [0305]	Eight-byte real: contains 2 8-byte real numbers. The first is the size of a database unit in user units. The second is the size of a database unit in meters. For example, if your library was created with the default units (user unit = 1 µm and 1000 database units per user unit), then the first number would be 0.001 and the second number would be 10^{-9}. Typically, the first number is less than 1, since you use more than 1 database unit per user unit. To calculate the size of a user unit in meters, divide the second number by the first.
ENDLIB [0400]	No data is present. This marks the end of a library.
BGNSTR [0502]	Two-byte signed integer: contains creation time and last modification time of a structure (in the same format as that of BGNLIB) and marks the beginning of a structure.
STRNAME [0606]	ASCII string: contains a string which is the structure name. A structure name may be up to 32 characters long. Legal characters are 'A' through 'Z', 'a' through 'z', '0' through '9', underscore, question mark, and the dollar sign, '$'.
ENDSTR [0700]	No data is present. This marks the end of a structure.
BOUNDARY [0800]	No data is present. This marks the beginning of a bounary element (polygon).
PATH [0900]	No data is present. This marks the beginning of a path element.
SREF [0A00]	No data is present. This marks the beginning of a structure reference element (a reference or "call" to another cell in the library).
AREF [0B00]	No data is present. This marks the beginning of an array reference element (an array of cells).
TEXT [0C00]	No data is present. This marks the beginning of a text element.

LAYER [0D02]	Two-byte signed integer: contains the layer number. The value must be from 0 to 63.
WIDTH [0F03]	Four-byte integer: contains the width of a path or text lines in database units. A negative value for width means that the width is absolute; i.e., I is not affected by the magnification factor of any parent reference. If omitted, zero is assumed.
XY [1003]	Four-byte signed integer: contains an array of XY co-ordinates in database units. Each X or Y coordinate is four bytes long.

Path and boundary elements may have up to 200 pairs of coordinates. A path must have at least 2, and a boundary at least 4 pairs of coordinates. The first and last point of a boundary must coincide.

A text or SREF element must have only one pair of co-ordinates.

An AREF has exactly three pairs of coordinates, which specify the orthogonal array lattice. In an AREF the first point locates a position which is displaced from the reference point by the inter-column spacing times the number of columns. The third point locates a position which is displaced from the reference point by the inter-row spacing times the number of rows.

A node may have from 1 to 50 pairs of coordinates. A box must have five pairs of coordinates with the first and last points coinciding.

ENDEL [1100]	No data is present. This marks the end of an element.
SNAME [1206]	ASCII string: contains the name of a referenced structure.
COLROW [1302]	Two-byte signed integers: the first 2 bytes contain the number of columns in the array. The third and fourth bytes contain the nunber of rows. Neither the number of columns nor the number of rows may exceed 32,767 (decimal) and both are positive.

NODE
[1500]

No data is present. This marks the beginning of a node.

TEXTTYPE
[1602]

Two-byte signed integer: contains the text type. The value of the text type must ge in the range of 0 to 63.

PRESENTATION
[1701]

Bit array: contains 2 bytes of bit flags for text presentation. Bits 10 and 11, taken together as a binary number, specify the font. Bits 12 and 13 specify the vertical presentaton (00 means top, 01 means middle, and 10 means bottom). Bits 0 through 9 are reserved for future use and must be cleared. If this record is omitted, then top-left justification and font 0 are assumed.

STRING
[1906]

ASCII String: contains a character string for text presentation, up to 512 characters long.

STRANS
[1A01]

Bit array: contains two bytes of bit flags for SREF, AREF, and text transformation. Bit 0 (leftmost) specifies reflecton. If it is set, then reflection about the X axis is applied before angular rotation. For AREFs, the entire array lattice is reflected, with the individual array elements riidly attached. Bit 13 flags absolute magnification. Bit 14 flags absolute angle. Bit 15 (rightmost) and all remaining bits are reserved for future use and must be cleared. If this record is omitted, then the element is assumed to have no reflection and its magnification and angle are assumed to be non-absolute.

MAG
[1B05]

Eight-byte real: contains a magnification factor. If omitted, a magnification of 1 is assumed.

ANGLE
[1C05]

Eight-byte real: contains the angular rotation factor, measured in degrees, counterclockwise. For an AREF, the angle rotates the entire array lattice (with the individual array elements regidly attached) about the array reference point. If this record is omitted, and algle of zero degrees is assumed.

REFLIBS
[1F06]

ASCII string: contains the names of the reference libraries. This record must be present if there are any reference libraries bound to the current library. The name for the first reference library starts at byte 0 and the name of the second library starts at byte 45 (decimal). The reference library names may include directory specifiers (separated with ":") and an extension (separated with "."). If either library is not named, its place is filled with nulls.

FONTS
[2006]

ASCII string: contains names of textfont definition files. This record must be present if any of the 4 fonts have a corresponding textfont definition file. This record must not be present if none of the fonts have a textfont file. The name of font 0 starts the record, followed by the remaining 3 fonts. Each name is 44 bytes long and is null if there is no corresponding textfont definition. Each name is padded with nulls if it is shorter than 44 bytes. The textfont definition file names may include directory specifiers (separated with ":") and an extension (separated with ".").

PATHTYPE
[2102]

Two-byte signed integer: contains a value of 0 for square-ended paths that end flush with their endpoints, 1 for round-ended paths, and 2 for square-ended paths that extend a half-width beyond their endpoints. Pathtype 4 signifies a path with variable square-end extensions (see BGNEXTN and ENDEXTN).

GENERATIONS
[2202]

Two-byte signed integer: contains a positve count of the number of copies of deleted or backed-up structures to retain. This number must be at least 2 and not more than 99. If the GENERATIONS record is not present, a value of 3 is assumed.

ATTRTABLE
[2306]

ASCII string: contains the name of the attribute definition file. This record is present only if there is an attribute definition file bound to the library. The attribute definition file name may include directory specifiers and an extension (see FONTS). Maximum size is 44 bytes.

EFLAGS
[2601]

Bit array: contains 2 bytes of bit flags. Bit 15 (rightmost) specifies template data. Bit 14 specifies external data (also referred to as "exterior" data). All other bits are currently unused and must be cleared to 0. If this record is omitted, then all bits are assumed to be 0. Further information about template data can be found in the *GDSII Reference Manual*. Information about external data can be found in the *CustomPlus User's Manual*.

NODETYPE
[2A02]

Two-byte signed integer: contains the node type. The value of the node type must be in the range of 0 to 63.

PROPATTR
[2B02]

Two-byte signed integer: contains the attribute number. The attribute number is an integer from 1 to 127. Attribute numbers 126 and 127 are reserved for the user integer and user string properties, which existed prior to Release 3.0.

PROPVALUE
[2C06]

ASCII string: contains the string value associated with the attribute named in the preceding PROPATTR record. Maximum length is 126 characters. The attribute-value pairs associated with any one element must all have distinct attribute numbers. Also, there is a limit on the total amount of property data that may be associated with any one element: the total length of all the strings, plus twice the number of attribute-value pairs, must not exceed 128 (or 512 of the element is an sref, aref, or node). For example, if a boundary element used a property attribute 2 with property value "metal", and property attribute 10 with property value "property", then the total amount of property data would be 18 bytes. This is 6 bytes for "metal" (odd length strings are padded with a null) plus 8 for "property" plus 2 times the 2 attributes (4) equals 18.

The following records are not supported by Stream Release 3.0:

BOX
[2D00]

No data is present. This marks the beginning of a box element.

BOXTYPE
[2E02]

Two-byte signed integer: contains the box type. The value of the boxtype must be in the range of 0 to 63.

PLEX
[2F03]

Four-byte signed integer: a unique positive number which is common to all elements of the plex to which this element belongs. The head of the plex is flagged by setting the seventh bit; therefore, plex numbers should be small enough to occupy only the rightmost 24 bits. If this record is omitted, then the element is not a plex member.

Plex numbers are not commonly used.

BGNEXTN
[3003]

Four-byte signed integer: applies to pathtype 4. Contains four bytes which specify in database units the extension of a path outline beyond the first point of the path. The value can be negative.

EXDEXTN
[3103]

Four-byte signed integer: Applies to pathtype 4. Contains four bytes which specify in database units the extension of a path outline beyond the last point of the path. The value can be negative.

MASK
[3706]

ASCII string: Required for Filtered format, and present only in Filtered Stream files. Contains the list of layers and data types included in the data file (usually as specified by the user when generating the Stream file). At least one MASK record must follow the FORMAT record. More than one MASK record may follow the FORMAT record. The last MASK record is followed by the ENDMASKS record. In the MASK list, data types are separated from the layers witha semicolon. Individual layers or data types are separated with a space. a range of layers or data types is specified with a dash. An example MASK list looks like this:

1 5-7 10 ; 0-63

ENDMASKS
[3800]

No data is present. This is required for Filtered format, and is present only in a Filtered Stream file. This terminates the MASK records. The ENDMASKS record must follow the last MASK record. ENDMASKS is immediately followed by the UNITS record.

LIBDIRSIZE [3902] Two-byte signed integer: contains the number of pages in the Library directory. This information is used only when reading the data into a new library. If this record is present, it should occur between the BGNLIB record and the LIBNAME record.

SRFNAME [3A06] ASCII string: contains the name of the Sticks Rules File, if one is bound to the library. This informationis used only when reading the data into a new library. If this record is present, it should occur between the BGNLIB and LIBNAME records.

LIBSECUR [3B02] Two-byte signed integer: contains an array of Access Control List (ACL) data. There may be from 1 to 32 ACL entries, each consisting of a group number, a user number, and access rights. This information is used only when reading the data into a new library. If this record is present, it should occur between the BGNLIB and LIBNAME records.

The following record types are either not used, not released, or are re-lated to tape formatting:

TEXTNODE	[1400]
SPACING	[18]
UINTEGER	[1D]
USTRING	[1E]
STYPTABLE	[2406]
STRTYPE	[2502]
ELKEY	[2703]
LINKTYPE	[28]
LINKKEYS	[29]
TAPENUM	[3202]
TAPECODE	[3302]
STRCLASS	[3401]
RESERVED	[3503]

2.9.5 Stream syntax in Bachus Naur representation

An element shown below in CAPITALS is the name of an actual record type. An element shown in lower case means that name can be further broken down in to a set of record types. The following table summarizes the Bachus Naur symbols:

Symbol	Meaning
: :	"Is composed of"
[ ]	An element which can occur zero or one time.
{ }	Choose one of the elements within the braces.
{ } *	The elements within the braces can occur zero or more times.
{ } +	The elements within braces must occur one or more times.
< >	These elements are further defined in the Stream syntax list.
|	"Or"

<stream format> ::= HEADER BGNLIB [LIBDIRSIZE] [SRFNAME] [libsecur] libname [reflibs] [fonts] [attrtable] [generations] [<FormatType>] UNITS {<structure>}* ENDLIB

<FormatType> ::= FORMAT | FORMAT {MASK}+ ENDMASKS

<structure> ::= BNGSTR STRNAME [STRCLASS] {<element>}* ENDSTR

<element> ::= {<boundary> | <path> | <SREF> | <AREF> | <text> | <node> | <box>} {<property>}* ENDEL

<boundary> ::= BOUNDARY [EFLAGS] [PLEX] LAYER DATATYPE XY

<path> ::= PATH [EFLAGS] [PLEX] LAYER DATATYPE [PATHTYPE] [WIDTH] [BGNEXTN] [ENDEXTN] XY

<SREF> ::= SREF [EFLAGS] [PLEX] SNAME [<strans>] XY

<AREF> ::= AREF [EFLAGS] [PLEX] SNAME [<strans>] COLROW XY

<text> ::= TEXT [EFLAGS] [PLEX] LAYER <textbody>

<node> ::= NODE [EFLAGS] [PLEX] LAYER NODETYPE XY

<box> ::= BOX [EFLAGS] [PLEX] LAYER BOXTYPE XY

<textbody> ::= TEXTTYPE [PRESENTATION] [PATHTYPE] [WIDTH] [<strans>] XY STRING

<strans> ::=   STRANS [MAG] [ANGLE]

<property> ::=   PROPATTR PROPVALUE

2.9.6 Example GDSII Stream file

The following is a dump of a minimal GDSII Stream file, consisting of just one polygon (boundary). The GDSII file was created with the program DW2000 from Design Workshop. The binary dump was created on a VAX with the VMS command DUMP. The hex numbers are read **backwards, from right to left**, with each pair of digits representing a byte. Reading the first line below, we see that the file begins with the bytes 00 06 00 02, telling us that the first record contains 6 bytes, that the first record is type 00 (the header), and that record contains data of type 02 (two-byte signed integer). The corresponding ASCII representation on the right is read from left to right.

```
02000200  60000201  1C000300  02000600  ............`....  000000
01000E00  02000200  60002500  01000E00  .....%.`........  000010
42494C45  4C504D41  58450602  12002500  .%....EXAMPLELIB  000020
413E0503  14000300  02220600  59524152  RARY.."........>A  000030
1C00545A  9BA02FB8  4439EFA7  C64B3789  .7KÆŞï9D˘/..ZT..  000040
60000000  01000E00  02000200  60000205  ...`..........`  000050
58450606  0C001100  01000E00  02000200  ..............EX  000060
0100020D  06000008  04000045  4C504D41  AMPLE..........  000070
0000F0D8  FFFF0310  2C000000  020E0600  ........,....Ø˘..  000080
FFFF204E  00001027  0000204E  00001027  '...N ..'...N ..  000090
0000F0D8  FFFFF0D8  FFFFF0D8  FFFFF0D8  Ø˘..Ø˘..Ø˘..Ø˘..  0000A0
00000004  04000007  04000011  04001027  '...............  0000B0
00000000  00000000  00000000  00000000  ................  0000C0
```

The following is an ASCII representation of this file created by the program SDUMP,[198] which translates the token numbers into the names listed in the previous section.

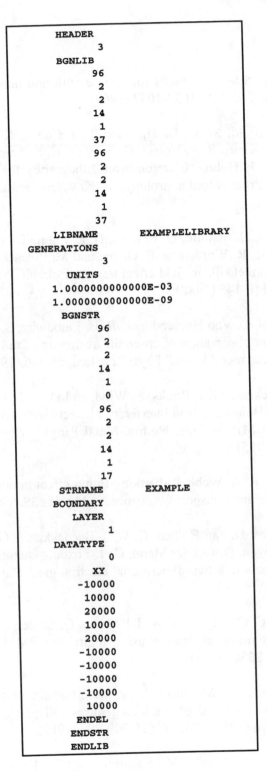

```
        HEADER
          3
        BGNLIB
          96
          2
          2
          14
          1
          37
          96
          2
          2
          14
          1
          37
        LIBNAME        EXAMPLELIBRARY
      GENERATIONS
          3
        UNITS
      1.0000000000000E-03
      1.0000000000000E-09
        BGNSTR
          96
          2
          2
          14
          1
          0
          96
          2
          2
          14
          1
          17
        STRNAME        EXAMPLE
        BOUNDARY
        LAYER
          1
        DATATYPE
          0
          XY
       -10000
        10000
        20000
        10000
        20000
       -10000
       -10000
       -10000
       -10000
        10000
        ENDEL
        ENDSTR
        ENDLIB
```

REFERENCES

1. M. Hatzakis, "Electron resists for microcircuit and mask production," *J. Electrochem. Soc.* **116**, 1033-1037 (1969).

2. M. G. Rosenfield, M. G. R. Thomson, P. J. Coane, K. T. Kwietniak, J. Keller, D. P. Klaus, R. P. Volant, C. R. Blair, K. S. Tremaine, T. H. Newman, and F. J. Hohn, "Electron-beam lithography for advanced device prototyping: Process tool metrology," *J. Vac. Sci. Technol.* **B11** (6), 2615-2620 (1993).

3. S. A. Rishton, H. Schmid, D. P. Kern, H. E. Luhn, T. H. P. Chang, G. A. Sai-Halasz, M. R. Wordeman, E. Ganin, and M. Polcari, "Lithography for ultrashort channel silicon field effect transistor circuits," *J. Vac. Sci. Technol.* **B6** (1), 140-145 (1988).

4. C. P. Umbach, C. Van Haesendonck, R. B. Laibowitz, S. Washburn, R. A. Webb, "Direct observation of ensemble averaging of the Aharonov-Bohm effect in normal metal loops," Phys. Rev. Lett. **56** 386 (1986).

5. V. Chandrasekhar, M. J. Rooks, S. Wind, and D. E. Prober, "Observation of Aharonov-Bohm Electron Interference Effects with Periods h/e and h/2e in Individual Micron-Size, Normal-Metal Rings," *Phys. Rev. Lett.* **55**, 1610-1613 (1985).

6. S. Washburn, R. A. Webb, "Aharonov-Bohm effect in normal metal quantum coherence and transport," Advances in Physics **35**, 375 (1986).

7. B. J. van Wees, H. van Houten, C. W. J. Beenakker, J. G. Williamson, L. P. Kouwenhoven, D. van der Marel, C. T. Foxon, "Quantized conductance of point contacts in a two-dimensional electron gas," Phys. Rev. Lett. **60**, 848 (1988).

8. M. J. Rooks, C. C. Eugster, J. A. del Alamo, G. Snider, E. Hu, "Split-gate electron waveguide fabrication using multilayer PMMA," J. Vac. Sci. Technol. B 9, 2856 (1991).

9. P. H. Woerlee, G. A. M. Hurkx, W. J. M. J. Josquin, and J. F. C. M. Verhoeven, "Novel method of producing ultrasmall platinum silicide gate electrodes," *Appl. Phys. Lett.* **47** (7), 700-702 (1985).

10. E. Anderson, V. Boegli, M. Schattenburg, D. Kern, and H. Smith, "Metrology of electron-beam lithography systems using holographically

produced reference samples," *J. Vac. Sci. Technol.* **B9** (6), 3606-3611 (1991).

11. R. Viswanathan, D. Seeger, A. Bright, T. Bucelot, A. Pomerene, K. Petrillo, P. Blauner, P. Agnello, J. Warlaumont, J. Conway, and D. Patel, "Fabrication of high performance 512K static-random access memories in 0.25 μm complementary metal-oxide semiconductor technology using x-ray lithography," *J. Vac. Sci. Technol.* **B11** (6), 2910-2919 (1993).

12. S. Y. Chou, H. I. Smith, and D. A. Antoniadis, "Sub-100-nm channel-length transistors fabricated using x-ray lithography," *J. Vac. Sci. Technol.* **B4** (1), 253-255 (1986).

13. P. W. Hawkes and E. Kasper, *Principles of Electron Optics,* Academic Press, London (1989).

14. P. Grivet, *Electron Optics*, Elsevier, Oxford, Pergamon imprint (1965).

15. E. Munro, "Numerical modelling of electron and ion optics on personal computers," *J. Vac. Sci. Technol.* **B8** (6), 1657-1665 (1990).

16. H. Boersch, "Experimentelle Bestimmung der Energieverteilung in Thermisch Ausgelosten Elektronenstrahlen," *Z. Phys.* **139**, 115-146 (1954).

17. M. Gesley, "Thermal field emission optics for nanolithography," *J. Appl. Phys.* **65** (3), 914-926 (1989).

18. T. H. P. Chang, "Proximity effect in electron beam lithography," *J. Vac. Sci. Technol.* **12**, 1271-1275 (1975).

19. D. F. Kyser and N. S. Viswanathan, "Monte Carlo simulation of spatially distributed beams in electron-beam lithography," *J. Vac. Sci. Technol.* **12**(6), 1305-1308 (1975).

20. M. Hatzakis, "Recent developments in electron-resist evaluation techniques," *J. Vac. Sci. Technol.* **12** (6), 1276-1279 (1975).

21. G. Brewer, ed., *Electron-Beam Technology in Microelectronic Fabrication,* Academic Press (1980).

22. R. Birkhoff, in *Handbuck der Physik,* E. Fluegge, ed., Springer, Berlin and New York, 53 (1958).

23.   K. Murata, D. Kyser, and C. Ting, "Monte Carlo simulations of fast secondary electron production in electron beam resists," *J. Appl. Phys.* **52**, 4396-4405 (1981).

24.   University of Califronia, Berkeley, Department of Electrical Engineering, Berkeley, CA USA.

25.   Leica Ltd., Cambridge, UK; USA: 708-405-0213, -0147 fax. UK: 44-223-411-411, -211-310 fax.

26.   Sigma-C GmbH, Rosenheimer Landstr. 74 D-85521 Ottobrunn Germany, 49 89 609 60 51.

27.   AISS GmbH, represented byTranscription Enterprises Limited, 101 Albright Way, Los Gatos, CA 95030. 408-866-1851, fax: 408-866-4839.

28.   S. A. Rishton and D. P. Kern, "Point exposure distribution measurements for proximity correction in electron beam lithography on a sub-100 nm scale," *J. Vac. Sci. Technol.* **B5** (1), 135-141 (1987).

29.   M. Rosenfield, S. Rishton, D. Kern, and D. Seeger, "A study of proximity effects at high electron-beam voltages for x-ray mask fabrication. 1. Additive mask processes," *J. Vac. Sci. Technol.* **B8** (6), 1763-1770 (1990).

30.   E. Kratschmer, "Verification of a proximity effect correction program in electron beam lithography," *J. Vac. Sci. Technol.* **19** (4), 1264-1268 (1981).

31.   K. K. Christenson, R. G. Viswanathan, and F. J. Hohn, "X-ray mask fogging by electrons backscattered beneath the membrane," *J. Vac. Sci. Technol.* **B8**(6), 1618-1623 (1990).

32.   Y. Yau, R. F. W. Pease, A. Iranmanesh, and K. Polasko, "Generation and applications of finely focused beams of low-energy electrons," *J. Vac. Sci. Technol.* **19**(4), 1048 (1981).

33.   M. A. McCord and T. H. Newman, "Low voltage, high resolution studies of electron beam resist exposure and proximity effect," *J. Vac. Sci. Technol.* **B10**(6), 3083-3087 (1992).

34.   M. Parikh, "Self-consistent proximity effect correction technique for resist exposure (SPECTRE)," *J. Vac. Sci. Technol.* **15**(3), 931-933 (1978).

35. H. Eisenmann, T. Waas, and H. Hartmann, "PROXECCO - Proximity effect correction by convolution," *J. Vac. Sci. Technol.* **B11** (6), 2741-2745 (1993).

36. K. Harafuji, A. Misaka, K. Kawakita, N. Nomura, H. Hamaguchi, and M. Kawamoto, "Proximity effect correction data processing system for electron beam lithography," *J. Vac. Sci. Technol.* **B10** (1), 133-142 (1992).

37. K. Cummings, R. Frye, E. Rietman, "Using a neural network to proximity correct patterns written with a Cambridge electron beam microfabricator 10.5 lithography system," *Appl. Phys. Lett.* **57**, 1431-1433 (1990).

38. J. Jacob, S. Lee, J. McMillan, and N. MacDonald, "Fast proximity effect correction: An extension of PYRAMID for circuit patterns of arbitrary size," *J. Vac. Sci. Technol.* **B10** (6), 3077-3082 (1992).

39. B. D. Cook, S.-Y. Lee, "Fast proximity effect correction: An extension of PYRAMID for thicker resists", J. Vac. Sci. Technol. **B11**, 2762 (1993).

40. G. Owen and P. Rissman, "Proximity effect correction for electron beam lithography by equalization of background dose," *J. Appl. Phys.* **54** (6), 3573-3581 (1983).

41. M. Gesley and M. A. McCord, "100 kV GHOST electron beam proximity correction on tungsten x-ray masks," *J. Vac. Sci. Technol.* **B12** (6), 3478-3482 (1994).

42. Y. Kuriyama, S. Moriya, S. Uchiyama, and N. Shimazu, "Proximity effect correction for x-ray mask fabrication," *Jpn. J. Appl. Phys.* **33**, 6983-6988 (1994).

43. T. Abe, S. Yamasaki, T. Yamaguchi, R. Yoshikawa, and T. Takigawa, "Representative Figure Method for Proximity Effect Correction [II]," *Jpn. J. Appl. Phys.* **30** (11), 2965-2969 (1991).

44. CAPROX, trademark of Sigma-C GmbH, Rosenheimer Landstr. 74 D-85521 Ottobrunn Germany, 49 89 609 60 51. Distributed by Raith GmbH, Hauert 18, D-44227 Dortmund, Germany (0231-97-50-000) or Raith USA, 6 Beech Rd, Islip, NY 11751, 516-224-1764, 516-224-2620 fax, 73164.1330@compuserve.com.

45. PROXECCO, distributed byTranscription Enterprises Limited, 101 Albright Way, Los Gatos, CA 95030. 408-866-1851, fax: 408-866-4839.

46. Y. Pati, A. Teolis, D. Park, R. Bass, K. Rhee, B. Bradie, and M. Peckerar, "An error measure for dose correction in e-beam nanolithography," *J. Vac. Sci. Technol.* **B8** (6), 1882-1888 (1990).

47. Raith GmbH, Hauert 18, D-44227 Dortmund, Germany (0231-97-50-000) or Raith USA, 6 Beech Rd, Islip, NY 11751, 516-224-1764, 516-224-2620 fax, 73164.1330@compuserve.com.

48. J.C. Nabity Lithography Systems, PO Box 5354, Bozeman, MT 59717 USA, (406-587-0848), jcnabity@aol.com.

49. Data Translation Inc., 800-525-8528.

50. J. C. Nabity, M. N. Wybourne, "A versatile pattern generator for high-resolution electron-beam lithography," Rev. Sci. Instrum. **60** (1) (1989).

51. The Leica SEM division and the Zeiss SEM/TEM division have merged to form a new, separate comany, Leo Electron Optics. US Address: One Zeiss Drive, Thornwood, NY 10594, 800-356-1090.

52. R. Kendall, S. Doran, E. Weissmann, "A servo guided X-Y-theta stage for electron-beam lithography," J. Vac. Sci. Technol. **B9**, 3019 (1991).

53. R. Innes, "Yaw compensation for an electron-beam lithography system", J. Vac. Sci. Technol. **B12**, 3580 (1994).

54. H. Ohta, T. Matsuzaka, N. Saitou, "New electron optical column with large field for nanometer e-beam lithography system", Proc. SPIE **2437** 185 (1995).

55. Jenoptik Technologie GmbH, Microfabrication Division, D-07739 Jena, Germany, 49-3641-653181 (voice) 49-3641-653654 (fax). The electron beam lithography division of Jenoptik has recently been acquired by Leica Ltd., Cambridge, UK, to form Leica Lithographie Systeme Jena GmbH; USA: 708-405-0213, -0147 fax. UK: 44-223-411-411, -211-310 fax.

56. J. Ingino, G. Owen, C. N. Berglund, R. Browning, R. F. W. Pease, "Workpiece charging in electron beam lithography," J. Vac. Sci. Technol. **B12** (3) 1367 (1994).

57. M. Gesley, F. Abboud, D. Colby, F. Raymond, S. Watson, "Electron beam column developments for submicron- and nanolithography," Jpn. J. Appl. Phys. **32** 5993 (1993).

58. M. Gesley, "MEBES IV thermal-field emission tandem optics for electron-beam lithography," J. Vac. Sci. Technol. **B9** (6) 2949 (1991).

59. H. Pearce-Percy, R. Prior, F. Abboud, A. Benveniste, L. Gasiorek, M. Lubin, F. Raymond, "Dynamic corrections in MEBES 4500," J. Vac. Sci. Technol. **B12** (6) 3393 (1994).

60. A. Murray, F. Abboud, F. Raymond, C. N. Berglund, "Feasibility study of new graybeam writing strategies for raster scan mask generation," J. Vac. Sci. Technol. **B11** (6) 2390 (1993).

61. Lepton Inc., Murray Hill NJ 07974, 908-771-9490.

62. D. M. Walker, D. C. Fowlis, S. M. Kugelmass, K. A. Murray, C. M. Rose, "Advanced mask and reticle generation using EBES4," Proc. SPIE **2322**, 56 (1994).

63. M. G. R. Thomson, R. Liu, R. J. Collier, H. T. Carroll, E. T. Doherty, R. G. Murray, "The EBES4 electron-beam column," J. Vac. Sci. Technol. **B5** (1) 53 (1987).

64. D. W. Peters, D. C. Fowlis, A. von Neida, C. M. Rose, H. A. Waggener, W. P. Wilson, "EBES4: Performance of a new e-beam reticle generator," SPIE vol. 1924, 193 (1993).

65. H. C. Pfeiffer, D. E. Davis, W. A. Enichen, M. S. Gordon, T. R. Groves, J. G. Hartley, R. J. Quickle, J. D. Rockrohr, W. Stickel, E. V. Weber, "EL-4, a new generation electron-beam lithography system," J. Vac. Sci. Technol. **B11** (6) 2332 (1993).

66. P. F. Petric, M. S. Gordon, J. Senesi, D. F. Haire, "EL-4 column and control," J. Vac. Sci. Technol. **B11** (6) 2309 (1993).

67. J. D. Rockrohr, R. Butsch, W. Enichen, M. S. Gordon, T. R. Groves, J. G. Hartley, H. C. Pfeiffer, "Performance of IBM's EL-4 e-beam lithography system", Proc. SPIE **2437** 160 (1995).

68. R. Kendall, S. Doran, E. Weissmann, "A servo guided X-Y-theta stage for electron-beam lithography," J. Vac. Sci. Technol. **B9**, 3019 (1991).

69. R. Innes, "Yaw compensation for an electron-beam lithography system", J. Vac. Sci. Technol. **B12** 3580 (1994).

70. H. Elsner, P. Hahmann, G. Dahm, H. W. P. Koops, "Multiple beam-shaping diaphragm for efficient exposure of gratings," J. Vac. Sci. Technol. **B11**(6) 2373 (1993).

71. K. Nakamura, T. Okino, S. Nakanoda, I. Kawamura, N. Goto, Y. Nakagawa, W. Thompson, M. Hassel Shearer, "An advanced electron beam lithography system for sub-half-micron ultra-large-scale production: the distortion corrector technology," J. Vac. Sci. Technol. **B8**(6) 1903 (1990).

72. T. Komagata, H. Takemura, N. Gotoh, K. Tanaka, "Development of EB lithography system for next generation photomasks," Proc. SPIE **2512**, 190 (1995).

73. H. C. Pfeiffer, "Projection exposure with variable axis immersion lenses:a high-throughput electron beam approach to "suboptical" lithography," Jpn. J. Appl. Phys. **34** 6658 (1995).

74. Y. Someda, H. Satoh, Y. Sohda, Y. Nakayama, N. Saitou, H. Itoh, M. Sasaki, "Electron-beam cell projection lithography: Its accuracy and its throughput," J. Vac. Sci. Technol. B12(6) 3399 (1994).

75. G. H. Jansen, "Coulomb interactions in particle beams", J. Vac. Sci. Technol. **B6** 1977 (1988).

76. K. Hattori, R. Yoshikawa, H. Wada, H. Kusakabe, T. Yamaguchi, S. Magoshi, A. Miyagaki, S. Yamasaki, T. Takigawa, M. Kanoh, S. Nishimura, H. Housai, S. Hashimoto, "Electron-beam direct writing system EX-8D employing character projection exposure method," J. Vac. Sci. Technol. **B11**(6) 2346 (1993).

77. K. Sakamoto, S. Fueki, S. Yamazaki, T. Abe, K. Kobayashi, H. Nishino, T. Satoh, A. Takemoto, A. Ookura, M. Oono, S. Sago, Y. Oae, A. Yamada, H. Yasuda, "Electron-beam block exposure system for a 256 M dynamic random access memory," J. Vac. Sci. Technol. **B11**(6) 2357 (1993).

78. A. Yamada, K. Sakamoto, S. Yamazaki, K. Kobayashi, S. Sago, M. Oono, H. Watanabe, H. Yasuda, "Deflector and correction coil calibrations inan electron beam block exposure system," J. Vac. Sci. Technol. **B12**(6) 3404 (1994).

79. M. Kawano, K. Mizuno, H. Yoda, Y. Sakitani, K. Andou, N. Saitou, "Continuous writing method for high speed electron-beam direct writing system HL-800D," J. Vac. Sci. Technol. **B11**(6) 2323 (1993).

80. G. H. Jansen, *Coulomb Interactions in Particle Beams* (Academic, Boston, 1990).

81. S. Berger, D. J. Eaglesham, R. C. Farrow, R. R. Freeman, J. S. Kraus, J. A. Liddle, "Particle-particle interaction effects in image projection lithography systems," J. Vac. Sci. Technol. **B11**(6) 2294 (1993).

82. Y. Someda, H. Satoh, Y. Sohda, Y. Nakayama, N. Saitou, H. Itoh, M. Sasaki, "Electron-beam cell projection lithography: Its accuracy and its throughput," J. Vac. Sci. Technol. **B12**(6) 3399 (1994).

83. Y. Nakayama, S. Okazaki, N. Saitou, H. Wakabayashi, "Eelctron-beam cell projection lithography: A new high-throughput electron-beam direct-writing technology using a specially tailored Si aperture," J. Vac. Sci. Technol. **B8** 1836 (1990).

84. J. A. Liddle, C. A. Volkert, "Stress-induced pattern-placement errors in thin membrane masks," J. Vac. Sci. Technol. **B12**(6) 3528 (1994).

85. H. P. W. Koops, J. Grob, *Springer Series in Optical Sciences: X-ray Microscopy* (Springer, Berlin, 1984) vol. 43.

86. S. D. Berger, J. M. Gibson, "New approach to projection-electron lithography with demonstrated 0.1μm linewidth," Appl. Phys. Lett. **57** (2) 153 (1990).

87. S. D. Berger, J. M. Gibson, R. M. Camarda, R. C. Farrow, H. A. Huggins, J. S. Kraus, "Projection electron-beam lithography: A new approach," J. Vac. Sci. Technol. **B9**(6) 2996 (1991).

88. J. A. Liddle, S. D. Berger, C. J. Biddick, M. I. Blankey, K. J. Bolan, S. W. Bowler, K. Brady, R. M. Camarda, W. F. Connely, A. Crorken, J. Custy, R. C. Farrow, J. A. Felker, L. A. Fetter, B. Freeman, L. R. Harriott, L. Hopkins, H. A. Huggins, C. S. Knurek, J. S. Kraus, D. A. Mixon, M. M. Mkrtchyan, A. E. Novembre, M. L. Peabody, W. M. Simpson, R. G. Tarascon, H. H. Wade, W. K. Waskiewicz, G. P. Watson, J. K. Williams, D. L. Windt, "The Scattering with Angular Limitation in Projection Electron-Beam Lithography (SCALPEL) System," Jpn. J. Appl. Phys. **34**, 6663 (1995).

89. J. A. Liddle, H. A. Huggins, S. D. Berger, J. M. Gibson, G. Weber, R. Kola, C. W. Jurgensen, "Mask fabrication for projection electron-beam lithography incorporating the SCALPEL technique," J. Vac. Sci. Technol. **B9**(6) 3000 (1991).

90. G. P. Watson, S. D. Berger, J. A. Liddle, W. K. Waskiewicz, "A background dose proximity effect correction technique for scattering with angular limitation projection electron lithography implemented in hardware", J. Vac. Sci. Technol. **B13**, 2504 (1995).

91. H. W. P. Koops, *Microcircuit Engineering 88* (North-Holland, New York, 1989) p.217.

92. G. E. Shedd and P. E. Russel, "The scanning tunneling microscope as a tool for nanofabrication," Nanotechnology **1**, 67 (1990).

93. N. C. MacDonald, W. Hofmann, L.-Y. Chen, J. H. Das, "Micro-machined electron gun arrays (MEGA)", Proc. SPIE **2522**, 220 (1995).

94. W. Hofmann, L.-Y. Chen, N. C. MacDonald, "Fabrication of integrated micromachined electron guns", J. Vac. Sci. Technol. **B13**, 2701 (1995).

95. N. Shimazu, K. Saito, M. Fujinami, "An approach to a high-throughput e-beam writing with a single-gun multiple-path system," Jpn. J. Appl. Phys. **34**, 6689 (1995).

96. T. H. P. Chang, D. P. Kern, L. P. Murray, "Arrayed miniature electron beam columns for high throughput sub-100 nm lithography", J. Vac. Sci. Technol. **B10**, 2743 (1992).

97. D. A. Crewe, D. C. Perng, S. E. Shoaf, A. D. Feinerman, "Micromachined electrostatic electron source", J. Vac. Sci. Technol. **B10**, 2754 (1992).

98. G. W. Jones, S. K. Jones, M. D. Walters, B. W. Dudley, "Microstructures for control of multiple ion or electron beams", IEEE Trans. Electr. Dev. **36**, 2686 (1989).

99. E. Kratschmer, H. S. Kim, M. G. R. Thomson, K. Y. Lee, S. A. Rishton, M. L. Yu, T. H. P. Chang, "Sub-49nm resolution 1 keV scanning tunneling microscope field-emission microcolumn," J. Vac. Sci. Technol. **B12**, 3503 (1994).

100. E. Kratschmer, H. S. Kim, M. G. R. Thomson, K. Y. Lee, S. A. Rishton, M. L. Yu, T. H. P. Chang, "An electron-beam microcolumn with improved resolution, beam current, and stability", J. Vac. Sci. Technol. **B13**, 2498 (1995).

101. Cadence Design Systems, 555 River Oaks Parkway, San Jose, CA (USA) 408-943-1234. See also http://www.cadence.com.

102. Mentor Graphics Corp. Gateway Marketing Center, P.O. Box 5050, Wilsonville, OR 97070. 800-547-3000, fax: 503-685-8001. E-mail: gendel@gateway.mentorg.com

103. Silvar Lisco, 703 E. Evelyn Av., Sunnyvale, CA 94086. 800-624-9978, 408-991-6000, fax: 408-737-9979.

104. Integrated Silicon Systems, P.O. Box 13665, Research Triangle Park, NC 27709. 800-422-3585.

105. Refer to the *Semiconductor International Buyer's Guide* issue for a list of other CAD vendors.

106. Design Workshop, 4226 St. John's, Suite 400 D. D. O. Quebec H9G 1X5, 514-696-4753, fax: 514-696-5351.

107. Tanner Research, 180 North Vinedo Av., Pasadena, CA 91107. 818-792-3000, fax: 818-792-0300.

108. DXF to GDSII conversion software is available from Artwork Conversion Software, 1320 Mission St. #5, Santa Cruz CA 95060 (408-426-6163.)

109. For information on ordering these programs and on the Berkeley Industrial Liaison Program, see http://www.eecs.berkeley.edu/ILP/Catalog/index.html

110. R. W. Hon, C. H. Sequin, *A Guide to LSI Implementation*, Second Edition, p.79. (XEROX Palo Alto Research Center, 3333 Coyote Rd., Palo Alto, CA 94304, 1980).

111. C. Mead, L. Conway, *Introduction to VLSI Systems* (Addison-Wesley, Reading MA 1980).

112. See http://info.broker.isi.edu/1/mosis

113. These rules provided by S. Reynolds, ISI (MOSIS) 4676 Admiralty Way, Marina del Rey, CA 90292.

114. Transcription Enterprises Limited, 101 Albright Way, Los Gatos, CA 95030. 408-866-1851, fax: 408-866-4839.

115.   SIGMA-C GmbH, Rosenheimer Landstr. 74, D-85521, Munich, Germany, phone 49-89-609-6051, fax 49-89-609-8112, caprox@sigma-c.de. U.S. distributor: Raith Co., 6 Beech Rd, Islip, NY 11751, 516-224-1764, 516-224-2620 fax, 73164.1330@compuserve.com.

116.   JEBCAD is sold by JEOL-USA, 111 Dearborn Rd, Peabody, MA 01960 (508-535-5900.) In Japan, JEOL Ltd., 1-2 Musashino 3-chome, Akishima Tokyo 196 (0425-42-2187.)

117.   Design Workshop, 4226 St. John's, Suite 400 D. D. O. Quebec H9G 1X5, 514-696-4753, fax: 514-696-5351.

118.   E. Reichmanis, L. F. Thompson, "Polymer materials for microlithography," in *Annual Review of Materials Science* vol. 17, R. A. Huggins, J. A. Giordmaine, J. B. Wachtman, Jr., eds. (Annual Reviews, Inc. Palo Alto, CA, 1987) p. 235.

119.   E. Reichmanis, A. E. Novembre, "Lithographic resist materials chemistry," in *Annual Review of Materials Science* vol. 23, R. A. Laudise, E. Snitzer, R. A. Huggins, J. A.Giordmaine, J. B. Wachtman, Jr., eds. (Annual Reviews, Inc. Palo Alto, CA) 1993, p. 11.

120.   C. Grant Willson, "Organic resist materials - theory and chemistry," in *Introduction to Microlithography*, L. F. Thompson, C. G. Willson, M. J. Bowden, eds., ACS Symposium Series 219 (American Chemical Society, Washington DC, 1983) p.87.

121.   *Materials for Microlithography - Radiation-Sensitive Polymers,* L. F. Thompson, D. G. Willson, J. M. J. Fréchet, eds., ACS Symposium Series 266 (American Chemical Society, Washington DC, 1984).

122.   C. G. Willson, "Organic Resist Materials", and L. F. Thompson, "Resist Processing", in *Introduction to Microlithography,Second Edition,* L. F. Thompson, C. G. Willson, M. J. Bowden, eds. (American Chemical Society, Washington DC, 1994).

123.   A. Weill, "The spin coating process mechanism," in *The Physics and Fabricaton of Microstructures and Microdevices*, M. J. Kelly, C. Weisbuch, eds., (Springer-Verlag, Berlin, 1986) p. 51.

124.   T. Tanaka, M. Morigami, and N. Atoda, "Mechanism of resist pattern collapse during development process," Jpn. J. Appl.Phys **32**, 6059 (1993).

125. The program SELID is available from Sigma-C GmbH, Rosenheimer Landstr. 74 D-85521 Ottobrunn Germany, 49 89 609 60 51.

126. T. E. Everhart, in *Materials in Microlithography*, L. F. Thompson et al., eds. (American Chemical Society, Washington DC 1984).

127. Gold etch solution type TFA from Transene Co., Rowley MA.

128. Chrome etch type CR-14 from Cyantek Corp., 3055 Osgood Ct., Fremont CA 94538.

129. M. Kurihara, M. Arai, H. Fujita, H. Moro-oka, Y. Takahashi, H. Sano, "Primary processes in e-beam and laser lithographies for phase-shift mask manufacturing II," SPIE vol. 1809, *12th Annual BACUS Symposium*, 50 (1992).

130. C. A.Kondek, L. C. Poli, "A submicron e-beam lithography process using an overcoating conducting polymer for the reduction of beam charging effects on lithium niobate and quartz," Proc. SPIE vol. 2194 p.366 (1994).

131. M. Angelopoulos, J. M. Shaw, K. Lee, W. Huang, M. Lecorre, M. Tissier, "Lithographic applications of conducting polymers," J. Vac. Sci. Technol. **B9**(6) 3428 (1991).

132. M. Angelopoulos, N. Patel, J. M. Shaw, N. C. Labianca, S. A. Rishton, "Water soluble conducting polyanilines: Applications in lithography," J. Vac. Sci. Technol. **B11**(6) 2794 (1993).

133. I. Haller, M. Hatzakis, R. Srinivasan, "High-resolution positive resists for electron-beam exposure," IBM J. Res. Develop. **12** 251 (1968).

134. M. Hatzakis, "Electron resists for microcircuit and mask production," J. Electrochem. Soc. **116** 1033 (1969).

135. PMMA vendors include: OCG Microelectronic Materials Inc., 5 Garret Mountain Plaza, West Paterson, NJ 07424, 800-222-4868. Microlithography Chemical Corp., 1254 Chestnut St. Newton, MA 02164 617-965-5511 617-965-5818 fax. Mead Chemical Co., 10750 County Rd. 2000, PO Box 748, Rolla, MO 65401. 314-364-8844.

136. G. H. Bernstein, D. A. Hill, "On the attainment of optimum developer parameters for PMMA resist," Superlattices and Microstructures **11** (2) 237 (1992).

137. B. P. Van der Gaag, A. Sherer, "Microfabrication below 10nm," Appl. Phys. Lett. **56** 481 (1990).

138. D. W. Keith, R. J. Soave, M. J. Rooks, "Free-standing gratings and lenses for atom optics," J. Vac. Sci. Technol. **B9** (6) 2846 (1991).

139. W. C. B. Peatman, P. A. D. Wood, D. Porterfield, T. W. Crowe, M. J. Rooks, "Quarter-micrometer GaAs Schottky barrier diode with high video responsivity at 118 μm," Appl. Phys. Lett. **61** 294 (1992).

140. R. C. Tiberio, G. A. Porkolab, M. J. Rooks, E. D. Wolf, R. J. Lang, A. D. G. Hall, "Facetless Bragg reflector surface-emitting AlGaAs/GaAs lasers fabricated by electron-beam lithography and chemically assisted ion-beam etching", J. Vac. Sci. Technol. **B9** 2842 (1991).

141. Note that this liftoff process allows the use of ultrasonic agitation because chrome sticks very well to silicon. The ultrasonic process causes lines of aluminum to peal off the surface. A common belief is that once the substrate is dry, the metal cannot be made to separate from the surface. This is not necessarily true. If the metal pattern adheres well to the substrate (e.g., Cr or Ti), then further ultrasonic agitation in the solvent may well continue the liftoff process and improve the yield of devices.

142. T. Tada, "Highly sensitive positive electron resists consisting of halogenated alkyl α-chloroacrylate series polymer materials," J. Electrochem. Soc. **130** 912 (1983).

143. Toray Marketing and Sales, 1875 S. Grant St., Suite 720, San Mateo, CA 94402. 415-341-7152. Toray Industries, 1-8-1 Mihama Urayasu Inc., Chiba, Japan.

144. K. Nakamura, S. L. Shy, C. C. Tuo, C. C. Huang, "Critical dimension control of poly-butene-sulfone resist in electron beam lithography," Jpn. J. Appl. Phys. **33**, 6989 (1994).

145. M. Widat-alla, A. Wong, D. Dameron, C. Fu, "Submicron e-beam process control," Semiconductor International (May 1988), p. 252.

146. Pre-spun mask plates are sold by Hoya Electronics Co., Ft. Lee, NJ.; Balzers Optical Co., Marlborough, MA; see the Semiconductor International Buyer's Guide for other vendors.

147. Mead Chemical Co., 10750 County Rd. 2000, PO Box 748, Rolla, MO 65401. 314-364-8844.

148. Nippon Zeon is represented in the US by Nagase California Corp., 710 Lakeway, Suite 135, Sunnyvale, CA 94086. 408-773-0700.

149. K. Kurihara, K. Iwadate, H. Namatsu, M. Nagase, H. Takenaka, K. Murase, "An electron beam nanolithography system and its application to Si nanofabrication," Jpn. J. Appl. Phys. **34** 6940 (1995).

150. T. Nishida, M. Notomi, R. Iga, T. Tamamura, "Quantum wire fabrication by e-beam lithographyusing high-resolution and high-sensitivity e-beam resist ZEP-520," Jpn. J. Appl. Phys. **31**, Pt. 1, no.12B, 4508 (1992).

151. J. Pacansky, R. J. Waltman, "Solid-state electron beam chemistry of mixtures of diazoketones in phenolic resins: AZ resists," J. Phys. Chem. **92** 4558 (1988).

152. Hoechst Celanese Corp, AZ Photoresist Products, 70 Meister Ave., Somerville, NJ 08876. 908-429-3500.

153. M. Kurihara, M. Komada, H. Moro-oka, N. Hayashi, H. Sano, "EBR900 processes in e-beam and laser beam lithographies for photomask production", Proc. SPIE **2437**, 240 (1995).

154. A. E. Novembre, R. G. Tarascon, O. Nalamasu, L. Fetter, K. J. Bolan, C. S. Knurek, "Electron-beam and x-ray lithographic characteristics of the optical resist ARCH", Proc. SPIE **2437**, 104 (1995).

155. OCG Microelectronic Materials Inc., 5 Garret Mountain Plaza, West Paterson, NJ 07424, 800-222-4868.

156. Shipley Inc., 455 Forest St., Marlboro, MA 01752. 800-343-3013.

157. D. Macintyre, S. Thoms, "High resolution electron beam lithography studies on Shipley chemically amplified DUV resists," presented at the MNE Conference, September 1996; to appear in *Micro- and Nano-engineering 96, Procedings of the International Conference on Micro- and Nano-engineering*, S. P. Beaumont ed., vol. 29.

158. E. Reichmanis, L. F. Thompson, "Polymer materials for microlithography," in *Annual Review of Materials Science*, v.17, R. A. Huggins, J. A. Giordmaine, J. B. Wachtman Jr., eds. (Annual Reviews, Palo Alto, 1987) p.238.

159. T. Yoshimura, Y. Nakayama, S. Okazaki, "Acid-diffusion effect on nano-fabrication in chemical amplification resist," J. Vac. Sci. Technol. **B10**(6) 2615 (1992).

160. E. A. Dobisz, C. R. K. Marrian, "Sub-30nm lithography in a negative electron beam resist with a vacuum scanning tunneling microscope," Appl. Phys. Lett. **58**(22) 2526 (1991).

161. A. Claßen, S. Kuhn, J. Straka, A. Forchel, "High voltage electron beam lithography of the resolution limits of SAL601 negative resist," Microelectronic Engineering **17** 21 (1992).

162. D. A. Mixon, A. E. Novembre, W. W. Tai, C. W. Jurgensen, J. Frackoviak, L. E. Trimble, R. R. Kola, G. K. Celler, "Patterning of x-ray masks using the negative-acting resist P(SI-CMS)," J. Vac. Sci. Technol. **B11**(6) 2834 (1993).

163. A. E. Novembre, D. A. Mixon, C. Pierrat, C. Knurek, M. Stohl, "Dry etch patterning of chrome on glass optical masks using P(SI-CMS) resist," Proc. SPIE **2087** 50 (1993).

164. C. W. Lo, W. K. Lo, M. J. Rooks, M. Isaacson, H. G. Craighead, A. E. Novembre, "Studies of 1 and 2 keV electron beam lithography using silicon containing P(SI-CMS) resist", J. Vac. Sci. Technol. **B13** 2980 (1995).

165. K. J. Stewart, M. Hatzakis, J. M. Shaw, D. E. Seeger, E. Neumann, "Simple negative resist for deep ultiraviolet, electron beam, and x-ray lithography", J. Vac. Sci. Technol. **B7** 1734 (1989).

166. K. G. Chiong, S. Wind. D. Seeger, "Exposure characteristics of high-resolution negative resists", J. Vac. Sci. Technol. **B8** 1447 (1990).

167. K. G. Chiong, F. J. Hohn, "Resist patterning for sub-quarter micron device fabrications", Proc. SPIE **1465** 221 (1991).

168. N. LaBianca, J. D. Gelorme, "High aspect ratio resist for thick film applications", Proc. SPIE **2438** 846 (1995).

169. W. Moreau, C. H. Ting, "High sensitivity positive electron resisit," US Patent 3934057, 1976.

170. S. Mackie, S. P. Beaumont, Solid State Technology **28** 117 (1985).

171. M. J. Rooks, C. C. Eugster, J. A. del Alamo, G. L. Snider, E. L. Hu, "Split-gate electron waveguide fabrication using multilayer poly(methyl methacrylate)," J. Vac. Sci. Technol. **B9**(6) 2856 (1991).

172. Microlithography Chemical Corp., 249 Pleasant St., Watertown, MA 02172. 617-926-3322, -2919 fax.

173. M. Hatzakis, "PMMA copolymers as high sensitivity electron resists," J. Vac. Sci. Technol. **16**(6) 1984 (1979). M. Hatzakis, "High sensitivity resist system for lift-off metallization," U.S. Patent No. 4024293 (1977).

174. P(MMA-MAA) and PMMA may be purchased from OCG Microelectronic Materials Inc., 5 Garret Mountain Plaza, West Paterson, NJ 07424, 800-222-4868; or from the Microlithography Chemical Corp., 249 Pleasant St., Watertown, MA 02172. 617-926-3322, -2919 fax.

175. R. E. Howard, E. L. Hu, L. D. Jackel, "Multilevel resist for lithography below 100nm," IEEE Trans. Electron. Dev. **ED-28**(11) 1378 (1981).

176. G. J. Dolan, "Offset masks for lift-off photoprocessing," Appl. Phys. Lett. **31**, 337 (1977).

177. R. E. Howard, D. E. Prober, "Nanometer-scale fabrication techniques," in *VLSI Electronics: Microstructure Science* vol. 5, (Academic Press, New York, 1982).

178. H. Takenaka, Y. Todokoro, "A PMMA/PMGI two layer resist system for stable lift-off processing," Proc. SPIE **1089** 132 (1989).

179. M. P. de Grandpre, D. A. Vidusek, M. W. Legenza, "A totally aqueous developable bilayer resist system," Proc. SPIE **539**, 103 (1985). M. W. Legenza, D. A. Vidusek, M. P. Grandpre, "A new class of bilevel and mono-level positive resist systems based on a chemically stable imide polymer," Proc. SPIE **539**, 250 (1985).

180. R. C. Tiberio, J. M. Limber, G. J. Galvin, E. D. Wolf, "Electron beam lithography and resist processing for the fabrication of T-gate structures," Proc. SPIE **1089**, 124 (1989).

181. A. N. Broers, "Micromachining by sputtering through a mask of contamination laid down by an electron beam," in *Proceedings of the First International Conference on Electron and Ion Beam Science and Technology*, R. Bakish, ed. (Wiley, New York, 1964) p.191.

182. R. Voss, R. B. Laibowitz, A. N. Broers, "Niobium nanobridge DC SQUID," Appl. Phys. Lett. **37** 656 (1980).

183. C. P. Umbach, S. Washburn, R. A. Webb, R. Koch, M. Bucci, A. N. Broers, R. B. Laibowitz, "Observation of the h/e Aharonov-Bohm interference effects in sub-micron diameter, normal metal rings," J. Vac. Sci. Technol. **B4** 383 (1986).

184. P. Mankiewich, H. G. Craighead, T. R. Harrison, A. Dayen, "High resolution electron beam lithography on CaF_2", Appl. Phys. Lett. **44** 468 (1984).

185. E. Kratschmer, M. Isaacson, "Nanostructure fabrication in metals, insulators, and semiconductors using self-developing metal inorganic resist," J. Vac. Sci. Technol. **B4**(1) 361 (1986).

186. M. Isaacson, A. Muray, "In situ vaporization of very low molecular weight resists using 1/2 nm diameter electron beams," J. Vac. Sci. Technol. **19,** 1117 (1981).

187. W. Langhenrich, A. Vescan, B. Spangenberg, H. Beneking, Microelectronics Engineering **17,** 287 (1992). W. Langhenrich, H. Beneking, Jpn. J. Appl. Phys. **32,** 6248 (1993).

188. J. Fujita, H. Watanabe, Y. Ochiai, S. Manako, J. S. Tsai, S. Matsui, "Sub-10 nm lithography and development properties of inorganic resist by scanning electron beams", J. Vac. Sci. Technol. **B13,** 2757 (1995).

189. D. R. Allee, X. D. Pan, A. N. Broers, C. P. Umbach, "ultra-high resolution electron beam patterning of SiO_2: A review," in *Science and Technology of Mesoscopic Structures*, S. Namba, C. Hanmaguchi, T. Ando, eds. (Springer-Verlag, Tokyo, 1991) p. 362.

190. M. J. Lercel, G. F. Redinbo, F. D. Pardo, M. Rooks, R. C. Tiberio, P. Simpson, H. G. Craighead, C. W. Sheen, A. N. Parikh, D. L. Allara, "Electron beam lithography with monolayers of alkylthiols and alkylsiloxanes," J. Vac. Sci. Technol. **B12**(6) 3663 (1994).

191. R. C. Tiberio, H. G. Craighead, M. Lercel, T. Lau, C. W. Sheen, D. L. Allara, "Self assembled monolayer electron beam resist on GaAs," Appl. Phys. Lett. **62,** 476 (1993).

192. S. W. J. Kuan, C. W. Frank, Y. H. Y. Lee, T. Eimori, D. R. Allee, R. F. W. Pease, R. Browning, "Ultrathin Poly(MMA) resist films for microlithography," J. Vac. Sci. Technol. **B7**, 1745 (1989).

193. M. Böttcher, L. Bauch, "Surface imaging by silylation for low voltage electron-beam lithography," J. Vac. Sci. Technol. B12, 3473 (1994).

194. C. Pierrat, S. Tedesco, F. Vinet, T. Mourier, M. Lerme, B. Dal'Zotto, J. C. Guibert, "PRIME process for deep UV and E-beam lithography", Microelectronic Engineering, **11**, 507 (1990).

195. C. Pierrat, "New model of polymer silylation: application to lithography", J. Vac. Sci. Technol. **B10**, 2581 (1992).

196. M. Irmscher, B. Höfflinger, R. Springer, "Comparative evaluation of chemically amplified resists for electron-beam top surface imaging use," J. Vac. Sci. Technol. **B12**, 3925 (1994).

197. Portions of the *GDSII Stream Format Manual*, Documentation No. B97E060, Feb. 1987, reprinted with permission of Cadence Design Systems, Inc., 555 River Oaks Parkway, San Jose, CA 95134. 408-943-1234. See also the web site http://www.cadence.com.

198. A useful set of GDSII utilities is available for the VMS operating system. This set includes programs for syntax checking, dumping to ASCII, building from ASCII, rotating and scaling cells, printing cell hierarchies, printing data extents, and displaying layer occupation. For purchase information contact the Cornell Nanofabrication Facility at 607-255-2329, or costello@cnf.cornell.edu.

CHAPTER 3

X-Ray Lithography

Franco Cerrina
University of Wisconsin

CONTENTS

3.1 Introduction

X-ray lithography was invented in the early 70s and, almost 25 years later, has not yet reached a manufacturing stage. Clearly, the cause of this delay is the explosive development of optical lithography. Optical lithography has however exhausted its initial "resolution reserve" (i.e., the distance from the diffraction limit) that it originally enjoyed: several generations of devices were manufactured with H-line and I-line tools, but only one or at most two will be fabricated using 248 nm (DUV) before the move to 193 nm become necessary; it is unlikely that 193 nm will be used beyond the 4 Gbit DRAM. After 193 nm the horizon is dark. In parallel, X-ray lithography has continued to evolve, trying to reach a continuously changing target. With the 100% accuracy of hindsight, we can say that it would have been better to concentrate the development efforts on the most advanced target (1 Gbit and beyond) rather than playing catch-up to the faster moving optical efforts. This notwithstanding, today the tooling for X-ray lithography has reached significant levels; it is the only technology that can support multi-generation of chips fabrication, all the way to the production of nanostructures.

There is little doubt that X-ray lithography is not well understood outside a small set of technologists, and our intent in this chapter is to explain the basis of the technology and the physical principles on which it rests, dispelling erroneous concepts and clarifying others. We will focus on the basic of the technology, discussing the sources, masks and image formation processes. We will not discuss in any great detail photoresist processes, since the thematic is common with other lithographies. For much of the detailed discussions that we were forced to leave out of the chapter, we point the interested reader to the extensive literature.

X-ray lithography is the main contender for sub-0.25 μm lithographies. The combination of high-resolution, large exposure window and high through-put satisfies all the requirements for a manufacturing process. In this chapter, we will present the physical basis of the technique, and its current implementation.

The X-rays were discovered just about 100 years ago, in 1895, by W. H. Roentgen, while experimenting on the nature of cathodic rays [1]. Later, Roentgen became the first recipient of the newly instituted Nobel prize for his discovery. The name reflects the uncertainty of the times about their nature, but Roentgen realized quickly the potential applications of the high penetrating power of the radiation. The applications of X-rays to medicine and industrial inspection are well known, but why is this radiation of interest in lithography? In short, because of their ability to define very high resolution images in thick materials. The resolution comes from the extremely short wavelength, of the order of $0.01 - 1.0$ nm, and the high penetration from the transparency of most materials in this region of the spectrum, where photons are only weakly coupled to condensed matter.

The use of X-rays in semiconductor lithography was first proposed by H.I. Smith in a seminal paper, [2]. Recently, E. Spiller has described the early history at IBM [3], while Alan Wilson reviewed the more recent developments [4]. X-ray lithography was proposed as a simple proximity imaging system, as shown in Fig. 3.1. The resist is exposed by the radiation transmitted by the transparent areas of the mask, illuminated with the X-rays generated by a source.

Perhaps because of this apparent simplicity, the development of X-ray

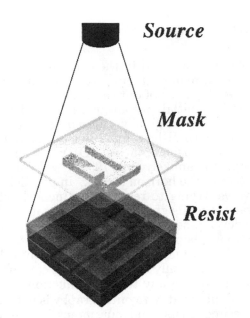

Figure 3.1: General arrangement of an X-ray lithography system.

lithography has been plagued by false starts and oversimplifications. It has been only recently, in the nineties, that the foundation of X-ray lithography has been put on a firm physical ground and, on that basis, a solid technology is being developed [7]. The prerequisites, i.e., powerful and reliable sources of X-rays have become available [5, page 439], while the properties of the interaction of the X-rays with materials both in the mask and in the resist are becoming progressively clearer [6, 8]. Masks with resolution approaching 0.14 μm over the field and with adequate placement accuracy are beginning to be available [9]. Large scale integrated circuits exposed with X-ray lithography have been demonstrated repeatedly, particularly at IBM [10], NTT [11], and NEC [12]. An example of 0.25 μm structures printed in a commercial resist is shown in Fig. 3.2.

The path to a complete process is emerging, and integrated solutions have been developed for small-scale demonstrations [14] as well as large area circuits [10] X-rays have also been applied to other imaging application as well. Chief is the area of micromachining, where X-rays are used for their penetrating power at very short wavelengths. Fig. 3.3 shows an example of high-aspect structures patterned in PMMA using X-rays [15]. In this process, a film or sheet of PMMA is exposed to harder X-rays with $\lambda = 0.05 - .1$ nm, so that the exposed area may be several millimeters, or even centimeters, thick. After development, the plastic matrix is used as a plating mold. The parts obtained by plating can be used directly as sub-assemblies in micromachines, or as printing molds for the production of very high resolution impressions [16, 17].

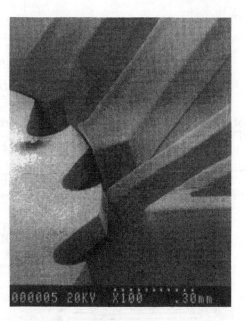

Figure 3.2: 0.25 μm SRAM patterns exposed in AZ-PF514, [13]

Figure 3.3: 500 μm thick nickel turbine and plated copper stator patterned using hard X-ray radiation [15]

3.1.1 X-ray properties

X-rays are electromagnetic waves, with a wavelength λ in the range $0.1 - 100$Å, with the subset $0.6 - 44$Å often called *soft X-rays*. Like all elementary excitations, they live a dual life as particles *(photons)* and waves. In general, it is convenient to think of them as photons when we consider energy related problems (absorption, statistics) and as waves in optical-like phenomena (diffraction, interference) [5, 18]. An equivalent approach is to consider X-rays propagating as waves and interacting as photons.

X-rays are generated in atoms when an electron from a higher atomic orbital decays into a core hole, leading to the emission of a photon of energy equal to the difference in energy between the two levels, i.e., by fluorescence [19]. Because of the competing Auger process, the fluorescence becomes efficient only for relatively heavy atoms ($Z > 28$ (Cu)). Another process, classical, discussed in detail below, is the emission of radiation by a (high-energy) accelerated electron, by direct coupling of the moving charge with the EM field [20, 5].

In the soft X-rays, the absorption process is dominated by the photoelectric effect [21]. At energies around 1-2 keV, the electric field of the radiation couples most efficiently with the atoms' inner electrons. The absorption of a photon results in the excitation of a photoelectron from a bound atomic state to the unbound continuum [22], thus yielding an absorption spectrum characterized by edges *(bound $\to$ continuum)* rather than lines *(bound $\to$ bound)*. This process is well known, and relatively simple formulae can describe the process, for instance in the framework of the scattering theory.

The interaction of an incoming field E_{in} leads to a scattered spherical field E_{out} as [21]:

$$E_{out} = E_{in} \frac{e^{ikR}}{R} (f_1 + i f_2) \tag{3.1}$$

where (f_1, f_2) are the *atomic scattering factors* that describe the fraction of the radiation of wavevector k scattered away from the incoming beam [23]. The first term, f_1, represents the reactive (dispersion) and the second, f_2, the dissipative (absorption) term. The atomic scattering factor depend only on the atom considered, not on its chemical or physical environment [21]. This simplifies the treatment of the interaction of materials and X-rays. In Fig. 3.4 we show the scattering factors for a series of materials of different compositions and densities.

A large set of compiled atomic scattering factors is available for all the atoms, from energies as low as 10-20 eV to several 100's of keV [24, 25, 26].

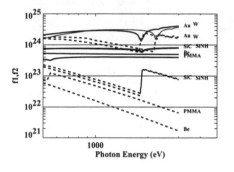

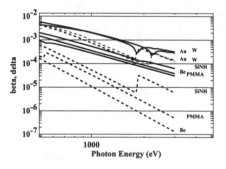

Figure 3.4: Average atomic and molecular scattering factors for several materials, times the material atomic density. Solid line – real part, dashed – imaginary part

Figure 3.5: Complex index of refraction for the same materials

Fig. 3.4 shows the typical behavior: when the X-ray energy becomes sufficient to ionize a new shell, a jump is observed in the imaginary part (dissipative term). As the energy increases beyond it, the strength of coupling of the interaction decreases with a law proportional to E^m, $m \approx -2.2$; the real (dispersive) part remains fairly constant, and converges to the value of Z (atomic number).

3.1.1.1 Optical Constants

The macroscopic response of a system is described by its dielectric function $\tilde{\epsilon}(\omega) = \epsilon_1(\omega) + \jmath\epsilon_2(\omega)$. It is easily computed from the atomic scattering factor (f_1, f_2) [24, 23] by summing the contributions of all the electrons in the material. From $\tilde{\epsilon}$ we can compute the complex index of refraction $\tilde{n} = n + \jmath k$, and from it reflectivity and absorption coefficient. Hence, for a single type of atom:

$$\tilde{\epsilon} = 1 - \alpha - \jmath\,\gamma \tag{3.2}$$
$$= 1 - K f_1 - \jmath\, K f_2 \tag{3.3}$$
$$K = \frac{e^2\lambda^2}{\pi\, m_0 c^2}\frac{N_a\rho}{A} \tag{3.4}$$

where N_a is Avogadro's number, A the atomic weight and ρ the density.

$$\tilde{n} = 1 - \delta - \jmath\beta = 1 - \alpha/2 + \jmath\,\gamma/2 \tag{3.5}$$

In Fig. 3.5 we show examples for materials often used in X-ray lithography.

It is important to notice how small the values of the optical constants are. Indeed, we typically have to write $n = 1 - \delta$, with $\delta \approx 10^{-3} - 10^{-5}$, while $\beta \approx 10^{-3} - 10^{-5}$ as well. Hence, the interaction between X-rays and materials can be considered as *"weak"*, since the radiation must propagate for hundreds of wavelengths before appreciable variation in the optical path build up.

In general, the strength of the interaction of a *material* with a beam of radiation depends on how many electrons are available for the interaction, i.e., on the *polarizability* of the material at that frequency, and $\beta \approx \lambda^{2.5}Z^2$. Hence, materials with a large number of electrons per cm^3 will interact more strongly than materials with fewer electrons. Materials of high density and containing elements of large atomic numbers (materials such as *Au, W, Pt, U, etc.*) will have large absorption compared to lighter ones (i.e., *H_2O, organic materials, Si, N_2, etc.*). Fig. 3.6 shows actual values for several important materials. The trend in the curves can be approximated by: [27]

$$\mu = 4\pi\beta/\lambda \approx \lambda^2 Z^2 \tag{3.6}$$

We see that the absorption coefficients are all around $0.1 - 1\mu m^{-1}$. This implies (a) that a shadow casting system must have absorbers several thousands of Å thick to achieve a sufficient contrast and (b) relatively large thicknesses of resist can be exposed uniformly. At the same time, the very low value of $\tilde{n}$ implies that materials are not good reflectors in the X-ray region, and this is indeed observed to be true: only at grazing angles is the reflectivity large enough to allow the fabrication of effective mirrors [21, 27].

3.1.2 Image Formation and Modeling

The simplest level is that depicted in Fig. 3.1, i.e., shadow-casting. If we neglect any diffraction effect, then the image modulation is simply given by the difference in transmission between absorbing and clear parts of the mask, as obtained from:

$$M = \frac{I_{clear} - I_{absorber}}{I_{clear} + I_{absorber}} = \frac{1 - T_{absorber}}{1 + T_{absorber}} = \frac{R - 1}{R + 1} \tag{3.7}$$

where T is the absorber transmission, $I_{absorber}/I_{clear}$, and is wavelength dependent. The contrast R is defined as the reciprocal of the absorber

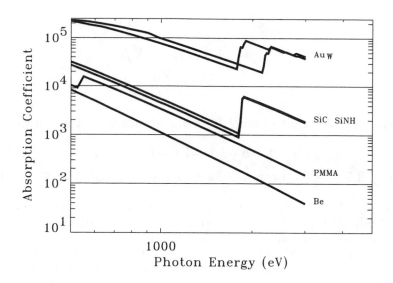

Figure 3.6: Absorption coefficients for the same set of materials, in μm^{-1}.

transmission. For a narrowband and an absorber of thickness t, Eq. 3.7 reduces to:

$$M = \frac{1 - e^{-\mu t}}{1 + e^{-\mu t}} \qquad (3.8)$$

while for an extended bandwidth it is necessary to average over the source power spectrum. In Fig. 3.8 we show the value of the contrast for the case of Au, for both a typical synchrotron source ($\lambda \approx 0.5 - 1.2$ Å) and for a plasma narrow band system ($\lambda \approx 12 - 15$ Å).

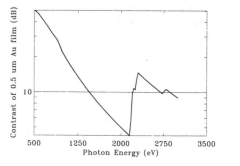 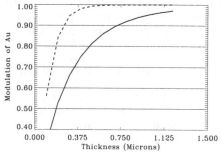

Figure 3.7: Monochromatic Contrast of 0.5 μm Au as a function of photon energy

Figure 3.8: Image modulation by an Au absorber of various thickness for synchrotron (solid) and plasma (dashed).

In general it is considered that it is necessary to have a modulation above 0.6 [28], i.e., $R > 4$, in order to have a good image definition in lithography. On that basis, it would appear that the thickness of the absorber must be limited to be above 0.25 μm. The modulation is excellent for the longer wavelengths of plasma sources.

We notice that because of the relatively weak absorption in the x-rays the absorber is physically thick, with $t \approx 300 - 500nm$ for high resolution ULSI, and up to several microns for deep lithography. We can specify a tolerance for the absorber uniformity over the mask. Assuming a 10% linewidth control budget, we can restrict the amount allocated to δt to maybe 1%. Hence,

$$\delta T / T = -\mu \delta t \leq 0.01 \qquad (3.9)$$

Since $\mu \approx 1 \ \mu m^{-1}$, δt should be of the order of 10nm or less across the mask field.

3.1.2.1 Cascaded systems

The image formation can be described by borrowing the framework of signal theory, i.e., by decomposing the system in a series of *cascaded* subsystems, whereby the output of each element provides the input of the next. This is particularly true with the X-rays, where the low value of reflectivity gives no return signal. Each element is controlled by a set of parameters, some of which are adjustable; these parameters control the behavior of the transfer function that specifies the characteristics of the output.

In general, we can study an optical system by solving the wave equation on planes orthogonal to the optical axis, with particular attention given to the boundaries between different media. The input of each element is the electric field $\tilde{E}(x, y; \omega)$ that describes the image at that plane, and the output is the result of the transformation due to the optical element. The same is true with X-rays, and the analysis is simplified by the impossibility of strong optics. Optically speaking, the X-ray lithography system is described by the incoming radiation (*illumination*), the image formation (*mask*), the propagation region (*gap*) and the absorbing material (*resist*).

In the case of a lithographic image, we must also distinguish among different types of images that are created on the way to the final developed resist. We define (in technical units):

1. **Mask output** – The field intensity, $I(x, y)$, immediately after the mask, mW/cm^2.

2. **Aerial image** – The intensity of the electric field $I(x, y)$ after propagation through the gap to the resist surface, before absorption. It is independent of the resist material and its units are mW/cm^2.

3. **Dose image** – The energy absorbed in the resist volume, $D(x, y, z)$, in mJ/cm^3.

4. **Latent image** – The distribution of the chemical species in the resist created by the absorption of X-rays.

5. **Developed image** – The distribution of material left on the substrate after the development of the exposed resist.

These images are formed sequentially and are incoherent (in the image formation sense) with each other. Each is formed as the result of a complex subprocess, which may or may not be linear. For instance, the formation of the latent image from the dose image is highly nonlinear.

In the discussion of image quality, it is also useful to distinguish between *mask pattern* and *target pattern*. The first is the physical implementation of the pattern in the mask, while the second is what has been specified by the device designer and what the lithographic process step is expected to deliver. The two are not necessarily the same, since various corrections and optimizations are often included in the mask pattern [29].

The first step is thus setting up of an accurate and verifiable model of image formation, capable of predicting the final image observed at the end of the X-ray system. The model can then be used to study the image properties and to determine the factors that affect its quality. The model can be used in particular to study the origin and extent of image biases, so that the mask pattern can be counterbiased to deliver the correct final image [30]. It is then possible to optimize the optical system design, in order to deliver an image with the required quality [31].

We now consider in some detail each step in the image formation.

3.1.2.2 Geometrical Image

Proximity imaging, in its simplest form, has been described above in term of shadow casting. In that case the mask is represented simply by a transmission function $T(x_m, y_m)$, and the aerial image (shadow) is given by

$$I(x, y) \;=\; T(x_m, y_m)\, I_0 e^{-\mu_0 g} \tag{3.10}$$

$$x \;=\; x_m + g\frac{R}{D} \tag{3.11}$$

$$y \;=\; y_m + g\frac{R}{D} \tag{3.12}$$

where (x_m, y_m) is the mask location, R the field position $(\sqrt{x_m^2 + y_m^2})$, D the source distance, μ_0 the aabsorption coefficient of the gap material, and g the gap separation between mask and wafer. Eq. 3.12 yields a faithful reproduction of the mask with a magnification term due to the source divergence. Alternatively, we can write the "corrected" location on the wafer (neglecting the resist thickness) is given as

$$x_m = x - g\frac{R}{D} \tag{3.13}$$

Notice that in this equation the magnification is $M = \partial x/\partial R = g/D$. In some cases it may be quite large, i.e., if $g = 20 \ \mu m$ and $D = 20 \ cm$ we have $M = 100 \ ppm$. This magnification is uniform and can be easily removed by including it in the mask layout, Eq. 3.13. However, variations in g will result directly in a local distortion of the pattern, so that for point sources the variations in g due to mask nonflatness or wafer topography must be carefully examined.

If we consider the source to have a finite extent $s(x, y)$ we then have to introduce a convolution with a blur function $h(x, y)$ that is obtained by scaling $s(x, y)$ to the projected coordinates [32, 6]:

$$h(x, y) = s(x\, g/D, y\, g/D) \tag{3.14}$$

$$I(x, y) = \int I(x', y') h(x - x', y - y')\, dx'\, dy' \tag{3.15}$$

Both terms (divergence and blur) can be unified using the concept of brightness [32, 6]. This analysis, based on simple geometrical optics, is adequate for a first pass, and for the study of the overlay contributions.

3.1.2.3 Fresnel Diffraction

In order to obtain the true image, it is necessary to use the full power of diffraction theory. Image formation in X-ray lithography appears to be deceptively simple. That we cannot neglect diffraction is immediately apparent if we consider that a periodic pattern of pitch p illuminated by λ will form a beam displaced by $\delta = g\ \lambda/p$. For $\lambda = 1\ nm$ and $g = 20\mu m$, $p = 250\ nm$, then $\delta = 80\ nm$, certainly not negligible. This simple example shows that diffraction effects, although present, are likely to be relatively modest since we are far removed from the diffraction limit. Proximity image formation is the subject of a large body of study. From a physical point of view, the textbook description is based on the Fresnel-Kirchhoff approximation (FKA) [34]. A more refined discussion was first introduced by A. Sommerfeld (edge diffraction field) [33, 34]; a detailed discussion is given in Ref. [35, 6]. The FKA assumes that the mask has zero thickness and disregards field continuity problems. If R is the distance between the observation point (x, y) and the mask coordinate (x', y') we can write:

$$
\begin{aligned}
\tilde{E}(x, y) &= \int \frac{e^{jkR}}{R} \tilde{E}_{in}(x', y') T(x', y')\, dx'\, dy' \\
&\approx \int \exp\left(j \frac{(x - x')^2 + (y - y')^2}{g\lambda} \right) \\
&\qquad \tilde{E}_{in}(x', y') T(x', y')\, dx'\, dy'
\end{aligned}
\tag{3.16}
$$

The first equation is nothing but the Huygens construction (superposition of spherical wavelets), and the second equation represents the FKA approximation. All the physical information is contained in the transmission function $T(x, y)$. The main advantage of the FKA is that it can be implemented very efficiently using FFT techniques, since Eq. 3.16 is essentially a convolution of the filtered field $T\ E_{in}$ with a propagation kernel, i.e., a simple product in the Fourier domain [6]. As in the case of optical imaging, the system is linear in the electric field and bi-linear in the intensity.

In Fig. 3.9 we show the intensity computed beyond the X-ray mask, for realistic values of the various parameters. Notice the strong modulation of the intensity, extending over tens of microns. Proximity diffraction does not act as a spatial frequency cutoff filter ($\lambda << d$), but rather produces a mixing of the phases that end up "confusing" the image at gap that are too large, as shown in Fig. 3.10. Since all the spatial frequencies up to very high values are transmitted, in proximity X-ray lithography the challenge is solely in the control of the phase of the transmitted fields.

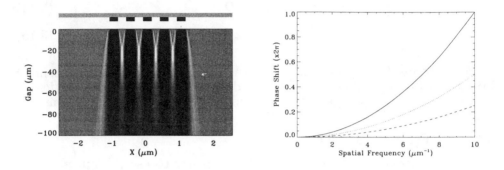

Figure 3.9: Intensity of the diffracted field beyond a 0.25 μm mask (exact calculation)

Figure 3.10: Phase error in proximity imaging, for various gaps. Solid line, 20 μm gap, next 10 and 5. $\lambda = 10\mathring{A}$.

The edge diffraction represents an important case and cannot be neglected a-priori [8, 6]. As mentioned above, the thickness of the absorber used in X-ray masks is of the order of several hundreds of X-ray wavelengths. The field is scattered in the interaction with the material, so that at the plane that corresponds to the mask end (mask exit plane) the field will be different than that created by a simple truncation. This effect has been analyzed in detail by considering the propagation of the radiation field into the absorber, and it can be best explained in terms of the edge fringing field first discussed by Sommerfeld. Briefly, the outgoing field at the mask exit plane can be thought of as being formed by two terms,

$$E_{out} = E_{FKA} + E_{Edge} = T_{Mask}E_{input} + E_{Edge} \qquad (3.17)$$

The first is the normal field as used in the Fresnel-Kirchhoff theory, i.e., a truncated wave, while the second has a cylindrical form. This phenomenon is beautifully illustrated in Fig. 3.11 and Fig. 3.12.

The edge field is of a dimensionality lower than the aperture field (line vs. area). Thus it will contain an extra $1/R$ term that explains its faster decay. Hence, at small gaps (large Fresnel numbers $f = W^2/\lambda g$, where W is the size of the diffracting feature) we cannot ignore the edge field [8]. At larger gaps (smaller Fresnel numbers, $f \leq \approx 12$), however, we can safely use the very efficient Kirchhoff approximation without introducing large errors.

3.1.2.4 Mask Transmission

At the lowest approximation, the mask is thought to be a flat binary filter $T(x, y) = \{0, 1\}$. In reality this is too simple an approximation, and the mask should be specified by a complex transmission (the electric field is a phasor) and by a three-dimensional material distribution. Several levels of approximations can be constructed.

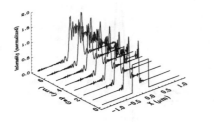

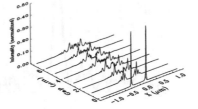

Figure 3.11: Intensity diffracted by an aperture in the Kirchhoff approximation (notice the sharp intensity transitions at the mask exit plane). The image is then computed at increasing distances from the mask

Figure 3.12: Edge field contribution, obtained by subtracting the data shown in Fig. 3.11 from a complete calculation carried out with the BPM method. The contribution is about 8% at the exit plane, nonnegligible. Notice the fast decay of the intensity at increasing gaps; at gaps $\geq 6\ \mu m$, the edge field has all but vanished.

In first approximation, we can still think the mask to be flat but use a form for the transmission of the absorbing parts that will include a phase shift. The contrast of the mask is always finite, so that the absorber is transmitting some radiation, i.e., acting like an attenuator rather than a complete absorber :

$$\tilde{T} = e^{\jmath\tilde{k}z} = e^{\jmath 2\pi(1-\delta+\jmath\beta)z/\lambda} = e^{-2\pi\beta/\lambda}e^{-\jmath 2\pi(1-\delta)z/\lambda} \qquad (3.18)$$

where z is the absorber thickness and $\tilde{k}$ is the complex wavevector. The field is phase shifted because of the different speed of the radiation in the absorber and in the clear parts of the mask. This phase shift contributes *positively* to the image quality, creating a sharper edge at the image boundary [36, 37].

At the next level of approximation, we can consider the mask topography since the material distribution affects the outgoing field. In particular, we should not treat the mask as a simple two-dimensional object because the sidewalls of the absorber may deviate from an ideal, flat and vertical wall. A sloped sidewall can be included in a two-dimensional (flat) mask model by having a grayscale mask, with a transmission that is changing smoothly near an edge rather than abruptly [6, 39]. Thus, we compute $T(x, y)$ as:

$$T(x, y) = e^{-\mu(x,y)\,z(x,y)}e^{-\jmath 2\pi(1-\delta)z(x,y)/\lambda} \qquad (3.19)$$

where both the absorber coefficient μ and the mask height z are local variables. This form of the transmission function then can be used in modeling the image formation using the FKA.

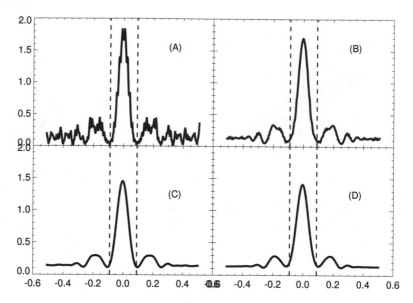

Figure 3.13: Image formation process, for a 0.18 μm feature. (a) monochromatic point source image, (b) polychromatic point source, (c) monochromatic extended source, (d) polychromatic extended source.

3.1.2.5 Computational Models

A detailed description of the image formation modeling has been published in [6], and here we will only summarize the main points. In the image formation, the most accepted method is an implementation of the Beam Propagation Method (BPM), whereby the system is divided in a series of cascaded *layers* [6, 8, 40]. The propagation within the absorber can be treated accurately and yields the line edge field predicted by Sommerfeld. The BPM is simple and can easily deal with nonuniform systems; it applies well to XRL because of the very low reflectivity at the layer boundaries. The illumination plays an important role, both in terms of spectral distribution and spatial coherence. The spectrum is taken into account by adding the intensities corresponding to each wavelength, after computing the monochromatic images with the BPM. This leads to an image that is spatially coherent, i.e., rich with interference fringes, Fig. 3.13. The finite extension of the source is treated by noticing that the Fresnel equations are linearly shift invariant, and thus the partially coherent problem can be reduced to a convolution with the source [32, 38]. The combination is shown in Fig. 3.13. [1]

Thus, the first step in the image formation is described by the diffraction of the incoming radiation through the absorber and then through the gap region. The aerial image (mJ/cm^2) is formed at the surface of the resist and represents the input to the chemical changes [41].

The dose image is formed when the X-rays are absorbed. As mentioned above, the main process is by photoelectron emission. From the knowledge of

---

[1] All the calculations presented in this and in the following figures refer to a 0.18 μm feature, illuminated with a filtered synchrotron spectrum and at a gap of 20 μm.

the aerial image spectral distribution and of the resist absorption coefficient, we can compute the local dose (in mJ/cm^3) [6]. The energy absorbed in the resist is further redistributed by the collisions of the two hot-electrons generated by the event [42, 43]. Using a *Local Continuous Slowing Down Approximation, LCSDA,* we can simulate the energy redistribution process. A blur function, similar to the case of Electron Beam Lithography (EBL), can be obtained by fitting a set of gaussians to the calculations, Fig. 3.14 [43]. These "blur functions" can be used by convolution with the dose image to obtain the real energy distribution in the resist material. The photon energy dependence is be taken into account by parameterization and leads to a simple description of the process that can be extended to include interfaces as well [44]. The main point of this analysis is that the effective range of the photoelectron blur is much shorter than previously thought. The reason is found in the threshold nature of the resist response, so that the part of the trajectory beyond the threshold energy doesn't play any role in the image formation. The traditional Gruen range [45] that estimates the stopping distance of the electron (i.e., the average distance from source point to where the electron reaches zero energy) cannot be used to estimate the resist response. Typically, the effective range is as short as 1/5 to 1/10 of the Gruen range. This notwithstanding, as the photon energy increases the effective range increases as well, as shown in Fig. 3.15.

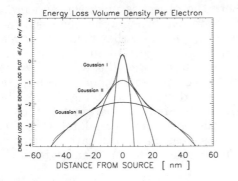
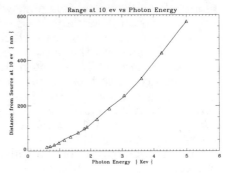
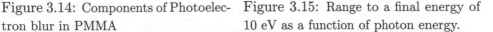

Figure 3.14: Components of Photoelectron blur in PMMA

Figure 3.15: Range to a final energy of 10 eV as a function of photon energy.

We can expect that each atomic species in the resist material will contribute two gaussians, resulting from the photoelectron and the Auger, respectively. More realistic energy deposition models (LCSDA) will somewhat alter this simple description [44] but will not change the main conclusion.

The energy deposited locally affects the chemical properties of the photoresist, creating the latent image. Many processes happen at this stage [41, 46]. The discussion of the resist response is not different from any other system, and we will not discuss it here.

3.1.3 Patterning Ability

The models of image formation in X-ray lithography have evolved considerably in the last few years, leading to a much improved understanding of the physical processes [6, 8, 40]. We can distinguish two main steps in the

X-ray lithography process, that is, aerial image and latent image formation. The first step depends only on the mask and the illumination system, while the second takes into account the properties of the recording material and its interaction with the radiation. Of course, the development process adds another term to the sequence, but we will not consider it here. In lithography, there are three main figures of merit (FOM) of the exposure process: *resolution, placement* and *throughput*. In terms of processing, the first two translate directly to critical dimension (CD) control and overlay. It is important to stop to consider the meaning of the various terms used in lithography, particularly "resolution," which is often used to compare different lithographic techniques and hence should receive special attention. When confronted with this issue, most turn to the use of various criteria based on the quality of *"imaging resolution,"* adopted originally for continuous-tone image forming systems, in astronomy, microscopy and photography [47]. *Their direct use is not correct in patterning*, because the goal of the technique is not that of achieving a faithful representation of an object (as, for instance, in microscopy) but rather that of defining a specific shape – the pattern. Exploiting the knowledge of the system optical transfer function (OTF), it is indeed possible to design masks that can be replicated beyond the traditional resolution of the optical system [48, 50, 49]. In doing this, we exploit the knowledge of the optical system and the freedom in designing an object whose image will form the pattern; this is very different from the case of more general imaging, where the object is unknown and no a-priori information is available. Indeed, in optical lithography phase shifting masks represent an extreme case of this principle, since they hold little direct similarity (fidelity) to the pattern exposed [51]. We thus must replace the concept of *imaging resolution* with that of *patterning ability*, expressing how close we can come to the realization of a given pattern under certain imaging conditions. In optical lithography, this is done by modifying the *Raleigh criterion* through the adjustment of the value of the k_1 parameter, that is, $\delta = k_1 \lambda/NA$ where $NA = sin(theta)$ is the optics' numerical aperture [51]. In proximity imaging, things have been less clear, and most have resorted to the standard application of the Raleigh imaging criterion that leads to the formula

$$\delta = k_1 \sqrt{\lambda g} \qquad (3.20)$$

where λ is the spectrum's average wavelength and g the distance between mask and wafer (gap). Traditionally, the prefactor k_1 has been assumed to be given by the Raleigh criterion, i.e., $k_1 = 1.22$. Recently, we have shown that it is possible to achieve large exposure latitude by properly designing the X-ray mask and the illumination system [6, 30, 39] beyond the value predicted by a k_1 of 1.22. If we use those results to obtain an empirical value for the prefactor in Eq. 3.20, we would use a value closer to 0.6. This is confirmed by experiment, which show that patterning can be performed with very high resolution at large gaps [8, 30, 40, 52]. Fig. 3.16 show results from NTT [53], and Fig. 3.17 a print from a phase shifting X-ray mask at CXrL.

The second component in image formation is that of the image blur introduced by the recording process [42, 43, 44]. This part is easier to handle, since it is incoherent and thus we can resort to the modulation transfer function (MTF) [32]. We can thus build a global *patterning ability*, considering both the optical capabilities and the photoelectron blur. Since they are inde-

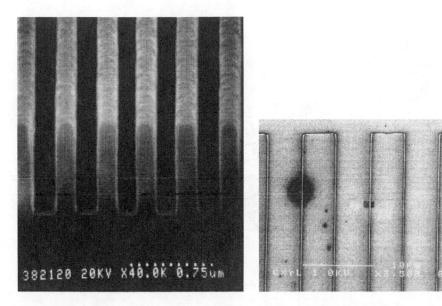

Figure 3.16: 0.15 μm printed at 40 μm gap.

Figure 3.17: X-ray phase shifting: sub-0.1 μm features printed at 20 μm.

pendent and incoherent we can add them in quadrature, yielding the result shown in Fig. 3.18. Clearly, *X-ray lithography can be used to pattern well into the sub-0.1 μm domain* [55, 44]; see the discussion on Page 53. We also notice that the patterning ability is a relatively weak function of the photon energy in the region between 1 and 2 keV. Since the dimension of interest for semiconductor processing in the next few generations (0.25 to 0.15 μm) are considerably larger than the patterning ability of X-ray lithography, the "resolution reserve" translates directly into large depth of focus and exposure tolerance. This updated model should be used to predict a first approximation of the patterning capabilities of X-ray lithography[44].

3.2 Implementation: System and Components

We now discuss in detail the various components of the exposure system, concentrating on the practical aspects and the current status.

3.2.1 X-ray Sources

Only two types of X-ray sources have enough power to be considered viable candidates for XRL-based manufacturing. They are based on plasmas [57, 58] or electron accelerators [5]. Many types of sources have been proposed over the past twenty years, but until now only the electron storage rings (ESR, often called symply synchrotrons) have reached a very mature stage of development.

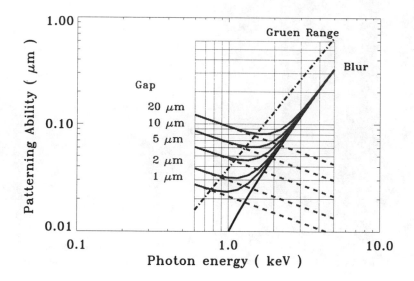

Figure 3.18: XRL patterning ability, for an optimized mask and exposure system. The diffraction is computed for $k = 0.7$, and the photoelectron range is that shown in Fig. 3.15.

3.2.1.1 Point sources

While there is no commercially available plasma X-ray source for XRL, several are under development [58]. The basic process of X-ray emission from a plasma source is the creation of a superthermal region of plasma, where highly ionized atoms emit soft X-rays because of atomic transition and blackbody radiation. The plasmas can be created in several ways, but we can distinguish between the laser and non laser sources. The image formation mechanism in a point source is not different from other type of sources [56]. A requirement specific to point sources is the need to collimate the beam of X-rays, requiring either a large distance or an optical collimator [58, 61]: the highly diverging beam may created sloped exposures at the edges of the field, seriously affecting the CD control.

Laser plasmas Laser plasma sources were brought close to fruition by Hampshire Instruments, using a Nd:YAG glass slab laser system, pumped with arrays of solid-state emitters [59, 60]. The radiation from the laser is focused in a diffraction limited spot of short time duration. The power deposition rate is much larger than the removal, and extremely high temperatures can be achieved. A plasma micro-ball is formed and, before it dissipates, emits X-rays. The conversion efficiency can be quite high, depending on the combination of laser radiation and material used. The targets range from metals such as Fe [59] to solid gas targets of Ne [60] for longer wavelengths. Debris is always formed and can be stopped by using a gas puff or a mechanical actuator. The main problem of plasma sources has been, until now, a low power output. While the power in the shot is very high, in lithography what matters is the CW power. Laser plasma sources typically deliver a few mW/cm^2 average in typical situations. Because of the appeal of the process several efforts are underway to improve the power level [58].

Dense plasma sources Another approach is based on the creation of the plasma by the collapse of an electrical discharge [62, 63, 64]. Several attempts in the past were unsuccessful in delivering a reliable source [58]. This type of system relies on the discharge of a large amount of energy in an ionized gas, causing a large increase in energy density and a superthermal plasma. The radiation mechanism is the same as is laser sources, but the power can be much larger. These sources have been plagued by stability and power supply problems, because of the requirement of rapid firing of the capacitor banks. Electrode erosion may also lead to instability in the discharge, and the large current often cause EM interference that may adversely affect delicate electronic systems. Nevertheless, sources such as the high-density plasma in development at SRL [62] may eventually challenge the synchrotrons.

3.2.1.2 Synchrotron Radiation Sources

Synchrotrons and Electron Storage Rings (ESR) are accelerators [65, 5] capable of producing stable beams of particles of very high energy, in the million (MeV) to billion (GeV) electronvolt range. The machines were invented for high energy and nuclear physics research but found very important applications in the spectroscopical area because of their intense emission in the ultraviolet and X-ray regions of the spectra [66], as illustrated in Fig. 3.19. The radiation generated by the source is filtered and delivered to the mask-wafer assembly, securely held by the stepper. Today, there are more than 45 rings installed at various laboratories in the world (for a complete review, see [67]) and the family is continuously growing.

Fig. 3.20 shows the IBM electron storage ring [96] Helios (manufactured by Oxford Instr., [69, 70] from which several beamlines [71] extract the radiation and direct it to the exposure stations. The general structure, including the injector, storage ring and beamlines, is quite general. With some impropriety, we commonly refer to ESRs as synchrotrons. Their basic structure are quite different but the properties of the radiation are the same – hence the terminology.

A closer view of a cluster of beamlines used to relay the radiation from the ring to the exposure stations is shown in Fig. 3.21, while Fig. 3.22 shows the visible part of the radiation extracted through an optical window. Several reviews of the properties of synchrotron radiation have been published [72]; here we will concentrate on the aspects relevant to the X-ray lithography process [5]. The radiation emitted by low-energy electrons captured in a circular path is dipole radiation [23]. This can be easily understood if we recognize that any circular motion can be decomposed in two linear oscillations with angular frequency ω. The two oscillations are exactly 90° out of phase. When the speed of the electrons becomes very large, approaching c (the speed of light), the Lorentz contraction sets in. The wavelength of the radiation will be blue shifted by the Doppler effect. In the treatment of syncrotron radiation, the energy of the electrons is measured by the relativistic parameter γ, defined in term of the electron velocity v, as:

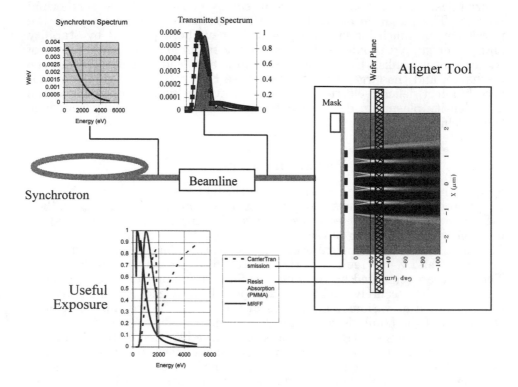

Figure 3.19: Schematic of an X-ray Lithography system based on a synchrotron. Notice the synchrotron, beamline and mask arrangement as well as the changes in the spectrum of the radiation.

ALF

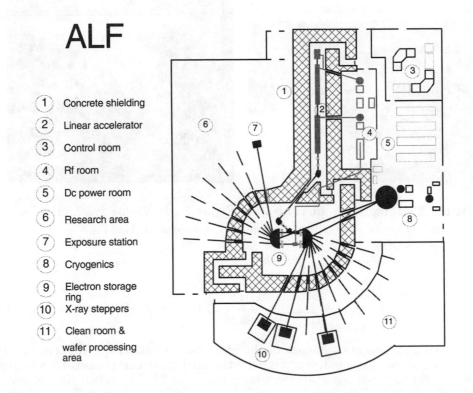

1. Concrete shielding
2. Linear accelerator
3. Control room
4. Rf room
5. Dc power room
6. Research area
7. Exposure station
8. Cryogenics
9. Electron storage ring
10. X-ray steppers
11. Clean room & wafer processing area

Figure 3.20: Basic accelerator structure in an XRL facility: IBM's Helios ring installed at ALF

Figure 3.21: Beamlines from Aladdin to CXrL's steppers. The source is visible in foreground.

Figure 3.22: Visible Synchrotron radiation emission at Aladdin

$$\gamma = \frac{1}{\sqrt{1 - \frac{v^2}{c^2}}} = \frac{E}{m_0 c^2} \approx 1957 \, E \, [GeV] \qquad (3.21)$$

An important consequence of the relativistic electron speed is the "folding" of the radiation pattern, whereby the angles are compressed by a factor of γ along the instantaneous electron velocity [5, 23] These two effects, combined, result in a profound change in the radiation spectrum, i.e., the way in which the power emitted is distributed among the different wavelengths.

A more formal analysis leads to the definition of a series of formulae useful for predicting the properties of syncrotron radiation. In practical units (with E in GeV, I in A and R in m):

$$
\begin{array}{llll}
\lambda_c \, (\text{Å}) & = & 5.59 \, R/E^3 & \text{Critical Wavelength} & (3.22) \\
\epsilon_c \, (\text{eV}) & = & 12,398 \, /\lambda_c & \text{Critical Energy} & (3.23) \\
P \, (kW) & = & 88.5 \, E^4 I/R & \text{Total Power} & (3.24) \\
P \, (kW) & = & 15.83 \, \epsilon_c E I & & (3.25) \\
P \, (W/mrad) & = & 14.1 \, E^4 I/R & \text{Power/mrad} & (3.26)
\end{array}
$$

An electron storage ring is essentially a very efficient X-ray antenna: the power radiated depends only on the radius, beam energy and current. It is

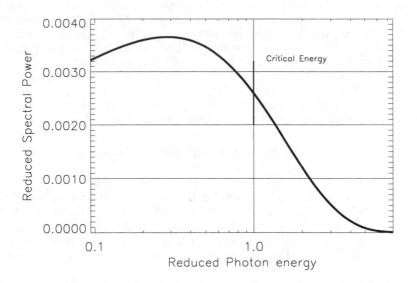

Figure 3.23: Universal radiation spectrum emitted by a storage ring. The photon energy is normalized to the critical energy ϵ_c and the spectral power (in W/eV to the product EI for a 1 $mrad$ angle. To obtain an actual spectrum, multiply the photon energy by ϵ_c and the power by EI.

important to note how it is possible to obtain the same radiation (both in spectrum and in total power) from two very different machines, as long as ϵ_c and the product EI are the same, as shown in the last equation.

Synchrotron source spectral characteristics A universal syncrotron radiation spectrum is shown in Fig. 3.23. Notice the broad range of energy covered by the radiation. For example, let us consider a machine with $E = 1\ GeV$, $R = 2\ m$ and $I = 0.1\ A$. Using Eq. 3.26 above, we can see that such a ring will have $\epsilon_c = 1109\ eV$, $\lambda_c = 11.18$Å will radiate $0.705\ W/mrad$. If we keep in mind that an exposure beamline will accept around 30 mrads of radiation, more than 21 W will be delivered to the beamline. Such a machine has a γ of 1957 and a critical wavelength of 11.18 Å. At $\lambda = \lambda_c$ the machine will radiate $1.6\ 10^{13}$ $photons/s$ in a 1% bandwidth centered at $\lambda = 11.18$ Å. It is important to notice that while the median energy is 11.18 Å, the spectrum extends well into the visible (and even infrared) on the longer wavelength side.

In summary, the spectral distribution of the radiation emitted by a synchrotron or by an electron storage ring is characterized by a very wide frequency distribution, extending from the infrared to the X rays. The power radiated is also large, making the electron storage ring a very efficient X-ray source.

Synchrotron source spatial characteristics Another important property of the radiation emitted by ESRs is its angle distribution. As mentioned above, the Lorentz transformation contracts the *"figure 8"* shape typical of the dipole pattern in a narrow beam of aperture $1/\gamma$. The motion of the electrons along the orbit makes the narrow cone "sweep" horizontally, thus leading to a very uniform horizontal distribution. In the vertical direction the beam is instead characterized by a distribution of width $\approx 1/\gamma$ *(rads)*, i.e., about a few *mm* high at the end of a typical beamline length $(10\ m)$. The strong chromaticity of the source is typical of the synchrotron radiation process, and in first approximation we can write that photons of wavelength λ will be emitted with a distribution of standard deviation given by [20, 73, 74]:

$$\theta = \frac{1}{\gamma} \left(\frac{\lambda}{\lambda_c} \right)^{0.435} \tag{3.27}$$

The narrow vertical emission angles have led to considering syncrotron radiation as being "collimated." This statement is an approximation, as shown above. It is however true that the opening angles are very small (a fraction of a milliradian), particularly in comparison with other sources. In the horizontal directions these effects are averaged out by the motion of the electrons.

We want to briefly mention the strong linear polarization of the radiation. To an observer located exactly in the orbit plane, the radiation would appear to be 100% polarized horizontally. By integrating over the full spectrum [5] we find that the degree of polarization can be obtained by noticing the ratio between parallel emission and perpendicular emission, yielding a polarization degree of exactly 75% [75]. Although polarization effects are important in spectroscopy they do not appear to play any significant role in XRL, so we will not consider the argument further.

The emission of radiation is a random process governed by Poisson statistics, controlled by the deterministic laws described above [18]. The results we have so far derived apply to the case of an isolated electron moving on the central orbit. The electrons stored in an ESR do not all have exactly the same parameters but rather are distributed in a random way around some average values. The distribution of the electron trajectories is governed by statistical laws, so that for example the distribution describing the deviation of the electrons from the standard orbit is to a very good approximation a gaussian function.

In a real machine, the photons are not generated exactly along the electron trajectory but rather at some random angle whose distribution is consistent with the radiation distribution [32, 76]. The radiation angles are always referred to each individual electron *instantaneous* orbit plane, so that if the electron orbit forms an angle α with the reference or central orbit at the emission point, then the radiation lobe will be oriented along the orbit at the same angle α. The radiation pattern will then be the convolution of the radiation fan and of the electron directions at that orbit location; if we approximate the radiation with a gaussian distribution as well we can make use of the fact that the convolution of two gaussians is still a gaussian with a standard deviation given by:

$$\sigma_T^2 = \sigma_R^2 + \sigma_{x'}^2 \tag{3.28}$$

where σ_R is the standard deviation of the radiation. The angle distribution used in Eq. 3.27 is thus augmented by the electrons' angular distribution. For machines of low beam energy, the photon energy spread normally dominates.

In conclusion, the physical aspect of the synchrotron radiation source is that of a gaussian distribution of point sources, with a distribution given by the electron beam size σ_x and with an angle aperture following Eq. 3.27.

Equivalent source   In order to achieve a uniform intensity at the mask position some form of beam scanning or expansion must be provided. This can be done by keeping the mask-wafer stationary and rastering the X-ray beam or by keeping the radiation fixed and mechanically scanning the mask-wafer assembly. Other schemes, based on oscillating the electron beam itself, are unlikely to gain widespread acceptance. From the point of view of XRL what is important is the penumbra that is created by the finite size of the electron beam. D, P being the distance from source and from the pole of the mirror, it can be shown [77] that the penumbra is obtained by:

$$\sigma_{g,x} \;=\; \frac{g}{D}\sigma_x \tag{3.29}$$

$$\sigma_{g,z} \;=\; \frac{g}{P}\sqrt{\sigma_z^2 + (D-P)^2\sigma_{z'}^2} \tag{3.30}$$

so that the scanning action introduces, quite unexpectedly, the effect of angle divergence as well as the beam size [6, 32]. The term under the square root correspond to the projected source size at the mirror location $(D-P)$; hence, the meaning of Eq. 3.30 is that the scanning mirror defines a new virtual source at its location of size $\sigma_{g,z}$. Thus, even a collimated beam $(\sigma_{z'} = 0)$ or a point source $(\sigma_z = 0)$ will introduce penumbral effects when scanned. The case of electron beam wobbling corresponds to having the mirror coincident with the source, so that $P = D$. Similarly, the scanning of the mask-wafer assembly leads to the same $D = P$ condition.

Summary   We can collect in a table some parameters that are typical of two machines, a large "research"(Aladdin, at the University of Wisconsin) and a "compact" one (Helios, at IBM's Advanced Lithography Facility), together with the performances that can be expected in both cases.

All electron accelerators of the synchrotron/ESR type share the same block structure (compare also Fig. 3.20),

| Source → Pre Accel → Injector→(Booster) → Machine. |

The choice between the different schemes is dictated by the issues of cost, efficiency, and reliability that are compatible with a given application. For XRL applications, reliability (i.e., uptime) is of the utmost importance. No rings have yet built and used in an industrial manufacturing environment; the lessons learned from the physics experience do indicate that all of those approaches are viable, provided that they are correctly implemented and that maintenance is regularly used. Storage rings are well behaved, predictable machines with few critical components: maintainability should be relatively easy to design in. Safety issues are also very important, since the injection subsystem is a region of intense radiation that must be fully contained.

Parameter	Units	Research	Compact
Beam Energy	GeV	1.0	700
γ		1957	1370
Radius	m	2.0	0.519
Magnetic Field	kG	16	45
Critical Energy	Å	11.4	8.45
Radiation Angle	mrads	0.5	0.8
Current	mA	100	200
σ_R	mrads	0.5	0.8
σ_x	mm	0.52	0.7
$\sigma_{x'}$	mrads	0.2	1.5
σ_z	mm	0.075	0.7
$\sigma_{z'}$	mrads	0.013	0.3
Penumbra (@ 15 m) x	nm	0.8	1.3
Penumbra z	nm	0.05	1.3
Power/mrad	W/mrad	0.7	1.3
Power (Beamline)	W	21	39

Table 3.1: Comparison between Aladdin, a research machine, and Helios, the superconducting storage ring built by Oxford Instr. Ltd.

Experience at several facilities based in industrial R&D labs have shown that superconducting accelerators perform reliably, with very high uptime (always in excess of 90%) [70, 78, 79].

The problem of delivering a high intensity radiation beam to the mask requires a careful engineering study in order to ensure cost-effectiveness, reliability and safety. The system used to accomplish this is called a *beamline*. It can be divided into subsystems performing the different tasks required for the overall functionality, that is, the relay of the X-rays generated by the ESR to the mask-wafer assembly.

3.2.1.3 Beam Delivery Systems

The discussion of the previous section indicates how the photon energy range between 1 and 2 keV provides the best resolution. For the case of synchrotron sources, we have quite a large freedom in designing the spectrum delivered to the mask-wafer system. What should we design for? We can answer this question by studying the deposition of power among the various components of an X-ray lithography system. We will use the case of Helios, as described in Table 3.1 The all-important part of the radiation emitted by a storage ring is that which finds its way into the photoresist; it is thus relevant to ask which fraction of the power delivered by a given beamline falls in that "window" [80]. We can compute this fraction by defining a filter to select the useful part of the radiation. This filter is shown in Fig. 3.24; a similar concept was introduced by Grobman [81]. We call it the *mask-resist filter function* (MRFF) [82]. If we consider a mask of transmission $T(E)$ and a resist of absorbance $A(E)$, we can write the filter as:

$$MRFF(E) = \frac{T_{mask}(E)A_{resist}(E)}{max(F(E))} \qquad (3.31)$$

where $max(F(E))$ is a normalization factor that guarantees a maximum value of 1 for $MRFF$.

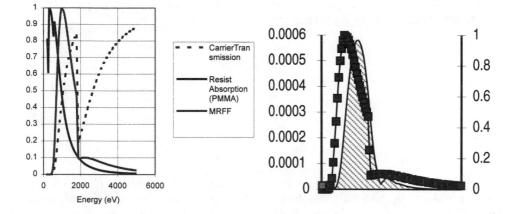

Figure 3.24: Definition of Mask-resist filter function

Figure 3.25: Overlap between beamline spectrum and MRFF

In Fig. 3.25 we show an example of beamline settings yielding a good overlap between the spectra transmitted and the MRFF changes. The degree of overlap can be used to optimize the beamline design.

In a beamline, mirrors act as a low-pass element, and filters as high-pass. Together, they define a spectral bandpass that, multiplied by the source spectrum, determines the delivered spectrum, as shown in Fig. 3.25. The filtering process can be used to improve the matching between the source and the mask-resist, but at the cost of rejecting power.

We now have all the ingredients to determine the best combination of elements in an X-ray lithography system. The absorbed power is directly related to the throughput and should always be maximized. This, in turn, requires the design of a system with a minimum number of reflections and a thin exit window. A good match between beamline and MRFF is highly desirable because it reduces the deposition of X-ray energy in unwanted parts of the system, particularly in the mask membrane and in the underlying substrate.

The beamline is the illumination system for the lithographic step. Its optical characteristics have been discussed in detail and are well understood. For a syncrotron radiation-based system, the beamline should collimate, filter and vertically scan a uniform beam. Using a specially designed mirror, it is possible to achieve all the functions with a single reflecting surface [83, 84, 85]. The exit window must be able to withstand the stress induced by the atmospheric pressure and be resistant to the large radiation dose deposited over its useful life. Trial-and-error has converged on the use of Be as the material and on a curved thin plate geometry [86]. This geometry maximizes the resistance to fracture, transforming the forces induced by the pressure differential in tensile stress [87]. For a window of radius r, a simple relation can be written connecting thickness and induced stress. In general, we want the internal stress to be a fraction of the ultimate tensile strength (UTS), which is material and fabrication dependent; this is shown in Eqn. 3.32, where t is the thickness of the membrane, P is the pressure, and σ_y is the UTS of the membrane. Since the transmission of the window, T, depends as well on the material thickness t, we can include all the terms together in a simple relation that depends only on the material properties (its UTS). Thus, selecting a stress no larger than 80% of the UTS and a transmission greater than 90%, we can write:

$$t \quad \geq \quad \frac{rP}{0.8\sigma_y} \tag{3.32}$$

$$T \quad = \quad e^{-\mu t} \geq 0.9 \tag{3.33}$$

$$\frac{\mu}{\sigma_y} \quad \leq 0.8\frac{ln(0.9)}{r\,P} \tag{3.34}$$

where t is the thickness of the window, r the bending radius, P the hydrostatic pressure, σ_y the ultimate tensile strength, μ the absorption coefficient and T the transmission in the X-rays [82]. The last equation shows us a relation between two quantities, on the left, which are only material related with those on the right, which are only design values. In a μ vs. σ_y plot, only those materials falling below the straight line can be considered satisfactory in the design. Such a plot is shown in Fig. 3.26, and various materials are indicated on it. Sharper radii of curvature allow thinner windows, and thus the absorption becomes less of an issue. Clearly, both Be and SiC are acceptable candidates. However, Be is a metal and as such can be bent easily in circular shapes; this is more difficult to achieve for ceramic-like materials such as SiC. Surprisingly, Si does not appear to be a good candidate. This is because the internal stress does not depend on the material Young's modulus but rather on the boundary condition of the membrane, and the safe stress is dependent on the UTS.

Finally, the hard radiation generated in the ring during injection must be fully contained for the operators' safety. A large body of literature exists on this subject [88] and we will not dwell on it, save for noticing that radiation safety is achieved by carefully placed shields located along the beamlines and by heavier, fixed shielding around the ring itself. The shielding along the beamlines is not needed because of the X-rays, that are efficiently stopped by the vacuum vessel walls, but rather because of the harder gamma rays that may be generated by the collision between the high energy electrons and the residual gas molecules. This bremsstrahlung may happen to be oriented along the beamline if by chance a region of (relatively) large gas trapping exists at the tangent point.

3.2.1.4 Synchrotrons and Lithography

A synchrotron-based X-ray lithography system comprises three main subsystems: the source, the beamline and the exposure station. We have described

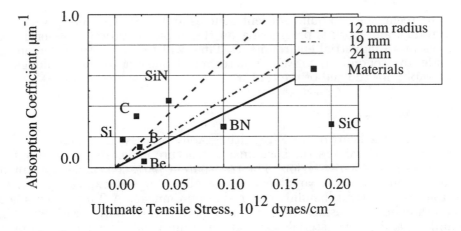

Figure 3.26: Absorption - stress plot for various materials and radii of curvature

the source and the beamline, while the exposure tools will be covered in a different section of this chapter. At this point we want to emphasize that the design of a successful lithography process requires a careful engineering systems study. There are many trade-offs that are possible, delivering performances that are approximately the same but with widely different costs. For example, beamline acceptance may be used to compensate for a low-power machine; modularity is essential in order to accommodate tools with different scanning requirements; and so on.

From a lithography point of view, there are two parameters that will determine the viability of the approach. Resolution is not one of them, at least until features of less than 0.1 μm will be needed. Power density, cost, and cost of ownership are. The economics have been addressed in several papers [89, 90, 91, 92]. The absolute cost is not a very significant figure of merit; cost effectiveness, often quantified as cost of ownership (COO), is a better figure. Cost-effectiveness must not be confused with net cost: to make a trivial example, a large bulldozer is certainly more expensive than a garden spade, but few would argue that the spade (or even many spades) is a more cost effective tool to build highways. Synchrotron based X-ray lithography is certainly expensive, but it is also capable of exceedingly favourable economies of scale.

A well-designed X-ray lithography syncrotron radiation system can easily deliver in excess of 50 mW/cm^2 over a field of 50 mm (H) by 25 mm (V) of collimated radiation in the lithographically useful exposure window. This makes it possible to achieve the goal of 1 s/field exposure; assuming an overhead of 1 s for stepping and alignment, an 8 inch wafer would be exposed in 1 min. Including wafer loading/unloading the system should be capable of delivering more than about 50 8 inch wafers per hour and will most likely be limited by the exposure tool overhead. For a manufacturing plant using 10 beamlines, this would be equivalent to $50 \times 20 \times 10 = 10,000$ wafer levels a day. A storage ring can support a very large silicon operation. Clearly, even a large initial capital cost can be amortized quickly by such a production rate [89, 90, 91, 92]. The real cost component on which the cost-effectiveness hinges is the mask, not the accelerator cost [92].

The technology to build efficient sources and beam transport systems is here today, and work is progressing on the implementation of several such sources in an industrial environment. Time will tell if the synchrotrons do provide an economical, as well as a technical, answer to the challenges of lithography for the end of the century.

3.2.2 Masks

The mask is the heart of the X-ray lithography process. It combines pattern and optical system, and the basic mask structure is dictated by the optical properties in the X-ray region. The transparent part, the carrier (or "membrane"), must be transparent enough to allow for fast exposures and yet be able to withstand handling and radiation damage. A 2 μm thick silicon membrane has a transmission of about 50% (see Table 3.3), for a typical SR beamline, and other materials have similar values [93]. In general, the membrane will be of the order of 1 to 2 μm thick and made of low-Z materials for high transmission. Since the exposure is a 1× process, the placement accuracy of the pattern must be within the bounds dictated by the error allocation budget. This implies that the membrane must be rigid enough so that no distortion of the pattern is induced by the handling and the exposure process. The membrane is too compliant to be self-supporting, and it must be fastened to some support that provides the necessary rigidity. This is the form of an outside frame of suitable shape, as shown in Fig. 3.27 and Fig. 3.28.

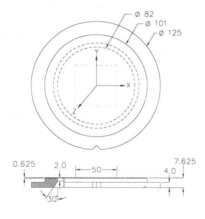

Figure 3.27: NIST X-ray mask structure. [95].

Figure 3.28: An X-ray mask fabricated at IBM for the US Lithography program (Notice the location of the kinematic mounts and the V-cut for the locating pin)[96]

In general, the mask is fabricated starting form a 3" or 4" silicon wafer, following the steps described in Table 3.2. The details may vary from implementation to implementation, but the fundamental structure remains con-

Step	Description	Process
1	Membrane deposition	CVD
2	Wafer bonding	Anodic bond
3	Membrane formation	KOH etch
4	Plating base deposition	Sputtering
5	E-beam patterning	EBL
6	Au absorber formation	Plating
7	Resist ashing	Plasma
8	Inspection	KLA SEMSPEC
9	Repair	Micrion FIB

Table 3.2: Example of a mask fabrication process (additive)

stant: a thin, uniform membrane is mounted on a structurally rigid holding frame. The pattern is applied on the membrane, and the mask is mounted on the exposure and processing tools using the frame.

3.2.2.1 Stress and Distortion

Before discussing the fabrication details, it is paramount to consider the mask as a mechanical structure. Since there is no active optical system in proximity XRL, the mask combines the pattern and the imaging system. The influence of the absorber topography on image formation has been described above, and here we will discuss instead the influence of the mask structure on the placement accuracy. The mask must be considered as a complete three dimensional mechanical system in order to correctly evaluate the issues concerning the pattern placement; because of the complexity of the system, Finite elements methods (FEMs) are best suited for the analysis [97, 98]. A simplified but illuminating discussion was presented by A. Yanof [99], in which he derived analytical expressions for out of plane distortions (OPD) and in plane distortions (IPD) in simple cases. For collimated, or near-collimated sources such as the synchrotrons, OPD has much less importance than IPD. Ku derives the following expression [100], that specializes to the formula found by Yanof [99]:

$$\delta = \sigma_a t_a \frac{(1 - \nu^2)(L - l)(t_a + t_m)l}{2t_m((E_m t_a + E_m t_m)l + E_a t_a(L - l))} \tag{3.35}$$

where δ is the displacement caused by a feature of length L, located at the center of a square mask of width L, σ_a the stress in the absorber of thickness t_a, E_s the substrate Young modulus, σ_s stress and t_s its thickness, ν is the Poisson ratio. Notice how the IPD is independent of the substrate stress. It is important to understand the various distortions and their physical origin in order to control and minimize their effect on the image. In Table 3.3 we list the most important materials properties, from the point of view of X-ray mask fabrication. In mask fabrication, the stability of the membrane to external forces (solidity) is also very important. The *burst*

Materials	Density	T	T	Y	UTS	α	κ	Damage
Units	g/cm^3	–	–	d/cm^2	d/cm^2	–	$W/cm/^\circ K$	ppm
Carriers		1μ	2μ					
SiNH	2.75	.56	.39	1.7	0.05	2.1	0.2	0.86
SiC	3.2	.57	.41	4.6	0.2	4.6	0.41	0.01
Si:B	2.33	.63	.46	1.7	0.07	3.7	1.6	0.04
BNH	1.63	.88	.80	1.7	0.1	2.25	0.8	20.0
Diam	3.52	.58	.47	10.5	0.02	3.5	6.55	0.17
Absorbers		.35	.7				–	
Au	19.3	0.14	0.05	.8		14.2	3.18	–
W	19.3	0.14	0.05	3.9		4.5	1.74	–
Ta	16.6	0.20	0.07	1.9		6.5		–

Table 3.3: *Mechanical properties of materials used in X-ray masks fabrication; T is the transmission of a 1 or 2 μm thick film for carriers, and 0.35 or .7 μm for absorbers; Young modulus and UTS in 10^{12} dynes/cm² units; α is the expansion coefficient in ppm/°K, and the damage is expressed as the IPD corresponding to a delivered dose of 10^4 J/cm². Adapted from [94]*

strength, i.e., the pressure required to burst the membrane is a good measurement of the strenght of the membrane. Obviously, it is desirable to maximize the burst strength while minimizing IPD and X-ray absoprtion. The first two requirements lead to a thicker substrate, while the transparency improves with thinner films. Hence, a compromise needs to be reached. A study of Table 3.3 shows clearly that SiC is the best material for a carrier or membrane.

3.2.2.2 Intrinsic Distortions

These refer to the errors that are observed in a mask that is held in a perfect way, i.e., with no external forces. While this clearly is an impossible case, it is nevertheless instructive. The sources of pattern errors can be several. To be specific, by "pattern error" we mean for the moment the pattern displacement relative to an ideal grid [100].

The first source of error is found in the finite accuracy of the pattern generator itself. The pattern will have some distortions which are tool-dependent. Today, two EBL tools stand as the best. One is the IBM in-house EL4+ system (a variable-shaped beam), the other the commercial LEICA 100 kV EBPG (a gaussian beam) [9]. The LEICA has 12 nm positioning accuracy over a 25mm field, obtained by a combination of a high-voltage beam, an accurate stage and a high-resolution laser interferometer. The mask writing is essentially a blind operation, since no previous marks are available on the membrane and no simple way exists to determine the position of the electron beam. A more sophisticated approach has been proposed, using a fiducial grid deposited on the mask to monitor the beam position [102]; alternatively, to reduce overlay errors caused by mix-and-match, alignment marks may also be deposited on the membrane prior to patterning (as done on direct write) with the tool used for the previous level (e.g., an optical stepper) [103]. Notice that this second approach relaxes the requirements on the patterning relative to an ideal cartesian grid in order to reduce the global overlay error.

The second source of error is the fabrication process; that stress would affect the pattern position accuracy quickly became clear [104, 105]. In both subtractive and additive processes, a pattern of metal is created on the membrane and because of the formation process, the metal is under some level of stress [106, 107, 108, 109, 110, 111]. For instance, a tensile stress in the absorber will cause a compressive stress in the membrane to balance the forces and reach equilibrium. Any changes in the stress distribution will create a corresponding strain field, i.e., a pattern displacement. An analysis of the distortions thus created shows a strong pattern specificity, precluding simple corrections at the mask writing level. A worst-case analysis may help in setting some boundaries; one such case is that of a mask half-covered with absorber. Clearly, the boundary at the center will undergo maximum displacement. Following Yanof's discussion [99], we can quickly conclude that

$$\frac{\sigma_a t_a}{2E_s t_s} \leq \frac{\delta}{W} \qquad (3.36)$$

for a mask of side $2W$ half-covered with an absorber, δ being now the maximum acceptable IPD. For a worst-case scenario based on a 5 nm budget component and a 25 mm SiC membrane, the stress in the absorber must be less than about 2×10^7 $dynes/cm^2$, roughly equivalent to 1/10 of the membrane stress. This value is achieveable by careful processing. The distortion is pattern-specific, since the patterning leads inevitably to a non uniform stress in the final structure, and hence to a non uniform strain. The long-range component of the strain field can be compensated relatively easily, but the local strains require a product-specific pattern correction [112]. Of course, the key is providing a low-stress absorber, thus minimizing unwanted displacements [100], and local corrections may provide a fall-back position. Nonuniform deposition (or etching) conditions can cause both pattern shifts and linewidth variations. For instance, a nonuniform temperature in the RIE of W absorber will cause CD variation because of changes in etching rates, and He-backside cooling is being explored to solve this problem – both during deposition and etching. The solution to all these problems is a controlled deposition and postprocessing to minimize the stress in the absorber.

3.2.2.3 Extrinsic Distortions

These refer to the distortions cause by external forces on an otherwise perfect mask [101, 113]. They include mostly mechanical (gripping forces), gravity and environmental (temperature) effects. The first can be minimized by an intelligent design of the gripping system which must not overconstrain the mask. Kinematical mounts have been proposed as one of the main approaches, in particular by IBM [113, 114]. A cassette is used to hold the mask in a minimum-stress configuration in both modern steppers (such as the Suss Model 2m, 3 and 4, Fig. 3.29) and EBL tools (such as the Leica, Fig. 3.30), so that any forces applied to the cassette are not transmitted to the mask itself [115]. Conversely, it has been proposed to used controlled external stresses to compensate for inherent distortions in the mask pattern cause by other factors, or to produce a controlled amount of linear magnification [?, 116]. A mask structure suitable for kinematic mounts is shown in Fig. 3.27.

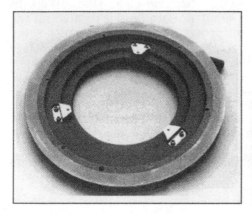

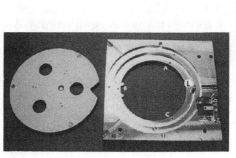

Figure 3.29: Mask cassette with kinematic mounts for Suss stepper; notice clips position (mask will be facing "down"

Figure 3.30: Leica EBL mask cassette (left,backplate); A, B and C refer to the kinematic mount locations, and L to the locating pin

The effect of mask orientation must be evaluated with care, because of the force of gravity; while this may sound surprising, error budgets with entries of a few nanometer make these effects far from negligible. For instance, the mask is typically held horizontally in the electron beam writer and in the inspection systems, while it is held vertically in the stepper tool [?]. In Fig. 3.31 we show the effect of the force of gravity on the mask (vastly magnified) held in a kinematic mount; a different distortion is produced if the mask is held horizontally. The gravity produces a deformation in the mask, so that the pattern is written not onto an ideal substrate but on a distorted one, leading to a placement error Δ_{ebl} relative to an ideal grid; when the mask is mounted vertical, we will have another error Δ_{xrl}. The total error is the *vectorial* sum of the two since both are deterministic, leading to a base error of:

$$\vec{\Delta}_{base} = \vec{\Delta}_{ebl} + \vec{\Delta}_{xrl} \qquad (3.37)$$

By using this approach, i.e., adding vectorially the distortions due to the mask flexing in the (horizontal) e-beam patterning tool with those of the mask in the (vertical) stepper, we obtain the result shown in Fig. 3.32, i.e., we can compensate for the distortions at each step. This is achieved by a careful design of the holding system so that the two errors balance out, pretty much like in the design of lenses [?].

While it is possible to generalize the approach to an arbitrary mask design, the response is strongly dependent on the particular frame cross section. A Finite Elements Model (FEM) study becomes necessary in determining the effects of a new type of frame. We notice that in this case we are not trying to minimize the absolute error at each stage but rather to adopt a *look ahead* strategy based on our knowledge of the system. The total remaining

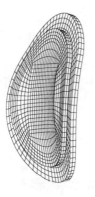

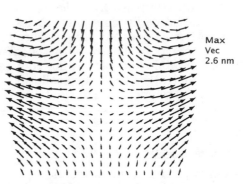

Figure 3.31: Distortion in X-ray mask frame under the effect of gravity force, magnified 50,000×

Figure 3.32: Market plot of compensated distortions in X-ray masks due to gravity; notice the systematic character and the small absolute value.

error is well below the limit for 0.1 μm lithography, and could be completely eliminated by mapping a correction in the EBL tool.

Temperature variations and gradients may introduce further distortions, because of the expansion coefficients of the materials forming the mask [117, 118, 119, 120]. Again, computer modeling is necessary for detailed maps of the distortion caused by mask heating [121], but we can draw some general conclusions from the properties of the materials. For uniform materials and temperature fields, the linear distortion is simply given by

$$\Delta X = X\alpha\Delta T, \ \alpha \approx 10^{-6} \tag{3.38}$$
$$X = 25\,mm, \ \Delta X \leq 5\,nm \to \Delta T \leq 0.2\,^{o}C \tag{3.39}$$

for most material, we need to keep the temperature uniformity to better than 0.1 ^{o}C to remain within the error budget. Additionally, the difference in expansion coefficient between absorber and carrier may introduce pattern dependent distortions. The same is true if the temperature field is nonuniform. In any case, the key is to reduce and minimize the temperature fields in the mask.

The origin of temperature variations comes from the energy absorbed from the X-rays beam. The membrane itself typically absorbs 30 to 50% of the radiation; the absorber may reach close to 100% for masks with high coverage. This corresponds to an input of several tens of mW/cm^2 and would cause a rapid heating on a system of such small thermal mass. In the case of a syncrotron radiation scanning beam, the moving hot spot will create a strain field that follows its motion [118]. When the beam is in a nonsymmetric position, the membrane expansion will lead to a change position because of the unequal forces. However, the mask membrane and wafer are held at the same temperature by the gas in the gap region [117]. The distance of a few tens of microns provides an efficient cooling path, capable of dissipating the input heat into the thermally massive wafer, provided that the exchange gas has a good thermal conductivity. Helium is typically chosen for this reason, and also to decrease X-ray losses; in a well-designed system the temperature rise can be easily contained to less than $0.1\ ^{\circ}C$. Experimental evidence supports the fact that temperature rise can be minimized.

A finer point has to with the *absolute* mask temperature, since it may be different in the patterning tool (e-beam) than in the stepper, or even in the metrology tool. Typically in a well-designed stepper it is possible to adjust the temperature to whatever value is necessary, providing the baseline correction.

3.2.2.4 Mask Fabrication Process

Details of the mask fabrication process have been reported several times. Here we will use as a guideline the process developed at CXrL [122], and then discuss those developed at other centers.

Membrane   The membrane is formed by depositing $SiNH$ by LPCVD on a batch of Si(100) wafers, with the LPCVD parameters adjusted to provide the required tensile stress $\sigma = 10^8 - 10^9 dynes/cm^2$ and high optical transparency ($T > 60\%$) [123]. The wafer is then bonded to the frame by using either an epoxy or an anodic bonding. This bonding operation is crucial because it may introduce and/or fix bowing and/or warping in the final mask [124]. It can also be used to *remove* bowing, by using for instance a glass ring with a smaller coefficient of expansion than the silicon wafer, and a moderate temperature bonding process. Glasses with expansion coefficients lower than (Hoya SD-1), equal to (SD-2) and larger than (SD-3) silicon exist commercially [125]. The back window, defining the membrane area, is defined by RIE after lithography; care must be exercised so that the sides of the square are properly aligned with the crystallographic directions. A KOH anisotropic etch follows to remove the substrate and free the membrane. A square membrane is preferred (unless one uses flip bonding [104, 126]) because no sharp features (ragged edge) should be present at the edge between Si and membrane; edge raggedness seriously affect the membrane strength. The membrane is then inspected for pinholes and other defects. One of the key parameters in the blank is the amount of bow, caused by the pull of the membrane onto the wafer and frame [124]. The final amount is determined by the torsional rigidity of the frame cross-section and by the membrane perimeter force. The membrane is flat to a fraction of a wavelength of visible light, but the bow in the frame can reach several microns; while it doesn't affect the image formation, it may create problems of interference with the wafer if the bow is too pronounced. A membrane $25 \times 25\ mm^2$, 2 μm thick and with a stress of $10^9\ dynes/cm^2$ may produce a rotation in the frame leading to a bow of tens of microns [124]. A flat mask blank can be achieved by controlling the stress distribution in the wafer and in the frame. The cross-section shown in Fig. 3.27 is stiff enough to reduce the bow to a fraction of a micron, if the process steps are designed correctly [124]. In the U.S., the activity has concentrated on the standard frame proposed by NIST. In Japan, no convergence exists yet, but several approaches are being tried.

Other materials used for membranes are *boron-doped silicon, [127], polysilicon [128], boron nitride, [129], diamond [130] and SiC [131]*. Of all these materials, SiC is currently the material of choice because it has the highest radiation resistance [132]. Over the course of a typical lifetime, the mask is exposed to huge amounts of radiation (in excess of $1.0 \cdot 10^6$ J/cm^2 for a 6 months lifetime) and must remain stable without any alterations in mechanical properties that may lead to pattern shifts. The absorbed energy may involve breaking of bonds, and the following de-excitation must return the material to the same initial ground state configuration. If strained bonds are present, the material may relax in a different configuration [133], leading to atomic motions and, ultimately, macroscopic pattern distortions. Silicon nitride damages quite easily, particularly if oxygen is present [134, 135]. So far, SiC fits the bill of the ideal absorber for ULSI lithography [132]; however, nonuniform irradiation of the membrane may lead to unacceptable displacements even in radiation-hard materials [136] so that aperturing may not be an acceptable practice.

Since an alignment system typically looks at the marks through the membrane, it is important that the optical transparency be good. It is also important that the reflectivity be low, so that the mark detection S/N is good. Surface roughness affects the alignment signal and should be minimized. The large index of refraction of SiC results in a large reflectivity so that an antireflection layer may be necessary to improve the alignment signal [137]. This is typically an indium-tin-oxide (ITO) layer.

The absorber must be patterned in a high-Z material in order to have a large density of electrons to absorb the X-rays. A look at the periodic table shows that Au, Pt, Ta and W are some of the highest density materials and thus suitable candidates, particularly because of their well known processing characteristics. For instance, U and Pb are not, because of inferior materials properties in terms of chemical and mechanical stability.

Absorber: Additive   Gold has been one of the most widely used materials for X-ray absorbers [122, 138, 139, 140, 141, 142, 143, 144]. Gold tends to etch isotropically so that it is difficult if not impossible to achieve sharp walls in a subtractive process, and it is used exclusively as a plating metal. It does not oxidize, it is chemically stable and it is easy to apply by electrolysis. A plating base is applied to the mask blank by evaporating a Cr/Au bilayer. An e-beam resist is spun on and patterned with a suitable tool. After exposure, the blank is developed. If a chemically amplified resist is used, particular care must be exercised to avoid temperature nonuniformities in the postbake step; the best results are achieved by using a specially designed chuck with a helium heat-exchanging medium [145]. After development, a light plasma etching may be necessary to ensure that the plating base is clean. The mask blank is then plated to the required thickness of Au. The plating operation is delicate, and several approaches have been developed. The thin plating base may result in nonuniform potential distribution because of resistance paths, and thus the plating may yield nonuniform thickness. A solution is to use large dummy pads in the plating area so that the pattern is a small change to the plating area. Another is to perform a calibration. In general, the plating must be configured properly to avoid absorber thickness variations. Both DC and pulsed plating techniques have been used, as well as flowing or stationary baths. Today commercial cyanide-free baths with Th brightener are used [122]; the brightener concentration can be used to control the metal stress.

After plating, the resist is removed and/or ashed. In the ashing, the mask must not heat above the Au phase transition, around 70 $^o C$, because of the resulting large pattern distortions [140]. An example of such a plating process is shown in Fig. 3.33, for the case of a 0.25 μm SRAM pattern fabricated by IBM for the U.S. Advanced Lithography Program (ALP); IBM's Advanced Mask Facility can be considered to be the Center with the widest knowledge of Au electroplating. Sometimes materials such as Ni are used for harder X-ray masks in special applications for diffractive X-ray optics [146].

Absorber: Subtractive There are more choices for absorber materials in the subtractive process. First, one must decide on which absorber to use. W is the most commonly used material [147, 148, 149, 150, 151, 152], but Ta is also widely used [153, 154]. W and Au have similar optical properties but are very different from a materials point of view. W can be deposited by physical evaporation, sputtering and CVD. W can be reactive ion etched (RIE) anisotropically, yielding very well controlled sidewalls [149], Fig. 3.34; this is not possible in Au. The same applies to Ta and other refractory materials. W films may end up with a very high built-in stress, as well as gas inclusions. Depending on the deposition conditions, one can have W films which are microcrystalline or have columnar structures [108]. The most important parameter is the stress, which can vary from compressive to highly tensile. As discussed above, σ_i should be as low as possible, and certainly below 10^8 $dynes/cm^2$. For normal W sputtering, the stress is a strong function of the Ar pressure; in-situ monitoring [155] may not be accurate enough, and postprocessing may be necessary to yield the required stress. Even more important that average stress values, nonuniformities in the deposition conditions may lead to nonuniform stress in the membrane, and later to distortions after the pattern etching [156]. Alloys of refractory metals are very interesting, and studied mostly in Japan [157]. These materials have an amorphous structure and produce low stress films over a very wide range of deposition conditions. Recently, Tantalum Silicon Nitride amorphous absorbers have been used for absorbers [158]. Back-side cooling may be necessary to avoid nonuniform etch rates [156], and the RIE conditions must be accurately tuned [149].

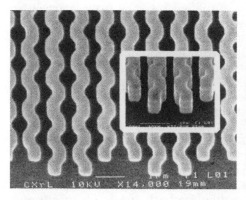 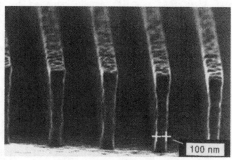

Figure 3.33: Advanced Lithography Program (ALP) X-ray mask fabricated at IBM. The inset shows a detailed side view of the topography in the mask.

Figure 3.34: W subtractive process mask fabricated at NRL's Nano Processing Facility

3.2.2.5 Pattern Inspection and Repair

The inspection of the X-ray masks includes the detection of defects, the measurement of the linewidth (LW) distribution and of the positional accuracy of the pattern. They are three different types of measurement. Repair [160] is applied when the number of defects is small enough to be economically feasible.

Manufacturing quality masks must be defect free. As usual, we define clear defects (missing absorber) and dark defects (extra absorber). KLA has recently developed a high-throughput inspection tool, the Semspec [159], that allows the inspection of the mask. Defects as small as 0.030 μm can be detected and their position marked. In parallel, Micrion has developed a sophisticated repair tool based on focused ion beams [161] that can read the SEMSPEC data and then repair the defect. A dark defect is eliminated by local sputtering, a clear by local deposition of W from an organometallic, followed by ion sputtering [160]. Some re-deposition may occur unless the process is optimized.

The LW is measured by top-down SEM, although TEM-like methods have been developed at NIST [162]; these measurements are particularly valuable because they yield more information on the absorber distribution. X-ray masks tend to have absorbers with some degree of slope, and it is important that the angle be known.

3.2.2.6 X-ray Mask Design Variations

Over the years, several designs of X-ray masks have been introduced, typically following an organization-specific approach. In the U.S., there is a general convergence to a mask process whereby the membrane is formed first, by back-etching the *Si* substrate, and then patterned *(Patterning After Etching, PAE)*. In Japan, instead, the membrane is formed after patterning the absorber *(Patterning Before Etching, PBE)*. At first, PAE may appear to be superior to PBE because unwanted displacements may be introduced in the masks by nonuniform absorber stress when the membrane is released. However, if the stress is controlled, the displacements may be minimized. In general, U.S. technology favors thinner wafers and wider glass rings to provide the required rigidity, while in Japan a thicker wafer is often used. Various types of mask structures have been described in the past, and here we will review briefly the main ones, referring to the literature for more in-depth discussion. In general, the mask must present a membrane surface as flat as possible (less than 1 wavelength of He-Ne), with no interference from other structures in the gap. This means that the remaining wafer and/or the glass frame must have either a convex shape or be recessed. The following groups have developed complete in-house mask fabrication processes:

Hewlett-Packard, USA We mention this early mask only for historic reasons; it was based on a BNH film that we now know is unstable under high doses of X-rays. After deposition, the BNH-coated wafer was bonded to a glass frame *on the BNH film side* and the wafer fully removed. This left a BNH film bonded to a glass ring. Both W and Au were used, and the activity led to the first studies in stress distribution and placement accuracy [104, 147].

IBM, USA The IBM mask fabrication process was set quite early [163], and is continually being refined. A recent IBM mask is shown in Fig. 3.28. The membrane is formed using a Si:B film [127], and a glass ring frame with a "lip," clearly visible in the photograph, where the kinematic mounts are set. The shape of the frame has evolved in the NIST format. Si:B is far from ideal, and the large number of defects and dislocations make alignment very difficult. IBM had to use off-board alignment windows, increasing the complexity and the overlay errors. Fortunately, the mask frame is stiff, and its biggest merit is that it could be adapted easily to kinematic mounts. Being conductive, the Si:B substrate is well suited for plating. The mask is easily adaptable to other absorbers, and today IBM is moving to a SiC based system. IBM has fabricated thousands of these masks, for product development as well as for research recently obtaining very encouraging results in terms of defect control ("perfect" masks have been achieved) and positioning accuracy [112].

Hampshire Instr., USA Although the company has disappeared, its mask should be mentioned because of its robust design. Based on a 3" wafer, it took advantage of the increased torsional rigidity of the glass ring. Both Au and W were used as absorbers [164]. The 3" format continues to be used by MIT, AT&T and Motorola.

AT&T, USA After an early process based on BNH, AT&T has developed an innovative mask technology, based on the use of polysilicon films deposited on quartz wafers. The quartz is then etched to define the membrane. Polysilicon is very strong, and thin films can be formed with good performances. Both W [165] and Au [144] absorbers, with a format suitable for the use of the point source stepper developed at Hampshire, now at AT&T.

Fraunhofer Institute, Germany This format defined the European standard, and was based on a 4" wafer and glass ring. It allowed for easy fabrication but was prone to distortions due to the (relatively) weak glass ring and the vacuum chucking used. Several membrane materials were used. Nevertheless, it was used for several years at Bessy. SiC was also used as a membrane material, together with W and Au as absorbers. Stress compensation was achieved by ion-implanting the absorber layer [166]. Unfortunately, recently the activity has been discontinued.

NTT, Japan This mask structure represents a sophisticated combination of a thick wafer and low-stress membrane (SiNH) with a Ta absorber. This structure has *a single mounting point*, where a drop of epoxy is used to mount the membrane blank on the glass frame (square). This elegantly eliminates distortions due to stress in the frame, but raises questions about stability during e-beam patterning. NTT has fabricated hundreds of these masks, and the process has been transferred to a commercial company. [167]

Mitsubishi Electric, Japan This mask is based on SiC, both for the membrane *and* for the frame. The design yields a very stiff system, together with the radiation resistance of SiC [168]. W:Ti alloys are used for the absorber [157].

MIT, USA The MIT group has produced several innovations in the course of their extensive research program. Here we mention in particular the *mesa-mask*, whereby the silicon wafer itself is recessed relative to the membrane plane. This design is particularly attractive for nanotechnology, and can be easily adapted to various configurations. Recently, a *flip-design* has also been introduced [126], to reduce the risk of bow and allow smaller gaps. Both Au and low-stress W are used as absorbers for the SiNH membranes [150].

NRL, USA The Naval Research Laboratory has a very active programs in advanced lithographies, and in particular in X-ray mask fabrication. Both W and Au absorbers are commonly used, but a large effort has been invested in the development of high-quality W patterning [149].

CXrL, USA The Center for X-ray Lithography process uses a gold absorber and a SiNH membrane to provide a low-cost high-quality mask. The main weakness is in the radiation damage of the membrane, although the process could be easily extended to SiC membranes. A NIST ring is used to keep a minimum-force condition during exposure in the e-beam system as well as in the X-ray tool. These masks are used for process characterization, for device fabrication [14] and for GaAs MMIC development [122].

L2M, France The group at Baigneux has developed an extensive mask technology, with a focus on nanostructures using both Au and W absorbers. The format is for use on the Suss 200/2 stepper installed at LURE [169].

IESS, Italy IESS, in Rome, has established a process based on gold on silicon nitride [143]. High-resolution Fresnel Zone Plates masks have been succesfully generated [146], as well as ULSI test structures. Masks for the study of image formation have been fabricated [30, 40].

Many other types of masks have been described, with a wide variety of results.

3.2.2.7 Global Mask Optimization

The short wavelengths used in X-ray lithography produce sharp images but may cause some problems because the diffraction phase shift causes unwanted features (similar to speckle, Fig. 3.13). In lithography, the resist acts as a hard-limiter (threshold detector) so that a change in exposure dose will result in a change in linewidth because of the finite slope at the threshold image position. If the slope is smooth, then the variation is predictable and monotonic. If it is not smooth, such as when high-frequency "ripples" are present, then the linewidth may change wantonly with dose, complicating process control. A low-frequency pass filter is needed in order to produce a smooth image, and a careful understanding of the process is needed in order to optimize the exposure conditions.

We have demonstrated in a series of papers [6, 31, 40, 170, 171] that the accurate image formation model depends on many factors such as the spectral bandwidth, the spatial coherence of the illumination, the mask modulation and phase shift, the proximity gap between mask and wafer, and the

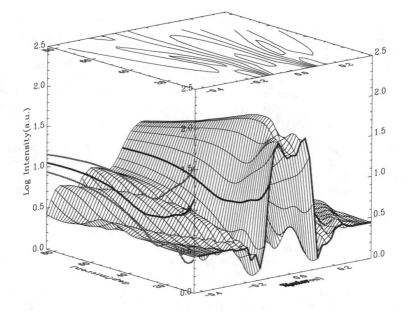

Figure 3.35: Evolution of the aerial image (intensity on z-axis vs. lateral position on front axis) with the gap (left horizontal axis, increasing from 10 to 80 μm). Notice the highlight of the intensity corresponding to the 0.25 μm line. The projection on the left panel of the 0.25 μm line and of the two lines corresponding to 0.275 and 0.225 μm defines the range of doses for which we have an acceptable linewidth, that is, the exposure window.

electron scattering in the photoresist. The role for the absorber is to provide the contrast and phase shift necessary to form the best image, i.e., to maintain image fidelity and linewidth control. By exposure window we define the region in the exposure-gap space (equivalent to exposure-defocus in optical lithography, [173]) where a given tolerance can be maintained, as shown in Fig. 3.35.

H.I. Smith remarked several years ago [36] that a π phase shift will produce an image with sharper "walls," i.e., better contrast. This is true, but we have found that the requirement for a π shift is only one of several needed to deliver a large exposure latitude. The combination of these factors can increase the exposure window greatly and at the same time make the mask fabrication process *easier*.

Absorber Optimization: Modeling of Walls Slope and Bias For the synchrotron spectrum generally used, the thickness of the mask absorber is usually in the range of 0.3 $\sim$ 0.5 μm to provide the contrast of $\sim$ 4 to 10, depending on the spectrum and absorber material. Simulation has shown [6, 8, 38], and recent experiments have confirmed [52], that thinner absorbers with a π phase shift give steeper image edge profiles. However, the π phase shift is usually accompanied by a lower contrast (<10) in the image, which could give rise to ghost features, particularly in two-dimensional structures. Typically, these ghost features are due to higher spatial frequency terms.

There are two ways to suppress ghosts and increase the exposure window

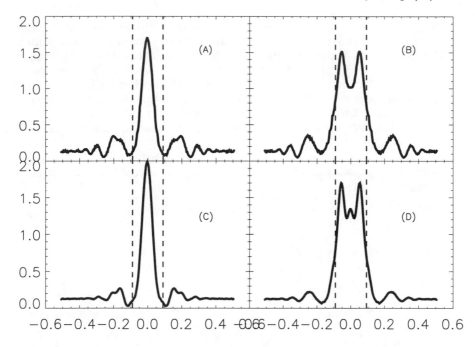

Figure 3.36: Image formation using BPM. (A), no bias or slope; (B) bias of 50 *nm*, (C) slope of 60 nm over a 450 nm thick absorber (about 8°), (D) slope and bias. The best case (sharper modulation and reduced ghosts) is obtained by the combination of the two.

at the same time, both based on the removal of the higher frequencies in the image. The first is to create a smooth modulation in the electric field transmitted by the mask rather than a sudden transition at the mask edges; the second is to reduce the spatial coherence.

The truncation in the electric field typical of an (ideal) binary absorber is never observed in practice. In the fabrication process of X-ray masks, it is very difficult if not impossible to obtain rectangular absorber profiles with square corners and exactly 90° sidewalls. In making the mask with an etching process, sloped sidewalls usually result and efforts are made to eliminate them. In a *Au* plating process, a rounded top and sometimes a foot are usually present [162]. But these "imperfections" produce a smoothly modulated field coming out of the mask and the very high spatial frequency components are removed, thus eliminating sidelobes and Poisson spots caused by field truncation [40]. We notice that the effect of the absorber variations is different from that of illumination (see below) because it acts on the field (i.e., before diffraction) rather than on the intensity (i.e., at the resist). In conclusion, an absorber with a profile formed by smooth corners and possibly a small edge slope will produce a better image [40]. From the manufacturing point of view, this also makes the requirements on mask fabrication process less difficult. Another important result of our activity was the clear definition of the need to introduce a bias in the mask features [30]. The effects of introducing these two factors (bias and slope) are illustrated in Fig. 3.36.

The "optimal" value for the bias, slope and thickness is not obvious.

Among other things, one should have a clear definition for the meaning of "optimal". It is possible to perform an efficient search for the combination of bias, slope and thickness that provides the best result – for instance, the widest exposure range for a set of different features with a defocus of 10 μm[39, 40], with the results shown in Fig. 3.37 and Fig. 3.38. Very interestingly, the modeling points out that a clear optimum thickness of absorber is found around 0.3-0.4 μm, while the bias has relatively less influence. This points out the need for a correct mask design in order to achieve the required image properties.

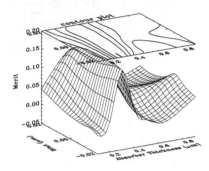

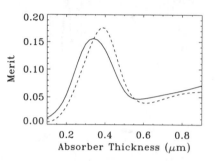

Figure 3.37: Figure of merit in function of bias and thickness of mask for a 0.18 μm feature; notice the small dependence on bias

Figure 3.38: FOM in function of thickness at optimum bias; notice the strong peak at $t \approx 350 nm$

Partial Coherence The effect of coherence in optical lithography is fairly well understood from a large amount of simulation and experimental work. In optical lithography, reduced spatial coherence gives a larger exposure window for the same CD and helps to eliminate standing waves formed in the resist by interference of coherent light. But in X-ray proximity lithography, it was believed that the penumbra due to a finite light source size was a limiting factor to resolution [172]. Our analysis has concluded exactly the opposite [6, 38, 30]. In X-ray lithography, a moderate amount of blur (reduced spatial coherence) is a positive factor in forming an image with a large exposure window (exposure latitude and depth of focus). Essentially, the diffraction process is a linearly shift invariant process, so that incoherent components can be convoluted with the diffracted image. The optimum blur size (3σ for a Gaussian source) should be around $1/3 \sim 1/2$ the feature size [30].

We have also shown that reduced spatial coherence has the same effect as any other blur sources such as vibration and photoelectron scattering in the resist [43, 44]. Hence, we can distinguish two types of blur that have an influence in the image formation: that due to the finite degree of spatial coherence of the source and that due to other effects. To be more specific, random vibrations and other perturbations in the optical setup happen on a time scale much longer than the propagation of the radiation so that they can be convolved with the intensity of the image. We notice that pure sinusoidal vibrations will affect adversely the linewidth, because of the shape of the time-dependent blur. Photoelectrons and chemical species diffusion,

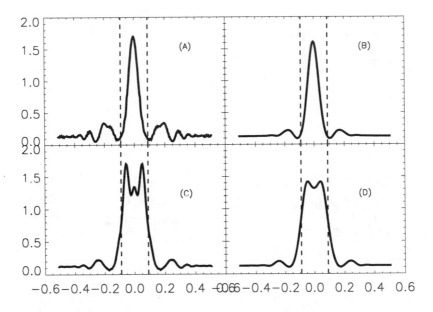

Figure 3.39: Effect of partial coherence in proximity X-ray lithography. Panel (a): no blur, bias or slope (but full syncrotron radiation source), (B) with 90 nm of blur, (C) slope and bias, (D) slope, bias and blur. The calculations refer to an 0.18 μm standard gold mask on a syncrotron radiation beamline. Similar results are also obtained for plasma sources.

which have a gaussian-like behaviour, play a similar role. All in all, these broadening processes are beneficial – within limits – in that they remove the highest spatial frequencies. These factors made X-ray lithography perform much better than previous simulation had predicted [173]. The results of these effects on the image formation are illustrated in Fig. 3.39.

As an experimental verification, we have performed a dose-gap matrix study [174] on the Suss XRS-200/1 at CXrL. Hence, a single 6" wafer will contain the whole map for a given resist processing conditions. The metrology was performed at Sematech on an AMRAY 1880FE SEM. The results, for a set of structures, are presented in Fig. 3.40. The exposure conditions were not the best, and the gold thickness was too large. That notwithstanding, the results show excellent linewidth control, particularly for the L/S case, as shown in the figure. The agreement between theory and experiment is also excellent. A similar experiment was recently performed on the Suss 200/2m of CXrL, using APEX-E resist [175].

3.2.2.8   Optimal Exposure Conditions

The conclusions of our activity can be summarized in a "recipe" for an optimized X-ray mask strategy. Table 3.4 illustrates our conclusions. The exposure window predicted, based on these parameters, is shown in Fig. 3.41. The values are not too different for the case of 0.18 μm process.

Fig. 3.41 shows how different features can be printed simultaneously with broad latitude. The results of our modeling efforts have been validated

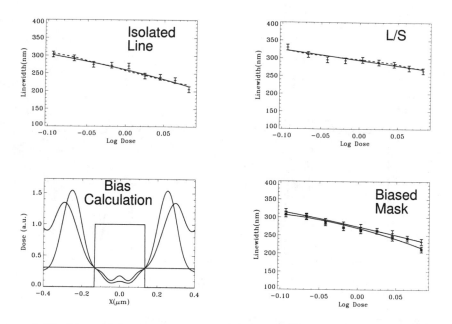

Figure 3.40: Linewidth variation studies for isolated and dense lines. The dotted curve shows the theory and the crosses the experimental data. The agreement between data and theory is excellent; noticed the shift .between isolated and l/s cases, and its elimination by biasing

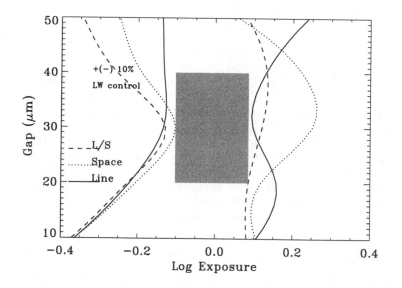

Figure 3.41: Exposure window for optimized mask at 0.25 μm. Notice the wide range of gaps (depth of focus) and the tolerance to exposure variations, in excess of 40%. Similar results can be achieved at 0.18 μm

Guidelines for Optimized
0.25 μm and 0.18 μm X-ray lithography mask

Parameter	0.25 μm	0.18 μm
Central Energy (nm)	≈ 0.83	same
Absorber Thickness (nm)	400	350
Sidewall slope	$\approx 56/400\ (8^\circ)$	same
Usable Proximity gap (μm)	5-40	5-30
Angular blur (mrads, 3σ)	4.5	4.5
Constant blur (nm, 3σ)	≈ 30	21
Mask bias (nm/edge)	30	20

Table 3.4: Optimized X-ray exposure parameters.

independently by MIT [8] and by other work at IBM. Recently, an IBM group has verified experimentally our predictions that a thinner Au mask should exhibit a larger process latitude than those with thicker Au. [52, 31] In general, similar recipes apply to the case of 0.18 μm lithography, although the optimization should be reevaluated for the specific conditions of interest.

3.2.2.9 Suppliers

Blanks can be obtained from Hoya Corp. [176], with various types of membranes and frames. In the U.S., MTC [177] supplies SiC films, and Nanostructures [178] has developed various types of X-ray mask processes. The main supplier of finished X-ray masks is certainly IBM's Advanced Mask Facility with NIST format and Au absorber [179]. Various degrees of certification are available. NTT spin-off NATC [180] also supplies finished Ta-absorber masks.

3.2.3 X-ray Steppers or Aligners

An X-ray aligner is a simplified version of an optical stepper. Simplified, because it does not need the large and complex imaging lens and because the mask is held in close proximity to the wafer, facilitates alignment. The X-ray stepper is based on a sophisticated mechanical stage, held vertical to match the pattern of the syncrotron radiation. Fig. 3.42 and Fig. 3.43 show the latest SAL stepper being commissioned at CXrL. Several companies have attempted the development of X-rays aligners for commercial production [164, 181]; currently SVGL [182], Canon [183], SHI [184] and SAL [115] are active in this area. Many more experimental aligners have been developed [185, 186, 187]. New tools are being developed (SAL, SVGL, Canon) at the time of writing, that are a radical departure from previous approaches. For instance, SAL is developing the new XRS-200/4, using a scanning beamline designed at CXrL.

Today, there is a consensus that the X-ray stepper should have a vertical stage to accommodate both the syncrotron radiation and point sources, in order to reduce development costs. In the case of point sources, barring

Figure 3.42: SAL Mod4 stepper during installation. The "butterfly" is the wafer stage, while the mask is held in the other granite frame.

Figure 3.43: Front view of SAL Mod4 stepper, without environmental chamber in place.

dramatic developments in average power delivered, a collimator is required to form a parallel beam of X-rays. From the point of view of the exposure, the characteristics of the stepper are no different from those of an optical tool:

		0.25 μm	0.18 μm
Exposure time	s	1	1
Step. & Align time	s	1	1
Wafer size	in	8	12
Exposure field	mm^2	50×50	50×50
Mask type		NIST	NIST
Gap	μm	50-10	50-5
Alignment	nm	30	20

Table 3.5: Specification of X-ray Steppers

The alignment systems are discussed below. Given a good alignment error detection system, the stepper must then implement the necessary fine motions. The mechanical stage performances and the feedback loop implemented are critical for the achievement of the required positioning accuracy. In order to maintain the error within the allocated budget entry, the sources of overlay error must be analyzed in detail, from the beamline to the mask stress and thermal changes. The mask conditions have a strong influence on the exposure conditions. Because of the high compliance of the membrane, small forces or perturbations may be enough to introduce unacceptable strains. Conversely, the judicious creation of strain fields can be used to reduce and/or eliminate long-range systematic distortions. The preferred mask design (in the U.S.) is the NIST format previously described. Kinematic mounts have been demonstrated to be capable of holding the mask with high accuracy and are being implemented in both the SVGL and SAL steppers. The temperature control of the environment around the mask is as critical as in many other processing steps. In general it is agreed that an *He* atmosphere is necessary in the region surrounding the exposure area. It is unclear if still environments are to be preferred to flow conditions. Karl Suss and now SAL [115] steppers use a flow of He directed along the X-ray path to the mask. Canon [183] is using a reduced pressure. Experience will tell which approach is more effective. Finally, the condenser optics in the beamline may affect the image placement because of non collimated beams, having higher order aberrations [85, 188]; these can be eliminated by using the correct condenser optics [85].

3.2.3.1 Alignment Error Detection System

The alignment system requirements of XRL are the same as those of optical lithography: the aerial image must be in registry with the previous level to about 1/3 of the CD. This value includes all contributions, and hence the aligner itself is allowed only a small fraction. For 250 nm CD, the overall error is about 80 nm. Of these, only 50 nm are allocated to the aligner tool and a correspondingly smaller fraction to the alignment system. The data accumulated to date indicate clearly that the overlay is mask limited, rather than stepper limited.

The detection of the alignment error can be based on imaging or non-imaging (interferometric) methods. Most systems have demonstrated errors of less than 15 nm, and some less than 10. The alignment strategy is very important. There is a consensus that at least 4 marks per field are necessary to allow for redundancy and robustness. Field-by-field, global and enhanced global are also implemented [189]. The mask mark should be located on the membrane itself, rather than outboard, in order to allow for measurement of distortions. An interesting approach has been proposed by SVGL, based on a mutual mark detection scheme between mask and wafer [190]. Imaging systems have been pioneered by Suss [191], with the development of the ALX-100. IBM used an imaging system in its early stepper at Brookhaven [4]. Intensity (nonimaging) systems have also been developed at AT&T, Micronix, Hampshire, CXrL (with the two-states aligner, [192]) and Canon. Imaging and intensity based systems are quite sensitive to the reflectivity of the membrane and marks on the wafer. Interferometric and Moiré systems have been developed at MIT [193], NTT [185], SORTEC [187], and others. Interferometric systems are susceptible to errors induced by multiple reflections but are well suited to remote operation because of the lack of short-focal length imaging optics.

After the error signal has been acquired, there are several strategies for compensating. A global alignment system is preferred in production because of shorter alignment time and the possibility of correcting for wafer field rotation and distortion, coupled with the possibility of accepting poor marks signals by extrapolating from the neighboring fields. With a magnification correction in place on the mask, it is also possible to compensate for trapezoidal distortions and reduce overlay error. Global systems capitalize on the short-term positioning accuracy of the stage, which is typically under laser interferometer control at $\lambda/1024$, i.e., $\approx 5 \ nm$. Particular care must be exercised to eliminate external influences on the laser optical path – in particular He admixture, since a 0.1% amount of He changes the path by a large amount.

Mark detection remains a critical test of the feasibility of alignment systems, and mask AR layers may be necessary to improve the S/N ratio for some types of masks. In this area, bright field systems are inferior to dark field or interferometric because of a lower S/N for marks on high-reflectivity substrates, although antireflection films can be used.

3.2.3.2 Suppliers

Of the current tools manufacturers, only SVGL, SAL, SHI and Canon are actively involved in developing exposure tools for 0.25 and 0.18 μm. In Japan, several companies have developed aligners, but mostly for in-house use. NTT has probably the most advanced steppers (SR-1 and SR-2). SORTEC and Toshiba have reported on their tools as well.

3.2.4 Resists

The development of X-ray resists has always lagged behind the development of optical and e-beam resists. In general, the large image modulation observed in XRL relaxes the requirements on the resists. Perhaps the most critical issue is that of resist speed. Most often, e-beam resists are also good X-ray resists since both are dose-imaging materials. Indeed, the mechanism of exposure is exactly the same in the two cases, with the difference that XRL uses an "internal" source of electrons (the photo- and Auger electrons) while EBL uses an external source (the beam of electrons). Of course, no bleaching and no standing waves are observed in X-ray resist images. After formation of the latent image, the process and evolution of the resist is exactly the same as that of normal photoresist [28]. Following the trend of optical resists, chemically amplified resists are widely used in XRL, mostly because of their high sensitivity [194, 195]. Because of the superior aerial image quality, X-ray resists are less burdened than their optical counterparts in delivering a sharp sidewall. Thick resists are also commonplace, with aspect ratios often exceeding 5, and with the limit being set by the material mechanical collapse rather than by the imaging [79]. Because of the excellent depth of focus of XRL, no surface imaging layer is necessary. Overall, the XRL resists process is much simpler than the corresponding optical or e-beam, and this accounts for a large fraction of the lower manufacturing cost of XRL. In XRL only the resist itself is necessary, with no bottom antireflection layer nor image enhancement processes. A topcoat may be necessary to avoid contamination of the acid from environmental pollutants but has no imaging role. In a word, the XRL resist process is simple: a single-layer resist coupled to a very sharp aerial image is a winning combination.

Because of the high modulation in X-ray lithography, the concept of critical MTF [28] loses some of its meaning. Indeed, as pointed out by Seeger [195] even low contrast resist materials can yield vertical walls. More critical is the background dissolution rate (i.e., the erosion of the unexposed material for a positive resist); hence, the "contrast" required is more that defined by the difference in development rate of the unexposed and fully exposed areas (rather than the more commonly used slope of the NRT, normalized remaining thickness, curve).

3.2.4.1 Resist Types

The main types of resist used today are chemically amplified. There are several in-house as well as commercial products.

AZ-PF 514 A positive material, widely used in XRL, with good sensitivity and resolution. Manufactured by Hoechst, it has excellent etch resistance, Fig. 3.44. [196]

APEX-E This IBM-developed DUV resist is now marketed by Shipley. It has very good resolution and exposure latitude, and performs very well in the X-ray region Fig. 3.46. [197]

SAL-605 This negative resist material is made by Shipley and performs very well to sub-100 nm dimensions, Fig. 3.45. [197]

PMMA It remains the material with the highest resolution and, unfortunately, poor etch resistance and sensitivity. Used mostly in mask making and advanced research. [198]

CANI Developed at NTT, it is a chemically amplified resist of high-resolution, Fig. 3.47 [199].

ZEP It is a metacrylate copolymer with excellent resolution and better etch resistance than PMMA. [200]

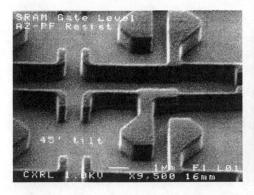 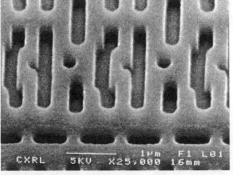

Figure 3.44: AZ-PF 514 Photoresist images   Figure 3.45: Images from SAL 605

3.3 Applications

XRL has been around for more than two decades, and has been applied to many types of problems in devices manufacturing. In general, we can divide the field in two broad areas which capitalize on different aspects: the first is more interested in high-resolution and high-volume, with typical applications being in the domain of ULSI devices for semiconductor processing. The second exploits instead the penetrating power, at modest resolutions, to fabricate exotic structures as those found in micromachining.

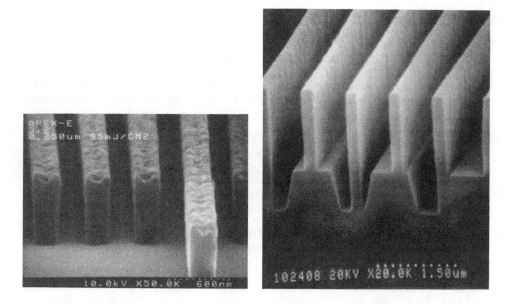

Figure 3.46: 0.25 μm APEX-E exposures obtained at IBM's ALF

Figure 3.47: High-resolution structures over large topography, exposed at NTT's SOR, using CANI

3.3.1 High Resolution Lithography

3.3.1.1 ULSI

By far, the main application in terms of commercial applications is in the area of semiconductor devices. As the dimension of the gate shrinks to less than 0.18 μm it is necessary to explore lithographic processes that are alternative to the optical mainstream. The realization of advanced devices was demonstrated very early on isolated devices [4], with sub 0.1 μm transistors fabricated at MIT as early as 1985 [201]. Large field 0.2 μm devices were routinely obtained at CXrL using several levels of X-rays exposures [202]. However, the achievement of large-area devices has been an elusive goal [10]. This is because of the requirement of low defect densities for multilevel circuits; defect-free masks are still difficult to manufacture. The lack of aligners with overlay better than $\approx 70 nm$ has also made demonstrations harder; indeed, it may be self defeating to try to demonstrate manufacturing capabilities without an adequate toolset. Given the lack of tools, one should then compare the technology in areas where meaningful experiments can be done. This is in process latitude and uniformity. For instance, IBM [203] and NTT [204] have both studied arrays of ring oscillators and demonstrated how the propagation delay distribution is much more clustered for XRL than for deep UV lithography. This is a direct result of the superior exposure latitude and process robustness of XRL. While these demonstrations used syncrotron radiation, advanced MOS structures have been fabricated with point sources at MIT [205] and AT&T [206]. Today, the most complex device has been manufactured at NTT, with 5 levels of XRL [11]. The largest area device record belongs to IBM with a 64 Mbit chip [10], and very recently Mitsubishi Electric has announced a 1 Gbit cell made with XRL [207]. The main message that emerges from the ULSI activity is that the gating factor is the defect density in the mask.

3.3.1.2 Nanolithography

MIT is clearly the leader in the application of XRL to nanolithography. Several types of quantum electron devices (QEDs) have been manufactured using the longer wavelength of point sources, and a microgap technique [208]. As shown in Fig. 3.18, the highest resolution is achieved at softer wavelengths and smaller gaps, although the dependence on the wavelength is quite weak [209]. An example of a *PRESTFET* gate is shown in Fig. 3.48. At the L2M center in France, 50 *nm* and smaller features have been repeatedly demonstrated for a wide variety of applications. An example of a 150nm gate-length HEMT is shown in Fig. 3.49 [210, 211]. XRL is the main candidate for the manufacturing of QED because of its parallel nature. An intriguing possibility is that of using phase shifting masks for defining extremely high-resolution lines Fig. 3.17. This technique, first introduced in optical lithography [51], has also been demonstrated in the X-ray domain, yielding structures as small as 400 Åwith good exposure latitude [55]. Since many QEDs rely on local very small features while the rest of the device is relatively sparse, X-ray lithography is a good candidate for the exposure of such structures.

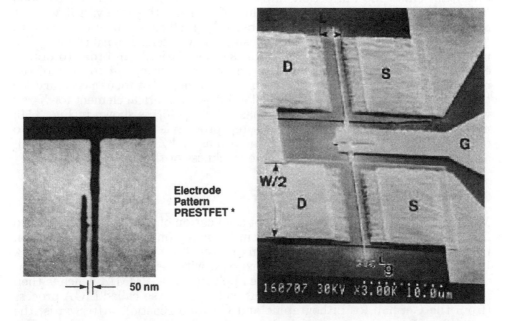

Figure 3.48: MIT's PRESTFET Transistor.   Figure 3.49: L2M's 150 nm gate HEMT.

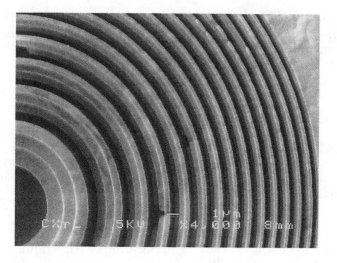

Figure 3.50: High efficiency Fresnel phase plate. The thinnest line is 0.1 μm. Notice the high aspect ratio and the multilevel structure.

3.3.1.3 Diffractive Optics

Diffractive optics are optical devices that capitalize on the ability of lithography to define arbitrary and well-controlled diffraction gratings. Typically, the pitch of gratings for the visible region is of the order of 200 to 1000 *lines/mm* since the diffraction limit for $\lambda = 0.5$ μm is reached at 4000 *lines/mm*. This is well within the limits of optical lithography, if it weren't for the efficiency issue. In order to obtain high efficiency it is necessary to "blaze" the grooves (lines) either by using subwavelength patterning (pulse modulation) or by controlled slope (phase modulation). In order to obtain efficient devices, sub-grooves need to be patterned from 1/4 to 1/16 of the main groove. XRL has the resolution and the depth of focus necessary for this kind of patterning. An example of a diffractive optical element for X-ray wavelengths (an optimized Fresnel phase plate, [146]) is shown in Fig. 3.50; notice the aligned exposures. A two-step process is often used in order to obtain a thick absorber (shifter), using a master FZP patterned with EBL as an X-ray mask [146] to increase the thickness of the exposed resist.

3.3.1.4 High Penetrating Power

Micromachining has become a very hot application of XRL; high aspect ratio lithography (HAL) was demonstrated in Germany on the Bonn Synchrotron [16]; it is based on the high penetrating power of very hard X-rays Fig. 3.3. Hundreds of microns of resist can be exposed with vertical profiles. The material almost exclusively used for these applications is PMMA, which is then used as a mold for the plating of various metals in the so-called LIGA process (from the German for Lithographic and Galvano Plastic). Afterwards, the metal form is used as a printing dye for transfer in other materials; in Europe, Microparts has emerged as a high-volume manufacturer of components based on the LIGA and stamping process, [212]).

At large thicknesses, the resist stress becomes very large and may lead to cracking. It is thus necessary to apply the resist in a very controlled way, with repeated annealing cycles to ensure a low-stress layer. Recently, a new process based on actual slabs of PMMA (Lucite) has been developed [17]. In a variation of the process first a layer of high viscosity and molecular weight PMMA is spun on the wafer. Then the slab is glued to this layer using a monomer solution as solvent. The exposure is performed normally, using filters to remove the softer part of the spectrum. The maximum thickness is determined by the absorption coefficient of PMMA, and by its crosslinking properties. When irradiated with very heavy doses, PMMA first decomposes to smaller weight fragments that may be repolymerized by further exposure. This leads to a three-dimensional linked material, hard to remove with any agent. The ratio between dissolution and crosslinking is about a factor $R = 10$ to 30. Since the *bottom* of the resist must be exposed to define the feature, the radiation absorbed at the top of the resist of thickness t is $\exp \mu t$ times that at the bottom. The mask must provide enough contrast not to expose the top of the (dark) areas, causing resist loss. This fixes the requirement on the absorber thickness, in general several microns or tens of microns [16]. The mask itself is much simpler than those found in ULSI XRL, with dimensions in the range of a few microns; often, a replica mask is formed with XRL to obtain a high-contrast absorber.

Mask and resist prepared, the exposure is a simple process, following the same lines as the ULSI proximity. Aspects ratios in excess of 100 can be achieved by using very hard X-rays. Stress control is perhaps the biggest challenge, because of the large changes in stress caused by the exposure process itself. Gas evolution may also be a problem in very thick slabs. After exposure, the exposed PMMA is developed in a high-contrast developer to maintain vertical structures. The next step is electroplating (for the LIGA), leading to the growth of a metal structure into the resist template. In this way many types of systems have been produced: embossing punches, magnetic motors, electrostatic motors, nozzles, filters and so on. Much activity is underway in Japan, the U.S. and Europe in this technology that has captured the imagination of scientists and laymen alike.

3.4 Status of X-ray lithography

The development of the X-ray lithography technology continues at a fast pace. The countries with the most advanced programs are the United States and Japan. Recently, both Korea and Taiwan have entered the field with programs based at their respective synchrotron sources, in Pohang and Hsinchu. Europe has fallen out of the development, with the exception of England where both LEICA and Oxford Instruments are located. European research is now concentrating on LIGA and MEMS, rather than on ULSI devices. In the U.S., the major industrial players are IBM, Loral, AT&T and Motorola, all members of the X-ray lithography Association [213, 214]. These companies focus on the use of the ALF facility at East Fishkill, and on the new SVGL X-ray stepper. AT&T is also interested in point sources, and continues to develop a system originally based on a stepper from Hampshire. Among the U.S. tools suppliers for X-ray lithography, only SVGL and SAL are left. SAL is involved in a development program with Sanders (now a Lockheed company) and CXrL, geared to the development of a new stepper (the SAL XRS-200/4) capable of operation with both point sources and synchrotrons. The companies involved in point source development are JAMAR and SRL. The last has developed a high-power high-density plasma source that will soon undergo lithographic tests [62]. JAMAR continue to develop a laser based plasma source [58]. The U.S. effort is led by ARPA, which funds and coordinate the activity [214].

Non industrial players include universities, DOD and DOE laboratories, in particular Sandia, Livermore and Lawrence Berkeley. DOE laboratories concentrate on EUV lithography. The Naval Research Laboratory in Washington, D.C., is a very active center of activities in mask fabrication and nanotechnology. Academic activity is centered at MIT, Wisconsin-Madison (CXrL) and Louisiana State (CAMD). MIT has the oldest program and focuses on point sources systems and nanolithography. CXrL uses the synchrotron source of the SRC for developing ULSI steppers and processes, while CAMD's activity concentrates on LIGA and microfabrication. The University of Florida has a strong program in EUV source development.

In Japan several companies are involved in XRL development. The main driver is certainly NTT (Atsugi), followed by Mitsubishi Electric Labs (MEL, Osaka). Many Japanese companies have developed their own accelerators and aligners, with very good results. Canon is developing a commercial stepper, with a prototype installed at MEL that has been recently used for manufacturing 1 Gbit cells (0.14 μm). The accelerator technology has been developed at SORTEC, whose machines have remarkable performances.

If we examine the status of the key components of XRL, we can reach the following conclusions.

Source This is a solved problem, with commercial sources available on the market. The challenge is now on cost reduction, and indeed the new rings from Sumitomo Heavy Industries (Aurora 2) and Oxford should see a lower sticker price. Technical issues solved.

Beamline Very efficient and cost effective beamlines have been designed. Technical issues solved.

Aligner Three new aligners are under test, all targeted to the 0.18 μm region, and beyond. No surprises are expected here, since the components (stage, alignment, etc.) have been demonstrated independently and are well understood.

Mask The mask fabrication has advanced in the area of patterning, where the 100 kV LEICA EBPG can be considered to be the ultimate tool available on the open market. Improvements in throughput, mounting and stress control are still needed. Inspection and repair have been demonstrated. It is still weak in the defect density area, mostly because of the lack of a commercial mask maker in the fabrication loop.

General infrastructure This too is lagging behind, mostly because of the contraction in investment on the part of some large end users which, in some case, have become competitors with tool and infrastructure suppliers. It is difficult to expect much development in this area until a clear commitment to manufacturing is seen.

Metrology Like in all other lithography areas, it is lagging behind. Metrology of the 1× mask is admittedly more difficult, but recent developments in the LMS2020 and in the Nikon 4I should come a long way in solving this common problem.

3.5 Conclusions

X-ray lithography has matured considerably. We now understand the physical processes in much greater detail; this understanding has led to simpler processes, such as in the case of X-ray mask absorbers. The development of a viable manufacturing technology is hampered by the lack of infrastructure. In a typical positive feedback loop, the lack of demand for XRL tools and processes keeps supplier companies from entering the market; in turn, the lack of tools makes end users more cautious about the technology. Perhaps the greatest obstacle to XRL is the obscurity of the practical aspects of the technology: only a small number of engineers has ever come in contact with it – and often the force of opinions is inversely proportional to the amount of knowledge. All in all, the number of people involved with XRL in the U.S., and in the world, is small and no commercial promoter exists. This is to be contrasted with the hundreds of engineers with experience in optical lithography, reinforced by a strong and well honed sales force. XRL can meet the 0.25 μm challenge as well as the forthcoming 0.18 μm generation, and can evolve easily to 0.15 μm. Indeed, it is the *only* technology for which the transition from generation to successive generation is incremental, rather than revolutionary. While it is clear that the end users do not have any great love for alternative lithographies, it is most likely around the 0.18 μm generation that they will have to leave behind the comfortable processes of optical lithographies. The simple fact is that XRL works and can deliver the required manufacturing processes. Time will tell if the equipment manufacturers can rise to the challenge, and if the market is ready to accept this new production tool.

3.6 Suggested Readings

Most of the papers related to X-ray lithography can be found in the following specialized conferences proceedings.

- **The International Conference on Electron, Ion and Photon Beams, Technology and Nanofabrication**, published in the Journal of Vacuum Science and Technology, usually the Nov/Dec. Issue.

- **Microprocess Conference Proceedings**, published in the *Japanese Journal of Applied Physics*

- **Proceedings of Micro- and Nano-Engineering**, published by Elsevier.

- **SPIE Symposia on X-ray, Electron and Ion Lithography**, published by SPIE.

Bibliography

[1] H.I. Smith, "100 Years of x-rays: impact on micro and nanofabrication", Journ. Vac. Sci. and Techn., **B13**, 2323 (1995)

[2] D. Spears and H.I. Smith, "High-resolution pattern replication using soft x-rays", Electr. Letters, **8**, 102 (1972)

[3] E. Spiller, "Early history of x-ray lithography at IBM", IBM Journ. Res. Devel., **37**, 287 (1993)

[4] A. Wilson, "X-ray lithography in IBM, 1980-1992, the development years", IBM Journ. Res. Devel., 302 (1993)

[5] W. Glendinning and F. Cerrina, "X-ray lithography", Handbook of VLSI Lithography, W.B. Glendinning, J.N. Helbert, Ed., Noyes (1991)

[6] J. Guo, F. Cerrina, "Modelling proximity lithography", IBM Journ. Res. Devel., **37**, 331 (1993)

[7] F. Cerrina, "Recent advances in X-ray Lithography", Jpn. Journ. of Appl. Phys., **31**, 4178 (1992)

[8] S.D. Hector, H.I. Smith, and M.L. Schattenburg, "Simultaneous optimization of spectrum,spatial coherence,gap,feature bias and absorber thickness in synchrotron-based x-ray lithography", Journ. Vac. Sci. and Techn.,**B11**, 2981 (1993).

[9] B.H. Koek, T. Chisholm, A.J. van Run and J. Romijn, "Sub-20nm Stitching and Overlay for Nano-Lithographic Applications", Jpn. Journ. of Appl. Phys., **33**, 6971 (1994)

[10] R. Dellaguardia, D. Puisto, R. Fair, L.W. Liebmann, T. Zell, D. Seeger, G.J. Collini, R. French, B.R. Vampatella, J.M. Rocque, S.C. Nash, A.C. Lamberti, F. Volkringer, J.M. Warlaumont, "Fabrication of a 64 Mbit DRAM using X-ray Lithography", SPIE Proc. 2144, 112 (1995)

[11] K. Deguchi, K. Miyoshi, H. Ban, T. Matsuda, T. Ohno, Y. Kado, "Fabrication of 0.2 μm large-scale integrated circuits using SR lithography", Journ. Vac. Sci. and Techn., **B13**, 3040 (1995)

[12] K. Fujii, T. Yoshihara, Y. Tanaka, K. Suzuki, T. Nakajima, T. Miyatake, E. Orita and K. Ito, "Applicability test for SR XRL in 64 Mbit DRAM fabrication process", Journ. Vac. Sci. and Techn., **B12**, 3949 (1994)

[13] A.A. Krasnoperova, S. Rhyner, E. Zhu, J.W. Taylor, F. Cerrina, and W. Waldo, "Modeling of a positive chemically amplified photoresist for x-ray lithography", SPIE Proceedings, **2194**, 198 (1994)

[14] R. Nachman, G. Chen, G. Wells, J. Wallace, H.H. Li, M. Reilly, A. Krasnoperova, P. Anderson, E. Brodsky, E. Ganin, S. Campbell, J. Taylor and F. Cerrina, "X-ray lithography processing at CXrL from beamline to quarter-micron NMOS devices", SPIE Proc., **2194**, 106 (1994)

[15] T. Wiegele, "Micro-turbo-generator design and fabrication", University of Wisconsin Thesis (1995)

[16] W. Ehrfeld and D. Munchmeyer, "Three-dimensional microfabrication using synchrotron radiation", Nucl. Instr.and Methods in Physics, **A303**, 523 (1991)

[17] H. Guckel, K. J. Skrobis, T. R. Christenson, J. Klein, "Micromechanics for actuators via deep x-ray lithography", SPIE Proc., **2194**, 2 (1994)

[18] F. Cerrina, "Ray tracing of x-ray optical systems: source models", SPIE Proc., **1140**, 330 (1989)

[19] N.A. Dyson, *X-rays in atomic and nuclear physics*, Longmanns Group Ltd., London (1973)

[20] A.A. Sokolov, I.M. Ternov, "Radiation from relativistic electrons", p. 82 and ff., Amer. Inst. of Phys., New York (1986)

[21] B.L. Henke, "Ultrasoft-x-ray reflection,refraction and production of photoelectrons(100-1000ev region)", Phys. Rev., **A6**, 94 (1972)

[22] G. S. Brown, S. Doniach, "The principles of x-ray absorption spectroscopy", *Synchrotron Radiation Research*, 353-383, H. Winick, Ed., Plenum (1980)

[23] W.D. Jackson, *Classical Electrodynamics*, John Wiley & Sons(1975)

[24] B.L. Henke, P. Lee, R.L. Shimabukuro T.J. Tanaka, and B.K. Fujikawa, "Low energy x-ray interaction coefficients:photoabsorption,scattering and reflection", *Atomic Data and Nuclear Data Tables*, **27**, 1 (1982)

[25] J.M. Auerbach and G. Tirsell, LLNL Rep. UCRL91230 (1984)

[26] D.T. Cromer, D. Liberman, "Anomalous dispersion calculations near to and on the long-wavelength side of an absorption edge", Acta Cryst., **A37**,267 (1981)

[27] A.G. Michette, *Optical systems for soft x-rays*, Plenum, London (1986)

[28] W.M. Moreau, *Semiconductor Lithography*, Plenum Press (1988)

[29] S. Turner, F. Cerrina, "Optimization of aerial image quality", Journ. Vac. Sci. and Techn., **B11**, 2446 (1993)

[30] J. Guo, Q. Leonard, F. Cerrina, E. DiFabrizio, L. Luciani, M. Gentili, and D. Gerold, "Experimental and theoretical study of image bias in XRL", Journ. Vac. Sci. and Techn., **B10**, 3150 (1992)

[31] F. Cerrina, J.Z.Y. Guo, S. Turner, L. Ocola, M. Khan and P. Anderson, "Image formation in x-ray lithography: process optimization", Microelectronic Engineering, **17**, 135 (1992)

[32] D. So, B. Lai, F. Cerrina, "The effect of beam emittances on x-ray lithography exposure line resolution", Journ. Vac. Sci. and Techn., **A5**, 1537 (1987)

[33] A. Sommerfeld, *Optics*, Academic Press, NY (1963)

[34] M. Born and E. Wolf, *Principles of Optics*, Pergamon (1980).

[35] Z.Y. Guo, G. Chen, M. Khan, V. White, S. Turner, P. Anderson and F. Cerrina, "Aerial image formation in XRL: The whole picture", Journ. Vac. Sci. and Techn., **B8**, 1551 (1990)

[36] Y.C. Ku, E.H. Anderson, M.L. Schattenburg, H.I. Smith, "Use of a pi-phase shifting x-ray mask to increase the intensity slope at feature edges", Journ. Vac. Sci. and Techn., **B6**, 150 (1988)

[37] Y. Yamakoshi, N. Atoda, K. Shimizu, T. Sato and Y. Shimizu, "High-resolution x-ray lithography using a phase mask", Appl. Optics, **25**, 928 (1986)

[38] Z.Y. Guo, F. Cerrina, "Verification of partially coherent light diffraction models in x-ray lithography", Journ. Vac. Sci. and Techn., **B9**, 3207 (1991)

[39] J. Xiao, M. Khan, R. Nachman, J. Wallace, Z. Chen and F. Cerrina, "Modeling image formation: application to mask optimization", Journ. Vac. Sci. and Techn., **B12**, 4038 (1994)

[40] M. Gentili, E. DiFabrizio, L. Grella, M. Baciocchi, L. Mastrogiacomo, R. Maggiora, J. Xiao and F. Cerrina, "Fabrication of controlled slope attenuated phase shift x-ray masks for 250 nm synchrotron lithography", Journ. Vac. Sci. and Techn., **B12**, 3954 (1994)

[41] A.A. Krasnoperova, M. Khan, W. Waldo and F. Cerrina, "Simulations of the development of negative chemically amplified photoresists", Proc. 10th Int. Conf. Photopolymers, 176-184 (1994)

[42] K. Murata, D.F. Kyser, Adv. Electr. Phys., **48**, 323 (1989)

[43] L.E. Ocola, F. Cerrina, "Parametric modeling of photoelectron effects in x-ray lithography", Journ. Vac. Sci. and Techn., **B11**, 2829 (1993)

[44] L.E. Ocola, F. Cerrina, "Parametric modeling at resist-substrate interfaces", Journ. Vac. Sci. and Techn., **B12**, 3986 (1994)

[45] T.E. Everhart, P.H. Hoff, "Determination of keV energy dissipation vs. penetration in solid materials", J. Appl. Phys., **42**, 5837 (1971)

[46] L. Capodieci, A. Krasnoperova, F. Cerrina, C. Lyons, C. Spence and K. Early, "Postexposure bake simulation for lithography process modeling", Journ. Vac. Sci. and Techn., **B13**, 2963 (1995)

[47] See, for instance, L. Levi, *Applied Optics*, Wiley (1964) for a complete discussion of various resolution criteria.

[48] G. Toraldo di Francia, "Resolving power and information", J. Opt. Soc. Amer., **45**, 497 (1955)

[49] A.W. Lohmann, D.P. Paris, "Superresolution for nonbirefringent objects", Appl. Optics, **3**, 1037 (1964)

[50] J.W. Goodman, *Introduction to Fourier Optics*, McGraw-Hill Book Company(1968).

[51] B. Lin, "The attenuated phase-shifting mask", Sol. State Techn., **35**,43(1992)

[52] M.M. McCord, A. Wagner, D. Seeger, "Effect of mask absorber thickness on x-ray exposure latitude and resolution", Journ. Vac. Sci. and Techn., **B11**, 2881 (1993)

[53] K. Deguchi, NTT, private communication(1995)

[54] B.V. Gnedenko, *Theory of Probability*, Chelsea, New York (1962)

[55] F. Cerrina, "The limits of patterning in X-ray Lithography", Mat. Res. Soc. Symp. Prco., **380**, 173 (1995)

[56] Z.Y. Guo, F. Cerrina, "Comparison of plasma and synchrotron sources", SPIE Proc., **1465**, 330 (1991)

[57] D.J. Nagel, R.R. Whitlock, J.R. Grieg, R.E. Pechacek, M.C. Peckerar, "Laser-plasma source for pulsed X-ray Lithography", SPIE Proc., **135**, 46 (1978)

[58] J. Maldonado, "Prospects for granular x-ray sources", SPIE Proc., **2523**, 2 (1995)

[59] J. Frackoviak, G.K. Celler, C.W. Jurgensen, R.R. Kola,A.E. Novembre L.E. Trimble, "Performance of the hampshire instruments model 5000 proximity x-ray stepper", SPIE Proc., **1924**, 258 (1993)

[60] D. A. Tichenor, G. D. Kubiak, M. E. Malinowski, R. H. Stulen, S. J. Haney, K. W. Berger, R. P. Nissen, G. A. Wilkerson, P. H. Paul, S. R. Birtola, P. S. Jin, R. W. Arling, A. K. Ray-Chaudhuri, W. C. Sweatt, W. W. Chow, J. E. Biorkholm, R.R. Freeman, M. D. Himel, A. A. MacDowell, D. M. Tennant, L. A. Fetter, 0. R. Wood II, W. K. Waskiewicz, D. L. White, D. L. Windt, T. E. Jewell, "Development of a laboratory extreme-ultraviolet lithography tool" SPIE Proc., **2194**, 95 (1994)

[61] D.G. Stearns, N.M. Ceglio, A.M. Hawriluk, R.S. Rosen, S.P. Vernon, "Multilayer optics for soft x-ray projection lithography: problems and prospects", SPIE Proc., **1465**, 80 (1991)

[62] R.R. Prasad, M. Krishnan, J. Mangano, P.A. Greene, N. Qi, "Neon dense plasma focus point x-ray source for sub-0.25 μm lithography", SPIE Proc., **2194**, 120 (1994)

[63] S.C. Plidden, M.R. Richter, D.A. Hammer, D.H. Kalantar, "1-kW x-pinch soft x-ray source", SPIE Proc., **2194**, 209 (1994)

[64] J. Chen, E. Panarella, B. Hilko, H. Chen, "Soft x-ray output from the spherical pinch plasma radiation source for microlithographic applications", SPIE Proc., **2194**, 231 (1994)

[65] W. Scharf, *Particle accelerators and their uses*, Harwood Academ. Publishers, New York (1986)

[66] G.Margaritondo, *Introduction to synchrotron radiation*, Oxford, New York (1988)

[67] H. Winick, "Overview of sychrotron radiation facilities outside the USA", in the *Proceedings of the 6th synchrotron radiation instrumentation conf.* Berkeley(1989) and Nucl. Instr.and Methods in Physics, **A291**, 487 (1990)

[68] Courtesy of J. Silverman, IBM East-Fishkill.

[69] C.N. Archie et al., "Installation and early operating experience with the helios compact SR X-ray source", Journ. Vac. Sci. and Techn., **B10**, 3224 (1992)

[70] D.E. Andrews, C.N. Archie, "Helios compact SR x-ray source: one year of operation at ALF", SPIE Proc., **1924**, 348 (1994)

[71] J.P. Silverman, C.N. Archie, J.M. Oberschmidt, and R.P. Rippstein, "Performance of a wide-field flux delivery system for synchrotron x-ray lithography", Journ. Vac. Sci. and Techn., **B11**, 2976 (1993)

[72] A very extensive treatment is found in Ref. [75], but the publication is hard to find. Excellent summaries are also presented by H. Winick in [73] and in Ref. [74]. A simplified treatment is found in Ref. [66].

[73] H. Winick, "Properties of synchrotron radiation", *Synchrotron Radiation Research*, p. 11, Plenum Press, New York (1980)

[74] S. Krinsky, "Characteristics of synchrotron radiation and of its sources", *Handbook on Synchrotron Radiation*, E. Koch, Ed., North Holland (1985)

[75] G.K. Green, Brookhaven National Laboratory Report BNL 50522, (1973)

[76] K. Chapman, B. Lai, F. Cerrina, "Modeling of undulator sources", Nucl. Instr. and Meth. in Physics Review, **A283**, 88-89 (1989)

[77] D. So, B. Lai, F. Cerrina, "A model for optimal beamline design for synchrotron radiation x-ray lithography", SPIE Proc., **773**, 30 (1987)

[78] T. Hosokawa,T. Kitayama, T. Hayasaka, et al., "NTT superconducting storage ring - super-ALIs", Rev. Sci. Instr., **60**, 1783 (1989)

[79] N. Atoda, "Progress of SR lithography – a path to 0.1 μm feature size", Proc. of Int. Conf. on Advanced Microelectronic Devices and Processing, Sendai (1994)

[80] R. Cole and F. Cerrina, "Novel toroidal mirror enhances x-ray lithography beamline at the Center for X-ray Lithography", SPIE Proceedings, **1465**, 111 (1991)

[81] W. Grobman, "Sychrotron radiation x-ray lithography", Synchrotron Radiation Research, H. Winick, Ed., Pergamon Press, NY(1985)

[82] M. Khan, L. Mohammad, L. Ocola, J. Xiao and F. Cerrina, "An updated system model for X-ray Lithography", Journ. Vac. Sci. and Techn., **B12**, 3930 (1994)

[83] J. Xiao and F. Cerrina, "Design of an aspheric mirror for synchrotron radiation x-ray lithgraphy beamline", Nucl. Instrum. and Methods in Physics Research, **A347**, 231 (1992)

[84] J. Xiao and F. Cerrina, "Effect of condenser mirror surface roughness on partially coherent image formation in X-ray Lithography", SPIE Proc., **2194**, 187 (1994)

[85] J. Xiao, F. Cerrina and R. Rippstein, "Novel single mirror condenser for X-ray Lithography beamlines", Journ. Vac. Sci. and Techn., **B12**, 4018 (1994)

[86] E. L. Brodsky, "The mechanical design of thin beryllium windows for synchrotron radiation", Nucl. Instr.and Methods in Physics, **A266**, 358 (1988)

[87] R. Engelstad, University of Wisconsin, private communication (1989)

[88] National Council on Radiation Protection and Measurements Rep. **39**(1971b).

[89] T. Kitayama, Mitsubishi Electric,private communication (1994)

[90] R. H. Hill, "The future costs of semiconductor lithography", Journ. Vac. Sci. and Techn., **B7**, 1387 (1989)

[91] A. Wilson, "X-ray lithography: can it be justified?", Solid State Tech., **29**, 249 (1986)

[92] K. Early, W. H. Arnold, "Cost of ownership for x-ray proximity lithography " SPIE Proc., **2194**, 22 (1994)

[93] G.K. Celler, J.R. Maldonado, "Materials aspects of X-ray Lithography", Mat. Res. Soc. Symp. Prco., **306**, (1993)

[94] J. Maldonado, "X-ray Lithography development at IBM", SPIE Proc., **1465**, 6(1991)

[95] U.S. SEMI Standard for X-ray Mask, SEMI document 2538

[96] Courtesy of J. Silverman, IBM East-Fishkill.

[97] P. Lenius, R. Engelstad, S. Palmer, E. Brodsky and F. Cerrina, "Mechanical Distortions of Support Frames For X-ray Lithography Masks", Journ. Vac. Sci. and Techn., **B8**, 1570 (1990)

[98] M.F. Laudon, D.L. Laird, R.L. Engelstad and F. Cerrina, "Mechanical response of x-ray masks", Jpn. Journ. of Appl. Phys., **32-1**, 5928 (1993)

[99] A.W. Yanof, D.J. Resnick, "X-ray mask distortions: process and pattern dependence", SPIE Proc., **632**, 118 (1986)

[100] Y.C. Ku, M.H. Lim, J.M. Carter, M.K. Mondol, A. Moel and H.I. Smith, "Correlation of in-plane and out-of-plane distortion in X-ray Lithography process", Journ. Vac. Sci. and Techn., **B10**, 3169 (1992)

[101] D. Laird and R. Engelstad, "Effect of Mounting Imperfections on X-ray Masks", SPIE Proc., **1465**, 134 (1991)

[102] H.I. Smith, S.D. Hector, M.L. Schattenburg and E.H. Anderson, "A new approach to high fidelity e-beam and ion-beam lithography based on an in situ global-fiducial grid", Journ. Vac. Sci. and Techn., **B9**, 2992 (1991)

[103] S. Palmer, Texas Instruments, private communication (1989)

[104] M. Karnezos, "Effect of stress on the stability of x-ray masks", Journ. Vac. Sci. and Techn., **B4**, 226 (1986)

[105] K.H. Muller, P. Tischer, W. Windbracke, "Influence of absorber stress on the precision of x-ray masks", Journ. Vac. Sci. and Techn., **B4**, 230 (1986)

[106] A. Yanof, "X-ray mask distortion from arbitrary integrated circuit patterns: closed-form and finite elements calculation", Journ. Vac. Sci. and Techn., **B9**, 3310 (1991)

[107] S. Nash, T.B. Faure, "X-ray mask process-induced distortion study", Journ. Vac. Sci. and Techn., **B9**, 3324 (1991)

[108] C.C. Fang, F. Jones, R.R. Kola, G.K. Celler, V. Prasad, "Stress and microstructure of sputter-deposited thin films: molecular dynamics simulations and experiment", Journ. Vac. Sci. and Techn., **B11**, 2947 (1993)

[109] S. Ohki and H. Yoshihara, "X-ray mask distortion analysis using the boundary method", Journ. Vac. Sci. and Techn., **B8**, 446 (1990)

[110] A. Moel, M. Itoh, S. Mitsui and Y. Gomei, "Mask distortion analysis for the fabrication of 1 Gbit DRAM by X-ray Lithography", Jpn. Journ. of Appl. Phys., **32**, 5947 (1993)

[111] D.J. Resnick, K.D. Cummings, W.J. Dauksher, H.T.H. Chen, G.M. Wells, W.A. Johnson, P.A. Seese, R. Engelstad and F. Cerrina, "The effect of aperturing on radiation damage induced pattern distrotion of X-ray masks", Symposium on Electrons,Ions and Photos Beams 1995, and Journ. Vac. Sci. and Techn.,**B13**,3046 (1995)

[112] S. Nash, T.B. Faure, J.P. Levin, D.M. Puisto, J.M. Rocque, K.R. Kimmel, M.A. McCord and R.G. Viswanathan, "High-accuracy defect-free x-ray mask technology", Jpn. Journ. of Appl. Phys., **V33**, 6878 (1994)

[113] D.L. Laird, M. Laudon and R.L. Engelstad, "Practical considerations in x-ray masks mounting methodology", Journ. Vac. Sci. and Techn., **B11**, 2953 (1993)

[114] A. Chen, S.N. Lalapet and J.R. Maldonado, "Elastic deformation of X-ray Lithography masks under external loading", Journ. Vac. Sci. and Techn.,**B9**, 3306 (1991)

[115] C.J. Progler, A.C. Chen, T.A. Gunther, P. Kaiser, K.A. Cooper, R.E. Hughlett, "Overlay performance of X-ray steppers in IBM's ALF" Journ. Vac. Sci. and Techn., **11**, 2887 (1993)

[116] A.C. Chen, J.P. Silverman, "Magnification correction for proximity X-ray Lithography", SPIE Proc. **2437**, 140 (1995).

[117] K. Heinrich, H. Betz, A. Heuberger, "Heating and temperature-induced distortions of silicon X-ray masks", Journ. Vac. Sci. and Techn., **B1**, 1352 (1983)

[118] Y. Vladimirskij, J. Maldonado, R. Fair, R. Acosta, O. Vladimiskij, R. Viswanathan, H. Voelker, F. Cerrina, G.M. Wells, M. Hansen, R. Nachman, "Thermal effects in SR irradiation of x-ray masks", Journ. Vac. Sci. and Techn., **B7**, 1657 (1989)

[119] A. Chiba, K. Okada, "Dynamic in-plane thermal distortions analysis of an x-ray mask for SR X-ray Lithography", Journ. Vac. Sci. and Techn., **B9**, 3275 (1991)

[120] K. Yamazaki, F. Satoh, K. Fujii, Y. Tanaka, T. Yoshihara, "Evaluation of temperature rise and thermal distortions of x-ray mask for sychrotron radiation lithography", Journ. Vac. Sci. and Techn., **B12**, 4028 (1994)

[121] E.A. Haytcher, R.L. Engelstad, N.M. Schnurr, "Finite element analysis of dynamical thermal distortions of an x-ray mask for SR lithography", SPIE Proc., **1671**, 347 (1992)

[122] G.M. Wells, M. Reilly, F. Moore, F. Cerrina and K. Yamazaki, "X-ray mask fabrication process", SPIE Proc., **2512**, 167 (1995)

[123] G.M. Wells, M. Reilly, R. Nachman, F. Cerrina, M.A. El-Khakani, M. Chaker, "Characterization of a silicon nitride mask membrane process", Mat. Res. Soc. Symp. Prco., **306**, 81 (1993)

[124] M. Laird, R. Engelstad, "Predicting out-of-plane distortions during x-ray masks fabrication", Micro and Nanocircuit Engineering (1995)

[125] T. Shoki, Y. Yamaguchi, H. Nagasawa, "Effect of anodic bonding temperature on mechanical distortion of SiC x-ray mask substrate", Jpn. Journ. of Appl. Phys., **31**, 4215 (1992)

[126] M.L. Schattenburg, N.A. Polce, H.I. Smith and R. Stein, "Fabrication of flip-bonded mesa masks for X-ray Lithography", Journ. Vac. Sci. and Techn., **B11**, 2906 (1993)

[127] C. Uzoh, J.R. Maldonado, J. Angilello, "Structural defects in B-doped Si substrates for x-ray masks", Journ. Vac. Sci. and Techn., **B5**, 266 (1987)

[128] L. Trimble, G.K. Celler, J. Frackoviak, A. Liddle, G.R. Weber, "Production of an X-ray mask blank for a point source stepper", SPIE Proc., **1671**, 317 (1992)

[129] S.S. Dana, J. Maldonado, "Low pressure chemical vapor deposition boro-hydronitride films and their use in x-ray masks", Journ. Vac. Sci. and Techn., **B4**, 235 (1986)

[130] B. Lochel, H.L. Huber, C.P. Klages, L. Schafer, A. Bluhm, "Diamond membrane based x-ray masks", Journ. Vac. Sci. and Techn.,**B10**, 3217 (1995)

[131] T. Shoki, H. Nagasawa, H. Kosuga, Y. Yamaguchi, N. Annaka, I. Amemiya, O. Nagarekawa, "Properties of thin SiC membranes for x-ray masks", SPIE Proc., **1924**, 450 (1993)

[132] P.A. Seese, K.D. Cummings, D.J. Resnick, J.P. Wallace, G.M. Wells and A.W. Yanof, "Accelerated radiation damage testing of x-ray mask membrane materials", SPIE Proc., **1924**, 457 (1993)

[133] K.H. Lee, S.A. Campbell, R. Nachman, M. Reilly and F. Cerrina, "X-ray damage in low temperature ultra-thin silicon dioxide", Appl. Phys. Lett., **61**, 1635 (1992)

[134] H. Okuyama, Y. Yamashita, K. Marumoto, H. Yabe, Y. Matsui, Y. Yamaguchi, T. Shoki, H. Nagasawa,"Synchronton irradiation stability of x-ray-masks utilizing stress-free W-Ti absorbers and SiC membranes", SPIE Proc., **2194**, 144 (1994)

[135] J. Ahn, K. Suzuki, S. Tsuboi, Y. Yamashita, "UHV ECR-CVD SiNx films for X-ray Lithography mask membrane: properties and radiation stability", Jpn. Journ. of Appl. Phys., **V33**, 6908 (1994)

[136] D.J. Resnick, K.D. Cummings, W.J. Dauksher, H.T.H. Chen, G.M. Wells, W.A. Johnson, P.A. Seese, R. Engelstad and F. Cerrina, "The effect of aperturing on radiation damage induced pattern distortion x-ray masks", 1995 Symposium on Electrons,Ions and Photos Beams and Journ. Vac. Sci. and Techn., **B13**, 3046 (1995)

[137] M. Reilly, P. Anderson, E. Brodsky, J. Wallace, Q. Leonard, C. Capasso, J.W. Taylor, F. Cerrina, W. Waldo, G. Chen, K. Yamazaki, K. Simon, "Performance of a modified Suss XRS-200/2m stepper at CXrL", Jpn. Journ. of Appl. Phys., **12B**, 6899 (1994)

[138] W. Windbracke, H. Betz, H.L. Huber, W. Pilz, S. Pongratz, "CD control in x-ray masks with electroplated Au absorbers", Microcircuit Engineering, **5**, 73 (1986)

[139] W. Chu, M. Schattenburg and H.I. Smith, "Low-stress gold electroplating for x-ray masks", Microcircuit Engineering, **11**, 223 (1990)

[140] C. Khan Malek, B. Kebabi, A. Charai, P. de la Houssaye, "Effect of thermal treatment on the mechanical and structural properties of Au thin films", Journ. Vac. Sci. and Techn., **B9**, 3329 (1991)

[141] W.A. Johnson, R.E. Acosta, B.S. Berry, W.C. Pritchet, D.J. Resnick, W.J. Dauksher, "Stress reduction of gold absorber patterns on x-ray masks", Journ. Vac. Sci. and Techn., **B10**, 3155 (1992)

[142] S.L. Chiu and R.E. Acosta, "Electrodeposition of low-stress Au for x-ray masks", Journ. Vac. Sci. and Techn., **B8**, 1589 (1990)

[143] M. Gentili, L. Grella, E. DiFabrizio, L. Luciani, M. Baciocchi, M. Figliomeni, R. Maggiora, L. Mastrogiacomo, F. Cerrina, "Development of electron-beam process for the fabrication of x-ray nanomasks", Journ. Vac. Sci. and Techn., **B11**, 2938 (1993)

[144] G.K. Celler, C. Biddick, J. Frackoviak, C.W. Jurgensen, R.R. Kola, A.E. Novembre, L.E. Trimble, D.M. Tennant, "X-ray mask development based on SiC membrane and Au absorber", Journ. Vac. Sci. and Techn., **B10**, 3186 (1995)

[145] D.J. Resnick, K.D. Cummings, W.A. Johnson, H.T.H. Chen, B. Choi and R.L. Engelstad, "Temperature uniformity across an x-ray mask membrane during resist PEB", Journ. Vac. Sci. and Techn., **B12**, 4033 (1994)

[146] A. Krasnoperova, J. Xiao, F. Cerrina, E. DiFabrizio, L. Luciani, M. Figliomeni, M. Gentili, W. Yun, B. Lai, E. Gluskin, "Fabrication of hard x-ray phase zone plates by X-ray Lithography", Journ. Vac. Sci. and Techn., **B11**, 2588 (1993)

[147] M. Karnezos, R. Ruby, B. Heflinger, H. Nakano, R. Jones, "Tungsten: an alternative to Au for x-ray masks", Journ. Vac. Sci. and Techn., **B5**, 283 (1987)

[148] M. Chaker, S. Boily, Y. Diawara, M.A. El-Khakani, E. Gat, A. Jean, H. LaFontaine, H. Pepin, J. Voyer, J.C. Kieffer, A.M. Haghiri-Gosnet, F.R. Ladan, M.F. Ravet, Y. Chen, F. Rousseaux, "X-ray mask development based on SiC membrane and W absorber", Journ. Vac. Sci. and Techn., **B10**, 3191 (1992)

[149] E.A. Dobisz, C.R. Eddy, J. Kosakowski, O.J. Glembocki, L.M. Shirey, K.W. Foster, W.P. Chu, K.W. Rhee, D.W. Park, C.R.K. Marrian, M.C. Peckerar, "Comparison of dry-etch approaches for tungsten patterning", SPIE Proc., **2194**, 178 (1994)

[150] Y.C. Ku, L.P. Ng, R. Carpenter, K. Lu, H.I. Smith, L.E. Haas, I. Plotnik, "In-situ stress monitoring and deposition of zero-stress W for X-ray masks", Journ. Vac. Sci. and Techn., **B9**, 3297 (1991)

[151] R.R. Kola, G.K. Celler, J. Frackoviak, C.W. Jurgensen, L.E. Trimble, "Stable low-stress absorber technology for sub-half-micron X-ray Lithography", Journ. Vac. Sci. and Techn., **B9**, 3301 (1991)

[152] M. Chaker, S. Boily, Y. Diawara, M.A. El Khakani, E. Gat, A. Jean, H. Lafontaine, H. Pepin, J. Voyer, J.C. Kieffer, A.M. Haghiri-Gosent, F.R. Ladan, M.F. Ravet, Y. Chen and F. Rousseaux, "X-ray mask development based on SiC membranes and W absorber", Journ. Vac. Sci. and Techn., **B10**, 3191 (1992)

[153] S. Ohki, M. Kakuchi, T. Matsuda, A. Ozawa, T. Ohkubo, M. Oda and H. Yoshihara, "Ta/SiN-structure x-ray masks for sub-half-micron LSIs", Jpn. Journ. of Appl. Phys., **28**, 2074 (1989)

[154] T. Yoshihara and K. Suzuki, "Sputtering of fibrous-structured low-stress Ta films for x-ray masks", Journ. Vac. Sci. and Techn., **B12**, 4001 (1994)

[155] M. Mondol, H. Li, G. Owen, H.I. Smith, "Uniform-stress tungsten on X-ray mask membranes via He-backside cooling", Journ. Vac. Sci. and Techn., **B12**, 4024 (1994)

316 / Cerrina

[156] M.F. Laudon, R. Engelstad, K. Thole, W.A. Johnson, D.J. Resnick, W.J. Dauksher, "Modeling of in-plane distortions due to variations in absorber stress", presented at Micro and Nanocircuit Engineering (1995)

[157] K. Marumoto, H. Yabe, S. Aya, K. Kise, Y. Matsui, "Total evaluation of W-Ti absorber for X-ray masks", SPIE Proc., **2194**, 221 (1994)

[158] W.J. Dauksher, D.J. Resnick, K.D. Cummings, J. Baker, R.B. Gregory, N.D. Theodore, J.A. Chan, M.A. Nicolet and J.S. Reid, "Method for fabricating a low-stress x-ray mask using annealable amorphous refractory compounds", Journ. Vac. Sci. and Techn., **B13**, 3103 (1994)

[159] The isntrument in question is the SEMSPEC, manufactured by KLA Corp., USA

[160] P.G. Blauner and J. Mauer, "X-ray mask repair", IBM Journ. Res. Devel., **37**, 421 (1993)

[161] D.K. Stewart, T. Olson, B. Ward, "0.25 μm x-ray mask repair with focused ion beams", SPIE Proc., **1924**, 98 (1994)

[162] M.T. Postek, J.R. Lowney, A.E. Vladar, W.J. Keery, E. Marx, R.D. Larrabee, "X-ray mask metrology: the development of linewidth standards for X-ray Lithography", SPIE Proc., **1924**, 435 (1993)

[163] R. Viswanathan, R.E. Acosta, D. Seeger, H. Voelker, A. Wilson, I. Babich, J. Maldonado, J. Warlaumont, O. Vladimirsky, F. Hohn, D. Crockett, R. Fair, "Fully scaled 0.5 μm MOS circuits by SR X-ray Lithography: mask fabrication and characterization", Journ. Vac. Sci. and Techn., **B6**, 2196 (1988)

[164] S.M. Preston, D.W. Peters, D.N. Tomes, "Preliminary testing results for a new x-ray stepper", SPIE Proc., **1089**, 164 (1989)

[165] G.K. Celler, C. Biddick, J. Fracoviak, C.W. Jurgensen, R.R. Kola, A.E. Novembre, L.E. Trimble, and D.M. Tennant, "Masks for X-ray Lithography with a point source stepper", Journ. Vac. Sci. and Techn.,**B10**, 3186 (1992)

[166] H. Betz, H.L. Huber, S. Pongratz, W. Rohrmoser, W. Windbracke, U. Mescheder, "Silicon x-ray masks: pattern placement and overlay accuracy", Microcircuit Engineering, **5**, 41 (1986)

[167] S. Ohki, M. Oda, M. Kakuchi and H. Yoshihara, "High-accuracy x-ray masks with sub-half-micron 1M-DRAM chips", Microcircuit Engineering, **13**, 251 (1991)

[168] H. Sumitani, K. Itoga, M.Inoue, H. Watanabe, N. Yamamoto, Y. Matsui, "Replicating characteristics by SR lithography", SPIE Proc., **2437**, 94 (1995).

[169] Y. Chen, F. Carcenac, F. Rousseaux, D. Decanini, M.F. Ravet, H. Launois, "Improvements of nanostructure patterning in SOR x-ray mask making", Jpn. Journ. of Appl. Phys., **V33**, 6923 (1994)

[170] J. Guo, F. Cerrina, "Absorber roughness effect in XRL image formation", SPIE Proc.,**1924**,382 (1993)

[171] J.Z.Y. Guo, F. Cerrina, "Optimization of partially coherent illumination in x-ray lithography", SPIE Proc., **1671**, 442 (1992)

[172] W. Grobman, *X-ray lithography*, Synchrotron Radiation Handbook, H. Koch, ed., North Holland (1984).

[173] B. J. Lin, "Methods to print optical images at low-k_1 factors", SPIE Proc., **1264**, 2 (1990)

[174] J.Z.Y. Guo, Q. Leonard, F. Cerrina, E. DiFabrizio, L. Luciani, M. Gentili and J. Frank, "Experimental study of aerial images in x-ray lithography", Journ. Vac. Sci. and Techn., **B11**, 2902 (1993)

[175] K. Early, D. Trindade, Q. Leonard, F. Cerrina, K. Simon, M. McCord and D. DeMay, "Resolution and components of critical dimension variation in x-ray lithography", SPIE Proc., **2437**, 62 (1995)

[176] Hoya Corp., 3-3-1 Musashino, Tokoyo, Akishima-shi, Japan

[177] Materials & Technologies Corp., Poughkeepsie, NY 12601

[178] Nanostructures, Santa Clara, CA 96051

[179] Advanced Mask Facility, Microelectronics Division, Essex Junction, VT 05452

[180] NTT Advanced Technology Corp., Atsugi-shi, Kanagawa, Japan

[181] W.T. Novak, "A lithography system for X-ray Lithography development", SPIE Proc., **393**, 106 (1983)

[182] P. Brikmeyer, SVGL, private communication (1994)

[183] Y. Fukuda, Canon Nanotechnology Div., private communication (1994)

[184] S. Hamada, K. Ito, T. Miyatake, F. Sato, K. Yamazaki, "22-nm overlay accuracy of synchrotron radiation stepper using an improved chromatic bifocus alignment system" SPIE Proc., **2194**, 73 (1994)

[185] M. Fukuda, M. Suzuki, M. Kanai, H. Tsuyuzaki, A. Shibayama and S. Ishihara, "Overlay accuracy of a SR stepper evaluated by two-mask double-exposure", Journ. Vac. Sci. and Techn., **B12**, 3256 (1994)

[186] R. Hirano, T. Higashiki, H. Nomura, O. Kuwabara, T. Nishizaka, and N. Uchida, "Evaluation of overlay accuracy for the x-ray stepper TOXS-1", Journ. Vac. Sci. and Techn., **B12**, 3247 (1994)

[187] K. Hoga, T. Itoh, S. Kusumoto, K. Araki, Y. Yasui, H. Takeuchi, S. Aoki, "Improvement of heterodyne alignment for x-ray steppers", Journ. Vac. Sci. and Techn., **B11**, 2179 (1993)

[188] J. Guo, J. Xiao, F. Cerrina, "Effects of illumination system aberrations on proximity x-ray lithography images", SPIE Proceedings, **1924**, 320 (1993)

[189] W. Waldo, Alignment systems in *Handbook of VLSI Lithography*, W.B. Glendinning, J.N. Helbert, Ed., Noyes(1991)

[190] M. Nelson, J. Kreuzer and G. Gallatin, "Design and test of a through-the-mask alignment sensor for a vertical stage x-ray aligner", Journ. Vac. Sci. and Techn., **B12**, 3251 (1994)

[191] R.E. Hughlett, K.A. Cooper, "Video-based alignment system for X-ray Lithography", SPIE Proc., **1465**, 100 (1991)

[192] G. Chen, J. Wallace, R. Nachman, G. Wells, D. Bodoh, P. Anderson, M. Reilly and F. Cerrina, "CXrL aligner: an experimental quarter-micron feature x-ray lithography system", Journ. Vac. Sci. and Techn., **B10**, 3229 (1992)

[193] E.E. Moon, P.N. Everett and H.I Smith, "Immunity to signal degradation by overlayers using a novel spatial-phase-matching alignment system", 1995 Symposium on Electrons,Ions and Photos Beamsand Journ. Vac. Sci. and Techn., **B13**, 2648 (1995)

[194] J.W. Taylor, C. Babcock, M. Sullivan, "Chemically amplified x-ray resists", ACS Symposium Series, **527**, 224 (1993)

[195] D. Seeger, "Resist materials and processes for X-ray Lithography", IBM Journ. Res. Devel., **37**, 435 (1993)

[196] Hoechst Celanese Corp., Conventry, RI 02816

[197] Shipley Co., Marlborough, MA

[198] OCG Microelectronics Materials AG (Switzerland).

[199] H. Ban, J. Nakamura, K. Deguchi and A. Tanaka, "High speed positive photoresist suitable for precise replication of sub-0.25 μm features", Journ. Vac. Sci. and Techn., **B12**, 3904 (1994)

[200] Nippon Zeon Co., Electronic Chemicals Dept., 2-6-1 Marunouchi, Chiyoda-ku, Tokyo 100 (Japan)

[201] S.Y. Chou, D.A. Antoniadis, and H.I. Smith, "Observation of electron velocity over-shoot in sub-100nm channel MOSFET's in silicon", IEEE Electr. Dev. Lett., **EDL-6**, 665 (1985)

[202] R. Nachman, G. Chen, G. Wells, J. Wallace, H.H. Li, M. Reilly, A. Krasnoperova, P. Anderson, E. Brodsky, E. Ganin, S. Campbell, J.W. Taylor and F. Cerrina, "X-ray lithography processing at CXrL: from beamline to 0.25 μm processing", SPIE Proc., **2194**, 106 (1994)

[203] R.Viswanathan, D. Seeger, A. Bright, T. Bucelot, A. Pomerene, K. Petrillo, P. Blauner, P. Agnello, J. Warlaumont, J. Conway, D. Patel, "Fabrication of high performance 512K SRAMs 0.25 μm CMOS technology using x-ray lithography", Journ. Vac. Sci. and Techn., **B8**, 2910 (1993)

[204] K. Deguchi, K. Miyoshi, H. Ban, T. Matsuda, I. Ohno, Y. Kado, "Fabrication of 0.2 μm LSI circuits using SR X-ray Lithography", 1995 Symposium on Electrons,Ions and Photos Beams and Journ. Vac. Sci. and Techn., **B13**, 3040 (1995)

[205] I.Y. Yang, H. Hu, L.T. Su, V.W. Wong, M. Burkhardt, E.E. Moon, J.M. Carter, D.A. Antoniadis, H.I. Smith, K.W. Rhee, W. Chu, "High-performance self-aligned sub-100 nm MOSFETs using X-ray Lithography", Journ. Vac. Sci. and Techn., **B12**, 4051 (1994)

[206] G.E. Rittenhouse, W.M. Mansfield, A. Kornblit,D.N. Tomes, R.A. Cirelli, F. Klemens, J. Frackoviak, G.K. Celler, "Sub-0.1 μm NMOS transistors fabricated using X-ray Lithography", SPIE Proc., **2437**, 126 (1995)

[207] Y. Nishioka, K. Shiozawa, T. Ohishi, K. Kanamoto, Y. Tokuda, H. Sumitami, S. Aya, H. Yabe, K. Itoga, T. Hifumi, K. Marumoto, T. Kuroiwa, T. Kawahara, K. Nishikawa, T. Oomori, T. Fujino, S. Yamamoto, S. Uzawa, M. Kimata, M. Nunoshita and H. Abe, "Gigabit scale DRAM cell with new simple $Ru/(Ba, Sr)TiO_3$ stacked capacitor using X-ray Lithography", IEEE Int. Electr. Dev. Meet., 1995, Washington

[208] W. Chu, C.C. Eugster, A. Moel, E.E. Moon, J.A. DelAlamo, H.I. Smith, M.L. Schattenburg, K.W. Rhee, M. Peckerar, M.R. Melloch, "Conductance quantization In a GaAs electron waveguide device fabricated by x-ray lithography", Journ. Vac. Sci. and Techn., **B10**, 2966 (1992)

[209] K. Early, M. Schattenburg, H.I. Smith, "Absence of resolution degradation in X-ray Lithography for wavelengths from 4.5 to 0.83 nm", Microcircuit Engineering, **11**, 317 (1990)

[210] Y. Chen, R.K. Kupka, F. Rousseaux, F. Carcenac, D. Decanini, M.F. Ravet, H. Launois, "50-nm X-ray Lithography using synchrotron radiation", Journ. Vac. Sci. and Techn., **B12**, 3959 (1994)

[211] F. Rousseaux, D. Decanini, F. Carcenac, E. Cambril, M.F. Ravet, C. Chappert, N. Bardou, B. Bartenlian, P. Veillet, "Study of large area high density magnetic dot arrays fabricated using SR-based X-ray Lithography", presented at 1995 Symposium on Electrons,Ions and Photos Beams and Journ. Vac. Sci. and Techn., **B13**, 2789 (1995)

[212] Microparts Gesellschaft fur Mikrostrukturetechnik mbH, Karlsruhe, Germany

[213] M. Peckerar, J. Maldonado, "X-ray lithography: an overview", Proc. IEEE, **81**, 1249 (1993)

[214] M. Peckerar, J. Maldonado, "The advanced lithography program – government role in x-ray development", Sol. State Techn., **37**, 44 (1994)

CHAPTER 4

Deep-UV Resist Technology:

The Evolution of Materials and Processes for 250-nm Lithography and Beyond

Robert D. Allen
IBM Almaden Research Center

Willard E. Conley
IBM Microelectronics Division

Roderick R. Kunz
MIT Lincoln Laboratory

CONTENTS

4.1 INTRODUCTION

Deep-UV lithography using 248- and 193-nm light will likely be the microlithography technology of choice for the manufacture of advanced memory and logic semiconductor devices for the next decade. Photoresists capable of exploiting these deep-UV wavelengths operate on an entirely different imaging mechanism than traditional positive photoresists. As a result, the preparation, formulation, process peculiarities, reflection suppression requirements, etching properties, etc., are substantially different from (traditional) novolac-based mid-UV resists. The payoff (and it is huge) will be the ability to manufacture devices using optical lithography for device generations perhaps as small as 0.14 μm. This chapter introduces the basic chemistry behind the DUV resists of today and discusses issues involved in the implementation of these new resist materials into manufacturing. Additionally, we describe the new work in the field of 193-nm photoresists, which many hope will be ready for manufacturing devices before the turn of the century. The issues concerning materials, process, reflectivity control, manufacturing facilities, quality control, photoresist manufacturing, and photoresist cost are discussed.

4.1.1 Evolutionary/Revolutionary Changes in Microlithography of the 1990s

The explosive growth in performance of semiconductor devices has been fueled in part by advances in microlithography and photoresist technology. The current generation of advanced microprocessors and DRAM memory chips have critical dimensions approaching 0.35 μm and are for the most part printed using novolac-based mid-UV photoresists. The science and technology of novolac/diazonaphthoquinone (DNQ) photoresists has recently been reviewed in great detail by R. Dammel in an SPIE monograph,[1] and is discussed in this chapter only for comparison purposes (to the often radically different DUV resists).

Novolac-based photoresists are available with "wavelength-limited" resolution (i.e., the ability to print images below the exposure wavelength) resulting from an enormous investment in tailoring novolac structure and molecular weight and improvements in DNQ inhibitors. The basic chemistry of the traditional DNQ/novolac resist is illustrated in Fig. 4.1. The novolac resin behaves as an inert, aqueous-base soluble matrix polymer, while the DNQ undergoes (noncatalytic) radiation-induced photochemistry (the Wolfe rearrangement), which converts the DNQ from a strong dissolution inhibitor to a mild dissolution promoter. A typical dissolution rate scheme for novolac/DNQ resists is presented in Fig. 4.2. This dissolution response is substantially different from the DUV resists discussed below. Another notable difference (there are so many!) is the relatively low contrast of traditional novolac/DNQ resists as compared to DUV formulations. Figure 4.3 shows a typical "contrast curve" of a positive mid-UV resist. Newer DUV resists introduced below show typically a factor of 2 to 3 contrast improvement over these more traditional materials. The message is, the performance gain in the transition from traditional (mid-UV) resists to newer DUV resists comes as a consequence of *both* wavelength considerations and contrast improvement.

The inhibition of novolac dissolution by DNQ compounds is a complex function of inhibitor structure and novolac molecular properties. Both the bonding structure (ortho-ortho, meta/para ratio, etc.) and the molecular weight characteristics profoundly impact the novolac/DNQ interactions and thus the lithography. The exposed resist film is baked, *not* to allow chemistry to occur (the chemistry occurs during the exposure), but to diffuse the standing waves resulting from the interference effects, which impart a (periodic) difference in exposure intensity as a function of film thickness. Recent work examining the detailed molecular weight influences on novolac dissolution and resist performance[2] and novolac/DNQ interactions[3] is helping to unveil the mysteries surrounding these complex, important, high-performance photoresists. The tremendous and continuing progress in novolac-based mid-UV resists, if repeated in the DUV resist technology described in this chapter, will produce ever-evolving, ever-improving photoresists well into the next century.

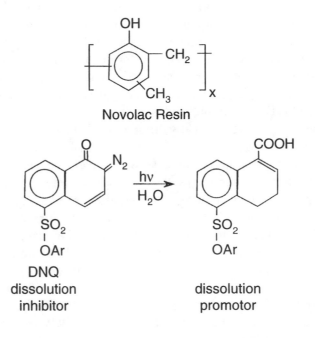

FIG. 4.1 Basic chemistry and materials used in novolac/diazonaphthoquinone mid-UV photoresist.

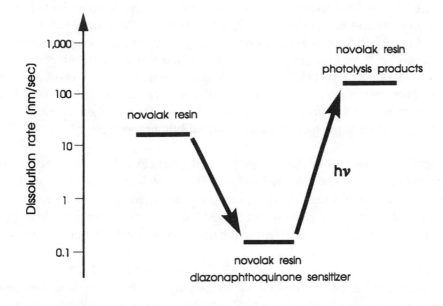

FIG. 4.2 (from Ref. 1)

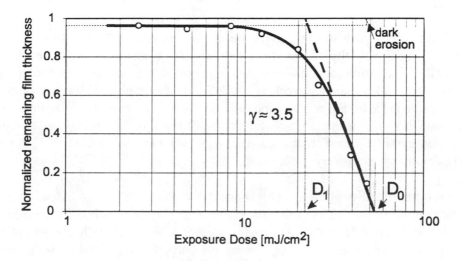

FIG. 4.3 (from Ref. 1)

The purpose of this chapter is to inform the reader (the photoresist user) of the current work in the research and development of new photoresist technology that will be in use in the near future (i.e., DUV lithography), including a broad background, reasons for and benefits of newer resist technology, problems and difficulties in developing, manufacturing, and processing these present and future materials, and an analysis of the photoresist technology that (we anticipate) will move us into the next century. This chapter is not an exhaustive review of photo-resist chemistry and in fact is not an exhaustive review of advanced (DUV) chemistry and materials. Several excellent reviews of that type have been published over the past few years, most notably by Reichmanis[4] and Willson and their coworkers.[5] A very recent status report by Ito[6] is a valuable addition to the DUV resist literature. This chapter attempts to provide insight into the activities in DUV photoresist R&D in the recent past and to provide a frame of reference for future developments that will be useful (we hope) to a broad spectrum of lithographers.

The next generation of devices (0.35 μm and smaller) will be produced with a mixture of optical lithographies including traditional 365-nm photolithography, and at shorter wavelengths (ca. 250 nm, deep-UV) combined with newer (chemically amplified) photoresists. It is likely that the industry conversion from i-line to DUV for critical levels will be completed in the transition to the 0.25-μm generation of devices.

While the move from g-line to i-line lithography was *evolutionary* in nature (using common resist technology, novolac/DNQ positive photoresist), the move to DUV is *revolutionary*, involving new exposure technologies, imaging chemistry (chemical amplification), and new materials (hydroxystyrene-based polymers). In light of this tremendous resist design change, it is not surprising that the revolution took so long in coming!

The technology path toward device generations beyond 0.25 μm (e.g., 1-Gbit DRAM) is currently the subject of much discussion. With the advent of new high-NA DUV exposure tools and the development of a new generation of high-performance DUV (single-layer) resists, 248-nm lithography will certainly play a critical role in the 0.25-μm generation and may prove to be robust well beyond. Add into the DUV mix the opportunities for phase-shifting mask technology and/or more complex multilayer resist schemes, and the window for critical-level DUV (248-nm) lithography opens up very wide, indeed.

ArF excimer (193-nm) lithography is another viable approach to extend optical lithography beyond 0.25 μm, but the resist technology developed for traditional DUV lithography is problematic at this deeper UV wavelength due to optical transparency considerations. We discuss the current thinking on 193-nm resist design toward the end of this chapter (Sec. 4.8) and describe the limited number of resist systems published to date.[7] Both single-layer and multilayer (dry) processes are covered, with emphasis on the materials issues involved in the design of resists for 193-nm lithography in contrast to traditional DUV materials. Additionally, the newest research in the area of 193-nm resist materials is covered.

Photoresists for 193-nm lithography have only recently (since about 1992) received attention from resist research groups. As a result, much progress needs to be made in a relatively short time (the next several years), when stepper manufacturers should have developmental/early manufacturing exposure tools ready for commercialization. The compression of the typical photoresist development cycle represents an enormous challenge for the photoresist industry.

4.1.2 Chemically Amplified Photoresists: A Brief History and Description

Key in the development of deep-UV lithography was the revolutionary work of Ito, Willson, and Frechet at IBM in the early 1980s.[8] These workers recognized very early that if DUV lithography was to become a reality, a new imaging mechanism with very high efficiency (fast photospeed) was required, due to the limited output of mercury lamps at 250 nm. The new resist design, termed *chemical amplification* (CA) involved the preparation of an acid-reactive polymer, formulated with an "onium salt" photoacid generator. The critical feature was that the acid-labile group (attached to the polymer) would react with the photogenerated acid in such a way that a new molecule of acid would be generated, thus beginning a catalytic cycle. Thus, a molecule of photogenerated acid might produce 500 to 1000 chemical reactions (deprotection/deblocking steps).[9] This design is illustrated in Fig. 4.4.

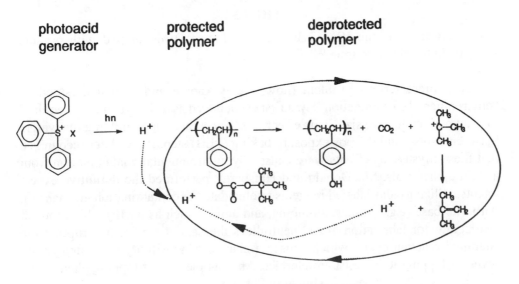

FIG. 4.4 Chemical amplification.

Concerted research in the early to mid 1980s uncovered much promise, as well as many problems, with this new mode of lithography. A variety of acid-catalyzed chemical reactions were incorporated into the basic resist design,[4,5,10] and CA resists were extended into negative-tone imaging.[11] The contrast of these new resists was extremely high compared with traditional novolac/DNQ positive photoresists. Problems initially arose, however, in gaining aqueous development in these early CA resists. As a result, the first commercial resist process implemented into

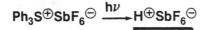

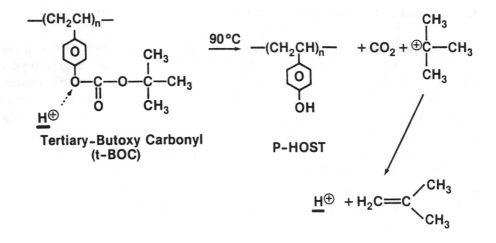

FIG. 4.5

manufacturing was the IBM TBOC resist (Fig. 4.5), developed with organic solvents in the negative-tone.[12]

An even more serious problem (now widely known and well understood as "environmental contamination") was first encountered as a change in dose required to print a given feature size. This effect was exacerbated upon holding the wafers between expose and the postexposure baking (PEB) step. It was later recognized that these mysterious effects were caused by environmental contamination from airborne chemicals. MacDonald and Hinsberg[13] performed the definitive experiments, utilizing radio-labeled reagents to elucidate the contamination mechanism. Airborne base (e.g., amines, ammonia, and amides such as NMP), ubiquitous in semiconductor fabrication areas, neutralizes the catalytic acid, interrupting the chemical reaction cycle, which causes a change in sensitivity and often (more severe) t-topping at the resist/air interface. An example of the t-topping phenomena produced be airborne bases is shown in Fig. 4.6.

Methods have been developed to address the environmental contamination problems of CA resists, including filtration of clean-room air (activated charcoal),[14] the addition of weak organic acid filters to wafer tracks,[15] use of resist top-coat (RTC),[16] and chemical treatments, such as additives to DUV resists.[17]

The second generation of CA resist technology was the result of a breakthrough in the development of aqueous developing positive-tone materials. IBM APEX resist[18] is the best known of this class of materials. AT&T developed a conceptually

15 min in filtered air　　　　　15 min in 10 ppb
　　　　　　　　　　　　　　　NMP before exposure

FIG. 4.6

different material (CAMP) that behaved similarly.[19] APEX is based on the recognition that (cleanly) aqueous developing positive-tone resist materials require a balance between hydrophobicity and hydrophilicity to be achieved. If a resist film is too hydrophobic, this will result in surface inhibition (at best) and undevelopable residue. Resist formulations that are too hydrophilic will result in unacceptably high (unexposed) erosion rates, which severely limits process latitudes.

By preparing copolymers of hydroxystyrene with a t-BOC protected hydroxystyrene, the IBM workers were able to achieve the desired hydrophobic/hydrophilic balance. This resist design has become the most commonly used methodology for positive DUV photoresists and is schematically illustrated in Fig. 4.7. In essence, a partially protected poly(hydroxystyrene) is formulated with a photoacid generator. Exposure produces catalytic quantities of acid. Postexposure baking drives the acid-catalyzed deprotection to produce a (hopefully) large differential dissolution.

APEX resist was the first (relatively) large-scale, commercially available DUV (positive-tone) photoresist. Several subgenerations of this resist were developed in the late-1980s and early 1990s, including dyed versions.[20] It is well suited for manufacturing 0.35-µm devices with DUV tools with moderate numerical apertures (e.g., SVGL Micrascan™ II, NA = 0.5). Clean (airborne) environments are critical for APEX performance, however, due to the contamination problems discussed above.

It is anticipated that new generations of DUV resist technology currently being explored and introduced (in the 1996 time frame) will offer improved process latitude for 0.25-µm (production) lithography, along with significant enhancement in environmental stability. The following discussion offers technical detail on the resist design principles that led to the new generation of DUV resist technology.

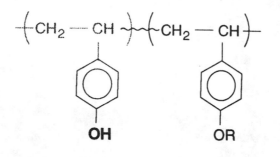

• Novolac isomer   • Acid Labile
• Hydrophilic
• High T_g
• Acidic
• Transparent (248 nm)
• Etch resistant

FIG. 4.7 Traditional (APEX) DUV resist design.

4.2 DUV (248-NM) RESIST MATERIALS

4.2.1 Newer Concepts in Positive Resists

Several recent publications have outlined the problems associated with early positive-tone chemically amplified resist systems. Since positive resists are more widely used, the majority of research has focused on their improvement. One of the first publications discussed how a series of sulfonium and iodonium photoacid generators (PAGs) might be considered to solve the problem of latent image stability and possible causes.[21] Subsequent authors have discussed the use of PAGs not only to produce a photoacid but to assist in the inhibition of the unexposed region.[22] Several authors have investigated the use of 4-DNQ photoactive compounds as PAGs for chemically amplified systems.[23] Several papers have discussed barrier layers to suppress or eliminate environmental contaminants, the addition of bases to prepoison the photoresist,[24] "suicide" bases to provide more stability,[25] and the concept of annealing the film to reduce the film density, which would assist in minimizing the contamination of the film surface.[26] More recently there have been several publications demonstrating materials that have little or no effect from postexposure bake or postexposure bake delay.[27] Much effort has been invested in gaining an increased knowledge of and control over this problem of environmental degradation (delay stability) in CA resists.[6]

The major factor in the substantial improvement in the performance and environmental stability of DUV resists involved a change in the chemistry of the protecting group. A divergence of resist design has recently occurred, leading to two completely different resist classes, each with its promises and problems. These new resists (once again based on hydroxystyrene copolymers) can be grouped by activation energy (reactivity of the protecting group to the photogenerated acid). Figure 4.8 shows the modern DUV resist "family tree."

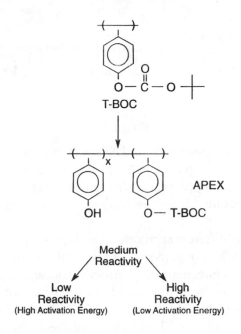

FIG. 4.8 DUV resist family tree.

The low-activation-energy resists have protecting/blocking groups that are extremely reactive to acid. In fact, published evidence indicates that the (acid-catalyzed) deprotection chemistry occurs during (or immediately after) exposure, before post-expose bake (PEB). This renders this critical delay period much less important. In fact, this type of resist is conceptually similar to the time-tested i-line (novolac/DNQ) chemistry in that the PEB is intended to diffuse standing waves, not to thermally activate the deprotection. Resists based on this design concept are typified by IBM's KRS,[27] Hoechst AZ's DX-46,[28] and most recently OCG's new positive resist, ARCH.[29]

The high-activation-energy resists have protecting groups with reactivity lower than that of APEX/TBOC-type resists. The lower activity allows for improvements in thermochemical stability. This is typified by IBM's ESCAP (Fig. 4.9) family of resists.[31] The use of lower-reactivity protecting groups allows for high-temperature bakes [both postapply bake (PAB) and PEB] to be performed. Ito and Hinsberg at IBM recognized this as an advantage, which led to *the annealing concept*, where the resist film is baked above the glass transition temperature (softening) T_g. At or

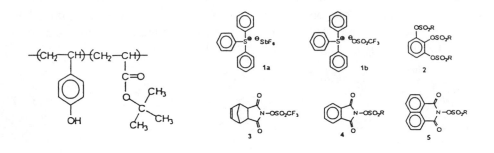

FIG. 4.9

above the T_g, the film compacts and residual stress and residual casting solvent are removed, resulting in a sharp drop in the permeation rate of unwanted chemicals into the film. This annealing concept is represented in Fig. 4.10. The environmental stability of the resist using this approach is greatly improved.

Each of these radically different approaches has its pros and cons. For example, the low-activation-energy/high-reactivity approach offers fantastic environmental properties and excellent imaging quality. Areas of concern are based on the lability of the protecting group (live by the sword, die by the sword), which may impact resist component manufacturability and shelf-life stability. Line slimming, which is thought to result from continued acid-catalyzed chemistry after PEB (but before the development step can be accomplished), appears to be an issue with these new resists. Conversely, the high-activation/low-reactivity resists should be shelf-stable

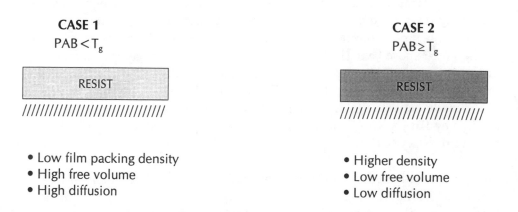

FIG. 4.10   The annealing concept.[30]

and manufacturable in a straightforward fashion by virtue of the robustness of the protecting groups used, but because the annealing temperature (PAB) is quite high for this resist design, choice of PAG is crucially important in terms of thermal stability. Similarly, controlling acid diffusion using high bake temperatures places a tremendous burden on the PAG.[31] Time will tell which of these very different resist designs will predominate in the next generation of high-performance DUV resists.

4.2.2 Negative DUV Resists

Solvent-developable negative photoresists that image swelled or DNQ image reversal systems were the only types of negative resists available for manufacturing until workers at Hitachi[32] developed nitrene crosslinking of hydroxystyrene polymers. Feely et al. first discussed an aqueous developable negative DUV resist in the mid 1980s.[11] Thackeray and coworkers at Shipley Company introduced various forms of their negative DUV resist based on Feely's initial studies.[33] Dammel et al. discussed a negative chemically amplified system for x-ray and electron beam lithography; several following publications discussed development of RAY-PN.[34]This development further escalated the research of negative resist with numerous publications from authors in Asia, North America, and Europe. There were several publications from workers at IBM discussing various types of negative resist systems for x-ray, electron beam, mid-UV, and DUV lithography,[35] most recently on the application of CGR 248 negative DUV resist to memory and logic devices for 500- to 300-nm lithography.[36] Holmes, Brunsvold and coworkers reported on the benefits of negative resists for phase shift, across-chip linewidth variation improvements, block levels for ion implantation, and little or no effect to environmental stability problems that have hampered positive CA resist systems.[37]

A serious problem with "microbridging" as the failure mechanism of negative DUV resists has been investigated.[38] Figure 4.11 shows an example of microbridging. It is thought that polymer molecular weight plays an important role in this failure mechanism. As device dimensions continue to shrink, the issue of resist pattern collapse has arisen from time to time. This problem is tone independent and can cause a serious the reduction of process latitude and the ability to print high aspect ratio imaging to satisfy RIE engineer's concerns. A recent publication from Tanaka and co-workers at SORTEC [39] addressed the issue of resist pattern collapse with high aspect ratio images by addressing the issue of rinse water surface tension. Reducing the surface tension of the rinse liquid increases the usable resolution and imaging was obtained with feature sizes of 0.15 μm with high aspect ratios. As device manufacturers begin to investigate 0.25-μm manufacturing, methods to enhance image resolution and process latitude such as this could become important.

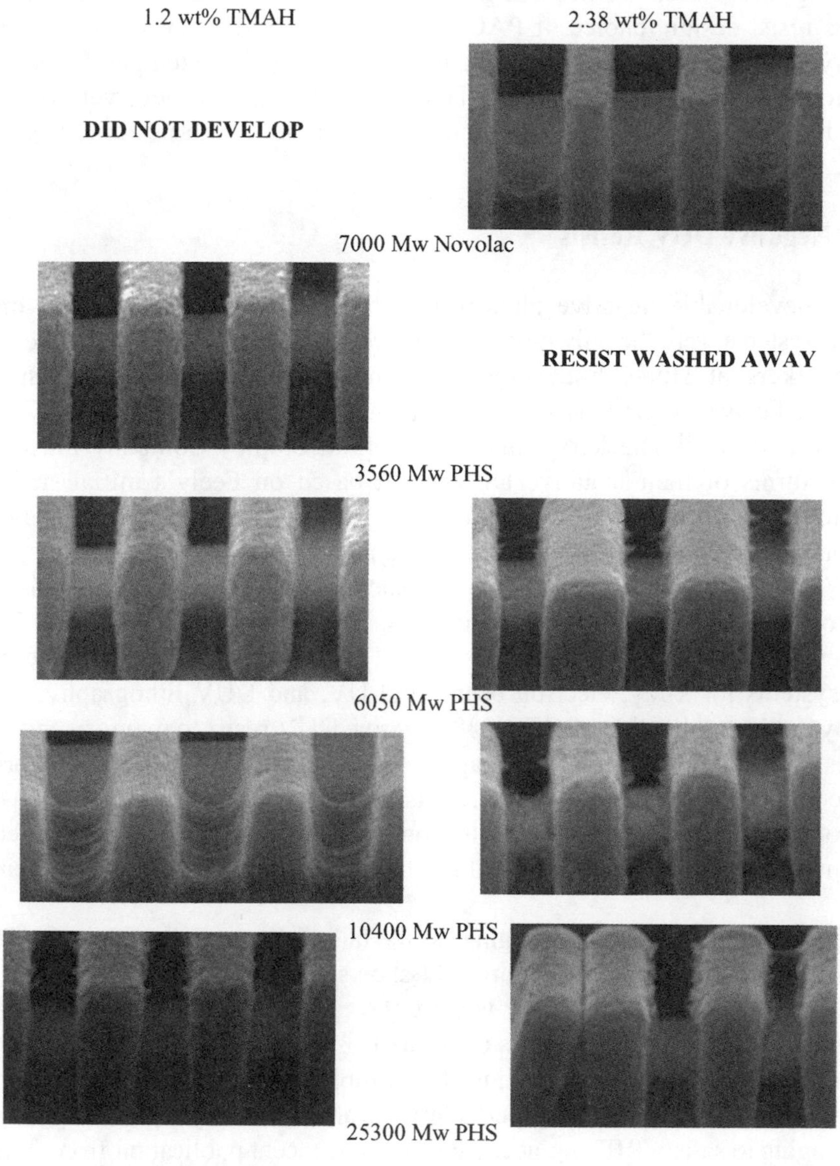

1.2 wt% TMAH

2.38 wt% TMAH

DID NOT DEVELOP

7000 Mw Novolac

RESIST WASHED AWAY

3560 Mw PHS

6050 Mw PHS

10400 Mw PHS

25300 Mw PHS

FIG. 4.11 Lithography experiment results: CA negative resists.

4.3 PROPERTIES OF DUV (248-NM) RESIST MATERIALS

4.3.1 Acid Diffusion in CA Resists

Implicit in leveraging the tremendous potential of DUV (CA) photoresists is the ability to understand, and then control, diffusion of the photoacid before, during, and after the PEB. The many materials and chemistry issues of photoactive component diffusion and thermochemical stability dominate DNQ chemistry, and the issues of photoacid diffusion and deprotection kinetics dominate chemically amplified systems. The issue of standing waves in DNQ/novolac resist and chemically amplified DUV systems is a direct result of diffusion or lack thereof. Issues regarding solvents[40] could be of even greater concern not only due to their effect on resist photospeed, coating quality, etc., but the residual components that could be acidic or acid. This could be of major concern especially with some of the more extremely acid-sensitive protecting groups. One difficulty in isolating acid-diffusion effects in CA resists is separating the acid-catalyzed chemistry (kinetics and mechanisms) from acid diffusion.[41] Another recently recognized phenomena is *acid evaporation* from the resist/air interface.[42] This tremendously complicates resist dissolution response in positive CA resists.

4.3.2 Environmental Stability; Divergence in DUV Resists

As positive-tone DUV materials continue to mature, problems such as environmental stability will become less of a concern. The annealing concept is a fine example of methodology in use today to stabilize CA resists. This concept involves densification of the film to reduce or eliminate the poisons entering the film surface that could neutralize the photoacid. Several authors have also discussed the use of low-E_a (activation energy) polymers such as ketal and acetal protected materials. These materials have demonstrated excellent delay stability and processability over substrates such as titanium nitride.

An excellent example of improved performance of the new generation of DUV resist technology resulted from a joint development effort between IBM and Shipley that has yielded a new material called UV2HS.[6, 43] This material has demonstrated superior PEB stability, as shown in Fig. 4.12, which shows imaging with time delay from 0 up to 7 hours. UV2HS resist was developed primarily for the Micrascan imaging tool, with UVIII being targeted for the excimer laser tools. The bulk polymer in these resists consists of a copolymer of 4-hydroxystyrene and t-butylacrylate. These resists have been designed to overcome the so-called delay problem associated with chemically amplified resists. Stability toward airborne base contamination is achieved by baking the resist film at unconventionally high temperatures in order to minimize the free volume. The high-temperature process is possible due to the high glass transition temperature of 150°C of the resin and much higher thermal deprotection temperature of 180°C. These high processing temperatures require the selection of a PAG that can tolerate these conditions along with a PAG that isn't volatile. Examples of bulky acids have been described by Ito and coworkers. Smaller acids such as methane sulfonic and trifluoromethane sulfonic acids are unsuitable acid generators due to their volatility out of phenolic resist films during the postexposure processes employed for UV2HS-type resists.

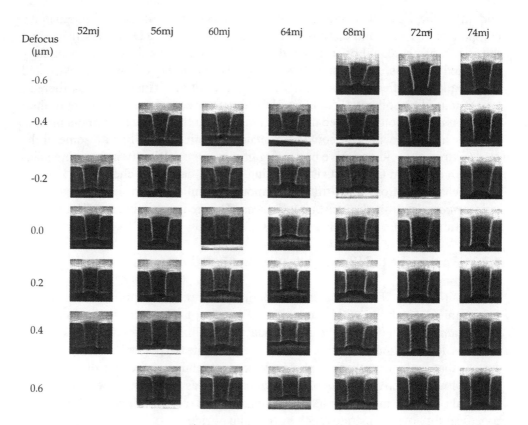

FIG. 4.12a

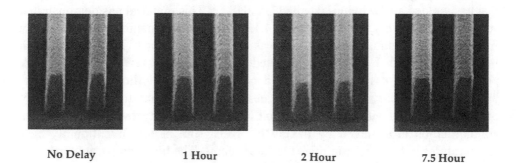

FIG. 4.12b

The annealing of the resist can benefit not only the environmental stability but also the lithographic performance due to reduction of acid diffusion achieved in conjunction with the use of a bulkier acid. What is important is not to eliminate but to control acid diffusion. A certain amount of diffusion is needed to achieve

NO DELAY

24HR DELAY

FIG. 4.13 Delay between PEB and development.

acceptable sensitivities and to reduce standing wave patterns on the resist sidewall. The robustness of these resists permits the manipulation of the bake conditions for optimal performance while the linewidth shift per °C of the PEB temperature with UV2HS (4–8 nm/°C) is only half of APEX (15 nm/°C), which is presumably due to the reduced free volume. Thus, thanks to these attributes, UV2HS demonstrates extraordinary resolution (k=0.40), opens small via holes at a dose only slightly higher than that for line/space patterns, and can print holes (Fig. 4.12) and bars to the same size at the same dose, which is extremely difficult. Since the PAB at 150°C provides an outstanding PEB stability of 18 hours, reduction of the PAB temperature by 10°C still maintains excellent resistance to contamination. In addition to the much reduced free volume in the UV2HS resist film, the incorporation of the acrylate structure is believed to be beneficial in limiting absorption of airborne contaminants.

Recently, authors have discussed in greater detail the use of ketal and acetal protecting groups for positive DUV chemically amplified photoresists.[27] Researchers at OCG Microelectronics[28] recently published data on studies conducted with acetal and ketal protected copolymer of poly(hydroxystyrene) (PHS). In this work they studied the stability of the protecting groups and measured the relative reaction rates of various copolymers. The general conclusions from these studies suggested that ketal-protected polymers have a much higher reaction rate than the acetal analogs and that storage stability would be a major problem with the ketal polymers because of their greater acid sensitivity.

The joint development effort between IBM and Shipley to further develop DUV resist technology has yielded a new material called KRS (27). This material has demonstrated excellent PEB stability (Fig. 4.13), which shows imaging with time delay from 0 to 24 hours. Here one can see that there is no t-topping or effects of this processing delay. Another attribute of this material is the insensitivity to PEB temperature. This is a major problem that the resist developer must be concerned about. Since hotplate temperatures can vary across the plate, minimizing or eliminating this effect is extremely valuable for 250-nm lithography. In Fig. 4.14, the IBM workers demonstrated PEB insensitivity of KRS with imaging from no PEB to 102°C PEB. Here we can see that no measurable change in linewidth is achieved.

4.3.3 Image Thermal Stability

We briefly mentioned the importance of thermal stability with our DNQ/novolac resists; this is of equal concern for DUV photoresists. The high-activation-energy resists (e.g.,ESCAP), which can be annealed above the T_g, have excellent thermal stability in comparison to the previous generation of resists (e.g., APEX), due principally to residual solvent effects. In any case, the thermal (image) stability of DUV resists should be superior to more traditional i-line systems due to the intrinsic T_g difference between novolac and PHS (see Table 1).

Table 4.1

POLYMER	MOLECULAR WEIGHT	GLASS TRANSITION (°C)
Novolac resin	Low, broad distribution	90–110
poly(hydroxystyrene)	Medium, narrow distribution	160–180

Negative resist materials have not enjoyed the same popularity as positive materials, but the advent of more process-friendly materials has created new applications. The concerns that govern positive PHS-based systems are also true for negative PHS systems. The literature has numerous examples of various types of crosslinkers that have been used, ranging from solvent developable materials to the most recent aqueous-developable, high-resolution systems. Figure 4.15 shows a micrograph of IBM's CGR 248 negative-tone DUV photoresist exposed on Micrascan II resolving

250 nm nested features. Negative resists currently satisfy a small need in the fabrication of semiconductors; increases in popularity will be unlikely unless these materials become more compatible with current process conditions, such as process bake temperatures and developer concentration. Today's negative resists suffer from the requirement fot weaker nonstandard developers, and as positive-tone systems continue to improve, negative usage could remain flat or decrease over time.

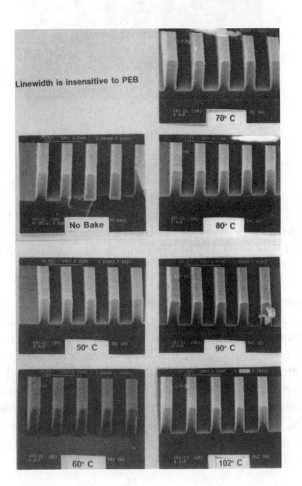

FIG. 4.14 SEM images of 0.4 μm L/S in KRS processed without PEB at 50, 60, 70, 80, 90, and 102°C respectively for 90 seconds.

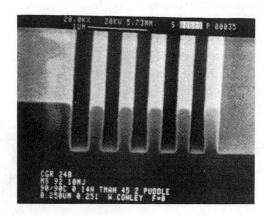

FIG. 4.15

4.3.4 Reflection Control

Reflective notching on highly reflective substrates and linewidth (critical dimensions, or CD) variations due to topography and film thickness nonuniformity have been a difficult problem for semiconductor manufacturers. The thin film interference (TFI) contribution to CD variation is a topic receiving much current attention.

The most common cure for these problems is to increase optical density of the photoresist by direct addition of an unbleachable highly absorbing dye to the undyed resist formulation. This approach is typified by the recent report by Plat et al.,from IBM, for positive-tone (novolac/DNQ) i-line resists.[44] The increased optical density of dyed resist provides reduction of the thin film swing effect and dampening of the reflected light without major changes to the manufacturing process. However, there are many problems associated with dyes directly added to the photoresist, such as particle generation, reduction of photospeed, thinning of the unexposed resist, degradation of the printed profiles, and general degradation of resist performance (resolution, DOF, and dose latitude).

Development of a high-performance dyed resist presents a constant challenge to the resist suppliers. The optimization of optical density of the resist for a particular application is necessary in order to minimize the effect of dyes on the resist performance. Generally, the bleachable absorbance, or Dill A factor, increases with increased concentration of photoactive component (PAC), while nonbleaching absorbance (Dill B) increases with addition of an absorbing dye. A large variation in a Dill A parameter is impractical due to photospeed constraints and limited solubility of commonly used DQ-type photoactive components. The increase in the nonbleaching absorbance (B) can be easily achieved by addition of a small amount of highly absorbing dye to the undyed photoresist.[44]

Chemically amplified DUV photoresist systems have a more difficult reflection suppression problem in that they are based on either copolymers or terpolymers of hydroxystyrene and are extremely transparent in DUV. Additionally, DUV PAGs typically do not bleach during exposure (with few exceptions).[23] Reflectivity control options in DUV resists include the addition of dye as a component in the resist formula as in the traditional i-line resist approach detailed above, or more typically the incorporation of a top antireflective coating and/or a bottom antireflective layer to the DUV resist process. Of course, any complexity added to the process is unwanted; however, reflections over topography must be combated.

As the optical density of photoresist materials increases, there is a significant reduction in process latitude and resolution. As dyed additives are considered, they must be chemically neutral and serve only to provide absorbance or to enhance the performance of the material. The use of diazonaphthoquinone as PAGs and in combination as a highly absorbing material to serve as an acid generator and dye has been discussed. For example, the use of DNQ/novolac resist for DUV lithography to suppress reflective notching has been investigated. The application of this "old" technology for addressing this problem is truly effective, with no environmental stability issues and great reflection control. However, sloped profiles, high imaging doses, and lack of optical density control from one resist lot to another make this system difficult to employ.

The introduction of Dyed APEX-E DUV photoresist by IBM has demonstrated significant improvements in reflectivity control with minimal reduction in process latitude and resolution.[20] Shown in Fig. 4.16 are features in standard APEX-E from .35 to .225 µm. Process latitude for these features increases with an increase in optical density up to an absorbance of 0.5. In Fig. 4.17 we have plotted Prolith‾ simulated swing curves for various optical densities of dyed APEX-E to demonstrate the reduction in swing vs increasing optical density. This work additionally investigated the ideal optical density and referenced work from previous authors that have discussed optimizing optical density around 0.42 per micron of nonbleachable absorbance.[45] Figure 4.18 is a plot of depth of focus for the dyed APEX-E resist vs increasing optical density. This plot clearly demonstrates that the optimum optical density to obtain the highest depth of focus is about 0.40 per micron. This material also demonstrated an interesting feature of minimizing dose impact with increasing optical density, which the authors attributed to the unique conjugated aromatics dyes, as shown in Fig. 4.19.

Top antireflective coating, introduced by Okazaki and coworkers at Hitachi, is extremely useful to suppress reflectivity.[45] The ideal material would have the square root of the refractive index of the photoresist to completely eliminate swing effects.[46] The use of top antireflective coatings is far more attractive than bottom antireflective layers (ARLs), since the present cost of these materials is lower than their bottom counterparts. However, top ARLs have been found to only reduce notching, not to completely eliminate it. The coupling of heavily dyed photoresist and top antireflective layer may provide enough reflection control to maintain critical dimension. The process complexity is significantly reduced vs bottom antireflective layers due to the materials removal during the development process, while the bottom counterpart is removed with a costly and complex additional reactive ion etch process.

Both top and bottom ARLs add the cost and complexity of additional process steps. Since current DUV photoresists are high in cost, the addition of an expensive ARL reduces the attractiveness of the technology. For ARL technology to be more attractive, the cost will need to be less than the cost of the photoresist and the benefit have much more impact than that of currently available materials. Future antireflective materials will need far greater optical density than current materials. Shown in Fig. 4.20 is a swing curve of undyed APEX-E over bare silicon and over 130 nm of bottom ARL, where a significant reduction in amplitude is realized. There should be no interaction between the photoresist and the ARL that will create a foot or an undercut at the resist/ARL interface. Figure 4.21 shows the undercutting profile over this bottom layer. Any defect created could transfer into the substrate and cause a number of unwanted defects that could significantly reduce device yield.

(a) 350 nm

(b) 300 nm

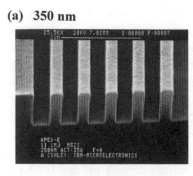

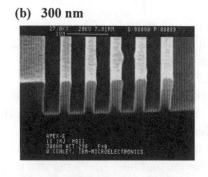

(c) 250 nm

(d) 225 nm

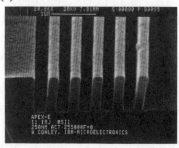

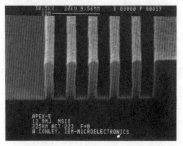

FIG. 4.16

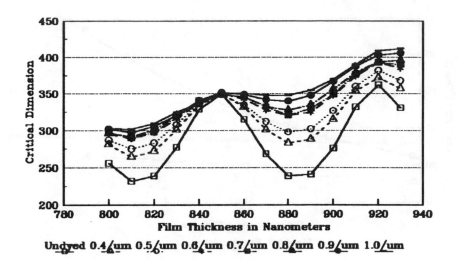

FIG. 4.17 APEX-E dyed and undyed CD vs thickness. Prolith simulated for various optical densities. MSII exposure system.

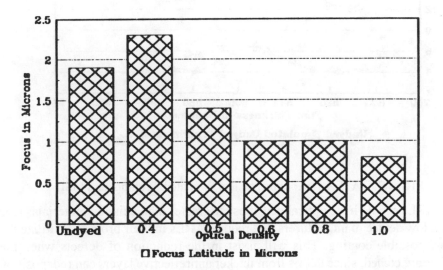

FIG. 4.18 Dyed APEX-E focus latitude for various ODs. MSII, double puddle overlap, 350nm 1/s.

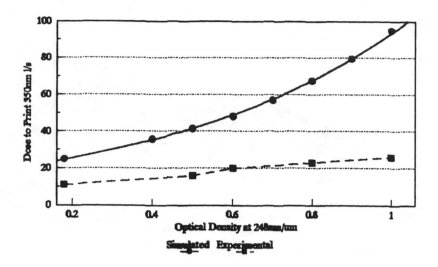

FIG. 4.19 APEX-E dyed resist dose to print. Prolith simulated and and actual for various ODs. MSII exposure system (350nm 1/s).

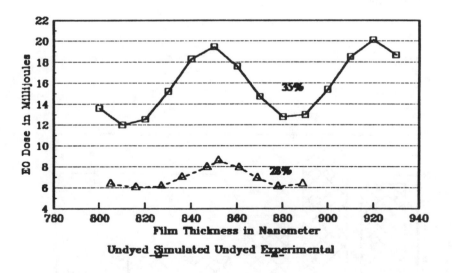

FIG. 4.20 APEX-E experimental and simulated. Prolith/2 and MSII.

Future ARLs should have optical densities far greater than existing materials. This would allow device manufacturers that require ARLs in their process to utilize the thinnest possible coating. This will assist in the reduction of defects when the materials are etched, since debris from thicker antireflective layers can redeposit on the device during long etch processes. This consequence of the etch process is a significant problem and can be a serious yield detractor. If the ARL etch rate could be 1.5 to 2 times higher than that of the photoresist, it would assist in minimizing defects. One possibility is to radically alter the design of organic ARL materials, as

Kunz and Allen showed for the development of an ARL well-suited for 193-nm lithography (see below).[47]

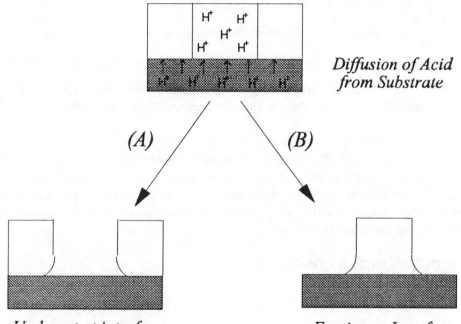

Diffusion of Acid from Substrate

(A) *(B)*

Undercut at interface
Positive Resist

Footing at Interface
Negative Resist

FIG. 4.21 Substrate (BARC) contamination.

4.3.5   Etch Resistance in DUV Resists

Etch selectivity with 850 nm of resist thickness and the etching of 1200 nm of oxide/nitride plus the addition of polymerizing gases (i.e., CHF_3) will lead to higher selectivity but also to tapered profiles. Resist footing is intolerable (undercut is undesirable but more manageable) at the resist/substrate interface. Increased thermal stability of the photoresist allows for higher etch rates for throughput considerations. The optical density of ARL materials is unfortunately inversely proportional to etch rate, which could lead to reticulation during etch (Fig. 4.22). Increased thermal stability can also assist in the reduction or elimination of striations or scalloping in contact holes, which are formed at the resist surface during etch and are directly transferred into the substrate. Hardening processes may be required but are difficult, considering the nature of the material and the additional cost contribution in process and floor space.

Polymer	Etch Rate Å/min
poly(methylmethacrylate)	11228
poly(methylmethacrylate) + 25% Novolac	9479
poly(anthrylmethacrylate) (100% anthracene)	4600
poly(vinylmethylether-co-maleic anhydride)	15763
poly(styrene-co-maleic anhydride)	7628
poly(vinylmethylether-co-maleicanhydride) (5% anthracene)	6000
APEX	7429
Spectralith	7246

FIG. 4.22 O_2 etch rates of polymer systems. Materials were etched on an ARIES system (MRC). Pressure: 2 mtorr; Flow: 20 sccm O_2; RF Power: 2500 Watts.

The creation of sidewall passivation layers requires the use of RIE residue strippers. These materials typically consist of volatile, basic reagents such as NMP, ethanolamine, or other hydroxyl amines that are harmful to the chemically amplified resist. The development of strippers that are more compatible with use of chemically amplified photoresists will require cooperative development between the stripper company and the photoresist company.

4.4 DUV PROCESS CONSIDERATIONS

4.4.1 Process Ideology

Development engineers need to alter the philosophy of photoresist processing with DUV materials and pass this new philosophy along to manufacturing personnel. The process of handling DUV photoresists is different in many ways from i-line type materials. The ability to batch process is currently not available, and integrated process clusters are a key to more successful process development in the semiconductor fabs. The ideology of designing and maintaining process sectors that are less contaminated by process chemicals that are harmful to chemically amplified photoresists may be a requirement. The introduction of ancillary materials that are compatible with chemically amplified photoresists will require further consideration in the development of processes. Along with traditional methods of reducing cost, increasing process yield and maintaining maximum throughput in the manufacturing of 0.25-µm (and smaller) devices will be essential.

The control of hotplate temperatures, dose, developer temperature, and resist thickness are items of more serious consideration than in the recent past, not just because it's DUV lithography but because it's 250-nm lithography! Extreme CD control of 0.05 µm was good for 0.50-µm lithography; however, for 0.25 µm this will need to be reduced further, to 0.025 µm (and that doesn't include etch bias!).Thickness control is also of increased importance, for several reasons. As dimensions shrink and the conversion to a shorter wavelength occurs with more transparent materials, we must remember that as wavelength decreases, swing amplitude and periodicity increase.[48] This physics contribution to our chemical process will require even tighter thickness control with these relatively transparent materials.

4.4.2 Positive-Tone Process Considerations

Next-generation photoresist requirements for 250-nm lithography will require >20% dose latitude and >1.4-μm depth of focus, linear resolution to 225 nm, and ultimate resolution to 200 nm, where k_1 is well below 0.5 and in fact approaches 0.4. However, next generation photospeed requirements are unclear, with several stepper manufacturers designing next-generation steppers or step-and-scan systems to be powered by excimer lasers. If this development is realized, then photospeed requirements could be relaxed slightly. However, if imaging doses become too high, issues such as lens distortion or damage with the more powerful lasers will need to be addressed. Slower speed DUV photoresists would provide additional (needed!) flexibility in the design of more process-robust, environmentally stable resist systems that could be processed over substrates that have typically resulted in footing or undercutting problems.

The use of standard developers is a must if DUV materials are to be a part of mix-and-match lithography, since i-line photoresists use 2.38% TMAH and the majority of current DUV materials require varying developer normality. The integration of the exposure tool and the process track are an absolute necessity for current DUV materials. This integration could eliminate surface contamination problems or line slimming because of delays in postexposure bake or development.

Stepper and track manufacturers are continuing to investigate solutions to air filtration and elimination of contaminants that either enter the process chamber or emanate due to sealants or other adhesives used in the build process. Recent publications have discussed methods to detect airborne contamination and the quantitative measurement of NMP along with ammonia in the process area.[49] The further development of methods to accurately detect airborne contamination and to quantify the actual measurements vs degradation or contamination of the resist image will require further research. Processing over TiN, BPSG, SiN, and other metal surfaces has caused various degrees of image footing, as shown in Fig. 4.23. This problem has been combated by the incorporation of a thin antireflective layer to act as a barrier layer; however, this additional process step and material requirement adds cost. If improvements in photoresist materials can solve the footing problem, perhaps this will translate into surface contamination reduction or elimination.

4.4.3 Negative-Tone Process Considerations

There are numerous benefits that current negative-tone DUV resist systems can deliver to improve device performance and process simplification. Improved

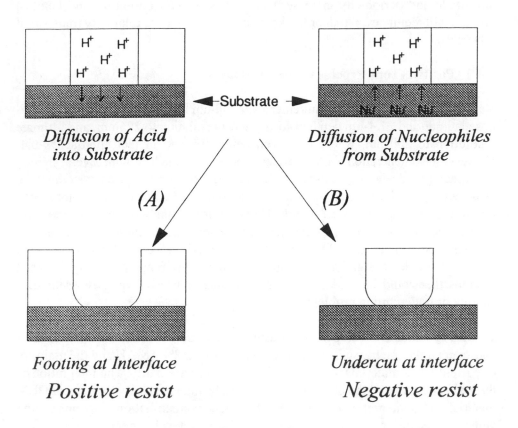

Diffusion of Acid
into Substrate

◄—Substrate—►

Diffusion of Nucleophiles
from Substrate

(A)

(B)

Footing at Interface

Positive resist

Undercut at interface

Negative resist

FIG. 4.23 Substrate contamination.

resolution and process performance have been attributes demonstrated by most recent systems. Workers at IBM have published several papers regarding CGR 248 negative DUV photoresist for improved linewidth vs PEB temperature and process yield improvements for 0.50-, 0.30-, and 0.25-μm phase shift mask lithography for advanced microprocessor applications.[50] Commercially available systems have fewer problems then their positive-tone counterparts. The issue of surface contamination is minimal vs. positive-tone materials; however, issues such as coating-bowl compatibility, microbridging, nonstandard developer normality, and image undercutting on TiN, BPSG, or SiN are issues that reduce the viability of the negative-tone technology. Process engineers are concerned with cross contamination with positive materials; separate coating bowls are unacceptable. Fabs are standardizing with 2.38% TMAH developer, and most commercially available negative resists are processed in lower-normality developers. The commonality of one developer for all processes eliminates logistical problems with additional developer lines, storage, etc., but is seriously challenging resist formulators.

4.4.4 Reflection Suppression Process

To maintain process latitude and resolution, and to minimize process complexity, the incorporation of a top ARL can be applied to a process to assist in controlling reflectivity. Top ARLs can minimize but not eliminate notching. The most effective means for suppression of reflective notching is the incorporation of a bottom ARL. Of course, there are a number of disadvantages that must be considered along with the advantages. First, current antireflection systems require vastly different processing conditions than most photoresists. These process conditions require semiconductor manufacturers to have additional hotplate stations and therefore add cost to tracks.

Suppressing reflective notching can significantly improve critical dimension linewidth control and in some cases is a necessity. Unfortunately, there are disadvantages to any additional coating process, especially if that process is significantly different in regard to the baking process. The requirement of removing the ARL with a separate reactive ion etch step creates another process step and adds to cost. However, if no additional process step is required and the substrate could be etched, this could make the technology more attractive.[51]

ARLs serve another purpose in the processing of DUV photoresists: as a barrier layer over substrates such as BPSG, TiN, SiN, and other metals. Positive systems today are not immune to contamination (resist foot) on these surfaces, and the ARL may have several uses and perhaps will be a necessity for device processing.[46]

4.5 FACILITIES CONSIDERATIONS

Unfortunately, various photoresist grades and tones, additional filtration units, hotplate stations, coating bowls, antireflective layers, and excimer lasers will require semiconductor manufacturers to construct additional fab space to accommodate these process tools. The development of filtration units by both stepper and track manufacturers will consume either process floor space or support areas. The use of ancillary materials with nonstandard process conditions will require additional hotplates along with separate spin apply bowls. If ARLs require reactive ion etch, then additional tooling will be required to accommodate these additional process levels. Future DUV exposure systems may require additional toxic process gases for laser usage and the necessary safeguards and space to accommodate the excimer laser. More tools equals more space and more cost. However, the payback is higher-performing devices!

4.6 MATERIALS MANUFACTURING AND QUALITY CONSIDERATIONS

4.6.1 Manufacturing Considerations

The manufacturing of DUV photoresist materials will bring a new set of rules to the synthesis of polymers, photoacid generators, additives, and the final formulation of the photoresist. New safeguards must now be considered to avoid cross contamination between various types of photoresists and starting materials. The level of reactor cleanliness must be reexamined to ensure that even parts-per-billion levels of cross contamination do not occur. The formulation of developers, strippers, and other ancillary materials will require that chemically amplified photoresist starting materials be synthesized and formulated in separate facilities. These concerns may affect the scale-up of starting materials with the necessity and additional cost of separate facilities, which may also be required to bottle and package materials to ensure no cross contamination.

Traditional methods of reducing metal contamination levels have been with filtration and the use of cationic, anionic, and a mixed bed of ion exchange resins. Each method and material serves a separate purpose; however, each brings with it a concern about polymer problems. There is an increasing level of activity to reduce the filtration pore size down to 0.05 μm from current 0.2-to 0.1-μm levels, which is driven by the need to reduce particle size for fabs that are developing 0.25-μm lithography. The old rule was that the particle size could be only ¼ of the size of the geometry.[52] This level of filtration could be extremely time consuming and, coupled with the cost of the filters and the time required, could drive the cost of producing the photoresist higher. In fact, it is possible that the cost of filtration alone could exceed the cost of the raw materials. The ¼ particle size to imaging geometry rule indicates that if 0.05 μm creates materials problems, what are the consequences at 0.18-μm or 0.15-μm geometries?

The use of ion exchange resins to reduce metal contaminants will require significant research; since all current polymer systems are acid sensitive, some are more acid labile than others. Some of these photoresist systems are in danger of degradation during processing with cationic exchange resins due to deprotection of the acid-labile materials during the attempted purification. This is also true for the more sensitive negative systems, which are acid hardening and could crosslink and thus reduce storage stability.

4.6.2 Quality Control Considerations

The production of materials for 0.25 μm and sub-0.25 μm and the issues related to quality control, along with testing accuracy, deserve some discussion. Semiconductor manufacturers are continuing to drive down the specification tolerances for acceptance of process chemicals. The future may require that purity-level specifications be in the parts-per-trillion range, and the requirement to expand the number of metals tested is increasing. In the previous subsection we discussed the level of filtration required to reduce particles; however, the ability to test to such levels will require suppliers to invest more capital in detection methods and cleaner manufacturing/testing and application facilities. The testing of incoming and outgoing materials may be more labor and capital intensive and continue to keep the cost of DUV materials higher than traditional high-resolution i-line photoresists. Semiconductor manufacturers continue to drive the level of metal contaminants down, even though no thorough study has been published correlating device yields with defects created from photoresist; that is independent of all other process steps. It is possible we could be overspecifying the materials and adding unnecessary cost to the photoresist!

The areas in quality control that differ from i-line to DUV may be in the purity of the PAGs and the ability to measure the photospeed of the resist. The necessity to accurately measure the amount of residual acid in a photoacid generating compound is far more critical with chemically amplified resist systems than with DNQ materials. The literature shows that various researchers have developed materials with extremely acid-labile copolymers and monomeric inhibitors. This could seriously impact shelf life and render useless the photoresist formulated with PAGs with extremely low levels of residual acid.

The ability to accurately test a photoresist that is 10 to 15 times more sensitive than conventional i-line materials creates a unique challenge. The goal is to develop a test that can accurately determine the photospeed of a resist with test errors similar to current novolac/DNQ materials. Recent literature suggests that test errors ranging from 0.2% to 0.4% are the current level that suppliers are working toward.[53] At IBM, the further development of cost-effective DUV photospeed testing has been given a high priority. The ability to accurately measure the photospeed of a photoresist material that has an E_0 dose of several millijoules is extremely difficult to perform with today's excimer based exposure systems. The requirement of accurately reproducing fine incremental dose control for the types of photoresist systems that are being tested is nearly impossible without a system such as Micrascan from SVG Lithography.

4.7 COST AND MARKET CONSIDERATIONS

Several publications have discussed the cost of ownership for DUV lithography. The common factor for each analysis is that DUV will and does cost more than current high-resolution i-line processes. These concerns will drive semiconductor manufacturers to pursue the extension of i-line technology or work diligently to reduce the cost of DUV via tools, materials, or the reduction of process steps. Recent models discuss the cost impact of tools, chemicals, service, operators, floor space, consumables, engineering, and maintenance on current i-line lithography. Tool cost, chemicals, service, and maintenance appear to be the major factors in the higher cost of DUV lithography. Within this set, the cost of chemicals—i.e., photoresist and ancillary materials—is higher than in any other category. Therefore, techniques such as resist minimization are continuously investigated and will continue to gain attention, because development and manufacturing fabs are currently using shot sizes ranging from 4 to 6 mls for 8-inch wafers and are working to reduce this by 50% to 75% to reduce the cost per level.

Moreau et al. discussed the ability to reduce photoresist shot size with various casting solvents for DNQ resist.[54] The paper stated that shot size reduction is affected by the viscosity of the formulation, the surface tension of the resists and substrate, and the evaporation rate of the resists. To achieve a minimum shot size, the resist formulation should have the lowest possible viscosity, lowest surface tension, and moderate evaporation rate. Deposition of a liquid resist on top of a liquid film can overcome surface forces and aid in shot size reduction. This is a technique the track manufacturers are offering, since they have demonstrated that a solvent pre-wet of the wafer surface can help to reduce resist shot size significantly and that a 1-ml shot for planar surfaces is possible.[54]

By the year 2000, many expect the usage of DUV photoresist to range between 35,000 and 50,000 gallons per year. Extremely large capital and material investment plus the lack of market acceptance/volume usage will keep photoresist cost high. As the i-line market developed, five years passed before it was 25% of the G-line market. Suppliers expect that it will be 9 or 10 years before the DUV market is 25% of the i-line market. At the year 2000, suppliers expect that the DUV market will be 10% of the i-line market, or 4% of the total market, making large investments in people and capital difficult.

4.8 193-NM LITHOGRAPHY

4.8.1 Introduction

ArF excimer (193-nm) lithography is a viable approach to extend optical lithography beyond 0.25 μm, but the resist technology developed for traditional DUV lithography is problematical at this deeper UV wavelength. We discuss the materials issues involved in the design of resists for 193 nm lithography with regard to optical properties, resolution, photospeed, and etch resistance.

The high-volume manufacture of semiconductors today involves the exclusive use of single-layer resist (SLR) processes, although potentially higher performance top-surface-imaged (TSI) and bilayer (BL) photoresist systems have long been proposed as alternatives. These resist processes, depicted in Fig. 4.24, may be distinguished

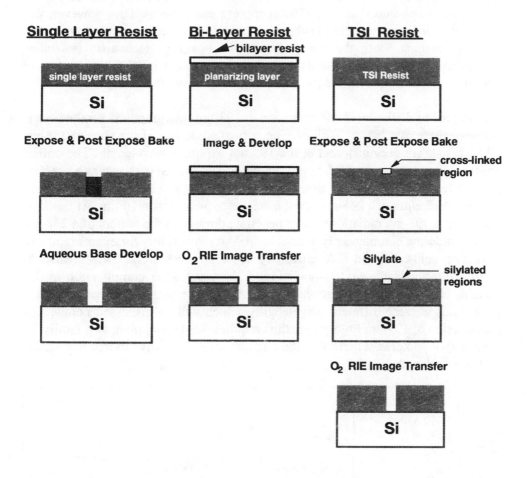

FIG. 4.24

by their complexity, with the number of process steps increasing considerably in the progression from SLR to BL and TSI approaches. As a consequence, the relative simplicity of the SLR approach has always been preferred, as long as the desired products are manufacturable using this time-tested approach. This in fact is the main driver to shorter wavelengths and new photoresist materials and imaging mechanisms. Put another way, the semiconductor industry will push optical lithography with SLR processes as far as possible. When the requirement for the extension of DUV lithography (either 248- or 193-nm) to performance limits that exceed the capability of SLRs, thin imaged resist systems (either in the form of TSI or BL) will likely emerge as important high-end manufacturing processes.

It would indeed be fortunate if the transition from 248- to 193-nm lithography could occur in the same (relatively smooth) manner that the transition from 436 to 365 nm occurred in the late 1980s. Many 436-nm (g-line) photoresists functioned effectively at 365 nm (i-line) because both the optical properties and the photochemistry of the DNQ/novolac photoresists were similar at these wavelengths. This, however, was not the case in the very difficult subsequent transition from 365-nm lithography to 248-nm imaging. Both the materials and the imaging mechanism (including photochemistry) were necessarily completely changed with the introduction of chemical amplification (in the form of acid-catalyzed deprotection).

The degree of difficulty in the transition to 193-nm lithography is somewhere in between these past two transitions. The optical properties of the current 248-nm DUV resists are very different at the 193-nm exposure wavelength. The optical absorbance of a typical p-hydroxystyren-based DUV resist is shown as a function of wavelength in Fig. 4.25. It should be noted that at a wavelength of 193 nm, the 0.3 absorbance/µm of a 248-nm resist increases to greater than 10/µm. At this high absorbance, 193-nm radiation cannot penetrate through to the bottom of a 248-nm resist, rendering commercially available DUV resists useless for exposure at 193 nm. The acid-catalyzed CA chemistry used in commercial DUV resists is completely applicable to 193-nm imaging, however. One complication in this imaging chemistry is the difference in chemical reaction pathway for generation of photoacids at these different wavelengths, which will be discussed below. The introduction of 193-nm lithography thus requires the development of a family (or families) of photoresist materials and chemistries specifically tailored to function at this wavelength.

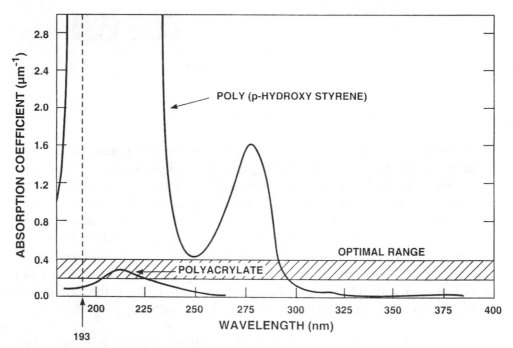

FIG. 4.25 Absorption of resins.

The successful introduction of 193-nm lithography requires the availability of a high-performance single-layer photoresist for a number of reasons: First, a simple lithographic process must be available to develop 193-nm DUV lithography exposure tools. Second, industry acceptance of a new lithography wavelength is likely to be most effective when SLR processing accompanies the new lithography. Finally, 193-nm lithography will probably replace 248-nm lithography when and if the replacement is economical and provides significant improvement in chip performance (or manufacturing yield).

Consequently, new 193-nm SLRs with high-performance imaging characteristics (resolution, depth of focus, process latitude, adhesion, sensitivity), plasma etch resistance equivalent or even superior to conventional DUV resists, and compatibility with industry standard processing chemicals (aqueous-base developers) must be developed.

A single-layer, chemically amplified resist based on 248-nm (DUV) resist chemistry is probably the most attractive candidate for a 193-nm lithography process, since it represents the most direct path from current manufacturing processes. Ideally, this would represent a drop-in replacement for the current generation of DUV resists and would exploit the considerable knowledge base that already exists for these systems. Unfortunately, the optical(transparency) requirements for these two systems are quite different.

Traditional DUV resists are based on hydroxystyrene polymers, phenolic resins with optical properties at 248 nm much improved over the structurally similar novolac resins. Hydroxystyrene polymers are extremely opaque at 193 nm, however. In fact, these resins are ideally suited for top-surface imaging at 193 nm, as will be discussed below. Consequently, a 193-nm single-layer resist with high-performance imaging characteristics (resolution, depth of focus, photospeed, adhesion, etc.), plasma etch resistance equivalent to conventional DUV resists, and compatibility with standard TMAH developers (0.14–0.26 N) is the challenge facing 193-nm resist designers.

The design of positive (single-layer) resists for 193-nm lithography is a significant challenge. This emerging field of photoresist research has recently been reviewed.[7] The imaging chemistry is quite similar to that practiced in traditional DUV lithography: photogeneration of a strong acid followed by acid-catalyzed deprotection to render the exposed regions of the film soluble in aqueous base. The differences in resist design between traditional (248 nm) and 193-nm lithography are related to matters of optical transparency, thus governing the selection of polymer family.

New polymer materials are required for 193-nm (single-layer) resists, with high optical transparency at the exposure wavelength combined with properties that hydroxystyrene polymers (and few others) possess:

- hydrophilicity (for good positive-tone development characteristics),
- high T_g (130–170°C), for good thermal properties and the latitude to perform higher postexposure bakes,
- aromatic rings in high concentration (for good etch resistance), and
- an easily blocked hydroxyl group (for incorporation of acid-cleavable functionality).

Acrylic polymers have been the most widely investigated polymer platform for 193-nm resist design because of their excellent optical transparency (see Fig. 4.25) and easily tailored structure. Many of the SLR approaches currently under development (based on current reports in the literature) are combinations of acrylic co- (or ter-, tetra-, etc.) polymers formulated with photoacid generators. Figure 4.26 shows the generic structure of an acrylic skeleton. The characteristics and advantages of acrylic CA resist systems are as follows:

- *Transparency at 193 nm:* The aliphatic nature of the polymer and the low extinction coefficient make for a very transparent polymer indeed. Figure 4.25 shows a comparison of the absorbance properties of poly(4-hydroxystyrene) (PHOST) with poly(methyl methacrylate) (PMMA).

- *Property diversity:* Substitution of the ester R group permits tremendous property flexibility. Additionally, copolymerization can dramatically expand the property spectrum. A virtually infinite variety of polymer characteristics can emanate from this family. Fortunately, the optical transparency at 193 nm is very high, providing a terrific platform on which to build a resist.

- *Ease of synthesis:* Acrylic polymers prepared for 193-nm single-layer resists are products of a "direct synthesis," unlike traditional DUV polymers, which are made via a two-, three- or even four-step process, adding significantly to the complexity and cost. Additionally, in another departure from DUV materials, acrylic polymers are synthesized without the need for acids or bases. In short, resist component manufacturability, an enormous impediment to the move to DUV technology, may not be nearly as problematic in 193-nm SLRs.

- *Plasma etch resistance:* Simple acrylics have etch rates (in aggressive plasma environments found in many semiconductor processes) approximately twice that of phenolic resists. Therefore, the polymer must be substantially modified to provide suitable etch resistance.

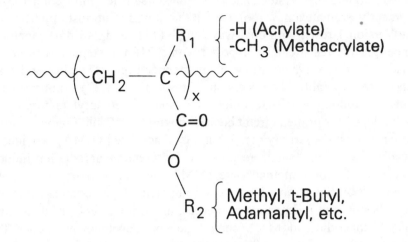

FIG. 4.26 Acrylic polymers.

Three key breakthroughs in the development of acrylic-based resist chemistry prior to the recent worldwide efforts toward the development of a 193-nm SLR were as follows:

1. Development of acid-catalyzed deprotection chemistry (CA) in the early 1980s, including acid-catalyzed reactions of esters of methacrylic acid by Ito and coworkers at IBM.[8]

2. Development of all-acrylic, aqueous-developing positive resists employing the methacrylate terpolymer concept, by IBM in the late 1980s.[55]

3. Recognition by workers at Fujitsu that plasma etch resistance and 193-nm transparency can reside simultaneously in an acrylic resin through incorporation of alicyclic components.[56]

Next we discuss the various approaches currently under investigation, most of which involve acrylic backbones. These approaches have been grouped according to basic resist properties.

4.8.2 193-nm Acrylic Single-Layer Resists

4.8.2.1 Tool evaluation resist

The first high-speed, high-resolution positive single-layer resist (IBM Version 1) was developed in 1993 by the IBM/MIT Lincoln Laboratory team.[57] The primary purposes of this resist was to evaluate the imaging performance of the SVGL prototype 193 step-and-scan system, and to demonstrate the feasibility of a 193-nm SLR. To properly evaluate a new tool, we felt that it was of utmost importance to design the Version 1 resist with dual-wavelength (193- and 248-nm) capability. With this design, the resist can be exposed with a 248-nm stepper with known characteristics. In this way, questions about stepper (optics) quality and resist performance are separable. The Version 1 resist (Fig. 4.27) comprises two components: an iodonium triflate onium salt and a methacrylate terpolymer originally developed for printed circuit board lithography.[55] Each monomer serves a separate function in the terpolymer. T-butyl methacrylate (TBMA) provides an acid-cleavable side group that is responsible for creating a radiation-induced solubility change. Methyl methacrylate (MMA) promotes hydrophilicity for photoinitiator solubility and positive-tone development characteristics while also improving adhesion and mechanical properties, and minimizing shrinkage after expose/bake. Methacrylic acid (MAA) controls aqueous development kinetics. This polymer is prepared in a single step from readily available, inexpensive components. By selecting the terpolymer composition and molecular weight, imaging properties(including dissolution properties, photospeed, contrast)can be altered to a significant extent.

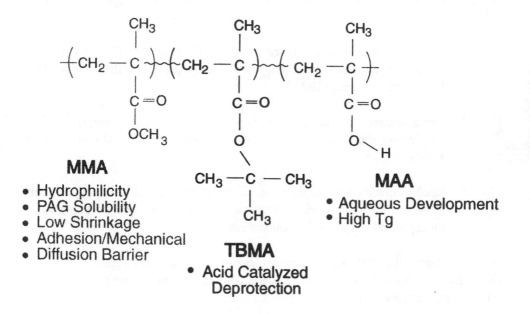

FIG. 4.27 IBM methacrylate terpolymer. (Version 1 resist).

Version 1A (initial formulation) resist was exercised extensively on the SVGL Micrascan 193 prototype. In fact, the resist was integral to the optical characterization of the tool and accelerated optimization of the prototype's optical characteristics. The highest resolution obtained at 193 nm (NA=0.5) for Version 1A was 0.22 μm in 0.75 μm of resist. Figure 4.28 shows 193-nm imaging of this resist using the SVGL prototype step-and-scan system.

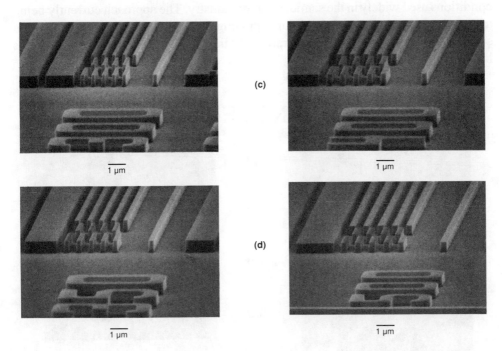

(a)

(c)

(b)

(d)

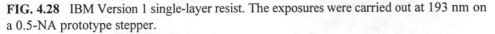

FIG. 4.28 IBM Version 1 single-layer resist. The exposures were carried out at 193 nm on a 0.5-NA prototype stepper.

An advanced version of this resist (Version 1B) was developed. Figure 4.29 shows the resolution capability in Version 1B exposed with a 248-nm stepper (with an NA=0.48) at dimensions of 0.30 μm and 0.35 μm, pictured through a dose range of 4.4 to 5.6 mJ/cm². Recent imaging results with 193-nm exposure (ISI ArF Microstepper) show a further gain in resolution to 0.14 μm. This photoresist has been extensively used recently in the development and evaluation of new 193-nm exposure tools.

It should be noted that most of the acrylic-based resists develop with base (hydroxide)concentrations significantly lower than is typically used with phenolic (248- or 365-nm) resists (0.01–0.05 N vs. the current production concentrations of 0.21–0.26 N). This difference could impact the introduction of this class of 193-nm resists into a manufacturing environment.

4.8.2.2 Etch resistant resists

The Version 1 resists described above have etch properties surprisingly similar to conventional DUV resists (e.g., IBM APEX-E) in CF₄-based (oxide) etch recipes. More aggressive etch chemistries (e.g., aluminum and polysilicon etching) demand substantial increases in etch resistance. For example, the etch rate of the Version 1 resist can be three times as high as that of a novolac-based resist under halogen etch conditions used widely in the semiconductor industry. The approach currently being used to impart etch resistance without degrading the optical transparency of 193-nm single-layer resists involves incorporation of alicyclic (aliphatic/cyclic) compounds into the polymer structure.

V1.0b 193-nm SINGLE-LAYER RESIST

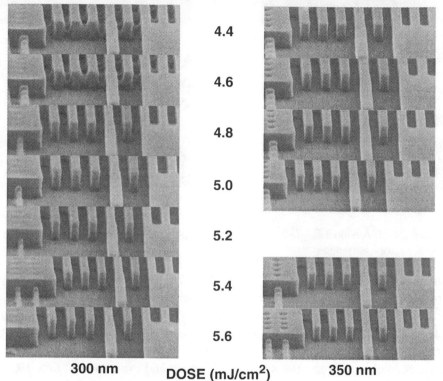

300 nm   DOSE (mJ/cm^2)   350 nm

- GCA XLS 0.48 NA DUV TOOL
- APEX-E DOSE-TO-SIZE AT E_{max} = 4.8 mJ/cm^2
- DEVELOPED IN 0.02 N TMAH

FIG. 4.29

The etch resistance of acrylic resists can be substantially improved by increasing the carbon content of the polymer via the incorporation of cyclic aliphatic functionality. For example, the etch rate of poly(isobornyl methacrylate) is less than half that of PMMA in a hydrogen bromide plasma. This approach was originally demonstrated by workers at Fujitsu.[56] It is based on earlier studies that correlated the etch rate with the carbon/hydrogen ratio of a series of polymers[58] and is currently the most commonly employed method used to develop etch-resistant, transparent 193-nm photoresists.

A more etch resistant two-component resist was recently described by workers at Fujitsu.[59] The resist features a hydrophobic copolymer (Fig. 4.30) of adamantyl methacrylate (ADMA) and 3-oxocyclohexyl methacrylate (OCM). The onium salt PAG used is triphenylsulfonium hexafluoroantimonate. The protecting group is thought to be more reactive than the t-butyl ester used in the IBM system, although with the "superacid" being generated upon exposure with this resist, it is not clear why this choice was made. Because of the hydrophobic nature of the Fujitsu copolymer, aqueous development is very difficult. Isopropyl alcohol is added to 0.26 N TMAH developer to aid in development. Substantial improvements in resist design and lithographic performance have been recently reported by Fujitsu (see below).

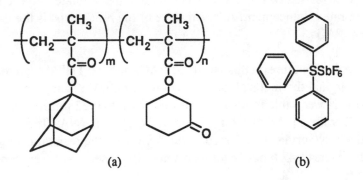

(a) (b)

FIG. 4.30 Resist structure of (a) 3-oxocyclohexyl methacrylate and adamantyl methacrylate copolymer (OCMA-AdMA), and (b) triphenylsulfonium hexafluoroantimonate (Ph_3SSbF_6).

The resist process uses a topcoat that apparently reduces t-topping. Photospeeds are highly dependent on the isopropyl alcohol content in the developer and are typically between 20 and 90 mJ/cm^2 with 248-nm exposure and somewhat faster at 193-nm (probably due to optical density). These workers were able to print 0.17-μm features (overexposed) with 0.4 μm of resist when exposed to 12 mJ/cm^2 on a Nikon prototype (NA=0.55) 193-nm exposure system, as shown in Fig. 31.

An aqueous-developing etch-resistant 193-nm resist was recently described by workers at NEC.[60] The resist material (Fig. 4.32) involves a terpolymer of methacrylic acid(adhesion/aqueous dissolution), tetrahydropyrannylmethacrylate (THPMA), an acid-labile protected monomer, and tricyclodecanyl acrylate

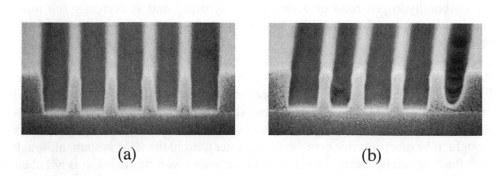

(a)   (b)

FIG. 4.31 Resist patterns of 0.17 μm lines and spaces printed by the ArF exposure system (NA=0.55), (a) with over-top coating, and (b) without over-top coating. A mixture containing 25wt% IPA is used as the developer. The exposure dose is (a) 12.0 mJ/cm², and (b) 13.8 mJ/cm².

(alicyclic, etch resistant). These authors claim that the acrylate a-hydrogen offers a dissolution rate enhancement. A key feature of the NEC resist is the use of nearly transparent PAGs.

The DUV resist literature on this family of PAGs and the THPMA protecting group suggests marginal thermal stability under current DUV resist process conditions. These workers were able to resolve features down to 0.22 μm in a 0.5-μm film at 50 mJ/cm2 with an ArF prototype exposure system (NA=0.55). The authors employed a poly(acrylic acid) topcoat and developed the resist/topcoat in 0.005 N TMAH. Etch rates 1.42 times faster than a novolac-based resist were reported (CF_4-RIE).

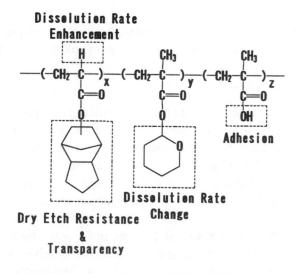

FIG. 4.32 Structural formula of Poly ($TCDA_x$-$THPMA_y$-MAA_z)

In summary, it is apparent from the two-component resist approaches published to date that high-resolution lithography using a 193-nm SLR is possible. It is equally likely that plasma etch resistance approaching DUV (aromatic) resists is feasible. Much more difficult is the development of a two-component resist formulation that possesses a combination of both imaging quality and plasma etch resistance. The lack of flexibility in a two-component approach, coupled with the (often competing) considerations of hydrophilic-hydrophobic balance, lithographic contrast, high T_g, thermal stability, and etch resistance, limits the options of the 193-nm resist designer.

In response to the design limitations of two-component resists, workers at IBM developed an alternative strategy to increase etch resistance.[61] This design involves the introduction of three-component resists consisting of methacrylate polymer slightly modified from Version 1, alicyclic dissolution inhibitor compound, and finally a photoacid generator. After evaluating large numbers of dissolution inhibitor compounds, it became apparent that a class of compounds (5-B steroids) was available (from natural sources) with very desirable properties:

- high solubility in resist and PGMEA
- strong dissolution inhibition
- high exposed dissolution rate
- 193-nm transparency
- moderating influence on T_g
- plasma etch resistance
- good thermal stability (>200°C)

This three-component approach (IBM Version 2) using steroidal dissolution/etch inhibitors offers flexibility for tuning resist performance and realizing increased etch resistance simultaneously. The etch resistance of Version 2 resist is far better than that of Version 1. The presence of alicyclics in both the polymer and dissolution inhibitor produced chlorine etch rates only slightly faster than those of novolac resins. Version 2 resist has achieved etch rates 1.2 times that of traditional DUV resists (APEX-E), in a resist formulation with good imaging quality. Figure 4.33 shows 0.5-μm features printed in 0.8 μm of Version 2 resist, exposed with a GCA XLS DUV stepper (NA=0.48) at 248 nm. This resist has been "scaled-up" and used extensively at MIT Lincoln Laboratory in a prototype device program.[62] The limitations include inadequate resolution (below 0.25 μm on the SVGL Microscan 193 stepper), the need for diluted developer (0.025–0.05N TMAH), and poor tolerance to aggressive semiconductor processes, such as ion implant.

Workers at Toshiba have recently described another methacrylate-based three-component resist where the increased etch resistance is provided by naphthalene functionality.[63] They have attempted to exploit the absorption shift that occurs on increasing the conjugation length of aromatic chromophores in the hope of identifying a phenol analog with a transparent window at 193 nm. A problem with this approach is that the most pronounced shift is observed for the longest wavelength absorption, while the higher-energy (shorter-wavelength) transitions are affected to a lesser extent. Substituent and matrix effects also play a role in the position and intensity of the absorption of the aromatic chromophore. The authors' three-component system is based on a tetrapolymer containing 5% naphthalene, and a dissolution inhibitor and PAG that are both naphthalene based. In practice, the amount of aromatic incorporation afforded is too low to impart significantly more etch resistance. A resolution limit of 0.16 μm at an imaging dose of 300 mJ/cm2 was reported.

PRELIMINARY RESULTS – V2.0 SINGLE-LAYER RESIST

RESIST: V2.0 – THREE-COMPONENT RESIST WITH IMPROVED ETCH RESISTANCE
POLYMER – 16% WEIGHT ISOBORNYL MOIETY
INHIBITOR – CHOLIC ACID ESTER DERIVATIVE
PHOTOACID – BIS (t-BUTYL PHENYL) IODONIUM TRIFLATE
TOTAL RESIST 32% WEIGHT ALICYCLIC CARBON

ETCH RESISTANCE: $1.2 \times$ NOVOLAC IN HIGH-DENSITY Cl_2 PLASMA (Helicon)
IMAGING: 0.48 NA DUV STEPPER

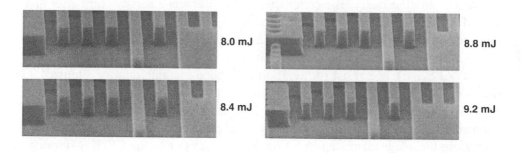

8.0 mJ

8.8 mJ

8.4 mJ

9.2 mJ

500-nm FEATURES

FIG. 4.33

4.8.3 Recent Advances in Materials for 193-nm Single-Layer Resists

In the past twelve to eighteen months (late 1995 through 1996) there has been a dramatic worldwide increase in the development activities in many areas of 193-nm resists. The following are references to some of this work.

4.8.3.1 New polymers

A cooperative effort of workers at IBM, University of Texas at Austin, and BF Goodrich has demonstrated that 193-nm resists can be developed using (nonacrylic) polymers based on the materials derived from the polymerization of cyclic olefins (e.g., norbornene derivatives).[64] The imaging properties do net yet approach those of photoresists based on acrylic polymers (see Fig. 4.34), but plasma etch rates lower than that of novolac have been demonstrated. These cycloaliphatic backbone polymers derived from the polymerization of cyclic olefin monomers have transparency (at 193 nm) comparable to that of acrylic polymers. The etch resistance of poly(norbornene) (PNB) is extremely high compared to that of alicyclic acrylic resist systems like the IBM Version 2 resist. In fact, this cycloaliphatic polymer is *more etch resistant than novolac,* when compared in an aggressive chlorine plasma. Table 4.2 shows the relative etch rate of a variety of resist materials (etch rate relative to novolac polymer in a chlorine plasma).

Table 4.2

Material	Etch Rate
IBM Version 1 Resist	2.2
IBM Version 2 Resist	1.4
APEX-E DUV Resist	1.2
Novolac Resin	1
Poly(Norbornene)	0.85

Workers at AT&T (Lucent Technologies) have published information on a new resist system that is effectively a cyclic olefin/acrylic hybrid prepared from the free radical alternating copolymerization of norbornene and maleic anhydride. Acrylate monomers (acrylic acid and t-butyl acrylate) are also included in the polymerization. Photoresists based on these novel materials show quite attractive resist performance.[65] Imaging properties improve dramatically when these hybrid polymers are formulated as three-component resists, using steroidal additives to improve the performance of these new materials, in much the same way as the IBM Version 2 resist.

FIG. 4.34 Contact print image (15mJ/cm² @ 254-nm) of cycloaliphatic polymer resist, containing a poly(norbornene-t-butyl ester) formulated with triphenylsulfonium hexafluoroantimonate. Feature sizes of 1.5 µm (top) and 1.25 µm (bottom) are shown.

4.8.3.2 New high-performance resists

Workers at Fujitsu,[66] NEC,[67] and Toshiba[68] have recently published information on new (second-generation) acrylic resists that combine good etch resistance and clearly improved imaging characteristics. Lithographic performance data on 193-nm steppers with these new resists remains to be demonstrated. The Fujitsu resist is particularly noteworthy in that it combines for the first time three key properties: etch resistance approaching DUV resists, high resolution, and compatibility with industry-standard 0.26-N TMAH developers.

4.8.4 Plasma Etching of 193-nm Resists

Although it seems clear that acrylic-based polymers are well suited for high-resolution imaging applications at exposure wavelengths down to 193 nm, the chemical durability of this class of polymers is, in general, too low to withstand the harsh plasma etch steps required to transfer the resist pattern into the underlying substrates. Since the first reports of a high-resolution 193-nm SLR in 1993, work has focused on developing formulations with improved etch resistance. These reports have primarily centered on the incorporation of alicyclic moieties into acrylic resist formulations as the means to improve their etch resistance, as was discussed above. However, the reported improvements in etch resistance were obtained using etch chemistries (e.g., Ar, CF_4, or HBr) that are not always representative of the most aggressive plasma etches used in semiconductor manufacturing. However, systematic etch durability studies have recently been conducted using a high-ion-density helicon plasma source on over 30 acrylic resin/inhibitor combinations incorporating 12 copolymers and 4 alicyclic dissolution inhibitors.[69] All the formulations were etched in aggressive chlorine plasmas (i.e., metal etch) under identical process conditions. The results were then compared to empirical polymer parameters. The first, reported by Ohnishi,[58] is the

$$\text{Ohnishi parameter} = N/(N_c - N_o) ,$$

where N, N_c, and N_o are the total number of atoms, number of carbon atoms, and number of oxygen atoms, respectively, per monomer.

The Ohnishi parameter was originally used as a means to develop a relationship between chemical structure and reactive ion etch rate for a high- energy (300–500 eV) ion bombardment, but it showed a poor relationship to chemical structure for etch rates in a downstream glow discharge, where the etching mechanisms are largely chemical in nature. Kunz[69] has introduced an additional parameter, called the ring parameter:

$$\text{Ring parameter} = \text{MCR/MTOT} ,$$

where MCR and MTOT are the mass of the resist existing as carbon atoms contained in a ring structure and the total resist mass, respectively. Of these two parameters, the ring parameter appears to be a better predictor of etch rate in a high-ion-density, low-ion-energy plasma. Figure 4.35 shows the experimental data summary for normalized etch rate versus the Ohnishi parameter [Fig. 4.35(a)] and the ring parameter [Fig. 4.35(b)]. The experimental data were obtained using materials ranging from PMMA on one extreme to tetrapolymers incorporating alicyclic methacrylates, as well as monomeric additives based on the steroid structures discussed above.[61] From these experiments, models can be developed to assist in the determination of optimal chemical structure of new 193-nm photoresists that can meet the needs of tomorrow's semiconductor manufacturing.

4.8.5 Reflectivity Control at 193 nm

The evolution from 365 to 248 nm has presented lithography engineers with new challenges caused by increased thin-film interference effects in resists during exposure. These challenges were the result of the inherently higher reflectivity that

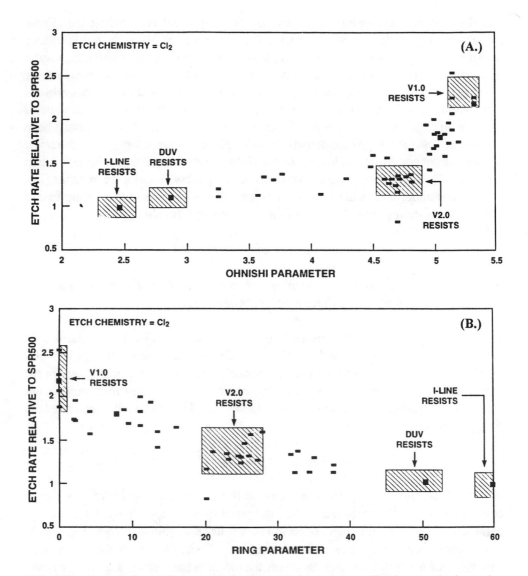

FIG. 4.35 (a) Experimentally measured polymer etch rates, normalized to standard novolac i-line resist, as a function of the Ohnishi parameter. The etch conditions in the helicon rf plasma source were: chlorine gas, source power of 2000 W, 75 W rf power applied to the wafer, 2 mTorr pressure, 100 sccm chlorine flow. (b) The same data as in (a), but now it is plotted vs the ring parameter.

occurs at shorter wavelengths and the development of new nonbleachable photoresists, and it has fueled the development of both top and bottom antireflective layer (ARL) technology for use at 248 nm. The extension of production photolithography to 193 nm will also require reflection suppression for certain lithography levels. Already, extensive experimental and modeling work has been done at 248 nm to better understand requirements, limitations, and performance of ARL technology.[70] This framework will have a direct bearing on developments at

193 nm, since much of this modeling can be generically applied to any exposure wavelength. The limitation at 193 nm, however, lies in knowledge of the optical properties of the materials in use, since few commercially available instruments can measure the complex refractive index at wavelengths <200 nm.

Most bottom ARLs designed for use at 248 nm are composed of strongly absorbing phenyl-containing polymeric systems. A recent survey of refractive indices of organic polymers at 193 nm[47] suggests that many aromatic systems have reflection coefficients at 193 nm that, although small (2%–5%), are several times higher than reflection coefficients for 248 nm. These increased reflectivities are caused by dispersion in the real part of the refractive index caused by very high absorptivities. For example, at 193 nm, novolac has a real component to the refractive index of only 1.36—a considerable mismatch from the photoresist's value of 1.68. This suggests that aromatic-based (248-nm) deep ultraviolet ARLs may not be ideally suited for use at 193 nm. However, prototype 193-nm ARLs have already been developed that incorporate a more transparent methacrylate functionality.[47] The refractive indices of 1.5 to 1.6 for these ARLs better match the real part of the refractive index of the 193-nm resist systems, which are also methacrylate based. From these early experiments, it seems as though 193-nm ARL development should be straightforward and limited only by the availability of precision measurement tools.

4.8.6 TSI and Multilayer Resists for 193-nm Lithography

Multilayer (bi- and trilevel) and top-surface-imaged resists have been the subject of sporadic interest over the last decade. The TSI and bilevel approaches rely on oxygen-plasma-based pattern transfer from a thin, silicon-rich imaging layer to a thick planarizing layer. An additional pattern transfer step is required in the trilevel scheme. The advantages of these approaches include improvement in the depth of focus, ability to print high-aspect-ratio features, and minimization of reflectivity and topography effects. Given the challenges inherent in the design of 193-nm SLRs (enumerated in detail above), and also as a consequence of recent advances in silylation tools and plasma etch systems, TSI and multilayer resist approaches have become very attractive candidates for early implementation of 193-nm lithography. In this section we describe 193-nm resist systems based on these approaches.

Since 193-nm resist development began in 1988, bilayer and TSI resist systems have been considered to be lower-risk approaches than single-layer resists, since many more resist chemistries are available when requirements on transparency are relaxed. Over this time, bilayer resist schemes using poly(silylmethacrylates),[71] poly(silanes),[72] poly(silynes),[73, 74] and poly(siloxanes)[75] have been reported. The most mature of these processes, however, involve TSI silylation resists employing a single-component poly(vinyl phenol) resin,[76] a far more simple process than TSI resists for 248-nm lithography, which require a 248-nm absorbing dye and rely on photoacid-catalyzed cross linking. In the silylation scheme, exposure directly crosslinks the polymer resin, thereby reducing the indiffusion rate of silicon-bearing reagents. These reagents, typically silyl amine derivatives, react with the phenolic hydroxyl group to form a silyl ether. The diffusion selectivity has a contrast of 1 to 2, which when combined with a highly selective oxygen-plasma dry etch step, can yield overall process contrast values in excess of 50. Figure 4.36 shows 200-nm dense and isolated lines exposed by a 0.5-NA 193-nm step-and-scan. Here, the silylating agent was dimethylsilyldimethylamine, and the silylation depth was 130 nm. The development was performed using a helicon RF oxygen plasma source. Using similar conditions, a process depth of focus of 1.0 um was obtained for 175-nm dense lines and resolution to 150 nm has been achieved.

Several areas of further improvement are required before this process meets all production specifications. Of these improvements, the photospeed of ~45 mJ/cm² must be improved. Preliminary work[77] using chemically amplified systems at 193 nm suggests that photospeeds of ~5 mJ/cm² are achievable, and work[78] at 248 nm indicates that acceptable imaging performance is achievable with chemically amplified silylation systems. The perceived weakness of the silylation approach stems largely from early pilot-line experience with TSI processes where the use of older generation equipment proved troublesome. Issues that will require experimental reverification on current tools before full acceptance can be realized are (1) cross-wafer and lot-to-lot silylation rate and oxygen etch uniformity, (2)

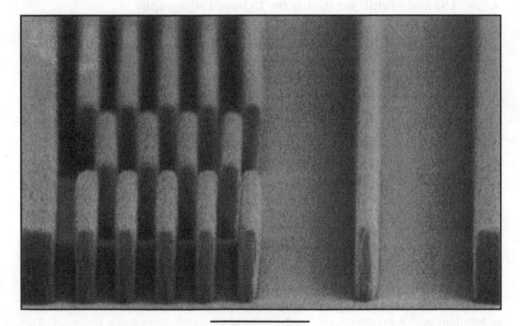

1 μm

FIG. 4.36 SEM of 0.20-μm silylated resist features. Resist 0.76-μm-thick PVP silylated with DMSDMA at 90° C and 25 Torr for 60 s was etched under optimized conditions.

control and minimization of line-edge roughness, and (3) yield as it relates to substrate damage and process-generated particulates. Given the current state of modern high-ion-density etchers and production-worthy silylation systems, the outlook for TSI processes to finally becoming accepted (at either 193 nm or 248 nm) looks promising, as lithographers look to all means for enhancing the resolution capabilities of their exposure tools.

Although the enabling technologies are clearly on the horizon for TSI, economic factors may prove to be the limiting factors for implementation. For these reasons, multilayer resist schemes must also be considered backup technologies should single-layer approaches fail to meet all imaging and etch-resistant requirements. Although a number of bilayer schemes have already been reported for use at 193 nm, even the most promising systems still suffer from major design flaws such as the need for nonaqueous developers, too low a glass transition, or limited shelf life. The development of these approaches into fully optimized processes will require a materials effort on scale with the SLR development programs. As an alternative, trilayer processes (using an intermediate hard-mask layer) may be possible for process steps possessing stringent etch requirements. In this type of process, a Version 1 type (see above) SLR system could be used to pattern a thin (~200 nm) oxide hard mask lying on top of an organic planarizer. The planarizer would then be patterned using an oxygen plasma etch and, together with the hard mask, would act as an etch mask for the actual device etch. Although the technology for such a trilayer resist process exists today, economics will again dictate whether this route is a viable one.

4.9 CONCLUSION

In this chapter we have discussed the revolutionary changes in resist materials required to extend the life of optical lithography beyond 0.25-µm devices. There are a great many factors and obstacles that DUV technology must overcome to be a viable manufacturing technology for the production of sub-0.35-m geometries. Reviewing the recent literature has given a broad overview of the obstacles that have been overcome and challenges that need to be addressed for DUV to become the industry workhorse for sub-300-nm lithography. Materials and process-related issues have been addressed for various aspects including positive and negative tone along with materials concerns for controlling reflectivity; we have also addressed issues pertaining to the facilities required to support the additional process tools and support systems required for DUV manufacturing along with new psychology regarding the manufacturing of DUV materials. Further discussion has been offered addressing issues related to the quality control and the more stringent controls required to manufacture photoresist materials that semiconductor manufacturers will require.

Finally, i-line will most likely take us to 300 nm, and some will attempt to push i-line lithography all the way to the 250-nm generation. The 250 nm (256-Mb) device generation is the point where DUV will begin to dominate and perhaps take us to 180 nm (1 Gb) early device generations as stepper manufacturers developer higher NA lens systems. This should be rapidly followed by the introduction of early 193-nm manufacturing, provided the nascent ArF stepper infrastructure and high-performance 193-nm single-layer resists become available in a timely fashion. Higher NA (DUV) brings more resolution but at the expense of depth of focus, which may accelerate the need for 193-nm lithography. Perhaps this is where top surface imaging or bilayer resist technology will enter to provide relief as device manufacturers search for more lithographic process latitude.

REFERENCES

1. R. R. Dammel, *Diazonaphthoquinone-based Resists,* SPIE Vol. TT11, SPIE, Bellingham WA (1993).
2. M. Hanabata, F. Oi, and A. Furuta, Proc. SPIE 1466, p. 132 (1991); R. D. Allen, K. J. R. Chen, and P. M. Gallagher-Wetmore, Proc. SPIE 2438, p. 250 (1995); P. C. Tsiartas, L. L. Simpson, A.Qin, C. G. Willson, R. D. Allen, V. J. Krukonis, and P. M. Gallagher-Wetmore, Proc. SPIE 2438, p. 261 (1995).
3. See collection of papers in Proc. SPIE 2438, Session 4: Novolac/DNQ Interactions, pp. 282 ff., (1995).
4. E. Reichmanis, F. Houlihan, O. Nalamasu, and T.Neenan, *Chem. Mater.* **3,** 394 (1991).
5. S. A. MacDonald, C. G. Willson, and J. M. J Frechet, *Acct. Chem. Res.* **27,** 151 (1994).
6. H. Ito, *Solid State Technol.*, pp. 164-173 (July 1996).
7. R. D. Allen, G. M. Wallraff, D. C. Hofer, R. R.Kunz, S. C. Palmateer, and M. W. Horn, "Photoresists for 193-nm Lithography," *Microlithography World,* 21 (Summer 1995).
8. H. Ito, C. G. Willson, Technical Papers of SPE RegionalTechnical Conference on Photopolymers, 331 (1982); H. Ito, C. G. Willson, J. M. J. Frechet, U.S. Patent 4,491,628 (1985).
9. D. McKean et al., *J. Polym. Sci.: Polym. Chem.*, **27,** 3927 (1989).
10. A. A. Lamola, C. R. Szmanda, and J. W. Thackeray, *Solid State Technol.*, **34**(8), 53 (1991).
11. W. Feely, J. Imhof, C. Stein, T. Fisher, and M. Legenza, *Polym. Engin. Sci.,* **16,** 1101(1986); W. Feely, Proc. SPIE 631, p. 48 (1986).
12. J. Maltabes et al., Proc. SPIE 1262, p. 2 (1990).
13. S. A. MacDonald et al., Proc. SPIE 1466, p. 2 (1991); W. D. Hinsberg et al., Proc SPIE 1925, p. 43 (1993).
14. S. A. MacDonald et al., Proc. SPIE 1466, p. 2 (1991).
15. T. Grafe, "Control and measurement of airborne ammonia, NMP and other bases in deep UV lithographytools," Donaldson Co., (Presented at SVGL

Users Group Meeting), Santa Clara, CA (March 10, 1996).

16. O. Nalamasu et al., *J. Photopolym. Sci. Technol.*, **4**, 299 (1991); O. Nalamasu et al., *J. Vac. Sci.Technol.*B, **10**(6), 2536 (1992).

17. H. Roschert et al., Proc. SPIE 1672, p. 33 (1992).

18. R. Wood, C. Lyons, J. Conway, and R.Mueller, Proc. KTI Inteface '88, p. 341 (1988).

19. R. Tarascon et al., *Polym. Eng. Sci.*, **29**, 850 (1989); O.Nalamasu et al., Proc. SPIE 1262, p. 32 (1990).

20. W. Conley et al., Proc. SPIE 2195, p. 461 (1994).

21. G. Schwartzkopf et al., Proc. SPIE 1466, p. 26 (1991).

22. T. Fedynyshyn, J. Thackeray, H. Georger, and M. Denison, *J. Vac. Sci. Technol.* B, **12**(6), 3888 (1994); T. Ueno, L. Schlegel, N. Hayashi, H. Shiraishi, and T. Iwayanagi, *Polym. Eng. Sci.*, **32,** 1512 (1992).

23. R. Hayase, Y. Onishi, H. Niki, N. Oyasato, and S. Hayase, *J. Electrochem. Soc.*, **141**(11), 3141 (November 1994).

24. Y. Kawai, A. Otaka, A. Tanaka, T. Matsuda, *Jpn. J. Appl. Phys.*, **33**, 7023 (1994); Y. Kawai et al., *J. Photopolym. Sci.. Technol.*, **8**(4), 535 (1995); K. Asakawa, T. Ushirogouchi, and M. Nakase, Proc. SPIE 2438, p. 563 (1995).

25. K. Przybilla et al., Proc. SPIE 1925, p. 76 (1993); S. Funato et al., *J. Photopolym. Sci. Technol.*, **8**(4), 543 (1995).

26. H. Ito et al., Proc. SPIE 1925, p. 65 (1993); H. Ito et al., *J. Photopolym. Sci. Technol.*, **7**, 433 (1994).

27. W. Huang, R. Kwong, A. Katnani, and M. Khojasteh, Proc. SPIE 2195, p. 37 (1994); C. Mertersdorf et al., Proc. SPIE 2438, p. 84 (1995).

28. See reference 17; also M. Padmanaban et al., Proc. SPIE 2195, p. 61 (1994).

29. O. Nalamasu et al., in *Microelectronics Technology: Polymers for Advanced Imaging and Packaging,* E. Reichmanis, C.Ober, S.MacDonald, T. Iwayanagi, and T. Nishikubo, Editors, ACS Symposium Series 614, pp. 4-20 (1995).

30. W. D. Hinsberg et al., Proc. SPIE 1925, p. 43 (1993); P. Paniez et al., Proc. SPIE 2195, p. 14 (1994).

31. H. Ito et al., Proc. SPIE 2438, p. 53 (1995).

32. T. Iwayanagi et al., *IEEE Trans. Electron. Devices*, **28,** 1306 (1981); S. Nonogaki et al., Proc. SPIE 539, p. 189 (1985).

33. J. Thackeray et al., Proc. SPIE 1086, p. 34 (1989).

34. C. Eckes et al., Proc. SPIE 1466, p. 394 (1991).

35. B. Reck et al., *Polym. Engin. Sci.*, **29**(14), 960 (1989); R. Allen, W. Conley and J. Gelorme, Proc. SPIE 1672, p. 513 (1992).

36. W. Conley et al., Proc. SPIE 1925, p. 120 (1993).

37. W. Brunsvold et al., Proc. SPIE 2195, p. 329 (1994).

38. J. Thackeray, G. Orsula, and M. Denison, Proc. SPIE 2195, p. 152 (1994); L. Linehan et al., Proc. SPIE 2438, p. 211 (1995).

39. T. Tanaka, M. Morigami, and N. Atoda, *Jpn. J. Appl. Phys.*, **32**(12B), 6059 (1993).

40. V. Rao, W. Hinsberg, C. Frank, and R. Pease, Proc. SPIE 1925, p. 538 (1993); C. Mack, D. DeWitt, B. Tsai, and G. Yetter, Proc. SPIE 2195, p. 584 (1994); V. Rao et al., Proc. SPIE 2195, p. 596 (1994); B. Beauchemin et al., Proc. SPIE 2195, p. 610 (1994).

41. G. Wallraff et al., Proc. SPIE 2438, p. 182 (1995); J. Hutchinson et al., Proc. SPIE 2438, p. 486(1995).

42. N. Kihara et al., J. Photopolym. Sci. Technol., 8(4), 561 (1995).

43. W. Conley et al., Proc. SPIE 2724, p. 34 (1996).

44. M. Plat et al., Proc. SPIE 2438, p. 272 (1995).

45. T. Tanaka, N. Hasegawa, H. Shiraishi, and S. Okazaki, J. Electrochem. Soc., 137, 3900 (1990).

46. T. A. Brunner, C. F. Lyons, and S. S. Miura, J. Vac. Sci. Technol. B, 9, 3418 (1991).

47. R. Kunz and R. Allen, Proc. SPIE 2195, p. 447 (1994).

48. T. Brunner, Proc. SPIE 1466, p. 61 (1991).

49. K. Dean, N. Thane, and J. Sturtevant, OCG Interface Proc., p. 199 (1994).

50. W.Conley et al., Proc. SPIE 1925, pp. 120–132 (1993); S. Holmes et al., Proc. Semicon/Kanasi-Kyoto Technology Seminar, pp. 85–96 (1993); M. Neisser et al., OCG Interface Proc., p. 161, (1994).

51. W. Moreau et al., Proc. SPIE 2195, p. 225 (1994).

52. W. Moreau, Semiconductor Lithography, Principles, Practices and Materials, Plenum Press (1988).

53. W. H. Ostrout, T. L. Brown, "Advanced Method for Determining Photoresist System Capability," Proc. SPIE 1926, p. 134 (1993).

54. W. Moreau et al., Proc. SPIE 2438, p. 646 (1995).

55. R. Allen, G. Wallraff, W. Hinsberg, and L. Simpson, J. Vac. Sci.Tech., B9(6), 3357 (1991).

56. Y. Kaimoto, K. Nozaki, S. Takechi, and N. Abe, Proc. SPIE 1672, p. 66 (1992).

57. R. R. Kunz, R. D. Allen, W. D.Hinsberg, and G. M. Wallraff, Proc. SPIE 1925, p. 167 (1993); R. D. Allen et al., J. Photopolym. Sci. Tech., 6(4), 575 (1993), R. D. Allen et al., "Methacrylate TerpolymerApproach in the Design of a Family of Chemically Amplified PositiveResists," Chap. 11 in ACS Symposium Series, No. 537, Polymers for Microelectronics, L. Thompson, C. G. Willson, and S. Tagawa (1994).

58. Y. Ohnishi, M. Mizuko, H. Gokan, and S. Fujiwara, J. Vac. Sci. Technol., 19(4), 1141 (1981); H. Gokan, S. Esho, and Y. Ohnishi, J. Electrochem. Soc., 130(1), 143 (1983)

59. M. Takahashi et al., Proc. SPIE 2438, p. 422 (1995).

60. K. Nakano, K. Maeda, S. Iwasa, T. Ohfuji, and E. Hasegawa, Proc. SPIE 2438, p. 433 (1995).

61. R. Allen et al., Proc. SPIE 2438, p. 474 (1995).

62. C. Keast et al., presentation at the1996 International Conference on 193-nm Lithography, Sematech, Colorado Springs, August 1996.

63. M. Nakase et al., Proc. SPIE 2438, p. 445 (1995).
64. R. D. Allen, R. Sooriyakumaran, Juliann Opitz, G. M. Wallraff, R. A. DiPietro, G. Breyta, D. C. Hofer, R. R. Kunz, S. Jayaraman, R. Shick, B.Goodall, U. Okoroanyanwu, and C. G. Willson, Proc. SPIE 2724, p. 334 (1996).
65. T. I. Wallow, F. M. Houlihan, O. Nalamasu, E. Chandross, T. X. Neenan, and E. Reichmanis, Proc. SPIE 2724, p. 355 (1996).
66. K. Nozaki, K. Watanabe, E. Yano, A. Kotachi, S. Takechi, and I. Hanyu, *J. Photopolym. Sci. Technol.*, **9**(3), 509 (1996).
67. S. Iwasa, K. Maeda, K. Nakano, T. Ohfuji, and E. Hasegawa, *J. Photopolym. Sci. Technol.*, **9**(3), 447 (1996).
68. N. Shida, T. Ushirogouchi, K. Asakawa, and M. Nakase, *J. Photopolym. Sci. Technol.*, **9**(3), 457 (1996).
69. R. Kunz et al., Proc. SPIE 2724, p. 365 (1996).
70. T. Ogawa et al., Proc. SPIE 1927, p. 263 (1993).
71. B. W. Smith et al., Proc. SPIE 2438, p. 504 (1995).
72. R. Kunz et al., *Jpn. J. Appl. Phys.*, **31**, 4327 (1992)
73. R. Kunz et al., J. Vac. Sci. Technol.B, **10**, 2554 (1992).
74. R. Kunz, presented at the IEEE Workshop, Santa Fe, NM, Oct. 1992.
75. M. Hartney et al., Proc. SPIE 1262, p. 119 (1990).
76. M. Harney et al., Proc. SPIE 270 (1993); D. Johnson and R. Kunz, *J. Photopolym. Sci Technol.*, **6**, 593 (1992); S. Palmateer et al., Proc. SPIE 2438, p. 455 (1995); K. Macda et al., Proc. SPIE 2438, p. 465 (1995).
77. D. LaTulipe et al., Proc. SPIE 2195, p. 372 (1994).
78. C. Garza, E. Soloweij, M. Boehm, *Polym Eng. Sci.*, **32**, 1600 (1992).

CHAPTER 5
Photomask Fabrication Procedures and Limitations

John G. Skinner
JGSA Inc.

Timothy R. Groves
IBM Semiconductor Research and Development Center

Anthony Novembre
Bell Laboratories, Lucent Technologies

Hans Pfeiffer
IBM Semiconductor Research and Development Center

Rajeev Singh
Intel Corporation, Technology and Manufacturing Group

CONTENTS

5.1 INTRODUCTION

This chapter is concerned with masks for optical lithography. Photomasks used in present-day semiconductor manufacturing are reduction reticles, where the pattern is formed in a chromium layer over a fused silica substrate. The pattern represents one level of an integrated circuit (IC) design. An optical stepper forms a four to five demagnified image of the reticle on the wafer. The reticle usually contains one or more identical circuit patterns plus wafer process test patterns and marks for aligning the reticle in the optical stepper. Typically, 20 to 25 different mask levels are required for a complete IC device.

A photomask is made by exposing, or writing, the required circuit pattern in a resist layer spun on top of the chromium layer. The resist layer is developed to form the required pattern in the resist layer, the chrome layer is then etched through the patterned areas in the resist, and the resist layer is then removed. The mask features have to be measured to insure that the etched pattern is correct to within some specified tolerances, and the pattern also has to be examined for integrity, including missing or extra chromium defects. The final mask is cleaned, and the front and back surfaces are then protected with a thin standoff membrane called a pellicle.

Photomasks have always been an important part of the lithographic process, but in recent years they have become a very critical part of lithography because of the difficulty in achieving the ever tighter pattern specifications. Mask tolerances for near-future IC designs are measured in tens of nanometers, defect sizes in hundreds of nanometers, and pattern data volume in gigabytes. Masks are becoming increasingly difficult to fabricate because specifications are reaching the electromechanical limitations. Mask pattern errors that were once considered insignificant, such as feature edge shifts caused by proximity effects and resist heating during pattern generation, and feature displacement errors caused by the gravitational effect on the mask substrate, are now becoming a significant part of the allowed photomask tolerances. This chapter reviews mask specifications, mask types, pattern generation, mask processing, inspection and repair, and metrology.

Mask technology made several major changes around 1990 with the introduction of 0.5-μm technology. Lithography was beginning to hit a barrier with the available lithography tools and innovative mask technologies that were started a decade earlier were investigated further. Thicker mask substrates were introduced to reduce mask distortion, and fused silica mask blanks were more in demand for better transmission at the shorter actinic wavelengths. The introduction of new mask technologies together with tighter mask specifications continue to make mask making a challenging and interesting field.

Photomasks have nine principal fabrication steps:

1. Substrate preparation
2. Pattern writing
3. Pattern processing
4. Metrology
5. Inspection for pattern integrity
6. Cleaning
7. Repair
8. Pellicle attachment
9. Final defect inspection.

This chapter is concerned with topics 1 through 6, the formation and verification of the mask pattern.

5.2 PHOTOMASKS, MASKS, AND RETICLES: TERMINOLOGY

Photomasks are used in conjunction with actinic radiation that can be "focused" with glass or fused-silica lenses, as compared with thin membrane masks used for x-ray lithography and stencil masks used for ion-beam lithography.

The terms masks and reticles are used interchangeably, but traditionally the term *mask* refers to a 1X patterned substrate that contains the entire wafer pattern and is used for contact printing with a wafer. The term *reticle* refers to a patterned substrate that contains one to about six copies of the chip pattern and is used with an optical projection system to image the pattern onto the wafer.

"Reticle" is derived from the Latin word meaning grids or nets, and originally referred to an open frame across which "marks" were formed, either from spider webs or fine metallic or plastic threads.[1] These were replaced with a glass disc in which the "marks" and "masking patterns" were etched into the glass surface or into an opaque layer coated on the glass surface; these were known as graticules. Based on this earlier terminology, the term *graticule* appears to be more appropriate, but reticle has become the established name.

5.3 PHOTOMASK SPECIFICATIONS

5.3.1 Mask Parameters

Photomasks are specified by three principal parameters: (1) pattern position accuracy, (2) feature size control, and (3) defect density. Recently, the "fidelity" of the photomask pattern has also become an item of concern, and while there is no clear specification for pattern fidelity, efforts are being made to "correct" the inherent feature shape errors caused by the mask writing and processing. Another parameter that is becoming a concern is the feature edge roughness, because it will eventually affect feature size measurements and the detection of defects.

5.3.1.1 Pattern position

Pattern positioning accuracy may refer to the pattern alignment between two or more critical mask levels of a given mask set, typically called the overlay accuracy, but more recently it is with respect to some "absolute grid." This grid is a reference that is used for multiple mask sets rather than just one mask set. At the present time, it is usually the grid of a given pattern generator or coordinate measuring machine. A two-dimensional standard on a 6″ square fused-silica substrate, 0.25″ thick, is available from the German National Bureau of Standards (the Physikalisch--Technische Bundesanstalt) through Leica Inc. USA.[2] The standard consists of a 29×29 array of crosses. The calibration consists of determining the total length of three selected rows and columns of cross marks, and the coordinates of 169 cross marks arranged in a square grid. The uncertainties of the measurements, given as twice the standard deviations, are

- Total lengths: 30 nm
- X/Y Coordinates: 35 nm.

The orthogonality and row straightness of a grid pattern can be checked independent of a standard, with the aid of a coordinate measuring machine or an e-beam pattern writer, by rotating the grid 90° to check orthogonality and 180° to check row straightness.[3] Linearity can be checked by displacing the grid to compare one section against another. Length can be measured relative to the laser interferometer used to position the stage in the pattern writing tool, but inherent machine errors usually limit the absolute accuracy of the writing tools. Absolute length measurement requires an external standard. Unfortunately, orthogonality, straightness, and linearity can only be checked on a macro basis, every centimeter for example, but the usefulness of a mask depends on the accuracy of these parameters on a micro scale. For this reason it is important to know and measure the machine errors that affect all mask parameters. (See Sec. 5.6.3 for more details.)

5.3.1.2 Critical dimension control

A *critical dimension* (CD) is one or more features defined by the circuit designer and/or wafer process engineer as being the most important with respect to dimensional control. The feature(s) may be oriented in the X or Y direction and may be clear or opaque. The three parameters that are important are:

1. the average value of a given critical dimension relative to the specified value,
2. the uniformity of the critical dimension, and
3. the linearity of feature sizes down to some lower limit.

The accuracy of the linewidth measuring tool and the cross-section profile of the features play an important role in determining the above parameters. These items are discussed in Sec. 5.8.

Each of the three parameters has a unique effect on the wafer processing. (1) The correct average feature size is important to maintain the wafer exposure level that gives maximum exposure latitude. Biasing the wafer exposure to compensate for a large deviation in the average mask CD reduces the available exposure latitude. Also, large variations from one mask level to another requires recalibrating the wafer exposure level. (2) The CD uniformity is important to yield maximum circuit performance. (3) The importance of feature size linearity is dependent on the circuit design.

5.3.1.3 Defects and pattern fidelity

Defects may be defined as extra or missing chromium film, etc., whereas pattern fidelity, or integrity, covers many issues such as incorrect pattern shape due to machine writing and processing limitations (see Secs. 5.4 and 5.9).

Pattern fidelity is also defined by the pixel size used to write the pattern; the smaller the writing pixel the better the pattern definition. The pixel size used to define the pattern is called the pattern address size. The circuit designer has to design the circuit feature sizes as multiples of the address size. The relation between the pattern address, the pattern location, and the pattern-writing time is a function of the architecture of the pattern writer (see Sec. 5.6.3.4).

5.3.2 Specifications: 1994 SIA Roadmap

The Semiconductor Industry Association (SIA)[4] periodically reviews the projected roadmap for integrated circuit technology and outlines critical specifications, process limitations, needed developments, schedules, etc. The latest projections (Oct. 1994) were established by extending historic trends for dynamic random access memory (DRAM) bit count by a factor of four every three years for the next fifteen years (to 2010). Table 5.1 lists the specifications and expected time schedule for critical mask levels out to 0.13-μm design rules, based on the latest SIA Roadmap. The mask requirements are for "critical mask levels at the defined years" and the "mask levels are assumed to be relatively small and difficult to produce." The SIA Roadmap is based on DRAMs being the driver for leading technology, with microprocessors following. However, the time gap between the technologies has been closed and microprocessors are expected to be the drivers by 2000.[5]

It will be noted in Table 5.1 that the mask specifications for 4X reticles is less than four times the 1X mask specifications. This is because the 1X tolerances are for x-ray masks and it is expected that the pattern fidelity transfer from the x-ray mask to the wafer will be better than that from an optical reticle to the wafer. This is because of the loss of the high spatial frequency components of the pattern through the stepper lens.

TABLE 5.1 Specification for the critical mask levels—SIA Roadmap.

Time schedule		1995		1998		2001		2004		
Design rule	(μm)	0.35		0.25		0.18		0.13		
Magnification		4	5	1	4	1	4	1	2.5	4
Image placement	(nm)	70	70	30	44	22	32	15	26	26
CD uniformity* (3 sigma)	(nm)	50 ↓ 35	50 ↓ 35	20	40 ↓ 25	15	30 ↓ 18	10	20 ↓ 13	20 ↓ 13
Mean to target*	(nm)	40 ↓ 20	40 ↓ 20	10	30 ↓ 15	7	20 ↓ 10	5	15 ↓ 8	15 ↓ 8
Defect size*	(nm)	280 ↓ 210	350 ↓ 262	50 ↓ 38	200 ↓ 150	36 ↓ 27	144 ↓ 108	26 ↓ 20	65 ↓ 40	104 ↓ 78
Minimum field size at 1X	(mm²)	22×22		26×26		26×30		26×36		
Data volume	(Gbytes)	0.5		2		8		32		

* Range of values reflects continuous improvements required by mask makers.

Optical lithography may be employed out to 130-nm design rules with the appropriate enhancements such as phase shift technology, oblique stepper illumination, and optical proximity correction. The application of optical lithography is projected as shown in Table 5.2:

TABLE 5.2 Application of optical lithography.

SCHEDULE			
PILOT PRODUCTION	*MASK REQUIRED	DESIGN RULE	LITHOGRAPHY
1995	1993	0.35 μm	I-line or I-line with optical enhancements
1998	1996	0.25 μm	I-line+ optical enhancements & 248 nm
2001	1999	0.18 μm	248 nm + optical enhancements & 193 nm
2004	2002	0.13 μm	193 nm with optical enhancements

* Mask schedule based on mask being available two years ahead of pilot production.

From the time schedule in Table 5.2, the required mask specifications in Table 5.1, and a knowledge of today's capabilities, it is easy to see why photomasks are a critical part of the lithographic process.

5.3.3 Combined Feature Placement and CD Specification

At the present time, the feature size and feature position are two separate specifications, as listed in Table 5.1, but on many circuits the critical parameter is the feature edge placement; the edge of one feature must be positioned to a given precision to the edge of another feature. On the photomask, the feature edge accuracy is a combination of the pattern-positioning accuracy and the critical dimension control (see Fig. 5.1). These two parameters are usually considered as independent parameters, but if one parameter becomes more difficult to achieve than the other it may be possible to readjust the tolerances appropriately. This is not an accepted policy at this time, but we believe it warrants further considerations.

5.4 MASK TYPES

5.4.1 Binary Masks

The mask pattern on a conventional photomask is a close replication of the circuit designer's pattern, with the possible addition of some wafer-processing bias, etched into an opaque film. The patterned area is either clear or opaque, hence the term *binary*. Limitations in the mask writing and processing remove sharp corners on rectangular figures, may cause some figures to have serrated edges, displace feature edges, and cause some level of nonuniformity of feature sizes across the mask. The mask pattern is further degraded by the limitations of the stepper lens and the wafer processing; altogether, the final wafer pattern may look slightly different from the original designer's pattern. The pattern difference increases as the dimensions of the mask features approach the resolution limit of the stepper lens. The relation between these parameters is given by

$$\text{resolution} = \frac{k_1 \times (\text{actinic wavelength})}{\text{numerical aperture of stepper lens}}.$$

For a typical wafer production line, $k_1 > 0.7$. For $k_1 < 0.7$, the CD performance of the stepper lens is nonlinear and the mask CD tolerances must be smaller to maintain the same mask contribution at the wafer. At $k_1 = 0.5$, the required mask CD uniformity is approximately half that for $k_1 = 0.7$.

Using the values in Table 5.2, where actinic wavelength = 248 nm, resolution = 180 nm, and NA = 0.6, requires $k_1 = 0.44$. This requires the CD tolerance to be even less than half that for $k_1 = 0.7$. The mask CD uniformity specification is eased if 193-nm actinic radiation is used for 180-nm lithography.

The typical deviations from the designer's pattern are generally small enough not to affect the quality of the circuit performance for >1-μm design rules, but as new technologies push IC designs below 0.5 μm, wafer pattern errors caused by the above-mentioned limitations require some mask pattern modifications to improve the pattern fidelity. The wafer pattern fidelity errors can be separated into two arbitrary divisions: those caused by reticle pattern errors caused by the presence, or lack, of surrounding features, and those caused by the resolution limit of the wafer stepper lens and the wafer processing. Reticle pattern errors due to the proximity of other features can be corrected by proximity effect corrections (PEC). This is essentially a mask maker's problem because it is dependent on the writing-tool characteristics. Pattern errors due to lens limitations and/or wafer process steps can be corrected, within limits, by applying pattern modifications to the reticle, which is known as optical proximity correction (OPC) or optical pattern correction. OPC is the wafer engineer or circuit designer's problem to correct, but the mask maker's problem to apply.

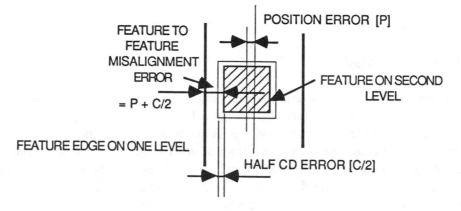

FIG. 5.1 Combined registration and CD specification.

The resolution for a given tool set can be improved by reducing k_1. This can be done by improved wafer processing and/or modifying the wafer stepper illumination (see Chap. 1). Another technique is by "destructive light interference" at the wafer produced by introducing a phase shift in different parts of the projected pattern image. This is achieved by means of phase shift masks (PSM) (see Chapt. 1).

5.4.2 Proximity Effect Correction Masks

Changes in pattern shapes may occur due to local pattern exposure variations that occur during the mask-writing process. In the case of the electron beam pattern gen

erator, the incident electrons on the mask are scattered by the resist and the mask substrate, and the backscattered electrons cause additional exposure on nearby features. The range of the backscattered electrons, which is of the order of 1 to 3 μm, is dependent on the energy of the incident electrons. The higher the energy, the greater the backscatter range. This additional exposure affects the size and shape of adjacent features. This effect can be minimized by a "GHOSTing" exposure,[6] where the mask substrate is exposed a second time with a relatively large-diameter beam at a low exposure level. The intend is to "fog" the mask resist with a dose exposure that approximately equals the dose due to the backscattered electrons. A more elegant, but also more complicated, technique is to calculate the effect of the back-scattered electrons at every point in the mask pattern and adjust the written pattern to compensate for the extra backscatter dose (see Sec. 5.6).

5.4.3 Optical Proximity Correction Masks

Optical proximity correction is a modification to the photomask pattern to compensate for changes in feature shape and size that occur during the pattern transfer from the mask to the wafer. These feature changes may be caused by extra exposure due to the presence of adjacent features, a limitation in the wafer stepper resolution, or a variation in the activity of a given wafer process step. OPC is considered by many as a necessity to increase the wafer process latitude for 0.35-μm design rules, and by some as a means to achieve sub-0.25-μm designs, together with other optical enhancements such as off-axis stepper illumination. There is also a need to include OPC correction in phase shift masks to maximize the benefit gained from PSM technology.

Some limited form of OPC has been in use for at least two or three decades. These pattern modifications were usually requested by a wafer engineer based on knowledge of a particular process step. In recent years, OPC has become more of a science than an art due to the introduction of several OPC software programs.[7] The OPC correction process consists of measuring several generic test patterns processed on a wafer and constructing a multilevel lookup table from the measured data. The final mask pattern may contain serifs at feature corners and incremental changes along the length of a feature. The degree of OPC pattern modification, measured by the number of added serifs and line jogs, depends on the required level of pattern correction at the wafer level. Typically, the OPC computation includes imaging and wafer process effects out to a range of ~2 μm from a given point on a feature edge, and this usually accounts for ~85% of the pattern fidelity error.

The writing time and complexity of an OPC increases considerably compared to an equivalent pattern without OPC modification. The writing address for a 4X OPC reticle may be 25 nm compared to 100 nm for a similar reticle without OPC correction, and the additional features, such as serifs and incremental line changes, may increase the mask data size by ~6X (see Fig. 5.2).

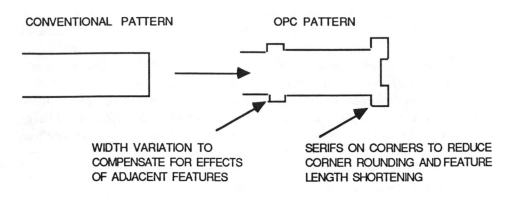

CONVENTIONAL PATTERN

OPC PATTERN

WIDTH VARIATION TO
COMPENSATE FOR EFFECTS
OF ADJACENT FEATURES

SERIFS ON CORNERS TO REDUCE
CORNER ROUNDING AND FEATURE
LENGTH SHORTENING

FIG. 5.2 Optical proximity correction pattern. Features are added to the primary pattern on the photomask to compensate for stepper lens limitations and wafer processing nonlinearities.

5.4.4 Phase Shift Masks

OPC masks correct pattern fidelity errors and increase the wafer process latitude, but the resolution improvement is minimal. A new mask technology, known as phase shift masks,[8] was introduced in 1982 that significantly improved the resolution of a given wafer stepper exposure tool. There are several PSM structures consisting of clear and opaque regions with a phase shift between certain clear regions. Figure 5.3, an alternating grating array, illustrates the details of a phase shift mask. There is a half-wave phase shift, at the wafer stepper wavelength, between the two different levels of the clear regions. This structure can be fabricated from a conventional quartz mask substrate by etching the recessed areas,[9] or the raised regions can be layer-deposited on the quartz surface and then etched away in the appropriated areas.[10]

Phase shift structures can be separated into "strong" and "weak" PSM technologies, based on the level of improved resolution. The alternating PSM structure is the "strongest" PSM technology and can improve the resolution of a given wafer stepper by ~40%,[11] but problems are encountered in merging the etched and nonetched clear regions. An abrupt transition from 0° to 180° produces unwanted features on the wafer. This can be avoided by making the full transition in three 60°,[9] or by etching a gradual transition region (see Fig. 5.4).[12]

PSM structures with opaque and clear regions are usually difficult to lay out and very difficult to fabricate without defects. Another PSM structure that provides less resolution improvement but is considerably easier to fabricate is the halftone or "embedded" PSM.

5.4.5 Halftone and Embedded Phase Shift Masks

The halftone PSM is a two-layer structure consisting of a "transparent" film with a phase shift of 180°, covered with a halftone film with a specified transmission between 5% to 10%, deposited on a fused-silica substrate. The mask pattern is formed in the combined film layer. The light passing through the combined film layers is reduced in intensity and is a half wave out of phase with the light transmitted through the clear regions.[13] This structure has been upgraded by combining the phase shift and attenuation into a single layer. This is called an embedded PSM structure (see Fig. 5.5).[14]

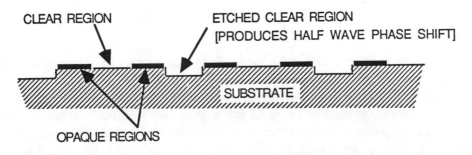

FIG. 5.3 Alternating phase shift mask structure.

One difficulty with attenuated PSM is that although the transmission through the embedded layer is only 5% to 10% at the stepper actinic wavelength, it can be as much as 30% to 60% at the stepper alignment wavelength. This can cause alignment problems on some steppers. Another problem is the overlap region of the exposed areas on the wafer around the mask pattern. These problems can be overcome by coating the phase-shifting layer with an opaque film, such as chrome, around the edge of the mask patterns. Another technique is to etch a high-resolution pattern around the stepper alignment marks and around the circuit patterns on the masks, which diffracts the light around the marks out of the alignment detector capture area.

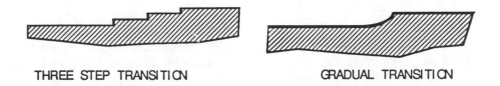

THREE STEP TRANSITION   GRADUAL TRANSITION

FIG. 5.4 Phase transitions.

Because of the fabrication difficulties with the strong PSM structures, the embedded PSM structure, together with off-axis wafer stepper illumination, is the first phase shift technology to be used in a wafer production mode of operation.

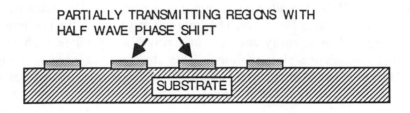

FIG. 5.5   Embedded phase shift mask.

It should be noted that although the strong PSM technologies have been available for more than a decade, they are not in general use in the U.S. The introduction of improved wafer stepper lenses and shorter-wavelength illumination sources, which extended the resolution limit to ~0.35 μm without any significant change in the binary mask, together with the difficulty in obtaining a defect-free PSM mask with a fast delivery time, delayed the introduction of phase shift technology into a wafer production line. The embedded PSM will undoubtedly hasten the general use of phase shift technology. Below 0.35 μm it will be necessary to employ one or more of these innovative mask technologies such as PSM or OPC.

5.5   PHOTOMASK SUBSTRATES

The standard-size substrate for leading edge photomasks is 6″ square by 0.25″ thick and is made of fused silica. The 6″ substrate is needed to accommodate the larger stepper lens field sizes for 2.5X to 5X masks, the thickness is required to minimize pattern placement errors due to substrate distortion, and the fused silica is needed for its low coefficient of expansion to minimize the effect of temperature variations, the higher transmission at the shorter actinic wavelengths, and the higher mechanical strength.

5.5.1   Substrate Size

The substrate thickness was increased to 0.25″ from the previous thicknesses of 0.09″ and 0.12″ to minimize pattern placement errors due to the gravitational sag of the substrate and the substrate distortion caused by the plate clamping mechanism. The problem arises because a substrate bends about the neutral plane of the plate, which is in the center of the plate's thickness, and one surface of the substrate is stretched and the other is compressed by any bending motion. The gravitational sag of a kinematically mounted 6″ × 0.25″ thick substrate (i.e. one with gravitational sag and no clamping distortion) is 0.62 μm, and the associated pattern placement error, relative to the same pattern on a flat substrate, is 40 nm.[15] The pattern error for a 0.12″ thick substrate is 80 nm. If the two opposite sides of the substrate were clamped in a horizontal plane, the error would be reduced to 5 nm for a 0.25″ plate and 10 nm for a 0.12″ plate. However, this requires the clamping surfaces to be perfectly flat and in a common plane. For a purely gravitational sag, the pattern placement error is only reduced a small amount by increasing the thickness of a 6″ substrate beyond 0.25″.

If the substrate distortion is uniform from plate to plate, the pattern placement error can be premeasured and corrected with the pattern generator software.[16] For example, if the 0.25″ thick plate were kinematically mounted (i.e., free-standing with no clamping error) on two edges, then a linear magnification compensation across one direction of the pattern area will reduce the pattern placement error to 9 nm. However, it is essentially impossible to have two edges of the substrate resting uniformly on two support rails. A true kinematic mount has three support points, but unfortunately the gravitational effect produces a deflection in two directions that requires a nonuniform magnification correction along two axes. This correction technology should be available in the next-generation pattern writers (see Sec. 5.6).

The method of clamping the substrate and the reproducibility of the clamping mechanism can also have a large effect on the flatness of the mask and the pattern placement accuracy. The errors are a significant portion of the allowed tolerance and must be taken into account when measuring the pattern placement accuracy.

The introduction of 0.25″ thick substrates initially caused an unexpected problem with the uniformity of feature sizes across the mask. Thinner substrates (0.09″) written with the same writing tool and developed with the same processing equipment had a smaller CD variation. While not clearly stated, it is believed that the problem was caused by larger-than-expected thermal nonuniformities throughout the substrate during the resist coating and prebake process. These nonuniformities caused the resist sensitivity to vary across the substrate, which in turn produced a larger-than-normal CD variation. Careful control of the prebake conditions has reduced this problem. Hoya, a manufacturer of photomask blanks, has recently proposed using a chill step to rapidly cool the substrate after prebake to further control thermal effects.[17] Recent evaluations of CD uniformity suggest that the cooling effect at the corners of a hot plate used for baking the substrates is still a significant contributor to CD variations.

The mask substrate size was increased from 5″ square to 6″ square around 1990 to accommodate the larger die sizes. Up to that time, the dominant demagnification for optical wafer steppers had been 5X, but with the introduction of larger die sizes and the desire to stay with ≤6″ square mask substrates, the industry is moving to 4X demagnification. For 0.35-μm designs, optical lithography may use 4X or 5X reticles. In order to minimize wafer costs, it is necessary to place two die on the photomask, therefore, two 0.35-μm die (22 mm × 22 mm) can fit on a 6″ reticle. A 7″ substrate is required for two 0.25 μm die (26 mm × 26 mm), and an even larger reticle is required for expected die size for 0.18-μm technology (26 mm × 30 mm).

The question of the next reticle size has been raised by SEMI/Sematech[18] and a task group from Japan.[19] The discussions are based on the expectations that the wafer exposure tool for ≤0.25 μm designs will be a step-and-scan projection system where a narrow field of view, ~2 mm wide, is scanned synchronously across the mask and the wafer (see Chap. 1). The length of the slit, which has a maximum length equal to the diameter of the lens field of view, is the width of the die, and the length of the slit travel will be the length of two die. This requires the mask substrate equal [(4 × die size) + 30 mm], where the 30 mm is the required clearance between the pattern and the edges of the substrate. This dictates the substrate be at least 8″ for the 0.18-μm technology, and 9″ for 0.13 μm. There is a need for a 6″ × 9″ substrate at the present time for the SVGL MicroScan Wafer Stepper,[20] but the industry agrees that the mask substrate should be symmetrical, i.e., square, to minimize problems with resist coating and resist developing, etc. This suggests the next size substrate should be 9″ square.

At a recent SEMI meeting[21] it was agreed that the next size reticle will be 230 mm square. The thickness has yet to be agreed upon. The proposed layout for the 230-mm reticle is shown in Fig. 5.6. The two-rail support is intended to keep the plate sag uniform in one direction, which will simplify the pattern error correction, and the close spacing and the location of the rails will help minimize the plate sag. The pattern placement error is a minimum if the plate supports are spaced with a separation of 0.577 of the plate length. These are known as the Airy support points.

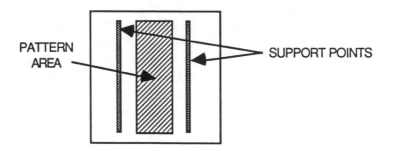

FIG. 5.6 Proposed layout for 9-inch masks.

5.5.2 Substrate Material

Photomasks for leading edge technology use fused-silica (SiO_2) substrates. The low expansion coefficient of fused silica minimizes the effects of any temperature variations during the mask-writing process or thermal effects caused by heat transfer from the stepper. However, even with fused-silica substrates a temperature change of 0.08°C will change the pattern placement accuracy by 10% of the allowed tolerance for 0.25-μm technology. Pattern generator environmental chambers are typically specified at ±~0.03 °C, which will maintain the pattern placement error, due to thermal effects, to <10% of the total allowed error, providing the mask substrate is allowed to reach thermal equilibrium before writing.

Fused silica replaced the low expansion borosilicate glasses for mask substrates not only because of the low expansion coefficient but also because of the higher transmission at the shorter wavelengths. The transmission of fused silica for 1-cm thickness is typically listed as about 90% at 248 nm wavelength, and about 85% at 193 nm, and the surface reflection is 8% and 10% respectively.[22] This is sufficient for mask substrates, but a large effort is underway to improve the quality of fused silica for stepper lenses. Recent transmission data[23] of selected samples of the improved material indicate the preradiation (see next paragraph) measured total loss due to scattering and absorption is 0.3% to 0.7% per cm and may be as low as ~0.15% per cm.

Fused silica is subject to laser-induced changes at 193 nm. These changes, typically referred to as "damage," include color center formation and compaction.[23] The color center formation causes a low level of fluorescence at about 400 nm wavelength, and the compaction causes a small change in the refractive index of the fused silica. These changes may limit the useful lifetime of the fused-silica lenses used in wafer exposure tools (see Chap. 1) but are not expected to affect the characteristics of a photomask, because the energy density of the stepper illumination is lower at the reticle than it is at the last lens element.

The advantages of quartz were initially dimmed for the mask maker because of the higher surface resistance of quartz, which gives rise to static surface charge that can readily damage a mask pattern, and the initial problem of adhesion of the traditional chrome film to the quartz surface. Mask makers and users have learned to live with the potential static charge problem, and the chrome film adhesion has been overcome.

5.5.3 Substrate Surface Quality and Flatness

Quartz is a relatively hard material and requires special efforts to achieve the required surface flatness and surface finish. The flatness of a typical mask substrate over the pattern area is ≤2 μm, and flatness of ≤1 μm is also available. As noted above, a deflection of 0.62 μm for a 0.25″ thick substrate produces a pattern placement error of 40 nm. A plate with a polished flatness of ~1.0 μm will mask the gravitational sag, and plate sag measurements to correct for this error will be meaningless.

The surface roughness of the polished blank is ~10Å, peak to valley.[24] The surface must be free from pits and microscratches that can cause defects on the finished mask.

5.5.4 Opaque Films

The traditional opaque film deposited on the photomask substrate is a varying compound of chromium, nitrogen, and oxygen, possibly plus other elements. The composition of the film varies through its depth to provide different characteristics. The film surface next to the substrate acts as a "glue" layer to give good adhesion to the quartz. The intermediate film composition is optimized to give maximum attenuation without undue film thickness. The top surface, which faces the stepper lens, is usually an antireflective film to minimize the reflection of any light that may be reflected back from the wafer surface. The required optical density of the opaque layer is 3.0 and the film thickness is ~100 nm.

As noted earlier, the introduction of quartz substrates was accompanied with the unexpected problem of "poor" adhesion of the chrome film. The problem showed as an increase in the number of pinholes in the chrome film after a high-pressure cleaning cycle or an ultrasonic cleaning process. This problem prompted the introduction of a different opaque film that had improved adhesion to quartz, demonstrated by fewer pinholes after the high-pressure or ultrasonic cleaning, and could be readily dry-etched with the then-available resists used for mask fabrication. The film was molybdenum silicide (MoSi) film.[25] Moly-silicide films are typically 80 to 100 nm thick, have an optical density of 3.0, and have an antireflection coating of MoSiO to reduce the surface reflection to ~15%.[26] Moly-silicide films are not in general use now but are finding applications in the embedded phase shift masks.

As noted above, any change in the bow of the mask will change the pattern placement. The plate bow can be produced by surface strains as well as by clamping errors. A common source of variable surface strain is the deposited chromium film. If the deposited film is stressed, due to the temperature of deposition, etc., the film stress will undoubtedly change after the mask has been exposed and processed. Changes in plate bow of the order of 1 μm are not unusual. Therefore, it is very important that there is minimal strain in the deposited film.

5.6 PATTERN GENERATION FOR MASKS: CHALLENGES AND PROJECTIONS

5.6.1 Introduction

Lithographic patterns printed on wafers are only as good as the mask or reticle used to print them. For this reason the pattern generation on the mask is a critical, limiting step. Because requirements on the mask patterning are more severe than those on the wafer, mask pattern generation will be continually challenged.

Commercially available pattern generators include laser writers and e-beam writers. Laser writers employ an array of focused laser beams. Each beam forms an independent writing probe with roughly Gaussian intensity profile. The beams are individually blanked and unblanked as the array is scanned in a raster fashion. E-beam mask writers are of two types: Gaussian beam and variable shaped beam (VSB). In Gaussian systems the beam is, again, blanked and unblanked as it is deflected across the mask substrate. In variable shaped beam systems, a rectangular spot of variable width and height is employed, and the spot size is dynamically changed as the pattern is written. The basic optics of these three systems are shown schematically in Fig. 5.7(a–c).

The optics of a Gaussian e-beam writer are shown schematically in Fig. 5.7(a). A point source, typically a thermal field emission tip, is magnified into a high-speed beam blanker. The beam is blanked by deflecting onto a blanking aperture. While unblanked, the beam passes through the aperture and is focused onto the writing surface. Deflection occurs just above the final lens. The writing substrate is clamped to an XY stage, which positions the substrate under the beam. The column depicted here is modeled after the MEBES 4500 column, manufactured and marketed by Etec Systems, Inc. of Hayward, California.[27]

The optics of a multiple Gaussian beam laser writer are shown schematically in Fig. 5.7(b). A 364-nm laser beam is split into 32 independent beams. These beams are scanned across the writing surface by a 24-facet rotating polygonal mirror. The beams are interweaved in both X and Y directions to average out mirror and stage displacement errors. In addition each beam has 17 gray beam exposure for variable dose. The system shown in the figure is modeled after the ALTA 3000 laser writer, manufactured and marketed by Etec Systems, Inc.[28]

The optics of a VSB electron beam column are shown schematically in Figure 5.7(c). The electron source, typically a single-crystal lanthanum hexaboride thermionic emitter, uniformly floods the first spot-shaping square aperture. The first condenser lens forms an image of this first aperture onto the second spot-shaping aperture. The spot-shaping deflector displaces this image, forming a composite shadow that is a rectangle of variable size and aspect ratio. This is demagnified and focused onto the writing surface. The main field deflector is immersed in the final projection lens and scans the beam in two transverse dimensions. The beam is blanked and unblanked in synchronism with the deflection to generate the pattern. The writing substrate is held on an XY stage, which moves in a step-and-repeat pattern as the individual fields are exposed. The system shown is modeled after the EL-3 column, developed by IBM, and has been used in their research, development, and manufacturing facilities for direct write and mask making from 1978 to the present time.[29]

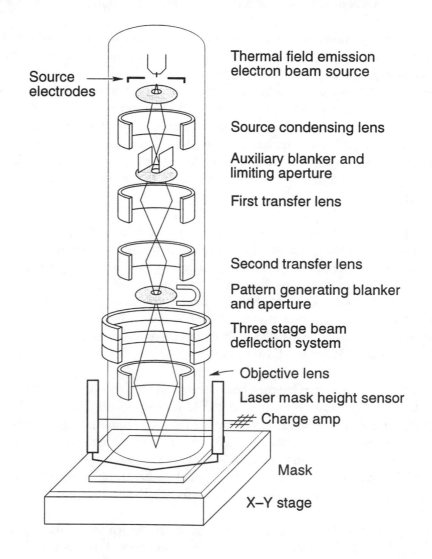

Source electrodes

Thermal field emission electron beam source

Source condensing lens

Auxiliary blanker and limiting aperture

First transfer lens

Second transfer lens

Pattern generating blanker and aperture

Three stage beam deflection system

Objective lens

Laser mask height sensor

Charge amp

Mask

X–Y stage

FIG. 5.7(a) Optics of a Gaussian e-beam pattern writer.

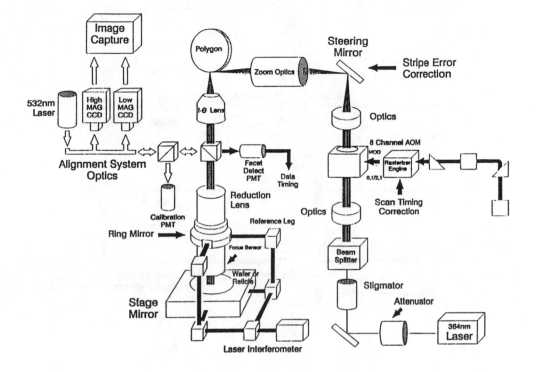

FIG. 5.7(b) Optics of a multiple Gaussian beam laser pattern writer.

Table 5.3 summarizes the use of the different machine architectures in several commercial pattern generators available today.

TABLE 5.3 Architecture of commercially available pattern generators.

SYSTEM	IBM EL-4	JEOL JJBX-7000	JENOPTIK ZBA-31H	LEICA VECTOR BEAM	LEPTON EBES 4	MEBES 4500	ALTA 3000
STAGE MOTION	Step and Scan	Step and Scan	Contin-uous	Step and Scan	Contin-uous	Contin-uous	Contin-uous
PATTERN PLACE-MENT	Vector Scan	Vector Scan	Vector Scan	Vector Scan	Vector Scan	Raster Scan	Raster Scan
PATTERN FILL	Shaped Beam	Shaped Beam	Shaped Beam	Single Gaussian Beam	Single Gaussian Beam	Single Gaussian Beam	Multi-Gaussian Beam

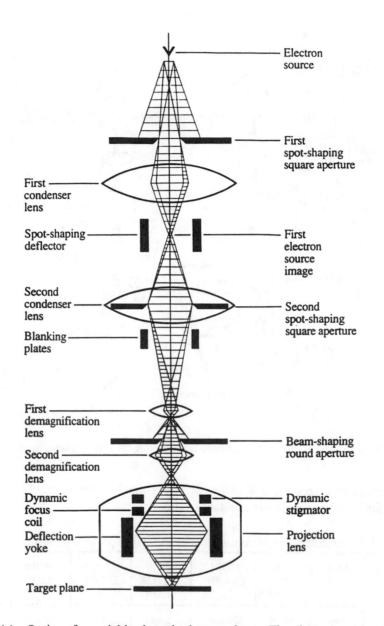

Electron source

First spot-shaping square aperture

First condenser lens

Spot-shaping deflector

First electron source image

Second condenser lens

Second spot-shaping square aperture

Blanking plates

First demagnification lens

Beam-shaping round aperture

Second demagnification lens

Dynamic focus coil

Dynamic stigmator

Deflection yoke

Projection lens

Target plane

FIG. 5.7(c) Optics of a variable shaped e-beam column. The electron source, typically a.single crystal lanthanum hexaboride thermionic emitter, uniformly floods the first spot-shaping square aperture. The first condenser lens forms an image of this first aperture onto the second spot-shaping aperture. The spot-shaping deflector displaces the image of the first aperture on the second aperture, forming a composite shadow that is a rectangle of variable size and aspect ratio. This is demagnified and focused onto the writing surface. The writing substrate is clamped on an XY stage, which moves in step-and-repeat fashion as the individual fields are exposed. The column shown is modeled after the EL-3 system developed by IBM.

A pattern is created by sequentially exposing elementary shapes that are "butted" or "stitched" together. This is shown in Fig. 5.8 for the cases of a Gaussian round beam raster scan system, and a variable shaped beam system.[30] The composite shape shown in the figure is built up of 10 scan lines of variable length for the Gaussian round beam system. The same composite shape is built up of two elementary shapes in the variable shaped beam system. In both examples the composite shape contains 120 pixels, where a pixel is a fundamental unit of resolution with diameter d. The number of spots or "flashes" is 120 for the Gaussian system, and two for the variable shaped beam system, where the flashes are exposed sequentially. This represents a difference of a factor-of-60 difference in flashes for this particular example. To obtain equivalent writing speed would mean the flash rate must be 60 times higher for the Gaussian beam system as compared with the shaped beam system.

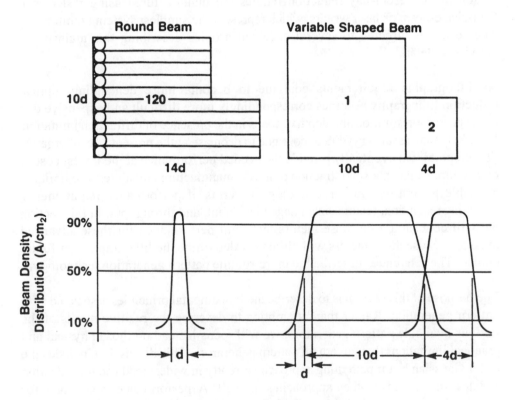

FIG. 5.8 Stitching of elementary shapes to form a composite pattern shape. The nature of these elementary shapes is defined by the writing strategy, of which two, a Gaussian round beam raster scan and a variable-shaped beam vector scan, are shown.

Pattern generators can be characterized in terms of their parallelism, or the number of pixels that can be simultaneously exposed. A Gaussian beam is capable of exposing one pixel at a time. Each beam of a multiple Gaussian beam array can expose one pixel, with all of the beams operating independently and simultaneously.

The parallelism is, thus, equal to the number of beams. A variable shaped beam can expose one or more pixels at a time, depending on the spot size. The writing speed, in units of area swept out per unit time, is directly proportional to the degree of parallelism. This becomes important for large patterns, which can require several hours to expose.

As noted in Sec. 5.2, most masks used in present-day semiconductor manufacturing are reduction reticles, where the pattern is formed in a chromium layer over a fused silica plate. An optical stepper forms a demagnified image of the reticle onto the wafer. This has the advantage that the pattern errors, including placement and image size errors, are demagnified as well. Relaxed performance requirements, together with the need to minimize cost, favor simplicity in the pattern generator. For this reason most present-day reduction reticles are manufactured using raster scan Gaussian e-beam or laser writers, which represent the simplest system architecture. The reader is directed to several excellent articles describing this mainstream activity in detail.[31,32]

As lithographic requirements continue to become more demanding, optical reduction lithography becomes correspondingly more difficult and expensive due to the need for resolution and depth of focus in the presence of diffraction limitation. It is likely that, assuming CDS to continue to decrease at the present rate, alternative lithographies will eventually provide the needed performance at the lowest cost of ownership. At most, a small fraction of masks manufactured today test the limits of available pattern generation technology in terms of performance requirements, including resolution, feature size, image placement, and throughput. In this section we will concentrate on those factors that limit performance for these advanced masks, because these factors will ultimately determine the lithographies of future choice. The references describe the more-routine pattern generation technology.

The purpose of this section is to describe the fundamental principles associated with pattern generation. Rather than attempting to describe the relative merits of the various available pattern generators, we will focus instead on those physical and practical limitations that govern all pattern generators. This discussion includes the raster Gaussian beam patterning approach currently in widespread use for reduction reticles and includes other approaches as well. A performance benchmark for present-day systems will be described. Finally, the trends of pattern generation in the future will be discussed, along with some predictions regarding performance.

5.6.2 Challenges

5.6.2.1 Performance parameters

Device patterns are enormously complex. An upper limit on the number of pattern features can be obtained by dividing the linear chip size by the CD, and squaring this number. This is of the order 2×10^{10} for a 1-Gb DRAM chip with 180 nm CD. For patterns to be written in a reasonable amount of time, the throughput must be adequate. This becomes a critical parameter as well. The definition of "reasonable time" depends on the application. It typically takes much longer to write a mask than to expose a wafer using that mask. This is acceptable, because the time required to write the mask can be amortized over all of the wafers printed by a given mask. For a mask-manufacturing facility, the "reasonable time" required to write a mask might be on the order of two or three hours.

Pattern placement accuracy and image size control require stability, whereas a high writing speed tends to compromise stability. For this reason, high throughput acts in opposition to accuracy and control. A pattern generator may be deemed good if it meets the requirements for control while still generating patterns in a reasonable time. Throughput is intimately connected with cost of ownership, since, in principle, it is possible to increase the overall throughput of a mask-making facility by simply adding more pattern generator systems. In summary, pattern placement accuracy, image size control, and throughput are the key performance parameters of any pattern generator.

Table 5.1 showed the specifications for 1X and 4X masks needed to expose the device levels. Although pattern placement accuracy and image size control are included, throughput is not. In the following we will assume the SIA target specifications to be the relevant performance targets.

5.6.2.2 Performance drivers

One estimate of writing speed, in units of area swept out per unit time, is the ratio of the total beam current divided by the incident dose in units of charge per unit area required to expose the resist. This represents the writing speed in the limit where no significant system overhead times exist. It is, therefore, an upper limit.

An example of a system overhead is the time the electronic requires to execute a flash. Another upper-limit estimate of writing speed can be derived from this, namely, the pattern area, divided by the total number of pixels, times the data rate in pixels per unit time. This quantity is implicitly proportional to the parallelism, or number of pixels exposed simultaneously, as discussed previously.

The smaller of these two estimates of writing speed is the limiting value, assuming no other system overheads are important. Which contribution dominates depends, among other things, on the writing strategy, Gaussian single-pixel vs variable shaped spot, for example. The size of the fundamental resolution element or pixel must scale with the CD in order for the resolution to remain a suitably small fraction of the CD. The throughput is inversely related to the number of spots or "flashes." This is shown in Fig. 5.9 for cases of a Gaussian single-pixel system and a variable shaped beam system. As the resolution is reduced by a factor of two the number of spots increases by a factor of four for a Gaussian single-pixel system, but does not increase for a variable shaped beam system. This implies that the throughput scales quite differently for the two writing strategies, with the advantage going to the variable shaped beam approach. This advantage becomes more pronounced as the CD becomes smaller. This assumes that neither system is limited by the brightness of the electron source. This is the case for CDs on the order of 50 nm or larger as measured on the reticle, which should include all useful circuit designs for some time to come.

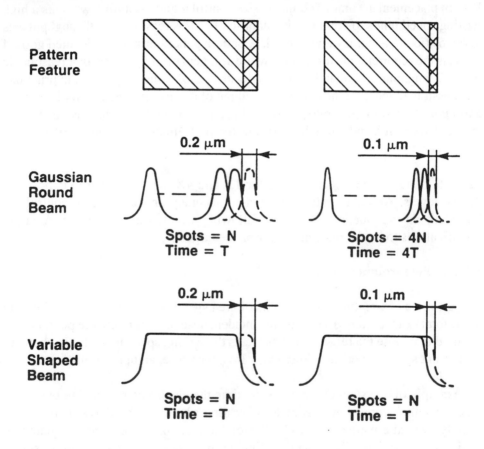

FIG. 5.9 Scaling law for the number of flashes as a function of resolution. The writing time is assumed here to scale with the number of spots.

An exception to the limitation of the single-pixel Gaussian beam, as described above, is the technique employed on the laser pattern generator ALTA 3000, where feature widths can be incrementally changed by a small fraction of one pixel by use of a staggered pixel writing scheme plus multiple gray levels of exposure. It has been proposed to adapt this same technology to the MEBES e-beam architecture.[33]

5.6.2.3 Physical limitations

As the beam current is increased, the resolution inevitably deteriorates, due to the stochastic Coulomb interaction between beam electrons occurring in the drift length of the column. The size of the blurring due to stochastic Coulomb interaction is intimately related to the pointlike nature of the beam electrons.[34] Because the electrons are, by assumption, initially randomly distributed in space, it follows that the vector separation between electrons is randomly distributed as well. The magnitude and direction of the Coulomb interaction depends directly on these vector spacings. The interaction takes place along the entire length of the optical path. It is strongest at crossovers, since the electron density in the transverse direction is highest here. The interaction is manifest by transverse displacement of the electrons in the writing plane, and by broadening of the energy distribution. Both of these result in blurring, the latter through chromatic aberration of lenses and deflectors. Because the electron density along the beam axis increases directly with current, the blurring increases monotonically with current as well.

One typically increases the beam current until the resolution is barely tolerable for the desired CD. At sufficiently low beam current (and writing speed), one can obtain any desired resolution. This is because the resolution of an electron beam is, for practical purposes, limitless. The fundamental limits of resolution occur in the realm of electron microscopy, which deals with dimensions three orders of magnitude smaller than those encountered in present-day lithography. The same cannot be said for laser writers, for which the resolution in present-day lithography is limited by diffraction. This represents a serious limitation on the extendibility of laser writers to future device generations. For this reason we concentrate on e-beam pattern generation in this article. In summary, a fundamental trade-off exists between resolution and throughput for e-beam writers.

Another class of fundamental limitations, called proximity effect, is imposed by the interaction of the electron beam with the mask substrate. Patterning is accomplished by the deposition of incident beam energy in the resist at the desired locations. The solubility of the resist in the developer is altered by this selective deposition. The dose and contrast depend on the specific resist/developer chemistry (see Sec. 5.7).

The beam passes through the resist layer and penetrates to some depth within the substrate. The stopping range of fast electrons in bulk matter is approximated by[35]

$$R(\mu m) = 0.04 \ (kV)^{1.75} / \ r(g/cm^2) \ ,$$

where kV = beam voltage in kilovolts, and r = density of the bulk matter. For bulk quartz at 10 kV the range is 0.9 μm. Since the resist is typically a few tenths of micrometers thick, it follows that the beam deposits only a faction of its energy in the resist. The remaining energy is deposited in the substrate as heat, or is backscattered. Through the 1.75 power law, the range is quite sensitive to the beam voltage.

The backscattered current emerges from an area with radius comparable with the range R. In the process the resist undergoes unwanted exposure over an area that is much larger than the size of the incident beam. To make matters worse, the amount of this unwanted exposure at any given point in the pattern varies according to the fraction of the adjacent area that is exposed. This depends on the specific pattern being written. This effect has been named proximity effect and manifests as a variation in the size of written pattern features.

Fortunately it is possible to predict the amount of unwanted dose for a given pattern. This forms the basis for two general correction techniques. In one approach, the incident dose is modified so that the resulting exposure dose, with backscattering, is uniform and correct at all points in the pattern. In the other approach, the size of pattern features is varied by an amount equal and opposite to the variation caused by the backscattering. Both of these approaches require modification of the pattern data.

Proximity effect can also be minimized by a "GHOSTing" exposure,[6] where the mask substrate is exposed a second time with a relatively large-diameter beam at a low exposure level. The intent is to "fog" the mask resist with a dose exposure that approximately equals the dose due to the backscattered electrons.

In addition to backscattering, the beam also undergoes forward scattering as it passes through the resist. Acting together with backscattering, the forward scattering causes the edges of pattern features to be significantly displaced from their design locations. This error is also amenable to correction. Forward scattering represents a blurring, or loss of resolution. The amount of blurring is typically 10–200 nm and is roughly inversely proportional to the beam voltage for a given resist thickness.

Much effort has been devoted over the last 25 years to correcting proximity effects.[36] As a result it has been demonstrated that correction to arbitrary precision is possible. The most accurate methods require computation to properly modify the pattern data. Because of ever-increasing data volumes, this is a nontrivial problem. A fundamental trade-off exists between the accuracy and the computation time,[37] the latter of which can be many hours, even using the most powerful computers. This computation can be performed prior to the writing step and, therefore, does not represent a direct limitation on throughput.

Most of the incident energy is deposited in the writing substrate as heat. This heat is, in turn, transferred to the resist layer. Depending on the chemical nature of the resist and the amount of heating, the resist sensitivity can be locally altered, again leading to unwanted image size variation. This effect is similar to backscatter proximity effect, but suffers from the additional complication that in addition to depending on the local pattern density, it depends on the speed and order in which the pattern is written. This is because heat diffusion in the substrate is time-dependent. The addition of the time variable makes it impractical to correct for beam heating in a way similar to backscatter proximity correction.

Because of the poor thermal conductivity of fused silica reticles, the temperature rise during e-beam exposure is relatively high. Present-day e-beam reticle writers operate at 10 kV and 1.2 $\mu C/cm^2$, for which the beam heating is negligible. Future reticles will require dry processable resists and higher resolution. The latter tends to drive the beam voltage higher. With resist sensitivity of 10 $\mu C/cm^2$, beam currents of 30 A/cm^2, and 50 kV e-beams, the temperature rise on bulk fused silica is 45 to 85°C, depending on pattern density.[38]

It is interesting to examine the scaling laws for beam heating as a function of beam voltage and resist sensitivity. The total energy per unit area deposited in a bulk substrate is equal to the beam voltage times the incident dose (in units of charge per unit area) required to expose the resist. As the beam voltage is increased, the resist layer becomes proportionally more "transparent" to electrons; consequently, the higher the beam voltage, the more dose that is needed. The energy per unit area deposited in the substrate is, therefore, proportional to the square of the beam voltage. It is also proportional to the incident dose needed to expose the resist.

Proximity x-ray lithography employs a membrane mask that is no more than a few microns thick. When writing these masks, backscatter proximity effect and beam heating are less important, because the beam goes through the membrane without significant backscattering or energy deposition. This presumes high beam voltage of at least 100 kV, which is desirable, if not required, for membrane masks. From the point of view of beam heating, bulk substrates require low beam voltage, or sensitive resist, or both. With regard to beam voltage, the requirements for membrane masks are fundamentally opposed to those of bulk masks.

The fundamental trade off between resolution and writing speed is shown in Fig. 5.10(a,b) for 1X and 4X masks, respectively, for high-volume manufacturing. The writing tool, the process, and the need for productivity each impose a limitation. Each of these limitations divides the diagram into two regions, one allowed, and the other forbidden, by the limitation. For example, the stochastic Coulomb interaction limits the tool performance to an allowed region above and to the left of the diagonal line. Similarly, the process limits performance to the region above the horizontal line. The need for productivity or throughput in manufacturing requires a minimum writing speed, independent of resolution. This restricts the useful area to the region to the right of the vertical line. The cross-hatched area is the common region that is permitted by all three limitations. This is the useful region of operation for manufacturing. The black dot in each figure represents the best allowable performance, which is the desired operating point. A given pattern generator may be deemed good if it operates near this point. The 1X and 4X reduction cases differ because of the differing resolution requirements on the reticle and the need to pattern differing areas in a given time.

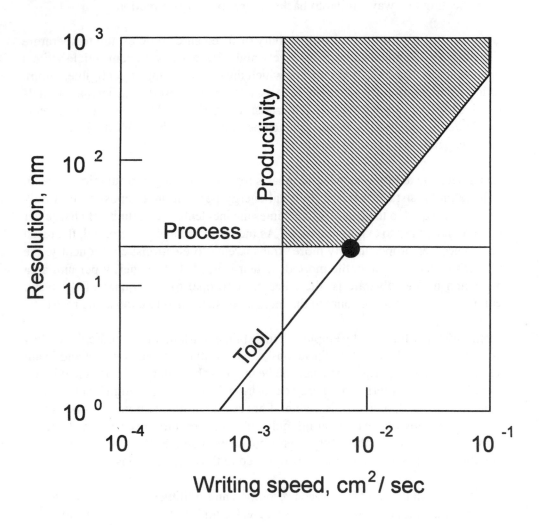

Fig. 5.10(a) Resolution and writing speed at 1X.

5.6.2.4 Practical limitations

Physical limitations, including stochastic Coulomb interaction, proximity effect, and beam heating, act to constrain the useful range of operating conditions for which resolution and throughput meet the necessary requirements. There are, in addition, several important practical limitations that deserve mention.

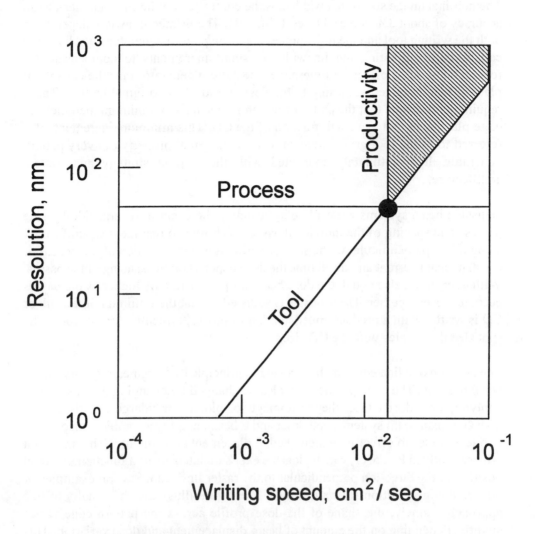

FIG. 5.10(b) Resolution and writing speed at 4X.

Pattern data originate from the device design and progress through conversion to numerical control (NC) data usable by the pattern generator, through inspection and repair of the final mask, through printing on the wafer. This process involves many steps, with diverse hardware, software, and personnel. Verification is required to insure that the data have not changed in ways that could compromise the pattern fidelity. Data volume increases with the number of pattern features in a chip. DRAM density increases by a factor of four every three years, leading to roughly a factor-of-four increase in data volume. One Gb DRAM chips with 180 nm CD will require a few GB of data. Data management represents a significant system problem and is an ongoing challenge.

The nominal image size or linewidth must be controlled on the finished mask to an accuracy of about 3% of the CD (see Table 5.1). The printed linewidth depends on both the writing tool and the resist processing. The pattern generator must have the capability of adjusting the linewidth in small increments to compensate for repeatable process offsets. Furthermore, the pattern generator must be capable of placing feature edges according to the design grid of the original pattern. These requirements dictate that the pattern generator must have a minimum increment of edge placement that is a small fraction of the CD. This minimum increment, also referred to as the tool grid increment, is a fundamental property of every pattern generator. It is intimately connected with the writing strategy and system architecture.

Gaussian beam systems write a line by scanning the beam in several side-by-side passes. The spacing of the individual passes is chosen so that the dose uniformity along a line perpendicular to the scan direction is a small fraction of the dose. For constant beam current and dwell time the dose depends on the spacing of the passes. With everything else equal, the dose becomes proportionally higher as the passes become closer together. The spot size is adjusted so that the minimum linewidth or CD is written with a predetermined number of passes, typically three or four. The spot size then scales with the CD size.

The position of a line edge can be changed in principle by incrementally varying the beam position. The edge position can also be changed by changing the dose, since applying more dose causes the resist image to "bloom" or enlarge slightly. Raster scan Gaussian beam systems, both laser and e-beam, employ a combination of these two techniques to effect incremental edge placement adjustment.[33] This is shown schematically in Fig. 5.11, which plots the exposure intensity as a function of lateral position in the direction perpendicular to the raster line scan. Several examples of dose and edge position shift are shown. The figure illustrates the penalty of this approach, namely, the shape of the dose profile across the pattern edge varies slightly, depending on the amount of beam displacement and dose variation. This causes the variation in edge placement resulting from process variations, such as resist development, to be nonuniform across the pattern. Despite these drawbacks, these techniques have been successfully applied to meet present-day requirements for linewidth control on 4X and 5X reticles.

Variable-shaped beam systems behave quite differently from Gaussian beam systems with respect to incremental edge placement adjustment. The electron optical column employs a shaping section, not present in Gaussian systems, as shown in Fig. 5.7(c). This allows the edge position to be varied independent of the centerline placement and dose. The minimum edge placement increment is equal to the shaping increment. Shaping capability must be designed into the electron column and data path, thus affecting the system complexity. A fundamental advantage over the Gaussian beam systems is that the dose profile across a pattern edge for a variable shaped beam is independent of the size of the edge placement increment. Edge dose profile is, therefore, uniform within the written pattern. It follows that small variations in the process affect all pattern features uniformly in variable shaped beam systems. Repeatable process offsets can be compensated in this case by applying a uniform bias to all pattern features. For this reason, a variable shaped beam system is preferred for applications that require very tight linewidth control, such as 1X masks and advanced NX reduction reticles.

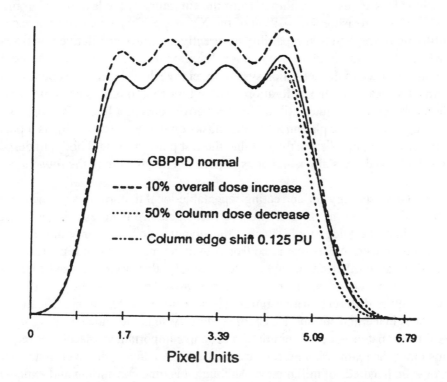

FIG. 5.11 Exposure intensity as a function of position perpendicular to scan direction for a raster Gaussian beam system. The right-hand edge of the intensity profile can be moved in small increments by varying the dose, the edge position, or both. The solid curve represents the normal Gray beam with per pixel deflection profile. The broken curves represent the profile with various corrections applied.

The pattern generator will be challenged to meet the ever-increasing demands for pattern placement, image size control, and throughput. This must be accomplished at reasonable cost. A fundamental trade-off exists between resolution and throughput, as determined by limitation on total beam current imposed by the stochastic Coulomb interaction in the drift length of the column. The critical supporting technologies of data management, resist, defect inspection and repair, and metrology represent practical limitations that, if not properly addressed, can seriously compromise the overall mask fabrication process.

All pattern generators create small errors in pattern placement and image quality. Some errors arise from the fundamental limitations discussed and are not correctable. Many errors are amenable to correction by feedback and calibration, however. Every successful pattern generator relies on a strategy of measurement and correction of pattern errors.

An important example of a repeatable, correctable placement error is local pattern distortion. This arises, for example, from the curvature of the laser interferometer mirrors used to measure the position of the XY stage. This error can be measured by utilizing the registration capability to directly measure the relative positions of the writing beam and marks placed on a calibration test plate. Once measured, the errors are fed back to the deflection system to remove them from subsequent written patterns. By measuring the calibration plate in four rotations, it is possible in principle to obtain an absolute calibration of all errors except a linear, isotropic scale factor, regardless of the placement of the masks on the test plate.[39] This is a purely geometric result and relies on the fact that the test plate, as a rigid body, undergoes only rotation and translation, and does not vary in height among the four views.

An alternative strategy for correcting repeatable local distortion is to write a test pattern and then measure the placement of written pattern features using a separate XY coordinate metrology tool. This has the advantage over the previous approach in that it eliminates errors that arise from the difference between registration and writing. It has another advantage as well, namely, that it can be used to correct pattern density dependent errors. Such errors arise because patterns of differing density require different writing times. These induce subtle, but significant local differences in thermal and surface charging duty cycle in the optical writing system. This form of pattern-specific emulation becomes important as placement requirements enter the realm where errors are measured in nanometers over distances of tens or even hundreds of millimeters. Although it is time consuming and expensive to perform this calibration for every unique pattern, there will always remain a fraction of critical masks that can benefit from this, and for which it is worthwhile, despite the cost.

Yet another strategy for correcting local distortions is to utilize the registration capability to detect the position of the writing probe relative to the actual writing surface at intervals during the writing. The most common and straightforward approach is to move the stage so that a registration mark is positioned under the beam. This mark must be rigidly positioned relative to the writing surface. It can either be permanently on the stage, to which the writing substrate is, presumably, rigidly clamped or, preferably, printed on the writing substrate prior to the writing step. This later is often undesirable, because the additional processing tends to introduce defects on the writing substrate that must be subsequently detected and repaired. After measuring the position of the writing beam relative to the mark, the measurement error is then used to correct the position of the writing probe. This approach requires that the registration errors be insignificant relative to the require placement accuracy and that no additional error is introduced in the process of moving the XY stage between the registration and the writing locations.

A variation of this approach is to utilize the writing beam to register to a grid that is printed in the resist layer prior to the writing steps. This technique, called global fiducial registration,[40] requires the grid errors to be insignificant relative to the placement accuracy required for the written pattern. It also requires that the exposure dose during the registration step not compromise the written image size significantly.

With membrane masks for proximity x-ray or projection e-beam lithography, it is possible to register to a grid that is permanently mounted beneath the membrane. This technique requires that the membrane be sufficiently transparent to the writing beam. Membrane masks suffer from process-induced distortion after the writing step. This tends to compromise all of the above correction approaches. Only pattern-specific emulations, described above, are useful here, assuming the process-induced distortion is repeatable across all membrane masks written with a given pattern.

None of the correction approaches described here is perfect in the sense of removing all errors in the written pattern. An uncorrectable error component, however small, is always present. For this component there is no substitute for a stable, repeatable pattern generator and process. It is preferable not only that the amount of drift and variation be small during the interval between calibrations, but that throughput be high enough to minimize drift and calibrations, but not so high that accuracy and stability are compromised.

5.6.3 Meeting the Challenges

5.6.3.1 Present state of the art

In assessing the capability of pattern generators to meet future challenges, it is logical to begin by surveying the present state of the art. There are a number of different mask writers in use today, employing the various approaches discussed earlier. Because of their unique characteristics, it would be impractical to survey all of them here. It is useful to concentrate on one example, preferably a system that has demonstrated the ability to meet the most advanced requirements. It is hoped that this will give the reader a feel for the characteristics that a state-of-the art pattern generator must have. Actual performance data will also be presented.

We will describe the EL-4 e-beam mask maker, designed by IBM.[41] This system is installed and running in the Advanced Mask Facility in Essex Junction, VT. It writes quartz reticles and masks for proximity x-ray lithography. X-ray masks are written at 1X, while quartz reticles are typically written at 2X to 5X magnification. X-ray masks, therefore, represent the more demanding application.

A block diagram of the system is shown in Fig. 5.12. The major subsystems include an electron column, with control electronics, an XY stage to move the mask under the beam, a digital pattern data path, system control computer hardware and software, and a mechanical system to control vacuum and mask loading. Most of these subsystems are present in some form in all pattern generators. They will be described in turn.

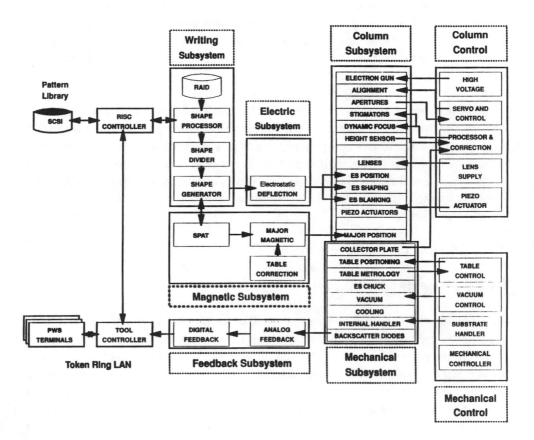

FIG. 5.12 Block diagram of the EL-4 variable-shaped beam system from IBM. The various subsystems include the pattern section, writing, electric, magnetic, feedback column, column control, mechanical, and mechanical control subsystems.

The electron column employs a variable shaped beam.[42] The focused electron probe consists of a rectangular spot for which the size can be varied independently in the X and Y dimensions between 0 and 2.0 μm, with 8 bits of shaping in each axis. In this process the current density is maintained at a constant value of 30 $Å/cm^2$. The electron source is single crystal LaB_6, with accelerating voltage of 75 kV.

The main field magnetic deflection is 2.1 mm, and the subfield electrostatic deflection is 37.5 μm. The magnetic deflection uses the variable axis immersion lens (VAIL) concept, shown schematically in Fig. 5.13.

A yoke in the plane of the final projection lens applies a transverse field synchronous with the magnetic predeflection. This transverse field precisely offsets the radial component of the projection lens field so that the magnetic symmetry axis of the lens is always centered on the beam axis. The deflection aberrations are thus effectively canceled. The beam intersects the writing surface at normal incidence, decoupling pattern placement from variations in height of the writing surface. The subfield deflection utilizes two coaxial dodecapoles, separated by a distance along the beam axis. This double deflection causes the beam to pivot about the back focal plane of the projection lens. In this way the subfield deflection maintains normal incidence on the writing surface as well. The total resulting deflection aberration at the edges of the 2.1-mm field is 18 nm. A dodecapole deflector has twelve electrodes, distributed around the circumference. By applying variable voltages to the individual plates, a uniform electrostatic deflection field is created, which is electrically rotatable. This permits addressing any position within the subfield with high speed and short settling time. The twelve plates are of equal width, with widths chosen in such a way that the field is highly uniform near the beam axis. This results in low third-order deflection aberration and excellent deflected resolution at the edges of the subfield. The advantage of this twelve-pole deflector over the typical eight- or twenty-pole deflectors is that only two bi-voltages are needed for the X and Y axes, respectively.

The XY stage is a servo-guided planar stage,[43] shown schematically in Fig. 5.14. The position is controlled in three degrees of freedom, X, Y, and theta (yaw). The remaining three degrees of freedom, Z, pitch, and roll, are constrained by the fact that the stage moves on a very flat surface. This represents a significant advantage over traditional rail-guided stages. Another advantage is the low mass, which permits higher acceleration, and faster response.

The digital pattern data path uses a Power PC to control the flow of pattern data. A 16 GB redundant array of inexpensive disks (RAID) stores the pattern data. The custom logic is contained in field programmable gate arrays (FPGA). The final stage in the data path is a shape generator, which translates the pattern data into shaped spots. The effective data rate is equal to the product of the spot transfer rate times

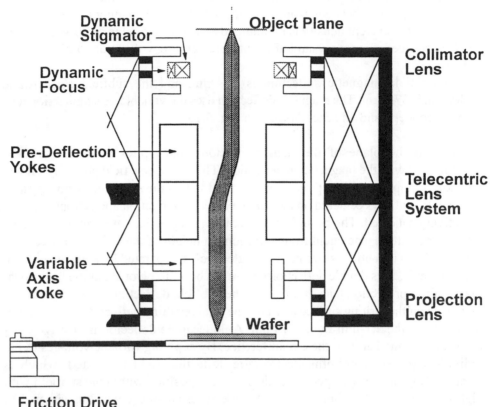

Dynamic Stigmator

Object Plane

Collimator Lens

Dynamic Focus

Pre-Deflection Yokes

Telecentric Lens System

Variable Axis Yoke

Wafer

Projection Lens

Friction Drive

FIG. 5.13 Variable axis immersion lens (VAIL), schematic. This is the final lens and deflection system in the EL-4 electron column. With deflection the beam is first displaced parallel to the optic axis prior to passing through the projection lens. The variable axis yoke is excited in synchronism with the deflection in such a way that the radial component of the projection lens field is precisely canceled. In this way the beam behaves as it would if it were on the central magnetic axis of the lens with no deflection This arrangement has the property that the deflection aberrations are very low. The immersion property also insures that the on-axis aberrations are low as well, resulting in uniformly good resolution throughout the deflection field. The beam impacts the writing surface at perpendicular incidence, thus decoupling pattern placement accuracy from small variations in the height of the writing surface.

the number of pixels in the shaped spot. The spot transfer rate is 20 MHz. A 0.25 × 2.0 μm shape contains 32 pixels, in which case the effective pixel rate is 640 MHz. Despite this high rate, the electronics are only required to run at the relatively modest spot transfer rate of 20 MHz.

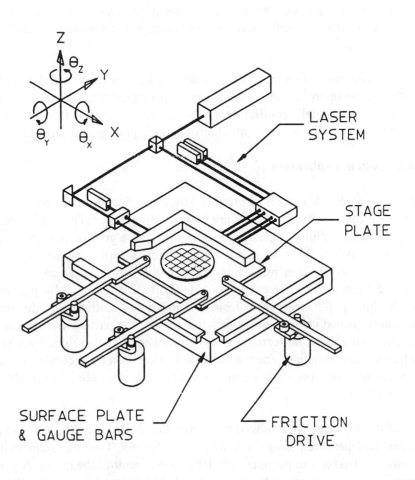

FIG. 5.14 Planar XY stage, schematic. The writing substrate is clamped to this stage in the EL-4 system. The stage slides on the surface plate, which is coated with a low friction material. Three laser beams measure x, y, and theta (yaw). The measurement information is fed back to correct the position. This arrangement has the advantage that the pitch and roll degrees of freedom are tightly constrained, resulting in high placement accuracy. It also has the advantage over more conventional stacked rail bearing stages that the moving mass is low, resulting in fast servo response.

The system control computer is a PS/2 Model 95, running under the OS/2 operating system. This permits multitasking, so that various calibration, data logging, and data analysis functions can be carried out in the background while the system is writing. The operator interface is point-and-click, and includes job deck capability to run consecutive exposures without operator intervention. All of the subsystems are connected via a LAN, which comprises the main path for handshake and interrupt functions of the various components. Calibration data can be collected, analyzed, and graphically displayed on-line using a custom APL2 program package. The system can be operated remotely via LAN. This allows an expert to run the system, and collect and analyze performance data resident at the system or from hundreds of miles away.

The mechanical subsystem loads and unloads masks under automated control. Two cassettes contain up to 10 masks each. These may be pumped independently and left in the system to thermally equilibrate. The mechanical system is run under separate PS/2 control, including the vacuum control, and all gauges and sensors.

5.6.3.2 System calibration

Calibration of EL-4 centers around a strategy of measuring and correcting repeatable errors. Deflection errors are removed by a process called Learn, where the registration capability of the system is used to scan a grid, permanently mounted on the stage. All of 56×56 subfields are independently calibrated, and linear distortions removed. This procedure takes about two minutes and is routinely performed before and after every mask is written. It can also be performed at intervals during the writing, if desired. During this calibration, the magnetic deflection is stepped in a serpentine path that mimics the writing. In this way a class of repeatable magnetic deflection errors is removed. These include hysteresis and yoke temperature variation. The magnetic deflection must be accurate to within the placement tolerance, presently 25 nm. Over the 2.1-mm deflection field this is of the order 1×10^{-5}.

Errors of the XY stage are removed by a procedure called emulation. A test pattern is written, and measured using the Leitz LMS-2000 tool. The measurement data are automatically fed back to the magnetic deflection to remove the errors. A variant of this process uses measurement targets embedded in the product pattern. This allows the possibility of removing errors that depend on local pattern density. These errors arise from thermal and charging influences in the column, and in the writing substrate itself.

The height of the writing surface is measured locally under the beam. The measurement data are used to automatically correct focus. Height-dependent linear field distortions are also correctable; however, this is normally not required because of the normal beam landing.

Focus and astigmatism are evaluated from written images, using test patterns included with every product mask. Once set up, these require only occasional and minimal adjustment.

The system is located in an environmental chamber. The temperature is controlled to ±0.05°C. Twenty temperature sensors are located at critical points around the system. The data are automatically logged. Good thermal management is essential, as the thermal expansion of the mask can easily compromise pattern placement if temperature is not well controlled.

5.6.3.3 Performance of the EL-4 system

As mentioned, the key performance parameters are pattern placement, image size control, and throughput. Absolute pattern placement is typically about 30 nm, with resist images measured on a Leitz LMS-2000 metrology tool. Placement data are shown for a typical large pattern in Fig. 5.15(a,b). Figure 5.15(a) shows the repeatability of two masks written one day apart. This has a 3-sigma value of 14 nm.

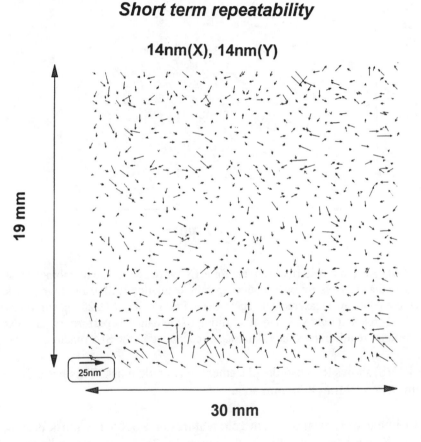

FIG. 5.15 (a)   Pattern placement repeatability for two x-ray masks written one day apart. The membrane masks were written using the EL-4 system, and the position of pattern features measured using a Leica Leitz LMS-2000 metrology tool. The errors are relative to the common average of the absolute placement errors for the two masks. The pattern is equivalent to a 256 Mb DRAM pattern.

Absolute Pattern Placement

21nm(X), 22nm(Y)

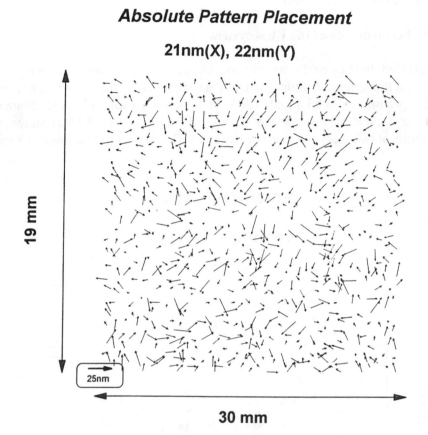

FIG. 5.15(b) Absolute pattern placement for a 1X x-ray mask. The membrane mask was written using the EL-4 system with product specific emulation (PSE), and four passes. The errors represent the raw measurement data from the Leica Leitz LMS-2000, with translation and rigid body rotation removed. No other fitting of the data was performed. This exceeds the performance requirements for 1X masks, as contained in the SIA roadmap.

Figure 5.15(b) shows the absolute placement of a single mask. This has a 3σ value of 22 nm over a 19 mm × 30 mm area.

An SEM image of typical 0.18-μm gold features on a 1X x-ray mask is shown in Fig. 5.16. Image size control is in the range 11–25 nm for these images. This performance is consistent with SIA targets.

5.6.3.4 Future challenges

For present-day mainstream pattern generation, the leading practical limitation is CD control. This is limited by the relatively coarse resolution of the Gaussian beam

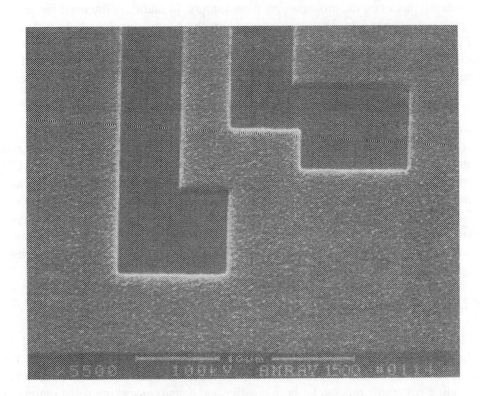

FIG. 5.16 1X mask for proximity x-ray lithography. Gold structures are 0.18 μm wide and 0.48 μm high. Pattern is gate conductor level for a high-density CMOS SRAM circuit.

mask writers in widespread use, together with the coarse tool grid on which incremental pattern edge placement is constrained in raster-based systems. Pattern placement is not a serious limitation here, because placement errors on reticles are reduced by 4X to 5X on the printed wafer, and because the placement stability of these writers is adequate for reduction reticles, with some effort. These limitations are not fundamental but are driven by a cost-performance trade-off. They are, to a large extent, an artifact of the prevalence of *N*X reduction lithography, which in the past has removed the performance burden from the reticle, and placed it in the optical stepper.

As mentioned, the resolution and tool grid increment are, for practical purposes, limitless with electron beams. Moreover, the cost of improving these parameters beyond what is commercially available today is not high. Therefore, it is anticipated that CD control and image quality will cease to be limiting factors with advanced mask makers. Pattern placement will become the practical limitation for 1X lithography.

Data management represents a key challenge. Ever-expanding data volumes constantly tax even the most powerful computers. In addition, the need for introducing corrections into the data requires algorithms to be developed. An example of this is OPC, which attempts to correct for diffraction in the optical stepper by building local subresolution pattern features into the mask. Since OPC will be needed for optical lithography below 0.25 μm, this represents an urgent problem.

5.6.4 Conclusion

We have seen that pattern generation for mask making is both exacting and complex. The key performance parameters of pattern placement accuracy and CD control scale with CD. Throughput, though not an overriding issue for mask making, cannot be ignored in the overall system design. The requirements, as detailed in the SIA Roadmap, become more difficult with every device generation. Systems exist that can meet the present requirements, with effort. A significant, ongoing investment will be needed to support the technology learning needed to meet future requirements, however.

5.7 MASK FABRICATION: RESISTS AND PROCESSING

This section reviews the process steps and chemical processes involved in converting an exposed resist-coated mask substrate to a patterned mask ready for inspection. A detailed discussion of the radiation induced chemistry and corresponding patterning process for both positive and negative working resist materials is presented. The chemical process is reviewed with a description of an e-beam resist commonly used for mask fabrication, namely PBS, and is then extended to include other resists that have a potential use for mask fabrication. Resist selection, capabilities, and limitations are also reviewed.

A radiation-sensitive material, referred to as a resist, is typically composed of an organic polymer and small molecule additives that serve to enhance the lithographic performance of the material. The action of the exposing radiation on the resist produces a change in the solubility of the exposed vs nonexposed regions of the resist film. This differential solubility is used during the development step to form the pattern in the resist. The opaque masking layer, typically a chromium film on the fused silica substrate, is etched through the resist pattern to form the final mask.

The ability to reliably and cost-effectively fabricate each circuit level pattern on a reticle is essential to the successful production of an integrated circuit. The required CD, specifications on CD uniformity, and line edge acuity may determine what exposure tool and fabrication process will be used for the production of a particular mask or reticle. The critical mask requirements given in Table 5.1 represent the guidelines for the type of exposure tool (electron-beam or laser-based) that will be used. Selection of the exposure tool and minimizing the overall exposure time will greatly influence the imaging material (resist) chosen. The final decision in the process flow is whether the chromium etch pattern transfer step will be based on "wet" or "dry" chemistry. This latter consideration can also influence the choice of the imaging material.[44] Process enhancements, such as process control techniques for the fabrication process, are currently in the development stage. However, it is expected these techniques will become mainstream if the mask CDs decrease to ~0.50 µm.[45, 46]

5.7.1 Selection of Exposure Tool

The presence of electron-beam and laser exposure tools in a given mask-making facility can result in having two vastly different resist processes active in the shop. The ideal situation is one in which one material is interchangeable between the two systems. In reality, no one material can meet the sensitivity requirements for both forms of exposure and to date there is no active development program to achieve this. The next section provides a description of how an imaging material for a given exposure tool is selected.

5.7.2 Imaging Material Considerations

The choice of the resist must be compatible with the exposure environment of the tool, and the dose requirement must enable high tool throughput and cost effectiveness.

Historically mask manufacturing was performed using photolithographic methods.[47] Optical pattern generators operating in either a contact print or a step-and-repeat mode used the same resist for the imaging material as used in silicon device fabrication. These "conventional" photoresists are comprised of a copolymer of a cresol and formaldehyde monomer, and a diazonaphthoquinone (DNQ) photoactive compound or sensitizer. The copolymers are referred to as novolacs and due to the acidic nature of the resin, they readily dissolve in an aqueous base solution. The insertion of the photoactive compound renders the resist film insoluble in the alkaline solution. Exposure to light converts the sensitizer to an acidic product and increases the solubility of the exposed resist areas in an alkaline solution, resulting in positive tone pattern formation. Figure 5.17 provides the structure and light-induced chemistry occurring in the novolac based resist.

The simplicity and cost effectiveness of the material and process represented attractive attributes for making masks this way. However, as time progressed, the enormous increase in pattern density, the decrease in CD size, and tighter CD uniformity specifications brought about the need to replace the optical-based exposure tool to one employing electrons.[48]

The resist requirements therefore changed dramatically and became a function of

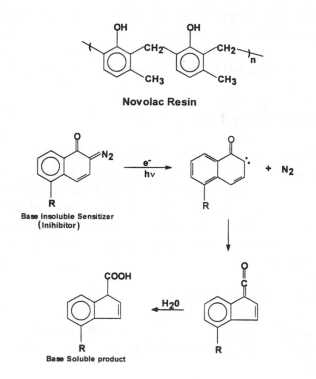

FIG. 5.17 Structure of novolac and sensitizer and radiation-induced chemistry leading to positive-tone image formation.

a mask-fabrication process based on an electron beam exposure. This process is depicted in Fig. 5.18.

The process depicted in Fig. 5.18 requires the resist characteristics listed in Table 5.4. These properties are essential to achieve consistent mask-manufacturing capability.

The properties listed in Table 5.4 effectively limit the classes of commercially available electron-beam and optically sensitive materials. For example, the resist coating shelf-life requirement may preclude the possibility of using resists based on chemical amplification.[49] The required lithographic performance of the resist will further reduce the number of suitable resists. Table 5.5 lists the lithographic performance and processing requirements necessary for the manufacturing of the 256 Mb DRAM generation reticles.

TABLE 5.4 Resist material properties requirements.

MATERIAL PROPERTY	REQUIREMENT
Resist solution shelf life	> 6 months
Batch to batch reproducibility	< 5% variation in Composition and Molecular Weight.
Resist film coating shelf life	> 3 months
Resist thermal properties	Glass transition temperature* $T_g > 80$ °CDecomposition temperature $T_d > 120$ °C
Wet etch Cr chemistry	No degradation or adhesion failure
Dry etch Cr chemistry	Min 1:1 selectivity in Cl_2-O_2 based plasmas**
Solubility	In environmentally safe spin coating solvents and aqueous base developers
Stripability	Removal in commercial amine based stripping solutions or O_2 , halogen based plasmas

*The glass transition temperature, Tg, is the temperature at which the resist changes from a glassy amorphous state to a rubbery state.
**The plasma, used for "dry-etching" the chromium film, may change according to the chromium film composition.

TABLE 5.5 Resist lithographic performance criteria.

LITHOGRAPHIC PROPERTY	REQUIREMENT
Sensitivity: e-beam*; laser	‾2.0 $\mu C/cm^2$ @ 10 keV; 100 mJ/cm^2 @ I =365 nm
Contrast (γ)	(4 for 85° resist wall profile
Resolution	± 0.50 μm

*The required e-beam resist sensitivity is a factor of the required writing rate of the pattern generator. A low resist sensitivity requires a higher exposure dose, and a faster writing rate requires a higher dose intensity. Together, these lead to increased resist heating and the possibility of a thermally induced CD variations (see Sec. 5.6).

The resist process latitude to achieve the required mask CD uniformity (±3%) also places limitations on the resist selection. The processing parameters that are investigated include exposure dose and development latitude characteristics, and the extent of film erosion in a positive resist of the nonexposed film regions. Allowance of a certain level of film erosion can improve the sensitivity of the resist but can have a profound affect on resist contrast, pattern profile, and pattern edge roughness.

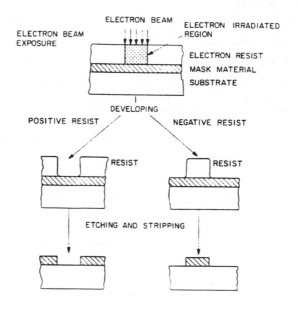

FIG. 5.18 Schematic of the mask fabrication process.

The standard resist exposure environment in an electron beam system is one of high vacuum ($\sim 10^{-6}$ Torr), and depending on the exposure dose and the e-beam accelerating voltage, there is the possibility the film will undergo a rise in temperature. The high-vacuum environment of the tool and the elevation in temperature require that for the resist to be compatible with the operation of this tool the following conditions must be met: the components of the resist and the radiolysis products must be nonvolatile at the maximum temperature (T_{max}) the mask will realize, and the onset temperature for decomposition (T_d) and the glass transition temperature (T_g) of the composite resist formulation must be greater than T_{max}.

Any material evolution from the film is cumulative and unacceptable. These undesirable volatile products can result in a performance degradation of the writing tool through contamination of the thermal emission source, and in surface charge buildup leading to pattern placement errors. Careful thermal analysis of the resist is therefore required as well as a fundamental understanding of the radiation-induced chemistry.

For laser-based exposure tools where the exposure wavelength is centered at 364 nm and operates under atmospheric conditions the likelihood of material evolution during exposure is reduced. As with any optical exposure system, one of the primary properties of the resist under consideration is its opacity at the exposure wavelength and, from a processing standpoint, its exposure dose. Unacceptably long exposure times may lead to a heat rise and therefore the same problems as discussed above may prevail.

5.7.3 Commercially Available Electron Beam Resists

5.7.3.1 Positive-acting resists

Throughout the evolution of the Etec MEBES, the basic resist sensitivity require-ment has remained essentially constant. With each MEBES generation, improve-ments in writing frequency, electron source emitter, and resolution have occurred. However the maximum allowable dose at 10 keV beam accelerating voltage in a single scan trace has remained in the 1–2 $\mu C/cm^2$ regime. As a consequence only a limited number of positive-acting resist materials are sensitive enough to operate in this exposure dose range. By far the primary resist in use over two decades is the alternating copolymer of 1-butene and sulfur dioxide (PBS).[50] The copolymer structure and electron-beam-induced chemistry are given in Fig. 5.19.

PBS is initially prepared as a polymer having an average molecular weight in the range of 5×10^5 to 1×10^6 g/mol. Upon exposure to electron-beam radiation PBS undergoes main chain scission via cleavage of the carbon–sulfur bond. This scission event is highly efficient in PBS, and the irradiated polymer exhibits a reduction in its molecular weight. The scission efficiency in PBS, or any polymer, is expressed as the number of scission events (G-value) for 100 eV of energy absorbed.[51] G(s) values provide a quantitative approach to identifying highly sensitive positive acting resists. This G-value for scission has been measured at 10 for PBS.[52] In comparison, methacrylate-based polymers such as poly(methyl methacrylate) (PMMA) have a much less efficient scission process and G(s) values are in the range of 1.5 to 4.0. The radiation-induced degradation mechanism in PMMA is represented in Fig. 5.20.[53]

Main chain scission in PMMA is initiated by cleavage of the main chain carbon to carbonyl carbon bond. The radical that is formed along the main chain undergoes a rapid rearrangement that results in the cleavage of the main chain and production of small molecule volatile products and radical species.[53,54]

A comparison of the mechanisms shown for PBS and PMMA reveals a secondary event occurring in the case of PBS. Figure 5.19 also depicts that spontaneous depolymerization can occur after the main chain scission event. This is, however, temperature dependent, and for PBS chain unzipping is observed at temperatures above 60°C.[55] This unzipping of the polymer chain will only occur if the localized heating of the film exceeds the above temperature during the exposure. This is

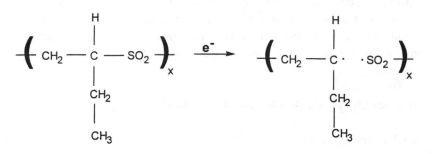

Initial high molecular weight   Lower molecular weight after irradiation

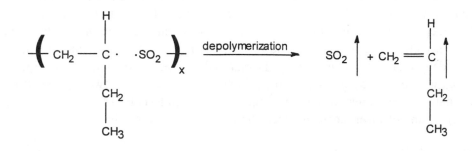

FIG. 5.19   Structure of repeat unit in PBS resist and electron-beam-induced reaction mechanism.

easily verified by the presence of a partial relief image in the film viewed immediately after exposure and represents the undesirable situation of evolving volatile products such as SO_2 in the exposure chamber of the tool. Therefore, an increase in the PBS exposure dose can lead to outgassing of the resist film while it is being exposed.

The exposure-induced reduction in molecular weight in PBS eventually leads to a total separation in the molecular weight distribution (MWD) of the exposed and nonexposed polymer. This separation, as depicted in Fig. 5.21, provides the basis for generation of a positive-tone relief image.

The difference in the MWD causes a difference in the solubility rate of the exposed versus non exposed film regions for a given solvent/developer. The difference in the thermodynamic characteristics of the developer chosen from that of the polymer also influences the ease of finding a solvent that for a specified development time will only act to dissolve the exposed portions of the film. For the case of PBS a two-component developer consisting of 5-methyl-2-hexanone and 2-pentanone is commonly used. The developer strength is modified by varying the proportions of the two components, and as shown in Table 5.6, increasing the proportion of 2-pentanone dramatically increases the development rate.

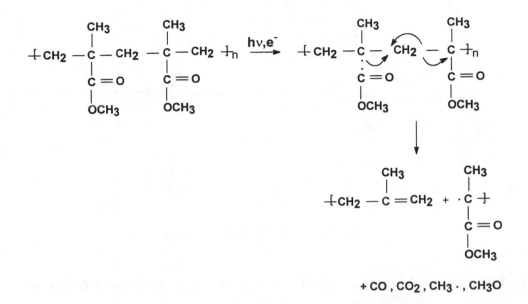

FIG. 5.20 Radiation-induced reaction mechanism in PMMA resist.

TABLE 5.6 Effect of developer composition on PBS dissolution rate.

DEVELOPER COMPOSITION (% 2-pentanone)	DEVELOPMENT RATE* (nm/sec)
30	15.8
20	11.7
10	6.1

* Dose = 1.4 $\mu C/cm^2$ @10 keV, Temp. = 20°C, RH = 35%

The reduction in 2-pentanone content in the developer provides for an improvement in PBS resist contrast and image profile. This in turn provides the capability of a higher level of pattern resolution to CDs below 0.50 μm. However, these improvements require longer development times and a greater use of these chemicals. The strong dependence of the dissolution rate of PBS on development composition and environmental conditions will be further discussed in a later processing section. A second major limitation to the use of PBS is that the chrome etch pattern transfer process is limited to using a wet chemical etchant. PBS exhibits extremely poor resistance to the O_2Cl_2-based plasma used in the dry etching of

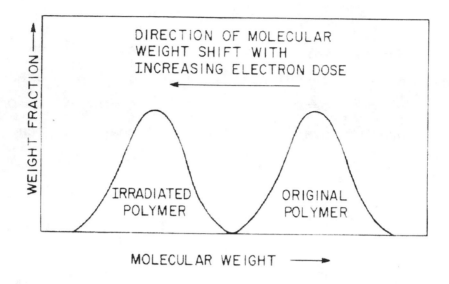

FIG. 5.21   Molecular weight distribution of a positive-acting resist as a function of irradiation dose.

chromium films. The need to progress to a dry etch pattern transfer process will be discussed in a section on processing of masks.

A number of alternative chain scission resists have been commercialized to provide reduced sensitivity to the development environmental conditions and slight improvement in the dry etching characteristics of PBS . The Toray EBR-9 family of resists are based on halogenated methacrylate homo and copolymers.[56] These resists exhibit higher G(s) values than PMMA and have a higher contrast than PBS, but require an exposure dose in excess of 5 μC/cm^2 at10 kV. As with PBS, the developing chemicals are all organic based, and issues relating to toxicity and flammability represent drawbacks to continued use of these materials. The improved dry-etching characteristics are marginal over what is required to minimize linewidth loss during the pattern transfer step.

Use of commercially available novolac-based resists, such as AZ-5206, provide a path for the removal of organic developing chemicals from the mask fabrication facility as well as maximizing the dry etch resistance for purely organic-based resists.[57] In these resists, conversion of the DNQ inhibitor to its aqueous base soluble product occurs via exposure to electrons and is consistent with the chemistry depicted in Fig. 5.17. The process is, however, plagued as being highly inefficient, and low DNQ conversion yields are observed. The highest practical sensitivities that have been achieved for these types of resists are of the order of 5 to 10 μC/cm^2 at 10 kV.[58] Improved sensitivity can be achieved by the use of higher-normality developers. Using this approach, up to 20% film loss in the nonexposed film regions can be observed and effectively compromises the measured contrast and image quality of the resist.

In addition, depletion of water from the film while it is present in the vacuum environment of the exposure tool leads to a delay in onset of development. Figure 5.22 shows development traces for a novolac-based resist in which the hold time in vacuum was varied. T_O and T_O' are delay times prior to the onset of development for substrates in which T_O' corresponds to a sample that was maintained twice as long as the sample in which the delay period is characterized by the onset for development at T_O. Increasing the time results in a longer development induction period. This variable induction period reduces CD control from mask to mask and in combination with poor sensitivity suggests resists of this type are unacceptable from a throughput and process-control standpoint.

Novolac-based resists do, however, provide the means of identifying a material that can act as an acceptable etch mask during dry etch patterning of the chrome layer. Selectivity of slightly >1.0 have been recorded and represent a marked improvement over what is observed for olefin sulfone and methacrylate-based polymers.

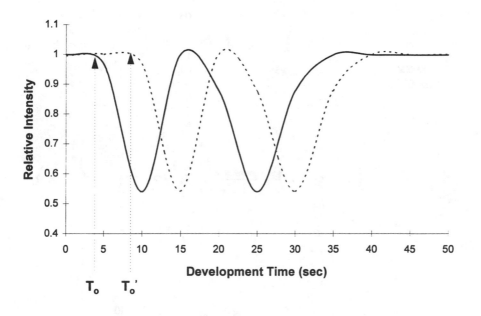

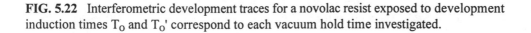

FIG. 5.22 Interferometric development traces for a novolac resist exposed to development induction times T_O and T_O' correspond to each vacuum hold time investigated.

The combination of the desirable etching resistant properties of novolac resists and the high sensitivity of olefin sulfone (PBS) resists has been achieved by blending the two polymers together.[59] The resist termed NPR consists of a cresol novolac resin and electron-beam-sensitive dissolution inhibitor. The inhibitor poly(2-methyl-1-pentene sulfone), though similar to PBS, is characterized by the depolymerization reaction (shown in Fig. 5.19) occurring at below room temperature. The radiation-induced chemistry and the path for the formation of a positive-tone relief image in NPR is given in Fig. 5.23.

At the proportions of the poly(2-methyl-1-pentene sulfone), PMPS, indicated in Fig. 5.23 the polymer acts as an effective dissolution inhibitor for dissolution of a novolac in an aqueous base solution. Electron-beam exposure results in the same initial carbon-sulfur bond cleavage as observed in PBS. However unlike PBS, the breakage of the carbon-sulfur bond in PMPS spontaneously leads to the scissioning of the majority of the remaining carbon-sulfur bonds in the polymer chain. The final monomeric products readily evolve from the film, leaving only the novolac resin remaining in the exposed film areas. These areas are in turn readily aqueous-base-soluble, leading to positive-tone pattern formation.

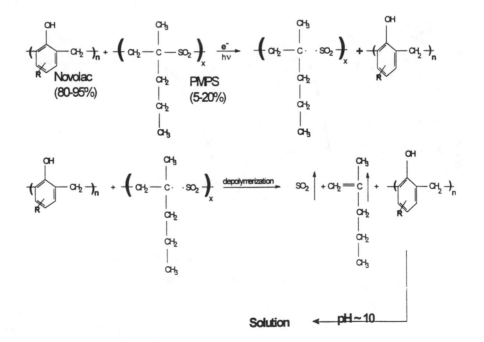

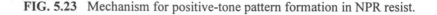

FIG. 5.23 Mechanism for positive-tone pattern formation in NPR resist.

The sensitivity of the resist is a function of the molecular weight of the PMPS added. Use of higher molecular weight results in higher sensitivity and values as low as 1.5 μC/cm² have been recorded.[60] Use of high molecular weight (>5 × 10⁵ g/mol) PMPS, however, reduces the compatibility of the blend, thereby reducing the image quality of the resultant pattern.

The use of an interruptive development process and PMPS having a MW < 5 × 10⁴ g/mol yielded the patterns shown in Fig. 5.24. Pattern dimensions of ~0.10 μm are shown, and the exposure dose used to generate patterns below 0.50 μm is in the 5–10 μC/cm² range.

Further improvement in the sensitivity of NPR, as well as minimizing the amount of outgassing from the film during exposure, must be realized before this material can be viewed as a viable alternative to PBS.

More recently high-resolution and high electron-beam sensitivity has been achieved with chemical-amplified resists that have been developed for use in deep-UV (λ = 248 nm) lithography. The observed high sensitivity is a result of an amplification of the initial radiation-induced event to produce a cascade of chemical events

0.20 μm Lines & Spaces

0.15 μm Lines & Spaces

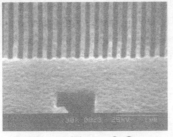

0.10 μm Lines & Spaces

FIG. 5.24 Resolution capability of NPR resist when exposed to electrons. Resist film thickness was 0.30 mm and a dose of 8 μC/cm² was used.

during, in most cases, a subsequent heating step. Sensitivities on the order of 1.0 $\mu C/cm^2$ and below have been recorded.[61] These resists also exhibit high contrast and the ability to print features having dimensions below 0.10 mm. Processing issues related to limited shelf life of the coated blank, the need to perform a postexposure bake step, and the sensitivity to the presence of basic airborne contaminants in the clean-room environment have precluded the use of these resists by mask manufacturers.[62] Of the issues identified, the greatest hurdle to use of these materials is the requirement of each mask house taking on the responsibility of coating the resist on the chrome-on-glass mask blank. Ownership of this process can be expensive and time consuming. These resists therefore are limited to applications involving patterning of silicon devices.

5.7.3.2 Negative-acting resists

Early interest in negative-acting resists was due to the inherent high sensitivity that could be achieved with these materials.[63] For a given resist, a range of sensitivities and contrast is achieved by adjusting the weight average MW and MWD of the polymer. For those resists composed of at least two monomers, alteration in composition can also affect the observed sensitivity of the resist. In addition, unlike the processing complexity associated with PBS, negative resists provided a reduction in the variables affecting line size and the process was typically without iteration. The initial family of negative working materials was based on the incorporation of an epoxy containing moiety.[64] The monomer of choice was glycidyl methacrylate, of which the homopolymer PGMA was offered as a commercially electron-beam resist for mask making. Copolymerization of glycidyl methacrylate with monomers such as ethyl acrylate (COP) and halogenated styrenes provided a variety of materials exhibiting varying sensitivity, resolution, and dry-etching resistance. The radiation-induced chemistry occurring in these resists is provided in Fig. 5.25.

The G-value for crosslinking $[G(x)]$ for resists containing an epoxide group has been measured as high as 10, and sensitivities at 10 keV of $<1\mu C/cm^2$ have been recorded. The high sensitivity in part is a function of the chain reaction of the ring-opening polymerization of the epoxide group present in these resists. From a sensitivity standpoint this reaction is advantageous; however, this reaction is observed to continue in the vacuum environment of the electron-beam exposure tool long after the resist has been exposed. This postirradiation continuation of the cross-linking reaction results in linewidth change as a function of the cure time in vacuum. For exposure times of >30 minutes, the measured resist linewidth uniformity is compromised by this continuation of the chain reaction. The growth in exposed resist film thickness as a function of cure time at increasing dose is shown in Fig. 5.26 for the resist consisting of 85 mole % GMA and 15% 3-chlorostyrene (3CLS).

The magnitude of the growth in film thickness/linewidth is largely a function of the process dose used. Higher doses minimize the effect, but at the consequence of

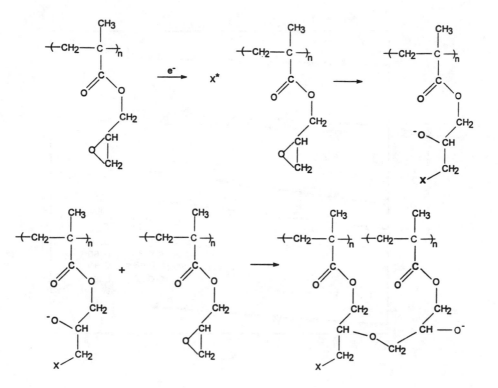

FIG. 5.25 Radiation-induced reaction mechanism for the cross-linking of epoxy-containing negative-acting resists.

throughput.[65] Elimination of the curing effect is accomplished through the use of resists based on styrene. Polystyrene is a weakly sensitive negative-acting resist, and incorporation of halogen-containing substituents produces a resist having a sensitivity in the 1–2 $\mu C/cm^2$ range. By far the majority of work has been with the homopolymer of chloromethylstyrene PCMS.[66] The absence of a continuation of cross-linking in vacuum is explained through the radiation-induced reaction mechanism given in Fig. 5.27. The mechanism shown is limited to the primary cross-linking reaction occurring with resists containing the chloromethyl functionality.

The inherent sensitivity of chlorinated and chloromethylated polystyrenes is due to the ease of cleavage of the carbon–chlorine bond present in these polymers. The radicals that are formed can abstract to produce additional radical species or combine with a radical of an adjacent chain, resulting in an insoluble network. The radical combination does not propagate, and there is no continuation of cross-linking after exposure. In addition to the lack of a curing reaction, the lithographic performance of styrene-based negative resists is greatly enhanced by

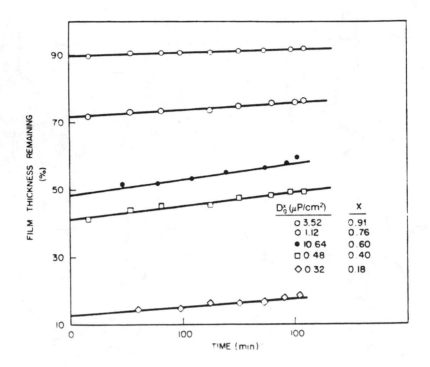

FIG. 5.26 Change in normalized film thickness remaining as a function of vacuum cure time at increasing dose. The value *x* represents the percentage of film thickness remaining at each dose investigated.

preparing materials exhibiting nearly mono-disperse MWD.[67] Contrast values as high as 3.0 have been reported for polymers having an MWD = 1.05.

Polystyrene-based resists also provide a significant improvement in plasma etching resistance over what has been measured for aliphatic polymers such as PGMA. Table 5.7 provides the etch rate for PGMA, poly(3-chlorostyrene), and copolymers of GMA and 3-CLS in a wet air plasma.

TABLE 5.7 Wet air plasma etch rate for selected negative-acting electron-beam resists.

RESIST	ETCH RATE (nm/min)
PGMA	28.5
P(GMA$_{0.50}$ CLS$_{0.50}$)	23.2
P(3-CLS)	19.5

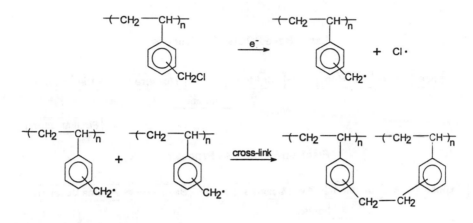

FIG. 5.27 Electron-beam-induced cross-linking reaction occurring on chloromethylstyrene-based negative acting resists.

5.7.4 Resist Processing Issues

5.7.4.1 Process sequence and its influence on CD control

Figure 5.28 is a process flow schematic for the patterning of a reticle using an electron-beam exposure tool/PBS resist and a laser-based tool/novolac (OCG 895i) resist.

The electron-beam resist process is more complicated than that required for the optically exposed novolac-based resist. The PBS process is characterized by a development step that is commonly performed in an iterative mode. This requires the process engineer to perform a first develop followed by a measurement of the critical feature or a linewidth control pattern on the reticle. The measured size dictates whether additional development time is required, and if so, the sequence is repeated until the desired line size at develop is achieved. This required sequence is a consequence of the effect the environmental conditions have on the development rate of the exposed PBS film areas. Figure 5.29 represents a plot of the PBS development rate as a function of percent relative humidity in the clean room at constant temperature. This data was obtained with a developer solution that contained no water and was stored in a nitrogen atmosphere to inhibit the absorption of water by the developer. A nonlinear dependency is observed for the typical operating condition of 35% to 40% RH. As can be inferred from the plot, slight changes in this range can produce significant changes in the critical feature size on the reticle.

Reduction in the sensitivity of the PBS development rate to the processing environment can be accomplished through some combination of increasing the

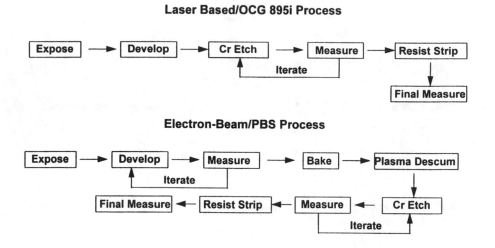

FIG. 5.28 Process flow for masks patterned using the laser-based and electron-beam exposure sources.

exposure dose and decreasing the developer concentration. The effect is to improve the contrast of the material, which in turn should provide an improvement in the development time latitude.

It has also been reported that PBS linewidth growth during development is a function of beam address size used during the exposure.[68] The linewidth growth may be further affected by the directionality of the feature. X-directed features have been shown to grow at a range that is different than Y-directed patterns; however, this effect is complicated by possible blanking errors in the e-beam deflector of the writing tool. The trend in line size as a function of beam address for a 2.5 μm Y-directed line is summarized in Fig. 5.30.

Extended development times also tend to reduce linewidth uniformity in that the development rate can be different in the center of the mask vs the edge. These radial effects present a major hurdle to achieving the proposed <~3% mask CD specification.

The second critical step in the PBS process is the plasma descum. A plasma descum is required to improve the line edge quality and reduce the number of opaque defects. Since PBS provides at best marginal dry-etching resistance quality, typical plasma treatment using O_2 or wet air must be optimized to provide a high level of control in the etch rate. Conditions providing etch rates of 25–35 Å/min. is most desirable from a control standpoint. Additionally, as a consequence of the poor dry-

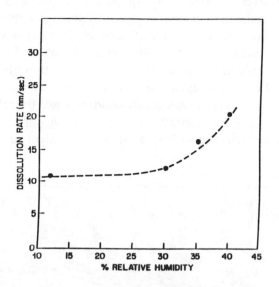

FIG. 5.29 Variation in PBS development rate as a function of % relative humidity in the mask-processing environment.

etching resistance of PBS, the chrome etch pattern transfer step must be carried out by wet-processing means. This isotropic pattern transfer process leads to under-cutting of the feature and is clearly depicted in Fig. 5.31.

Processes of this type ultimately limit the level of linewidth control and practical resolution that can be achieved in mask making. In contrast, the mask-patterning process associated with the use of the laser-based optical tool is simple and does not require a postdevelop bake and plasma descum step. Iteration, as in the case of PBS processes, is performed only at wet chemical chrome etch step. The use of a novolac-based resist such as the OCG 895i resist provides an avenue for the etch step to be performed using a plasma-based process. However, issues that must be addressed with the optical writer are the same as those experienced in silicon wafer processing. Resist absorption properties and thin film interference effects require that the coating of the resist be highly uniform.

5.7.4.2 Process control techniques

Elimination of the need to perform the development and wet chemical etching step in an iterative fashion can be achieved by attempting to control the process in real time.[69,70] Various methods have been proposed in which the goal is to reproducibly control the development/formation of the critical feature on a reticle so that the line-size uniformity can be improved and mask-to-mask variations are minimized. This becomes an essential requirement for the successful manufacturing of reticles for

64 and 256 Mbit generation devices. The general approach has been to adapt an apparatus onto the development/etching equipment that provides for the real-time monitoring of the change in the size of a mask feature. The commonly monitored signal is one that follows the change in the thickness of the exposed film regions as a function of development time. The sinusoidal trace is obtained from the reflected light intensity being collected by a sensor and produces a thin film interferometric trace. The number of extremes depend on the thickness of the resist film, the wavelength of light used to monitor the development process, and the refractive index of the resist at the monitoring wavelength. An example of an apparatus that has been adapted onto a commercially available spray/spin development machine is displayed in Fig. 5.32.

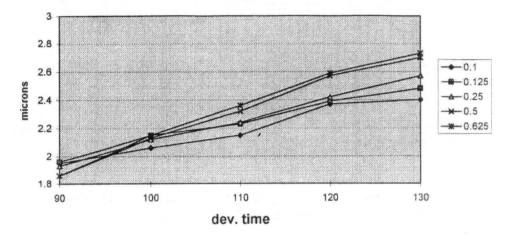

FIG. 5.30 Effect of beam address and feature orientation on PBS linewidth growth during development. The nominal feature size = 2.5 μm.

The apparatus depicted in Fig. 5.32 requires that a dedicated area of the mask be used to monitor the development of the exposed regions of the resist film. This is necessitated by the limitation on the amount of exposure area on any given mask level. Approximately 10% of the primary mask pattern area is required to obtain a signal that can be resolved above the noise threshold of the system. In the example of a contact or window level mask, the exposed area can be well below this 10% requirement. As a consequence, resist development rate monitoring pads (DMP) are located outside of the prime real estate on the mask and do not interfere with the optical stepper or defect inspection registration marks. The pads are represented in Fig. 5.32 and are located approximated 15 mm in from the edge of the mask and are 3.5 mm × 3.5 mm in size. The sensing head is located over one of the DMPs and collects the signal in a continuous mode. A reference point in the development-rate-monitoring trace is chosen to reproducibly identify the end point of development. The choice is varied, and for the system represented in Fig. 5.32 it is the penultimate

FIG. 5.31 Cross section of a 1-µm line patterned into the chromium layer of the photomask using a wet chemical pattern transfer process.

extremes of the interferomteric trace. An example trace identifying the reference or control point is given in Fig. 5.33.

The time to reach the control point T_R represents the variable from mask to mask, and this time essentially contains the history or whatever variation from the standard process has gone into patterning the particular mask. Additionally, a secondary control point could be instituted into the trace if, for a given resist development process, there is an induction time to development. As was shown in Fig. 5.18 this is common for novolac-based resists and this point can be simply the initial part of the trace in which the slope is not equal to zero.

The combination of this point with T_R provides the accurate determination of the development time for the given mask. An overdevelopment period, denoted T_O for a given mask level, is used to enable a particular geometry and feature size to be within 50 nm of the target size after the process development step. T_O is based on a percentage of T_R and is calculated by first generating a series of calibration curves of CD vs percent overdevelopment for a particular type of CD. For example, there

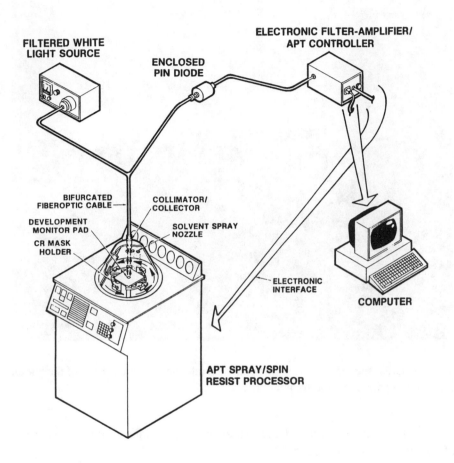

FIG. 5.32 Schematic of an apparatus used to actively control the development step of the mask fabrication process.

would exist a calibration curve for a 2.0-μm vertical isolated line and 2.0-μm contact hole, etc.

The apparatus in Fig. 5.32 is also used as a controller of the development process. Once T_R is determined and T_O calculated, the total development time of $T_R + T_O$ is performed. Upon completion of this time a signal is then sent to the development machine via the PC interface to terminate the spraying of the development chemicals and to switch on the rinsing chemicals. Table 5.8 provides results of masks processed using this controlling technology.

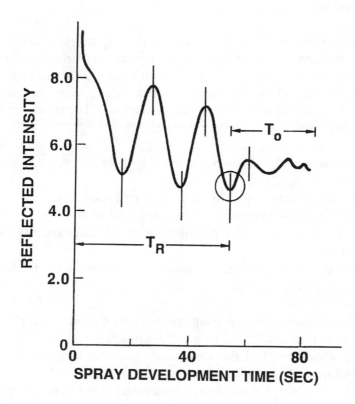

FIG. 5.33 Interferometric development trace identifying the reference point and over-development period used in controlling the development process.

TABLE 5.8 Critical dimension control performance for 1X and 5X zero-biased 5×5 masks.

MASK TYPE	CD (μm)	CD Target Size (μm)	POD*	RH/Temp (%/°C)	T_R (sec)	Avg. CD (μm)	3σ(nm)	Range (nm)
5x	5.25^	5	45	38/21	33	4.94	30	50
5x	5.25^	5	48	34/21	36	4.98	50	50
1x	1.5+	1.35	63	45/19	32	1.32	46	40
1x	1.5+	1.35	63	35/22	36	1.31	29	20

* POD = percent overdevelopment
^ Contact hole geometry
+ Isolated trench

5.7.5 Conclusions

The mask specifications on CD control, including mean-to-target and CD uniformity, plus the requirement for zero defects, place very stringent requirements on the resist development and the opaque film etch steps. The need to meet health and safety requirements limits the available chemical processes, and the constant attempt to reduce costs places further process limitations. The process requirements have been defined as:

- Reasonably high sensitivity—to avoid long writing times and/or thermal resist heating.
- Zero bias process—to achieve the required resolution.
- High contrast—to achieve the required resolution.
- Does not liberate volatile products—to avoid contaminating the e-beam column.
- Reasonable low etch rate compared with that of the opaque film.
- Aqueous base developer.

Two alternatives to current mask-making processes are chemically amplified resists and dry-developed resists.[71,72,73] Chemically amplified resists are sensitive to environmental conditions and the timing between exposure and the required postbake. The limitations are being minimized, but the best solution is to spin-coat the resist just prior to exposure and then automate the expose-to-bake step. The dry-develop resists are also dry deposited and again require an in-line process to deposit the resist film, expose, then dry develop. Both these enhanced resist processes will require a new level of automation that is not currently used by mask manufacturers.

5.8 METROLOGY

Three parameters are measured on a photomask: feature placement, feature size, and defect density. Feature placement is measured relative to some established grid and involves measurements over distances of centimeters, and feature sizes are local measurements over micrometers. Feature-size measurements require determining the average CD, relative to some standard, and the CD variation about the average value. The location and sizing of defects is discussed in Sec. 5.9.

The accuracy of the measuring tool affects the measured values of the mask parameters. This tool error can be minimized by multiple measurements of the same parameter, but this is time consuming. While repeated measurements are advisable for some level of improved performance, the measuring tool accuracy should be some "small" portion of the required parameter tolerance. This allowed measuring tool error is called the "gauge" spec. The recommended gauge specification is 10% of the allowed parameter tolerance.[74] Feature measurement on a 4X reticle for 0.25-

µm technology requires a tool with a repeatability of 2.5 nm (3σ) and a precision of 1.5 nm, and for feature position a tool with a repeatability of 4.4 nm (3σ). Today's measuring tools do not meet this requirement. If we assume that the mask error and the measuring-tool error both follow a normal distribution, then if the measuring tool error is 30% of the mask specification the measurement error is

$$1 - (1^2 + 0.3^2)^{1/2} = 1 - 1.045 = 4.5\% \ .$$

This has been accepted as a permissible error. For 4X reticles for 0.18-µm technology, a CD-measuring tool is required with a repeatability of $(0.3 \times 18 \text{ nm}) = 5.4$ nm (3σ), and for the feature placement a measuring machine with a repeatability of $(0.3 \times 32 \text{ nm}) = 9$ nm (3σ). Available confocal linewidth-measuring tools have a repeatability of ~6 nm (3σ)[75] and pattern-positioning tools have a repeatability of 5 to 7 nm (3σ).[77, 78] This means the tools are available for 4X reticles for 0.18-µm technology, but not without some limitations.

5.8.1 Pattern Placement

Pattern placement is measured on a coordinate-measuring machine that consists of a precision stage, whose position is measured with a laser interferometer, and an "alignment" microscope for locating the required feature.[76] The alignment microscope can also used to measure the feature size. Two commercial systems in use in today's mask shops are the Leica LMS IPRO[77] and the Nikon 5i.[78] These systems typically can handle up to 7″ substrates, with 9″ capability available as an option. The measurement range is 203 mm × 203 mm. The measurement repeatability is 5 to 7 nm (3σ) and the nominal accuracy is 12 nm (3σ). As noted in Sec. 5.3.1.1, a two-dimensional standard is available for calibrating coordinate-measuring machines.

The same precaution must be used to avoid nonrepeatable mask distortion during measurement, as is used during writing. The advanced coordinate-measuring machines support the mask on three support points, measure the sag of the mask, and compensate the measured values accordingly. However, as noted in Sec. 5.12, the flatness of the substrate has a significant effect on this measurement. Stage mirror bow errors are measured and compensated for in software. Temperature, atmospheric air pressure, and air circulation are all carefully controlled and again compensated for, as needed.

5.8.2 Feature Measurement

For pattern placement, the measured distance is from the center of one feature to the center of another feature. The cross section of the feature, Fig. 5.34, is not important, providing it is symmetrical. For the feature-width measurement, however, the feature profile is very important for several reasons. What region of

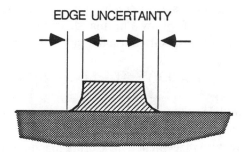

FIG. 5.34 Cross section of mask feature. The actual cross section depends on the chrome layer composition and the method of etching the chrome film.

the profile determines the projected image width on the wafer and how can the width at that region be measured? Since the chrome film thickness is only ~100 nm, which is much less than the depth of focus of a standard microscope, a microscope "views an integrated image."

Simulation of a 50-nm notch in the side wall of a bilayer MoSiON embedded phase shift film showed that the projected image is smaller than the same feature without the notch.[79]

The feature profile can be determined with an atomic force microscope, but it is usually sufficient to measure the width at the top and bottom of the feature to establish that the edge slope meets some acceptable limit. The feature width can be measured optically with a confocal microscope or a Mirau interferometer microscope.[80] The confocal microscope[81] employs a diffraction-limited illumination beam that is reflected back from the sample and is transmitted through a diffraction-limited aperture (see Fig. 5.35).

When the sample surface is in the focal plane of the microscope, essentially all the reflected light gets back through the limiting aperture and to the detector. A vertical displacement of the sample changes the focal point of the illuminating beam, and a smaller portion of the beam intensity gets through to the detector. The object is viewed by scanning the sample under the beam and correlating the beam intensity with the sample position. The vertical position of the sample is then changed, usually by means of a piezoelectric crystal, and the sample scan repeated. The depth resolution of the confocal microscope is not sufficient to measure the feature profile, but it is sufficient to separate and measure the feature width at the bottom and top of the feature.

The Mirau microscope (see Fig. 5.36) has a secondary mirror to form a Twyman-Green interferometer in conjunction with the sample. The mirror may be

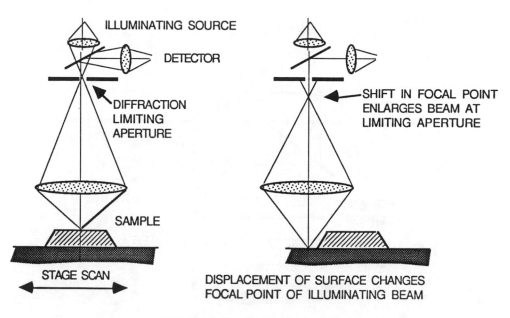

FIG. 5.35 Confocal microscope.

a semitransparent reference mirror and a beam splitter mounted below the microscope objective,[82] or an auxiliary mirror and beam splitter mounted above the objective.

The secondary mirror forms a reference plane with the focal plane of the microscope. Any part of the sample in the image plane appears bright in the detector. By varying the vertical position of the sample with a piezoelectric crystal, the width at different heights of the sample can be measured.

The advantage of the limited depth of focus of both these systems is the improved measurement repeatability. Both systems have 3σ repeatability ~6 nm compared with at least twice that value for conventional microscopes.

The major problem with nonvertical feature walls is the difficulty in defining a linewidth standard. The present NIST linewidth standards[83] have a listed width uncertainty of 50 nm due to the feature edge slope. This is unacceptable for leading-edge designs. Unfortunately, even the requirements of the feature to be used as a standard differ between the National Physical Laboratories (UK), and the NIST (US). The SIA specification for the mean feature width to the specified linewidth is 35 nm, reducing to 15 nm over a "few" years. This requires a linewidth standard with a calibrated width of ~12 nm, reducing to ~5 nm. This level of accuracy requires vertical walls. Unfortunately, wet etching is, and possibly will continue to

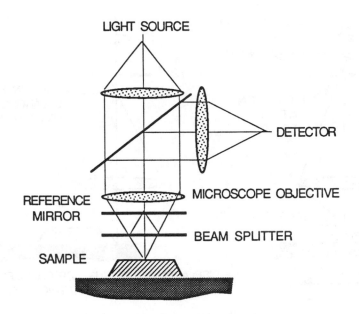

LIGHT SOURCE

DETECTOR

REFERENCE
MIRROR

MICROSCOPE OBJECTIVE

BEAM SPLITTER

SAMPLE

FIG. 5.36 Mirau correlation microscope.

be, a mainstay in mask fabrication, and this process yields nonvertical walls. If the standard has vertical walls, can it be used to calibrate features with nonvertical walls? The two solutions are to use the dry-etch process, which requires overcoming the higher defect density and associated pattern loading effects, or to devise a method of using a standard with nonvertical walls.

There are two approaches to using nonvertical wall standards. NIST has proposed using a lens projection system that emulates the wafer stepper lens and measuring the width of the image. The proposal is based on the recognition that a mask feature is viewed very differently by the mask maker and the mask user. The mask maker uses a microscope that typically has an illumination wavelength of ~500 nm and the lens numerical aperture ~0.9. The mask is "viewed" by the stepper lens with a illumination wavelength ≤365 nm and a numerical aperture of ~0.10. The profile of the feature formed in the wafer resist is an integrated effect of the mask feature profile, the stepper lens, the nonlinearity of the wafer resist, and the resist-processing conditions. The nonlinearity encountered in the wafer resist can be emulated by a "filter" in the image detector.

Another approach is to use the scatterometer developed at the University of New Mexico.[84] The scatterometer measures the intensity distribution of the diffracted and transmitted light. Results to date show that the measurements made over a series of grating are in the "middle" of measurements made with a number of different measuring tools. The instrument measures only angles and relative light intensities, and needs no special precalibration. The difficulty with the scatterometer is that it can only measure the width of a feature within a ~2 mm size grating. Typically the grating can be 1-μm to 2-μm lines and spaces. A proposal is to include a 2-mm-square grating on each mask, calibrate the features within the grating with the scatterometer, then use the calibrated features to calibrate the tool used for measuring the features across the mask. The advantage of this method is that the calibrated features within the grating will have the same profile as the primary pattern.

An important in-process measurement is to determine the feature size before stripping the resist. The feature size is usually left slightly oversize after the first etching step, measured, then etched again to achieve the required feature width. Etching the chrome film through the resist pattern requires undercutting the resist edge slightly, to clear out the base of the chrome feature. In the case of a wet etch process this may amount ~100 nm per edge. Although this undercut and other processing biases are usually allowed for in the data preparation for writing the mask, it is usually advisable to measure the feature width with the resist film in place so that a re-etch can be done if needed. The resist overhang prevents achieving an accurate CD measurement from the resist side of the mask. A process developed by SiScan is to measure the feature through the back of the substrate. This requires special microscope objectives corrected for spherical aberrations caused by viewing through the mask substrate.

5.9 MASK DEFECTS AND INSPECTION

A key requirement of mask making, which makes their manufacture different from semiconductor wafer lithography processes, is to meet the criteria of "zero defects."

5.9.1 Printability of Defects

Any anomaly in the mask pattern or substrate can be called a defect. However, due to reasons related to costs, defects on a mask will be classified as such if they are likely to print on a wafer during the mask image transfer and cause a failure in semiconductor chip performance. Here "failure" implies either total failure of functionality or any functionality below the required operating specification of the chip. On the lithographic level, a defect is classified as a "printable" defect if it causes greater than ±10% variation from the required size of the feature being printed. Therefore, the criteria of "zero defects" implies no defects above the minimum defect specification.

5.9.2 Hard vs Soft Defects

Mask makers classify all mask defects into two categories: "soft" and "hard." A soft defect is defined as any defect that can be removed by a cleaning process. By contrast, a hard defect is one that *cannot* be removed by a cleaning process. Particles, contamination, residue, stains, etc., on the chrome/quartz would be called soft defects. Missing or extra features in the chrome/absorber/phase shifter, pinholes, quartz pits, etc., would be called hard defects. All further discussions in this section will be related to hard defects. Within hard defects, the main discussion will be based on conventional, binary masks. Defects for PSMs have been addressed separately.

5.9.3 Types of Hard Defects

The classic hard defects are the missing and extra features generated during a mask-patterning process, i.e., pinholes, pinspots, intrusions, corner defects, missing features, etc. (see Fig. 5.37). Defects in the exposed quartz area of the mask, i.e., scratches, pits, and bubbles, would also be classified as hard defects. These defects have lithographic behavior similar to PSM defects.

There are also defect types related to the transmission in the clear/opaque areas, i.e., either semitransparent defects in the clear areas or transmission errors in the opaque absorber (Fig. 5.37). The next level of defect types are those resulting in errors from errors in the original mask data tape and also mask misprocessing. These involve misplacement and missizing of geometries, as depicted in Fig. 5.38.

Finally, gradual CD variations across the masks and edge quality of features, i.e., line edge roughness, are defect types that are not currently being detected and flagged but are likely to be specified in the future.

5.9.4 PSM Defects

The hard defects in the absorber, described above, also apply to the absorber film of phase shift masks. Furthermore, for transmissive absorbers, there is an additional defect type relating to the phase-shifting requirement. Then there is a new category of defects, i.e., phase-shifter defects, that occur in the quartz/clear areas due to the additional processing required to create the phase shift "layer." Some of these (there can be many more depending on the PSM type and the manner in which the mask is manufactured) are illustrated in Fig. 5.39. Note that a key aspect of PSM defects is that their lithographic performance may be different from standard opaque defects, and therefore the criteria for their rejection will differ, based on their printability.

5.9.5 Minimum Defect Requirement

At each generation of semiconductor lithography, a minimum defect size, based on minimum linewidth, has been identified. These are listed in the SIA Roadmap shown in Table 5.1.

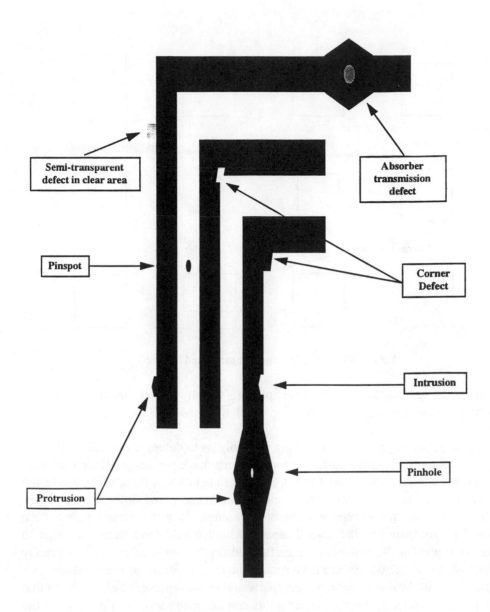

FIG. 5.37 Classic hard defects.

5.9.6 Inspection of Defects

Mask makers have to guarantee the masks they ship are "defect free." In order to do this they will typically inspect at a level below the minimum defect size requirement of the customer and, if they have the luxury, with a guard-band above the rated capability of the inspection tool. The defect inspection requirement,

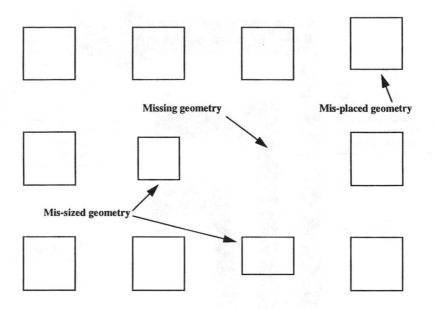

FIG. 5.38 "Misplacement & mis-sized" defects.

therefore, for inspection equipment suppliers is usually higher than the defect size requirement of the mask generation for which their tool will be used.

The task of the inspection tool, putting it simply, is to detect if there is light or no light being transmitted at a particular location on a mask per design. If we start with the premise that every die on a multiple-die reticle is good, then, using die-to-die inspection, any two dies should match exactly when compared simultaneously. Any differences found would represent defects. For single-die reticles, the reticle pattern would be compared to the chip database, which would have been modified to closely represent the resultant image formed on the mask after the lithography process. Assuming there were no errors in the modifications of the database, any differences in the die-to-database comparisons would represent defects. Now that it can detect defects, the only other requirement for the tool would be to do the inspections quickly and reliably. This is where the challenges arise.

A key aspect of the inspection tool that impacts its "comparison efficiency" is the algorithm used for difference detection. The designs of these algorithms are based on many factors, e.g., inspection light wavelength, pixel size, light sensor efficiency, analog/digital conversion speed and efficiency, stage movement resolution and rate, data conversion (from ideal layout to actual chrome mask), data conversion rate, rate of data transfer, etc. Plus, there are other factors related to the quality of the mask itself, e.g., reflectivity of the absorber film, line edge quality, etc., that also need to be considered in the algorithm development. Finally, the advent of PSMs has added another level of complexity to mask inspection algorithms. Ultimately all of these determine the accuracy, resolution, and speed of the defect detection of the inspection machine. As mask feature sizes and the minimum defect limits have reduced, the task has become more difficult—especially for die-to-database inspection.

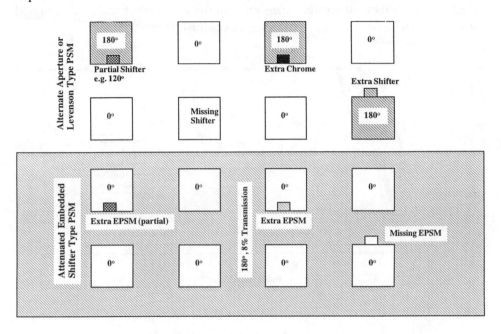

FIG. 5.39 Phase layer defects.

The ability of an inspection tool to accurately and completely detect the known defects on a standard defect mask determines its defect capture efficiency. A sample defect capture rate curve for a particular tool tested at a particular sensitivity setting is shown in Fig. 5.40.

Based on this result, the defect detection capability of this tool can be qualified as follows:

DEFECT TYPE	MINIMUM DEFECT SIZE DETECTED (μm)	DEFECT SIZE AT 100% CAPTURE EFFICIENCY
Pinhole	0.4 μm	0.5 μm
Pinspot	0.3 μm	0.5 μm
Protrusion	0.4 μm	0.6 μm
Intrusion	0.4 μm	0.6 μm
Corner	0.4 μm	0.6 μm

Mask makers certify their inspection tool capability on a regular basis to insure that the machine sensitivity has not changed due to lamp degradation or drifts in the system electronic hardware. The certification of an inspection tool is performed using a standard defect mask. One such standard defect mask is the Verimask™ (a trademark of DuPont Photomasks, Inc.). Verimask™ standards are available that cover all defect types shown in Figs. 5.37, 5.38, and 5.39 (except semitransparent defects), with defects of varying sizes on patterns with orthogonal, 45°, and odd-angle lines. There are no industry standards yet for PSMs.

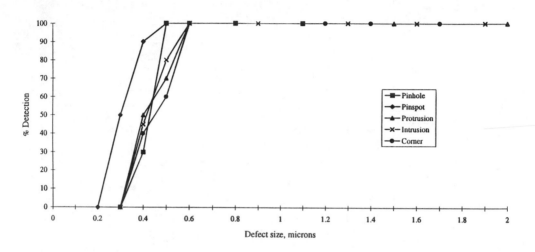

FIG. 5.40 Sample defect capture rate.

Ultimately, inspection tools used for mask defect detection are required to detect defects in a cost effective manner. In determining the cost of inspection, the major factors that impact the costs are data preparation time, number of inspections necessary, time of each inspection, defect review, classification, and disposition times. Inspection time is primarily driven by the sensitivity setting, i.e., the inspection pixel size. The smaller the pixel size used, the longer it takes to raster the mask; there is, as well, a higher probability of false defects. "False" defects are differences detected that are below the defect specification requirements for the mask being inspected. A high false defect detection rate impacts the inspection time and the time for defect review, classification, and disposition—and therefore cost.

Table 5.9 lists the latest generation of inspection tools being used by the industry for the manufacture of masks.

TABLE 5.9 Latest generation of defect inspection tools.

INSPECTION TOOL	SUPPLIER	INSPECTION MODES		MIN. DEFECT SENSITIVITY FOR (d/d) MODE (μm)
		Die-to-Die (d/d)	Die-to-Database (d/db)	
KLA 239 e	KLA Instruments Corp., USA	Y	Y	0.35
KLA 239HR	KLA Instruments Corp., USA	Y	Y	0.25
KLA 331	KLA Instruments Corp., USA	Y	Y	0.2
KLA 351	KLA Instruments Corp., USA	Y	Y	0.2
RT 810	Orbot Instruments Ltd., Israel	N	Y	0.35
R 8000	Orbot Instruments Ltd., Israel	N	Y	0.25
9 MD 83 SR	Lasertec Corp., Japan	Y	Y	0.3

5.10 MASK PROCESSING AND AUTOMATION

The principal steps in fabricating a photomask are straightforward. Load data in the pattern generator, load plate, write mask, remove mask, process mask, measure, and inspect for defects. However, the difference between a finished mask and a junk piece of glass may be as follows:

- Inspecting for particles on the substrate before and after writing the plate.
- Avoiding heat sources that may change the temperature of the substrate by >0.03 °C.
- Assuring the substrate is grounded correctly in an e-beam pattern writer.
- Assuring the substrate is held securely in the cassette without distorting the flatness of the substrate.
- Knowing the radial distribution of the process sprays are correct (see Sec. 5.11.1).
- Filtering the developing solutions and having them at the correct temperature.
- Assuring ahead of time that the mask-making data allows for the required process bias in the feature widths.
- Measuring enough points to achieve the required customer statistics.
- Being sure you are not using Company A's linewidth standard for Company B's product.
- Knowing what is an acceptable pattern infidelity.
- Being sure the mask number on the glass matches the pattern on the mask.
- Shipping the finished mask in the correct type of container.

The basic steps in fabricating a photomask are indeed straightforward, but at times very tedious and complex. As technology goes to smaller design rules, mask specifications get tighter, and mask substrates get larger, there will probably be a need to automate more of the mask fabrication steps to reduce the number of tedious tasks currently performed by operators.

Equipment performance can be monitored by the quality of the daily product, via the use of test patterns on every mask, and the use of periodic test masks. The use of statistical process control (SPC) together with a data-collecting system can help predict required equipment adjustments and some near-term failures, such as the end of useful life of light sources. With the cost of capital equipment being so high, it is essential a shop runs 24 hours a day, seven days a week; this requires expert operators to be available 168 hours a week for all critical manual operations, but the presence of an operator typically increases the local particle count, and the increase in mask sizes make masks very difficult to handle. The solution may be an increase of automation to pass the masks from one piece of equipment to another.

Wafer process lines use automation extensively. Mask making is more dependent on operator expertise than wafer processing, but automation is generally limited to individual pieces of equipment.

Several fully automated process lines have been developed that accept an exposed plate and completely process and measure the mask. The Siemens automated line[85] includes:

- Inspection of incoming blank-material
- Loading and unloading blanks into MEBES cassettes
- Loading and unloading plate cassettes into MEBES
- Complete PBS processing for different substrate sizes, including CD measurement
- Apply pellicle, all necessary glass sizes and thicknesses and pellicle types.

The Siemens auto-line halved the number of scrapped plates, reduced the number of defects, and doubled the number of zero-defect plates without repair (from 15% to 30%). The work flow in the Siemens mask shop is scheduled with a powerful software module real-time-event distributed scheduling (ReDS) to meet customer requirements. Potential bottlenecks that may prevent masks being shipped on time are detected by ReDS, and the customer is warned of potential delays.

For automation to be effective, the individual components must be reliable, easy to service, and compatible with computer control and data transfer. A new level of operator is required to provide rapid service when required. Automation is not without problems, but it may eventually be as common in the mask shop as it is in a wafer line.

5.11 PHOTOMASK CLEANING

The topic of photomask cleaning has traditionally not been of major concern in the photomask industry. This is because the task has been accomplished fairly successfully with the use of processes and technologies developed in the semiconductor wafer industry—where the requirements have been much more stringent. However, the situation is now changing.

The task of cleaning is to remove resist, particles, and other contamination from the substrate surface to a specified level of cleanliness, without damaging the surface. In the wafer industry, this task has been accomplished primarily by use of liquid-phase or wet-cleaning methods—and these methods have been easily transferred to mask making.

5.11.1 Requirements

The similarities in the requirements of mask making and wafer cleaning are summarized in Table 5.10.

TABLE 5.10 Comparison of wafer and mask-cleaning requirements by types of contaminants to be removed.

Wafer cleaning –removal requirement[86]	Mask cleaning requirement	Comments (relevance to masks)
(1) Gross and trace organics	Yes	Primary concerns are transmission loss and phase error
(2) Particles and particulates	Yes	
(3) Physically and chemically adsorbed ions	No	Unless they cause transmission loss and phase error
(4) Deposited and adsorbed metals	No	Unless they cause transmission loss and phase error
(5) Native and thin film oxides	No	Unless they cause transmission loss and phase error
(6) Impurities absorbed or entrapped by oxides	No	Unless they cause transmission loss and phase error

As can be seen from Table 5.10, the wafer-cleaning requirements related to removal of gross and trace organics, e.g., resists, oils, etc., and particles are those most relevant to mask cleaning. The other requirements are specific to wafers and would only apply to masks if the particular type of contamination caused loss of transmission through the quartz. This table exemplifies the larger nature of the task for semiconductor wafer manufacturers, which therefore explains the higher level of research activity occurring in wafer cleaning.

In Table 5.11, a closer look is taken at some of the specific characteristics of the two substrates being cleaned.

TABLE 5.11 Special characteristics of masks and wafers with respect to cleaning.

Characteristic	Wafers	Masks	Comments
Surface	Si, SiO$_2$, Si$_x$N$_y$, Al$_2$O$_3$, etc.	SiO$_2$, Cr$_x$O$_y$, Cr$_x$N$_y$O$_z$, Mo$_x$N$_y$O$_z$	
Surface topography	Severe, up to 1 μm	Mild, up to 0.15 μm	May be more severe for phase shift masks, i.e., up to 0.5 μm
Shape	Round	Square, Rectangular	
Minimum defect size	0.12 μm for 64 Mb DRAM generation[87]	0.7 μm for 64Mb DRAM generation[3]	Because of 4X or 5X magnification of features on mask
Defect density	60/m^2	Zero for active area of mask	Mask active area can be up to 13 cm x 13 cm

A discussion of each of the characteristics listed in Table 5.11 follows:

Surface The silicon or silicon dioxide surface. As the wafer and the mask surfaces are of similar chemical composition, the cleaning methods employed for the wafer are quite easily transferable to the quartz mask. There is, though, a difference: the wafer-cleaning methods have been developed not only to remove particles and organic contaminants but also to remove adsorbed ions and metals and to prevent redeposition of ionic contaminants and particles.[86]

The metal surface The aluminum and chromium oxide films deposited on the wafer and masks respectively are very different in their chemical reactivity. New films, i.e., molybdenum-based and chromium nitrides, are also coming into use for masks. Solutions used for cleaning an aluminum-coated wafer surface are typically solvents, since acids and bases easily attack the aluminum film.[86, 89] For mask makers, as aqueous cleaner is generally preferred to solvents, and since chromium oxide has a good resistance to acids and bases,[89] most of the wet-cleaning methods developed for the silicon dioxide surface have also been used successfully for the chromium/chromium-oxide-coated quartz mask surface.

Surface topography The mask surface does not have as severe a topography as a patterned silicon wafer surface.

Shape Where the cleaning methods involve immersion processes, there is no significant impact of the shape of the substrate. But if single substrate spin/spray processing is involved (which is becoming more common in mask cleaning), then the liquid flow patterns over the corners of a square/rectangular mask make the task of cleaning more difficult.

Minimum defect size The requirement for wafer cleaning is substantially tighter than for masks. This is primarily due to the fact that masks are manufactured with a 4X or 5X magnification factor, and after demagnification in an exposure tool, the impact of any defect is smaller by that factor. But with the advent of phase shift masks and a reduction of the k_1 factor in semiconductor lithography,[90] there can be an argument for the particle specifications on masks to be of the same level as that on the wafer.

Defect density This is a key difference for mask making that can make the task much more difficult in the future—the requirement of zero defects (larger than some specified dimension). The impact of this requirement has not been a significant problem in the past. But as minimum allowable defect sizes decrease, larger forces are required for defect removal, which makes this a major issue in mask cleaning.

This analysis shows that there are some specific requirements for masks—chromium oxide coating, square/rectangular shape, and zero defects, that make the task of mask cleaning different in some ways from wafer cleaning. Also, the possibility that the minimum defect size specifications for mask cleaning can be of the same magnitude as for wafers makes this task as challenging in some ways as wafer cleaning.

5.11.2 Adhesion Mechanisms

To be able to determine the appropriate strategy for cleaning it is necessary to understand the mechanisms by which contaminants adhere to a surface. There are both physical and chemical interactions occurring between a surface and a foreign material when the two come into contact.

There are numerous references in literature on the this topic.[91-96] The forces most likely to come into play are as follows:

Van der Waals forces These are forces of attraction between surfaces that arise from the electronic polarization of the atoms and molecules occurring when two materials come into proximity of each other. For particles on substrates, these forces are proportional to the diameter of the particle adhering to the surface.[91, 92, 95]

Electrostatic forces Two types of electrostatic forces come into play: The electrostatic image force is a Coulombic attraction between two surfaces that arises due to the bulk excess charges present on the surfaces. For particles on substrates, this force is proportional to the square of the diameter of the particle adhering to a surface.[92, 93, 95] The electrostatic double-layer force comes into play when two different materials, on coming into contact, develop a contact potential that is caused by the differences between their local energy states and work functions. For particles on substrates, this force is proportional to the diameter of the particle adhering to the surface.[91–93, 95]

Capillary forces These forces arise when a liquid film forms by capillary action between a particle and a substrate. Capillary force is proportional to the diameter of the particle and the liquid surface tension.[91, 96]

Adsorption and chemisorption Adsorption is a phenomenon by which molecules of a foreign material enter the internal pores of a surface and are accessible for selective combination to the substrate material. This selective action is most pronounced in a monomolecular layer next to the solid surface.[97] The forces of attraction are typically physical in nature, i.e., Van der Waals, etc.[98] In certain cases, the physical phenomenon is followed by a true reaction or chemical bonding; it is then called chemisorption.[97] These types of adhesive forces are more likely to occur on rough surfaces, which is not generally true for masks. Therefore, chemisorption has not been given further consideration.

Diffusion This is a mechanism by which a contaminant will penetrate a surface due to a concentration gradient in the bulk material[97] and therefore be "adhered" in the substrate. Again, this is a mechanism not relevant to masks. Therefore, this type of force has not been given further consideration.

Chemical bonding This is a phenomenon that is material specific, i.e., dependent on the chemical structure of the particle and substrate material it interacts with.[92–96] The solution to this adhesion problem will therefore be specific. Therefore, this topic has not been given further consideration.

In general, the primary adhesive forces of concern for mask surfaces and contaminants are the Van der Waals forces, electrostatic forces, and capillary forces.[93, 96] Between the two types of electrostatic forces, the electrostatic double-layer contact force is more important for particles smaller than 5 μm[95] (when the particles are of a material different from the substrate).

As mentioned above, the magnitudes of these adhesive forces of primary concern are a linear function of the size of the particle that needs to be removed from a surface. Therefore, as the particle diameter decreases, these forces will decrease. But, it has also been determined that as the particle size decreases, the ability to apply or transfer forces to it for its removal decreases as a nonlinear and stronger function of the particle diameter.[91, 92, 99] Therefore, the task of removing particles becomes more difficult as the particle size decreases, in spite of the adhesive forces having become smaller.

5.11.3 Development of Cleaning Methods

The main factors to consider in the development of any cleaning method are:
- to effectively overcome the forces in play between the substrate and the different foreign materials present on its surface,
- to prevent deposition and re-deposition of contaminants,
- to prevent damage to the substrate, and
- to prevent safety and environmental hazards.

Considering each of these factors individually:

- **Overcoming adhesive forces** This can be done by either providing a force that negates or diminishes the adhesive forces in play, or by applying an external force on the contaminant that overcomes the adhesive forces by sliding, rolling, or lifting off the particle.[92]

Methods to negate or diminish the adhesive forces:
1. Dissolve the foreign material in a solvent or aqueous solution.[92, 95]
2. Immerse in a solvent or aqueous solution to negate capillary forces and reduce Van der Waals forces.[92, 95]
3. Some solvents and solutions neutralize charges on particles thus reducing electrostatic forces.[92]
4. Chemical reaction, e.g. oxidation, will negate the adhesive forces on a contaminant by breaking down its chemical structure.
5. Etch the substrate surface to dislodge the foreign material.
6. Use laser energy to evaporate any liquid that is holding a particle by capillary action.[91, 106]
7. Surfactants reduce the interfacial tension between a solution and a substrate, thereby increasing wettability of the surface. This increased wettability improves the ability of the solution to dissolve contaminants and to hydrodynamically remove particles. A word of caution: at high concentrations surfactant molecules combine to form aggregates called micelles. These carbonaceous micelles are of submicron size and can be a source of added particles.[91, 100]

Methods to apply an external force:
1. Vibration action on the particle using megasonic or ultrasonic power to cause particle removal by liftoff.[91]
2. Electrostatic attraction of charged particles to cause liftoff. Not very effective for removal of small particles.[99]
3. Spinning the substrate to provide a centrifugal force that causes the particle to slide or roll off.[91]
4. Fluid flow across the substrate to provide a drag force on the particle, causing it to slide or roll off.[91]
5. Direct application of mechanical force, as with a brush or jets of high-pressure DI water or dry ice.[91, 94]

• **Preventing deposition and redeposition** A surface can potentially be contaminated by the cleaning method and equipment used. This is possible in two ways: (a) by exposing the surface to contaminants in the chemicals, equipment, and environment used for the cleaning or (b) by making the surface highly susceptible to mechanisms by which redeposition is promoted after cleaning.

Requirements to prevent deposition:
1. Pure and clean chemicals that do not contain chemical impurities and particles.
2. Construction materials of processing and handling equipment that do not particulate and leach impurities during use.
3. Point-of-use filtration with chemically compatible and stable filters.
4. Appropriate-class clean room with proper clean-room practices.
5. Proper preventive maintenance practices on equipment.

Methods to prevent redeposition:
1. Eliminate electrostatic charge on the surface.
2. Eliminate ionic charge on the surface.
3. Hydrogen-terminated silicon on the silicon dioxide surface that results from an HF etch makes the surface extremely susceptible to hydrocarbon contamination.[101] Therefore HF-last operations are considered a source of particle redeposition.[94]
4. Particles tend to deposit on a hydrophilic surface from the drying of the film that remained on the surface. Also, hydrophobic surfaces, like the hydrogen-terminated silicon surface, have a greater tendency to attract particles. The mechanism for this is not well understood.[91]
5. Zeta potential is a measure of the electrostatic double-layer repulsion forces created around surfaces in a liquid medium. An understanding of the zeta potential of surfaces in solutions is necessary to minimize particle redeposition. Zeta potential of particles is typically negative in alkaline solutions and becomes less negative to positive in acids. The silicon dioxide surface in most solutions is known to be negative.[91, 94]

6. Oxide passivation of the surface helps to minimize particle redeposition.[101]

A matter of key concern in mask making is the need to handle a mask after it has been cleaned to perform the pelliclization operation. This requirement has to be met without recontaminating the cleaned surface.

• **Preventing damage to the substrate** As it becomes more and more difficult to clean the substrate, the methods being used are becoming more and more aggressive. The following are known methods by which a substrate can be damaged during cleaning.[91, 100]

1. Etching/chemical attack
2. Megasonic energy
3. Static damage due to DI water
4. Plasma damage
5. Damage by mechanical force
6. Laser damage
7. Misaligned or dirty brush

• **Preventing safety and environmental hazards** The cleaning methods used should not cause safety hazards to the manufacturing personnel. Also, generation of toxic wastes and environmental hazards should be minimized, if not eliminated completely.

• **Cost effectiveness** A good cleaning method needs to minimize costs, i.e., capital cost, installation and start-up costs, operating costs such as chemical storage, handling and consumption, throughput, and maintenance, and disposal costs.

Ultimately, the choice of a cleaning method for a particular application will depend on a decision table analysis of the above-mentioned five factors.

On reviewing the development and use of cleaning methods in the past, one can understand why wet-cleaning methods have been preferred to dry-cleaning methods. Historically, the factors given key consideration were overcoming the adhesion forces, prevention of substrate damage, and cost. And it is evident from the discussions above that the advantages of wet-cleaning methods are that they
- provide many more opportunities to negate or diminish many of the adhesion mechanisms, i.e., Van der Waals forces, capillary forces
- provide more opportunities to apply external forces on particles, i.e., drag force from liquid flow, megasonic
- are less likely to damage the substrate
- are less expensive.

Finally, wet-cleaning methods are easily applicable to the removal of particles and contamination above 1 µm in size. In the future, as cleaning specifications become tighter, and the other factors such as deposition/redeposition, safety, and environmental requirements become more important, it is likely that gas or vapor phase cleaning methods may become more prevalent. The advantages of dry-cleaning methods are[105] that they:

- are less susceptible to redeposit contaminants on the surface
- provide improved process uniformity
- reduce use, handling, and disposal of hazardous chemicals
- allow more versatility in the range of process variables
- allow sequential, in-situ processing.

The other advantage is that water rinsing and the subsequent problems with drying, which are common to the liquid phase cleaning methods, are not encountered when using dry-cleaning methods.

5.11.4 Common Wet-Cleaning Equipment for Photomasks

In the mask-making industry, batch immersion processes were once most popular, as they were very cost effective. However, in the last three to four years, with the tightening of the cleaning specifications, spin/spray processes are becoming the norm. Some of the equipment suppliers widely used in the mask industry today include:

US based	- Fairchild (previously APT-Convac), SSEC
Europe based	- Hamatech (subsidiary of Steag)
Japan based	- MTC, WACOM, Clesen (previously Kuresen).

5.11.5 Common Wet-Cleaning Methods for Photomasks

Wet-cleaning technology has evolved from use of different approaches such as organic solvents, brush scrubbing, hot acid dips, and UV cleaning, to a few proven and reliable cleaning methods. Some of the hydrogen-peroxide-based wet-cleaning methods, first developed by Werner Kern[102] at RCA in 1965 for the wafer-cleaning industry, are commonly used in photomask cleaning. The chemistries involved are given in Table 5.12

TABLE 5.12 RCA cleaning chemistries.

CHEMICALS	PRIMARY APPLICATION
H_2SO_4 - H_2O_2	Removing heavy organics
H_2O - H_2O_2 - NH_4OH (SC-1)	Removing light organic residues and particles

The effect of each of these chemicals is discussed below:

H_2SO_4 -H_2O_2 (4:1)	This solution of sulfuric acid and hydrogen peroxide is also referred to as "piranha clean." This is mainly used to remove heavy organic materials, like resist and other organic contamination. It works as an oxidant and attacks the hydrocarbons.[86, 101]
H_2O - H_2O_2 -NH_4OH (5:1:1)	Also called RCA Standard Clean 1 (SC-1), this was designed to remove trace organic impurities on surfaces by the solvating action of the NH_4OH and the oxidation capability of the H_2O_2. Ammonium hydroxide can also serve as a complexant for many metallic contaminants. From research it was determined that the mechanism happening in the solution is that the peroxide oxidizes the surface while the ammonium hydroxide dissolves this chemical oxide. This sequential growth and etching of the surface helps to perform the cleaning action and removal of particles. However, the microetching occurring during 5:1:1 SC-1 clean is fairly nonuniform—resulting in microroughening of the substrate surface.[86,101] Recent research[104] reveals that lowering the NH_4OH concentration ratio to 0.01 –0.25 greatly reduces the microroughening while retaining the particle removal efficiency of the SC-1 clean.

The effects of variations in temperature, concentration, sequence of cleaning steps, application in immersion/spray/vapor phase systems, etc., and the resulting cleaning efficiencies and impact on the silicon and silicon dioxide surfaces have been studied extensively for the SC-1 clean.[98] However, there is very little published literature available on the piranha clean and cleaning of metal thin film surfaces.

5.11.6 Conclusions

Mask makers have so far been successful in adapting cleaning chemistries and methods that have been developed and optimized for cleaning the silicon wafer surface. The liquid phase cleaning technologies and peroxide-based processes have worked very well for the photomask requirements in the past. However, the job for mask makers pertains to cleaning the quartz substrate and the chromium/chromium oxide coating on it.

To further improve mask-cleaning technologies, the specific gaps pertaining to the needs of mask makers need to be addressed. These include

- understanding of the adhesion mechanisms on the chromium film. (The mask makers have a dilemma here. Particles on the chromium surface do not interfere with the transmission performance of the mask. However, if they exist and are detected, no semiconductor fab would like to risk the particle moving into a clear area at some later stage and therefore will want it removed.)

- understanding of the effects of cleaning chemistries on the quartz and chromium substrates
- understanding of the zeta potentials of particle and mask surfaces, hydrophobicity/ hydrophilicity of the mask surface after each process step, chemical and ionic components of particle and mask surface interactions, etc.

There is also a lack of clear understanding of the specifications for cleaning needed on masks to be used in the sub-half-micron lithography regime, especially when deep-UV wavelength exposure sources are used and low k_1 factors come into play.

To improve mask-cleaning efficiencies, work needs to be done to fill in the gaps identified above. Also, the biggest challenge for the mask makers in the future will be to accomplish removal of sub-half-micron defects to zero defect levels on a large area of a square/rectangular substrate.

5.12 MASK ERROR ANALYSIS

The correct analysis of mask errors is very important for today's short delivery times and tight specifications. Unfortunately, a given error may have several sources and often only an on-the-spot operator or engineer can determine the source of the problem. This section considers some of the errors that may be observed and the possible corrective action.

5.12.1 Pattern Placement Errors Across Mask

As noted earlier, even small distortions in the flatness of the mask substrate can produce pattern placement errors that are significant compared to the specifications. Therefore, one must determine first whether apparent placement errors are true errors or measuring errors.

If the error is not in the measurement, it has to be due to the pattern writer. There are typically about ten independent sources of pattern placement errors in any type of e-beam pattern generator. Machine error sources and the associated pattern error are listed in Table 5.13.

5.12.2 CD Variation Across a Mask

It is important to minimize variations in the critical dimension across a mask. Systematic errors must be eliminated and random errors should be analyzed for possible reduction or elimination. The following is a method for analyzing and reducing systematic errors.[107]

TABLE 5.13 Possible pattern errors and the associated mask error.

ERROR SOURCE	ASSOCIATED PATTERN ERROR
Stage interferometer error	X and/or Y pattern displacement
Beam scan length	Y pattern displacement Y feature width errors
Beam scan linearity	Y pattern displacement
Writing window distortion	X and/or Y pattern displacement
Reregistration error	X and/or Y pattern displacement
Substrate and/or cassette movement relative to stage	X and/or Y pattern displacement
Substrate/cassette clamping distortion	X and/or Y pattern displacement
X/Y Feature uniformity	Feature width errors
Tilted substrate	Pattern displacement Feature width errors
Substrate charging	Pattern displacement

First, measure the CD variations after each process step (i.e., develop, postbake, descum, etch, and resist strip), then separate the CD variations at each process step into linear and radial variations. This is done by systematically subtracting a linear factor from the measured mask data to reduce the CD variation to a minimum. Then subtract a radial term to reduce the CD variation to a random distribution. The linear variation across the plate may be due to the substrate or the writing system. This can be verified by writing a test pattern on the plate, rotating the plate in the cassette, and writing a second series of test patterns. If the processed CD variations of the two test patterns are in the same direction, the problem is due to the substrate; this may be caused by nonuniform prebake conditions. Radial CD variations may be due to spray nonuniformities in the developing and/or etching stations, or due to the resist-coating process such as a radial film thickness variation or variations in the temperature of the hot plate used for baking the resist. If the problem exists with plates from different suppliers, then it is probably the nonuniformity of the spray pattern.

Recent evaluations[108] indicate that a source of CD nonuniformity is the nonuniform temperature bake after coating the substrate with resist. The nonuniformity is due to the cooling effect at the four corners of the hot plate used for baking the resist-coated plates.

Some other unwanted effects are listed in Table 5.14.

Table 5.14 Possible causes of feature size variations.

EFFECT	PROBABLE CAUSE	POSSIBLE SOLUTION
Varying exposure and associated CD variation.	Proximity of adjacent features.	Proximity effect correction.
Varying feature width at writing scan boundary.	Feature located on writing beam scan boundary.	Correct beam scan length.
Varying feature width in X and Y directions.	Timing error on blanking modulator.	Adjust timing on the beam blanking modulator.
Systematic CD variation either radially or side to side.	Misadjusted developer station.	Adjust nozzle positions, and possibly adjust solution flow in developer station.
Incomplete resist development and/or etching of chrome film. Poor feature edge quality and large CD variation.	Incorrect chemical bias.	Adjust allowed processing bias.
Large CD variation and/or missing features.	Poorly resolved features.	Use higher-resolution resist and/or process.

5.13 SUMMARY

One intention of this chapter is to consider possible limitations and future directions to achieve the required SIA mask specifications. The four principal mask parameters, pattern placement, feature size control, image quality, and zero defects above some specified size, are all pushing at the limits of today's capabilities. Economics is also requiring a higher throughput, consistent with achieving the required mask parameters. The performance of the pattern writer affects all four mask parameters as well as throughput. The flatness of the mask substrate and the stress in the chrome film affect the writing and measurement accuracy of the pattern placement. The resist sensitivity and process technology affect throughput, linewidth control, image quality, and the quest for zero defects. Inspection for defects and metrology will affect throughput more than it does today because the detection of smaller defects requires a smaller inspection pixel and more gray levels, while CD and pattern placement metrology will require more measurements to verify the overall mask quality. Finally, the removal of hard defects, by focused ion beam or laser oblation, followed by the final clean to remove soft defects will continue to challenge the tool manufacturer and the mask maker.

The raster-scan Gaussian beam pattern writers have proven to be a cost-effective solution for pattern generation on 4X and 5X reticles for the last 15 years. These systems have undergone significant improvements over the past few years and will undoubtedly be improved further in the years ahead. However, indications exist that the requirements for 4X reticles for 0.25-μm technology are challenging the limits of pattern writers.[110] Two significant challenges, linewidth control and image quality, are limited by two factors in the raster scan systems: (1) In order to obtain a reasonable throughput it is necessary to increase the beam current, which in turn degrades resolution. (2) To obtain an adequately fine effective tool grid it is necessary to implement gray beam techniques, which degrade the exposure profile at pattern edges. These problems are not fundamental, but are peculiar to raster-based Gaussian systems. Variable shaped beam systems do not suffer from these limitations and show promise for providing a cost-effective solution for future, demanding products.

Several fundamental physical limitations govern the choice of the accelerating voltage for e-beam pattern generators. The stochastic Coulomb interaction between the beam electrons in the drift length of the column places an upper limit to the usable writing current for a given resolution. The effect diminishes with increasing beam voltage. Beam induced substrate heating causes the resist temperature to change, in turn affecting resist sensitivity and feature size control. The energy per unit area deposited in the writing substrate is given as the product of dose times beam voltage. The dose required to expose the resist is proportional to beam voltage, owing to the increased transparency of the resist layer at higher voltages. Heating thus depends on the square of the voltage. This suggests there is an optimum choice of voltage: as high as possible for good resolution and noise immunity, but not so high that the beam induced heating compromises linewidth control. Beam induced heating favors fast resists, such as chemically amplified resists.

A common misconception exists that increasing beam voltage leads to a reduction of writing speed, due to the increased dose requirement. In fact, one can increase the writing current with increasing voltage without penalty, thus recovering the writing speed.

E-beam proximity effect determines the difference in width between nested and isolated lines. Proximity effect is governed by the backscatter range of the electrons, which is proportional to the 1.75 power of the beam voltage. Proximity effect is most difficult to correct when the backscatter range is comparable with the linewidth. For example, quarter-micron devices require 1-μm lines an a 4X reduction reticle. The backscatter range for 10 kV electrons incident on bulk fused silica is also 1.0 μm, making this a poor choice of beam voltage from the point of view of proximity effect correction. At 50 kV, the backscatter range is ~10 μm; therefore with small feature widths and high voltage, variations in pattern density become small over distances comparable with the range of proximity effect. In this limit it becomes a simple matter to correct proximity effect.

The preeminence of optical reduction lithography during the last 15 years has greatly alleviated the burden on pattern generation for reticles. As a result, the development of production tooling for pattern generation has failed to keep pace with the ever-increasing demands on lithographic performance on the wafer.

Wafer lithography has been characterized by escalating tool cost, shrinking process latitude, and limited extendibility of existing tool sets to new generations. These factors tend to predict increasing cost of ownership in the lithography sector. This trend is offset by the fact that pattern generation technology has continued to progress in applications outside of mainstream manufacturing, such as x-ray, 1X projection, and direct-write e-beam lithographies. Although not yet in widespread use, the needed pattern generation technology has already demonstrated technological feasibility over the limited pattern area required for 1X lithography. Pattern generation technology is experiencing a rate of change that is unprecedented, even by the standard of the past 15 years.

To avoid measurement errors due to the profile of the mask features and to minimize the required process bias it will be necessary to use an anisotropic etch process, i.e., a dry-etch process, to transfer the resist pattern into the chrome,. Chemically amplified resists provide the necessary etch resistance for dry-etching the chrome film, as well as a high e-beam sensitivity to minimize the resist heating during the pattern writing. Despite the difficulties encountered with chemically amplified resists, a recent result yielded a resist CD uniformity of 12 nm (3σ) across the 110 mm × 110 mm pattern area, with near-vertical resist feature profiles. After dry etching, the CD uniformity with a 1 μm L/S pattern was 16 nm (3σ).[111]

The final pattern errors due to writing and processing will probably be limited by the "noise" in the many parameters involved, with no single major error source. There are, however, a number of elements that can be used to estimate some possible limitation to pattern placement. For example, feature placement accuracy may be estimated to a first approximation by (a) the pattern correction accuracy claimed by the OPC software, (b) the estimated temperature control, and also (c) the distortion of the mask substrate before, during, and after mask manufacture. An approximation of these errors are:

- The recent OPC model, by TVT, fits experimental data to within 4 nm (rms) at the wafer level. This is equivalent to 16 nm on a 4X reticle. This is 36% of the 0.25-μm tolerance and 50% of the 0.18-μm tolerance.[109]

- A substrate temperature change of 0.05°C, during the write time, produces
  - 2.6-nm placement error, over 104-mm pattern, which is 6% of the required pattern placement accuracy for 0.25-μm technology.
  - 5-nm placement error, over 198-mm pattern, which is 12% of the required pattern placement accuracy for 0.25-μm technology.

- The available data suggests the pattern placement error due to gravitational sag and the substrate clamping method may be as small as 5 nm or as large as 40 nm. The gravitational sag at the center of a kinematically supported 6"× 6" × 0.25" thick substrate is 0.62 μm, and the corresponding pattern placement error across the writing area is 40 nm (see Sec. 5.6). Although this error is theoretically reduced to 5 nm if the two sides of the substrate are clamped in a perfectly flat plane, experimental results indicate the sag is not reduced as much as expected.[15]

If the pattern placement error, due to the plate sag, is corrected for by measuring the plate "sag," then the polished flatness of the substrate is very important. For example, consider a substrate that is polished flat to within 0.62 μm of a flat plane when held vertically (i.e., with no gravitational sag). If the substrate is now held horizontally and clamped in such a manner to make the surface flat, the clamping would actually introduce a pattern placement error of 40 nm. At present, the typical high-quality substrate is polished flat to within 1 μm, therefore any attempt to correct the pattern placement error by "bending" the plate flat, to eliminate the 0.62-μm gravitational sag, will be fruitless. Likewise, any attempt to measure the sag of a "nonflat" plate with the intention of correcting for the plate sag in software will not yield meaningful results. The plate support method suggested for the 9" substrate (see Fig. 5.6) will minimize the plate distortion and the associated pattern placement error. The error is a minimum if the two plate supports are separated 0.577 of the plate length, assuming there are no clamping errors.

One final point. A mask is typically used by aligning two reference points on the mask with marching targets in the optical wafer stepper or on the wafer. However, the best pattern placement on the mask, measured over many points on the device pattern, may be relative to some grid that is offset relative to the mask alignment patterns. This is because of the pattern placement error of the alignment marks. Aligning the device pattern relative to the mask alignment patterns may increase the placement error by as much as 40% compared with doing a multipoint fit with the device pattern. This error can be minimized by determining the required displacement of the stepper alignment patterns on the mask to maximize the pattern placement accuracy, and then providing a means of off-setting the mask by the prescribed amount within the stepper.

REFERENCES

1. F. Twyman, *Prism and Lens Making,* Hilger & Watts Ltd, Hilger Division, London, NW 1, UK.
2. Calibrated by Physikalisch-Technische Bundesanstalt, Germany. Available in the USA through Leica Inc., Allendale, NJ 07401.
3. M. R. Raugh, Proc. SPIE 480, Paper 21 (1984).

4. SIA 1994 Roadmap, Semiconductor Industry Association, San Jose, CA 95129.
5. K. Brown, Keynote speaker, OCG Conference, San Diego (1995).
6. Andrew Murray and Robert L. Dean, "Proximity effect correction on MEBES for 1X mask fabrication," Proc. SPIE 1496, 171–179 (1990).
7. W. Oberdan et al., "Automatic optical proximity correction—a rules based approach," Proc. SPIE 2197, 278–293 (1994); J. P. Stirniman and M. L. Rieger, "Fast proximity correction with zone sampling," Proc. SPIE 2197, 294–301 (1994); SIGMA-C USA, Sunnyvale, CA 94087; SIGNAMASK, Berkeley, CA 94708.
8. M. D. Levenson, N. S. Viswanathan, and R. A. Simpson, "Improving resolution in photolithography with a phase shifting mask," *IEEE Trans. Electron Devices* **ED-29**(12), 1812–1846 (December 1982).
9. R. Ferguson et al., "Etched quartz fabrication issues for 0.25 µm phase shifted DRAM application," EIPB '93, San Diego, CA. 1993.
10. Y. Takahashi et al., "Primary process in e-beam and laser lithographies for phase shift mask manufacturing," Proc. SPIE 1674, 216–229 (1992).
11. Chapter 1 of this volume.
12. T. Saito et al., "Continuous phase transistions fabricated by subtractive process," in *13th BACUS Symposium on Photomask Technology and Management '93*, Proc. SPIE 2087, Paper 46 (1993).
13. B. J. Lin, "The attenuated phase-shifting mask," *Solid State Technology*, 43–47 (January 1992).
14. F. D. Kalk et al., "Attenuated phase shifting photomasks fabricated from Cr-based embedded shifter blanks," in *Photomask and X-Ray Mask Technology*, Proc. SPIE 2254 (1994).
15. M. Zander, "Flexure of photomasks in manufacture and applications," in *13th BACUS Symposium on Photomask Technology and Management '93,* Proc. SPIE 2087, Paper 08 (1993).
16. J. DeWitt et al., "MEBES 4000 optimization and characterization of MEBES 4500," in *14th Annual Symposium Photomask Technology and Management,* Proc. SPIE 2322, 79–91 (1994).
17. H. Kobayashi, K. Asakawa, and Y. Yokoya, "OCG-895i photomask blanks enhancements for the laser reticle writer," OCG Microlithography Seminar Interface '95, 229–239 (1995).
18. SEMATECH Open Meeting, Austin, Texas (February 8, 1995).
19. Y. Todokrk and H. Morimoto, "Next generation mask strategy-accuracy and mask size survey," in *15th BACUS Symposium on Photomask Technology and Management '95,* Proc. SPIE 2621, 216–237 (1995).
20. H. Sewell, "Step and scan: the maturing technology," Proc. SPIE 2440, 49–60 (1995).
21. SEMI Standards Task Force, "Next size reticle," San Francisco (July 15, 1996). Also SEMI Reticle Mtg, San Jose, CA (September 1996).
22. Corning Inc., Advanced Materials Department, Corning, NY 14831.

23. Second International Symposium on 193 nm Lithography, Colorado Springs, Colorado (July 1996).

24. Private communication, Hoya Electronics Corp., USA.

25. Y. Watakabe et al., "High performance VLSI photomask with a silicide film," *J. Vacuum Sci. Technol.* B **4**(4), 841–844 (July/August 1986).

26. A. Shigetomi et al., "High performance VLSI photomask with a molybdenum silicide film," *Microelectronic Engineering* **13**, 73–86 (1991).

27. ETEC Systems, Inc., 26460 Corporate Ave., Hayward, CA 94545.

28. See Ref. 27.

29. J. G. Hartley, T. R. Groves, and H. C. Pfeiffer, "Performance of the EL-3+ maskmaker," *J. Vacuum Sci. Technol.* B **9**(6), 3015–3018 (1991).

30. H. C. Pfeiffer, "The mask making challenge," *Opt. Eng.* **26**(4), 325–329 (1987).

31. J. DeWitt, J. Watson, D. Alexander, A. Cook, L. Gasiorek, M. Mayse, B. Naber, W. Phillips, and C. Sauer, "Investigation of MEBES 4500 composite system performance," Proc. SPIE 2621, 19–26 (1995).

32. R. Aprile, "Comparison of 'state of the art' lithography tools," in *Photomask Technology and Management,* Proc. SPIE 2087, 10–15 (1993).

33. A. Murray, F. Abboud, F. Raymond, and C. N. Berglund, "Address Data reduction and performance of graybeam writing strategies for raster scan mask generation," *J. Vacuum Sci. Technol.* B **12**(6), 3465–3472 (1994).

34. G. H. Jansen, *Coulomb Interactions in Particle Beams*, Advances in Electronics and Electron Physics, Supplement 21, P.W. Hawkes, ed., Academic Press, San Diego (1990).

35. A. E. Greun, Zeitschrift für Naturforschung, Teil A 12, 89 (1957).

36. G. Owen, "Method for Proximity effect correction in electron lithography," *J. Vacuum Sci. Technol.* B **8**(6), 1889–1892 (1990).

37. T. R. Groves, "Efficiency of electron beam proximity effect correction," *J. Vacuum Sci. Technol.* B **11**(6), 2746–2753 (1990).

38. T. R. Groves, "Theory of beam-induced substrate heating," *J. Vacuum Sci. Technol.* B **14**(6) (1996).

39. M. R. Raugh, Precision Engineering 7 (1985).

40. J. Ferrera, V. V. Wong, S. Rishton, V. Boegli, E. H. Anderson, D. P. Kern, H. I. Smith, "Spatial phase locked electron beam lithography, initial test results," *J. Vacuum Sci. Technol.* B **11**(6), 2342–2345 (1993).

41. H. C. Pfeiffer, D. E. Davis, W. A. Enichen, M. S. Gordon, T. R. Groves, J. G. Hartley, R. J. Quickle, J. D. Rockrohr, W. Stickel, and E. V. Weber, "EL-4, a new generation electron beam lithography system," *J. Vacuum Sci. Technol.* B **11**(6), 2332–2341 (1993).

42. P. F. Petric, M. S. Gordon, J. J. Senesi, and D. F. Haire, "EL-4 column and control," *J. Vacuum Sci. Technol.*, 2309–2314 (1993).

43. R. A. Kendall, S. K. Doran, and E. Weissmann, "A servo guided x-y-theta stage for electron beam lithography," *J. Vac. Sci. Technol.* B **9**(6), 3019–3023 (1991).

44. P. Buck and B. Grenon, "Photomask technology and development," Proc. SPIE 2087, 56 (1993).
45. P. Burggraaf, Semiconductor International 12(12), 72 (1989).
46. R. E. Novembre, R. G. Tarascon, L. F. Thompson, W. T. Tang, C. O. Tange, R. A. Bostic and D. H. Ahn, in *12th Annual Bacus Symposium,* Proc. SPIE 1809, 76 (1992).
47. A. J. O'Malley, *Solid State Technol.* **18**, 40 (1975).
48. J. P. Ballantyne, Chap. 5 in *Electron-Beam Technology in Microelectronic Fabrication,* G.R. Brewer, ed., Academic Press (1980).
49. H. Ito and C. G. Willson, Proc. SPIE 771, 11 (1987).
50. C. G. Willson, Chap. 3 in *Introduction to Microlithography,* 1st Edition, L. F. Thompson, C. G. Willson, and M. J. Bowden, editors, ACS Symposium Series 219, American Chemical Society, Washington D.C. (1983).
51. A. Charlesby, *Atomic Radiation and Polymers,* Pergamon Press, NY (1960).
52. J. Brown and J. O'Donnell, *Macromolecules* **3**, 265 (1970).
53. W. M. Moreau, *Semiconductor Lithography Principles, Practices and Materials*, 1st Edition, Plenum Press, New York (1988).
54. H. Hiroaka, *Macromolecules* **9**, 359 (1976).
55. T. Bowmer and J. O'Donnell, Polymer 22, 71 (1981).
56. T. Tada, *J. Electrochem. Soc.* **126**(11), 1829 (1979).
57. M. Bowen, *J. Polymer Sci.* **26**, 1424 (1981).
58. J. Shaw and M. Hatzakis, *J. Electrochem. Soc.* **126**, 2026 (1979).
59. M. J. Bowden, L. F. Thompson, S. R. Farenholtz, and E. M. Dorries, *J. Electrochem. Soc.* **128**, 1304 (1981).
60. M. J. Bowden, D. L. Allara, W. I. Vroom, J. Frackoviak, L. C. Kelly, and D. R. Falcone, *Polym. Electr.* **15** (1984).
61. Z. C. H. Tan, T. Stivers, H. Lem, N. DiGiacomo, and D. Wood, *J. Vac. Sci. Technol.* B **13**(6), 2569 (1995).
62. W. D. Hinsberg, S. A. MacDonald, N. J. Clecak, and C. D. Snyder, Proc. SPIE 1672, 24 (1992).
63. L. F. Thompson, E. D. Feit, and R. D. Heidenreich, *Polym. Eng. Sci.* **14**(7), 529 (1974).
64. Y. Taniguchi, Y. Hatano, H. Shiraishi, S. Horigome, S. Nonogaki, K. Naraolta, *Jpn. J. Appl. Phys.* **18**, 1143 (1979).
65. A. E. Novembre, M. J. Bowden, *Polym. Eng. Sci.* **23**(17), 975 (1983).
66. H. S. Choong and F. J. Kahn, *J. Vac. Sci. Technol.* **19**(4), 1121 (1981).
67. M. A. Hartney, R. G. Tarascon, and A. E. Novembre, *J. Vac. Sci. Technol.* B **3**, 30 (1985).
68. C. Braun, M. Stohl, A. Novembre, and F. Peiffer, Proc. SPIE 2621 (1995).
69. F. Rodriquez, P. Krasicky, and R. Gruele, *Solid State Technol.* **5**, 125 (1985).
70. A. E. Novembre, W. T. Tang, and P. Hsieh, Proc. SPIE 1087, 460 (1989).
71. M. A. Hartney, R. R. Kunz, D. J. Ehrlich, and D. Shaver, Proc. SPIE 1292, 119 (1990).
72. J. P. W. Schellekens and R. J. Visser, Proc. SPIE 1086, 220 (1989).

73. S. Tedesco et al., Proc. SPIE 1264, 144 (1990).
74. FORD Q-101 Quality System Standard, 1987 Edition.
75. R. E. Colgan and H. M. Marchman, "Optimizing 0.25 µm Lithography using confocal microscopy," in *14th BACUS Symposium on Photomask Technology and Management,* Proc. SPIE 2322, 336–343 (1994).
76. R. M. Silver, J. Potzick, and R. D. Larrabee, "Overlay measurements and standards," Proc. SPIE 2439, 262–272 (1995).
77. Leica LMS IPRO. Leica Inc., Allendale, NJ 07401.
78. K. Kodama and E. Matsubara, "Measuring system XY-5i," Proc. SPIE 2439, 144–149 (1995).
79. G. Dao et al., "248 nm DUV MoSiON embedded phase shift mask for 0.25 µm lithography," in *Photomask and X-Ray Technology II,* Proc. SPIE 2512, 319 (1995).
80. F. C. Chang, G. S. Kino, and W. R. Studenmund, Proc. SPIE 2196, 35–46 (1994).
81. I. R. Smith et al., "Photoresist metrology using confocal scanning microscopy," in *KTI Microelectronics Seminar, Interface '87,* 141–152 (1987).
82. S. S. C. Chim and G. S. Kino, in *Integrated Circuit Metrology, Inspection, and Process Control VI,* Proc. SPIE 1673 (1992).
83. J. Potzick, "Re-evaluation of the accuracy of NIST photomask linewidth standards," Proc. SPIE 2439, 232–242 (1995).
84. S. M. Gaspar Wilson et al., "Metrology of etched quartz and chrome embedded phase shift gratings using scatterometry," Proc. SPIE 2439, 479–494 (1995).
85. A. Oelmann and G. Unger, "Results from the First fully automated PBS process and pelliclization," in *13th Annual Symposium Photomask Technology and Management,* Proc. SPIE 2087, 57 (1993).
86. W. Kern, in *Handbook of Semiconductor Wafer Cleaning Technology,* W. Kern, editor, 3–67, Noyes Publications, Park Ridge, NJ (1993).
87. SIA Semiconductor Technology—Workshop Working Group Reports, Surface Preparation Requirements (Table 18), p. 78, SIA (1993).
88. S. Daugherty, "64- to 256-megabit reticle generation: technology requirements and approaches," SPIE Critical Review CR51, 213–224 (1993).
89. *CRC Handbook of Metal Etchants,* P. Walker and W.H. Tarn, editors, CRC Press Inc., Boca Raton, FL (1991).
90. F. C. Lo et al., "The ever increasing role of mask technology in deep submicron lithography," Proc. SPIE 2254 (1994).
91. V. B. Menon and R. P. Donovan, in *Handbook of Semiconductor Wafer Cleaning Technology,* W. Kern, editor, 379–432, Noyes Publications, Park Ridge, NJ (1993).
92. M. B. Ranade, *Aerosol Science and Technology* 7, 161–176 (1987).
93. R. A. Bowling, in *Particles on Surfaces 1: Detection, Adhesion, and Removal,* K. L. Mittal, editor, 129–142, Plenum Press, New York (1988).

94. R.P. Donovan and V.B. Menon, in *Handbook of Semiconductor Wafer Cleaning Technology*, W. Kern, editor, 152–197, Noyes Publications, Park Ridge, NJ (1993).
95. A. Khilnani, in *Particles on Surfaces 1: Detection, Adhesion, and Removal*, K.L. Mittal, editor, 17–35, Plenum Press, New York (1988).
96. M.B. Ranade et al., in *Particles on Surfaces 1: Detection, Adhesion, and Removal*, K.L. Mittal, editor, 179–191, Plenum Press, New York (1988).
97. *Chemical Engineering Handbook*, R.H. Perry and C.H. Chilton, editors, McGraw-Hill Book Company, New York (1963).
98. D.C. Burkman et al., in *Handbook of Semiconductor Wafer Cleaning Technology*, W. Kern, editor, 111–151, Noyes Publications, Park Ridge, NJ (1993).
99. D.W. Cooper et al., in *Particles on Surfaces 1: Detection, Adhesion, and Removal*, K.L. Mittal, editor, 339–349, Plenum Press, New York (1988).
100. J. Bardina, in *Particles on Surfaces 1: Detection, Adhesion, and Removal*, K.L. Mittal, editor, 329–338, Plenum Press, NY (1988).
101. G.S. Higashi and J.C. Chabal, in *Handbook of Semiconductor Wafer Cleaning Technology*, W. Ken, editor, 433–496, Noyes Publications, Park Ridge, NJ (1993).
102. W. Kern and D.A. Puotinen, RCA Review 31, 187–206 (1970).
103. P. Burggraaf, Semiconductor International, 86–90 (June 1994).
104. E. Kamieniecki and G. Foggiato, in *Handbook of Semiconductor Wafer Cleaning Technology*, W. Kern, editor, 497–536, Noyes Publications, Park Ridge, NJ (1993).
105. B.E. Deal and C.R. Helms, in *Handbook of Semiconductor Wafer Cleaning Technology*, W. Kern, editor, 274–339, Noyes Publications, Park Ridge, NJ (1993).
106. A.C., Engelsberg, "Removal of surface contamination using a dry, laser-assisted process," in Proceedings of Microcontamination Conference (1993).
107. M.D. Cerio, "Methods of error source identification and process optimization for photomask manufacturing," Proc. SPIE 2512 (1995).
108. Robert Dean and Charles Sauer, "Further work in optimizing PBS," in *15th Annual Symposium Photomask Technology and Management*, Proc. SPIE 2621, 386–398 (1995).
109. O.W. Otto, J.G. Garofalo, and R.C. Henderson, "Simplified rule generation for automated rules-based optical enhancement," in *15th Annual Symposium Photomask Technology and Management*, Proc. SPIE 2621, 577–587 (1995).
110. Larry Weins and Wayne Smith, "Benchmark study of mask writer lithography systems," Proc. SPIE 2884 (1996).
111. M. Katsumata, H. Kawahira, M. Sugawara, and S. Nozawa, "Chemically amplified resist process for 0.25 μm generation photomasks," in *16th Annual Symposium Photomask Technology and Management*, Proc. SPIE 2884 (1996).

CHAPTER 6
Metrology Methods in Photolithography

Laurie J. Lauchlan
IBM SSD

Diana Nyyssonen
IBM Microelectronics

Neal Sullivan
DEC

CONTENTS

6.1 INTRODUCTION

The photolithography process consists of aligning and optically transferring the pattern from a reticle onto a partially processed wafer that has been coated with photoresist. The resist is then developed and the resist image transferred into the underlying material by a chemical and/or thermal step (i.e., dry/wet etch, sinter, implant, etc.). This process may be repeated 15 to 25 times in the coarse of building a complex integrated circuit. The process steps themselves may distort the wafers, causing differences between wafers and lots. In order for the devices to function properly, it is necessary to ensure that levels are precisely aligned to one another, i.e., overlay, and to control the critical dimensions, i.e., image size, within each patterned layer.

The proper determination of image size and registration represent two of the primary goals of metrology as they are applied to the manufacture of integrated circuits. There is increasing concern that current metrology tools are inadequate for 250-nm device production. Historically, the accepted method of determining the acceptable contribution of metrology tooling is the Gauge-Maker's Rule. Applying this rule to 250-nm linewidths places the acceptable critical dimension (CD) tolerance at 25 nm with a corresponding uncertainty of 2.5 nm for the line-width metrology. This strict performance specification dictates the wafer-level metrology performance required. In order to achieve this high level of performance from metrology equipment, it will be necessary to pay attention to all aspects of the measurement process. In the following sections, we will discuss three of the principle areas of concern facing metrologists today: the meaning and application of standards, and a review of overlay and linewidth metrology methods.

The photolithography community has long been divided as to the necessity and applicability of measurement standards. There are many reasons for this division, but they basically fall into two camps. On the one hand, as long as integrated circuits can be built that meet the desired performance specification, the engineering community will not worry whether the values of critical features are "absolutely" correct, i.e., NIST traceable. On the other hand, in order to manufacture these devices economically, it is necessary to correctly estimate the manufacturing tolerances required to produce a given product. For 250-nm lithography processes, we will find that measurement errors are becoming nonnegligible components of the manufacturing process tolerances. This will necessitate changes not only in how measurement tools are corrected, i.e., calibrated, but in how they are evaluated as well. For this reason, we begin our chapter on metrology with a discussion of standards and artifacts.

6.2 STANDARDS AND ARTIFACTS

There is increasing concern that current CD and overlay (OL) metrology tools are inadequate for 250-nm and smaller device production. This discussion attempts to define some of the problems and discuss what is being done and can be done to meet the increasingly demanding metrology needs. At 250-nm linewidths, the CD tolerance is 25 nm with a 2.5-nm uncertainty budget on the metrology.[1] Similarly, the OL tolerance is 75 nm with a 7.5-nm metrology requirement.

The interpretation of what these metrology requirements actually mean depends on the assumptions made in the error analysis model used. A common assumption is that reproducibility is the most important contributor and that it can be characterized by the precision to tolerance (P/T) ratio. As shown in Fig. 6.1(a), measurement of a dimension D on a lot of wafers may produce a normal or Gaussian distribution centered about the mean. If nothing is known about the accuracy or precision of the metrology, no conclusion can be drawn about whether the part dimension is within the tolerance limits even though it may appear to be.

It is usual to assume some model of the metrology that includes both a measurement bias and tool precision. Dynamic precision of the tool is determined by repeat measurements at a single site or group of sites on a single wafer taken over a period of several days or more. The effect of tool precision is to broaden the measured distribution of part dimensions as shown in Fig. 6.1(b). The industry response to this has been to demand tighter and tighter precision from their metrology tools, for example, demanding 5 nm or less 3-sigma for CD tools so that precision becomes a non-issue.

Better precision, in some cases, has been achieved by a reduction in the sensitivity of the tool. For example, in the case of OL tools some manufacturers and users have reduced the objective numerical aperture (NA) to 0.5. The accompanying loss of resolution guarantees better precision, but the price paid is inability to identify edge detail such as asymmetry or edge roughness, which would affect the accuracy of the measurement. At an NA of 0.5, the Airy disk diameter of the objective lens is approximately 1.2 μm. Any edge detail on the order of 1/6 of this diameter or less than 0.2 μm will not be seen in the image of the OL target. This means that the tool will measure what would be considered grossly asymmetric structures as shown in Fig. 6.2 with no indication of a possible error in the measurement.

In general, bias in the metrology is either assumed to be negligible or constant as in the metrology model used by Kudva and Potter.[2] An offset is assumed to have been responsible for part of the mean shift as shown in Fig. 6.1(c), resulting in the good product being classified as bad (alpha error) and/or the bad product being classified as good (beta error).

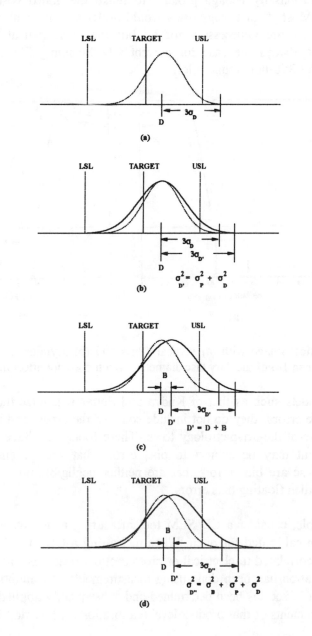

Fig. 6.1 The effect of precision and bias on a measured dimension D of a part: (a) No error, (b) with precision error only, (c) with precision and constant bias errors, and (d) with precision and variable bias errors both normally distributed.

The response to this type of error is to determine an offset or bias and correct the measurements. This response is cost effective only if the offset is significant enough to misclassify enough product to make the added cost of calibration worthwhile. Most IC metrology users would prefer systems with small offsets so that corrections are unnecessary for all but the most critical levels. This is, however, not always the case for current CD Scanning Electron Microscope (SEM) tools at 350-nm ground rules.

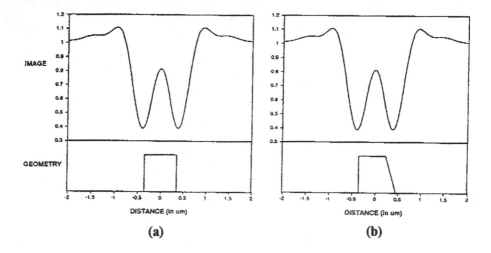

Fig. 6.2 Optical image with (a) vertical edges and (b) asymmetric edges. The edge width is less than 1/6 of the Airy disc diameter, which does not affect the image profile.

Although models such as that of Kudva and Potter recognize the cost of alpha- and beta-type errors, they do not include some of the errors and associated costs found in state-of-the-art metrology tools. These tools may have extremely good P/T ratios but may be subject to bias errors that are extremely difficult to quantify. These are bias errors that are neither negligible nor constant and are sometimes called floating bias errors.

As an example, consider a CD SEM tool measuring resist linewidth on a wafer that has been calibrated for pitch. When measuring a CD, it is found to have an offset when correlated to electrical or cross section measurements. Let us assume that a correlation has been done using measurements at a number of sites on a wafer and an offset has been determined and subsequently applied as a correction to all measurements of that product level (calibration for linewidth).

There are many sources of error that could affect subsequent measurements. A particular group of such errors occur because some aspect of the sample changes and affects the measurement, thereby introducing an error. For CD SEMs, one of these is edge geometry. In general, a CD SEM tool cannot measure a particular width such as "bottom" width (see Fig. 6.3). The tool typically measures some weighted average (top to bottom). When these measurements are correlated to a cross section measurement of "bottom" width or to electrical linewidth, the offset correction used in subsequent measurements is valid only as long as the resist edge profile does not change.

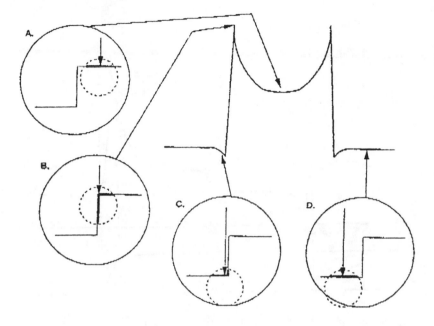

Fig. 6.3 SEM image of a line object showing the relationship of beam interaction volume (dotted circle) to the resulting image; secondary electrons generated at the intersection with the surface (heavy line) form the image.

A change in the resist edge geometry will cause an error in the measurement with or without changing the characteristic image waveform. That is, the changed geometry will result in a different correlation and different offset correction (see Fig. 6.4). The resulting measurement error may vary from chip to chip (e.g., with focus or dose changes) or wafer to wafer (e.g., with processing differences such as the position of the wafer in the carrier) or lot to lot (e.g., resist poisoning or contamination) and may vary from a few nanometers to 50 nm or more. In any case, whether the wafer mean or lot mean is used for dispositioning, the resulting CD measurement bias may be neither negligible nor constant. These types of errors are largest in process development and least in a well-controlled manufacturing line. The practice of measuring dynamic precision using a single wafer (characteristic of a particular process level) gives us no handle on the magnitude of these wafer-dependent errors, nor does calibration of the offset based on a single cross section or even a correlation based on an average of many sites on a single wafer. The only way to determine the magnitude and probability distribution for this kind of floating bias is to pull wafers from each lot at random and compare the measurements to those of a more accurate measurement process. From this data a mean bias can be determined as well as its probability distribution, variance, and standard deviation.

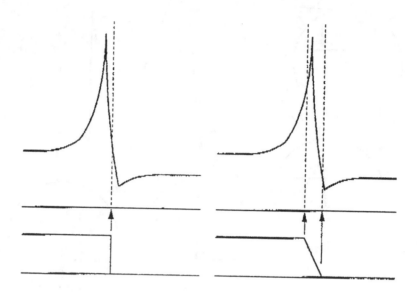

Fig. 6.4 Change of edge geometry with resulting change in correlation of image to bottom of physical edge.

Using this measured distribution of the variable bias, we can extend the model of Kudva and Potter[2] and recompute the cost of alpha- and beta-type errors. If we assume a normal distribution of the variable bias with a mean and sigma (σ_B) and assume that linewidth calibration has corrected for the mean bias, the effect on the measurement tool is to produce an error distribution with (σ_D) as shown in Fig. 6.1(d). That is, the tool will behave as though it had a poorer P/T ratio and the user will be unaware of it if he has measured only dynamic precision.

For example, assume a CD SEM with 10-nm precision ($3\sigma_P$) with a measured mean bias of 30 nm that has been corrected by an offset, but which actually varies by ± 15 nm ($3\sigma_B$). The tool will actually have an effective precision of 18 nm ($3\sigma_{EP}$) with the associated increased costs of alpha- and beta-type errors. If we initially had assumed a P/T ratio of 0.3 and a process capability (C_P) of 1.0, the tool would actually behave as though it had a P/T ratio of 0.5 and would have a metrology cost due to misclassification approximately 2 to 3 times that initially assumed (see Table 2 of reference 2).[2] The conclusion drawn here is that variable bias should be of as much concern to the user as precision. For CD, these bias errors will vary with process level. Different etch tools are also known to produce different profiles, so that CD SEM tool offsets may vary with etch tool and produce a bimodal distribution when two different etch tools are used on a single process level and measured by the same tool.

We have used edge profile as an example of variable bias for CD SEMs. There are other sources of such errors, including resist or layer thickness, properties of the layer and substrate materials, charging, etc. The magnitude of the bias, its probability distribution, mean, and 3-sigma are generally not known except in special cases due to the difficulty and cost of acquiring this information. In order to push total uncertainties to 2.5 nm and below, this issue will have to be addressed, its magnitude quantified, and a roadmap for improvement of calibration methods and metrology tools developed.

Similar types of errors occur in OL measurement where the tool-induced shift (TIS), which is supposed to be a characteristic of the measurement tool only, is found to vary with the wafer level being measured or even from wafer to wafer for the same feature.[3] However, we still have no practical way to measure lot-to-lot bias. Although variability of TIS is an indicator of the error, it is not an accurate measure of bias. A more accurate method of measurement than commercially available overlay tools is badly needed. One such method is currently under development at NIST.[4]

6.2.1 Calibration: A Reality Check

Given these precision and bias errors, how do we get the confidence we need in CD and OL metrology? One frequently heard response is the need for calibration standards. At the other extreme, is the statement that we don't need standards because we are building devices that work! The truth lies somewhere in between. Good standards do make the metrology job easier, reduce costs, and eliminate endless arguments and uncertainty, especially in process development. Knowing exactly how good a CD measurement is allows us to identify meaningful measurement differences and to fine-tune processes to produce devices that meet specifications with a minimum of reworking and lost time.

The role of standards has been debated over many decades. Standards originated at a time when technology was less complicated and knowing whether your grocer's scale was measuring weight properly (with weights traceable to NIST) was relatively straightforward. The initial purpose of standards was as a check on measurement equipment. A simple answer to the question does it or doesn't it measure accurately within the tolerance needed for a particular task was in many cases achievable. The idea of using standards to calibrate, i.e., to correct a measuring system, is a far more complex issue made more difficult by the high level of complexity and sophistication found in current dimensional metrology tools.

When dealing with a class of measurement tools that are all of similar design and the sources of measurement problems are known, a properly designed standard can be used to check that the measurements are accurate to the level desired. When the measurement systems differ significantly in design, as is the case for CD and OL tools, the usefulness of the standard is more problematic. For example, because of differences in software in OL tools, one system may measure targets at one level more easily and with better TIS than at another level, while a system from another manufacturer may do just the opposite.

Thus, the design of the standard and choice of materials may lead the user to erroneously consider one system better than another. Or the standard may differ significantly from product wafers such that it gives no indication of errors that may occur. This is the case with the current VLSI Overlay Standard.[5] It is a patterned silicon surface equivalent to a single-layer overlay structure and does not characterize the major source of OL measurement error, which is due to comparison of targets at two different levels on the wafer separated by a distance that exceeds the depth of focus of the imaging system. A single-layer standard is still a useful initial check on the tool to assure that the tool is working properly, but it does not tell us the complete story about potential sources of bias errors. Similarly, existing standards cannot be used to characterize variable bias errors. The standard is usually designed to allow accurate and repeatable measurements (i.e., with high contrast and vertical edges) and so inherently cannot evaluate bias errors that occur due to changes in the material characteristics (e.g., edge slope).

Larrabee[6] has frequently recommended use of precision standards. To quote him, "In-house control specimens (often referred to as 'golden standards') can act as interim precision standards but, if and only if, they closely match the specimens to be measured and are known to be stable with time." He does not address the major concern of users, which is what do you do if this "golden standard" is accidentally dropped or otherwise destroyed. A new standard selected from current product may differ physically from the earlier one, and its assigned value may have no tie to that of the broken standard. In addition, this type of precision standard cannot evaluate bias errors that result from changes in materials and structures. The answer to these problems is that, somehow, we have to have a better system or better measurement method that, even though it may be slower and more costly than a production measurement system, can provide the long-term stability and better accuracy that is needed.

We, thereby, introduce the idea of a Reference Measurement System that can be used to compare and quantify errors in less-costly and faster tools used in production. This system must be known to be, from basic physics, more accurate and more precise than tools used in the manufacturing line. Although we would like it to be an order of magnitude or more accurate, it does not have to be; it only needs to be significantly better to be of value. Many of us would like this "golden" system to reside at NIST, but the reality is that for many measurements, including CD and OL, at present it does not. Nor should it; a standard produced by NIST provides only a single or at best a few checkpoints for the wide variety of wafer levels and materials that are routinely measured in production. Every facility needs its own in-house, accessible Reference System to track bias in its metrology tools.

In-house, we already have systems that serve this role; for example, cross section CD SEMs and increasingly noncontact AFM systems. Sometimes, we simply choose our "best" or most stable CD or OL tool and relate all measurements to it. In one case, an average of several tools of different manufacture became the "golden" system (not recommended by metrologists). As an industry, we need to recognize that with the complex metrology and numerous sources of error found in fab tools, a "golden" tool has to exist and we can help production and reduce costs if such a tool is accessible to metrology users whether it resides in an in-house standards lab or in a manufacturing facility. We need to make use of the best system available to us now, while supporting the development of more accurate, faster and lower cost Reference Systems.

This paradigm requires acknowledgment that a Reference System is required and that it is needed to track and control production metrology tools. When recognized standards are available, they can provide a checkpoint on the Reference System. But it, in turn should provide measurements on product that will allow tracking of metrology errors and uncertainty. As a result, intelligent decisions can be made regarding the significance of measurement differences, how often corrections are required for production tools, and when changes in features may compromise measurements. In many fabrication facilities these tools do exist and to varying degrees supply the base measurements needed for stable metrology tools in production.

6.2.2 When No Standards Exist

Current metrology tools that measure pitch, CD, and OL, as already noted, have few standards and/or generally accepted methods of producing reference measurements. What do we currently do, what could we do better, and what do we need to better control metrology tools and wafer fabrication processes?

6.2.2.1 Pitch

In CD and OL tools, the most accurate measurement possible is pitch. This is a relatively simple extension of length measurement to microdimensions. Yet the common practice of using a single pitch measurement to "calibrate" tools (e.g., CD SEMs) as often as necessary is counter to the original intent of a standard. The measurement system should be stable and accurate enough to independently measure length to the desired accuracy using internal adjustments only. Then the pitch standard is used as a check on accuracy after internal adjustment is completed. The standard is used to indicate when the tool is "out of control" and requires remedial action. This is true of overlay tools that use an internally mounted scale to determine distance. It is not true of most CD SEMs which recommend the use of a permanently mounted CD standard to "calibrate" the system daily by supplying a single multiplicative constant in the software that is used to correct measurements.

Some of the problems that can result from using a single pitch measurement to adjust the system are illustrated in Fig. 6.5 and include effects of precision of the adjustment, assumption of linear slope equal to one, and a nonzero intercept. However, there remains the question that once the adjustment has been made, how is it known that the system will measure accurately above or below or even at the dimension(s) used for adjustment. It is common practice to omit checking the accuracy after "calibration."

Improved pitch measurement is possible, using an internally referenced line scale, interferometry, or similar method. Where such methods are not available, such as in CD SEMs, it is possible to do multipoint calibration with existing standards, using the Mandel method to fit the data to a straight line.[7] Mandel's method recognizes the existence of errors in the standard as well as in the measurement system, and the resulting linear fit will result in more stable and accurate pitch measurements (see Fig. 6.6).

The importance of calibration was recently reviewed by researchers from NIST. In the process of developing a new low-voltage SEM magnification calibration reference standard, NIST conducted an interlaboratory study to test the state-of-the-art of current SEM instrumentation from 0.2 to 3000 μm pitch dimensions and to determine the suitability of the new sample design.[8] Pitch calibration was determined to be necessary from this study. Furthermore, instrumental drift requires that SEMs must be checked, adjusted and rechecked periodically.

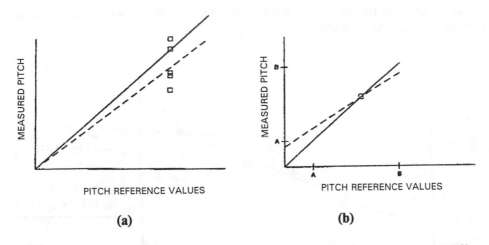

Fig. 6.5 Problems from use of a single pitch measurement to adjust a system: (a) Effect of precision and (b) linear curve with slope not equal to one and the effect of a nonzero intercept. The solid line is the assumed calibration curve after adjustment, and the dashed line is the true curve.

Current pitch standards are available with uncertainties of 3 nm (95% confidence level) at 1-μm dimensions.[9] Implementing an improved method becomes a software issue that can be performed off-line with existing systems and could be incorporated into the tool software if necessary. Again, we note that daily readjustment or calibration (recomputing of a calibration constant) can actually make dynamic precision worse. The goal should be adjustment of the system only if the measurement on a standard lies outside the control limits established for the tool and for an in-control tool should be a rare occurrence. In a CD SEM, this is best achieved by operating the system at a fixed magnification.

With reduction in ground rules and feature sizes, there is a need for a pitch standard accurate to < 2.5 nm at smaller dimensions. The SEM 200-nm pitch standard (used in the aforementioned interlaboratory study) is under development at NIST.[10] Currently, for ground rules with pitch < 1 μm, it is common practice to assume that the pitch generated by a good optical lithography tool is accurate and to use the pitch generated by this tool on product wafers as a calibration artifact. Thus, within most companies a de facto Reference System or source of pitch standards below 1 μm exists.

However, there are differences between steppers that may become significant at 250-nm ground rules. It is useful within a company or facility to designate the most stable, highest-precision tool as the Reference System. If two such tools exist and a comparison shows that any differences between the two are statistically insignificant, then the two can be compared regularly and tracked to catch possible drift or other source of error.

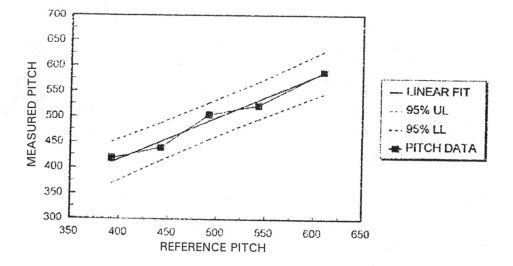

Fig. 6.6 Mandel method used to do a linear fit to measured pitch data. Measured data is fit using Eqs. 12.41–12.46 found in reference 7 and confidence limits are calculated using Eq. 12.35.

The problem with using stepper-generated pitch artifacts is that while they may be sufficiently accurate for many applications, for critical measurements they may not be. The uncertainty of the pitch is limited to the uncertainty of the mask used in the stepper and to the field-dependent aberrations of the stepper and depends on accurate magnification calibration of the stepper. In addition, pitch calibration at the center of the field is likely to be more accurate than pitch measured in the kerf (i.e., scribe line).

In advanced technology applications at the smallest ground rules where no standards exist, an in-house Reference System needs be established. Interferometry-based tools with 200-nm or better imaging resolution (either optical or SEM) are the most likely candidates. Optical systems with high NA (>.9) and a wavelength of 450 nm or lower would be satisfactory, as would phase imaging.[11] SEM systems with interferometric stages already exist.[12,13]

6.2.2.2 Linewidth

It is well known that pitch calibration alone does not guarantee accurate CD measurement.[14] Pitch is unique in that, as long as the image profiles of the lines cancel (Fig. 6.7). The lines may be poorly resolved, even asymmetric, and still produce accurate pitch. For CD measurement, however, edge detection (for SEM or optical or scanning probe microscopes) is the major problem. Differences exist between electrical and optical linewidth measurements, between SEM and optical, and between SEM cross-section measurements and atomic force scanning microscopes.[15,16] Each type of measurement involves interaction of a probe with the line structure, and in most cases the differences cannot be resolved due to lack of understanding of the physical processes involved. Hence, no linewidth standards or established methods of measurement currently exist for wafers.

Development of critical dimension standards for the semiconductor industry has been limited to date to those standards for which adequate analytical models exist for line edge determination.[17] A mask (chrome-on-glass) critical dimension standard, with uncertainty of 0.05 μm, was developed at NIST using the partial-coherence theory of optical imaging, for the case of transmitted light.[18] In this case, modeling was required to determine the correspondence of the actual edge location to its representation in the intensity profile. In general such models must account for both the physical interaction of the probe (e.g., optical, E-beam, AFM, etc.) with the sample and the effects introduced by the imaging equipment itself.

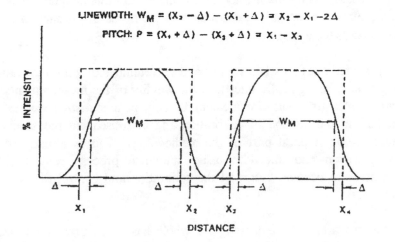

LINEWIDTH: $W_M = (X_2 - \Delta) - (X_1 + \Delta) = X_2 - X_1 - 2\Delta$

PITCH: $P = (X_1 + \Delta) - (X_3 + \Delta) = X_1 - X_3$

Fig. 6.7 The effect of an edge detection error (Δ) on pitch and linewidth measurements.

More recently, workers at NIST have demonstrated the feasibility of using Monte Carlo modeling for the development of an X-ray mask standard using the SEM in transmission mode.[19] Development of mask standards has moved at a much faster pace than wafer standards for two primary reasons. First, the modeling was performed for the case of transmitted light (electrons), and second, the materials used in mask fabrication are fairly standardized. Modeling the measurement system in transmission greatly simplifies the modeling compared to the case of reflected light. Further, the fact that photomasks typically consist of chrome on glass (optical) or a gold absorber on a chrome-polyimide-silicon support membrane (x-ray) requires that only a single set of materials interaction with the probe be accounted for in the modeling. CD-SEMS image in "reflected mode" (i.e., nontransmission) for CD wafer metrology applications, thereby seriously complicating the modeling effort. In addition, the materials measured for wafer CD metrology are many and varied, which presents severe challenges for modeling.

As a result, one must conclude that the expectation of a NIST standard for generalized wafer CD calibration is unreasonable. Further, accurate measurement would require that the IC manufacturing facility develop in-house standards for each of the various materials to be measured. To achieve this, it would be necessary to mimic the standards development process at NIST by requiring the incorporation of simulation and equipment modeling capability for manufacturing measurement tooling, in conjunction with the development of an in-house Reference Measurement System. While the complete level of certification achieved with a NIST standard is not feasible or even possible at a typical manufacturing facility, it is possible, using the methods recommended in this chapter, to achieve levels of uncertainty that are adequate to address the needs of production control for 0.25-μm process technologies.

Larrabee has pointed out that when we demand uncertainties of 2.5 nm or less, we are counting atoms; he rightfully questions the validity of the measurements.[20] In fact, nanometer measurements are meaningful only in a statistical sense. This means that both in the method of calibration and in subsequent measurements, statistics must be an integral part of the methodology. This is already true of determining precision and linewidth using statistical process control (SPC). However, it needs to become a part of the calibration and control of metrology tools as well.

For linewidth on wafers, the industry as a whole has relied upon cross-section SEM as a Reference System. The reason for its wide acceptance is that complex beam interactions with the substrate are eliminated and the cross-sectioned line becomes an ideal object with vertical edges composed of a single material. For such cases, we have some understanding from modeling of how to make a meaningful measurement, as shown in Fig. 6.8. However, if the line that is cross-sectioned is not straight, or the cross section is not perpendicular to the line, the line edge will not be vertical and the improved uncertainty will be lost.

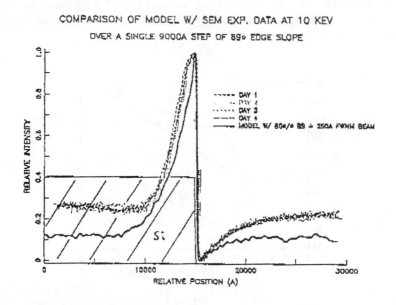

COMPARISON OF MODEL W/ SEM EXP. DATA AT 10 KEV

OVER A SINGLE 9000A STEP OF 89° EDGE SLOPE

Fig. 6.8 SEM images of a vertical silicon edge with results from simulation for comparison (courtesy of Bill Banke[21]).

There are several drawbacks to this method. It is destructive and highly dependent on the quality of the cross section. Only one point on a line is sampled, which may not be representative of the samples taken for the top-down measurements. In addition, because of cost, typically only a few cross sections are done and as a result do not provide the statistical averaging needed at nanometer dimensions. There is also concern about proximity effects for nested features approaching 250-nm linewidths. It is estimated that when done carefully, the method can provide measurements with an uncertainty < 10 nm.

Figure 6.9 plots two sets of measurements made top-down on two SEMs; one is a Reference System known to be more accurate by correlation to cross section; the other is a production tool. Both SEMs show precision of approximately the same magnitude. When the data is combined to form a calibration curve using the Mandel method,[7] the resulting linear fit is shown in Fig. 6.9(b). The slope of this curve is not equal to one and, contrary to one's initial assumption, is not due to inaccurate pitch calibration; pitch measures within 1% of nominal.

Unless the measured wafer is sacrificed to cross section, a two-step calibration process must be used. One set of wafers is measured both in cross section and top down on the primary calibration system, to generate the first calibration curve. Then identical uncut wafers are measured on both the primary and the second system to generate the calibration curve for the second system. For example, the

This two-step process further increases the uncertainty of the measurements. For the system in Fig. 6.9(b), the uncertainty is estimated to be 10 nm at 200-nm dimensions.

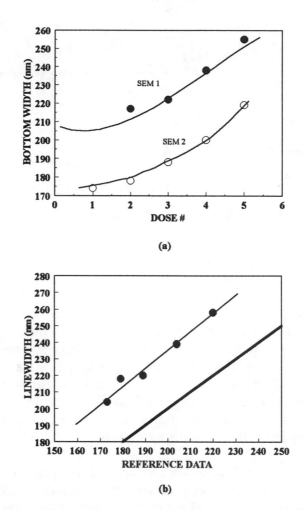

Fig. 6.9 SEM cross section calibration: (a) SEM measurement data for two SEMs, one used to calibrate the other; (b) the resulting calibration curve. The heavy line is the ideal calibration curve with slope of one.

Because of the destructive nature and the high cost of this type of calibration, there is increasing support for use of scanning probe microscopy, in particular atomic force microscopy (AFM) to provide profiling for cross section measurements. The issues relating to such a measurement are discussed in Sec. 6.4.4. In the noncontact, attractive mode with a proper tip, it is possible to get a profile of lines and some trenches (the latter limited by tip dimensions). The advantages are that profiling can be done nondestructively and, although slow, can provide more measurements on a single feature and single wafer than SEM cross section. Therefore, statistical averaging can be done meaningfully.

This method has more potential than reality at the present time. Systems are only beginning to be adapted for production use and need further development in positioning accuracy and pattern recognition to find sites and repeatably measure features of interest. The chief drawback has been and continues to be tip manufacture and characterization, in both size and uniformity (tip to tip). Tips have evolved from early parabolic tungsten tips to etched silicon tips (tapered and boot-shaped) and ion beam grown tips (needle-shaped), as shown in Fig. 6.10. In use, tip wear, tip flexing, and characterization of tip geometry have limited the uncertainty of the measurements.[22]

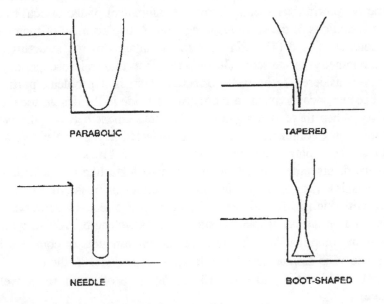

Fig. 6.10 Tip shapes used in AFM metrology.

A third reference method is correlation to electrical measurements of finished devices. This method is available in only a few cases where, for example, the electrical measurement speed of a device can be directly correlated to a CD measurement such as gate length. It is assumed that if the device meets specification, the CD measurement must have been correct. The uncertainty of this method is unknown, since many other variables in the process may have had some influence on the outcome. However, it does supply an average offset for linewidth calibration. The magnitude and distribution of any remaining variable bias is generally unknown. None of the above Reference Systems have proved wholly satisfactory. There is a need for an advanced Reference System that will satisfy the measurement requirements for 250-nm linewidths and below.

6.2.2.3 Overlay

For overlay, with lines of different materials at different levels (Z heights) on a wafer, the measurement is no longer a simple pitch measurement. The features for which pitch is being measured may produce different images due to different materials as well as different Z positions. At this time, there is no accepted Reference System that meets the needs of 250-nm ground rules requiring 7.5-nm uncertainty for overlay. This arises, in part, from tooling considerations, as well as material and geometric variations from the overlay structures themselves. As a result, the uncertainty components of an overlay measurement are difficult to deconvolve.

There are four major sources of error in optical overlay measurements; (1) asymmetric aberrations such as coma, (2) alignment of the optical system, (3) X and Y translations of the stage resulting from Z displacement of the two structures being compared, and (4) differences in the images of the structures, especially when asymmetry is present. Despite the fact that microscope objectives are generally considered to be well corrected, there is a problem, particularly with coma, because measurements are not always made in the design focal plane of the objective. When there is a significant Z displacement between the two levels of structures being compared, focus may be selected at a single plane, and coma in the out of focus plane may be significant. Fig. 6.11(a) illustrates the computed center-line displacement as a function of focus for a line image with 1/4 wave of coma (classic). Fig. 6.11(b) illustrates the effect of misalignment of the optical system, with Fig. 6.11(c) showing the combined effect. In addition, when white light is used, variation of focus position with wavelength, $\Delta Z/\Delta\lambda$, guarantees that "white light" illumination will introduce varying amounts of coma for any related focal plane. The $\Delta Z/\Delta\lambda$ may be as large as 1 μm, and in the example shown in Fig. 6.11 could easily introduce 10 nm or more of coma for a well-corrected objective.

In this example, the resulting overlay error for this structure will depend on the actual focal plane used in the measurement. One way of eliminating this source of error is referred to as the dual-focus technique, which makes two center-line measurements separately, one for each level structure, before taking the difference. This method assumes that there is no lateral motion of the focus mechanism and that in each case the measurement is made in the design focal plane of the objective, neither of which may be true. For a more detailed analysis of sources of OL error see Sec. 6.3.

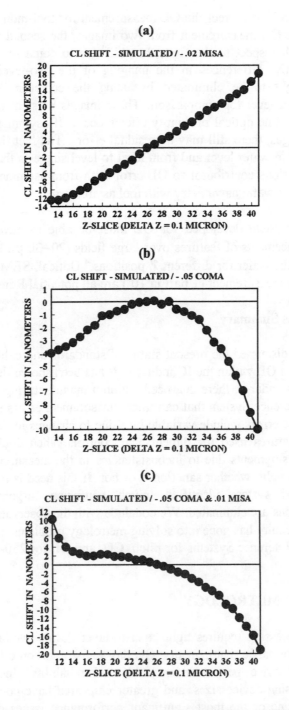

(a)

CL SHIFT - SIMULATED / - .02 MISA

(b)

CL SHIFT - SIMULATED / - .05 COMA

(c)

CL SHIFT - SIMULATED / - .05 COMA & .01 MISA

Fig. 6.11 Computed centerline displacement of a phase grating (180 nm SiO_2/Si) as a function of focus: (a) illustrates the effect of misalignment of the optical system, (b) illustrates the effect of 1/4 wave of coma (classic), and (c) the combined effect of coma and misalignment.

Another approach is to correct the OL measurement for tool-induced shift (TIS).[3] By averaging the OL measurement from two images, the second with the sample rotated 180° with respect to the first, the shift due to coma or misalignment is eliminated.[3] Lastly, differences in the imaging of the two-level structures are frequently thought to be eliminated by using the centerline of bars-in-bars structures at each level for comparison. These images must also be far enough apart laterally that no optical proximity effects occur. If asymmetry is present in any of these images, there still may be residual errors. The fact that TIS is found to vary both within wafer level and from level to level suggests that asymmetry in the wafer is a serious contributor to OL error both from asymmetric images and from interaction of wafer asymmetry with tool asymmetry.

Any Reference System developed for OL must be able to measure accurately nanometer displacements of features over large fields (20–30 μm) with no lateral displacement of the wafer for different Z positions.[4] Optical, SEM and AFM with accurate distance measurement (1 part in 10^5) are all potential Reference Systems.

6.2.3 Standards Summary

This section has discussed the present status of standards and calibration systems for pitch, CD, and OL within the IC industry. It has been shown that in each case, with or without standards, there is a need within a manufacturing or development facility for a Reference System that can track measurement errors in production or other tools. These errors include variable bias due to changes in the materials and geometries of features used for dimensional control. When doubt arises about accuracy of measurements due to inconsistencies in the measurements, a Reference System is sought, whether sanctioned or not. If this need is recognized and a well-characterized system identified, much confusion, argument, and bad metrology decisions are eliminated. We conclude with the observation that while a lot of human ingenuity has gone into solving metrology problems, there is need for development of Reference Systems for pitch, CD, and OL for 250-nm and smaller lithography.

6.3 OVERLAY METROLOGY

Lithographic processing requires tight layer-to-layer device overlay tolerances to meet device performance requirements. Overlay registration on critical layers can directly impact device performance, yield, and reliability. Increasing circuit densities, decreasing device sizes, and greater chip area have conspired to make pattern overlay one of the most significant performance issues during development of state-of-the-art semiconductor process technology. Figure 6.12 shows the evolution of device overlay and overlay measurement tolerances against successive generations of DRAM technology. As can be seen, current device overlay tolerance is in the 100–200 nm region with a corresponding overlay measurement tolerance (total uncertainty) of 10–20 nm, based on a 10% gauge capability.

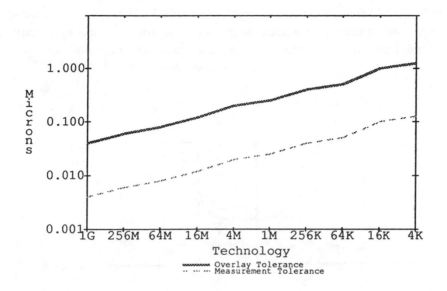

Fig. 6.12 Device overlay tolerance.

The total device overlay tolerance specification is derived in conjunction with the design rules for a given technology. Detailed analyses of the device performance requirements against manufacturing tolerances are undertaken. Typically all sources of error from the mask level through the measurement uncertainty are combined either linearly (most conservative approach) or in quadrature (assuming Gaussian error distributions) and used to derive the total overlay tolerance specification. Often, in the case of new process technologies, the tolerances are derived initially from a historical basis. In this case the measurement uncertainty budget becomes more of a goal than a performance specification.

The total process overlay budget must encompass all sources of error found in the photolithography process, including stepper stage and lens variations, resist application and develop-induced variations, substrate nonuniformities, and measurement errors. The overlay measurement tool is called upon to both quantify the magnitude of error derived from each of the sources and to verify the reduction of this error as process improvement activities take hold. It is critical that the contribution of the measurement tool to the total error, with respect to both precision and accuracy, be minimized and well controlled for all process levels. Minimizing the measurement error is the most important task facing the metrology engineer and is the focus of the following discussions.

Overlay error is defined as the planar distance from the center of the substrate (n level) target to the center of the resist defined (n+1 level) target. Overlay measurement involves the determination of the centerline of each structure along both the X and Y axis. As shown in Fig. 6.13, the centerline determination utilizes the symmetry around the structure center. Locations X_1 and X_2 shown in the figure represent the measurement-defined edge locations, while δ_1 and δ_2 account for deviations from the "true" edge locations, e.g., the error associated with the edge definition.

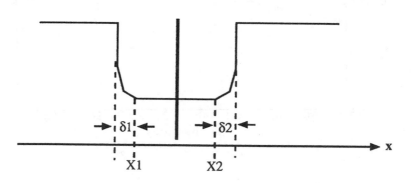

Fig. 6.13 Centerline determination (after Starikov).[23]

The benefit of this centerline symmetry is that the error associated with edge determination will tend to cancel from each side of the structure in the centerline calculation as shown in Eq. 6.1. If true mirror symmetry (e.g., $\delta_1 = \delta_2$), were present, then the errors associated with the edge determination would cancel exactly. In a linewidth determination for this same structure the error associated with each edge combines in an additive fashion, as shown in Eq. 6.2. Device overlay is defined as the difference between centerlines, CL(substrate) - CL(resist). The uncertainty in the overlay measurement, as will be shown, is a complex function of tool, target, and process interactions and is difficult to quantify.

$$CL = \frac{(X_1 - \delta_1) + (X_2 + \delta_2)}{2} \qquad (6.1)$$

$$LW = (X_1 - \delta_1) - (X_2 + \delta_2) \qquad (6.2)$$

The evolution of overlay measurement technology has followed the overlay requirements dictated by the development of process technology. Registration marks have evolved from simple, manual vernier structures that consist of pairs of bilayer, interleaved comb structures (Fig. 6.14). Such a measurement technology, while simple and efficient, was highly susceptible to operator-induced measurement error. Other technologies utilized for overlay measurement include electrical test, SEM, and optical microscopy. Of these, optical microscopy has become the dominant measurement technique due to its inherent advantage over both electrical and SEM techniques in a manufacturing environment.

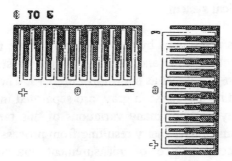

Fig. 6.14 Optical overlay verniers.

SEM-based overlay techniques are limited by the capabilities of the SEM at the low accelerating voltages required for nondestructive testing. At these voltages the SEM can image only the surface layer. Overlay registration often requires measurements of layers that are separated by thousands of angstroms. To accomplish this, SEMs require a special mask level that allow a subsequent etch step to remove the intermediate level and enable the placement of the resist level directly on the substrate (target) level. This additional process step adds complexity and cost to the process and is often not an option in a manufacturing environment. Electrical overlay measurements also occupy only a niche in overlay measurement. These measurements are limited by the requirement that the overlay be measured in a conducting film. Electrical overlay measurements can be obtained only following etch processing, where rework and direct closed-loop feedback to the photo-processing module are not as effective.

Optical overlay measurement systems have evolved a great deal from the simple optical microscope used for vernier measurements. The most advanced optical overlay measurement tools are fully automated and incorporate wafer handling, pattern recognition, image processing, and data processing hardware. The optics and illumination systems of the simple microscope, which had worked well for previous technology generations, are no longer suitable to meet the requirements of submicron semiconductor process technologies.

The inherent tool-limited measurement accuracy, commonly referred to as tool-induced shift (TIS),[3] has become a larger percentage of the measurement error. TIS is quantified by measuring the same feature at 0° and 180° (physical) rotations with respect to the tool. TIS is simply one half the sum of the measurements from each orientation. TIS, as will be shown, is a complex function of the optical alignment, mode of illumination, and aberrations of the optics in the system. Elimination of TIS often requires a full-scale assessment and redesign of the optical and illumination path to minimize complexity and improve the alignment of the total optical system.

Measurement accuracy is determined by the interaction of the measurement tool with both the measurement structure and the process. The most common overlay test structure is the box-in-box target as shown in Fig. 6.15. The boxes defined in Fig. 6.15, substrate in white and resist in gray, are separated in the Z dimension many times by a third layer. While many variations of this target exist, all are susceptible to several forms of asymmetry resulting from process interactions with target design. This second source of measurement inaccuracy is termed wafer-induced shift (WIS).[3] Unlike TIS, WIS is invariant with respect to rotation and, as a result, is not as easy to identify. In many cases, it is not possible to quantify WIS until subsequent etch processing is performed and the overlay structure remeasured. WIS must be addressed in both the measurement algorithms themselves and the overlay target design. Process-induced errors are a complex function of the measurement tool, the measurement algorithm, and the target design.

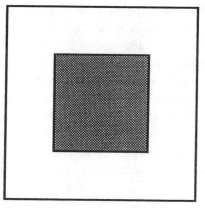

Fig. 6.15 Typical automated overlay measurement target.

The final component of the measurement contribution to the overlay error comes from the measurement precision of the tool. This component is quantified by performing repeated measurements of a sample to obtain a statistically significant quantification of the measurement variation. It is then, from the process engineer's standpoint, that these sources of measurement error combine to constitute the total measurement uncertainty. Minimization of all contributions to this measurement error is a primary function of the metrology engineer. As process technology continues to advance, the overlay measurement tolerance will continue to shrink. The capability of measurement technology to keep pace with the advancing process technology will be put to a severe test in the near future and will require a concerted effort to minimize all controllable error sources.

6.3.1 Measurement System Overview

In early semiconductor processing, layer-to-layer overlay registration requirements were not as close to the limitations of measurement technology. As shown in Fig. 6.12, a 2-μm CMOS process required total device overlay control of nearly 1 μm, for which the measurement error budget is nearly 200 nm. Overlay requirements for these technologies were adequately addressed by operator interpretation of optical vernier patterns (Fig. 6.14) with a standard bright-light optical inspection microscope.

It was not until process technology crossed the 1 μm minimum geometry threshold that automated overlay metrology was required. The need for automation is driven by the requirement for high-throughput, high-precision measurements. Automated measurement equipment can perform at rates exceeding 50 wafers per hour as compared to a maximum manual measurement capability of 10 to 15 wafers per hour. Additionally, the operator influence on the measurement outcome can easily be greater than 50 nm (3 sigma precision), a number that is three times larger than the total measurement error budget for a half-micron process. The accuracy and precision of manual vernier measurement is limited both by operator-to-operator differences and by the achievable resolution of vernier structures, which are typically limited to less than 50-nm minimum gradations. The analogous limitation for the automated overlay measurement tool is related to the measurement magnification and the resultant pixel size; e.g., for a magnification of 2000× and a 512×512 pixel image each micron will be represented by more than 10 pixels, and by using subpixel interpolation algorithms, edge definition to less than 10 nm is routinely achieved.

The optical overlay measurement tool can be broken down into five primary components. This oversimplification serves the discussion by allowing emphasis on those areas of the equipment critical for measurement. These major assemblies of an optical overlay tool are wafer handling, optical components, image processing, pattern recognition, and the system CPU.

6.3.1.1 Wafer handling

The wafer-handling subsystem is primarily responsible for the throughput of the tool. The wafer handler must be designed to minimize sources of particulate contamination. In the handler operation the wafer is subjected to prealignment for wafer center determination. Then the flat or notch of the wafer is detected via optical or other noncontact means, and positioned for measurement. The accuracy of the operation is gross by the standards of the fine alignment performed under the microscope but is critical for system throughput. The presence of unwanted rotation on the final wafer position will result in algorithmic difficulties and inaccurate measurements.

It is of critical importance that the stage be flat and that the wafer transfer to the stage is firm and repeatable with respect to location and flatness. The wafer stage must also be capable of supporting the full 200-mm wafer, since any exposed wafer edges are subject to bow and resulting measurement error. The stage must also be driven in X and Y in such a way as to provide the most accurate wafer translation so that high throughput can be maintained with as little target search as possible. In the case of "absolute" registration measurements, the stage must be accurate to less than 50 nm.

The stage Z drive must be accurate and repeatable for all Z motions, at all points on the 200-mm wafer stage. The stage must also be well behaved in the X-Y plane during Z motion. Any translations in X or Y which result from Z motion will contribute to measurement error. The worst-case measurement scenario for this Z-axis-induced translational error is during a dual-focus measurement sequence. In this measurement mode, which is used for structures exhibiting topographies larger than 2 μm in the axis perpendicular to the wafer (Z), centerline estimates from the substrate (target) and resist levels are taken at two separate focal planes. Any translation in X and Y that takes place during the move between focal planes will show up as measurement error (TIS). Finally, since it is usually the stage Z drive that limits the repeatability of the focusing subsystem, it must have minimum step increments that are less than required by the optics.

6.3.1.2 Optical components

The optical components of an overlay measurement tool consist of a microscope assembly, a camera, and an illumination system. Overlay system design has seen the most sweeping changes in the area of optics. The first measurement systems relied on standard, multiuse laboratory microscopes that were integrated into an automated measurement system. As process technology has advanced, performance requirements for overlay measurements have necessitated optical subsystems specifically designed for dedicated overlay applications. Overlay measurement error budgets of less than 20 nm have forced investigations of tool-induced asymmetries in the acquired image.[3,24,25,26,27,28]

The most common measurement error attributed to tool asymmetries is TIS. From a system optics perspective, illumination irregularities, optical alignment, and lens aberrations contribute most significantly to TIS. Optical design and manufacture for overlay measurement instruments must critically account for all optical components. The complexity of the optical path must also be reduced from that of a standard inspection microscope. Each optical element can contribute to misalignment, since each additional optical interface contained within the design could degrade the system performance. The most critical optical component for measurement accuracy is the final objective lens, with respect to both alignment and residual aberrations. Accurate measurement requires that the optical system alignment be achieved and maintained over time.

6.3.1.2.1 Lens effects

Aberrations in the objective (final) lens will contribute significantly to systematic tool-induced measurement error. Figure 6.16[29,30,31] schematically depicts the effects of spherical (symmetric) and comatic (asymmetric) lens aberrations on the image plane of a point source. Figure 6.16(a) shows a front view of the lens. Each of the labeled points represents a specific location on the lens from which a ray trace will be used to determine the image of a point source at the image plane for the comatic aberrations. The dashed circular lines represent the lens locations of groups of ray traces used to depict the effect of spherical aberrations.

As can be seen from Fig. 6.16(b), the spherical aberrations are symmetric in the plane of the image. From an overlay measurement perspective, where centerline determination of an object located within the center of the microscope's field of view is the critical determination, the effects of spherical aberration (assuming central object placement) will cancel exactly. This will hold true for all focal positions, since the error will exactly cancel across a centered measurement target in the centerline determination. This example may be further generalized to account for all symmetric aberrations, assuming central placement of the measurement structure within the field of view.[25] Figure 6.16(c) shows the effect of comatic aberrations on a point source image. The asymmetric nature of this aberration is readily apparent and will result in an incorrect assessment of centerline for a symmetrically placed overlay measurement structure. It is also important for the case of the comatic aberration to consider the interactions with sample topography and focus.

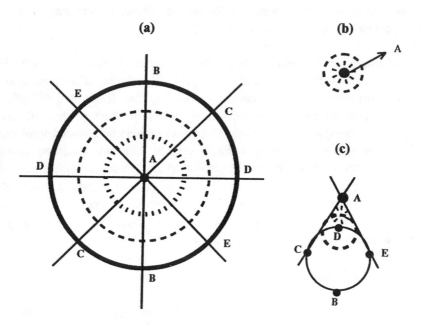

Fig. 6.16 The geometrical image of a point source (a) on the lens, (b) in the presence of spherical aberrations, and (c) comatic aberrations.

Kirk[32] has shown that brightfield optical systems are particularly susceptible to coma, and in the presence of 0.6λ of coma the range of error in edge location determination can be as large as 100 nm, over 1 µm of defocus. From an overlay measurement perspective, for a single focal plane and single-layer structure, it is expected that the asymmetric aberrations will not contribute significantly to the measurement error, since image effects due to the aberrations will affect each part of the target (inner and outer) in a like manner, canceling most centerline estimation error in the overlay calculation (CL_1 - CL_2). This will also be true, though to a reduced extent, for measurement of a dual-layer structure which measures each layer in focus (each measurement conducted at a separate focal plane). Measurement error resulting from tool factors in the dual-focus measurement mode would most likely result[27] from translations of the optical axis (due to unwanted X or Y stage motions), as viewed from the reference frame of the target, during focus.

The measurement instance most vulnerable to the effects of comatic aberrations is the single-focus measurement on a dual-layer sample; generally the standard measurement technique for all samples with less than 2-µm range in the Z direction. For it is in this instance that the two layers are found at two different focus positions and will see different effects from the comatic aberration, resulting in centerline estimates for each layer with different contributions from the comatic aberration. Simulations of comatic aberrations and their impact on overlay registration measurements have predicted[25] that errors of 35 nm (0.1 wavelengths of coma) to 84 nm[3,26] (0.2 wavelengths of coma) can result. Since these aberrations interact with focus[24] in ways that impact the overlay measurement error, it is essential to account for anticipated focus/sample interactions that may result from system design and measurement configuration.

Lens testing to monitor and maintain optics quality include Strehl ratio[33] testing (ratio of light intensity at the peak of the diffraction pattern of an aberration-free image), star testing (qualitative assessment of comatic aberrations by observing direction of coma flares from a diffraction limited array), and measurement of the optical transfer function (OTF).[25]

6.3.1.2.2   Resolution and depth of field

Other attributes of the optics subsystem that can impact measurement accuracy are resolution and depth of field (DOF). The common link between these two characteristics is the NA of the final objective. The NA of a lens is defined[34] as the light-gathering power of the objective. The relationship between NA, resolution, and depth of field is described by the Rayleigh criteria, as shown in Eqs. 6.3 and 6.4.

$$\text{Resolution} = k_1 * \left(\frac{\lambda}{\text{NA}} \right) \qquad\qquad (6.3)$$

$$\text{Depth of Field} = k_2 * \left(\frac{\lambda}{\text{NA}^2} \right) \qquad\qquad (6.4)$$

From these relationships it is evident that increased resolution comes at the expense of depth of field. As noted, it is often the case that in order to minimize systematic measurement errors due to stage effects, measurements are performed at one focal position. Large depth of field, then, is a very desirable quantity, from a theoretical consideration, to achieve measurement accuracy. It is fortunate that the image dimensions commonly utilized in overlay target design (10–20 μm) do not require the highest-resolution objectives and are consistent with larger depth-of-field objective designs.

6.3.1.2.3   Illumination

Illumination is perhaps the most critical part of the entire optical subsystem when it comes to measurement accuracy. Illumination will directly impact sensitivity to process variations and, if not properly aligned, can result in large increases in TIS. The illumination bandwidth must be chosen such that it will not expose photoresist, not be susceptible to diffraction phenomena, provide the required resolution, and maintain a stable output (intensity) over time. Typically broadband (450–650 nm) partially coherent[35] light is set up in Kohler[36] illumination in which the lamp filament is imaged on and completely fills the back focal plane of the objective. The aperture of the illuminator must be centered on the optical axis for symmetrical illumination.

The importance of symmetric (telecentric) illumination for the objective is derived from the centerline estimation process by which overlay measurements are generally performed. Figure 6.17 demonstrates the lateral image shift [25,27,37] that can result from an illumination axis that is not perpendicular to the wafer surface. The observed image can be seen to shift as the focal plane is changed. The worst case for measurement error occurs when the image optical path difference between layers is maximized. This is the case shown in Fig. 6.17, where the focal plane is at the substrate. It may also be noted that there is a focus position from which the optical path lengths from the two layers are equivalent. In this case the error resulting from the geometric image shift will be exactly compensated in each layer.

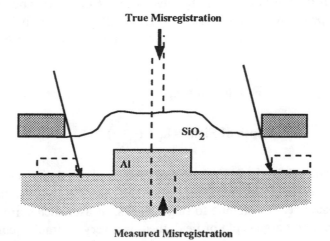

Fig. 6.17 The effect of illumination alignment on TIS.

Several authors have reported the results of simulations and actual tests of the effect on TIS when the illuminator stop is intentionally misaligned. Starikov[3] reports the results of a simulation in which the aperture was misaligned from NA=$\pm$ 0.42 to a range of NA from -0.28 to +0.56. A 29-nm measurement error (TIS) was induced in the substrate centerline determination. In other studies[27,37] adjustments were made to the illumination aperture and beamsplitter in an overlay measurement tool to cause a controlled misalignment. The effect on measurement, which was dependent upon the sample measured, was to induce a TIS that was a function of both the difference in optical path length between the layers and the angle of the incident light. Therefore the degree to which the illumination path is aligned is a strong determination for measurement accuracy and is sensitive to focus position during measurement. In addition to illumination alignment, the uniformity of illumination also will affect the measurement accuracy by causing shadowing effects that are a function of structure orientation and location. In many cases a diffuser[25] is put into the optical path to ensure even illumination.

6.3.1.2.4 Focus methods

Focus of the measurement system on the overlay target prior to measurement has been shown to negatively interact with tool optical asymmetries. The characteristics of the image, from a measurement perspective, are also very sensitive to changes in focus. Most overlay measurement systems have a measurement depth of field in the range of 0.6–1.2 µm, which is generally smaller than the range of topographies encountered. As a result there does not exist one focus position for which a typical two-layer structure is fully in focus; some parts of the structure will be out of focus.

However, it has been noted[28] that improvements in precision result when device overlay is determined from a single-focus position, in which centerline estimates from both layers (substrate and resist) are taken from a single focal plane. This is due to the elimination of Z axis changes and associated X and Y stage translations between measurements. These precision improvements, however, must be weighed against the potential loss of accuracy due to lens aberrations and illumination misalignment, which have been shown to be most sensitive for the case of the single-focal-plane measurement. Typically, focusing is performed by varying the Z distance between the final objective and the wafer surface and performing a signal contrast analysis of the signals gathered at each Z increment.

Image-contrast-based algorithms[38] rely on the sensitivity of the image signal intensity profile to changes in focus. These algorithms take advantage of the fact that the steepest image edge slope is found at the position of best focus. While this method is highly susceptible to substrate optical variations and thick film effects, which can limit the accuracy of the determined focus position to 200 nm, it is often the method most amenable to an automated measurement system. Refinements to this approach as reported by Kirk,[25] who used a harmonic analysis of a series of images through focus to demonstrate that maximum amplitude is not achieved by all spatial frequencies at "critical focus." In fact, only the higher-order harmonics were shown to be capable of producing a reliable focus.

Several alternative approaches to defining and maintaining focus have been utilized, including laser-based wafer-to-lens distance monitoring, aerial image contrast, and interferometric or confocal methods. A typical laser-based focus system[39] uses collimated near infrared laser light that is focused onto the sample. The light is reflected back across the sample through the same optical path and projected onto a split diode detector. The differential diode detector registers proper focus when the intensity on both halves of the detector is equal. Further improvements to the image contrast method have been proposed by using a graticule (aerial image of controlled contrast)[24] as the basis for the profile intensity optimization. Placement of the graticule near the microscope aperture, so that the shadow is imaged onto the substrate, results in the sharpest focus for the graticule at the substrate and a definite focus reference position.

Interferometric and confocal techniques,[40] due to a greatly decreased effective DOF and improved resolution, are capable of determining focus by scanning through a Z range and analyzing the resultant phase signal intensity information contained within a very narrow (< 0.2 μm) region as a function of Z (focus position). The narrow focal plane region excludes information from out-of-focus surfaces and ignores diffracted signals.

Most of the focus strategies discussed above have the capability to determine focus to a repeatability of < 250 nm (3 sigma), well within the depth of field of the measurement optics. As a result each is capable of providing adequate accuracy for focus determination and control.

6.3.1.3 Camera, image processing, and pattern recognition

The camera, usually a CCD sensor array, is mounted in the optical path and is responsible for gathering the image information used for measurement. Cameras chosen for overlay measurement applications are highly linear over the entire image area and provide an extremely stable response over time (years). Cameras are typically run very conservatively (low gain); all image processing is done within the specialized image processing hardware on the acquired gray-scale optical profile. Cameras must also be symmetric with respect to detection sensitivity across the image field and must not suffer from any image distortions or electronic noise.

The image processing components are fed directly from the camera and are responsible for all image processing functions. The image processor performs an analog-to-digital signal conversion and any required linear signal processing. The image processor is also responsible for all image quality enhancements, including frame averaging contrast/brightness adjustments, image filtering, line averaging during measurement, and normalization. The pattern recognition subsystem receives images from the image processor and "learns" selected images (model) for identification during automated measurement routines. The pattern recognition system performs a pixel-level electronic registration of subsequent image data against the stored model. The registration of the image results in a final positioning of the overlay test structure for measurement. The pattern recognition system must be capable of placing the measurement target to a repeatability of within 0.1 μm to attain the required levels of precision and minimize the impact of optical/illumination errors resulting from off-axis target placement. Typically pattern recognition is accomplished using an image correlation algorithm (using image gray-scale information) or an edge detection algorithm to identify images. A unique, nonrepetitive feature is usually required for successful "learning" to take place.

6.3.2 Data Analysis: Accuracy and Precision

Precision, accuracy, uncertainty, and traceability must be addressed during machine setup. Precision can be defined as the "repeatability of a measurement result."[41] Several assumptions concerning sample, test duration, test structure design, measurement data distribution, sampling plan, and machine operations allowed to occur between measurement passes must be clearly thought out and defined for each precision study performed. Accuracy is unambiguously defined as "the correctness of a measurement result."[41] The implications behind such a definition as reported by Larrabee and Postek are:[41]

1. There is universal agreement about the nature and definition of the quantity being measured.

2. There is a valid way to measure that quantity.

3. There exists some basic standard of comparison for that quantity that has been calibrated to within a known deviation from its true value.

4. There exists a valid calibration and operational procedure for the user's instrument that will effectively measure the standard and the user's specimens in the same way for all significant aspects of the measurement.

Each of the these attributes of accuracy must be assessed for each overlay measurement to be performed. Lack of adherence to one or all of the preceding caveats does not invalidate the measurement or reduce its usefulness; instead it defines the ways in which the data may (and may not!) be used. In the case of overlay measurement, only caveat 1 is realistically achievable, since there are no traceable standards and there are several methods for overlay measurement, each with differing degrees of validity. *Uncertainty* is defined as the "measure of the maximum expected error of measurement."[41] Thus, the error associated with overlay measurements is a combination of both accuracy and precision components. Mathematically, uncertainty (U) will be rigorously defined as[38]

$$U = E_O + 3 \times \left(s_o^2 + s_c^2 + s_m^2 \right)^{\frac{1}{2}} , \qquad (6.5)$$

where E_o is defined as the systematic error associated with the calibration of the standard, s_o is the standard deviation of the calibration standard measurement, s_c is the machine precision associated with the measurement of the calibration standard by the machine to be calibrated, and s_m is the standard deviation of the calibrated instrument as used for routine sample measurements.

Due to the lack of a basic overlay standard, absolute determination of uncertainty for overlay is not possible. However, in terms of what is known about overlay measurement, it is possible to create a working estimate of the uncertainty of an overlay measurement that accounts for equipment-induced inaccuracies. This definition can also be generalized and extended to include other sources of measurement error.[26] For this discussion only tool errors will be specifically considered. This working estimate of uncertainty will assume that all errors associated with pixel-to-micron calibration will cancel in the final overlay determination; CL(substrate) - CL(resist). Then, allowing for the symmetries associated with centerline determination itself, it is possible to rewrite Eq. 6.5 to read:[42]

$$U = TIS + 3 \times \left(s_{TIS}^2 + s_P^2 \right)^{\frac{1}{2}} , \qquad (6.6)$$

where TIS is the mean TIS obtained, s_{TIS} is the 1 sigma estimate of the variation of the TIS calibration, and s_p is the 1 sigma estimate of the instrument during routine sample measurement. The primary assumption herein is that the variation in TIS, s_{TIS}, accounts for the contributions of the interaction of process and target asymmetries with the measurement system.

6.3.2.1 Precision

The precision of an overlay measurement tool must be carefully qualified with regard to all aspects of the test. The value of precision will vary with test conditions. For example, the precision of a test that is performed by keeping a single sample under the objective and merely reacquiring and measuring the overlay will be far better than a test that is conducted over several days, where the wafer is cycled from a cassette to the measurement point during each measurement cycle. The first test is an example of what is often termed *static* or *short-term* precision, whereas the second is called *dynamic* or *long-term* precision. The short-term test will only assess the repeatability of the measurement algorithm, the illumination, and the image acquisition components of the overlay tool. The dynamic test incorporates all system activities leading up to the measurement and presents a more complete picture of the measurement capability. Often a designed experiment is used to partition measurement attributes in order to determine the largest contributors to measurement imprecision. The results of such a study for a given system and sample type combination can be directed at improving the tool precision through measurement, equipment, and target setup.

There are several graphical tools that can assist in the determination of areas of imprecision. By looking at data in a time or spatial sequence, it is possible to determine the sample contributions to imprecision at a point in time or at a stage location where the system produces anomalous measurements. Figure 6.18 looks at overlay measurement precision as a function of position on the wafer. The data collected in Fig. 6.18 is representative of static precision. System functions that occurred between measurements were limited to focus, image acquisition, and measurement. In this particular instance it is apparent that two of the die locations on the wafer have much poorer precision than the other three sites. Further analysis of the low-precision sites found that they were located in an area of the wafer that exhibited lower contrast due to the planarization process. The lower contrast at these sites made consistent measurements more difficult, due to the poorly defined structure edge.

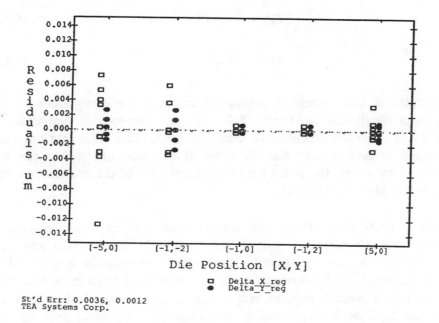

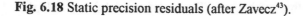

St'd Err: 0.0036, 0.0012
TEA Systems Corp.

Fig. 6.18 Static precision residuals (after Zavecz[43]).

6.3.2.2 Accuracy: tool-induced shift

Accuracy of overlay measurements has not received the same level of attention as accuracy for linewidth measurements. In the past, this may have been due to the larger total error budget (total uncertainty) enjoyed by overlay measurements, or the belief that overlay measurements are inherently accurate due to cancellation of errors and the underlying symmetry of centerline determination. While in large part this assumption holds, there are error sources unique to optical overlay measurement tools that arise as a direct consequence of the measurement symmetry.

TIS is the tool error component of overlay measurement that contributes most significantly to the accuracy of device overlay measurements. TIS is a complex function of the optical setup of the tool. TIS is typically analyzed by performing overlay measurements at a minimum of two wafer (flat or notch) orientations that are 180° apart. It is preferable to obtain data at all four orientations (0°, 90°, 180° and 270°) so that other calibration effects can be analyzed, e.g., pixel scale and nonlinear errors, but it is possible to quantify TIS with the minimum two orientations. In its simplest form, TIS is solved for by taking the algebraic mean of overlay measurements obtained at the first orientation, 0°, and the measurement values obtained at the 180° orientation for both X and Y registration. This is shown in Eqs. 6.7 and 6.8.

$$TIS_x = \frac{X_0 + X_{180}}{2} \tag{6.7}$$

$$TIS_y = \frac{Y_0 + Y_{180}}{2} \tag{6.8}$$

The strengths of this method of calculation are that it is straightforward, it demonstrates clearly any site-to-site, die-to-die, or wafer-to-wafer variation in TIS, and it is not overly dependent on sample plan heuristics. This data can also be analyzed in such a way that all terms of the total uncertainty can be incorporated and solved for, and a test of the distribution of the total uncertainty can be performed on the TIS data.

However, magnification effects and nonlinearities are not easily observed. Measurement nonlinearities, in general, occur when the measurement tool has reached some fundamental physical limitation (optical resonances in small CDs and SEM charging).[44] Other nonlinearities are the result of changes in the behavior of a measurement system's response over a range of process variations (a function of structure type) or improper tool or test setup. An alternative approach to solving for TIS is to perform a linear regression[45] of the first orientation (0°) against the second (180°). From this method the TIS is taken to be one half the intercept term of the regression. Using regression for TIS calculation can easily demonstrate both magnification and nonlinear effects but cannot easily partition data by site/wafer/lot. Regression is also very sensitive to test design. In fact, too small a misregistration sample range will result in poor confidence in the coefficients obtained. While such a predictor is often a valuable assessment of the validity of the test, it is not always feasible in practice to have a large range of misregistration.

Figure 6.19 and Table 6.1 represent the results of a TIS analysis conducted on a typical data set using both the regression and mean TIS calculation approaches. As can be seen, it is possible to obtain somewhat different results depending on the calculation approach used. In the example, both approaches give comparable values of TIS. However, analysis of the data using the algebraic TIS calculation demonstrates that there is systematic variation in TIS across the wafer, which is most likely, although not explicitly, highlighted by the large slope uncertainty in the regression model.

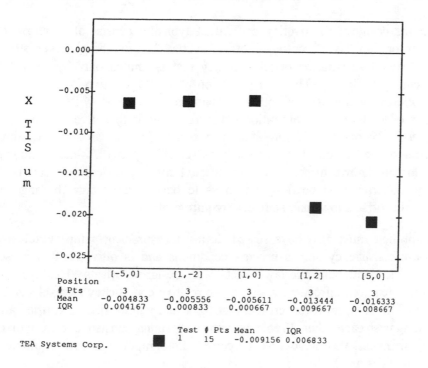

Fig. 6.19 Results from algebraic TIS calculation.[43]

Table 6.1 Regression results for TIS calculation (units in μm).

X Slope	X Slope (3σ)	X Intercept	X Intercept (3σ)
1.095	0.394	0.0103	0.0052

Typically, for a well-designed test and a robust set of overlay measurement targets, the two methods of TIS calculation should agree with respect to both mean TIS and variation of TIS. As shown by the example, it is instructive to use both methods so that the most comprehensive determination of the impact of the TIS error on the final reported measurement can be made. Additionally, it is not enough to simply quantify TIS for a single sample, since TIS is such a complex function of the tool setup and, as will be shown, process variations. TIS must be evaluated for each layer of the process with differing stack heights and contrast mechanisms, since each of these factors have the potential to result in differing TIS and differing levels of error in the (uncalibrated) overlay measurement value.

Test patterns designed for overlay calibration typically consist of a matrix of single-layer or dual-layer[46] designed offset overlay targets. Single-layer structures will have a registration offset accuracy that is limited only by the mask writer accuracy («10 nm). The misregistration offset for the two-layer designed offset structures[47] will consist of both wafer patterning and mask write errors. The advantages of the two-layer calibration structure design is that it can be made to mimic exactly the optical and film stack properties of actual product, including any process-induced asymmetries. The single-layer offset pattern, while containing more accurately known offsets, will not account for any process effects. Both calibration patterns must be designed so as to have a large enough range of misregistration offsets to satisfy statistical requirements.

Pixel calibration must also be assessed during measurement setup. Pixel size shows more consistency across process conditions and is often verified or set using a reticle containing overlay targets with designed offsets. This type of calibration is required infrequently, due to the relative consistency of pixel size as opposed to TIS. It is, however, critical that the x and y pixel size be verified and monitored against each other. Problems with maintaining consistent x vs. y pixel size will manifest themselves as perceived lithographic pattern equipment nonorthogonalities and higher-order distortions.

Overlay measurement accuracy errors, if not recognized and accounted for, can impact applications such as stepper setup and can produce false systematic stepper errors. Zavecz[45] reports that the introduction of systematic measurement errors into a simulated stepper setup data set results in the transposition of those errors to the stepper. TIS shows up in the translational term of the modeled stepper systematic errors, and pixel scale directly modifies the grid and field magnification (scale) coefficients. If these errors are not properly ascribed to the metrology tool as their source, they will end up as a stepper input correction that further degrades product overlay performance. It is important to accurately quantify and assign the measurement tool's TIS error component of the measured value and minimize its contribution via a hardware modification (e.g., optical alignment/collimation) or software calibration.

6.3.3 Process and Tool Interactions: Target Asymmetry and Measurement Algorithms

Careful design of all hardware components is critical for the minimization of overlay measurement error. The measurement asymmetry due to the measurement equipment's optical and wafer-stage misalignment can be determined by performing overlay measurements at $0°$ and $180°$ orientations on the same sample (TIS). However, the magnitude of the hardware error or TIS can also be observed to interact with the process conditions, achieving different values for different resist thicknesses, substrate surface roughness, substrate topographies, and structure designs.[37]

Figures 6.20 and 6.21 depict the TIS performance of two different overlay measurement systems over six different process layers in the X and Y directions, respectively. The substrates shown include metals, polysilicon, field oxide, and multilayer dielectrics; the resist thicknesses shown range from 0.5 µm to greater than 2 µm. The nominal wafer shown in Figs. 6.20 and 6.21 is a single-layer, (1 µm) resist on field oxide, with "designed" overlay-offset, frame-in-frame targets covering a 125-nm total offset range. Calculated TIS for this sample shows, as expected, minimum values for both measurement instruments, relative to the other process levels. This is due to the planar nature of the structure, which permits measurements at a single focal plane (both inner and outer frame in focus).

If TIS were independent of the process effects, one would expect to be able to predict a measurement instrument's behavior on all process layers, given an adequate understanding of the response in the ideal, minimum TIS instance. As can be seen from the figures, while it is generally true that minimum TIS results in best overall performance, it is also clear that there is a complex interaction between the various measurement schemes (optics, algorithms, illumination, etc.) represented by the different equipment and TIS performance on a given substrate. It is also notable that the trends observed for each piece of equipment are not the same in both X and Y directions.

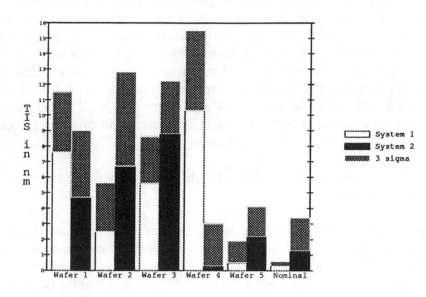

Fig. 6.20 X TIS.

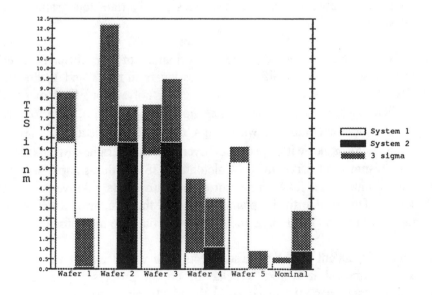

Fig. 6.21 Y TIS.

This data re-emphasizes the observations made during the discussion of the system optics. TIS is a *very* complex function of choice of focus method (e.g., single vs. double grab), optical configuration, and substrate material. It is not possible to accurately predict the TIS response for a given instrument on a new substrate based only upon an empirical understanding of the instrument's performance on other materials. It is therefore essential to first align and optimize the optical hardware and then to characterize the TIS performance of the instrument on *each* of the resist combinations that are to be measured.

Figure 6.22 shows the interactions of focus and TIS. The data for each of the systems was collected in the single focus measurement mode using Wafer 3 from Figs. 6.20 and 6.21 which has a resist thickness > 1.5 µm. The X axis of Fig. 6.22 shows the deviation from best focus (0.0) in microns, while the Y axis is the value obtained for TIS using an algebraic calculation. The X TIS behavior of both systems is typical of what would be expected: a region of focus that corresponds to a stable operating point across the DOF of the system optics.[37] The Y TIS as shown in Fig. 6.22, however, is not so well behaved. Across the DOF there does not appear to be a stable (TIS) point for either system. The anomalous behavior in Y TIS, corroborated by the two measurement tools, is believed to be due to an interaction with a target asymmetry. Starikov[26] is shown that much of the variation observed in TIS across a wafer can be accounted for by local variations in target symmetry. In fact, it is conceivable that the variations in TIS as shown in Figs. 6.21 and 6.22, both within and across wafers, is due to an interaction of the tool asymmetry (TIS) and the process asymmetry. Even across a single wafer, process conditions are interacting with the fundamental tool setup in different ways at different wafer locations, producing measurement errors of differing magnitudes.

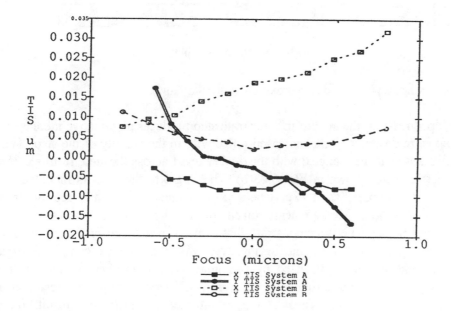

Fig. 6.22 TIS vs. focus.

6.3.3.1 Process asymmetries

Process asymmetries have become a significant fraction of the observed measurement error. Systematic measurement errors that exist even in the case of a fully optimized measurement tool (optimized with respect to TIS) are often referred to as wafer-induced shift. These errors arise from asymmetry in the overlay measurement target itself. Target asymmetry is usually the result of proximity effects or asymmetric film deposition conditions. Target asymmetry that results from proximity effects[3] is shown in Fig. 6.23.

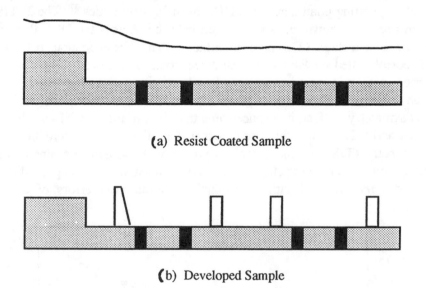

(a) Resist Coated Sample

(b) Developed Sample

Fig. 6.23 Mark asymmetry due to proximity (after Starikov[3]).

The sloped resist in the area of the substrate step produces an effective increase in the exposure dose due to the increased reflectivity in the vicinity of the step. This effective dose change, coupled with the thicker resist across the step, results in the asymmetric target shown in Fig. 6.23(b). The target to the right is included for comparison purposes, as it experiences no proximity effects and represents a well-formed target. Measurements taken by Starikov et al.[26] on structures exhibiting this asymmetry demonstrate that a mean overlay difference of 70 nm, along the axis affected by the asymmetry, was observed as compared to a structure that was not influenced by proximity effects. Further simulations by Starikov show that the introduction of even a 0.5% resist slope can result in a 23-nm measurement error. Tanaka[48] reports a target asymmetry that results from uneven deposition of a film. In the instance cited, a metal sputter deposition process produces film thicknesses that are dependent on topography and orientation with respect to the incident sputtered atoms. This results in a "snowdrift" effect within the box-in-box overlay target, whereby the metal is piled up in one side of the overlay target. This effect is shown schematically in Fig. 6.24.

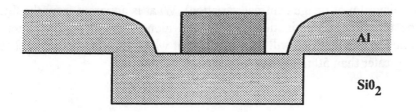

Fig. 6.24 Target asymmetry due to uneven film deposition.[48]

From an overlay measurement perspective, the snowdrift effect results in an inaccurate measurement due to an incorrect estimate of the "true" SiO_2 target centerline. The measurement inaccuracy that results from this type of target asymmetry can be greater than 100 nm (3 sigma) when comparing measurements taken after etch processing with standard overlay measurements taken after the develop process.[48]

The interaction between the process (WIS) asymmetries and the measurement tool asymmetries (TIS) for a given measurement tool is significant and complex. Kirk[32] has shown that tool-induced measurement errors in brightfield alignment systems, which in the study were shown to be very sensitive to comatic aberrations, can interact with the process topography in an additive fashion. Minimization of TIS alone cannot ensure accurate measurements, since even an overlay tool fully optimized with respect to TIS will not accurately measure an asymmetric target. It is necessary to be able to separately characterize the effects of TIS and WIS. The WIS component of inaccuracy must be addressed through measurement target design optimization and measurement algorithm development. These areas are becoming critical for accurate overlay measurement in sub-0.5-μm technologies where all controllable, systematic measurement error sources must be investigated and minimized.

6.3.3.2 Measurement algorithms

The threshold algorithm is increasingly running into accuracy difficulties resulting from the complex film stacks utilized in state-of-the-art processes. Given a lack of symmetry within the target of the type described in Figs. 6.23 and 6.24, the basic threshold algorithm is prone to systematic measurement error. The determination of a specified threshold on the image waveform, corresponding to a point on the edge of the overlay target, will not match reality. Figure 6.25 schematically shows a signal waveform obtained from the substrate portion of an overlay target. X1, X2, and X3 depict potential 50% threshold points on the curve. The asymmetry shown could be the result of tool (e.g., minor illumination misalignment) or mark sources (grainy film or image contrast variation). What is apparent from the figure is that the threshold determination is uncertain along the left edge. Centerline determination in this instance will be a function of which threshold point is chosen and errors greater than 50 nm can be expected to result.

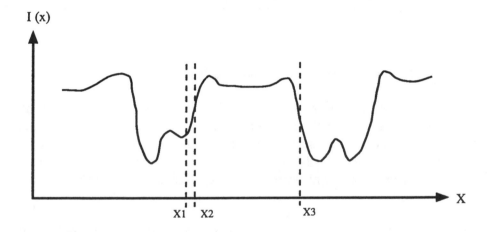

Fig. 6.25 Optical image intensity profile (after Kirk[25]).

It is becoming essential that more than a simple interpretation of the signal waveform for a percentage of the threshold be performed. The measurement algorithm must begin to take advantage of a-priori knowledge of the overlay target[23] and the physics of the subsequent interactions with the films and resist layers. The relationship between the location of the edge as determined by the measurement tool, and the true edge location, is critical for accurate measurements. All measurement algorithms, in one way or another, compare the centerline (or the intersection of two orthogonal centerlines) of a substrate layer to the photoresist layer. In addition, certain a-priori facts are known about the "ideal" target that can be used to interpret the manifestation of that target for a given substrate and target design. Use of such facts in the measurement process can lead to improved measurement accuracy, even in the absence of certified standards.

Starikov et al.[23] describe an algorithm that takes advantage of the inherent symmetry and redundancies present in overlay target design. Since target designs follow standard dimension practices, such information can be used to assess any asymmetries resulting exclusively from tool or process. Figure 6.26 illustrates the manner in which symmetry and redundancy considerations can aid measurement interpretation. By comparing the three separate estimates of centerline in Fig. 6.26, CL_1, CL_2, and CL_3, according to the redundancy relationship (RED) and the symmetry relationship (ASY) it is possible to determine which overlay measurement results contain a high degree of error due to target asymmetry. The RED assessment looks at two separate estimates of target centerline associated with different target edges. Errors in the RED assessment will result from random effects (film grains, voids, precipitates, and loss of edge contrast due to advanced planarization), target design errors, and target distortions. The RED assessment for a perfect target will be zero, due to the target design.

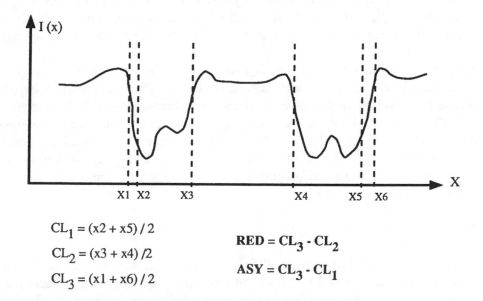

$$CL_1 = (x2 + x5) / 2$$
$$CL_2 = (x3 + x4) / 2$$
$$CL_3 = (x1 + x6) / 2$$

$$RED = CL_3 - CL_2$$
$$ASY = CL_3 - CL_1$$

Fig. 6.26 Target symmetry and redundancy (after Starikov[23]).

The ASY calculation compares two centerlines determined from different points along the same edge. By comparing these two centerlines, any effects due to variations in edge formation across the target will be highlighted.[26] ASY errors will result from process-induced asymmetries similar to those discussed in Sec. 6.3.3.1. These figures of merit (ASY, RED) can be used in a target optimization study to determine process-induced inaccuracies (WIS) for a given target design and process sequence. In a production mode this information can be compared against specified control values and used for data culling.[26] This information can also be used to assess the impact of process-induced asymmetry on the alignment marks used by lithography tools[23] since the same factors will induce similar asymmetric behavior.

Another algorithm that has been proposed to address process variations is a modified intercorrelation algorithm.[49] This algorithm uses all of the data contained within a defined region encompassing the actual edge, rather than just a single threshold point along the maximum slope. In this way there is an advantage for accuracy, since the whole waveform will be used for centerpoint determination and will tend to average over any low-contrast segments or sections of the waveform containing noise (e.g., due to metal grains). Reducing the sensitivity of the measurement tool to low contrast, grainy, or other poor signal-to-noise factors improves measurement accuracy over simple threshold techniques. A further, unexplored, benefit of this technique is use of the intercorrelation data to judge the symmetry of the target itself. In a well-formed highly symmetric target, the output of the intercorrelation is expected to be well behaved. Based on this, it is conceivable that a library of intercorrelation responses to varied target designs over all substrate and resist combinations could be used for target design optimization much in the same way that the ASY and RED responses can be used.

Finally, of critical importance to all measurement algorithms is final-measurement waveform construction and edge determination. Pixel interpolation beyond simple two-pixel interpolations is required, given that edges need to be known to better than 10 nm. As stated previously, typical minimum pixel size in measurement mode is on the order of 100 nm. At a minimum, multipixel regression interpolations should be utilized for edge determination and optimized for minimum error.[24] Advanced proposals have been made for improvements, and work is ongoing in this area.[50]

6.3.3.3 Overlay target design

Another avenue used to address WIS is overlay target design. This process is intimately connected to the measurement algorithm, tool setup, and process conditions. Examples of typical box-in-box targets are shown in Fig. 6.27. Typical target dimensions are 10–15 µm for the inner box and 20–30 µm for the outer box. Within each of the designs shown in Fig. 6.27 are a number of possible combinations of segmented frame outer, solid box inner, and so forth, which must be considered in relation to the amount of information derived from the structure, relative to the specific process layer.

When evaluating target designs, in addition to the differing edge profiles, it is also important to optimize the contrasts achievable with each of the structures by evaluating different phases of the inner and outer boxes. Substrate trench structures will provide, in many instances, vastly different contrast than mesa structures. In addition, the choice of contrast "tone" of the structure (e.g., resist inner box/substrate outer box vs. substrate inner/resist outer) will produce different measurement results. Typically the resist defined-box provides well-defined edges with little asymmetry (in the absence of proximity effects), whereas the substrate-defined box is subject to process variations (grainy materials, reflectivity variations, etc.). The best structures tend to be those that provide the greatest amount of redundant information for processing by the measurement algorithm, similar to those structures shown in Figs. 6.27(b) and 6.27(c). The structure of Fig. 6.27(c) has an additional advantage over the other structures shown in that the missing corners make this structure less susceptible to resist flow or viscosity-related buildup in sharp box corners.

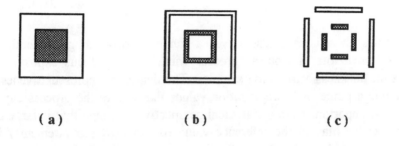

(a)   (b)   (c)

Fig. 6.27 Typical automated overlay target designs: (a) pad-in-pad, (b) frame-in-frame, and (c) segmented-frame.

6.3.4 Overlay Discussion

While it is beyond the scope of this section to rigorously define overlay measurement statistics, it is essential that the characteristics of a sound measurement setup be discussed, both in term of the physics and statistics of the problem and framed in the context of new process technologies. The basic question of overlay calibration and equipment optimization is one of minimizing TIS, WIS, and pixel-scale error through rigorous measurement evaluations. In the case of TIS and pixel-scale, repeated measurements at two wafer orientations are required, whereas overlay measurements over a range of structure designs (processed as identically as possible) should be performed for WIS. While different quantities are being measured, the basic statistical design can be the same in all cases.

6.3.4.1 Measurement optimization

Several factors are of critical importance when establishing a measurement optimization experiment: the range (misregistration) of the measurements evaluated, the homogeneity[51] of the variance over the defined measurement range, the partitioning of variance within the design, the degrees of freedom associated with each of the variance components, the representativeness of the samples to be used, and the level of understanding of all equipment and process assumptions. A brief discussion of each of these areas follows.

It is important when performing a tool optimization study that the samples chosen be representative of the process to be monitored. As a result, it is desirable to use material that has been through full processing to the point of the overlay measurements. Partial or short-loop lots will not contain the same topography nor appear optically similar as the fully processed material. To optimize on material that is not as close to product as possible is to run a real risk of improper tool setup.

The samples chosen must also cover a range of processing, preferably from several different lots or from a "process window" (e.g., focus exposure matrix and/or film thickness variations) samples. The degree of representativeness must also include a range of misregistration values that covers the process capability. This is also important from a statistical perspective. A range that is large enough to minimize the bias in the reference value (0° oriented measurements) for the linear regression[52] must be chosen for the experiment. Typically a number on the order of 25 times the measurement precision (1 sigma) will result in a bias in the estimate of the reference value that is less than 1%.

The experimental design must also address the analysis of variance and contain a statistically significant sampling plan. The way in which certain system or process factors contribute to the measurement variance and the best grouping for extraction in the analysis must be considered. Measurement precision is most often divided into three components: static, dynamic, and long term. In designing an overlay optimization experiment, the measurement equipment processes associated with each of these areas of precision must be defined. For example, within the rubric of static precision, illumination adjustment (contrast/brightness), focus, image acquisition, and measurement functions are often included. Dynamic and long-term precision measurements typically incorporate stage, long-term illumination variations, wafer placement, and pattern recognition accuracy.

The ability to assign causes to observed variance facilitates measurement optimization by allowing focused corrective action to take place. The test design needs also to assess the statistical behavior of the measurement system over the entire range of misregistration in the sample. The measurement statistics need to be robust enough to estimate the precision at each of the interval sampling points to determine the homogeneity of variance.[51] Poor homogeneity of variance, since it disturbs the calibration in a localized point in the range of measurements, can be misinterpreted as a systematic accuracy issue and may result in miscalibration of the measurement tool.

The final component of the measurement setup program involves test structure design. Unfortunately, there does not exist a single structure design that is optimized for all layers within a given process. It is often necessary to evaluate several target designs for precision and accuracy. An assessment of the process influences occurring for each measurement is critical and must lead to overlay target designs that address specific process-induced accuracy issues. Identification of the specific process asymmetry can also lead to a target design to address the accuracy issues. For example, the process asymmetry shown in Fig. 6.24, called the snowdrift effect, was resolved by a complex data analysis routine that used modeled stepper lens performance over the entire wafer to remove measurement outliers. An alternative approach would be to redesign the overlay target to incorporate symmetry and redundancy, similar to the design of Fig. 6.27(c), and utilize a measurement algorithm that was capable of assessing and reporting target asymmetries at individual measurement sites.

6.3.4.2 Current issues in overlay registration measurement

The increasing density of DRAM and CMOS logic, with both increased demands on a photolithography process whose process window has decreased (smaller DOF, larger field size, etc.) and an ever-increasing number of interconnect levels that are required to meet the needs of these advanced processes, has given rise to new planarization technologies. Most prominent among these new planarization technologies is chemical mechanical polishing (CMP). CMP, as the name suggests, is a combination of chemical etch and mechanical polishing and it is currently used for planarization on polysilicon, oxide, tungsten line, tungsten stud, and shallow-trench isolation levels, with more levels anticipated in the future.[53] The impact to overlay metrology with this process is a sharply reduced target contrast, as the planarization process produces wafers with less topography and potentially new forms of target asymmetry that result from the interaction of the CMP process with specific target designs.

The photos in Fig. 6.28 show the extreme loss of edge contrast that can occur in CMP. In order to successfully design targets that can survive the CMP process with a reasonable contrast remaining, it is necessary to understand the weaknesses of the CMP process and use these to one's advantage. CMP is most effective[53] when polishing small isolated structures, whereas with wide features (due to rounding/dishing) and dense areas (due to erosion) CMP is much less effective. Moreover maximizing substrate contrast through the use of mesas or trench in the target structure can make a difference on specific layers. As CMP becomes further integrated into process planarization technology it will necessitate a complete rethinking of approaches to stepper alignment and overlay measurement with respect to both target design and measurement algorithms.

Fig. 6.28 Effect of CMP on nonoptimized overlay targets.

6.3.4.3 Process control

The purpose of overlay measurement in the process sequence is to monitor the performance of the photolithographic alignment process. Typically, wafers are sampled from each lot to statistically assess the overlay performance of the lot. Overlay measurement data is also used in the setup of the lithography process, in terms of initial (and periodic) optimization of step-and-repeat lithography systems (steppers) and, during the process development stage, for setting stepper parameters on a lot-by-lot basis using send-ahead or pilot wafers. As discussed, overlay measurement errors can be attributed incorrectly to the stepper and result in poor overlay performance.

Using the overlay measurement system in a process control function requires a keen understanding of the measurement results, the associated uncertainty, and their implications for other process equipment. Specifically, the relationship between sampling plan and data distributional assumptions must be well understood. Assumptions of normally distributed overlay data must be carefully examined, as it has been shown, based upon overlay errors experienced with steppers, that this assumption is often not applicable.[54] The distribution of overlay measurement data is a function of the sampling plan. If, for example, data points are taken from two different points within a stepper field, the overlay data will be a bimodal distribution. This is a direct result of the stepper lens effects (rotation, magnification, and higher-order terms) that are a function of position within the field. Use of overlay measurement data to disposition product must account for sampling assumptions.

Finally, in a production environment where many systems are used to measure product, and dedication of measurement systems to specific layers is highly undesirable, it is essential that systems be properly matched to ensure consistency across all overlay tools. Methods for matching overlay metrology tools include corrected-mean (CM) matching, regression of raw data across systems,[55] golden standard matching, and mean deviation[56] matching. Another option proposed for accurate manufacturing measurements without requiring as extensive a measurement characterization and equipment/target optimization is simply to measure at both 0° and 180° orientations and calculate a corrected-mean value for the overlay registration[24] where

$$CM_x = \frac{X_0 - X_{180}}{2} \qquad (6.09)$$

$$CM_y = \frac{Y_0 - Y_{180}}{2} . \qquad (6.10)$$

In theory, one would expect to do away with the nuisance of all tool inaccuracies merely at the expense of throughput. However, recent data[28] indicate that this is not necessarily the case; the CM value was observed to be a function of the tool setup and its interaction with the process and target.

6.4 LINEWIDTH METROLOGY

Just as lithographic advancements have tightened layer-to-layer overlay tolerances, decreasing linewidth tolerances on "critical" levels (e.g., gate lengths) have come to place increasingly greater demands on linewidth measurement equipment. This is readily apparent when one considers the evolution of device gate lengths and their associated measurement tolerances against successive generations of DRAM technology. As shown in Fig. 6.29, current device gate lengths are in the 300–400 nm region, with a corresponding image size tolerance (i.e., total uncertainty) of 30–40 nm. This requires a tool with 3–4 nm uncertainty, based upon a 10% gauge capability. The uncertainty represents the total process linewidth budget, including all sources of variations in the imaging process, i.e., pattern proximity effects (iso-dense biases), stepper variations (focus variations/lens aberrations, etc.), resist image size changes caused by apply and/or develop nonuniformities, together with the uncertainties contributed by the measurement tooling.

The linewidth measurement tool is expected to both quantify the magnitude of linewidth variations generated from each source, as well as verify their reduction when process improvements are implemented. Historically, optical linewidth metrology was successfully applied when critical geometries exceeded 1 μm. However, as feature sizes approached 1 μm, it was soon learned that the size and shape of a structure itself affected the precision and accuracy of the optical measurement tool. This was a painful lesson for lithographers and metrologists alike.[6,14]

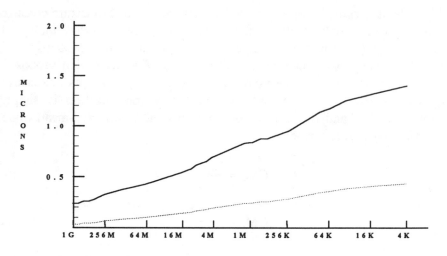

Fig. 6.29 Device linewidth progression.

Nevertheless, this situation provided a unique learning opportunity for the semiconductor industry. As metrologists and photolithographers scrambled to understand how diffraction and thin film effects influenced submicron optical image formation and linewidth measurements,[14] it became apparent that the "width" of a lithographic feature was subject to interpretation. This is illustrated in Fig. 6.30(a). Lithographers use the term linewidth unambiguously as the dimension of the resist at the resist/substrate interface (see Chap. 1 of this handbook). As simple as this sounds, it is not necessarily clear how to extract this information given the information available in most measurement systems, e.g., an optical intensity distribution, a SEM electron micrograph, or a scanning probe waveform. This situation has been documented in the SEMI standard linewidth definition, which accounts for width variations as a function of both measurement technique as well as position on a lithographic feature, and is illustrated in Fig. 6.30(b).[57]

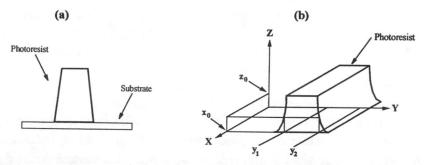

Fig. 6.30 (a) Idealized resist image; (b) SEMI standard definition for linewidth. Linewidth is given by the quantity $(y_2 - y_1)$ defined at the distance x_0 along the resist line and at height z_0.

In optical lithography, it is the lithographer's responsibility to produce resist images that meet specified manufacturing tolerances by transferring patterns from photomasks onto resist-coated substrates via light intensity distributions with ill-defined edges (i.e., aerial images; see Chap. 1 of this volume). Likewise, it is the metrologist's responsibility to determine whether these resist profiles do, in fact, meet the manufacturing tolerances by trying to correlate ill-defined intensity profiles with the true edges of the resist pattern. Thus, lithographers and metrologists have the same goal but work at the problem from different ends. Metrologists must determine where on a measurement profile/waveform correspond a feature's physical edges, e.g., the bottom. This problem is illustrated in Fig. 6.31. Herein lies the fundamental goal of linewidth metrology: to be able to measure the critical aspect of a structure in spite of possible changes in the measurement feature, e.g., width, thickness, sidewall slope, and substrate stack. Insofar as these parameters can affect measurements, it is critical that the metrologist be familiar with the lithography process as well as the process architecture in order to determine what processing factors might influence the metrology tooling.

Optical

SEM (I)

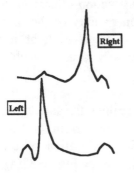

SEM (II)

AFM

Fig. 6.31 Measurement waveforms.

In the following sections, we will review different measurement techniques which have been applied to ever more complex photolithography problems, i.e., optical, electrical, and/or electron beam linewidth metrologies. In the case of scanning probe techniques, the application of the available instrumentation for serious "metrology" is only just beginning. For this reason, we will only briefly review the state-of-the-art of this approach.

6.4.1 Optical Linewidth Metrology

Although the measurement precision required on critical process levels of advanced semiconductor devices exceeds the measurement capability of most optical measurement systems today, there remain other applications, e.g., reticles, thin film heads, or flat panel displays, for which optical linewidth metrology is the technique of choice. This is based in part on technical requirements such as linewidths larger than .8 µm with patterned features less than ¼ λ, and in part by cost considerations. If an optical instrument can provide the required measurement capability, then it may offer a more cost effective solution than e-beam metrology. Therefore, the need remains to understand what factors affect this technique and how good measurements may be achieved. In addition, a firm grasp of the limitations of optical linewidth metrology provides a framework for understanding the limits and applicability of other linewidth measurement techniques. The following portion of this section will explore the application of optical metrology to the measurement of dimensions near 1.0 µm.

6.4.1.1 Diffraction and scattering effects on optical profiles

In the following discussion, we will concern ourselves with understanding how the determination of image size by optical metrology is affected by the diffraction of light. Optical diffraction is not a new concept and has been discussed in the context of imaging in Chap. 1[58] of this handbook. Diffraction effects in optical metrology become very important when the desired uncertainty of the measurement approaches, or becomes smaller than, the wavelength of light used to measure the feature.[59] Thus, diffraction effects can be very large in visible-light submicron optical metrology (i.e., 400 to 800 nm).

The complexity of the problem can be appreciated by recalling the simple example of Fraunhofer diffraction, i.e., the diffraction observed from an isolated transmissive slit in an otherwise opaque plane.[60] The orientation of the slit, together with its associated diffraction pattern, is shown in Fig. 6.32(a). The variation in light intensity as one might observe on a projection screen, together with its associated intensity distribution, is shown in Fig. 6.32 (b). As can be seen, there is no discernible "edge" to be found in this intensity distribution.

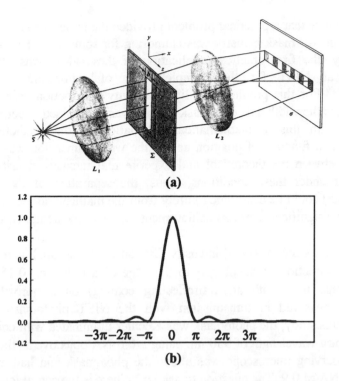

(a)

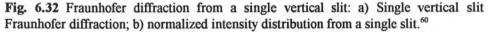

(b)

Fig. 6.32 Fraunhofer diffraction from a single vertical slit: a) Single vertical slit Fraunhofer diffraction; b) normalized intensity distribution from a single slit.[60]

For this simple example, the intensity distribution is given by Eq. 6.11:

$$I \propto \left(\frac{\sin z}{z}\right)^2 \qquad\qquad (6.11)$$

| Maximum Intensity | : | $z = 0$ |
| Minimum Intensities | : | $z = \pm\,\pi,\,\pm\,2\pi,\,\pm\,3\pi,\,...$ |

where z is defined to be equal to $\frac{\lambda \sin\theta}{D}$, λ is the wavelength of light illuminating the slit, D is the width of the slit, and θ is the angular displacement from the slit to a point of observation in the projection plane.

The intensity distribution of an edge as obtained with a microscope is further complicated by the components comprising the microscope (e.g., the coherence of illumination[25,61] and the numerical objective of the observing objective).[25,62] When these elements are added, the observed intensity distribution may or may not have the form described by Eq. 6.11.

A method for handling these complex situations was suggested by several researchers over the years.[14,18,61,63-68] However, it was not until the late seventies and early eighties that a practical problem provided the impetus to test these ideas. The semiconductor mask industry was clamoring for some sort of mask standard. Consequently, the first practical application of these ideas was attempted by researchers at NIST for the "simple" task of calibrating a photomask standard.[18,61,69] For this problem, the light intensity distribution as a function of position across the width axis of a mask feature was determined experimentally. In conjunction with this, a mathematical calculation of the predicted intensity distribution as a function of position across the width dimension was performed. Agreement between the theoretical and experimental intensity distributions was demonstrated under these conditions. Thus, the separation of the edges (i.e., physical width) could be determined purely from the mathematical calculation. To understand the significance of this achievement, consider the actual case.

Figure 6.33 shows the measured intensity distribution (solid line) as a function of position in a direction perpendicular to the edge of a thin (i.e., 0.15 μm thick) opaque chrome line (with an antireflecting coating) on a borosilicate glass substrate as observed in transmission with the NIST photomask calibration system.[18] In this study, the photomask was Koehler illuminated with coherent light from below at a wavelength of 530 nm using a 0.6-NA objective as the condenser lens. The observing microscope was above the photomask and had an objective lens with an NA of 0.9. The question to ask is: "Where is the actual location of the line edge on the measured intensity profile?" In the theoretical model, the chrome edge was assumed to have a vertical edge geometry. All other edges of lines and spaces on the mask were sufficiently removed from the edge of interest that overlap from the other diffraction patterns was assumed negligible. The open circles in Fig. 6.33 show the results of a theoretical calculation by D. Nyyssonen[18] of the diffraction pattern expected for the photomask specimen illuminated with the NIST photomask calibration system described above.[18] The theoretical curve has been normalized to 100% and has been shifted horizontally to best fit the experimental data. The fit is quite good, and since the position of the edge in the theoretical model is known, the edge on the theoretical curve can be taken to coincide with the physical edge of the line on the measured profile. For this example, the physical edge corresponds to the 25% point on the experimental profile.

For the example illustrated in Fig. 6.33, it is important to understand how the components of the microscope may affect the calibration threshold point. R. Larrabee[59] describes the situation in the following way. The NIST photomask calibration system is coherent over the effective area seen by the microscope. For this reason, if the opaque chrome line were to cover half the area seen by the microscope, then the electric field would be cut down by a factor of two and the corresponding intensity ($\propto E^2$) would be reduced by a factor of four. On the other hand, if the illuminating light were incoherent over this area, then the intensities (rather than the electric fields) would add and the edge would be at the 50% point. If the coherence of the light were unknown, the actual edge would lie somewhere between the 25% and 50% threshold points. For the current example (Fig. 6.33), the edge-location uncertainty due to a partially coherent light source could contribute as much as .08 μm per edge of uncertainty in the position of the line edge, depending on the degree of partial coherence. Note the linewidth uncertainty for this example would be twice this value.

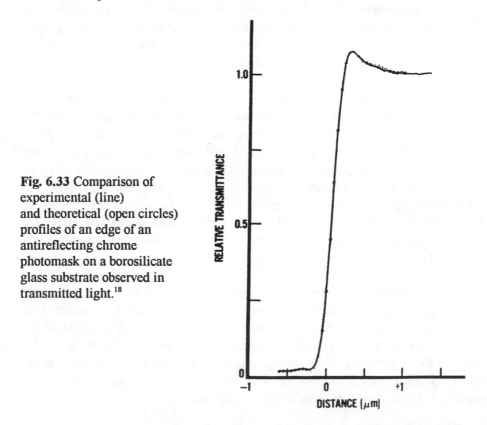

Fig. 6.33 Comparison of experimental (line) and theoretical (open circles) profiles of an edge of an antireflecting chrome photomask on a borosilicate glass substrate observed in transmitted light.[18]

The ability to determine an "accurate" linewidth on photomask features, i.e., opaque lines and clear spaces, was a tremendous advancement for the lithography community. It became possible for users to calibrate their own optical measurement systems using these NIST traceable standards. In order for the calibration to be valid it was not necessary to use the same experimental setup as the NIST system. Instead, it was sufficient for the user to calibrate the system using an arbitrary method of edge detection, i.e., not necessarily "correct." However, to be able to perform calibrated measurements on other photomasks it was necessary for these samples to behave optically like the NIST photomask standard, i.e., antireflecting chrome on glass. Insofar as the mask or reticle specimen deviates optically from the NIST standard, the uncertainty in the measured values will increase.[70]

As much as this work was a boon for the mask-making industry, it opened a Pandora's box for the lithography community. Once these methods had been applied successfully to calibrating photomask measurement systems, it was natural to try and extend this work to the more difficult problem of determining the "real" width of actual resist structures imaged on the complex topography created on silicon surfaces. It is fair to say that the ensuing exchanges between the standards community and the lithography community were not always constructive. The complexities of generating "standards" in the same fashion as the photomask standard were not always appreciated by the lithography community, who were themselves "compelled" to fabricate these structures whether or not they knew the precise dimensions.

The technical complexities cited by the standards community to generate a general-purpose linewidth standard were well founded. It was known that the intensity distribution in optical diffraction patterns is dependent on several characteristics of a feature, e.g., the optical constants n and k as well as the physical width and thickness of a structure.[71,72] In addition, the optical components and characteristics of the measurement system[25,73-77] can also affect the observed intensity distribution, e.g., the illumination optics,[25,62,68] as well as, different types of lens aberrations.[25] Consequently, the available thin film photomask standards could not be used to calibrate systems used to measure features patterned in thick layers, i.e., resist.[78]

Since most applications of interest to lithographers (and therefore metrologists) involve features that would be considered thick, it is useful to consider how thickness and edge profile affect the image intensity profile. An example of the effects of specimen geometry on the image profile is shown in Fig. 6.34. In this example,[14] the dependence of the image profile was investigated by varying the physical parameters of a 6-μm-wide and 0.6-μm-thick polysilicon line on an unpatterned silicon dioxide film on a silicon substrate.

The theoretical intensity profiles are shown superimposed on the polysilicon edges in Fig. 6.34. These profiles were computed for the case of imaging in reflected light using narrow cone-angle illumination (0.17 NA) with a wavelength of 530 nm and an imaging objective with NA of 0.85. It is instructive to compare these intensity profiles with those obtained for the photomask standard (Fig. 6.33). The calculated width of the polysilicon intensity profiles can be seen to be much larger than for the photomask case. In addition, the shapes are significantly more complex, i.e., more maxima and minima. The complexities of this intensity profile arise from additional optical effects introduced because of the thickness of the polysilicon within the film stack.[79] When a feature is thick,[78] the interaction of the phase of the electromagnetic fields at the feature interfaces become more complex and, in fact, significantly affect the magnitude and shape of the optical intensity profiles. For this reason, the term "diffraction" is reserved for thin, one-dimensional, features and the term "scattering" is reserved for thick, two- or three-dimensional, features.

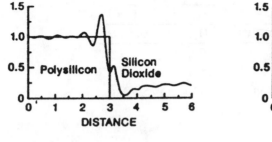

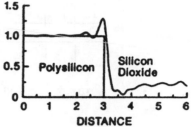

(a) Vertical poly-Si edge
poly-Si thickness = 6000 nm
SiO₂ thickness = 1050 nm

(b) Vertical poly-Si edge
poly-Si thickness = 6500 nm
SiO₂ thickness = 1050 nm

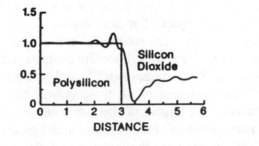

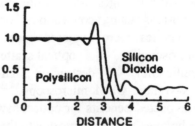

(c) Vertical poly-Si edge
poly-Si thickness = 6000 nm
SiO₂ thickness = 1250 nm

(d) Non-vertical poly-Si edge
poly-Si thickness = 6000 nm
SiO₂ thickness = 1050 nm

Fig. 6.34 Effect of specimen geometry on calculated image intensity profiles.[14]

The origin of the phase change may be understood by using simple thin film interference arguments (actual calculations of the intensity profiles are obviously more complicated). The physical origin of the phase changes are shown in Fig. 6.35. As can be seen, the phase of transmitted or reflected light from a structure with varying refractive index (i.e., thin film stack of various film types) varies with both the wavelength and thickness of the film. The intensity of the light (whether transmitted or reflected) is now a function of both the wavelength of the incident light as well as the thickness of the individual films in the film stack.

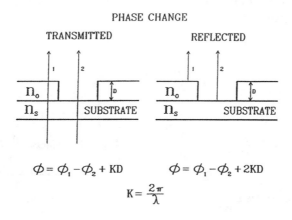

$$\phi = \phi_1 - \phi_2 + KD \qquad \phi = \phi_1 - \phi_2 + 2KD$$

$$K = \frac{2\pi}{\lambda}$$

Fig. 6.35 Optical phase change at thin film boundaries.

Scattering in combination with thin film interference effects have significantly limited the general application of optical linewidth systems in the submicron region. In the development of submicron processes (as was actively pursued in the 1980s), it became nearly impossible to deconvolute image profile changes created by width changes from profile changes created by film thickness or resist profile variations using conventional optical techniques. For this reason alternative techniques were explored such as confocal and coherence probe linewidth microscopies,[75,76,77,80] optical scatterometry,[81,82] electrical linewidth methods (where applicable),[83,84,85] destructive high-voltage SEM microscopy[86,87] and, eventually, low-voltage SEM microscopy.[88,89,90] The outcome of all these efforts has had unexpected benefits for optical metrology. The pursuit of alternative measurement techniques provided impetus to the manufacturers of optical tools to design better, more sophisticated instruments.[24,91,92,93] In addition, the metrologist learned how and when to better apply optical microscopy. The improvements in optical overlay discussed previously in this chapter illustrate this.

6.4.1.2 System and material effects on optical waveforms

In the previous section, optical diffraction (one-dimensional effect) and optical scattering (two- or three-dimensional effects) were shown to substantially influence linewidth intensity profiles. This section is intended to illustrate how optical system parameters and the material properties of the specimen affect optical intensity profiles and therefore contribute to linewidth measurement errors.

As stated in the preceding section, optical intensity profiles exhibit complex forms at or below 1-μm feature sizes. Several parameters of the optical system and the specimen materials can affect the shape of the intensity profile. Some of these parameters are listed in Table 6.2. Many of the system parameters listed in Table 6.2 have already been discussed in the context of overlay metrology. It is not surprising that both optical linewidth and overlay metrology require the control and characterization of many of the same system parameters.

Table 6.2 System/material parameters affecting image profile.

System	Material
Illumination	Reflectance
Spectral Bandpass	Optical Constants, n & k
Resolution	Film Thickness
Coherence	Texture/Grain Size
Aberrations	
Focus	
Flare	

Kirk[25] has modeled and studied the effects of illumination, lens aberrations, and focus offset on image profiles in order to quantify the contributions of these parameters on dimensional measurement errors. To simplify his calculations, Kirk used a simple test object, a chrome photomask, which was assumed to create an image profile with zero intensity over the width of the line and maximum intensity (e.g., unity) outside the line. Assuming an aberration-free lens, Kirk calculated the position of the edge positions of the line as a function of defocus and illumination coherency. This is shown in Fig. 6.36.

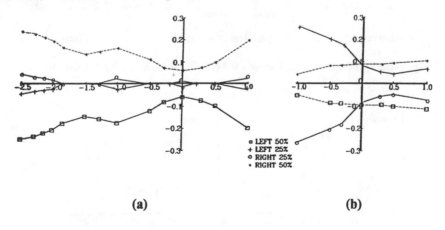

(a)   (b)

Fig. 6.36 Edge position errors (Y) as a function of defocus (X): (a) coherent illumination and (b) partially coherent ($s=\frac{2}{3}$) illumination (taken from Ref. 6.25).

With coherent illumination [Fig. 6.36(a)], the "true" edge of the line object lies at the 25% threshold. It remains at this position for defocus values as large as 1 μm. However, when the illumination is partially coherent [Fig. 6.36(b)], the 25% edge position now changes with defocus. As shown, however, the edge position determined from the 50% threshold does not change with defocus. Thus, in a practical measurement system with some degree of partial coherence, the best measurement repeatability would not necessarily be obtained using the most "accurate" edge threshold setting, i.e., 25%.

Spherical aberration, coma, and astigmatism are responsible for a lack of "sharpness" in an image.[94] The effect of these aberrations on image profile was modeled and experimentally verified by Kirk.[25] The results of some of this work are shown in Fig. 6.37. A theoretical image profile computed for a measurement system with no aberrations is compared with profiles computed for a system with spherical and comatic aberrations. Spherical aberrations reduce the amplitude of the fringes at the edges of the line, while coma produces asymmetry in the fringes of the edge profiles.

(a)

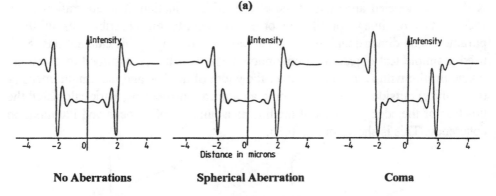

Theoretical profiles calculated using a 0.9 NA Objective and a 0.2 NA Condenser

(b)

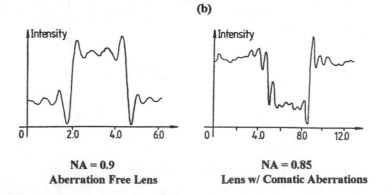

Fig. 6.37 Image profiles of an iron oxide line.[25]

These aberrations affect the determination of image size in different ways. Spherical aberrations decrease the slope of the measurement image profile. For measurement systems that use a steepest slope focusing algorithm, this can lead to poor measurement repeatability. For the case of comatic aberrations, the presence of asymmetric image profiles can lead to significant problems if the edge detection algorithms are not versatile enough to handle the asymmetries present in the profile (see following discussion).

The observation of asymmetric profiles, whether the result of a physical asymmetry or as a result of comatic aberrations, illustrate the importance of identifying a reproducible point on an image profile in order to perform repeatable measurements. The most common methods used in optical metrology are illustrated in Fig. 6.38. As can be seen, using a fixed-threshold edge detection method on an asymmetric profile could result in large measurement errors, depending on the severity of the asymmetry observed in the image profile.[95]

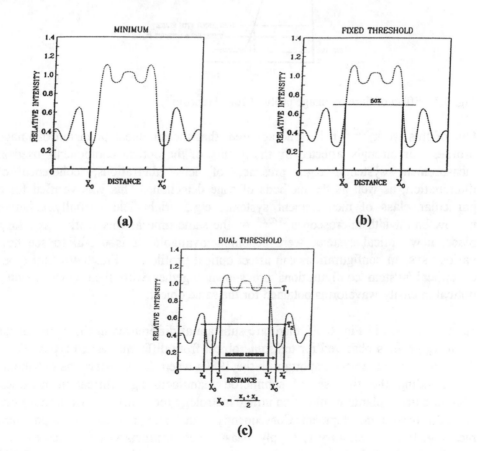

Fig. 6.38 Edge detection methods: (a) Minimum, (b) fixed threshold, and (c) dual threshold (from Troccolo[95]).

For image size determination, it is also necessary to include a discussion of flare. For the purpose of this discussion, flare is defined as any non-image-forming light that adds an incoherent background to the illumination system. Flare may or not be significant, depending on the edge detection method used to determine image size. If an absolute threshold algorithm is used to determine the position of the edge, then detected edge position will shift in the presence of flare, causing an error in the determination of linewidth. This is illustrated in Fig. 6.39.

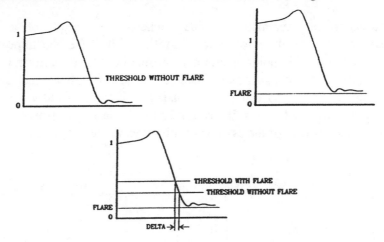

Fig. 6.39 Effect of flare on image profile (from Troccolo[95]).

Out of this work[25,62,68] it became apparent that measurement precision of image profiles was strongly affected by the quality of the optical components used in measurement systems, e.g., presence of lens aberrations, coherence of illumination, as well as the methods of edge detection. This was verified for a particular class of measurement systems, e.g., bright-field partially coherent narrow bandwidth microscopes.[25,62,68] At the same time as this work was taking place, new optical systems were becoming available. It is useful to see how various system configurations can affect optical profiles. In Fig. 6.40, four types of optical system configurations are shown together with their corresponding optical intensity waveforms obtained for the same object.

As can be seen in Fig. 6.40, there are substantial differences in the shape of the intensity profiles observed for the *same* object from different system types. These differences became very important in the 1980s when the industry was expanding and building the first set of submicron manufacturing fabrication facilities. Manufacturing plants would often utilize metrology tools different from those that had been used in development. Consequently, transferring a process took on a new meaning. It was necessary to supply measurement artifacts when processes were transferred. It is little wonder that the industry was clamoring for an NIST traceable optical linewidth standard.

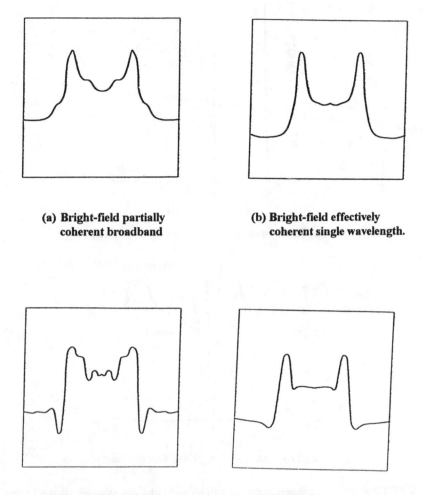

(a) Bright-field partially
coherent broadband

(b) Bright-field effectively
coherent single wavelength.

(c) Focused laser beam

(d) Confocal

Fig. 6.40 Optical responses of different system types on the same measurement artifact (from Troccolo[95]).

Unfortunately, different tool sets were not the only problems facing optical metrologists. Specimen thin film variations were also observed to have profound effects on measurement waveforms. This is not surprising, given what we have seen in the previous section. However, different systems were found to produce significantly different optical responses. An example of these effects is illustrated for the case of two optical systems, bright-field and confocal, in Fig. 6.41.[95] From this example, it is clear that measurement artifacts could only be applicable if they were reproduced over the acceptable process range and characterized on both the reference or development system and the manufacturing measurement system.

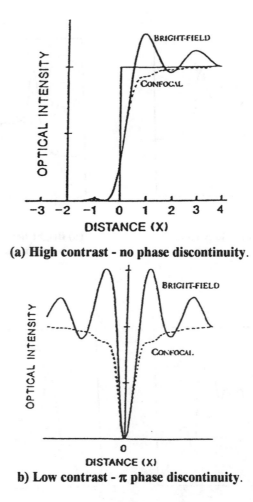

(a) High contrast - no phase discontinuity.

b) Low contrast - π phase discontinuity.

Fig. 6.41 Thin film waveforms for bright-field and confocal systems (from Troccolo[95]).

Given the variation of optical waveforms with both the specimen and the measurement system, it is not surprising that the semiconductor industry searched for alternative measurement techniques to either support or replace the optical measurement tools which were in use in the mid to late 1980s. There was much interest generated in more sophisticated optical techniques such as confocal and coherence probe. However, the confidence level on any optical measurement tool would always be tainted by the experiences of the 1980s. This loss in confidence was fueled in large part by the availability of commercial electrical linewidth test equipment.[96] Electrical linewidth metrology provided the means to directly correlate linewidth measurement test structures (gate lengths) with actual device performance (speed). Electrical linewidth metrology provided metrologists the opportunity to investigate alternative measurement techniques for process control and characterization.

6.4.2 Electrical Linewidth Metrology

"Four-point probe electrical linewidth measurement has been an important technique for submicron lithography because of its inherent high precision and is, perhaps, the only means to detect the subtle linewidth changes caused by various system or processing components."[97] The main drawback of electrical metrology is the slow turnaround time required to etch, strip, and probe the wafers. Although it does not give direct information about the photolithography step itself, electrical metrology is broadly utilized in semiconductor manufacturing to diagnose, characterize, optimize, and control semiconductor equipment and processes. The SEMI standard definition[57] of linewidth (Fig. 6.30) documents that the *physical* width of a feature is defined as the separation of edges given by the quantity (y_2-y_1), at a specified distance along the feature (x_o), and at a specified height (z_o). In contrast, the electrical width is defined as the effective conductive path width of a patterned, conducting film whose length is typically much larger than its width. Ideally, the *physical* linewidth of a line with perfectly vertical edges, uniform thickness, and uniform conductivity would equal the *electrical* linewidth measurement.[98]

The electrical test structure is designed to provide very precise and relatively fast determination of the average electrical linewidth (W in Fig. 6.42) of a conductive film using either an automated test system or a manual prober. A typical pattern consists of two types of four-point Kelvin structures: a van der Pauw sheet resistor and one or more bridge resistors. An example of the cross-bridge test structure is shown in Fig. 6.42. W is the electrical linewidth determined from this structure by measuring the sheet resistance of the van der Pauw resistor and the resistance of the bridge resistor of known length. As Linholm et al.[104] have discussed, current is forced through a section of the test structure and the potential drop between two points along that section of length is measured.

For the structure illustrated in Fig. 6.42, Linholm et al.[104] have shown that the electrical linewidth is given by:

$$W = \frac{R_s L}{R_b},\tag{6.12}$$

where W is the electrical bridge width, R_s the measured sheet resistance, L the design length of the bridge resistor, and R_b the measured bridge resistance. The sheet resistance is calculated using:

$$R_s = \frac{\pi}{2}\left\{ \frac{|V_{25}^+| + |V_{25}^-| + |V_{23}^+| + |V_{23}^-|}{4I} \right\}\tag{6.13}$$

where V^+_{25}, and V^-_{25} are the voltage drops measured across pads 2 and 5 when current I is alternately forced between pads 3 and 4. V^+_{23} and V^-_{23} are the voltage drops measured across pads 2 and 3 when current I is alternately applied across pads 4 and 5.

The bridge resistance is calculated using:

$$R_b = \frac{|V_b^+| + |V_b^-|}{2I_b} \qquad (6.14)$$

where V^+_b and V^-_b are the voltage drops measured across pads 5 and 6 when current I_b is forced alternatively between pads 1 and 3.[98]

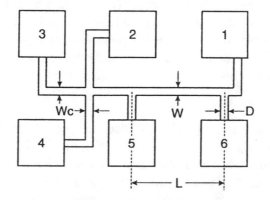

Fig. 6.42 A cross-bridge resistor electrical linewidth test pattern.[104]

In general, no significant current passes through the voltage taps, and therefore the measurements are not affected by resistive losses due to the interconnects or the points of contact. However, electrical test structures are often used to evaluate the resolution limits of the imaging systems and other process steps. In this case, there may be problems in resolving orthogonal lines due to lens aberrations or other imaging nonlinearities. In the sub-500-nm regime, it is not uncommon to resolve linewidth structures but fail to resolve the voltage tap connections. This may be eliminated by increasing the width of the voltage tap. Troccolo et al.[85] have shown that errors higher than the resolution of the measurement system will not be introduced by increasing the tap width by 20% for bridge structures with design widths less than 0.70 μm and lengths greater than 100 times the design width.

The magnitude of the currents used will be individually determined based on the resistivity of the material forming the test structure. The current should be sufficiently small to avoid heating effects. To determine if Joule heating is present, the forcing current is halved in a resistance measurement of the van der Pauw. If no significant change in the resistance is detected, then the original current is acceptable.[99] All resistors must be ohmic in the current ranges used. At the same time, the current must be large enough to induce a significant potential drop across the length of the sheet traversed. Shorter lengths will require larger currents.

"Cross-bridge resistor test structures should be designed such that the overall measurement uncertainty is limited by the electrical test system, not the design of the test structure."[98] The general design criteria for a cross-bridge resistor test structure are given in Table 6.3.

Table 6.3 General criteria for a cross-bridge resistor test structure.

Parameter	Value	Restrictions
W_C : Width of Van der Pauw	≥ 10 μm	
D : Width of Taps W : Width of Bridge	$D \leq 1.2\,W$ $D \leq W$	$W \leq 0.70$ μm and $L \geq 100\,W$ $L \leq 100\,W$
L : Length of Bridge	$L \geq 15\,W$	or 80 μm, whichever is larger

Variations of the basic design of the cross-bridge test structure have provided valuable process and tooling information. As one example, the bridge resistors can be surrounded by dummy lines to emulate the process bias due to proximity effects. On submicron structures containing densely packed dummy lines, the interconnects should hook up to the bridge resistors in a perpendicular fashion to prevent bridging, as shown in Fig. 6.43.

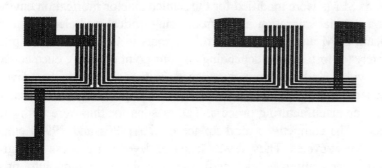

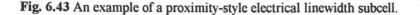

Fig. 6.43 An example of a proximity-style electrical linewidth subcell.

The pads can be arranged in a 1 × *n* layout for scribeline (or kerf) placement. This layout is illustrated in Fig. 6.44.

Fig. 6.44 An example of a 1 × *n* scribe line structure.

Contact size can be difficult to measure using any of the known techniques. B. Lin et al. introduced a technique for measuring average contact hole size by projecting these contacts onto metal lines and using an anisotropic etch to remove the exposed metal.[97] The technique was later verified by Lindsay et al. and the results showed good correlation to SEM.[100]

Electrical linewidth metrology continues to be an invaluable tool to semiconductor process engineers for numerous applications. The ability to use electrical test structures for process debug of critical tools (e.g., steppers) on critical process levels (e.g., gate layers) enabled the lithography community to progress past the measurement limits imposed by the available optical metrology tooling. Nevertheless, electrical linewidth metrology has limitations. It cannot provide a measurement of the physical width of an arbitrary structure and is usually limited to conductive process levels.

Therefore, as SEMs were modified for the semiconductor fabrication environment, there was a gradual migration away from using optical and electrical metrology tools to using SEM and electrical metrology tools in the development pilot lines. Unfortunately (or fortunately, depending on your point of view), once conventional optical linewidth methods were abandoned in the semiconductor development lines, it became nearly impossible to introduce an optical linewidth measurement system into the manufacturing process. The reasons for this were both pragmatic and political. The competitive marketplace of the 1980s and 1990s emphasized short development cycles. Thus, it was faster to develop a process on a given tool set (measurement equipment included) and transfer everything into production (pragmatic reason) than to delay a product introduction while development and manufacturing tried to deconvolute equipment/process interactions (political reason) on new equipment.

6.4.3 SEM Linewidth Metrology

In the last 10 years, SEM CD metrology has advanced significantly. It has progressed from a manual and destructive "reference" technique to become the standard CD measurement technique used today in high-volume wafer "manufacturing." Perhaps naively, optical metrologists were initially attracted to SEMs for the improved image that could be achieved with the short "electron" wavelengths. As an example, when an electron is accelerated across 1000 eV, the electron wavelength is only 0.4 Å. As a result, the limit of resolution of an electron microscope is potentially better than that of an optical microscope. More than any other parameter resolution was what motivated metrologists to pursue electron imaging. In this section, we will discuss what was required for the industry, both users and SEM manufacturers, to make the transition from optical-based measurement equipment to in-line metrology CD SEMs.

It should be recognized that SEM CD metrology is not a measurement panacea. Just as with optical microscopy, it is possible to be misled when SEMs are applied without understanding the limits and sensitivities of the instruments as well as the interactions of the electron beam with the sample. For these reasons, it is useful to compare the SEM and optical microscopes. As shown in Table 6.4,[101] the principle imaging advantages that the SEM offers over conventional optical techniques are high resolution and a large depth of field (i.e., focus). The resolution advantage as well as the elemental analysis capability clearly outweigh the less serious disadvantages, i.e., high vacuum and lower throughput, when SEMs were first introduced into semiconductor analytical labs. The clarity displayed in an electron micrograph is what led the semiconductor industry to embrace the SEM for inspection and metrology applications. However, the disadvantages, such as specimen/beam interactions, electron beam damage, and possible beam induced contamination, clearly gated the introduction of the SEM into the semiconductor manufacturing line for general applications, i.e., wafer-level inspection, or in process CD determination.

The practical resolution advantage offered by an SEM is affected by several factors, some of which are determined by system design (chromatic dispersion and spherical aberration), some by beam specimen interaction, and some by the sample interaction volume contribution to the detected electron signal. A schematic of a typical SEM is shown in Fig. 6.45. As can be seen, the microscope is made up of a source, apertures, condenser, and final lenses. The Weynelt cylinder encircling the source, focuses the electrons from the source to a point with a minimum cross-sectional dimension d (the crossover dimension). The electromagnetic lenses demagnify the crossover dimension as the electrons are accelerated down the column until the electron probe is formed and imaged on the sample. The electron probe is then raster scanned across the sample. Note that the spatial resolution of the microscope cannot be better than the diameter of the electron probe. As noted above, the sample interaction volume that contributes to the detected electron signal can further degrade the spatial resolution.

Table 6.4 SEM microscopy advantages and disadvantages over optical microscopy (from Postek[59]).

SEM Advantages	SEM Disadvantages
Potentially high resolution (0.8 - 200 nm)	High vacuum requirement
Excellent Depth of Focus	Lower throughput
Flexible Viewing Angle (0 to 60 degree Tilt)	Electron/Beam Sample Interactions
Readily Interpreted Image	Potential for Sample Charging
Elemental Analysis Capability	Lack of a Traceable Linewidth Standard
Minimum Diffraction Effects	Possible Beam-Induced Contamination
	Possible Damage due to electron beam

The high resolution observed in SEM micrographs can be misleading. As Postek[102] has observed, "The SEM image which is reproduced in the electron micrograph is an electronic representation of the sample, and only that. From even before the moment the electron comes into proximity to the sample, biases are introduced that can seriously influence any quantitative measurements made in this instrument, biases that had little or no effect on the original uses of the SEM (i.e., imaging or elemental analysis)." Thus, the successful application of SEM technology to linewidth metrology will depend on several factors: an understanding and appreciation of the SEM instrumentation, recognition that the sample and the electron beam interact in ways that may affect the accuracy or reproducibility of the measurements, and finally, a precise understanding of what needs to be measured.

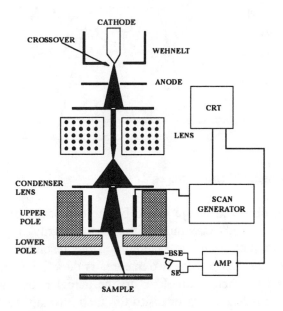

Fig. 6.45 Cross-sectional schematic of an SEM.

6.4.3.1 SEM Resolution/sharpness

Fully automated or semiautomated SEMs are currently in place on semiconductor manufacturing floors. In order for an automated instrument to routinely measure 250–500 nm features, the working resolution of the instrument should be 5–10 nm for accelerating voltages between 500 and 1000 eV. To meet these aggressive measurement goals, there have been significant improvements in the SEM instrumentation. Since resolution improvement has been one of the primary goals of SEM manufacturers over the last decade, it seems prudent to begin with a discussion of how resolution is evaluated in an SEM.

Many of the criteria used today to estimate SEM resolution have been carried over from optical microscopy. This was quite natural, given that SEM images were first recorded using optical film processing. In optical systems, the resolving power of an instrument is a measure of the ability to separate the images of two neighboring object points. For diffraction-limited optics — i.e., no aberrations — each image may be represented by a diffraction pattern. As the objects are brought closer together and these patterns overlap, it becomes more difficult to detect the presence of two images, i.e., separate the intensity maxima of each image. This situation is illustrated in Fig. 6.46.

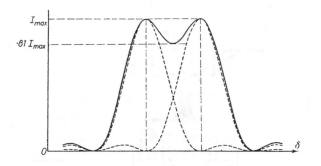

Fig. 6.46 Overlapping intensity fringes.[94]

In practice, the observed image would be the composite of both fringe patterns exhibiting a broad bright maximum with a central "gray" region. Quantifying the "gray" region is tantamount to determining the resolution. The limit down to which the "eye" can detect these two maxima is therefore somewhat arbitrary and may be affected by factors unrelated to the actual optics, e.g., image contrast (ratio of maximum to minima intensity for each intensity pattern), which can be affected by the object and the detectors, as well as the method of image processing. In order to be able to compare systems, optical microscopy has traditionally used the Rayleigh criterion to define the resolution of an optical system. According to this criterion, two images may be regarded as just resolved when the principle maximum of one corresponds to the first minimum of the other. This is illustrated in Fig. 6.46. For this example, the intensity distributions are of the form:[103]

$$\frac{I(\delta_i)}{I\max} = \left(\frac{\sin\left(\frac{\delta_i}{2}\right)}{\frac{\delta_i}{2}} \right)^2 \, , \qquad (6.15)$$

where $i = 1, 2$. The minimum contrast that can be "resolved" for this example is given by:

$$\frac{I_{\Delta\delta/2}}{I_{\max}} = \frac{8}{\pi^2} \sim 0.81 \, . \qquad (6.16)$$

The minimum distance $\Delta\delta$ between the two intensity peaks is then said to be the resolution of the optical system.

In principle, the optical criteria used to determine resolution can be easily applied to an electron microscope. Microscopists define the resolution of an SEM to be the minimum distance that an operator can distinguish between two "effectively" resolved points on an image. In practice, this can actually be very difficult to determine because the instrument, sample, operating conditions, and operator all contribute to the "achievable" resolution.

In order to minimize the dependence of the sample, it is common practice to evaluate resolution in an SEM by using a sample with intrinsically high image contrast. A typical sample will consist of gold particles evaporated on a carbon substrate. This sample gives excellent image contrast due to the high chemical contrast between gold (Z_{Au} = 79) and carbon (Z_C = 6); a typical micrograph is reproduced in Fig. 6.47. The minimum separation is measured between the particles of gold on an electron micrograph. The separation is converted to the actual distance in accordance with the recorded magnification. The minimum distance determined in this fashion is generally regarded as the SEM "resolution."

There are several problems with this technique. It is usually performed manually, directly on electron micrographs. Since it is known to be subjective in nature, it may be performed only during scheduled maintenance of the equipment. The problem with this situation is that the functional resolution of an SEM is not necessarily constant. Instrumental factors such as contamination within the column and beam misalignment, as well as tip degradation, can result in a loss of resolution. External factors such as vibration and electromagnetic stray fields, not to mention operator inexperience, can also degrade resolution. Moreover, resolution as defined here is an inadequate metrology metric because it fails to address image astigmatism. This characteristic is particularly relevant, since astigmatism tends to "wash out" edge definition. This is of critical importance in linewidth metrology, where the sole purpose is to determine the distance between two "edges." If the edges are not clearly defined, one can lose measurement precision, not to mention accuracy. For these reasons, SEM imaging performance cannot be adequately benchmarked using this type of subjective criteria. Therefore, there is a need to address the metric currently in place to evaluate SEM resolution by both users and equipment suppliers.

Recently, possible solutions to this problem have been discussed within the context of digital signal processing. Digital image processors have provided the means to perform image processing on electron micrographs. Although digital enhancements are not new for SEM microscopists, the application of new Fourier transform techniques for quantitative analysis is relatively new[104–107] When a Fourier transform is performed on an electron micrograph, one converts the information contained within the micrograph from the 2D contrast space to the 2D spatial frequency domain. Image "quality" may now be evaluated quantitatively. The low spatial frequency response is indicative of slow contrast variations on the micrograph, and the high spatial frequency response corresponds to "details" within the electron micrograph.[106,107,108] This approach is being investigated for several applications and may provide a more quantitative method to estimate both SEM resolution and astigmatism.[104-108]

As an example of this technique, Martin et al.[106] have performed Fourier transforms on resolution micrographs taken from gold-on-carbon specimens (see Fig. 6.47). The results of a Fourier transform analysis on a digitized micrograph are reproduced in Fig. 6.48(a). In this example, a minimum contrast may be defined to establish the spatial cutoff frequency (resolution) of the micrograph. When this technique is applied over the entire area of the micrograph, a contour shape of the resolution may be determined and astigmatism may be evaluated by comparing the X and Y cutoff spatial frequencies [Fig. 6.48 (b)].

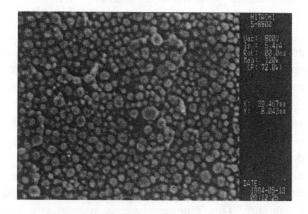

Fig. 6.47 Electron micrograph of typical SEM resolution target (Au/C).[106]

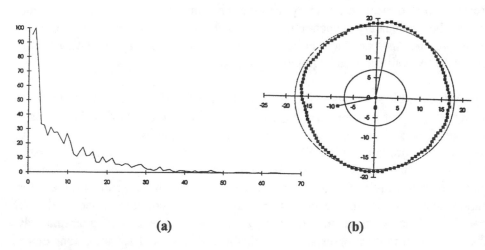

(a)   (b)

Fig. 6.48 Fourier transform of resolution electron micrograph.[106]

Finally, Postek et al.[107,109] have taken leadership positions to promote improved methods of instrument characterization for semiconductor applications. Realizing the need for improved testing and benchmarking procedures, Postek has developed a Fourier transform method to check and optimize the focus and astigmatism parameters of an SEM on a regular basis. Although not yet generally available, techniques such as those proposed by Postek will be required to ensure measurement capability is available to evaluate processes for 250-nm design rules. In conclusion, resolution is a critical instrument parameter for metrology applications that require continued monitoring throughout the working life of the instruments.

6.4.3.2 CD SEM System overview

The CD SEM used in high-volume wafer manufacturing today did not exist 10 years ago. Significant improvements in the areas of electron sources, lens designs, electron detectors, digital imaging, and system automation have provided the means to move SEMs out of the analytical laboratory and onto the manufacturing floors to produce state-of-the-art integrated circuits. These system improvements have not been achieved without an increase in cost to the end user. The increase in tool costs are not unexpected. Unfortunately, the expected improvements in performance may not be readily achieved if the user has not selected the appropriate technology for his/her application. It is necessary for users to make informed decisions when selecting equipment, and this is especially true for systems that are application specific such as we find with CD SEMs. In this section, the system components available in today's SEMs are reviewed in the context of their application to "nondestructive" low-voltage linewidth metrology (i.e., ~300–3000 eV).

6.4.3.2.1 Electron sources

The scanning electron microscope derives its name largely from its function. By scanning a finely focused electron beam over a well-defined rectangular area in a raster pattern, a microscopic image of a specimen may be obtained. As was previously described, an SEM is made up of a series of subsystems consisting of an electron source (gun), lenses, and detector(s). In addition, an SEM requires a means of viewing the "electron" image, such as a CRT, as well as specialized electronics designed for each subsystem. Unlike an optical microscope, the performance of an SEM is largely determined by its electron source. The current density of the incident electron beam determines to first order the magnitude of the emitted electrons (the electron signal). The size of the incident beam (i.e., the probe diameter) determines the limiting resolution of the electron microscope. In an SEM, both of these parameters, beam current and probe diameter, are largely determined by the electron source. For this reason, SEM types are distinguished principally by the type of electron source used.[110,111,112]

The main types of electron sources together with some of their operating requirements and performance characteristics are listed in Table 6.5.[113,114] The first SEMs used thermionic emission sources.[115] For these instruments, the electrons were obtained by heating a filament to a high temperature. In this process, at sufficiently high temperatures, a certain percentage of the electrons become sufficiently energetic to escape the source. Originally tungsten hairpin wires were used. However, the brightness (current density per unit solid angle) and probe diameter that could be achieved with this source limited the signal to noise ratio and resolution.

Table 6.5 Comparison of different electron source operating requirements and performance characteristics (From Postek[113,114]).

	Tungsten	LaB$_6$	CFE	TFE
Source Brightness (A-cm^{-2} Str^{-1})	10^6	10^7	10^9	10^8-10^9
Emitting Surface Area (μm^2)	$\gg 1$	> 1	0.02	0.2
Source Diameter (nm)	$> 10^4$	$> 10^3$	3 - 5	15 - 25
Energy Spread (eV)	1 - 3	1 - 1.5	0.2 - 0.3	0.3 - 1.0
Source Temperature (°K)	2900	1800	300	1800
Work Function (eV)	4.5	2.6	4.5	2.8
Operating Vacuum (Pa)	10^{-4}	10^{-6}	10^{-9} - 10^{-10}	10^{-8} - 10^{-9}
Typical Service Life (hours)	40 - 100	1,000	$\gg 2,000$	$> 2,000$

For this reason, alternative cathode materials were sought. The discovery of lanthanum hexaboride (LaB$_6$), with higher brightness (an order of magnitude higher than tungsten) and a smaller probe size significantly advanced the state-of-the-art of thermionic emission SEMs.[116] However, the technology improvement came at an increased cost. It was more difficult and, therefore, more costly to manufacture the LaB$_6$ cathodes. In addition, to prevent "poisoning" of the LaB$_6$ cathodes during normal operation of the SEM, higher gun vacuums were required with associated improvements (and increased costs) in the gun-pumping systems.[117] Thermionic emission systems continue to serve the electronics industry well in analytical applications. However, they were not sufficient for CD metrology applications, since they provided only moderate resolution at the low operating voltages required for nondestructive resist measurement and inspection. It was not until the introduction of commercially available SEMs that incorporated field emission sources that the semiconductor industry could pursue and embrace in-line SEM technology.[118–130]

As shown in Table 6.5, the cold cathode field emission source (CFE) and the thermally assisted field emission source (TFE) are the two basic types of field emission sources in use today. For field emitters, the cathode is in the form of a rod with a very sharp point at one end (typically on the order of 100-nm diameter, though nanometer-sized field emission tips are in development).[131] When the cathode is held at a negative potential relative to the anode, the electric field is so strong ($>10^7$ Vcm^{-1}) that the electrons can leave the tip without requiring additional thermal energy. The usual cathode material used in field emission tips is tungsten. This is because the high electric fields present at the tip create large mechanical stresses on the cathode and only very strong materials can withstand these forces without failing. The brightness potentially available from these field emission sources is several orders of magnitude larger than the thermionic sources (see Table 6.5). The cold field emission gun and column first designed by Crewe is reproduced in Fig. 6.49.[118]

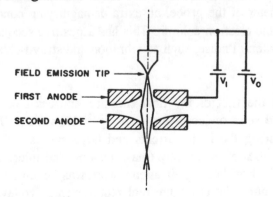

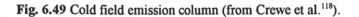

Fig. 6.49 Cold field emission column (from Crewe et al.[118]).

In order to maintain the field emission process, a good vacuum is required. The reasons for this are quite straightforward. Any contamination that is present on the cathode surface reduces the emission efficiency. For pressures on the order of 10 μPa, a monolayer of atoms can form on the tip in less than 10 seconds.[132] Consequently, in order to maintain the emission from the tip, a very good vacuum is required (better than 10 nPa). Even under these conditions, molecules will land on the tip from time to time, causing fluctuations in the emission current. Eventually, the whole tip will be covered and the emission will become very unstable. Under these conditions it is necessary to clean the tip. This is accomplished by rapidly heating the tip to a high temperature (~ 2300 K) for a few seconds in order to drive the molecules off and leave a clean surface. This situation describes what is required to operate the CFE SEMs. It is necessary to "flash" the CFE tips every 4 to 8 hours to maintain a usable emission current.

Alternatively, the field emission tip can be heated continuously so that the free molecules do not "stick" to the tip (should they come in contact with the tip they are immediately re-evaporated). These thermally heated field emitters do not require the periodic cleaning as do the CFEs and thus provide a more stable emission current during continuous operation. Zirconiated/tungsten <100> (ZrO) is the preferred material for thermal field emission cathodes because of its lower work function (see Table 6.5). In addition, the vacuum requirements are on the order of an order of magnitude less for the TFEs than are required for CFEs. Nevertheless, the typical service life of the TFEs is, in general, less than is achieved with the CFEs. This is the penalty one pays for operating the TFE tips at elevated temperatures.

The column designs are slightly different for CFEs and TFEs sources. The CFE has a sufficiently small probe diameter that it does not require a condenser lens in the electron column to demagnify the source (see Table 6.5). TFE sources have a probe diameter that is approximately a factor of 5 larger than the CFE sources. In order to reduce the diameter of the probe, an extra demagnifying condenser lens must be introduced into the electron column. This has a positive secondary effect since it further reduces external noise, such as vibration and stray fields, that may affect the source.[113]

Historically, CFEs were the first of the field emission sources to be designed, built, and applied toward semiconductor problems.[118-121] However, CFE sources were incapable of producing the high currents and large spot sizes needed for other applications, e.g., e-beam lithography (mask making and microfabrication) and analytical microscopy. For these applications, work was begun on alternative sources with relaxed vacuum and environmental requirements. Today, there are commercial instruments available that utilize either CFE or TFE sources. Both types of instruments have been successfully applied in numerous metrology applications. Although the electron source is a fundamental parameter of any SEM, it unfortunately does not solely determine whether the instrument will work best for a particular application. Unless the service life of the tip is the only relevant parameter, it is not possible to distinguish metrology instruments based on only the design of the source. For this, we must proceed further down the column.

6.4.3.2.2 Electron lens and detector designs

The design and optimization of CD SEMs have been driven by the specific application presented for quick and reliable measuring of critical dimensions patterned on large specimens (wafer diameters of 200 mm and soon to be 300 mm) in user-defined positions. These conditions have imposed constraints not only on the design of SEM wafer stages but also on the design of the SEM's final lens and electron detectors. It is well known that the highest "resolution"[133] is obtained in an SEM at the shortest working distance (the smallest distance between the final lens and the sample). At the time that full-wafer SEM were first introduced onto the semiconductor manufacturing floors, the final lens technology was predominately "pinhole" type used in combination with an Everhart-Thornley (ET) detector.[134] Examples of these first systems are illustrated in Fig. 6.50.

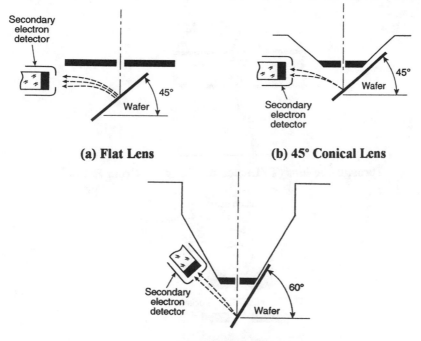

(a) Flat Lens   (b) 45° Conical Lens

(c) 60° Conical Lens

Fig. 6.50 Early pin-hole lens design with Everhart-Thornley detector.[109]

Since the first in-line SEMs were often dual purpose having both metrology and inspection capabilities, the working distance was still limited by the constraints imposed by the detectors for inspection applications, i.e., tilted specimens. To address this limitation, two improvements in lens design have been introduced: (1) through-the-lens (TTL) electron detection and (2) extended field (immersion) lens technology.

Through-the-lens refers to the fact that signal electrons are transported back through the focusing lens and collected above the final lens.[108,135,136,137] As shown in Fig. 6.51(a), by opening up the bore of the final lens, the electron detector could be placed in the area above the final lens. This offers two improvements for in-line wafer applications, regardless of wafer tilt or wafer size: the working distance can be reduced and the collection efficiency increased, both resulting in improved image resolution. A modification of the standard lens, the so-called inverted snorkel lens, enables a large sample to be immersed in the external field of the lens. One type of these immersion lenses is illustrated in Fig. 6.47(b). This lens offered several advantages: it could be incorporated into SEMs using "conventional" (i.e., below lens) SE detector positioning or it could be used in conjunction with two detectors, a dual-imaging mode. In the latter design, detectors could be positioned both above and below the final lens and the imaging mode determined by the particular application.

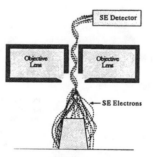

(a) Through-the-lens (TTL) electron detection (from Reilly[138]).

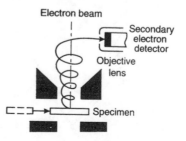

(b) Immersion lens (from Postek[109]).

Fig. 6.51 Lens designs.

In parallel with advances in lens designs, it has been necessary to improve electron detection capabilities for advanced in-line SEM metrology applications. The performance of the ET detector degrades significantly for the preferred SEM operating conditions (~600 eV and ≤10 pA) required for nondestructive measurements on semiconductor resist structures. In addition, the positioning of the ET detector can lead to alignment problems between the detector and the measurement sample. For linewidth applications this is critical, since misalignment can lead to asymmetric signal collection and thus distort the resulting linewidth intensity profiles. These problems resulted in the suggestion of a number of alternative solutions by different workers.

Dual[139,140] and quad[141,142] secondary ET detectors have been proposed and implemented. These designs improved imaging over the single ET detector by increasing the collection efficiency but did not necessarily improve the metrology situation. Instead of aligning one detector, it was now necessary to align and match two (or four) detectors. Alternatively, microchannel plate (MCP) detectors have been proposed by several workers.[143,144,145] These detectors can be placed above the sample (in close proximity to the final lens) or above the final lens, enabling through-the-lens detection thereby eliminating problems associated with sample/detector alignment on the primary electron axis.[109,143] These arrangements are depicted in Fig. 6.52. This configuration has also allowed the pursuit of alternative modes of image formation.

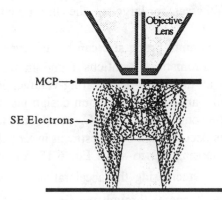

Fig. 6.52 MCP electron detection (from Reilly[138]).

6.4.3.2.3 Improved electron gun designs

As linewidth geometries have pushed into the deep submicron regime, CD SEM designs have pursued improvements in image resolution through the development of systems that incorporate low accelerating voltages and low total electron dose.

The low energy of the primary beam in current state-of-the-art CD SEMs de-emphasizes the resolution degradation produced by the interaction volume of the electrons with the sample (see Sec. 6.4.3.2.5). Under these conditions, resolution becomes primarily dominated by the incident beam probe size. In general, the probe size is given by Eq. 6.17.[117]

$$d_{eff}^2 = d_0^2 + d_d^2 + d_s^2 + d_c^2 \qquad (6.17)$$

where d_{eff} is the (effective probe size). For a system that utilizes field emission for the electron source, the aberration disk d_0, which represents the geometrical diameter (virtual spot size), is negligible due to the high brightness and small virtual source size of field emission technology. Of the remaining components [d_d (the diffraction disc)], d_s (the spherical aberration disc), and d_c (the chromatic aberration disc)], only the diffraction and chromatic terms contribute significantly to probe size. Thus, Eq. 6.17 can be simplified to:

$$d_{eff}^2 = \left[\frac{4I_P}{\pi^2\beta} + (0.6\lambda)^2\right]\alpha_P^{-2} + \left[C_c \frac{\Delta E}{E} \alpha_P\right]^2 , \qquad (6.18)$$

where I_P is the probe current, β is the gun brightness, λ is the DeBroglie wavelength, α_P is the electron probe aperture, E is the electron energy, ΔE is the energy spread of the gun (0.2 to 2 eV) , and C_C is the chromatic aberration constant. Equation 6.18 shows that when α_P is small, diffraction effects will dominate the probe size. Whereas, for large α_P, the chromatic term dominates.

From this consideration, minimum probe size can be achieved for a defined α_P and is dominated by the chromatic aberrations found in the final lens. Thus, achieving high resolution at these low-accelerating-voltage, low-dose operating conditions in the state-of-the-art CD SEM column design has been accomplished through minimization of the contribution of the C_C component to the system probe size. By and large, this has occurred not by investment in lens design optimization to minimize C_C but by choosing α_P to solve Eq. 6.18 for minimum d_{eff} while minimizing the $\frac{\Delta E}{E}$ term. By increasing the accelerating voltage of the primary beam (5 keV to 8 keV) as it passes through the final lens, $\frac{\Delta E}{E}$ can be reduced significantly and the resulting electron probe size becomes diffraction limited. After the incident beam has passed through the final lens, the beam is decelerated, either by an electrostatic lens just below the final pole piece of the magnetic lens[136] or by a decelerating potential applied to the wafer itself through the stage/chuck assembly,[146,147] to a "landing energy" in the 400 to 800 eV range. A schematic example of this column design is depicted in Fig. 6.53.

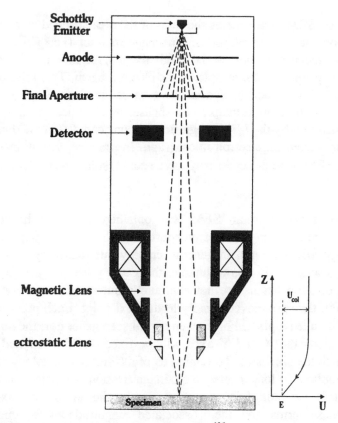

Fig. 6.53 New electron column schematic view.[136]

Due to the electrostatic retarding field that serves to push the signal electrons (SE or BSE) into the bore of the magnetic lens, the extremely strong final lens magnetic fields required to focus the energetic beams, and the extremely short working distance required to optimize the α_p for minimum probe size, signal detection has also required significant advancements in state-of-the-art CD SEMs. The standard ET detector placed at the level of the final lens or even in a TTL configuration, relying on a field grid to draw the low-energy secondaries to the detector, is no longer possible since the signal electrons will be accelerated to the full accelerating potential of the column very shortly after emission from the sample. Most advanced CD SEMs now place the detector above the lens and rely on deflection of the signal electrons either via electrostatic field of the final lens[136] or via a beam deflection field (E × B or Wein Filter) located above the final lens assembly[148] to push the signal electrons off the electron optical axis to the detector while having no effect on the primary beam. Such designs have resulted in tremendous improvements in signal collection efficiency, thereby allowing further reductions in total electron dose.

6.4.3.2.4 Magnification/field of view

Magnification in an SEM is defined as the ratio of the area scanned on a specimen with the corresponding area displayed on an image monitor (i.e., CRT display). Since the area of the image monitor is fixed, the magnification is changed by adjusting the area scanned on the sample by the electron beam. This relationship is illustrated in Table 6.6, which compares the area sampled on a specimen as a function of magnification. In principle, by measuring distances on an electron micrograph and dividing by the SEM magnification, one can determine dimensions of relevant features from an electron micrograph. In practice, this is made more difficult because SEMs do not necessarily have equal resolution (i.e., μm/pixel) in both axes.

Historically, the display from an SEM was obtained by a synchronous and continuous movement between the electron source and the electron display beams. Modern metrology SEMs do not, in general, employ this technology but instead use digital scan generators. In this technique, the x and y velocity components of the synchronous electron beams are not continuous but remain zero for a finite period of time until they are moved rapidly to the next point. Each point at which the beam dwells is called a pixel. For a given digital scan generator, the number of available pixels along the X and Y axes will be fixed. As the scanned area is changed, so too will the pixel size. To remind us of the digital technology currently in place, most suppliers no longer refer to the magnification but instead to the field of view (determined by multiplying the pixel size by the number of pixels in a digital scan). These terms and their associated magnitude as a function of "magnification" are included in Table 6.6 for completeness.

Table 6.6 Area sampled on a specimen as a function of magnification (or field of view).

	Analog Convention	Digital Convention	
Magnification Factor	**Area on Sample***	**Pixel Size***	**Field of View****
10	1 cm²	10 μm	1 cm
100	1 mm²	1 μm	1 mm
1,000	100 μm²	0.1 μm (1000 A)	100 μm
10,000	10 μm²	0.01 μm (100 A)	10 μm
100,000	1 μm²	1 μm (10 A)	1 μm

Assuming : *10 cm by 10 cm display / ** 1000 by 1000 pixel scan generator

Accurate SEM measurements require an accurate determination of the SEM magnification (or field of view). Currently, the only certified standard for the accurate calibration of SEM magnification is the NIST SRM 484.[149] This standard is composed of gold lines separated by layers of nickel providing a series of pitch structures ranging from 0.5 to 50 μm. This standard has served the industry well, but low voltage requirements in semiconductor applications have established the need for a new low-voltage SEM magnification standard. Postek has discussed the design, development, and manufacture of a new low-voltage magnification standard.[8,150] The SRM 2090 has a gradation of nominal pitch calibrations from 0.20 μm to 3 mm and has been designed for both high and low accelerating voltages. When this standard is issued, it will significantly advance the system calibration for advanced applications.

The design of the new magnification standard reflects the improvements in SEM technology as well as the increased sophistication of the end user. The artwork for the prototype of the NIST SRM2090 is shown Fig. 6.54. As can be seen, the new standard will allow calibration of magnification in both the X and Y axes without physically rotating the standard. This design reflects the goal of contemporary SEM metrologists to calibrate the X and Y pixels of the digital image measurement system rather than the SEM system "magnification." As was shown in the NIST Interlaboratory SEM study, the aspect ratio of the SEM display *may* not be square, i.e., the X and Y pixels are not equal, so it is important to calibrate both axes independently.[8] Note that calibration of the digital image system will allow relatively correct digital measurements but cannot be used to correct the micrographs obtained directly from the SEM.

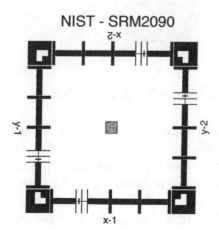

Fig. 6.54 SRM 2090 (from Postek[149]).

6.4.3.2.5 Electron image formation

Electron detection has been discussed primarily in the context of optimizing image formation. This is not unreasonable, since the quality of the SEM image is what first attracted the attention of semiconductor users. However, with the commercialization of MCP detector technologies, we have inadvertently opened Pandora's box. This is because the MCP detector can be designed to operate in either a secondary or backscattered detection mode. To appreciate the differences will require a brief digression on electron image formation.

Although secondary electron imaging has historically been the principal method used for critical dimension measurements, the behavior of these electrons with respect to image formation is not yet well understood. This is due to the complexity of secondary electron generation. The secondary and backscattered electron components used in image formation are depicted in Fig. 6.55.[88]

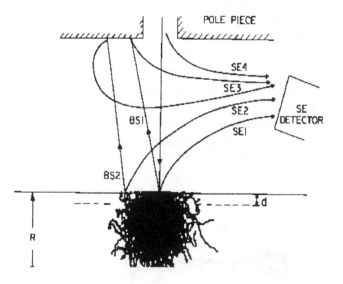

Fig. 6.55 SE and BSE formation (after Postek and Joy[88]).

The total secondary electron signal results from the combination of four unique secondary electron generation mechanisms. The SE1 electrons are generated during inelastic collisions between the primary electron beam and the sample. As little as 10% of the total SE signal is formed by the SE1 electrons, which comprise the highest resolution component of the total SE signal. The remaining 90% of the SE signal results from the SE2 (~ 30%) and SE3 (~60%) electron generation processes.[131] The SE2 component is generated by inelastic collisions between the backscattered electrons and the sample as the BSEs traverse the interaction volume within the SE escape depth (BSE1). These electrons are emitted within a larger distance from the initial beam interaction point than the SE1 and represent a lower "resolution" signal. The SE3 component results from the interaction of the backscattered electrons (BSE1 or BSE2) with the internal components of the SEM (e.g., pole piece, chamber walls). The SE4 component can be eliminated by proper column design and is generally not a factor in well-designed systems.

The energy distribution of secondary electrons is centered around 3–5 eV with $\geq$ 90% of the electrons having energies less than 10 eV.[132] The sample information contained in an SE image corresponds to SEs emitted from a sample depth that is ~3 times the mean secondary escape depth, d (see Fig. 6.55).[151] In addition, because of their low energy, secondaries are susceptible to sample charging. Since all SEs possess the same energy, it is difficult to separate the SE1 signal (which carries the edge component metrology information) from the SE2/SE3 components. As a result, the total secondary electron signal is used for CD measurement purposes. Since the composite SE measurement signal arises from numerous components, potential measurement errors can result (see below).

In contrast to secondary electrons, backscattered electrons have a much higher energy, which can range from 50% to ~100% of the primary electron energy. The BSE1 signal results from elastic scattering near the surface of the sample and provides high-resolution surface information. The other component of the backscattered signal (BSE2) results from multiple elastic and inelastic collisions. Since backscattered electrons have higher energies it is possible to filter out the SE signals by biasing the front face of the detector (< -50 eV) and thus image only BSEs. Wells[152] has demonstrated that BSEs may be screened by energy and has demonstrated extremely high resolution imaging with low-loss electrons (LLEs). The backscattered signal has a reduced sensitivity to sample charging due to its higher energy; however, BSEs must be collected by a detector that has "line of sight" placement with respect to the BSE signal generation. This can complicate the system design and potentially degrade the signal-to-noise ratio. However, this potential has not been lost on designers of new CD SEMs.[153]

Given the differences in origin between backscattered and secondary electrons, it should come as no surprise that there have been variations reported between measurements made using backscattered and secondary electron signals.[87,154,155] It is observed that the measurements of structures made using the backscattered signal are smaller and show better precision than those obtained using the secondary electron signal. These results are significant for two reasons. On the one hand, they suggest the potential for improved measurement precision for CD SEMs using BSE detection. On the other hand, they illustrate the complexities involved in electron image formation and how electron detection methods may contribute to linewidth measurement errors in SEM metrology.

6.4.3.3 Specimen effects

Charging of the specimen is a general problem for SEM microscopy. Nonconductive samples accumulate charge when exposed to an electron beam. This charge can distort the electron image and, if high enough, can lead to specimen damage from thermal or radiation effects. The general approach is to coat the sample with a conductive layer and, thereby, dissipate the charge. Unfortunately, this is not often an option in semiconductor applications.

The most common method used today to minimize specimen charging is to reduce both the accelerating voltage and the incident beam current. If one plots the normalized emission (total number of backscattered (η) and secondary electrons (δ) emitted) as a function of incident beam energy, a characteristic curve is obtained.[156] This curve is illustrated in Fig. 6.56. For insulators, a region exists for which the number of emitted electrons exceed the number of incident electrons, i.e., ($\eta + \delta$) > 1. This region is defined by two values of the incident energy, E_1 and E_2, which are referred to as the first and second crossover points. E_1 is typically a few hundred volts, 200 to 300 eV, while E_2 may vary over the range 500–5000 eV.

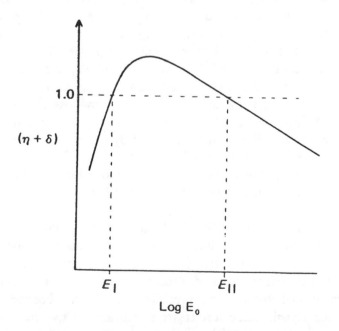

Fig. 6.56 Total electron emission as a function of incident beam energy E_0.[157]

There are significant differences in the electron image for these accelerating conditions. If the incident electron energy falls below E_1, then fewer electrons leave the sample than enter it. This will result in the buildup of a negative charge.

This charge lowers the effective energy of the incident beam and further reduces the electron emission. The negative charging can actually increase SE yield due to a decrease in the landing energy. This is often why negative charging appears bright in the SEM image. Eventually, a sufficient negative charge is built up so that the incident beam is deflected completely, i.e., no image. For energies between E_1 and E_2, the specimen is charged positively. This positive charge acts to decrease the number of secondary electrons emitted (δ), since the low-energy secondaries are attracted back to the sample. As a result of these competing processes, the effective value of the normalized emission becomes 1 and a state of dynamic equilibrium is set up in the presence of a small positive surface charge on the specimen. When $E > E_2$, negative charging will again occur. However, in this case, the effective energy of the incident beam will be reduced until it reaches E_2 and a quasiequilibrium is established. From a practical standpoint, this equilibrium does not produce stable imaging, since E_2 is found to vary greatly over the specimen surface and variations in surface conduction can perturb the image over time.[158]

Since each material has its own characteristic emission curve, it is possible (probable) that specimen charging will not be entirely eliminated for real semiconductor metrology applications. Nevertheless, the low accelerating voltage range available on modern CD SEMs should allow users to minimize specimen charging for most situations. More recently, Monahan and Marchman have reported on the control of sample charging by means of dose control as well as accelerating voltage.[152]

The effect of topographic contrast and its affect on accuracy in CD SEM metrology has been previously reported.[90,159] The electron emission intensity is known to be inversely proportional to the cosine of the solid angle between the surface normal and the incident beam.[160] This effect causes sloped edges to appear "bright" and can have serious consequences in linewidth metrology. The increased emission can distort image intensity profiles and lead to errors in the measured image size. The magnitude of the error will vary with several factors: image size (as features become smaller, the edges contribute a proportionally larger percentage of the image information), edge slope, as well as edge detection algorithms. This situation is illustrated in Fig. 6.57.

Although topographic contrast may introduce complications in the accurate determination of image size, it can also provide some advantages. The fact that the angular distribution of SE emission is concentrated in a direction normal to the emitting surface has been creatively exploited by one SEM manufacturer to distinguish between lines and spaces in the electron image.[140,141]

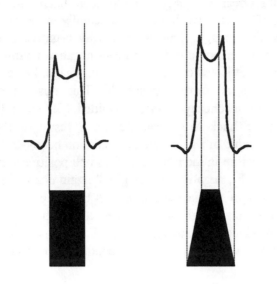

Fig. 6.57 The effect of topographic contrast on electron intensity profiles.

6.4.3.4 Automation

The automation of CD SEMs has proceeded somewhat nonlinearly since their introduction to semiconductor manufacturing lines in the mid 1980s. The initial automation offered was minimal and did not extend much beyond the selection and programming of measurement/inspection sites as well as the selection of a measurement algorithm. Given the state-of-the-art "minicomputers" available at this time, this was considered leading edge. Nevertheless, the industry did not advance significantly beyond this stage for several more years to come.

Progress has been incremental because full automation has required some fundamental modifications to the traditional SEM platform. The functional requirements for system automation are listed in Table 6.7. However, even with the appropriate subsystems in place, successful automated measurements may still be achieved only after extensive hands-on engineering efforts. An example of the problems that can be encountered when bringing automated metrology on-line was reported by Chain and Baaklini who discussed their experiences in automating a particular CD SEM.[161]

Nadler-Niv and Halavee first enumerated these system requirements for automated CD metrology in 1988.[139] It is important to realize that these functions do not operate independent of one another. Measurement automation requires the successful execution of each operation in order to advance to the next step. Without strict attention to each function, automation cannot be achieved. As an example, it is important to understand what type of feature will optimize target capture in the pattern recognition system. The answer to this question is not obvious, and recommended targets do, in fact, differ substantially between CD SEM suppliers (one supplier prefers horizontal/vertical edges for pattern recognition while another prefers diagonal edges).[162]

Table 6.7 Automation requirements to perform automated metrology.

Functional Requirements for SEM Automation
Accuracy of Wafer Alignment
Autofocus Capability
Robust Pattern Recognition (i.e., not confused by in-spec process variations)
Stable Image Formation
Digital Column Design
Digital Signal Processing
Low Voltage / High Resolution Imaging
Automated Measurement Algorithms

Most automated CD SEMs today use both optical and e-beam pattern recognition systems. The optical system is used because it has been found to provide very high capture rates at global alignment (~100 %). However, the successful application of an all e-beam pattern recognition system for automated CD metrology has been described by Chain and Baaklini.[161] In this application, a 100% capture rate was achieved for wafer global alignment by incorporating a scribe line autofocus pattern recognition structure.[163] The design of this unique structure enabled the SEM to automatically determine the mid range of wafer focus and simultaneously function as a unique feature used for global wafer alignment.

Image stability and resolution have been achieved by incorporating TFE sources into automated CD SEMs. This is not to say that automation with CFE sources has not been achieved.[161] However, this has been achieved at some cost, requiring additional engineering debug and extra tool maintenance. It is not clear whether most manufacturing lines could support this additional overhead.

Although the depth of focus of an SEM is considerable, it turns out that 8" wafers when not held down by vacuum can also exhibit sufficient warpage so as to be considered nonnegligible in an SEM.[164] For these reasons, it may be necessary to perform an e-beam autofocus more often than is normally recommended by the system suppliers. If there is a focus problem, the best situation will occur when the image is sufficiently out of focus to cause a measurement failure. The user can be flagged and the root cause investigated. The worst situation will occur when the measurement pattern recognition does not fail and a measurement is taken at a sufficiently wrong focus level to introduce measurement error. These may or may not result in wafer misprocessing — i.e., an α *or* β type error — but are usually difficult to trace unless they occur often or conveniently when engineering is present.

The final topic of interest in system automation is the selection of the measurement algorithm. Various methods, e.g., threshold,[89,90,138,165,166] peak to peak,[89,165] maximum slope,[89,166] linear regression,[138,163,167] and contact hole algorithms[159] have been applied to the measurement of SEM intensity profiles. The precision and accuracy (as compared to SEM cross sections) of these algorithms have been found to vary with system parameters as well as specimen variations. These results are summarized in Table 6.8.

Table 6.8 Beam/specimen parameters which can affect measurement precision.[141]

Algorithm	Observations
Thresholds	Sensitive to: Measurement focus; Resist Charging Loading effects (i.e., image proximity) Resist Sidewall Slope
Peak to Peak	Highly sensitive : System defocus Less sensitive : Beam Charging
Maximum Slope	Sensitive to image polarity (i.e., lines/spaces)
Linear Regression	No documented problems
Contact Hole	Problems with Signal Detection

The present level of automation available in today's CD SEMs is a testament to both the users and the suppliers. These instruments provide the requisite measurement capability for the current design rules in high-volume manufacture, i.e., 400 to 500 nm. Unfortunately, there is mounting evidence of measurement problems for the next generation of devices requiring 250–350 nm design rules.[16,168,169,170] This situation is reminiscent of what metrologists faced a decade ago with optical microscopy. However, this is not the same industry as it was 10 years ago. Metrology, as practiced today, is far more sophisticated and disciplined. Metrologists are acutely aware of the potential problems in linewidth measurements.

6.4.4 Advanced Probe Linewidth Metrology

Scanning probe microscopes are important new metrology tools which will complement rather than replace current optical and electron beam metrology equipment. This complementary nature has been demonstrated in recent work where force microscopy has been used to support and calibrate CD SEMs used in semiconductor development lines.[16,171,172,173] In lieu of an SEM linewidth standard, the potential to characterize linewidth profiles using scanning probe techniques has generated tremendous interest in these instruments.

Although the enthusiasm is fantastic, it is worthwhile to place scanning probe microscopy in perspective. Optical microscopy has been in use over the last 300 years and remains in wide use today. Scanning electron microscopy, on the other hand, has only been around for the last 30 years. The SEM is not as simple and requires special handling and sometimes specimen preparation, but for some applications it is the technique of choice. Now enter the scanning probe microscopies. It has been 25 years since researchers at the National Bureau of Standards first reported that the lateral resolution of profilometers could be significantly improved by using electron tunneling techniques in order to avoid hard contact with the sample surface.[174,175] In spite of this announcement, the potential for atomic resolution was not appreciated until the first scanning probe microscope, the tunneling microscope (STM), was announced and demonstrated by Binnig et al. 10 years later (1982).[176] It will take time to understand the nuances of these new techniques, and although great effort has been expended, 15 years is not a lot of time. As we shall see, it is necessary to understand the technical intricacies of the scanning motions as well as the ways in which the sample and the probe interact.

The basic concept on which the scanning probe techniques operate is the controlled scanning of a "probe tip" across a sample surface. By controlling the probe-to-sample spacing and monitoring a characteristic of the tip [e.g., current (STM), X, Y, Z deflection (AFM)] with a feedback control loop, topographic images of the sample may be obtained. The manner in which the tip-to-sample spacing is monitored or controlled is determined by the type of scanning microscope. Binnig's original STM relies on electron tunneling between a conductive probe tip brought into close proximity to a conductive sample (i.e., the electron tunneling distance of the sample: on the order of 1 to 2 nm). When a bias voltage is applied between the tip and the sample, a small current (on the order of

1 nA) will flow to or from the tip to the sample. Since the tunneling current density decays exponentially with distance between the tip and the probe, a feedback loop that maintains a constant current will control the physical separation between the tip and the sample. In this way, the tip tracks topographic changes of the sample. This is depicted in Fig. 6.58.

Fig. 6.58 The tip-to-sample spacing as controlled in an STM.

An image is obtained by rastering the tip across the field of view (FOV) and plotting the change in the tip position as it traverses the surface. The STM has very high vertical resolution (sub-angstrom) because of the exponential dependence of the tunneling current with separation distance. The lateral resolution is also of the same order because the current emitted from the tip occurs only from a small atomic cluster of atoms near the apex of the tip. Unfortunately, the STM has limited applications, for it is restricted to conductive samples. Widespread applications of the scanning probe technique awaited the arrival of the atomic force microscope (AFM).

The AFM was first described by Binnig et al. in 1986 and allowed scanning techniques to be applied generally.[177] As is known today, all surfaces exert atomic forces that vary with distance from the atomic surface and fall into two categories, attractive long range and repulsive short range. This is illustrated in Fig. 6.59. Binnig took into account the presence of these forces to develop a technique in which a sharp probe is brought close enough (within angstroms) to the surface of a sample to detect these forces. It is the objective of the AFM to maintain a constant force (and therefore distance) between the probe tip and the surface. When the distance between the tip and the surface is regulated with repulsive forces, the AFM is said to be using contact-mode scanning. On the other hand, when the feedback control is regulated with the longer-range (e.g., several *nanometers*) attractive force, the method is described as noncontact scanning. The primary advantage of the AFM is that it can be used on nonconductive (e.g., resist) samples.

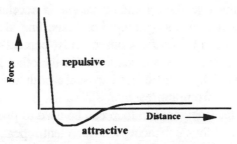

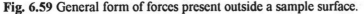

Fig. 6.59 General form of forces present outside a sample surface.

Some of the first applications of an AFM for characterizing features relevant to the semiconductor industry appeared soon after the AFM was first described.[178,179,180] The system described by Martin et al. is illustrated in Fig. 6.60.[173] This drawing serves to describe the basic operation of a metrological AFM. The tip is mounted on a cantilever beam whose motion may be monitored (the Z motion is monitored optically in the example shown in Fig. 6.60). In order for the forces to be sensed, the cantilever is vibrated at its mechanical resonance frequency and brought toward the sample. As the tip approaches the surface of the sample, the surface forces either dampen the amplitude or change the resonance frequency of the cantilever. In either case, the recorded change is used as the input signal for the feedback loop (see Fig. 6.60), which is used to regulate the tip to sample separation. In this way, the AFM, like the STM, obtains an image by rastering the tip across a portion of the sample and plotting the change in the tip position as it traverses the surface.

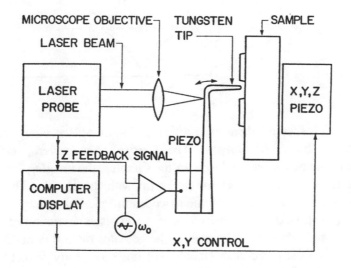

Fig. 6.60 Block diagram for an atomic force microscope (from Y. Martin et al [173]).

As described above, when repulsive forces are used to regulate the motion, the AFM is said to be operating in a contact-mode. Since the tip and surface are in such close proximity, damage to the tip and/or sample is accelerated. Moreover, when there is a sudden change in topography, i.e., resist line, deformation of the tip is difficult to avoid and can be severe. Consequently, contact-mode AFM does not work well in linewidth metrology applications. On the other hand, the longer-range noncontact-mode imaging can be used to sense both the vertical surface forces and the lateral forces from approaching sidewalls. Since the tip-to-sample separation is greater, the feedback control loop can be used to preserve the tip and avoid tip-sample encounters, thus enhancing measurement repeatability. For these reasons, noncontact-mode AFM is better suited for characterizing features with high aspect ratios as are encountered in many linewidth metrology applications. How faithfully a probe microscope reproduces surface topography depends strongly on the size and the shape of the probe tip. The imaging of undercut or re-entrant sidewalls requires the design of a probe tip that would allow the overhang to be reached. One such tip, the three-point tip described by Nyyssonnen et al., is reproduced in Fig. 6.61.[181]

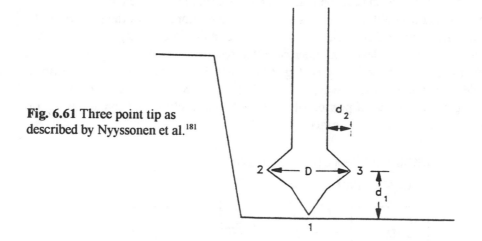

Fig. 6.61 Three point tip as described by Nyyssonen et al.[181]

As with other microscopies, there exist artifacts that affect the interpretation and measurement of the AFM images. These arise from several effects: the nonlinearities of the actuators used to scan the samples, the imaging (or scanning) algorithms, and the interactions of the probe tip with the sample. The actuators that are used in most scanning probe microscopes are piezoelectric tubes that use a $Pb(Zr,Ti)O_3$ compound.[22,182] This class of materials is ferromagnetic and exhibits hysteresis and creep.[183,184] Therefore, the actuator motion is not linear with applied voltage and the piezo scan signals will not necessarily reflect the actual position coordinates of the probe tip. Consequently, a plot of surface height versus uncorrected piezo scan signal will produce a curved or warped image. Independent monitoring of the actuator motions can correct these errors and increase the precision and repeatability of a measurement.[173,181] Capacitive and optical position sensing of the actuators have been reported.[22,181,185,186] However, in many systems, the piezo-driving voltage is altered to follow a low-order polynomial in order to linearize the motion. Attempting to address motion nonlinearities using a pre-defined driving algorithm for the actuator motions will be inadequate for dimensional metrology applications.

The problem created with scanning algorithms is not unique to the AFM, for it is present in both optical and scanning electron images. The nature of the problem is illustrated in Fig. 6.62. As shown, data is acquired along the scan axis in equal increments, regardless of topography. In an SEM micrograph, these distances would be determined by dividing the FOV by the number of pixels in a particular scan generator, i.e., 512, 1024, etc. In an AFM, however, the fixed, linear scanning algorithm is particularly vexing since it would be unable to detect abrupt changes in topography, i.e., a resist sidewall. For this reason, more complex algorithms are required that can readily respond to changing topography and therefore allow both forward and reverse motions. Nyyssonnen et al. were the first to report on a system that supported such 2D motion.[181]

Later, Martin and Wickramasinghe described a method that provided servo-controlled motion in both the lateral (X) and vertical (Z) scan directions, i.e., two dimensions, using a boot-shaped tip design.[187,188] In this design, 2D servoing is achieved through the use of a digitally controlled feedback loop for both piezos (with separate feedback for X and Z). Instead of making evenly spaced steps in the X and Y scan directions, the servoing follows along the contour of the surface. The scan direction is determined by detecting the change in surface slope and is continually modified to stay parallel to the sample surface. In this way, the data are not collected at fixed x intervals but are acquired at controlled intervals along the surface contour itself. This offers a significant advantage because one can increase the number of points collected when imaging sidewalls. For this reason, real CD widths may be determined at each Z-location using the 2D algorithm. A comparison of the 1D and 2D AFM scanning methods is reproduced in Fig. 6.62.[187]

Of perhaps greater concern in AFM imaging is the interaction of the probe tip with the sample. How faithfully the tip traverses a particular surface topography depends strongly on the size and shape of the probe tip. In profiling the complex structures often encountered in semiconductor processing, some features of the structure may be inaccessible to the probe tip (see Fig. 6.63), while other surfaces may be distorted.[22] These situations have been described by Griffiths et al. and are illustrated in Fig. 6.63.[22]

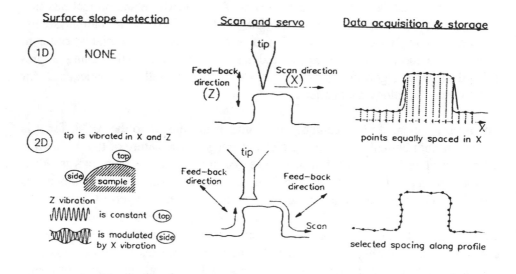

Fig. 6.62 The effect of scanning algorithm on AFM imaging (from Martin and Wickramasinghe).[187]

For samples with high aspect ratios (as is the case in linewidth metrology), several problems arise. Depending on the size and shape of the tip, one may or may not be able to access the areas of interest of the feature. This is illustrated in Fig. 6.63, where the tip shape does not permit access to the bottom of the trench, or the tip may not be able to reproduce all the angles of interest in a particular feature. In general, however, one wants very sharp probe tips in order to scan areas having abrupt surface changes such as resist sidewalls.[187,189,190] Unfortunately, most commercially available probes are conical or pyramidal in shape with rounded apexes. As is shown in Fig. 6.63, when these probes are used to scan features with high aspect ratios, the features may or may not be properly imaged. These situations illustrate the importance of probe tip geometry in AFM imaging.

Although scanning probe microscopy has been generally available for some time, it was the commercialization of Martin and Wickramasinghe's 2D AFM that prompted the general metrology community to seriously pursue CD AFM for linewidth applications.[191] This has allowed the semiconductor community to begin characterizing the CD AFM technique on a complete spectrum of real world problems.[16,170,192,193] This pursuit has identified a number of issues that need to be addressed in order for CD AFM technology to be widely used to characterize and measure the critical aspects of semiconductor features such as width, sidewall slope, and height.

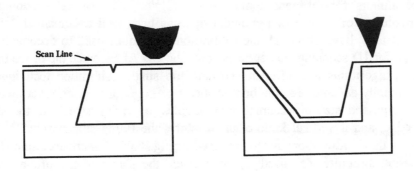

Fig. 6.63 Scan distortions caused by probe-sample interactions (from Griffiths et al.[22]).

The distortion of the scanned image by the interaction of the tip and sample complicates the determination of the physical width of a scanned feature and requires that the tip size and shape be characterized. Consider the case of the boot-shaped tip scanned across a lithographic feature using the 2D servoing algorithm described above. This situation is depicted in Fig. 6.64.[170,187] Observe how the scanned width appears wider than the actual width by the physical width of the probe. For this reason, it is essential that the probe tip width be determined accurately, since this value must be subtracted from all lateral measurements made by the CD AFM.

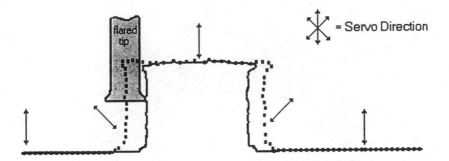

CD-AFM operating with flared tip and variable (2-dimensional) feedback

Fig. 6.64 Effect of probe width on a scanned image.[170]

Tip characterization is nontrivial, and different schemes have been described by several authors for different applications.[16,22,170,180,181,192] For 1D scanning, the tip-convolution of conical or pyramidal tips has been well documented.[194,195] As Wilder et al.[193] have observed, the calibration techniques used to deconvolve the tip shape for 1D scanning cannot be used for the CD AFM, since the probe tip and scanning algorithms are different. To date, no single calibration technique has been generally accepted for the boot-shaped tip. Lagerquist et al. determined the probe shape and size by scanning two samples, a grating structure, the silicon Nano-edge, and a vertical knife edge structure, the flared silicon ridge.[16,196] The results of these calibrations are reproduced in Fig. 6.65. For the example shown, the grating structure was used to characterize the shape of the probe and the effective width of the tip was determined by scanning the knife edge structure. Similarly, the calibration of the boot-shaped tip by a narrow (~350 nm) isolated etched silicon line with sloped sidewalls was reported by Wilder et al.[193]

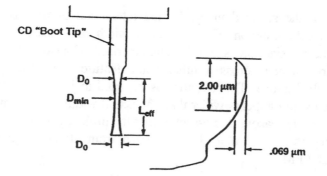

(a) The measurement of probe shape by scanning a grating structure.

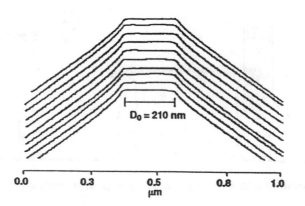

(b) The determination of tip width using a knife edge structure.

Fig. 6.65 The determination of shape and tip width using special structures (from Lagerquist et al.[16]).

Since tips degrade with use, it is necessary to check the tip width frequently. Wilder et al. have reported the variation of the boot-shaped tip width as function of tip use.[193] These results are reproduce in Fig. 6.66. The variation in the CD AFM image over time and use illustrates the need to monitor the tip width regularly. However, the adjustment of tip width should be considered carefully. Lagerquist et al. reported that tip width was adjusted only when shifts of a certain magnitude were observed using a running-average technique.[16]

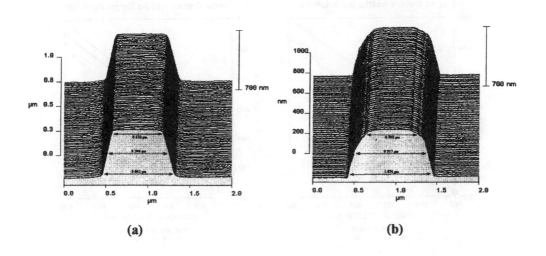

(a)   (b)

Fig. 6.66 The effect of tip aging on the CD AFM image of a calibration silicon artifact: (a) Fresh tip, and (b) after the same tip had been used for more than 20,000 μm of scanning (from Wilder et al.[193]).

Current-generation CD AFMs are capable of providing accurate profile information on features of various shapes and sizes. Since current low-voltage metrology SEMs are designed for top down imaging, this can limit the accuracy of their response when variable or re-entrant sidewalls are present.[168] With the versatile scanning capabilities of the CD AFM, it is desirable to characterize such features nondestructively and in so doing allow in-line CD SEMs to be calibrated on these complex structures. The feasibility of this approach is being pursued by a variety of workers and has been recently described.[16,192,193] Marchman characterized the relationship between AFM and SEM measurements for both DUV resist trenches and etched oxide trenches.[192] These results are reproduced in Fig. 6.67. For this situation, the measurement difference (SEM-AFM BOT) was constant for both isolated and dense structures. Consequently, a constant offset term could be added to the measurement algorithm, thereby improving the "accuracy" of the measurement. This relationship was not observed for the oxide features. The metrology bias from grouped to isolated trenches was a function of feature size and was attributed to charging. Consequently, the difference between the SEM and AFM was no longer just an offset. A linear correction, i.e., multiplicative and offset factors, is required to calibrate SEM linewidths.

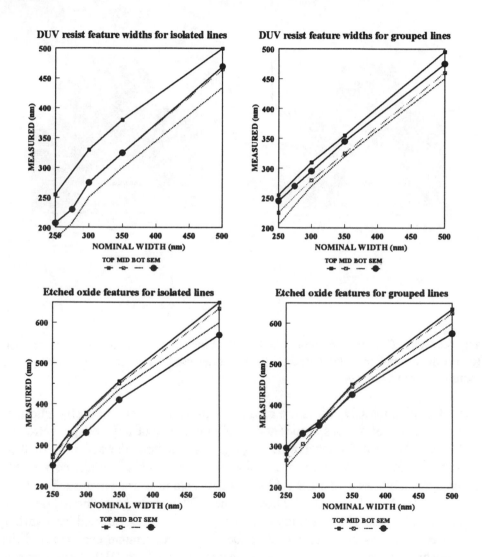

Fig. 6.67 SEM calibration using an AFM (from Marchman[192]).

The evaluation of the CD AFM by the semiconductor metrology community, however, has identified a number of issues that will need to be addressed in order for the CD AFM technology to be widely used. The issues that CD AFMs must address are complicated by the high level of automation currently available for most in-line metrology tools. The industry will be reluctant to give up features available in current tooling unless no measurement alternatives can be identified. Thus, in order for AFM linewidth metrology to be widely embraced, it will probably have to meet some (if not all) of the high standards that the low-voltage metrology SEMs currently maintain. The issues that are not yet resolved are listed in Table 6.9.

Table 6.9 CD AFM limitations.

Parameter	Problem
Throughput	**5 to 10 times slower than current SEM metrology tools.**
Navigation	**Metrology users require intrawafer positioning capability; correlation of large scale (macropositioning) and small scale (microscanning) is required.**
Automation	**Recipe generation must be straightforward. Host interface capability is required.**
Tip calibration & tip stability	**Need criteria, standard technique, and standard calibration artifact to monitor tip life.**

6.4.5 Linewidth Discussion

The determination of "linewidth" presupposes an understanding of which particular aspect of a feature needs to be measured (see Fig. 6.30). In integrated circuit manufacturing, this should not be taken for granted, since features are not necessarily simple and pattern transfer techniques may require that "linewidths" other than the lithographer's linewidth (i.e., width at the resist substrate interface) be monitored to control the full manufacturing process. As an example, a highly anisotropic plasma etching process would require that the top of a resist structure be controlled. On the other hand, if the etching process is isotropic, monitoring the bottom of the resist may be a better way to "control" the process. The determination of linewidth is further complicated by the possible change of the resist structure during process, as might occur when etch processes have nonnegligible lateral erosion rates. These effects are illustrated in Fig. 6.68.

These issues are becoming more important as devices shrink, because all process tolerances must be reduced. Variations in iso/dense etch biases below 400 nm are difficult to resolve when similar effects have been attributed to SEM-based metrology biases.[168,192] For reasons such as these, sub-350-nm process development will require the simultaneous characterization of both the process and the metrology equipment, i.e., low-voltage SEM and noncontact AFM. Since this is expected to be an ongoing process, it leads naturally to the last topic.

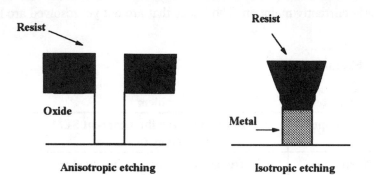

Fig. 6.68 Variation of linewidth with pattern transfer technique.

6.5 A 250-nm METROLOGY STRATEGY

From early 250-nm processing done by Thane et al. at SEMATECH, one may conclude that metrologists have their work cut out for them.[168] This follows in part from Thane's verbal observations that "the perceived process latitudes have always been dependent on the techniques used to measure the imaged resist pattern. The accepted CD for a given resist image is not only dependent on the metrology tool but also on how the engineer decides to use the measured data." So much for measurement confidence. On the other hand, the 250-nm resist SEM cross sections shown by Thane support this conclusion. These are reproduced in Fig. 6.69. The CD measurement at the top of a negative resist line is found to be 14% wider than a CD measurement at the resist base. The utilization of negative resist is associated with the emergence of reentrant resist profiles at 250 nm. This has significantly complicated the wafer-level linewidth metrology.

In order to deconvolve the process biases present at 250 nm, a modification of the current manufacturing measurement strategy will be required. The increase in throughput and improvement in measurement precision of both optical overlay and top-down SEM linewidth systems have dominated measurement tooling strategies for the last decade. Incremental process changes as occurred between 1000 nm and 500 nm could be well characterized and process latitudes controlled using existing reference techniques such as destructive transmission electron microscopy or cross-sectional SEM. Unfortunately, 250-nm lithography does not represent an incremental process shift. Lithography tooling, resist materials, and processing, as well as pattern transfer techniques, are *all* changing. For this reason, linewidth systems should not be expected to automatically provide the requisite measurement tolerances. Instead, the process complexities and tolerances present for 250-nm imaging will require a full characterization of the critical dimension image in three dimensions. This linewidth strategy is currently in place within GRESSI (Grenoble Submicron Silicon consortium).[187]

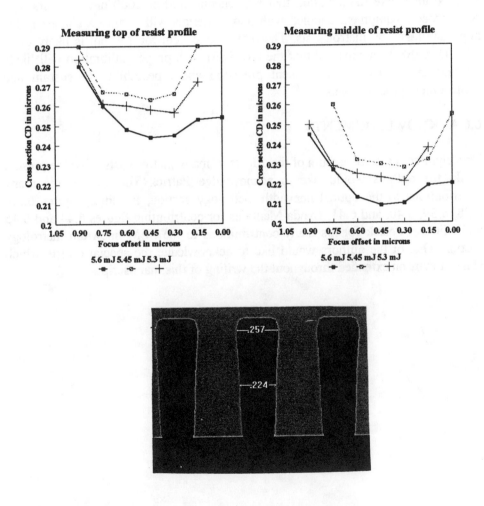

Fig. 6.69 The apparent variation in process latitude depends on what section of the resist profile is used for measurement (from Thane et al.[168]).

With the commercial availability of CD AFM, this system may necessarily evolve as many users' de facto "reference system" described in the first section of this chapter. The need for this system (or a system like this) is demonstrated by the presence of linewidth measurement problems using conventional top-down SEM imaging in predevelopment 250-nm lithography tests.[168] Similar problems in overlay metrology are to be expected, since the shallower depth of focus of the imaging equipment coupled with larger wafer sizes will place greater requirements on the planarization processes. The lateral, as well as the Z-height information, of the AFM image will be useful to characterize the OL structures.

The situation we face as 250-nm processing comes onto the development pilot lines is not unlike the situation that was encountered at 1000 nm 10 years ago. New process equipment coupled with new materials will require a change in the approach taken by metrologists. The reference system for 250 nm is likely to be the CD AFM. It is expected that the CD SEM with proper calibration will likely provide the requisite measurement precision and repeatability to sustain and control most process levels.

6.6 ACKNOWLEDGMENTS

This chapter is a compendium of the efforts of many metrologists. For the present work, the authors would like to acknowledge Patrick Toccolo for his many contributions to the optical linewidth metrology section, including figures 6.35, 6.38, 6.39, 6.40, and 6.41; Lynda Mantalas for contributing figures 6.44 and 6.45 and Michael Postek for his many contributions to the SEM linewidth metrology section. One of the authors would like to acknowledge the special efforts which Harry Levinson extended throughout the writing of this manuscript.

1. SIA, San Jose, CA, *The National Technology Roadmap for Semiconductors*.
2. S.M. Kudva and R.W. Potter, "Cost analysis and risk assessment for metrology applications," Proc. SPIE Vol. 1673, 2–13 (1992).
3. D.J. Coleman et al., "On the accuracy of overlay measurements: tool and mark asymmetry effects," Proc. SPIE Vol. 1261, 139–161 (1990).
4. R.M. Silver, J.E. Potzick, and R.D. Larrabee, "Overlay measurements and standards," Proc. SPIE Vol 2439, 262–272 (1995).
5. Available from VLSI Standards Inc., San Jose, CA.
6. R.D. Larrabee and M.T. Postek, "Parameters characterizing the measurement of a critical dimension," *Handbook of Critical Dimension Metrology and Process Control*, K.M. Monahan (ed.) (SPIE, Bellingham, WA), 2–24 (1993).
7. J. Mandel, *The Statistical Analysis of Experimental Data*, (Dover Publications, Toronto, Canada), 278 (1984).
8. M.T. Postek, A.E. Vladar, S.N. Jones and W.J. Keery, "Interlaboratory Study on the Lithographically Produced Scanning Electron Microscope Magnification Standard Prototype," J. Res. of Nat. Inst. Stan. & Tech., Vol. 98 No. 4, 447–467 (1993).
9. J. Potzick, "Re-evaluation of the Accuracy of NIST Photomask Linewidth Standards," Proc. SPIE Vol. 2439, 232–242 (1995).
10. B.L. Newell, M.T. Postek, and J.P. van der Ziel, "Performance of the prototype NIST SRM 2090A SEM magnification standard in a low accelerating voltage SEM," Proc. SPIE Vol. 2439, 383–390 (1995).
11. D. Nyyssonen, J. Seligson, and I. Mazor, "Phase image metrology with a modified coherence probe microscope," Proc. SPIE Vol. 1926, 299–310 (1993).
12. M.T. Postek, "SEM-based Metrological Electron Microscope System and New Prototype SEM Magnification Standard," Scanning Microscopy 3(4), 1087–1099 (1989).
13. Custom CD SEMS with interferometric stages are available from several manufacturers.
14. D. Nyyssonen and R.D. Larrabee, "Submicrometer Linewidth Metrology in the Optical Microscope," J. Res. Nat'l Bur. Stds., 92, 187–203 (1987).
15. R.A. Allen, et al., "Comparisons of Measured Linewidths of Sub-Micrometer Lines Using Optical, Electrical and SEM Metrologies," SPIE Vol. 1926, 34–43 (1993).
16. M. Lagerquest, W. Bither and R. Brouillette, "Improving SEM linewidth metrology by two dimensional scanning force microscopy," Proc. SPIE Vol. 2725, 494–503 (1996).
17. J. Potzick, "Accuracy in integrated circuit dimensional measurements," *Handbook of Critical Dimension Metrology and Process Control*, K.M. Monahan (ed.) (SPIE, Bellingham, WA), 120–132 (1993).
18. W. Bullis and D. Nyyssonen, "Optical linewidth measurements on photomasks and wafers," *VLSI Electronics: Microstructure Science, Semicon-*

ductor Microlithography, N.G. Einspruch (ed.) (Academic Press, New York, NY), Vol. 3, Chap. 7, 301–345 (1982).

19. M.T Postek, J.R. Lowney, A.E. Vladar, W.J. Keery, E. Marx, and R.D. Larrabee, "X-ray mask Metrology: The development of linewidth standards for x-ray lithography," Proc. SPIE Vol. 1926, 435–449 (1993).

20. R.D. Larrabee private communication to D. Nyyssonen.

21. Bill Banke, IBM Microelectronics, Burlington VT.

22. J.E. Griffiths and D.A. Grigg, "Dimensional metrology with scanning probe microscopes," J. Appl. Phys. 74 (9), R83–R109 (1993).

23. A. Starikov, et al., "Use of apriori information in centerline estimation," Proc KTI Microlitography Seminar, 277–294 (1991).

24. N. Smith and R. Gale, "Advances in optical metrology for the 1990s," Proc. SPIE Vol. 1261, 104–113 (1990).

25. C. Kirk, "A study of the instrumental errors in linewidth and registration measurements made with an optical microscope," Proc. SPIE Vol. 775, 51–59 (1987).

26. A. Starikov et al., "Accuracy of overlay measurements: tool and mark asymmetry effects," *Optical Engineering*, Vol. 31, No. 6, 1298–1312 (1992).

27. P. Troccolo, N. Smith, and T. Zantow, "Tool and mark design factors that influence optical overlay measurement errors," Proc. SPIE Vol. 1673, 148–156 (1992).

28. N. Smith et al., "Minimizing optical overlay measurement errors," Proc. SPIE Vol. 1926, 450–462 (1993).

29. D. C. O'Shea, *Elements of Modern Optical Design*, (John Wiley and Sons, New York, NY), 185 (1985).

30. M. Born and E. Wolf, *Principles of Optics*, 2nd Edition (The Macmillan Co., New York, NY), 211-218 and 473-480 (1964).

31. E. Hecht and A. Zajac, *Optics* (Addison-Wesley, Reading, MA), 175–186 (1979).

32. C.P. Kirk, "Theoretical models for the optical alignment of wafer steppers," Proc. SPIE Vol. 772, 134–141 (1987).

33. M. Born and E. Wolf, *Principles of Optics*, 2nd Edition, (The Macmillan Co., New York, NY), 461–462 (1964).

34. E. Hecht and A. Zajac, *Optics* (Addison-Wesley, Reading, MA), 309ff, (1979).

35. E. Hecht and A. Zajac, *Optics* (Addison-Wesley, Reading, MA), 214ff and 425ff (1979).

36. M. Born and E. Wolf, *Principles of Optics*, 2nd Edition (The Macmillan Co., New York, NY), 522, (1964).

37. A. Kawai, et al., "Dependence of offset error on overlay mark structures in overlay measurement," Jpn. J. Appl. Phys., Vol 31, 385–390 (1992).

38. C. Kirk and A. Gurnell, "Modeling optical linewidth measurement techniques in order to improve precision and accuracy," Proc. SPIE Vol. 565, 62–70 (1985).

39. W. Vollirath, "Microscope Objectives for Semiconductor Technology," *Optical Microlithography and Metrology for Microcircuit Fabrication*, M. Lacombat and S. Wittekoek (eds.), 1138, 166-171 (1989).

40. M. Davidson, et al., "An application of interference microscopy to integrated circuit inspection and metrology," Proc. SPIE Vol. 775, 233-247 (1987).

41. R.D. Larrabee and M. Postek, "Precision, accuracy, uncertainty and their application to submicrometer dimensional metrology," Solid State Electronics, Vol. 36, No. 5, 673-684 (1993).

42. A. Gurnell. Private communication.

43. TSTATS Software Users Maunal, Rev V6.0, 1993.

44. M.P. Davidson, K.M. Monahan and R.J. Monteverde, "Linearity of coherence probe metrology: simulation and experiment," Proc. SPIE Vol. 1464, 155-176 (1991).

45. T. Zavecz, "Lithographic overlay measurement precision and calibration and their effect on pattern registration optimization," Proc. SPIE Vol. 1673, 191-202 (1992).

46. H.S. Besser, "Characterization of a one-layer overlay standard," Proc. SPIE Vol. 1673, 381-391 (1992).

47. D. Corliss, Patent Pend (S/N - 07 / 723,170)

48. Y. Tanaka, et al., "New methodology of optimizing optical overlay measurement," Proc. SPIE Vol. 1926, 429-439 (1993).

49. O. Hignette, et al., "A new signal processing method for overlay and grid characterization measurements," *Microelectronic Engineering:6*, 637-643 (1987).

50. S.C. Douglas, "A frequency-domain subpixel position estimation algorithm for overlay measurement," Proc. SPIE Vol. 1926, 402-411 (1993).

51. R.R. Hershey and T.E. Zavecz, "Expert system for performing measurement system characterization," Proc. SPIE Vol. 1673, 568-579 (1992).

52. R.R. Hershey and T. E. Zavecz, "Figure of merit for calibration and comparison of linewidth measurement instruments," Proc. SPIE Vol. 1464, 22-34 (1991).

53. H. Landis et al., "Integration of chemical-mechanical polishing into CMOS integrated circuit manufacturing," *Thin Solid Films* Vol 220, 1-7 (1992).

54. R.M. Booth et al., "A statistical approach to quality control of non-normal lithography overlay distributions," IBM J. Res. Develop, Vol. 36, No. 5, 1992.

55. A. Engelsberg and D. Leach, "Overlay process control for 16-Mb DRAM manufacturing," Proc. SPIE Vol. 1673, 177-190 (1992).

56. M. Merrill, S.Y. Lee, Y.N. Kim, Y.S. Jung, and J.M. Lee, "Misregistration metrology tool matching in a one megabit production environment," Proc. SPIE Vol. 1673, 203-212 (1992).

57. "Specification for metrology pattern cells for integrated circuit manufacture." SEMI Book of Standards, Micropatterning Volume, 91-109 (1992).

58. See the section entitled, Optical considerations, in Chapter 1 of this handbook.

59. R. Larrabee, L. Linholm and M. Postek, "Microlithography Metrology," *Handbook of VLSI Microlithography*, W. B. Glendinning and J.N. Helbert (eds.) (Noyes Publications, Park Ridge, NJ), Chapter 3, 148–237 (1991).

60. See any general textbook on optics, e.g., E. Hecht and A. Zajac, *Optics*, (Addison-Wesley, Reading, MA), Chapter 10, 329–392 (1979); or M. Born and E. Wolf, *Principles of Optics*, (Pergamon Press, New York, NY), Chapter 8, 369–457 (1959).

61. D. Nyyssonen, "Spatial Coherence: The key to accurate optical micrometrology," Proc. SPIE Vol. 194, 34–44 (1979).

62. D. Nyyssonen, "Linewidth measurement with an optical microscope: effect of operating conditions on the image profile," Appl. Optics *16*, 2223–2230 (1977).

63. W.N. Charman, "Some experimental measurements of diffraction images in low resolution microscopy," J. Opt. Soc. Am. *53*, 410–414 (1963).

64. W.N. Charman, "Diffraction images of circular objects in high resolution microscopy," J. Opt. Soc. Am. *53*, 415–419 (1963).

65. B.M. Watrasiewicz, "Image formation in optical microscopy at high numerical aperture," Opt. Acta *12*, 167–176 (1965).

66. R.E. Kinzly, "Investigations of the influence of the degree of coherence upon the images of edge objects," J. Opt. Soc. Am. *55*, 1002–1007 (1965).

67. B.M Watrasiewicz, "Theoretical calculations of straight edges in partially coherent light," Proc. R. Soc. London Ser. *A 293*, 391–400 (1966).

68. C.P. Kirk, "Aberration Effects in an optical measuring microscope," Appl. Optics *26*, 3417–3424 (1987).

69. D.A. Swyt, "An NIST physical standard for the calibration of photomask linewidth measuring systems," Proc. SPIE Vol. 129, 98–105 (1978).

70. Mask uncertainty = 50 nm

71. D. Nyyssonen and C.P. Kirk, "Modeling the Optical Microscope Images of Thick Layers for the Purpose of Linewidth Measurement," Proc. SPIE Vol. 538, 179–187 (1985).

72. D. Nyyssonen, "Design of an Optical Linewidth Standard Reference Material for Wafers," Proc. SPIE Vol. 342, 27–34 (1982).

73. D. Nyyssonen, "Focused-beam vs. Conventional Bright-field Microscopy for Integrated Circuit Metrology," Proc. SPIE Vol. 565, (1985).

74. P.C.D. Hobbs, R.L. Jungerman, and G.S. Kino, "A Phase sensitive Scanning Optical Microscope," Proc. SPIE Vol. 565, 71–80 (1985).

75. S.D. Bennett, E.A. Peltzer, J. McCall, R. DeRosa, and I.R. Smith, "Confocal Metrology at 325 nm," Proc. SPIE Vol. 921, 85–91 (1988).

76. I. Smith, "Confocal Scanning Laser Microscopy: 3 Dimensional Metrology for Submicron Lithography," Solid State Technology, July (1986).

77. M. Davidson, K. Kaufman, and I. Mazor, "First results of a product utilizing coherence probe imaging for wafer inspection," Proc. SPIE Vol. 921, 100–114 (1988).

78. For the purpose of this discussion, the term "thick" applies to any film exceeding 1/4 the wavelength of the illuminating waverlength.

79. In this example, the illuminating wavelength was 5300 Å. The polysilicon thickness used clearly exceeded the 1/4 wavelength thickness criteria. The silicon dioxide film, however, can still be considered "thin" for this example.

80. S.S. Chim and G.S. Kino, "Submicrometer dimensional measurements with optical microscopy," Proc. SPIE Vol. 1673, 26–35 (1992).

81. Z.R. Hateb, S.L. Prins, J.R. McNeil, S.S.H. Naqvi, "16 MB DRAM trench depth characterization using dome scatterometry," Proc. SPIE Vol. 2196, 2–13 (1994).

82. M.R. Murnane, C.J. Raymond, Z.R. Hatab, S.S.H. Naqvi, and J.R. McNeil, "Developed Photoresist metrology using scatterometry," Proc. SPIE Vol. 2196, 47–59 (1994).

83. T.F. Hasan, D.S. Perloff, and C.L. Mallory, "Test Vehicles for the Measurement and Analysis of VLSI Lithographic and Etching Parameters," Semiconductor Silicon/1981, H.R. Huff, R.J. Kriegler, and Y. Takeishi (eds.), (The Electrochem. Soc. Inc.), 868–881 (1981).

84. D. Yen, L.W. Linholm, and M.G. Buchler, "A Cross-Bridge Test Structure for Evaluating the Linewidth Uniformity of an Integrated Circuit Lithography System," J. Electrochem. Soc. Vol. 129, 2313–2318 (1982).

85. P. Troccolo, L. Mantalas, R. Allen, and L. Linholm, "Extending Electrical Measurements to the 0.5 μm Regime," Proc. SPIE Vol. 1464, 90–103 (1991).

86. S. Jensen, "Quantitative Submicron Linewidth Determination Using Electron Microscopy," Proc. SPIE Vol. 275, 100–108 (1981).

87. M.T. Postek, W.J. Kerry, and R.D. Larrabee, "The Relationship between Accelerating Voltage and Electron Detection to Linewidth Measurement in an SEM," Scanning Vol. 10, 10–18 (1988).

88. M.T. Postek, and D.C. Joy, "Submicron Microelectronics Dimensional Metrology: Scanning Electron Microscopy," J. of R. of NBS Vol. 92, #3 (1987).

89. M.G. Rosenfield, "Analysis of linewidth measurement techniques using the low voltage SEM," Proc. SPIE Vol. 775, 70–79 (1987).

90. B. Singh, and W.H. Arnold, "Linewidth Measurements by Low-Voltage SEM," Proc. SPIE Vol. 921, 16–21 (1988).

91. F.C. Chang, G.S. Kino, and W.R. Studenmund, "Development of a deep-UV Mireau correlation microscope," Proc. SPIE Vol. 2196, 35–46 (1994).

92. D. Nyyssonen, J. Seligson, and I. Mazor, "Phase image metrology with a modified coherence probe microscope," Proc. SPIE Vol. 1926, 299 (1993).

93. Y.Xu, "Effects of dissimilar materials on submicron linewidth measurements from phase measurements," Proc. SPIE Vol. 2196, 24–34 (1994).

94. See any general textbook on optics, e.g., E. Hecht, and A. Zajac, *Optics*, (Addison-Wesley, Reading, MA), Chapter 6, 167–194 (1979), or M. Born and E. Wolf, *Principles of Optics*, (Pergamon Press, New York), Chapter 5, 202–231 (1959).

95. P. Troccolo, "Submicron Metrology," BACUS Symposium (1989).

96. T.F. Hasan and D. S. Perloff, "Automated Electrical Measurement Techniques to Control VLSI Linewidth, Resistivity, and Registration," Test and Measurement World, 5, 78–90 (1985).

97. B.J. Lin, J.A. Underhill, D. Sundling, and B. Peck, "Electrical Measurement of Submicrometer Contact Holes," Proc. SPIE Vol. 921, 164–169 (1988).

98. L.W. Linholm, R.A. Allen and M.W. Cresswell, "Microelectronic test structures for feature placement and electrical linewidth metrology," *Handbook of Critical Dimension Metrology and Process Control*, K.M. Monahan (ed.) (SPIE, Bellingham, WA) 91-118 (1993).

99. H. Schafft, "Standard Test Method for Determining the Average Width and Cross-Sectional Area of a Straight, Thin-Film Metal Line," ASTM Standard, Committee F-1, Doc. 209142, American Society for Testing Materials, 1916 Race Street, Philadelphia, PA.

100. T. Lindsay, K. Orvek, and R. Mumaw, "0.5 Micrometer Contact Measurement and Characterization," Proc. SPIE Vol. 1261 104–118 (1990).

101. R. Larrabee, L. Linholm and M. Postek, "Microlithography Metrology," *Handbook of VLSI Microlithography*, W. B. Glendinning and J.N. Helbert (eds.) (Noyes Publications, Park Ridge, NJ), Chapter 3, 175 (1991).

102. M. Postek, "Critical Issues in Scanning Electron Microscopy," J. Res. of Nat. Inst. Stan. & Tech., Vol. 99 No. 5, 641–671 (1993).

103. M. Born and E. Wolf, *Principles of Optics*, (Pergamon Press, New York, NY), Chapter 7, 332 (1959).

104. T.A. Dodson and D.C. Joy, "Fast Fourier Transform Techniques for measuring SEM resolution," Proceedings of the XIIth International Congress for Electron Microscopy, San Francisco Press, 406–407 (1990).

105. J. Bentley, N.D. Evans and E.A. Kenik, "Resolution Measurements for Scanning Electron Microscope," Proceedings of the XIIIth International Congress for Electron Microscopy, Paris, July (1994).

106. H. Martin, P. Perret, C. Desplat, and P. Reisse, "New approach in scanning electron microscope resolution evalution," Proc. SPIE Vol. 2439, 310–318 (1995).

107. M.T. Postek and A.E. Vladar, "SEM performance evaluation using the sharpness criterion," Proc. SPIE Vol. 2725, 505–514 (1996).

108. D.C. Joy, "Limits of SEM Resolution," Hitachi Instrument News, 1, 16–19 (1995).

109. M.T. Postek, "Scanning Electron Microscope Metrology," *Handbook of Critical Dimension Metrology and Process Control*, K.M. Monahan (ed.) (SPIE, Bellingham, WA) 70 (1993).

110. The exception to this rule is the recent introduction of a new type of SEM: the Environmental SEM. This technology has been introduced to look at samples generally prone to charging. Charging is minimized by viewing the specimens in an environment which surpresses charging, i.e., neutalizes charge build up on the specimen. This is new to the overall SEM field and especially new to the field of SEM metrology. See references 110 and 111 for further information.

111. G.D. Danilatos, "Introduction to the ESEM Instrument," Microsc. Res. and Tech., *25*, 354–361 (1993).

112. G.D. Danilatos, "Foundations of Environmental scanning electron microscopy," Adv. Electronics and Electron Physics, *71*, 109–250 (1988).

113. R. Larrabee, L. Linholm, and M. Postek, "Microlithography Metrology," *Handbook of VLSI Microlithography*, W. B. Glendinning and J.N. Helbert (eds.) (Noyes Publications, Park Ridge, NJ), Chapter 3, 178 (1991).

114. M.T. Postek, "Scanning Electron Microscope Metrology," *Handbook of Critical Dimension Metrology and Process Control*, K.M. Monahan (ed.) (SPIE, Bellingham, WA) 46 (1993).

115. V.K. Zworykin, J. Hillier, and R.L. Snyder, "A scanning electron microscope," ASTM Bulletin, *117*, 15–23 (1942).

116. J.M. Lafferty, J. Appl. Physics, *22*, 299 (1951).

117. J.I. Goldstein, D.E. Newbury, P. Echlin, D.C. Joy, C. Fiori, and E. Lifshin, *Scanning Electron Microscopy and X-Ray Microanalysis* (Plenum Press, New York, NY), Chap.2, (1984).

118. A.V. Crewe, D.N Eggenberger, J. Wall, and L.M. Welter, "Electron gun using a field emission source," Rev. Sci. Instr., *39*, 576–583 (1968).

119. A.V. Crewe and M. Isaacson, "The use of field emission tips in a scanning electron microscope," Proc. EMSA, G.W. Bailey (ed.) (San Francisco Press, San Francisco, CA), 359–360 (1968).

120. A.V. Crewe, M. Isaacson, and D. Johnson, "A simple scanning electron microscope," Rev. Sci. Instr., *40*, 241–246, (1969).

121. A.V. Crewe, D. Johnson, and M. Isaacson, "An electron gun scanning microscope," Proc. EMSA, G.W. Bailey (ed.) (San Francisco Press, San Francisco, CA), 360–361 (1968).

122. J. Orloff, "Thermal field emission for low voltage scanning electron microscopy," J. Microsc., 140 (3), 303–311 (1985).

123. L. Reimer, "Image Formation in Low Voltage Scanning Electron Microscopy," *SPIE Tutorial Text TT 12*, 144 (1993).

124. L.W. Swanson, "Field emission source optics," Electron Optical Systems, SEM Inc., AMF O'Hare, Il 60666, 137–147 (1984).

125. L.W. Swanson, "Comparative study of the zirconiated and built-up W thermal field cathode," J. Vac. Sci., 12(6), 1228–1233, (1975).

126. D.W. Tuggle and S.G. Watson, "A low voltage field emission column with a Shottky emitter," Proc. EMSA, G.W. Bailey (ed.) (San Francisco Press, San Francisco, CA), 455–457 (1984).

127. D.W. Tuggle, L.W. Swanson, and J. Orloff, "Application of a thermal field emitter source for high resolution, high current e-beam microprobes," J. Vac. Sci., 16(6), 1699–1703 (1979).

128. D.W. Tuggle, J.Z. Li, and L.W. Swanson, "Point cathodes for use in virtual source electron optics," J. Microsc., 140 (3), 293–301 (1985).

129. K.D. Van Der Mask, "Field emission, developments and possibilities," J. Microsc., 130 (3), 309–324 (1983).

130. L.M. Welter, "Application of a field emission source to SEM," *Principles and Technologies of Scanning Electron Microscopy*, M. A. Hayat (ed), 195–220 (1975).

131. M.R. Sheinfein, W. Qian, and J.C.H. Spence, "Brightness measurements of nanometer-sized field-emission tips," MSA Proceedings, G.W. Bailey and C.L. Reider (eds.), 632–633 (1993).

132. J.I. Goldstein, D.E. Newbury, P. Echlin, D.C. Joy, C. Fiori, and E. Lifshin, *Scanning Electron Microscopy and X-ray Microanalysis* (Plenum Press, New York, NY) (1981).

133. In this context, "resolution" is an ill-defined term.

134. The detector most commonly used in scanning electron microscopy is the scintillator-photomultiplier system developed by Everhart and Thornly in 1960. T.E. Everhart and R.F.M. Thornley, Rev. Sci. Instr., *37*, 246–248 (1960).

135. P. Kruitt, "Magnetic through-the-lens detection in electron microscopy and spectroscopy," Adv. Optical and Electron Microsc., *12*, 93–137 (1991).

136. S.R. Rogers, "New CD SEM Technology for 0.25 μm Production," Proc. SPIE Vol. 2439, 353–362 (1995).

137. K.M. Monahan, G. Toro-Lira, and M. Davidson, "A new low voltage SEM technology for imaging and metrology of submicrometer contact holes and other high aspect-ratio structures," Proc. SPIE Vol. 1926, 336–346 (1993).

138. T.W. Reilly, "Metrology algorithms for machine matching in different CD SEM configurations," Proc. SPIE Vol. 1673, 48–56 (1992).

139. T. Suganuma, "Measurement of surface topography using SEM with two secondary electron detectors," J. Elec. Microsc., *35* (1), 9–18 (1985).

140. B. Volbert and L. Reimer, "Advantages of two opposite Everhart-Thornley detector on SEM," SEM/1986/IV SEM Inc., (AMF O'Hare Chicago, IL), 1–10 (1986).

141. I. Nadler-Niv and U. Halavee, "Technologies for automated in-process E-beam metrology," Proc. SPIE Vol. 921, 2–15 (1988).

142. K. Harris, I. Nadler-Niv, and D. Levy, "Innate accuracy of a novel E-beam system," KTI Microelectronics Seminar Proc. San Diego, CA, 49–63 (1988).

143. M.T. Postek, W.J. Keery, and N.V. Frederick, "Low-Profile high efficiency micro-channelplate detector system for scanning electron applications," Rev. Sci. Instr., *61(6)*, 1648–1657 (1990).

144. M.T. Postek, W.J. Keery, and N.V. Frederick, "Development of a low-profile microchannel-plate electron detector system for SEM imaging and metrology," Scanning, *12*, I-27–28 (1990).

145. P.E. Russell and J.F. Mancuso, "Microchannel-plate detector for low voltage scanning electron microscopes," J. Microsc. *140*, 323–330 (1985).

146. L. S. Hordon, B. B. Boyer, and R. F. W. Pease, "Improved retarding field optics via image outside field," J. Vac. Sci. Technol. B13 (3), 826 (May/June 1995).

147. M. Ezumi, T. Otaka, H. Mori, and H. Todokoro, "Development of critical dimension measurement electron microscope for ULSI (S-8000 series)," Proc. SPIE Vol. 2725, 105–113 (1996).

148. M. Sato, H. Todokoro, T. Suzuki, and M. Yamada, "A snorkel type conical objective lens with E cross B field for detecting secondary electrons," Proc. SPIE Vol 2014, 17 (1993).

149. NIST Standard SRM 484, Office of Standard Reference Materials, National Institute of Standards and Technology, Room 204, Building 202, Gaithersburg, MD 20899.

150. M.T. Postek, "Scanning electron microscope-based metrological electron microscope and new prototype SEM magnification standard," Scanning Microscopy, 3(4) 1087–1099 (1989).

151. D.C. Joy, "Beam Interactions, contrast and resolution in the SEM," J. Microsc., *136*, (Pt 2) 241–258 (1984).

152. O.C. Wells, "Low-loss image for surface scanning electron microscope," Appl. Phys. Lett. *19*, 232–235 (1971).

153. K.M. Monahan, et al., "Benchmarking multi-mode CD-SEM metrology to 180 nm," Proc. SPIE Vol. 2439, 480–493 (1995).

154. M.T. Postek, "Electron detection modes and their relationship to linewidth measurement in the scanning electron microscope," Proceeding Joint Annual Meeting EMSA/MAS G. W. Bailey (ed.), 646–649 (1986).

155. N.T. Sullivan and R.M. Newcomb, "Critical Dimension measurement in the CD SEM: A comparison of Backscattered vs. secondary electrons," Proc. SPIE Vol. 2196, 118–127 (1994).

156. M. Knoll, "Aufladepotential und Sekundäremission elektronenbestrahlter Körper," Z. Tech. Phys., *16*, 467–475 (1935).

157. J.I. Goldstein, D.E. Newbury, P. Echlin, D.C. Joy, C. Fiori, and E. Lifshin, *Scanning Electron Microscopy and X-Ray Microanalysis* (Plenum Press, New York, NY), 466 (1984).

158. C.W. Oatley, *The Scanning Electron Microscope, Part I, The Instrument* (Cambridge University Press, Cambridge) (1972).

159. R.L. Van Asselt, H. Becker, and E.C. Douglas, "Development of accurate linewidth measurement techniques to support fabrication of devices with structures of 1 μm and less," Proc. SPIE Vol. 921, 22–32 (1988).

160. J.I. Goldstein, D.E. Newbury, P. Echlin, D.C. Joy, C. Fiori, and E. Lifshin, *Scanning Electron Microscopy and X-Ray Microanalysis* (Plenum Press, New York, NY), Chap. 4 (1984).

161. E.E. Chain and E.P. Baaklini, "Automated CD measurements with the Hitachi S-6280," Proc. SPIE Vol. 2439, 319–324 (1995).

162. Private communication to L. Lauchlan.

163. E.P. Baaklini and T.W. Reilly, unpublished.

164. H. Levinson, private communication.

165. D.K. Atwood and D.C. Joy, "Improved accuracy for SEM linewidth," Proc. SPIE Vol. 775, 159–165 (1987).

166. J.F. Mancuso and S. Erasmus, "Edge recognition in SEM Metrology," Proc. SPIE Vol. 565, 196–204 (1985).

167. M. Miyoshi, M. Kanoh, H. Yamaji, and K. Okumura, "A precise and automatic very large scale integrated circuit pattern linewidth measurement method using a scanning electron microscope," J. Va. Sci. Technol. B 4(2), 493–499 (1986).

168. N. Thane, R. Savage II, D. Stark, and L. Hollifield, "Reaching for 0.25 um with current available tool set," Proc. SPIE Vol. 2439, 89–103 (1995).

169. D. Nyyssonen, "Collection of low energy secondary electrons and imaging in a low-voltage SEM," Proc. SPIE Vol. 2725, 562–571 (1996).

170. G. Vachet and M. Young, "Critical dimension atomic force microscopy for 0.25 um process development," Proc. SPIE Vol. 2725, 555–561 (1996).

171. M. Davidson and N. Sullivan, "Monte Carlo simulation for CD SEM calibration and algorithm development," Proc. SPIE Vol. 2439, 334–344 (1995).

172. J. Schnier, T.H. McWaid, R. Dixson, V.W. Tsai, J.S. Villarrubia, E.D. Williams, and E. Fu, "Progress on accurate metrology of pitch, height, roughness, and width artifacts using an Atomic Force Microscope," Proc. SPIE Vol. 2439, 401–415 (1995).

173. Y. Martin and H. K. Wickramasinghe, "Toward accurate metrology with scanning force microscopies," J. Va. Sci. Technol. B 13(6), Nov. (1995).

174. R. Young, J. Ward, and F. Scire, Phys. Rev. Lett. *27*, 922 (1971).

175. R. Young, J. Ward, and F. Scire, Rev. Sci. Instrum. 43, 999 (1972).

176. G. Binnig, H. Rorher, C. Gerber, and E. Weibel, Phys. Rev Lett. *49*, 57 (1982).

177. G. Binnig, C.F. Quate, and Ch. Gerber, "Atomic Force Microscope," Phys. Rev Lett. *56*, 930–933 (1986).

178. Y. Martin, C.C. Williams, and H. K. Wickramasinghe, J. Appl. Phys. *61*, 4723 (1987).

179. Y. Martin, D.W. Abraham, and H. K. Wickramasinghe, Appl. Phys. Lett. *52*, 1103 (1988).

180. P.C.D. Hobbs and H. K. Wickramasinghe, "Metrology with an Atomic Force Microscope," Proc. SPIE Vol. 921, 146–150 (1988).

181. D. Nyyssonen, L. Landstein, and E. Coombs, "Two-dimensional Atomic force Microprobe Trench Metrology System," J. Vac. Sci. Technol. B9, 3612–3616 (1991).

182. D.A. Grigg, J.E. Griffith, G.P. Kochanski, M.J. Vasile, and P.E. Russell, "Scanning Probe Metrology," Proc. SPIE Vol. 1673, 557–567 (1992).

183. O. Nishikawa, M. Tomitori, and A. Minakuchi, Surf. Sci. *181*, 210 (1987).

184. L. Libioulle, A. Ronda, M. Taborelli, and J.M. Gilles, J. Vac. Sci. Technol. B9, 655 (1991).

185. J.E. Griffith, G.L. Miller, C.A. Green, D.A. Grigg, and P.E. Russell, "A scanning tunneling microscope with a capacitance-based position monitor," J. Vac. Sci. Technol. B8, 2023–2027 (1990).

186. R.C. Barrett and C.F. Quate, "Optical scan-correction system applied to atomic force microscopes," Rev. Sci. Instrum. *62*, 1393–1399 (1991).

187. Y. Martin and H. K. Wickramasinghe, "Method for imaging sidewalls by atomic force microscopy," Appl. Phys. Lett. *64* (19), 2498–2500 (1994).

188. O. Wolter, Th. Bayer, and J. Greschner, J. Vac. Sci. Technol. B9, 1353 (1991).

189. H.M. Marchman, J.E. Griffith, J.Z.Y. Guo, J. Frackoviak, and G.K. Celler, J. Vac. Sci. Technol. *B12*, 3585 (1994).

190. M.R. Rogers and F.D. Yashar, "Recent developments in atomic force microscopy applicable to integrated circuit metrology," Proc. SPIE Vol. 1673, 544–551 (1992).

191. The CDSXM developed at IBM is currently marketed by Veeco Instruments as the Dektak SXM.

192. H.M. Marchman, "Nanometer-scale dimensional metrology with noncontact atomic force microscopy," Proc. SPIE Vol. 2725, 527–539 (1996).

193. K. Wilder, B. Singh, and W.H. Arnold, "Sub-0.35-micron critical dimension metrology using atomic force microscopy," Proc. SPIE Vol. 2725, 540–554 (1996).

194. L. Montelius and J.O. Tegenfeldt, "Direct observation of the tip shape in scanning probe microscopy," Appl. Phys. Lett. 62 (21), 2628–2630 (1993).

195. F. Atamny and A. Baiker, "Direct imaging of the tip shape by AFM," Surf. Sci. *323*, 314–318 (1995).

196. Both characterization samples are available from IBM GMTC, Sindelfingen, Germany.

CHAPTER 7
Optical Lithography Modeling

Andrew R. Neureuther
University of California at Berkeley

Chris A. Mack
Finle Technologies, Inc.

CONTENTS

7.1 INTRODUCTION

7.1.1 Overview

Optical lithography is rapidly becoming very complex. As the limits of resolution are pushed to achieve feature sizes on the order of the wavelength of light or smaller, many phenomena must be understood. Technology innovation to extend optical lithography is also introducing many implementation options that must be assessed. Together, the increased concern for physical effects and the introduction of innovations have greatly increased the number of parameters whose effects must be characterized and balanced. Modeling offers a solid foundation for efficient characterization and a way to systematically quantify relationships and quickly investigate new innovations. The ultimate test is, of course, producing the desired features on product wafers. Yet a little time spent in understanding the models or in making a few simulation runs at a computer terminal can make working in the fabrication facility much more effective. It is also true that observational feedback from the fabrication facility can make modeling and simulation much more effective.

This chapter is designed to provide information about modeling and simulation at four distinct levels. It begins with an overview of the phases and nature of modeling and simulation. Then the underlying basic physical models and phenomena of optical imaging, substrate interactions, and resist dissolution are considered. The usefulness of modeling and simulation in concert with conventional characterization methods for determining the practical performance of lithography is then illustrated. Uses of modeling and simulation in assessing technology innovations in materials, exposure tools, masks, etc., are considered. Finally, a summary of available simulators is provided.

7.1.2 Overview of the Modeling and Simulation Process

Optical lithography modeling began in the early 1970s when Rick Dill started an effort at IBM Yorktown Heights Research Center to describe the basic steps of the lithography process with mathematical equations. At a time when lithography was considered a true art, such an approach was met with skepticism. The results of their pioneering work were published in a landmark series of papers in 1975,[1-4] now referred to as the "Dill papers." These papers not only gave birth to the field of lithography modeling, they represented the first serious attempt to describe lithography not as an art, but as a science. These papers presented a simple model for image formation with incoherent illumination, the first order kinetic "Dill model" of exposure, and an empirical model for development coupled with a cell algorithm for photoresist profile calculation. The Dill papers are still the most referenced works in the body of lithography literature.

Out of the Dill effort came a follow-on research program at the University of California at Berkeley. In 1979 this group presented the first result of their effort, the lithography modeling program SAMPLE.[5] SAMPLE improved the state of the art in lithography modeling by adding partial coherence to the image calculations and by replacing the cell algorithm for dissolution calculations with the string algorithm. But, more importantly, SAMPLE was made available to the lithography community. For the first time, researchers in the field could use modeling as a tool to help understand and improve their lithography processes.

In 1985 the model PROLITH (the Positive Resist Optical LITHography model)[6] was introduced. This model added an analytical expression for the standing wave intensity in the resist, a prebake model, a kinetic model for resist development (now known as the Mack model), and the first model for contact and proximity printing. PROLITH was also the first lithography model to run on a personal computer (the IBM PC), making lithography modeling accessible to all lithographers, from advanced researchers to process development engineers to manufacturing engineers.

In the years since this early work, modeling has become an essential tool for lithography research and development and is quickly becoming an important manufacturing tool as well. Many university, industry and commercial efforts have advanced the state of the art in lithography simulation and, at the same time, advanced our basic understanding of lithography.

Lithographic models must simulate the basic steps of image formation, resist exposure, post-exposure bake diffusion, and development to obtain a final resist profile. Figure 7.1 shows a basic schematic of the calculation steps required for lithography modeling. More details on these models can be found in subsequent sections. Of course, there can be many parts to a detailed blaock diagram of lithography simulation. Bake steps can cause chemical reactions which affect the final development rate behavior. Simulators frequently include some sort of "metrology" model which measures the width and sidewall angle of the resist profile. Also, analysis of the results (such as determining the depth of focus) can be a built-in part of the lithography simulator.

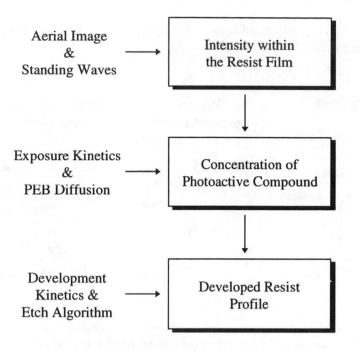

FIG. 7.1   Flow diagram of a lithography model.

7.2 MODELS FOR IMAGING

7.2.1 Exposure Systems and Key Parameters

Three basic methods of exposing layout patterns on masks and wafers are shown in Fig. 7.2. One approach is to form a basic elemental beam spot pattern with particles such as electrons or with a Gaussian optical beam. A scanning apparatus is then used to synthesize the desired pattern. The scanning can be raster type, with an array of predetermined locations and an on/off beam blanking, or vector type, in which only the exposed regions are scanned. Massively parallel exposure systems (where a large area of many patterns is exposed at one time) are classified as shadow (contact or proximity printing) or image forming (projection printing), as shown in Fig. 7.2. Optical and x-ray lithography are good examples of this "parallel" exposure approach and they involve fundamentally similar electromagnetic diffraction phenomena, even though the wavelengths used can differ by several orders of magnitude. In true contact printing the mask and wafer make intimate contact, while in proximity printing a small gap of height z is introduced to prevent damage to the mask. Within projection printing a mask is illuminated and the transmitted light (or for electrons or ions, the particle beam) is collected by a lens and imaged at the wafer plane. Very advanced lenses are required to transfer the massive amount of information in a layout pattern for a chip.

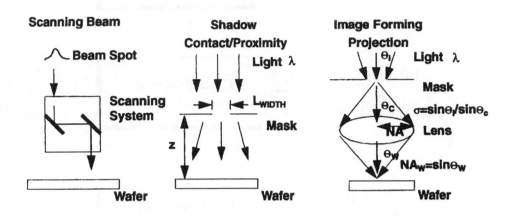

FIG. 7.2   Optical lithography configurations for printing and the fundamental parameters which describe them and determine resolution.

In scanning beam systems the exposure of the resist material can be characterized through the energy deposited along the particle trajectory. A simple first order model for topography-free surfaces is to consider the incident beam to have a Gaussian profile given by

$$I(r) = e^{-\frac{r^2}{2\beta^2}} \tag{7.1}$$

Here r is the radical distance from the center of the spot and β describes the standard deviation. The full width half maximum of this beam is 2.35β and is an indication of the typical spacing possible between scanned locations. The resolution, L_{width}, is

$$L_{width} = k_b\,\beta \tag{7.2}$$

where k_b is a constant representing the normalized resolution and is about 2.35 depending on the writing process utilized.

Optical contact printing with ultraviolet (UV) light has been used in production for linewidths down to 2.0 μm. The resolution is dependent on the separation between mask and wafer, and if intimate contact between the mask and wafer is maintained features in the 0.35 μm range can be produced.[7] The mask-to-wafer separation results in an inherent resolution limitation due to diffraction spreading such that the resolution is

$$L_{width} \approx \sqrt{k_c \lambda z} \qquad (7.3)$$

Here λ is the exposure wavelength, z is the mask-to-wafer separation as shown in Fig. 7.2 and the constant k_c is a constant around 1.6 that represents the normalized resolution and depends on the process utilized. The square root behavior is a consequence of the Fresnel theory of diffraction, which is required to model the near field region just below the mask opening.[8,9] As an example, the last generation of devices proximity printed by IBM was at 4 μm using UV light near 0.4 μm and a separation of 25 μm and corresponds to a k_c of 1.6. In x-ray lithography the wavelength is 400 times shorter (1 nm) and feature sizes below 0.2 μm are produced. To facilitate proximity printing the illumination is limited to a narrow set of angles (3°) and the gap variation Δz must be controlled to a few microns. While nearly perfect contact printing ($z \approx 0$) is possible, it has given way to proximity printing ($z > 0$), which avoids the potential of damage to the mask. While it may not be possible to mechanically preset the separation z to some desired accuracy, it is possible after the exposure to determine accurately the separation during the exposure by inspecting the resulting images.[10] A series of square open contacts works best for this purpose and the image changes (according to Fresnel diffraction) from square with ballooned corners ($\Delta v > 4.0$), to cloverleaf ($\Delta v \sim 3.5$), square again ($\Delta v \sim 3.0$), diamond ($\Delta v \sim 2.5$), and finally circular ($\Delta v < 2.0$) as the separation is increased.[11] Here, Δv is given by

$$\Delta v = L_{width} \sqrt{2 / (\lambda z)} \qquad (7.4)$$

High volume production is now accomplished with optical projection printing. The UV wavelengths of 436 nm and 365 nm, which are the g-line and i-line of mercury, the deep ultraviolet (DUV) wavelengths of 248 nm and 193 nm of KrF and ArF lasers, and the extreme ultraviolet (EUV) wavelengths near 13 nm and below are being considered. In projection printing, demagnification factors of 1X to 10X are common with 4X and 5X becoming more popular as a good trade-off between mask area and minimum mask feature size. Ring field systems, in which an image for a thin arc region is scanned across the chip, operate on quite similar optical principles to conventional step-and-repeat projection systems but put the focusing "power" in their reflective mirrors rather than the refractive lenses. A variety of optical system approaches have been envisioned, as can be found in articles by Bruning[12] and Markle.[13]

The working resolution of a projection system is given by[14]

$$L_{width} = k_1 \frac{\lambda}{NA} \qquad (7.5)$$

Here NA is the numerical aperture (the sine of the acceptance angle of the lens as viewed from the wafer shown in Fig. 7.2) and k_1 is a constant representing the normalized resolution that decreases from 0.8 to 0.5 as the resist, exposure tool, and mask technologies are improved. Several criteria for defining resolution are compared in Fig. 7.3. The well-known Rayleigh resolution criteria for resolving adjacent stars with a telescope corresponds to allowing the peak intensity of an adjacent star to fall into the null of the image of the first star, giving $k_1 = 0.61$. The fringe spacing of interfering coherent laser beams at an angle corresponds to $k_1 = 0.5$. Large angle illumination with many incident rays is known as incoherent illumination and can produce modulation for even smaller feature sizes. The modulation produced by incoherent illumination goes to zero for features at $k_1 = 0.25$ and smaller.

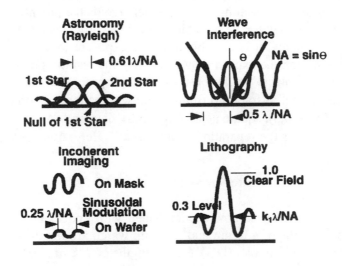

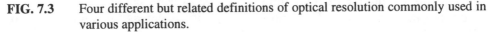

FIG. 7.3   Four different but related definitions of optical resolution commonly used in various applications.

The working total focus range (TFR) is

$$TFR = k_2 \frac{\lambda}{NA^2} \qquad (7.6)$$

where k_2 is a constant that describes the normalized TFR and is about 1.0, with a typical range of 0.6 to 1.5, depending on the resist, exposure tool, and mask technologies. The Rayleigh depth of focus,[15] which corresponds to a quarter wave phase error between the central (principal) ray and the extreme (marginal) ray, corresponds to $k_2 = 0.5$ in moving either toward or away from the lens (Fig. 7.4). Thus the Rayleigh defocus criteria predicts a total focus range of $k_2 = 1.0$. As an example, a 248 nm DUV projection printer with NA = 0.5 having $k_1 = 0.7$ and $k_2 = 1$ gives $L_{width} = 0.35$ µm and $TFR = 0.99$ µm.

Equations (7.5) and (7.6) are best interpreted as scaling equations in which k_1 and k_2 are the normalized resolution and TFR. It is often convenient to describe general lithographic trends using these two dimensionless variables rather than the actual resolution or TFR. One should be very cautious in using these two equations to predict resolution or TFR as a function of wavelength or numerical aperture. It is not appropriate to consider k_1 and k_2 as just "process dependent constants" rather than the variables they are.

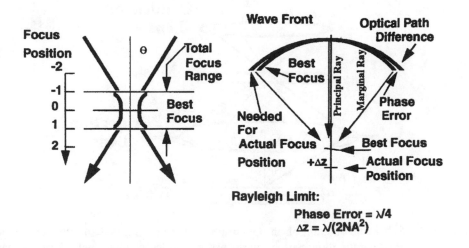

FIG. 7.4   Behavior of the optical image through best focus and the quarter wavelength phase error introduced by Rayleigh to define the focal depth.

7.2.2 Images in Projection Printing

An optical system for projection printing is shown in Fig. 7.5. The optical system consists of illumination, mask diffraction, and lens collection. Its operation is best understood by following rays through the system having various angles. The illumination can be viewed as consisting of a set of plane waves that strike the mask at various angles. The integrated circuit pattern on the mask diffracts these plane waves in the illumination into all possible visible angles. However, only rays at angles that fall within the acceptance angle cone of the projection lens (outer heavy line) continue toward the wafer. The greater thickness of the lens at its center slows down (adds phase) to the axial (principal) ray such that the rays at the edge of the lens (marginal rays) that are bent toward the axis can catch up and get sufficiently ahead (phase advance) to focus at the wafer. In an ideal lens this causes the broad diffraction from a pinhole to form a spherical wavefront converging toward the focal point on the wafer plane. The projection lens is thus acting like a low-pass filter with respect to ray angles. The thin-film stack on the wafer also influences the coupling and focusing of energy in the lithographic material and is, in essence, part of the optical system.

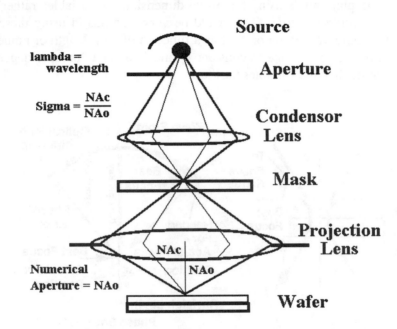

FIG. 7.5   Schematic diagram of an optical projection printing system optical column showing the illumination, mask, wafer, and lenses along with the rays that determine the numerical aperture (NA) and partial coherence (σ).

Optical system designers improve the imaging for lithographic purposes by tuning the illumination so that the system is neither fully coherent nor incoherent. The basic phenomena which is exploited is to make light illuminating adjacent pixels sufficiently coherent that the overlapping image tails of these two pixels will aid each other in synthesizing the composite image. There is a limit, though, to the amount of coherence that is useful, as too much coherence produces very noticeable ringing effects. Although each ray in the illumination system does not have any temporal coherence with any other ray, the desired local spatial coherence can be generated by restricting the degree that rays can be off-axis compared to the angles accepted by the lens. With restricted off-axis angles, the signal passing through adjacent pinholes due to each ray will have only a small phase difference. When this phase difference $\Delta\phi$ is less than

$\pi/2$ for every ray in the illumination the signals will be nearly identical at each instant of time. Partial coherence will be discussed in more detail shortly and is extensively described in Born and Wolf.[15]

The above discussion has been in terms of the numerical aperture seen at the wafer (NA_w) and has assumed a mask size equal to the wafer size (1X). Lenses are often described in terms of the F_{lens} number which describes the total power for bending rays on both the mask and wafer sides of a lens. The F_{lens} is the focal length divided by the diameter of the lens and is approximately $1/2NA_{lens}$. For a system with demagnification factor M (larger than 1) the sines of the angles at the mask are reduced by a factor M and

$$F_{lens} = \frac{0.5}{NA_{lens}\left(1 + \dfrac{1}{M}\right)} \tag{7.7}$$

$$NA_w = NA_{lens}\left(1 + \frac{1}{M}\right) \tag{7.8}$$

Imaging with incoherent illumination ($\sigma > 2$) is much easier to analyze because linear transform theory can be used on the intensity (rather than electric field). The approach used is to expand the mask intensity into sinusoids, weight the amplitudes of these components by the Modulation Transfer Function (MTF), and sum up the intensity at the image plane. A plot of the MTF is shown in Fig. 7.6. The MTF is unity for low spatial frequencies and decreases to zero for equal line and space patterns of width $0.25\lambda/NA_w$. A rule of thumb before partial coherence was well understood was that projection printers should operate at feature sizes corresponding to a 60% MTF. For example, at a wavelength of 436 nm and with a lens of $NA_w = 0.28$, a 1.22 μm line and 1.22 μm space pattern has a normalized spatial frequency of 0.62, which would produce an MTF of 0.6 or 60%.

We can see how the 60% MTF rule-of-thumb can be misleading by looking at another image quality metric called the contrast. The image contrast, C_{IMG}, is given in terms of the global intensity minimum and maximum by

$$C_{IMG} = \left(\frac{I_{max} - I_{min}}{I_{max} + I_{min}}\right) \leq 1 \tag{7.9}$$

as illustrated in Fig. 7.6. The contrast in the actual image is much higher than the MTF appears to indicate. First, the fact that the mask is an on-off square wave instead of a sinusoid increases the modulation at the wafer to 76%. More importantly, significant improvements in the image of square wave mask patterns are produced by the coherent image interaction effects from using partially coherent illumination. Specifically at a σ of 0.7 the resist at the wafer sees a 95% contrast for 1.22 μm equal lines and spaces. The MTF curve is also misleading in that it suggests a gradual decrease in contrast with decreasing feature size. In practice, when partial coherence is used to boost the contrast for large feature sizes a very abrupt falloff in image quality occurs for features smaller than $k_1 = 0.5$, as can be seen by the $\sigma = 0.3$ and 0.7 curves in Fig. 7.6.

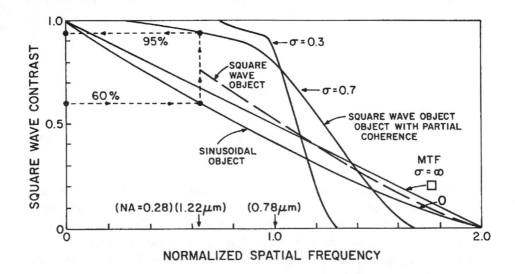

FIG. 7.6   Modulation Transfer Function (MTF) for a sinusoidal object as a function of the cycles per mm normalized to the highest frequency passed for coherent illumination. The contrast for imaging an equal line and space on-off pattern with partial coherence is also shown.[16]

The use of partial coherence can also improve image quality due to intra-feature interaction effects. The image of a simple knife edge shown in Fig. 7.7 is improved in several ways. The smoothest profile occurs for a sigma of infinity, which corresponds to a fully incoherent imaging system. For this case, linear transform theory (i.e., the MTF) can be used directly on the intensity and the image converges to half the discontinuity, or 0.5, at the knife edge. Introducing partial coherence by reducing sigma increases the edge slope and creates an overshoot in the peak intensity. More importantly, the intensity near the line edge is decreased and even takes on a minimum value there. This minimum plays a critical role in determining linewidth fidelity and profile quality. If sigma is reduced to as low as 0.2 the overshoot becomes excessive and extends laterally, indicating that interference (proximity) effects between adjacent features are likely. For this reason, values of sigma of 0.3 to 0.7 are used with 0.5 being typical. Note that as coherence is introduced the image converges to $0.5^2 = 0.25$ at the edge, as expected from using linear transform theory to calculate the electric field and then taking the square to get intensity.

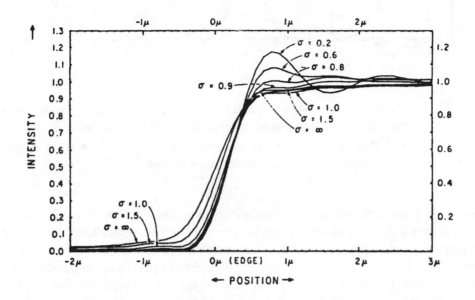

FIG. 7.7   Aerial image intensity versus horizontal distance for a knife-edge pattern at a wavelength (λ) of 0.436 μm and a numerical aperture of 0.28 with various degrees of partial coherence (σ).[16]

An important but not well recognized consequence of the use of partial coherence is that intensities add nonlinearly and superposition of intensity no longer holds. An illustration of the failure of superposition is shown in Fig. 7.8. Figure 7.8(a) shows aerial image intensities for a $0.3\,\lambda$ /NA opening with period $1.6\,\lambda$ /NA. Similarly, Fig. 7.8(b) shows a $0.5\,\lambda$ /NA opening with the same period. When the mask patterns are combined, one would expect to see a $0.8\,\lambda$ /NA opening with period $1.6\,\lambda$ /NA, as shown in Fig. 7.8(c). However, (a) and (b) do not add up to the image in (c), indicating that when partial coherence is present, superposition cannot be used.

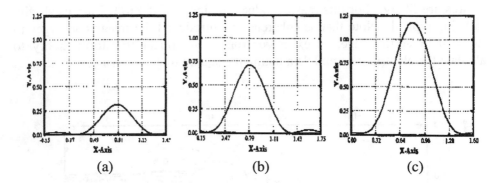

(a)   (b)   (c)

FIG. 7.8   Aerial image intensities for (a) 0.3 and (b) $0.5\,\lambda$ /NA openings with period $1.6\,\lambda$ /NA which when added together do not produce the intensity of (c) a $0.8\,\lambda$ /NA opening with period $1.6\,\lambda$ /NA, indicating that when partial coherence is present superposition cannot be used.[16]

Images of three different feature types are compared in Fig. 7.9. For this feature comparison a standard normalization is introduced. This normalization consists of setting λ = 0.5 and NA_w = 0.5. This makes the line size in microns equal to the numerical value of k_1 and the defocus in microns equal to k_2. Henceforth we will also refer to NA_w as NA. The first feature type is a periodic 0.8 line and 0.8 space pattern. Its image is the solid curve which is low on the left and high on the right. It is symmetrical about both edges of the plot and only half of the total period needs to be shown. An isolated line, shown as the dashed curve, is nearly but not quite similar, since it is not symmetrical about the right border of the graph. The isolated space shown as the dotted line is bright on the left side. In Fig. 7.9 the mask edge is at 0.4 and all of the images have an intensity of about 0.25 at the edge, as expected from the effects of partial coherence.

The optical images of integrated circuit (IC) patterns degrade with increasing focal error (defocus) or lens imperfections (aberrations). Aberrations are nothing more than optical path differences (OPD) across the lens. The OPD can be expanded in a Taylor series or a set of orthogonal functions (such as Zernike polynomials). Defocus for low NA lenses corresponds to a phase error that increases as the square of the distance of the ray from the center of the lens. Examples of focus effects are shown in Fig. 7.10. Note that the image appears to pivot about a point inside the bright area of the image, known as the image isofocal point. A more complete discussion of aberrations and their effects on image quality can be found in Born and Wolf.[15]

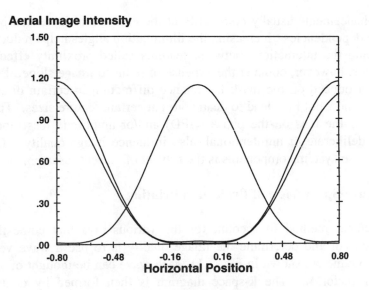

FIG. 7.9   Aerial image intensity for isolated line, isolated space, and equal line and space patterns versus horizontal distance.

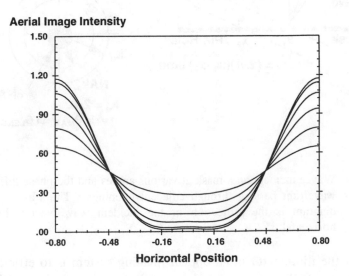

FIG. 7.10   Aerial image intensity of a $0.8\lambda/NA$ line and space pattern as focus is changed in steps of 0.4 Rayleigh units.

The illumination, mask diffraction, lens filtering, and thin-film environment all interplay together, and a systems-level approach must be used to understand how images of IC layout-type patterns might be selectively enhanced. Unfortunately, these enhancements usually come only at the expense of increased pattern size and type dependencies. Increasing the illumination angles helps reduce coherent ringing and the interactions between features called proximity effects. These advantages, however, come at the expense of reduced image slope. Phase shifts can be introduced on the mask to enhance diffraction in certain directions and suppress others but may lead to dead zones at certain feature sizes. The filtering qualities of the lens on the phase (OPD) and/or amplitude loss (apodization), whether deliberate or unintentional, also influence image quality. The role of each of these system components is the subject of current research.

7.2.3 Wave-Space View of Projection Printing

A convenient method to account for the various rays that come through an optical system and determine the image quality is to use a wave vector or **k**-space diagram. As shown in Fig. 7.11, each wave can be thought of as having a direction vector **k**. The **k**-space diagram is then formed by taking the ray projection on the x-y plane.

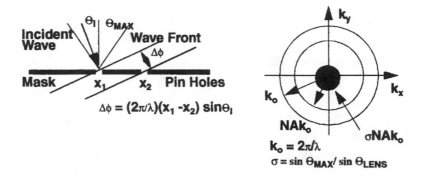

FIG. 7.11   Waves incident on a mask at various angles and the phase difference of the wavefront passing through adjacent openings. The wave vector **k**-space diagram is the projection of the incident wave vector direction on a horizontal.

The job of the illuminator in an optical imaging system is to efficiently collect light and to create the same set of plane waves for all points across the mask. In the conventional optical projection printing system shown in Fig. 7.11, the circular shape is a cone filled with incoherently related light rays illuminating the mask. The illuminator, or condenser, provides these rays and is ideally set up such that a well focused image of the shape of the illumination ray angle pattern is formed in what is known as the entrance pupil or **k**-space entrance plane of the lens (this is called Köhler illumination). Providing coherently unrelated illumination rays is very important and special techniques are used to insure that each ray illuminating the mask has no coherence in its relation to any other illumination ray. The fact that the ray angles are restricted to within a cone, however, has important consequences in that it introduces local spatial coherence on the mask and the system is said to use partial coherence. The ratio of the radius of the illumination cone to the acceptance cone of the lens is known as the partial coherence factor σ. The creation of this spatial coherence from the illumination will be discussed in detail after the **k**-space view of the entire optical system has been completed.

The IC pattern on the mask diffracts the plane waves in the illumination into many possible ray angles (or **k**-space locations). For purposes of illustration it is convenient to assume a pattern that is periodic. Consider first a single plane wave at angle Θ_I, as shown in Fig. 7.12, for a mask periodic in the x direction with period P_x. Light from the incident wave emerges through each of the holes, producing a spatially periodic pattern in x. This periodic signal gives rise to a set of plane waves leaving the mask and propagating downward toward the lens.

Each new wave has the property that its angle satisfies the Bragg condition; namely, the path differences between rays emerging from adjacent pinholes are integer multiples of a wavelength. The new waves, called diffracted orders, are numbered according to this path difference. The larger the period the smaller the angle between adjacent diffracted orders. In **k**-space the angles correspond exactly to periodic distances inversely proportional to the period in x or y. The magnitudes and phases of the scattered rays are, of course, usually all different and they may even change when the incident illumination angle changes, although it is usually assumed that they do not. Making the assumption that they are independent of the angle of incidence allows the role of the mask to be simply interpreted as that of making shifted and weighted copies of the pattern of the illumination. This gives rise to the array of circular shapes shown in Fig. 7.12.

The function of the lens is to collect the diffracted rays and reorient them in angle and phase such that the rays reconverge at the image location and resynthesize a good approximation of the mask signal at the wafer. For lithography purposes the ability of a lens system to capture a broad cone angle of diffracted rays is characterized by its numerical aperture NA ($= \sin\theta$) of the largest angle of a ray coming from the lens to the wafer. The lens thus acts like a low pass filter in **k**-space with an abrupt cutoff. Assuming there is no phase error or magnitude change in collecting and refocusing these rays, the lens is said to be diffraction limited. A **k**-space view of the role of the illumination, mask, and imaging lens is depicted in Fig. 7.13.

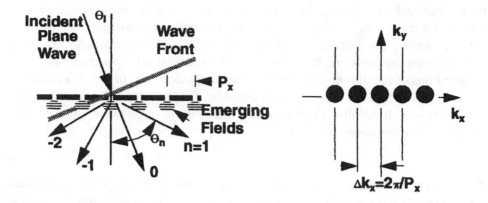

FIG. 7.12   Diffraction of an incident wave in passing through a periodic mask into various diffraction orders.   In **k**-space this corresponds into periodic replication of the illumination **k**-vectors at a spacing of $2\pi/P_x$.

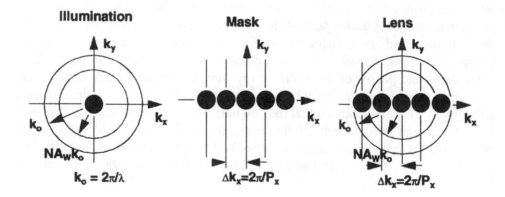

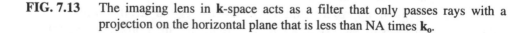

FIG. 7.13   The imaging lens in **k**-space acts as a filter that only passes rays with a projection on the horizontal plane that is less than NA times $\mathbf{k_o}$.

The **k**-space spectrum view is the foundation for formulating rigorous mathematical descriptions of the imaging.   The **k**-space view combines the illumination, mask diffraction, and lens collection effects.   Formally, the illumination can be viewed as providing a spectrum of plane waves with electric fields $W(\mathbf{k})$, which are then diffracted by the mask in direction **k´** by multiplying by $M(\mathbf{k´},\mathbf{k})$.   The lens collection is represented by integrating over waves in **k´**-

space and multiplying each by its transfer function $L(\mathbf{k}')$. To expose lithographic materials, work must be done on molecules that is proportional to the time-average of the square of the electric field, called the intensity. This intensity is proportional to the electric field times its complex conjugate. For the special case where the electric field and intensity for a large clear area are both normalized to unity, this time-average is simply $I = EE^*$.

Computing the intensity by taking one incident ray at a time, finding its diffraction pattern, and summing the collected fields is known as the Abbe method of imaging (also called the extended source method). This approach is used in many imaging programs. A second formulation is due to Hopkins,[17,15] who observed that if the complex mask diffraction efficiencies are not a function of the illumination angle then the integration over the illumination can be carried out before summing over the diffraction angles accepted by the lens. An advantage in Hopkins' approach is that the illumination integral is independent of the mask pattern and may be precalculated for a fixed optical exposure tool. Both approaches are found in simulation programs for lithography.

7.2.4 Images at the Small Feature Limit

The delta function system response due to a small pinhole mask pattern is called the point spread function. Light from a pinhole diffracts outwardly with equal amplitude in all directions and uniformly fills the entire collection cone in $\mathbf{k}$-space. The electric field is the inverse Fourier transform of the circular disk of plane waves. The intensity is this function times its complex conjugate (to give time average) and is the well known Airy function.[15]

$$I(x) = I_o \left[\frac{2J_1(v)}{v} \right]^2 \qquad (7.10)$$

$$v = 2\pi \left(\frac{NA}{\lambda} \right) x \qquad (7.11)$$

This function is plotted Fig. 7.14. Two important properties are that the null occurs at $0.61\lambda/NA$ as Rayleigh observed, and at $0.75\lambda/NA$ the first major sidelobe is at a maximum value of 0.044. The response to a delta function in one dimension (a thin line) is known as the line spread function also shown in Fig. 7.14. Its pattern also shows nulls and sidelobes at slightly smaller values of λ/NA.

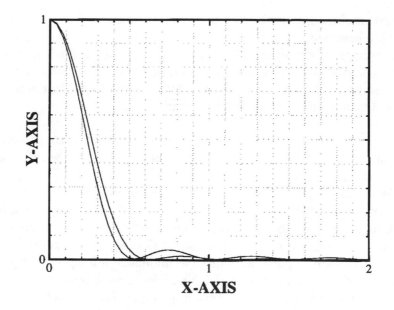

FIG. 7.14 Relative aerial image intensity versus normalized horizontal distance for a small mask opening as imaged by a diffraction limited system (outer curve). The normalized image of a small line opening is shown for comparison (inner curve).

An interesting relationship occurs between the point spread function and the mutual coherence due to the fact that they are both created by limiting maximum ray directions in the optical system. Both the resolution and the distance over which the illumination is strongly coherent are inversely proportional to the ray angle. Restricting the illumination to ray angles in **k**-space so that they are a factor of $\sigma < 1$ smaller makes the coherent distance across the mask sufficiently high that the resolution or pixel size is several times smaller. This means that interactions between adjacent pixels will basically be coherent. It is interesting that this partial coherence effect on the image is not a property of the illumination alone but rather the relative degree to which the illumination fills the acceptance cone of the lens.

The small feature limit of optical lithography is very useful for understanding differences in effects of feature types and mask polarity, the quality of the optical system, as well as the tendency for defects to print. Here "small" means features that are smaller than the resolution of the projection printer. As expected, these patterns give images that have a variation with horizontal distance given by the basic point or line spread function described earlier. It is, however, the behavior of their intensity versus their size that gives the insight.

The two basic feature types for the two polarities produce the four cases shown in Fig. 7.15.

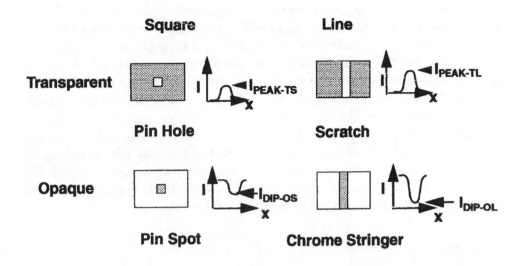

FIG. 7.15   Four basic types of small transparent/opaque square/line features in lithography. A simple perturbational model can be used to find the peak intensity or dip in intensity associated with each type.

Consider first the case of a small opening and a normalization such that a large clear area produces an electric field $E = 1$ and intensity $EE^* = 1$. Because the pinhole is small, any light coming through will be correlated. If, in addition, the pinhole also transmits the incident illumination waves without loss, the shape of the pinhole can be replaced by a square pinhole of side d that has equivalent area. Since the pinhole is smaller than a resolvable feature the electric field at the wafer contributed by the pinhole, δE, is proportional to its area. The image peak intensity will be proportional to $\delta E \delta E^*$, which goes as the area squared (or dimension[4]) and is given to good accuracy by[18]

$$I_{PEAK-TS} = 8.5\left(\frac{d}{\lambda / NA}\right)^4 \tag{7.12}$$

Intensities for the other three feature types can be related to the intensity $I_{PEAKS-TS}$ of the square pinhole (transparent square). A square pinhole can be formed by taking the product of two perpendicular lines. For this reason it should be expected that with reasonable coherence in the illumination, the peak intensity $I_{PEAKS-TL}$ for a scratch (transparent line) is the square root of that for the transparent square and increases with the width squared.

$$I_{PEAK-TL} = \sqrt{I_{PEAK-TS}} \qquad (7.13)$$

The opaque cases follow from a perturbation argument applied to the local field of a clear area. The electric field as a function of position on the mask for a pin spot is that for a clear mask minus the electric field for a pinhole. The electric field at the wafer in the vicinity of the pin spot image location is thus one minus the electric field of the pinhole, or 1-δE. The intensity for a pin spot is then (1-δE)(1-δE)* and is approximately 1-2δE. This of course means that the intensity will dip by 2δE at the location of the pin spot. For an opaque line a similar argument allows the electric field at the mask to be the clear field minus the transparent line. The field at the wafer and resulting intensity follow and can be expressed in terms of the peak intensity for a transparent square so that the effects of all four feature types can be quantitatively determined from substituting their dimension d into the expression for $I_{PEAKS-TS}$.

$$I_{DIP-TS} = 1 - 2\sqrt{I_{PEAK-TS}} \qquad (7.14)$$

$$I_{DIP-TL} = \left(1 - 2\left(\sqrt[4]{I_{PEAK-TS}}\right)\right) \qquad (7.15)$$

Fig. 7.16 shows the actual intensity behavior for feature sizes normalized in terms of λ/NA. For $d = 0.25\lambda/NA$ the above formulas predict that the peak intensities for the transparent case are 0.03 and 0.18 for a square and a line respectively. The corresponding opaque intensity in the dip should be 0.64 and 0.15 for the square and the line. The values in Fig. 7.16 indicate that these simple rules of thumb apply, although the error for the opaque line is considerable because the perturbation is 85% of the clear field intensity. Also notice that to achieve a peak intensity of 0.75 of the clear field intensity, square contacts must be about $0.7\lambda/NA$ while lines can be as small as $0.6\lambda/NA$. For this reason, lithographers often oversize contacts compared to lines. This, unfortunately, may not help the packing density much. With regard to defects, Fig. 7.16 shows that the requirement of a worst case 30% intensity through a clear defect means a square defect should be less than $0.45\lambda/NA$ and a scratch should be less than $0.3\lambda/NA$ wide. For opaque defects, where the intensity is to stay above 70%, the square and line should be less than 0.25 and $0.07\lambda/NA$ respectively. Unfortunately, defects typically coherently interact with features. As a result even tighter specifications are required on defects in the vicinity of features. For example, defects on the order of one third, and in some critical array cases one quarter, of the feature size L_W must be avoided. Incidentally, the assessment of the printability of these defects can be calculated using extensions of the perturbational model.

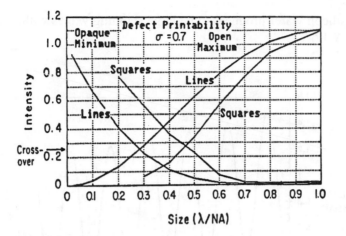

FIG. 7.16 Peak intensity for small transparent squares and lines and minimum intensity for small opaque squares and lines as a function of their size in λ/NA.[19]

7.3 MODELS FOR WAFER/RESIST INTERACTIONS

The reaction of the resist to the incident aerial image depends on the details of the intensity produced within the resist by the image, the bleaching of the resist during exposure, and the dissolution of the resist material during development.

7.3.1 Substrate interactions

The substrate in all forms of lithography contributes reflected photons or particles that usually degrade the image. In particle beam systems it is typically a broad background glow that is described by the backscattered term in the dual Gaussian beam model. In optical systems, the reflected light can interfere with incoming light, causing many interesting, and usually unwanted, effects.

For optical lithography in the UV and DUV regions substrate reflections are very severe and contribute standing wave effects. Even if there is no coherence between individual photons, each photon wave packet runs over its own tail as it reflects from the substrate. As long as the phase $\phi(t)$ in the plane wave expression vaires sufficiently slowly so that the phase does not change over about twice the resist thickness, distinct nulls will appear in the resist. The null to null spacing is $\lambda/(2n_{RESIST-R})$ where $n_{RESIST-R}$ is the real part of the refractive index for the resist. The ratio of the exposure in the constructive and destructive regions can be a factor of 8. The interference phenomena also makes the energy coupled into the resist a strong function of resist thickness. Standing wave effects are a major concern in optical lithography. A simple process of resist on oxide on a silicon wafer produces quite large standing waves within the resist, as seen in Fig. 7.17.

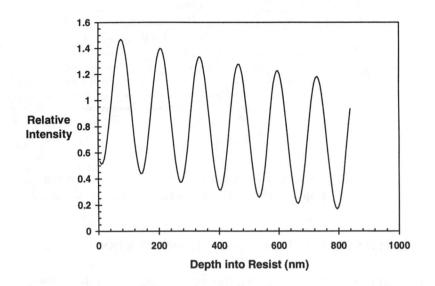

FIG. 7.17   Standing wave pattern of the resist on oxide on silicon due to interference between incoming and reflected waves.[20]

The relative exposure of the resist in the maxima and minima depends on the reflection coefficient seen by the wave in the resist material upon striking the substrate.   This reflection coefficient (the ratio of the reflected to incident electric fields) can be calculated from the complex refractive indices of the resist n_{RESIST} and substrate n_{SUB} as

$$\rho = \frac{\left(n_{RESIST} - n_{SUB}\right)}{\left(n_{RESIST} + n_{SUB}\right)} \qquad (7.16)$$

Note that the intensity reflectivity (a more easily measured quantity) is the square of the magnitude of the reflection coefficient.   The reflection coefficient of silicon seen from within the resist can now be calculated from the refractive indices of the resist media and silicon substrate.   At a wavelength of 436 nm typical values are $n_{RESIST} = 1.68$ and $n_{SUB} = 4.82 - i0.117$.   This gives

$$\rho_{inside} = \frac{\left(1.68 - \left(4.82 - i0.117\right)\right)}{\left(1.68 + \left(4.82 - i0.117\right)\right)} = 0.48 \angle 178.9 \qquad (7.17)$$

(Note that the phasar notation of magnitude and angle found in Eg. 7.17 is simply the polar representation of a complex number, in this case the reflection coefficient). The maximum and minimum of the electric fields can be estimated by neglecting the gradual decay of the waves as they propagate to and from the substrate surface. The maximum and minimum normalized to an incident wave in the resist with unit amplitude are

$$E_{max} = 1+|\rho| = 1.48; \; E_{min} = 1-|\rho| = 0.52 \qquad (7.18)$$

The energy deposited per unit volume is proportional to the square of the electric field times the extinction constant. Again normalizing to the exposure produced by a unit incident wave gives

$$I_{max} = (1+|\rho|)^2 = 2.19; \; and \; I_{min} = (1-|\rho|)^2 = 0.27 \qquad (7.19)$$

The ratio of I_{max} to I_{min} is 8.10 and shows that the exposure in the constructive interference nodes is about eight times larger than it is in the destructive interference nodes. The lateral intensity variation in the image must help overpower this vertical variation in the resist. The vertical variation of exposure can be viewed as a vertical contrast of

$$C_{VERTICAL} = \frac{(I_{max} - I_{min})}{(I_{max} + I_{min})} = \frac{(2.19 - 0.27)}{(2.19 + 0.27)} = 0.78 \qquad (7.20)$$

Given a vertical contrast of 0.78 it is no wonder that as a rule a horizontal image contrast of 0.8 to 0.9 is required in printing with conventional resist on silicon. The phase of the reflection coefficient describes the positioning of the nodes with respect to the substrate. A phase of about 180°, as occurs for silicon, produces an exposure minimum at the substrate. Aluminum has an index of $n_{Al} = 0.468 - i4.84$, which gives $\rho_{Al} = 0.942 \angle 142$. The larger magnitude of the reflection coefficient on aluminum produces an extremely low exposure in the destructive nulls, especially near the surface where the attenuation during propagation to and from the substrate has little effect. Fortunately, the 38° phase shift moves the positions of the destructive nulls about $\lambda_M/16$ away from the interface so the resist at the substrate interface receives more than the minimum exposure and is more easily removed during development.

A second major problem in optical lithography is that the periodic change in the resist reflectivity with resist thickness produces periodic variations in the efficiency with which energy is coupled into the resist layer. This variation with resist thickness is known as the *swing curve*. For substrates with a high real part of the refractive index, the resist can act like a matching layer, and a thin-film stack thickness of an odd multiple of quarter wavelengths is desired. An even multiple will result in poor coupling. Thus a quarter wavelength change in the resist coating thickness can be equivalent to a large change in exposure intensity as depicted in Fig. 7.18. For conventional resist on silicon, the energy coupled into the resist changes by about 50% for a 65 nm resist thickness change. For this reason coating thickness is typically controlled to within 5 - 10 nm. Unfortunately, topography on the wafer results in local resist thickness variations, which in turn produce coupling variations and linewidth control problems.

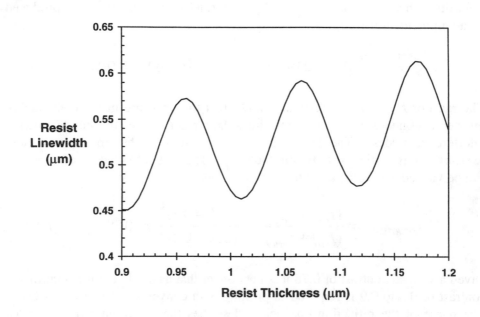

FIG. 7.18   Swing curve of changing resist linewidth with photoresist thickness caused by thin-film interference effects.

The extremes of the coupling of energy into the resist coated substrate can be calculated through a simple procedure if the attenuation in the resist is ignored. To avoid the complex algebra associated with complex indices of refraction, simply assume that the reflection coefficient of the same magnitude is produced by a replacement material with an equivalent but only real refractive index n_{EQ} underneath the resist. When coated with resist material of an odd or even number of quarter thicknesses, this equivalent material will produce effective refractive indices in air respectively of

$$n_{STACK} = \frac{(1 - |\rho|)}{(1 + |\rho|)} n_{RESIST} = \frac{n_{RESIST}^2}{n_{EQ}}$$

$$n_{STACK} = \frac{(1 + |\rho|)}{(1 - |\rho|)} n_{RESIST} = n_{EQ} \tag{7.21}$$

A simple interpretation is that the resist is acting like a quarter wave coating material for the equivalent substrate in the odd multiple case, and for the even multiple case it has no effect. The corresponding reflectivity of the resist coated layer seen in air for odd and even coating thickness will thus be

$$R_{STACK} = [\rho_{AIR}]^2 = \left[\frac{1 - n_{STACK}}{1 + n_{STACK}}\right]^2 \tag{7.22}$$

For the example refractive indices of silicon and resist used previously, $n_{EQ} = 4.78$ (very close to the real part of the true refractive index for silicon), $n_{STACK} = 0.59$ and 4.78, and $R_{STACK} = 0.43$ and 0.07. This means that for the odd multiples of a quarter wavelength coating 93% of the light is coupled into the thin-film stack, while for the even case only 53% is coupled in the thin-film stack. The variation in energy coupled between the two resist thickness cases corresponds to an equivalent exposure variation of a factor of 1.6. This variation in coupled energy, while moderated somewhat by attenuation/bleaching in the resist and minor dissolution differences in starting from a constructive/destructive interference position, is the main contributor to the variation in linewidth seen in the swing curve. It should be noted that regardless of the resist thickness and the resultant energy coupling, the relative maximum and minimum exposures within the resist are always the same as predicted above by the vertical contrast $C_{VERTICAL}$.

7.3.2 Exposure and the Latent Image

The kinetics of photoresist exposure is intimately tied to the phenomenon of absorption. The discussion below begins with a description of absorption, followed by the chemical kinetics of exposure.

The phenomenon of absorption can be viewed on a macroscopic or a microscopic scale. On the macro level, absorption is described by the familiar Lambert and Beer's laws, which gives a linear relationship between absorbance and path length times the concentration of the absorbing species. On the micro level, a photon is absorbed by an atom or molecule, promoting an electron to a higher energy state. Both methods of analysis yield useful information for describing the effects of light on a photoresist.

The basic law of absorption is empirical. It was first expressed by Lambert in differential form as

$$\frac{dI}{dz} = -\alpha I \tag{7.23}$$

where I is the intensity of light traveling in the z-direction through a medium, and α is the absorption coefficient of the medium and has units of inverse length. In a homogeneous medium (i.e., α is not a function of z), Eq. (7.23) may be integrated to yield

$$I(z) = I_0 \exp(-\alpha z) \tag{7.24}$$

where z is the distance the light has traveled through the medium and I_0 is the intensity at $z = 0$. If the medium is inhomogeneous, Eq. (7.24) becomes

$$I(z) = I_0 \exp(-Abs(z)) \tag{7.25}$$

where

$$Abs(z) = \int_0^z \alpha(z')dz' = \textit{the absorbance}$$

In 1852 Beer showed that for dilute solutions the absorption coefficient is proportional to the concentration of the absorbing species in the solution.

$$\alpha_{solution} = ac \tag{7.26}$$

where a is the molar absorption coefficient, given by $a = \alpha MW/\rho$; MW is the molecular weight; ρ is the density; and c is the concentration. The stipulation that the solution be dilute expresses a fundamental limitation of Beer's law. At high concentrations, where absorbing molecules are close together, the absorption of a photon by one molecule may be affected by a nearby molecule.[21] Since this interaction is concentration dependent, it causes deviation from the linear relation (7.26). Also, an apparent deviation from Beer's law occurs if the index of refraction changes appreciably with concentration. Thus, the validity of Beer's law should always be verified. For an N component homogeneous solid, the overall absorption coefficient becomes

$$\alpha_T = \sum_{j=1}^{N} a_j c_j \tag{7.27}$$

We will now apply the concepts of macroscopic absorption to a typical positive photoresist. A diazonaphthoquinone positive photoresist (such as AZ1350J) is made up of four major components: a base resin R that gives the resist its structural properties, a photoactive compound M (abbreviated PAC), exposure products P generated by the reaction of M with ultraviolet light, and a solvent S. Although photoresist drying during prebake is intended to drive off solvents, thermal studies have shown that a resist may contain 10% solvent after a 30 min. 100°C prebake.[22,23] The absorption coefficient α is then

$$\alpha = a_M M + a_p P + a_R R + a_s S \tag{7.28}$$

If M_o is the initial PAC concentration (i.e., with no UV exposure), the stoichiometry of the exposure reaction gives

$$P = M_o - M \tag{7.29}$$

Eq. (7.28) may be rewritten as[2]

$$\alpha = Am + B \tag{7.30}$$

where $\quad A = \left(a_M - a_p \right) M_o$
$$B = a_p M_o + a_R R + a_s S$$
$$m = M / M_o$$

A and B are called the bleachable and nonbleachable absorption coefficients, respectively, and make up the first two Dill photoresist parameters.[2]

The quantities A and B are experimentally measurable[2] and can be easily related to typical resist absorbance curves, measured using a UV spectrophotometer. When the resist is fully exposed, $M = 0$ and

$$\alpha_{exposed} = B \tag{7.31}$$

Similarly, when the resist is unexposed, $m = 1\left(M = M_o \right)$ and

$$\alpha_{unexposed} = A + B \tag{7.32}$$

From this A may be found by

$$A = \alpha_{unexposed} - \alpha_{exposed} \qquad (7.33)$$

Thus, $A(\lambda)$ and $B(\lambda)$ may be determined from the UV absorbance curves of unexposed and completely exposed resist (Fig. 7.19).

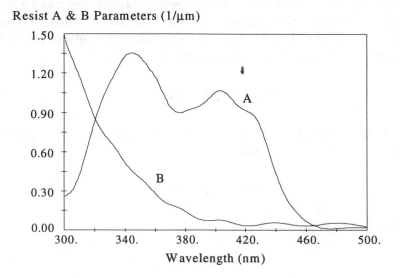

FIG. 7.19 Resist parameters A and B as a function of wavelength, measured with a UV spectrophotometer.

On a microscopic level, the absorption process can be thought of as photons being absorbed by an atom or molecule causing an outer electron to be promoted to a higher energy state. This phenomenon is especially important for the photoactive compound since it is the absorption of UV light that leads to the chemical conversion of M to P.

$$M \xrightarrow{\;UV\;} P \qquad (7.34)$$

This concept is stated in the first law of photochemistry: only the light that is absorbed by a molecule can be effective in producing photochemical change in the molecule. The actual chemistry of diazonaphthoquinone exposure is given below.

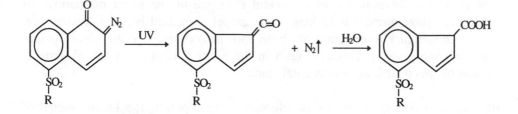

The kinetics of this exposure reaction are first order in photoactive compound M and in photon flux:

$$\frac{dm}{dt} = -CIm \qquad (7.35)$$

where C is the standard exposure rate constant and the third Dill photoresist parameter. If the intensity within the photoresist film remains constant during exposure, the exposure rate equation can be easily solved to give

$$M = M_o \exp(-CIt) \qquad (7.36)$$

A solution to the exposure rate equation is simple if the intensity within the resist is constant throughout the exposure. However, this is generally not the case. In fact, many resists *bleach* upon exposure, i.e., they become more transparent as the photoactive compound M is converted to product P. This corresponds to a positive value of A, as seen, for example, in Fig. 7.19. Since the intensity varies as a function of exposure time, this variation must be known in order to solve the exposure rate equation.[24]

7.3.3 Development

An overall positive resist processing model requires a mathematical representation of the development process. Early work took the form of empirical fits to development rate data as a function of exposure.[2,25] The model formulated below begins on a more fundamental level, with a postulated reaction mechanism which then leads to a development rate equation.[26,27] The rate constants involved can be determined by comparison with experimental data. Deviations from the expected development rates have been reported under certain conditions at the surface of the resist. This effect, called the *surface induction* or *surface inhibition*, can be related empirically to the expected development rate, i.e., to the bulk development rate as predicted by a kinetic model.

Unfortunately, fundamental experimental evidence of the exact mechanism of photoresist development is lacking. The model presented below is reasonable, and the resultant rate equation has been shown to describe actual development rates extremely well. However faith in the exact details of the mechanism is limited by this dearth of fundamental studies.

In order to derive an analytical development rate expression, a kinetic model of the development process will be used. This approach involves proposing a reasonable mechanism for the development reaction and then applying standard kinetics to this mechanism in order to derive a rate equation. We shall assume that the development of a diazo-type positive photoresist involves three processes: diffusion of developer from the bulk solution to the surface of the resist, reaction of the developer with the resist, and diffusion of the product back into the solution. For this analysis, we shall assume that the last step, diffusion of the dissolved resist into solution, occurs very quickly so that this step may be ignored. Let us now look at the first two steps in the proposed mechanism. The diffusion of developer to the resist surface can be described with the simple diffusion rate equation

$$r_D = k_D(D - D_S)$$
(7.37)

where r_D = rate of diffusion of the developer to the resist surface, D = bulk developer concentration, D_S = developer concentration at the resist surface, and k_D = rate constant.

We shall now propose a mechanism for the reaction of developer with the resist. The resist is composed of large macromolecules of resin R along with a photoactive compound M, which converts to product P upon exposure to UV light. The resin is quite soluble in the developer solution, but the presence of the PAC (photoactive compound) acts as an inhibitor to dissolution, making the development rate very slow. The product P, however, is very soluble in developer, enhancing the dissolution rate of the resin. Let us assume that n molecules of product P react with the developer to dissolve a resin molecule. The rate of the reaction is

$$r_R = k_R D_S P^n$$
(7.38)

where r_R = the rate of reaction of the developer with the resist and k_R = rate constant. (Note that the mechanism shown in Eq. (7.38) is the same as the "polyphotolysis" model described by Trefonas and Daniels[28]). From the stoichiometry of the exposure reaction,

$$P = M_o - M$$
(7.39)

where M_o is the initial PAC concentration (i.e., before exposure).

The two steps outlined above are in series, i.e., one reaction follows the other. Thus, the two steps will come to a steady state such that

$$-r_R = r_D = r \qquad (7.40)$$

Equating the rate equations, one can solve for D_S, and eliminate it from the overall rate equation, giving

$$r = \frac{k_D k_R D P^n}{k_D + k_R P^n} \qquad (7.41)$$

Using Eq. (7.39) and letting $m = M/M_o$, the relative PAC concentration, Eq. (7.41) becomes

$$r = \frac{k_D D(1-m)^n}{k_D / k_R M_o^n + (1-m)^n} \qquad (7.42)$$

When $m = 1$ (resist unexposed), the rate is zero. When $m = 0$ (resist completely exposed), the rate is equal to r_{max} where

$$r_{max} = \frac{k_D D}{k_D / k_R M_o^n + 1} \qquad (7.43)$$

If we define a constant a such that

$$a = k_D / k_R M_o^n \qquad (7.44)$$

the rate equation becomes

$$r = r_{max} \frac{(a+1)(1-m)^n}{a + (1-m)^n} \qquad (7.45)$$

Note that the simplifying constant a describes the rate constant of diffusion relative to the surface reaction rate constant. A large value of a will mean that diffusion is very fast, and thus less important, compared to the fastest surface reaction (for completely exposed resist).

There are three constants that must be determined experimentally, a, n, and r_{max}. The constant a can be stated in a more physically meaningful form as follows. A characteristic of some experimental rate data is an inflection point in the curve at about $m = 0.2$-0.7. The point of inflection can be calculated by letting

$$\frac{d^2 r}{dm^2} = 0$$

giving

$$a = \frac{(n+1)}{(n-1)}(1-m_{TH})^n \qquad (7.46)$$

where m_{TH} is the value of m at the inflection point, called the threshold PAC concentration.

This model does not take into account the finite dissolution rate of unexposed resist (r_{min}). One approach is simply to add this term to Eq. (7.45), giving

$$r = r_{max}\frac{(a+1)(1-m)^n}{a+(1-m)^n} + r_{min} \qquad (7.47)$$

This approach assumes that the mechanism of development of the unexposed resist is independent of the above-proposed development mechanism. In other words, there is a finite dissolution of resin that occurs by a mechanism that is independent of the presence of exposed PAC.

Consider the case when the diffusion rate constant is large compared to the surface reaction rate constant. If $a \gg 1$, the development rate Eq. (7.47) will become

$$r = r_{max}(1-m)^n + r_{min} \qquad (7.48)$$

The interpretation of a as a function of the threshold PAC concentration m_{TH} given by Eq. (7.46) means that a very large a would correspond to a large negative value of m_{TH}. In other words, if the surface reaction is very slow compared to the mass transport of developer to the surface, there will be no inflection point in the development rate data and Eq. (7.48) will apply. It is quite apparent that Eq. (7.48) could be derived directly from Eq. (7.38) if the diffusion step were ignored.

The kinetic model given above predicts the development rate of the resist as a function of the photoactive compound concentration remaining after the resist has been exposed to UV light. There are, however, other parameters that are known to affect the development rate, but which were not included in this model. The most notable deviation from the kinetic theory is the surface inhibition effect. The inhibition, or surface induction, effect is a decrease in the expected development rate at the surface of the resist.[29-31] Thus, this effect is a function of the depth into the resist and requires a new description of development rate.

Several factors have been found to contribute to the surface inhibition effect. High temperature baking of the photoresist has been found to produce surface inhibition and is thought to cause oxidation of the resist at the resist surface.[29-31] In particular, prebaking the photoresist may cause this reduced development rate phenomenon.[29,31] Alternatively, the induction effect may be the result of reduced solvent content near the resist surface. Of course, the degree to which this effect is observed depends on the prebake time and temperature. Finally, surface inhibition can be induced with the use of surfactants in the developer.

An empirical model can be used to describe the positional dependence of the development rate. If we assume that the development rate near the surface of the resist exponentially approaches the bulk development rate, the rate as a function of depth, $r(z)$, is

$$r(z) = r_B\left(1-(1-r_o)e^{-\beta_1 z}\right) \qquad (7.49)$$

where r_B = bulk development rate, r_o = development rate at the surface of the resist relative to r_B, and β_1 = an empirical constant. The induction effect has been found to take place over a depth of about 150 nm,[29,31] taken as the point at which the deviation from the bulk development rate has been reduced to 25% of its value at the surface of the resist (i.e., $\exp(-\beta_1 z) = 0.25$). Thus, an appropriate value β_1 is about $10\mu m^{-1}$.

The models presented above are certainly not the only acceptable description of positive photoresist development. In particular, the model of Kim has met with good success.[31,32]

7.4 ROLE OF SIMULATION IN PRACTICAL CHARACTERIZATION

Characterizing and understanding a lithography process is an important part of using processes effectively in manufacturing. Unfortunately, the requisite characterization experiments are expensive and time-consuming. Simulation provides an ideal tool to enhance experimental characterization and improve the understanding of a given process. The ease and flexibility of modeling tools also allow for quick process changes leading to optimization studies. The sections below describe how lithography simulation can be used to characterize a lithography process and understand the complex interaction among parameters.

7.4.1 Focus Effects

The effect of focus on a projection lithography system (such as a stepper) is a critical part of understanding and controlling a lithographic process. This section will address the importance of focus by providing definitions of the *process window* and *depth of focus* (DOF). Simulation proves an invaluable tool for predicting focus effects, generating process windows, and determining realistic values for the DOF.

In general, DOF can be thought of as the range of focus errors that a process can tolerate and still give acceptable lithographic results. Of course, the key to a good definition of DOF is in defining what is meant by tolerable. A change in focus results in two major changes to the final lithographic result: the photoresist profile changes and the sensitivity of the process to other processing errors is increased. Typically, photoresist profiles are described using three parameters: the linewidth (or critical dimension, CD), the sidewall angle, and the final resist thickness. The variation of these parameters with focus can be readily determined for any given set of conditions. The second effect of defocus is significantly harder to quantify: as an image goes out of focus, the process becomes more sensitive to other processing errors such as exposure dose and develop time. Of these secondary process errors, the most important is exposure.

Since the effect of focus is dependent on exposure, the only way to judge the response of the process to focus is to simultaneously vary both focus and exposure in what is known as a *focus-exposure matrix*. Fig. 7.20 shows a typical example of the output of a focus-exposure matrix using linewidth as the response (sidewall angle and resist loss can also be plotted in the same way) in what is called a Bossung plot.[33] Of course, one output as a function of two inputs can be plotted in several different ways. For example, the Bossung curves could also be plotted as exposure latitude curves (linewidth versus exposure) for different focus settings. Probably the most useful way to plot this two-dimensional data set is a contour plot – contours of constant linewidth versus focus and exposure (Fig. 7.21).

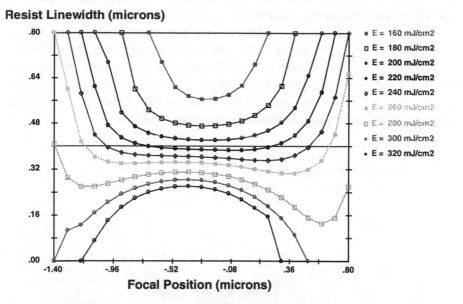

FIG. 7.20 Example of the effect of focus and exposure on the resulting resist linewidth. Focal position is defined as zero at the top of the resist with a negative focal position indicating that the plane of focus is inside the resist.

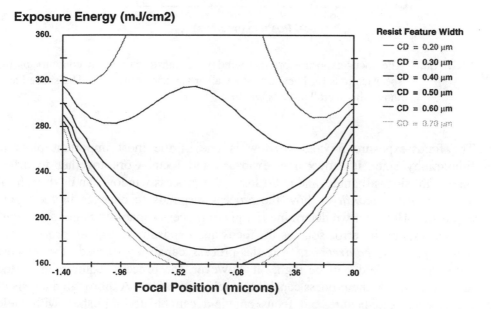

FIG. 7.21 Displaying the data from a focus-exposure matrix in an alternate form, contours of constant CD versus focus and exposure.

The contour plot form of data visualization is especially useful for establishing the limits of exposure and focus that allow the final image to meet certain specifications. Rather than plotting all of the contours of constant CD, one could plot only the two CDs corresponding to the outer limits of acceptability – the CD specifications. Because of the nature of a contour plot, other variables can also be plotted on the same graph. Fig. 7.22 shows an example of plotting contours of CD (nominal ±10%), 80° sidewall angle, and 10% resist loss all on the same graph. The result is a *process window* – the region of focus and exposure that keeps the final resist profile within all three specifications.

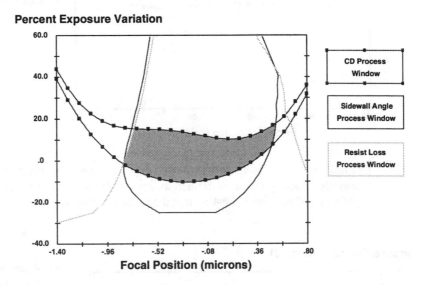

FIG. 7.22 The focus-exposure process window is constructed from contours of the specifications for linewidth, sidewall angle, and resist loss. The shaded area shows the overall process window.

The focus-exposure process window is one of the most important plots in lithography since it shows how exposure and focus work together to affect linewidth, sidewall angle, and resist loss. The process window can be thought of as a *process capability* – how the process responds to changes in focus and exposure. How can we determine if a given process capability is good enough? An analysis of the error sources for focus and exposure in a given process will give a *process requirement*.[34] If the process capability exceeds the process requirements, yield will be high. If, however, the process requirement is too large to fit inside the process capability, yield will suffer. A thorough analysis of the effects of exposure and focus on yield can be accomplished with yield modeling,[35] but a simpler analysis can be used to derive a number for depth of focus.

What is the maximum range of focus and exposure (that is, the maximum process requirement) that can fit inside the process window? A simple way to investigate this question is to graphically represent errors in focus and exposure as a rectangle on the same plot as the process window. The width of the rectangle represents the built-in focus errors of the processes, and the height represents the built-in dose errors. The problem then becomes one of finding the maximum rectangle that fits inside the process window. However, there is no one answer to this question. There are many possible rectangles of different widths and heights that are "maximum", i.e., they cannot be made larger in either direction without extending beyond the process window. (Note that the concept of a "maximum area" is meaningless here.) Each maximum rectangle represents one possible trade-off between tolerance to focus errors and tolerance to exposure errors. Larger DOF can be obtained if exposure errors are minimized. Likewise, exposure latitude can be improved if focus errors are small. The result is a very important trade-off between exposure latitude and DOF.

If all focus and exposure errors were systematic, then the proper graphical representation of those errors would be a rectangle. The width and height would represent the total ranges of the respective errors. If, however, the errors were randomly distributed, then a probability distribution function would be needed to describe them. For the completely random case, a Gaussian distribution with standard deviations in exposure and focus is used to describe the probability of a given error. In order to graphically represent the errors of focus and exposure, one should describe a surface of constant probability of occurrence. All errors in focus and exposure inside the surface would have a probability of occurring that is greater than the established cutoff. What is the shape of such a surface? For fixed systematic errors, the shape is a rectangle. For a Gaussian distribution, the surface is an ellipse. If one wishes to describe a "three-sigma" surface, the result would be an ellipse with major and minor axes equal to the three-sigma errors in focus and exposure.

Using either a rectangle for systematic errors or an ellipse for random errors, the size of the errors that can be tolerated for a given process window can be determined. Taking the rectangle as an example, one can find the maximum rectangle that will fit inside the processes window. Fig. 7.23 shows an analysis of the process window where every maximum rectangle is determined and its height (the exposure latitude) plotted versus its width (depth of focus). Likewise, assuming random errors in focus and exposure, every maximum ellipse that fits inside the processes window can be determined. The horizontal width of the ellipse would represent a three-sigma error in focus, while the vertical height of the ellipse would give a three-sigma error in exposure. Plotting the height versus the width of all the maximum ellipses gives the second curve of exposure latitude versus DOF in Fig. 7.23.

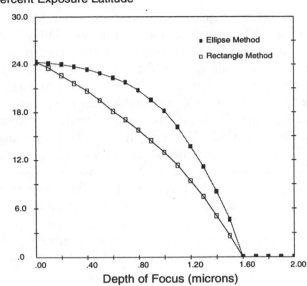

Percent Exposure Latitude

Depth of Focus (microns)

FIG. 7.23 The process window of Fig. 7.22 is analyzed by fitting all the maximum rectangles and all the maximum ellipses, then plotting their height (exposure latitude) versus their width (depth of focus).

The exposure latitude versus DOF curves of Fig. 7.23 provide the most concise representation of the coupled effects of focus and exposure on the lithography process. Each point on the exposure latitude - DOF curve is one possible operating point for the process. The user must decide how to balance the trade-off between DOF and exposure latitude. One approach is to define a minimum acceptable exposure latitude, and then operate at this point; this has the effect of maximizing the DOF of the process. In fact, this approach allows for the definition of a single value for the DOF of a given feature for a given process. The depth of focus of a feature can be defined as *the range of focus that keeps the resist profile of a given feature within all specifications (linewidth, sidewall angle, and resist loss) over a specified exposure range.* For the example given in Fig. 7.23, a minimum acceptable exposure latitude of 15%, in addition to the other profile specifications, would lead to the following depth of focus results:

$$DOF \ (rectangle) = 0.85 \ \mu m$$

$$DOF \ (ellipse) = 1.14 \ \mu m$$

$$DOF \ (average) = 1.00 \ \mu m$$

As one might expect, systematic errors in focus and exposure are more problematic than random errors, leading to a smaller DOF. Most actual processes would have a combination of systematic and random errors. Thus, one might expect the rectangle analysis to give a pessimistic value for the DOF, and the ellipse method to give an optimistic view of DOF. The average value of the two will be a more realistic number in most cases.

All of the above graphs were generated using simulation, which can not only generate focus-exposure data, but can analyze the data to determine the process window and DOF as well.

7.4.2 Aerial Image Optimization

Full optimization of a lithographic process requires thorough and time consuming calculations of many effects. One simplified approach to this optimization problem is to perform limited calculations at one point in space, say at a point in the resist corresponding to the nominal line edge, and try to optimize certain important properties of the resist feature at this point. Such a point optimization method, by its very nature, is somewhat limiting, since any interesting and important lithographic effects that occur elsewhere in the photoresist are not accounted for. However, if the point used is of interest (such as the nominal line edge) and the method used has physical significance, the results can be very useful.

To further simplify the analysis of a lithographic process, it is highly desirable to separate the effects of the lithographic tool from those of the photoresist process. This can be done with reasonable accuracy only if the interaction of the tool (i.e., the aerial image) with the photoresist is known. Consider an aerial image of relative intensity $I(x)$, where x is the horizontal position (i.e., in the plane of the wafer and mask) and is zero at the center of a symmetric mask feature. The aerial image exposes the photoresist to produce some chemical distribution $m(x)$ within the resist. This distribution is called the *latent image*. Many important properties of the lithographic process, such as exposure latitude and development latitude, are a function of the gradient of the latent image $\partial m / \partial x$. Larger gradients result in improved process latitude. By taking the derivative of Eq. (7.36), it can be shown that the latent image gradient is related to the aerial image by[36]

$$\frac{\partial m}{\partial x} = m \ln(m) \frac{\partial \ln I}{\partial x} \qquad (7.50)$$

where the logarithmic slope of the aerial image is often called simply the *log-slope*. The development properties of the photoresist translate the latent image gradient into a development gradient, which then allows for the generation of a photoresist image. Optimum photoresist image quality is obtained with a large development rate gradient. A lumped parameter called the photoresist contrast, γ, can be defined that relates the aerial image and the development rate r:

$$\frac{\partial \ln r}{\partial x} = \gamma \frac{\partial \ln I}{\partial x} \tag{7.51}$$

Eq. (7.51) is called the *lithographic imaging equation* and shows in a concise form how a gradient in aerial image intensity results in a solubility differential in photoresist. The development rate gradient is maximized by higher resist contrast and by a larger log-slope of the aerial image.

The above discussion clearly indicates that the aerial image log-slope is a logical metric by which to judge the quality of the aerial image. In particular, the image log-slope, when normalized by multiplication with the feature width, is directly proportional to exposure latitude expressed as a percent change in exposure to give a percent change in linewidth. This normalized image log-slope (*NILS*) is given by

$$NILS = w \frac{\partial \ln I}{\partial x} \tag{7.52}$$

This metric was first discussed by Levenson et al.,[37] and later in a related form by Levinson and Arnold,[38,39] before being explored to great extent by this author.[40-43,34] Simulation allows for the simple calculation of the NILS and thus its convenient use as an aerial image metric.

The well known effect of defocus on the aerial image is shown in Fig. 7.10. Both the edge slope of the image and the center intensity decrease with defocus, and the intensity at the mask edge remains nearly constant or increases slightly. To compare aerial images using the log-slope, one must pick an x value to use. An obvious choice is the mask edge (or more correctly, the nominal feature edge). Thus, all subsequent reference to the slope of the log-aerial image will be at the nominal feature edge. Now the effect of defocus on the aerial image can be expressed by plotting log-slope as a function of defocus, as shown in Fig. 7.24. The log-slope defocus curve has proven to be a powerful tool for understanding focus effects.

Some useful information can be obtained from a plot of log-slope versus defocus. As was previously discussed, exposure latitude varies directly with the log-slope of the image. Thus, a minimum acceptable exposure latitude specification translates directly into a minimum acceptable value of the NILS. The log-slope defocus curve can then be used to give a maximum defocus to keep the process within this specification. If, for example, the minimum acceptable normalized log-slope of a given process was determined to be 3.5, the maximum defocus of 0.5 μm lines and spaces on a 0.53 NA i-line stepper would be, from Fig. 7.24, about ±0.8 μm. This gives a practical definition of the depth of focus that separates the effects of the aerial image and the photoresist process. The printer determines the shape of the log-slope defocus curve, and the process determines the range of operation (i.e., the minimum NILS value). If the minimum log-slope needed was 6, one would conclude from Fig. 7.24 that this printer could not adequately resolve 0.5 μm lines and spaces. Thus, resolution can also be determined from a log-slope defocus curve.

Normalized Image Log-Slope

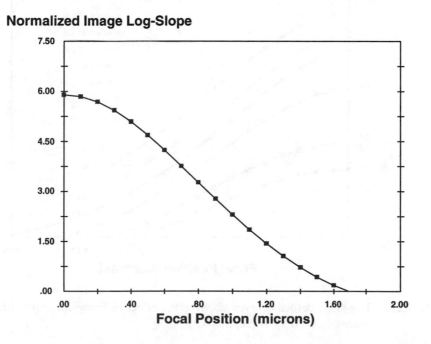

FIG. 7.24 An example of the log-slope defocus curve (0.5 μm lines and spaces, NA = 0.53, i-line, σ = 0.5).

To define resolution, consider Fig. 7.25, which shows the effect of feature size on the log-slope defocus curve. If, for example, a particular photoresist process requires a NILS of 3.8, one can see that the 0.4 μm features will be resolved only when in perfect focus, the 0.5μm features will have a DOF of ±0.7 μm, and the 0.6 μm features will have a DOF of ±0.9 μm. Obviously, the DOF is extremely sensitive to feature size, a fact that is not evident in the common Rayleigh definition. Since DOF is a strong function of feature size, it is logical that resolution is a function of the required DOF. Thus, in the situation shown in Fig. 7.25, if the minimum acceptable DOF is ±0.8 μm and the required NILS is 3.8, the practical resolution is about 0.55μm for equal lines and spaces. Resolution and DOF cannot be independently defined, but rather are interdependent. To summarize, DOF can be defined as the range of focus that keeps the log-slope above some specification for a given feature. Resolution can be defined as the smallest feature that keeps the log-slope above some specification over a given range of focus.

Normalized Image Log-Slope

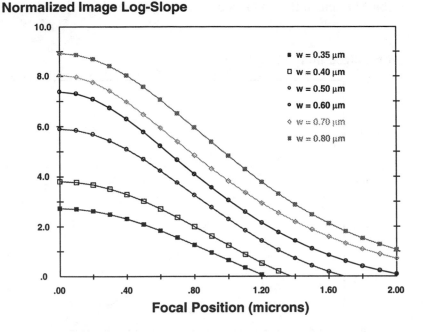

FIG. 7.25   Log-slope defocus curves showing the effect of linewidth (equal lines and spaces, NA = 0.53, i-line, σ = 0.5).

The key to the above image-based definitions for resolution and DOF is the linear correlation between the NILS and exposure latitude. But in order to make quantitative estimates, one must have a reasonable estimate for the minimum acceptable normalized log-slope. How is such an estimate obtained? By measuring a focus exposure matrix, one can obtain an experimental plot of exposure latitude (EL) versus defocus (exposure latitude being defined as the range of exposure which keeps the linewidth within specification, divided by the nominal exposure, and multiplied by 100%). This can be repeated for many different feature types and sizes, if desired. By comparing such experimental data with the log-slope defocus curves as in Fig. 7.25, a correlation between NILS and exposure latitude can be obtained. For example, one might find that data and simulated NILS are empirically correlated by the simple expression

$$EL = 8.1(NILS - 1.1) \qquad (7.53)$$

Eq. (7.53) in and of itself leads to very revealing interpretations. First, note that in this example a NILS of at least 1.1 must be used before an image in photoresist is obtained even at one exposure level. Above a NILS of 1.1, each increment in NILS adds 8.1% exposure latitude. Finally, if a minimum required exposure latitude is specified for a process, this value will translate directly into a minimum required NILS. For example, if an EL of 20% is required, the NILS that just achieves this level is 3.6. Thus, all images with a NILS in excess of 3.6 would be considered acceptable from an exposure latitude point of view. Correlations like Eq. (7.53) are very process dependent. However, for a given process, such a correlation allows imaging parameters to be studied by simply examining the log-slope defocus behavior.

Many image-related parameters can be easily studied using the log-slope defocus curves. The differences between imaging dense and isolated features, or lines versus contacts, for example, can be examined. The log-slope defocus approach has been used to optimize the numerical aperture and partial coherence of a stepper,[44] examine the differences between imaging in positive and negative tone resist[45] and study the advantages of off-axis illumination.[46]

Although defocus is strictly an optical phenomenon, the photoresist plays a significant role in determining the effects of defocus. As one might imagine, a better photoresist will provide greater depth of focus. In light of the above description of defocus using log-slope defocus curves, the photoresist impacts the DOF by changing the minimum acceptable log-slope specification. A better photoresist will have a lower log-slope specification, resulting in a greater usable focus range. This relationship between the photoresist and the log-slope specification is determined experimentally as described above by measuring exposure latitude versus defocus. In general, the resulting correlation between the NILS and the exposure latitude is given by

$$EL = \alpha \ (NILS - \beta) \qquad (7.54)$$

where β is the minimum NILS required to given any image at all in photoresist, and α is the percent increase in exposure latitude per unit increase in $NILS$. Thus, to a first degree, the effect of the photoresist on depth of focus can be characterized by the two parameters α and β.

Consider for a moment an ideal, infinite contrast photoresist. For such a case, the slope of the exposure latitude curve will be exactly 2/NILS.[47] Thus, using a typical linewidth specification of ±10%, an infinite contrast resist would make

$\alpha=10$ and $\beta=0$. The quality of a photoresist with respect to focus and exposure latitude can be judged by how close α and β are to these ideals.

7.4.3 Lumped Parameter Model

Typically, lithography models make every attempt to describe physical phenomena as accurately as possible. However, in some circumstances speed is more important than accuracy. If a model is reasonably correct and fast, many interesting applications are possible. With this trade-off in mind, the *lumped parameter model* (LPM) was developed.[47-49]

The mathematical description of the resist process incorporated in the lumped parameter model uses a simple photographic model relating development time to exposure, while the aerial image simulation is derived from the standard optical parameters of the lithographic tool. A very simple development rate model is used based on the assumption of a constant contrast. The photoresist contrast γ is defined theoretically as[50]

$$\gamma \equiv \frac{d \ln r}{d \ln E} \qquad (7.55)$$

where r is the resulting development rate from an exposure of E. Note that the base e definition of contrast is used here. If the contrast is assumed constant over the range of energies of interest, Eq. (7.55) can be integrated to give a very simple expression for development rate. In order to evaluate the constant of integration, let us pick a convenient point of evaluation. Let E_o be the energy required to just clear the photoresist in the allotted development time, t_{dev}, and let r_o be the development rate that results from an exposure of this amount. Carrying out the integration gives

$$r(x,z) = r_o \left[\frac{E}{E_o} \right]^\gamma \qquad (7.56)$$

Another component of the lumped parameter model is the effective resist thickness, D_{eff}, defined as

$$D_{eff} = \int_0^D \left[\frac{E(z)}{E(D)} \right]^{-\gamma} dz \qquad (7.57)$$

where D is the actual resist thickness. The effective resist thickness weights the actual resist thickness by a term related to the change in the development rate from the top of the resist film ($z=0$) to the bottom ($z=D$).

Eq. (7.56) is an extremely simple model relating development rate to exposure energy based on the assumption of a constant resist contrast. In order to use this expression, we will use a phenomenological explanation for the development process. This explanation is based on the assumption that development occurs in two steps: a vertical development to a depth z, followed by a lateral development to position x (measured from the center of the mask feature)[51] as shown in Fig. 7.26.

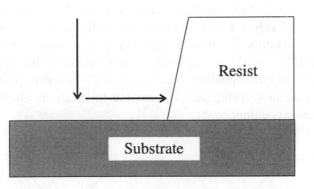

FIG. 7.26 Illustration of segmented development: development proceeds first vertically, then horizontally, to the final resist sidewall.

A development ray, which traces out the path of development, starts at the point $(x_o, 0)$ and proceeds vertically until a depth z is reached such that the resist to the side of the ray has been exposed more than the resist below the ray. At this point the development will begin horizontally. The time needed to develop in both vertical and horizontal directions, t_z and t_x respectively, can be computed from Eq. (7.56). The sum of these two segment times must equal the total development time. From the resulting equation, one can derive the integral form of the lumped parameter model:

$$\frac{E(x)}{E(0)} = \left[1 + \frac{1}{\gamma D_{eff}} \int_0^x \left(\frac{I(x')}{I(0)} \right)^{-\gamma} dx' \right]^{\frac{1}{\gamma}} \quad (7.58)$$

where $E(x)$ is the dose required to produce a feature of width $2x$, and $I(x)$ is the aerial image. Using this equation, one can generate a normalized CD versus exposure curve by knowing the image intensity, the effective resist thickness, D_{eff}, and the contrast, γ.

7.4.4 Modified Illumination

Conventional illumination schemes use a very simple shape for the light source: a uniform circle that results in a uniform cone of light striking the mask. Of course, many other shapes are possible, as shown in Fig. 7.27. Annular illumination, in which the central portion of the cone of light is blocked, was first proposed by Mack[52] and Fehrs et al.[53] Quadrupole illumination, which replaces a single circularly symmetric disk of light with four disks at right angles to each other, was proposed by Noguchi and coworkers at Canon,[54] Shiraishi and coworkers at Nikon,[55] and Tounai and coworkers at NEC.[56] All of these schemes have been called *off-axis illumination*, although the term is somewhat of a misnomer. Conventional partially coherent illumination includes a range of angles of incidence on the mask, both on-axis (normally incident) and off-axis (obliquely incident) illumination. Thus, the use of the term off-axis illumination to describe these new techniques is intended to mean an illumination scheme without any on-axis components.

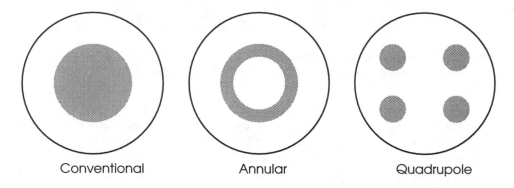

| Conventional | Annular | Quadrupole |

FIG. 7.27 Examples of various illumination shapes. The outer dark circle represents the objective lens aperture; the inner gray shapes are the illumination.

There are two major benefits in using off-axis illumination: resolution enhancement and depth of focus improvement. Although resolution enhancement may seem like sufficient justification to pursue the use of off-axis illumination, it is the impact of illumination angle on DOF that provides the more important benefits. Defocus causes a phase error for each diffraction order taht is proportional to the square of the radial position within the pupil. When the higher diffraction orders are out of phase relative to the zero order, the diffraction orders do not add properly when combining to form the aerial image, resulting in reduced image quality. For the case of on-axis illumination, the zero order is in the center of the aperture and thus will undergo no phase error due to defocus. The first orders, however, will have a phase error proportional to the square of their radial distance from the center of the aperture. The result will be a phase difference between the first and the zero orders, resulting in image quality degradation. Now consider the case of off-axis illumination where the size of the feature and the angle of incidence of the illumination are properly matched such that the zero order and one of the first orders are exactly the same distance from the center of the pupil. Thus, even though the orders will still have phase errors due to defocus, the errors will be the same and the relative phase difference between the zero and the first order will be zero. Such a situation will produce an image that is relatively immune to defocus, resulting in significant improvement in DOF. Note, however, that one would expect that the in focus performance for such a situation would be reduced, since only one of the first diffraction orders is being used to form the image.

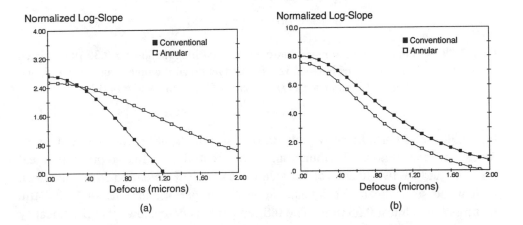

FIG. 7.28   Comparison of annular and conventional illumination for (a) 0.35 μm lines and spaces and (b) 0.7 μm lines and spaces with i-line, NA = 0.52. The conventional illuminator uses σ = 0.5, and the annular illuminator has a center σ of 0.5 with a width of 0.1.

To verify the improvement with off-axis illumination the log-slope defocus curve can be used to assess the quality of the aerial image. As an example, Fig. 7.28(a) compares log-slope defocus curves for conventional and annular illuminators (high NILS means a better quality image). As can be seen, the behavior is exactly as expected. The performance of the annular illumination system is worse than the conventional illuminator in focus, but is significantly better when out of focus by more than about 0.4 μm. The log-slope defocus curve can also be used to investigate the worst case feature size. In this case, 0.7 μm lines and spaces perform worse for all values of defocus with annular illumination compared to conventional illumination, as seen in Fig. 7.28(b). Fig. 7.29 shows that the annular illumination has a similar effect on isolated lines.

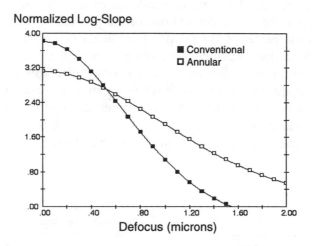

FIG. 7.29 Comparison of annular and conventional illumination for 0.35 μm isolated lines with i-line, NA = 0.52. The conventional illuminator uses σ = 0.5, and the annular illuminator has a center σ of 0.5 with a width of 0.1.

Fig. 7.30 shows mask linearity plots for 0.35 μm line/space patterns in and out of focus for conventional illumination. The central 45° line represents ideal linearity and the two lines to either side indicate ±10% deviation from this ideal. As can be seen, the in focus case shows linear performance down to 0.4 μm features (and almost 0.35 μm). The 0.8 μm defocus case shows that the linearity is somewhat reduced for all features and the 0.4 μm features have just gone out of specification. Fig. 7.30(b) shows the results for annular illumination optimized for 0.35 μm features. Note that the linearity of the smallest features is essentially the same as for conventional illumination. However, note that the 0.7 μm features have gone out of linearity, indicating that these features have less DOF than the smaller features!

Fig. 7.30 points out one of the problems of off-axis illumination. To see the benefits, Fig. 7.31 shows the resulting focus-exposure process windows for conventional and annular illumination for 0.35μm lines and spaces. In focus, the two illuminators show about the same exposure latitude. However, when out of focus the annular illuminator shows far less isofocal bias (the upward bending of the process window) and greater exposure latitude. As a result, the annular illumination case will show significantly greater DOF.

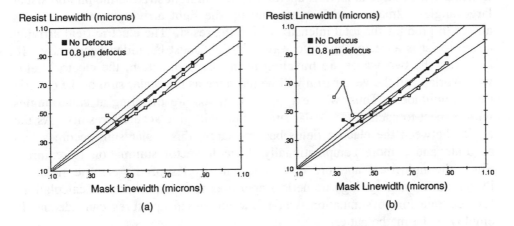

FIG. 7.30 Comparison of mask linearity for (a) conventional illumination and (b) annular illumination for equal lines and spaces with i-line, NA = 0.52. The conventional illuminator uses σ = 0.5, and the annular illuminator has a center σ of 0.5 with a width of 0.1.

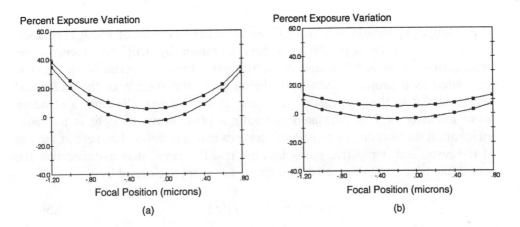

FIG. 7.31 Comparison of the focus-exposure process windows for (a) conventional illumination and (b) annular illumination for 0.35 μm equal lines and spaces with i-line, NA = 0.52. The conventional illuminator uses σ = 0.5, and the annular illuminator has a center σ of 0.5 with a width of 0.1. Contours are for ±10% linewidth specifications.

7.4.5 High Numerical Aperture Effects

Imaging with a large numerical aperture results in light striking the photoresist at large angles. Image models must sum up the light arriving at these various angles to produce the total intensity within the resist. The electric field arriving at the resist is a vector, and two waves of light must be added vectorially. If, however, the two waves are traveling in the same direction, the electric fields will overlap and the vector sum will be the same as the scalar sum of the electric field amplitudes. Thus, whenever electric fields are traveling at small angles with respect to each other, scalar summation will give accurate results. As the angles between the electric fields become large, this scalar sum becomes less accurate and a more computationally difficult vector summation is required. Since the numerical aperture determines the angles of the electric fields in lithographic imaging, high numerical apertures may require vector calculations for accurate image simulation, while low numerical apertures can adequately employ scalar mathematics.

It is not sufficient to describe a lithography imaging model as being simply "scalar" or "vector." Within these broad categories there are a number of different approximations that can be made to simplify the calculations. Several of these approximations are described below.

7.4.5.1 Zero order scalar model

The lithography simulator SAMPLE[5] and the 1985 version of PROLITH[6] used the simple imaging approximation first proposed by Dill[4] to calculate the propagation of an aerial image in photoresist. First, an aerial image $I_i(x)$ is calculated as if projected into air (x being along the surface of the wafer and perpendicular to the propagation direction of the image). Second, a standing wave intensity $I_s(z)$ is calculated assuming a plane wave of light is normally incident on the photoresist coated substrate (where z is defined as zero at the top of the resist and is positive going into the resist). Then, it is assumed that the actual intensity within the resist film $I(x,z)$ can be approximated by

$$I(x,z) \approx I_i(x)I_s(z) \tag{7.59}$$

For very low numerical apertures and reasonably thin photoresists, these approximations are valid. They begin to fail when the aerial image changes as it propagates through the resist (i.e., it defocuses) or when the light entering the resist is appreciably non-normal. This zero order model is also called the vertical propagation model.

7.4.5.2 First order correction to the scalar model

The first attempt to correct one of the deficiencies of the zero order model was made by Mack,[40,41] and later by Bernard.[57] The aerial image, while propagating through the resist, is continuously changing focus. Thus, even in air, the aerial image is a function of both x and z. An aerial image simulator calculates images as a function of x and the distance from the plane of best focus, δ. Letting δ_o be the defocus distance of the image at the top of the photoresist, the defocus within the photoresist at any position z is given by

$$\delta(z) = \delta_o + \frac{z}{n} \tag{7.60}$$

where n is the real part of the index of refraction of the photoresist. The intensity within the resist is then given by

$$I(x,z) = I_i(x,\delta(z))\,I_s(z) \tag{7.61}$$

Here the assumption of normally incident plane waves is still used when calculating the standing wave intensity.

7.4.5.3 Second order correction to the scalar model

The light propagating through the resist can be thought of as various plane waves traveling through the resist in different directions. Consider first the propagation of the light in the absence of diffraction by a mask pattern (that is, exposure of the resist by a large open area). The spatial dimensions of the light source determine the characteristics of the light entering the photoresist. For the simple case of a coherent point source of illumination centered on the optical axis, the light traveling into the photoresist would be the normally incident plane wave used in the calculations presented above. The standing wave intensity within the resist can be determined analytically[20] as the square of the magnitude of the electric field given by

$$E(z) = \frac{\tau_{12}E_I\left(e^{-i2\pi n_2 z/\lambda} + \rho_{23}\tau_D^2\, e^{i2\pi n_2 z/\lambda}\right)}{1+\rho_{12}\rho_{23}\tau_D^2} \tag{7.62}$$

where the subscripts 1, 2, and 3 refer to air, the photoresist and the substrate, respectively, D is the resist thickness, E_I is the incident electrical field, λ is the wavelength, and where complex index of refraction of film j: $\qquad\qquad n_j = n_j - i\kappa_j$

transmission coefficient from i to j: $\qquad\qquad \tau_{ij} = \dfrac{2n_i}{n_i + n_j}$

reflection coefficient from i to j: $\rho_{ij} = \dfrac{n_i - n_j}{n_i + n_j}$

internal transmittance of the resist: $\qquad\qquad \tau_D = e^{-i2\pi n_2 D/\lambda}$

The above expression can be easily modified for the case of non-normally incident plane waves. Suppose a plane wave is incident on the resist film at some angle θ_1. The angle of the plane wave inside the resist will be θ_2 as determined from Snell's law. An analysis of the propagation of this plane wave within the resist will give an expression similar to Eq. (7.62) but with the position z replaced with $z\cos\theta_2$.

$$E(z,\theta_2) = \frac{\tau_{12}(\theta_2)E_I\left(e^{-i2\pi n_2 z\cos\theta_2/\lambda} + \rho_{23}(\theta_2)\tau_D^2(\theta_2)e^{i2\pi n_2 z\cos\theta_2/\lambda}\right)}{1 + \rho_{12}(\theta_2)\rho_{23}(\theta_2)\tau_D^2(\theta_2)}$$

(7.63)

The transmission and reflection coefficients are now functions of the angle of incidence and are given by the Fresnel formulas. They are also a function of the polarization of the incident light. For the typical unpolarized case, the light entering the resist will become polarized (but only slightly). Thus, a separate standing wave can be calculated for each polarization and the resulting intensities summed to give the total intensity. A similar approach was taken by Bernard and Urbach.[58] Notice that by using a more accurate scalar model, vector effects such as polarization within the resist film still come into play.

By calculating the standing wave intensity at one incident angle θ_1 to give $I_s(z,\theta_1)$, the full standing wave intensity can be determined by integrating over all angles. Each incident angle comes from a given point in the illumination source, so that integration over angles is the same as integration over the source. Thus, the effect of partial coherence on the standing waves is accounted for. Note that the effect of the non-normal incidence is included only with respect to the zero order light (the light which is not diffracted by the mask).

7.4.5.4 Full scalar model

The above method for calculating the image intensity within the resist still makes the assumption of separability: that an aerial image and a standing wave intensity can be calculated independently and then multiplied together to give the total intensity. This assumption is not required. Instead, one could calculate the full $I(x,z)$ at once, making only the standard scalar approximation. The formation of the image can be described as the summation of plane waves. For coherent illumination, each diffraction order gives one plane wave propagating into the resist. Interference between the zero order and the higher orders produces the desired image. Each point in the illumination source will produce another image that will add incoherently (i.e., intensities will add) to give the total image. Eq. (7.63) describes the propagation of a plane wave in a stratified media at any arbitrary angle. By applying this equation to each diffraction order, an exact scalar representation of the full intensity within the resist is obtained.

7.4.5.5 Other issues in scalar modeling

Besides the basic modeling approaches described above, there are two issues that apply to any model. First, the effects of defocus are taken into account by describing defocus as a phase error at the pupil plane. Essentially, if the curvature of the wavefront exiting the objective lens pupil is such that it focuses in the wrong place (i.e., not where you want it), one can consider the wavefront curvature to be wrong. Simple geometry then relates the optical path difference of the actual wavefront from the desired wavefront as a function of the angle of the light exiting the lens, θ.

$$OPD(\theta) = \delta(1 - \cos\theta) \qquad (7.64)$$

Computation of the imaging usually involves a change in variables where the main variable used is $\sin\theta$. Thus, the cosine adds some algebraic complexity to the calculations. For this reason, it is common in optics texts to simplify the OPD function for small angles (i.e., low numerical apertures):

$$OPD(\theta) = \delta(1 - \cos\theta) \approx \frac{\delta}{2}\sin^2\theta \qquad (7.65)$$

Again, the approximation is not necessary and is only made to simplify the resulting equations.

Reduction in the imaging system adds an interesting complication. Light entering the objective lens will leave the lens with no loss in energy (the lossless lens assumption). However, if there is reduction in the lens, the *intensity* of the light entering will be different from that leaving since the intensity is the energy spread over a changing area. The result is a radiometric correction well known in optics[59] and first applied to lithography by Cole and Barouch.[60]

7.4.5.6 Vector model

Light is an electromagnetic wave that can be described by time-varying electric and magnetic field vectors. In lithography, the materials used are generally non-magnetic so that only the electric field is of interest. The electric field vector is described by its three vector components. Maxwell's equations, sometimes put into the form of the wave equation, govern the propagation of the electric field vector. The *scalar approximation* assumes that each of the three components of the electric field vector can be treated separately as scalar quantities and that each scalar electric field component must individually satisfy the wave equation.

The scalar approximation is commonly used throughout optics and is known to be accurate under many conditions. There is one simple situation, however, in which the scalar approximation is not adequate. Consider the interference of two plane waves traveling past each other. If each plane wave is treated as a vector, they will interfere only if there is some overlap in their electric field vectors. If the vectors are parallel, there will be complete interference. If, however, their electric fields are at right angles to each other there will be no interference. The scalar approximation essentially assumes that the electric field vectors are always parallel and will always give complete interference. This situation comes into play in lithography when considering the propagation of plane waves traveling through the resist at large angles. Reflection of a plane wave from the substrate will give two plane waves traveling past each other at some angle. For large angles, the scalar approximation may fail to account for these vector effects.[61]

7.5 MODELING TECHNOLOGY INNOVATIONS

In pushing the limits of lithography, many technology innovations are being introduced in projection printers, masks, resists, wafer topography, and coating. Figure 7.32 shows results from an interesting case study of simultaneously employing several emerging technology options to lithography at 250 nm feature sizes.[62] In this study rim phase-shift masks, annular illumination, a high (0.53) NA lens, and a *t*-BOC type resist (IBM APEX-E) were used. Simulation was first used to bias the features and size the 180° phase-shift rims in a mask to produce 220 nm lines with various pitches. The masks were then exposed and gave the phenomenal resist profiles seen in these SEMs over a wide depth of focus. These resist profiles have exceptionally high aspect ratios, show no sign of standing wave effects, and suggest that optical lithography should be quite capable of producing the 250 nm features required for the 256MB generation. There is, however, the very important issue of establishing sufficient understanding to make this a manufacturable process. In particular, it is evident from the SEMs that the process needs to be further tuned to eliminate the 60 nm difference in linewidth that occurred for the two pitches shown. In this rather complex environment, simulation can be very helpful in identifying likely causes for the difference and in suggesting quantitative adjustments for compensation. Simulation in this case shows that the difference in the aerial images for the two feature types accounts for about 10 nm, and the difference due to interactions with the substrate including annular illumination was an additional 12 nm. Most of the remaining 38 nm is associated with the properties of the DUV *t*-BOC resist material (which will be discussed below).

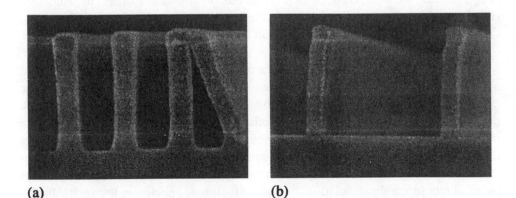

(a) (b)

FIG. 7.32 SEM photographs of (a) 0.22 μm dense lines and spaces and (b) 0.22 μm lines a pitch of 1.08 μm in 1.0 μm IBM APEX-E resist made on a 4X GCA DUV stepper with 0.53 NA and 0.6-0.7 partial coherence.[62]

Modeling and simulation can play a very important role in both technology development and for establishing manufacturability. When developing innovations the technologists must assess many different options. This technology development must be done rapidly and with minimal expense. Also, these innovations are not without their downside, which must be investigated as well. In establishing manufacturability these innovations introduce new parameters, and often the parameter space to be investigated becomes exceedingly large due to the tendency of many of the innovations to introduce feature size and type effects. Clearly, with such a large parameter space simulation can play an important role in identifying and characterizing the central effects and their associated trade-offs. This section gives a number of examples of the use of simulation in exploring advanced aspects of technology innovation issues.

7.5.1 Resist Post-Exposure Bake and Chemical Amplification

In designing resist systems chemists have used to advantage the fact that physical changes and chemical reactions can be driven by elevating the temperature of the resist for a fixed period of time in a post-exposure bake (PEB) process. For conventional novolac resists, when the PEB temperature is higher than the prebake temperature standing wave effects are reduced.[63] An alternative chemical reaction path can produce image reversal,[64,65] and activation of cross-linking is used in modern negative type resists with chemical amplification.[66-69] The PEB is also used to drive catalytic reactions in chemically amplified resists. For example, in the case of t-BOC based resist systems the acid (H^+) removes a protecting t-BOC side group leaving an OH which promotes the dissolution of the polymer in aqueous developers. The t-BOC group then further disassociates and releases an acid (H^+), which proceeds to initiate further deprotection reactions and appears to diffuse during the process. The t-BOC resist chemistry, characterization methods, and models can be found in Refs. 69-81.

Many of these physical changes and chemical reactions require a movement of some species, hence diffusion simultaneously plays a role in the reaction. In the case of t-BOC resists, for example, the acid must move to find new sites to deprotect. This movement might be facilitated by the very changes taking place in the resist material and thus may be thought of as a form of local concentration dependent diffusion. Hinsberg et al. have suggested that the acid mobility in t-BOC resists may increase exponentially with the free volume created by gaseous by-products in the disassociation.[72] The diffusion of species in the resist during PEB is critically important both to the improvement of line-edge profile quality due to removal of standing wave effects and to the undesired effect of creating a linewidth bias through the encroachment from the exposed to the unexposed areas. Both of these effects are particularly noticeable in t-BOC resists, which produce exceptionally vertical resist profiles and yet also have an undesired line undersizing or bias at best focus and exposure. A telltale sign of this bias is the rapid loss of linewidth of *t*-BOC resist features with additional PEB time.[73,78]

To model post-exposure bake effects in resists, the chemical reactions and diffusion of species must be considered simultaneously. For *t*-BOC based materials several reaction models and diffusion models are currently being researched. For illustration here the reaction equation of Ferguson et al.[71] is used in which the concentration of active (deprotected) sites C_{as} is driven by the availability of protected sites (1-C_{as}) and the acid concentration C_a raised to a power *m*. The acid concentration gets its initial distribution from the distribution of the cumulative energy deposited during exposure. During the PEB the acid decreases with time according to its concentration in the second rate equation due to possible acid loss mechanisms such as neutralization or time sharing with the deprotected material.

$$\frac{d}{dt}C_{as} = K_1\left(1 - C_{as}\right)C_a^m \tag{7.66}$$

$$\frac{d}{dt}C_a = K_2 C_a \tag{7.67}$$

Thus, in the absence of diffusion, these differential equations can be solved analytically[71] leading to an expression for the concentration of activated sites given by

$$C_{as}(t) = C_{CS}\left(1 - \exp\left[\left(-C_{ao}^m\left(\frac{K_1}{mK_2}\right)\right)1 - e^{-mK_2 t}\right]\right) \tag{7.68}$$

Here C_{CS} is a constant representing the maximum total deprotection, which in the simulation here is normalized to unity, and C_{ao} is the initial photo-generated acid concentration.

To apply these equations the acid concentration is first calculated from the exposure-bleaching process. Since the acid concentration cannot be directly observed, exactly how the acid is produced on exposure has yet to be definitively modeled. Generally, it is assumed that the rate of acid production is proportional to the cumulative density of energy deposited in the resist in J/cm^3. This is first calculated using the standard ABC exposure-bleaching model. For IBM APEX-E resist, these parameters have been measured to be *A* is -0.01 µm^{-1}, *B* is 0.37 µm^{-1} and *C* is 0.0042 cm^2/mJ, respectively.[74] The negative *A* indicates that a darkening reaction is taking place, evident only at extremely high doses (above 250 mJ/cm^2), and this may be associated with the base resin rather than the acid.

Nonetheless, for typical working doses of about 15 mJ/cm^2 the amount of photodarkening of the resist and the acid generation are practically directly proportional. To avoid the introduction of yet another constant, the acid concentration C_a is set equal to $(1-M)$, where M is found in the usual manner. This normalization is implicit in the values of the reaction rates given above. The use of a scaling factor would propagate through the reaction rate parameters.

The closed form solution of the rate equations can be used for determining the extent of the deprotection as a function of PEB conditions in the absence of diffusion. Thus, the equation is suitable for modeling large, uniformly exposed areas from an initial uniform acid concentration. The rate parameters K_1 and K_2 are a function of bake temperature and show a typical Arrhenius-type behavior for temperatures sufficiently removed from the glass transition temperature of the resist. Typical values for the rate parameters at 90°C for IBM APEX-E resist are K_1 = 2.0/sec, K_2 = 0.0033/sec, m = 1.8.[78] The reciprocal of these rate constants indicates that time to complete a deprotection event is about 0.5 seconds while the effective lifetime for the acid is several minutes. Note that if there is no loss mechanism ($K_2 = 0$) the reaction in the limit of long PEB times results in complete deprotection. These relationships are very useful for interpreting results of large area exposures on matched substrates that have uniform concentrations of species.

For nonuniform horizontal or vertical initial concentrations of species, effects of movement of the species in any of the various resist materials during the PEB can be modeled by adding terms to the right hand side of the reaction rate equations that reflect the change in supply due to diffusion. In the case of t-BOC resist in which the acid moves, the acid loss equation above is replaced by an equation that describes the change in the local acid concentration as a function of both diffusion and local loss mechanisms:

$$\frac{dC_a}{dt} = \frac{d}{dx}\left(D(C_{as})\frac{dC_a}{dx}\right) - K_2 C_a \qquad (7.69)$$

The new term with spatial derivatives (in one dimension for convenience) describes the movement of the acid C_a by means of a diffusivity $D(C_{as})$ that is dependent on the concentration of deprotected (active) sites C_{as}. Several models have been used to express this dependence on C_{as} and include relationships such as constant (Fickean) for concentration independent effects (D_o), linear to reflect possible use of the deprotected sites as stepping stones ($D = D_o + D_1 C_{as}$), and exponential to account for the free volume effects $\left(D = D_o e^{wC_{as}}\right)$. Typical values for the exponential model are D_o = 5.6 x 10^{-6} µm^2/min and w = 5.8.[78] These concentration dependent diffusion effects make resist modeling as complex as the modeling of the diffusion of impurities in the silicon substrate in forming localized device regions. The use of the exponential form requires careful attention to accuracy in simulation and special algorithms have been developed for this purpose.[77] Also, the introduction of spatial movement requires that boundary conditions be imposed at all edges of the simulation domain. Symmetry of the exposure is often used to give a zero flow condition in the lateral directions. At the top and bottom of the resist, boundary conditions that describe interactions with the external environment and substrate materials are needed. These boundary conditions have yet to be characterized experimentally and could range from simple recombination velocities to extensions for additional species entering the resist and participating in the reactions.

The dissolution rate of the *t*-BOC resist system is believed to be a property of only the level of deprotection of the polymer. This relationship can be determined from the measurements by Eib et al.[74] for APEX-E of the dissolution rate versus exposure and a transformation based on large area exposures as modeled by Eq. 7.68. Results for the dissolution rate versus the deprotection level from applying this transformation are shown in Fig. 7.33. A C_{as} of 0.17 or below is required to have a top loss less than 0.1 μm in a 1 minute development process. Extensive deprotection of over 70% of the sites ($C_{as} = 0.7$) is required to remove a 1 μm film in 1 minute. The slope and critical deprotection values of this characteristic curve are a function of the processing conditions, nature of the polymer or copolymer, etc.

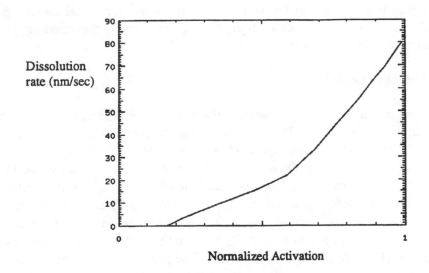

FIG. 7.33   Dissolution rate of IBM APEX-E resist as a function of the normalized level of deprotection as calculated by transformation of the measurement of Eib et al.[74]

The quantitative parameters in these models are determined through special experiments and simulation is often required to help deconvolve effects. FTIR measurements versus PEB conditions for various exposures of large, uniformly exposed areas are used to isolate reaction from diffusion and find K_1, K_2 and the energy density to acid concentration conversion factor. The parameters in the diffusion model are then found assuming the just-established reaction rates. Here FTIR measurements of patterned wafers versus PEB time, linewidth versus PEB time, and measuring thickness remaining versus PEB time of an unexposed wafer baked top-to-top with an exposed wafer[78] have been used. In all three cases there is some evidence that a type-II diffusion mechanism is present in IBM APEX-E which results in the acid moving as a sharp front at a velocity of about 30 nm/min.

Simulation can provide additional understanding of the role of the PEB process and resist system in relation to the exposure tool, mask, and illumination. We return to reconsider the SEM profiles of Fig. 7.32. Figure 7.34 shows the acid concentration produced by the initial exposure of the 0.22 µm equal lines and spaces and the 0.22 µm line with a pitch of 1.08 µm on a silicon substrate. Here the lens was focused one third of the way into the resist material. The strong standing waves at DUV are clearly evident. The simultaneous reaction and diffusion in the PEB process converts these profiles to the contour plots for the concentration of activated sites shown in Fig. 7.34. Here, the very high gradients in C_a in the vertical direction immediately produce a similar rapid vertical variation in C_{as} and together they cause the standing waves to bridge vertically. The horizontal push is much greater at the destructive interference locations where there is a gradient in the vertical as well as the horizontal directions to drive the increase in the concentration of acid. As a result, the standing waves reduce rapidly with PEB time.

7.5.2 Defect Printability

One of the major concerns in assessing the tendency of defects to print is their interactions with features. This has been characterized using a perturbational method.[83] Fig. 7.35 shows image contours for an isolated defect and the same defect near a feature. Here the standard normalized optical system parameters of $\lambda = 0.5$, NA = 0.5, and $\sigma = 0.5$ are used and the mask dimensions are in units of λ/NA = 1. The defect is a square of 0.25λ/NA on a side. The feature is a 0.8λ/NA line. When isolated, the defect produces an image whose size is independent of its shape and is given by the point spread function for the optical system (an Airy disk 1.2λ/NA in diameter). The peak intensity is independent of the defect shape and goes as the square of the area. For a square defect, this is the fourth power of the defect size and is only 3% of the clear field intensity for the 0.25λ/NA defect size. However, when this defect is adjacent to a feature, a considerable bulge is produced on the line edge. The intensity at the position of the nominal mask line edge is 30% of the clear field value but at the location of this defect rises to more than 45%. This sudden increase in the impact of the defect is due to the defect interacting with the feature in a highly coherent manner. This corresponds to adding the electric field of the defect (square root of intensity) to the electric field of the feature. The length of this bulge is determined by the diameter of the point spread function (1.2λ/NA) and the width is proportional to the area of the defect. The relationship of defect size to yield has been characterized by Wiley and Reynolds.[84]

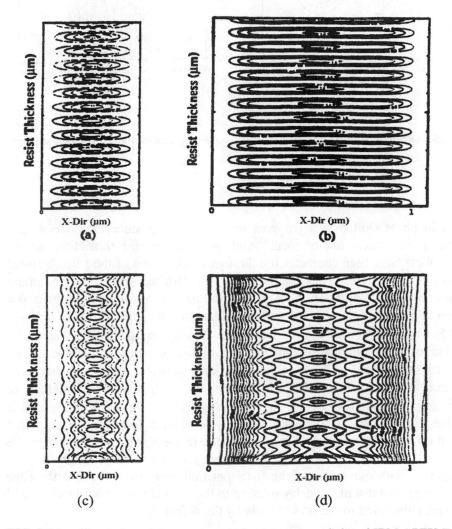

FIG. 7.34   Simulation of the exposure and post-exposure bake of IBM APEX-E resist: contour plots of the aerial image intensity for (a) 0.22 µm line and space, (b) 0.22 µm line with 1.08 µm pitch, and activated normalized activation level after post-exposure bake for (c) 0.22 µm line and space, (d) 0.22 µm line with 1.08 µm pitch.[82]

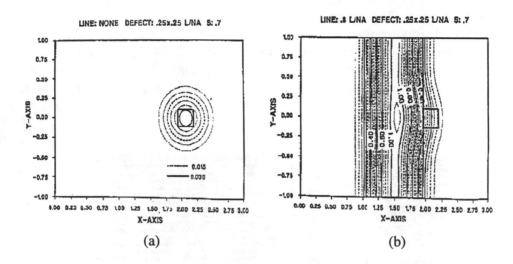

(a)   (b)

FIG. 7.35   Printability of a defect 0.25 λ /NA when (a) isolated and (b) when adjacent to a 0.8 λ /NA line.[83]

Defects in phase-shift masks are even more interesting since their worst case printability may occur out of focus[85] and more severe size restrictions apply. These defects have been characterized through an extension of the perturbational method. Figure 7.36, from the work of Wantanabe, shows the minimum intensity beneath a defect that is 100% transmitting with various phase shifts as a function of focal position. This graph was generated for λ = 0.5, NA = 0.5, and σ = 0.5. Defects of phase 180° cause the largest dip and in fact should be restricted to having only half the area of their chrome counterparts. Defects with phases of other than 180° cause the additional problem that they may pass inspection at best focus and fail when out of focus. The curve for defects of 120° in Fig. 7.36 is an example. In focus it is well above the 30% clear field level but goes as low as the 10% intensity level at a focus of 0.4μm. The explanation for this effect is that the small defect emphasizes the use of the edges of the lens while the large clear background creates a single on-axis central ray. With movement of the focal position away from the lens the edge rays undergo an extra phase delay relative to the central ray, which tends to add to the phase imparted to the rays initially by the defect.[86]

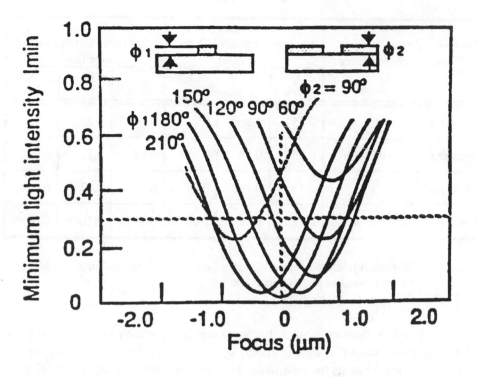

FIG. 7.36 Minimum intensity level produced by a phase defect of various phases as a function of the focus position.[85]

A study of the printability of defects in an attenuating phase-shift mask technology can be used to illustrate how the minimum defect size depends on location and relative phase.[86] Fig. 7.37 was made for a 10% transmitting attenuating type phase-shift mask. The attenuator had a 30° phase shift and the additional 150° phase shift was accomplished by etching the substrate. Optical systems were considered with $\lambda = 0.365$, NA = 0.5 or 0.6, and $\sigma = 0.5$. The features were 0.35 µm lines corresponding to 0.5 to 0.58λ/NA and contacts were oversize at 0.5 µm, or 0.69 to 0.82λ/NA. Both pinholes in the attenuating layer (with 30° phase shift) and pin spots (with 150° phase shift and no attenuation) were considered. The criteria for printability are that less than a 10% linewidth variation occur and that isolated defects have intensities less than 30% of the clear field.

Missing or Extra Absorber 180° phase 10% Att. Error	Pinhole Minimum Printable Defect Size (μm)			Pinspot Minimum Printable Defect Size (μm)		
	$\frac{\lambda}{NA}$	NA = 0.5	NA = 0.6	$\frac{\lambda}{NA}$	NA = 0.5	NA = 0.6
Isolated defect	0.35	0.26	0.21	0.34	0.25	0.21
Defect near an isolated line	0.25	0.18	0.15	0.15	0.11	0.09
Defect in an array	0.13	0.10	0.08	0.10	0.07	0.06
Defect in or near a contact	0.13	0.10	0.08	0.14	0.10	0.09

FIG. 7.37 Maximum tolerable defect size as a function of defect polarity and location for an attenuated phase-shift mask.[86]

There are five basic rules of thumb for assessing the tendency of defects to print and nearly all of them can be illustrated by Fig. 7.37. For convenience in use with other optical systems the minimum defect sizes will be described in λ/NA.

- Isolated defects should be less than about $0.35\lambda/NA$ or half of the minimum feature size (0.35 and 0.34 pinholes and pinspots).
- Defects near an isolated feature of nearly similar phase should be less than $0.25\lambda/NA$ or one third the feature size (0.25 pinhole).
- Defects near an isolated feature of nearly opposite phase should be less than $0.18\lambda/NA$ or one quarter of the feature size (0.15 pinspot).
- In arrays of contacts where marginal image quality occurs, defects should be an additional $\sqrt{2}$ smaller. (0.25 pinhole reduces to 0.13, and 0.15 pinspot reduces to 0.10).
- The use of resolution enhancement techniques that introduce feature size and type effects can degrade images of certain feature types, and hence if these degraded features are used, defects should be an additional $\sqrt{2}$ smaller.[86,87]

7.5.3 Exposure Tool Tuning and Imperfections

One of the major concerns in characterizing a projection printing tool is the extent to which tuning of the tools or residual minor nonidealities can affect performance. The leverage of the illumination distribution is clear from the earlier discussion of off-axis illumination. Phase-shift masks are a related means of tuning the tool performance. The addition of in-lens filters can also improve certain features such as contacts and their depth of focus.[88,89] These modifications also tend to interact with imperfections or aberrations in the lens. Tuning the exposure tools with or without these enhancements so the performance is equally good across the field is particularly important in addressing across chip linewidth uniformity issues.[90] Simulation can play an important role in that each of the imperfections or modifications can be assessed individually or collectively. Simulation can also assist in working backward from optical system performance observations, such as SEMs and electrical test data, to identify and address sources of nonidealities to improve process latitude and working depth of focus and make them uniform across the field.

7.5.4 Mask Transmission and Edge Effects

When modeling projection printing, a vertical-ray propagation model at the mask plane is typically assumed. This model treats the mask as a locally uniform thin-film stack and assumes that the electric field just under the mask is that from a vertically incident plane wave transmitted through the thin-film stack. For mask openings, this results in a nonphysical square wave modulation of the emerging field. For small mask openings and large vertical steps in phase-shift masks, a more detailed characterization of the emerging electric field may be necessary. Actual fields have tapered off-on transitions, have polarization dependent effects due to different boundary conditions for the electric and magnetic fields, and can have lateral cross-mixing in passing through the mask. Simulation has been used to characterize the extent to which these effects are present.[91-93] The effects are small in conventional masks, even for S-shaped chrome sidewalls, and are only of concern as mask openings near a wavelength in size. These effects are more than twice as pronounced for contacts than lines. However, a significant loss of intensity due to edge effects in phase-shift masks was predicted by simulation and then verified by experiment.[92] These edge effects occurred in trench etched masks where light emerging into the air across the bottom of the trench spread laterally in propagating and re-entered the substrate through the sidewall of the trench.

7.5.5 Substrate Topography

An additional concern in maintaining linewidth control is the unwanted lateral reflections from nearby topographic features. Scattering from topography can result in a sudden linewidth change or reflective notching. This is extremely important in process technologies where features must be defined between existing topographical features, such as in defining an electrical strap to a trench between two polysilicon lines.[94] It is also important even in defining simple polysilicon gate structures as they cross slightly nonplanar bowl-like dips in active areas.[95] These problems are treated in part by modifying the materials in the imaging process, such as by adding dye to the resist, using bottom antireflection coatings, or top antireflection coatings. Simulation can assist in sorting through the multitude of approaches and their combinations to address topography scattering issues.

7.5.6 3D Process Modeling

Important three-dimensional phenomena occur in lithography and pattern transfer processes. During exposure the aerial image is pattern and focus dependent due to the different focus behavior of small and large features and proximity effects between features and defects. In resist dissolution the etch fronts penetrate vertically and expand laterally, developing rapidly in highly exposed areas and expanding laterally into more slowly developing regions. In pattern transfer, etchant species visibility is pattern and depth dependent, as evident in reactive ion etch lag. These phenomena are critically important in dense layouts and in assessing the printability of defects. In some cases it may be possible to find planes of symmetry that include the most rapidly developing front, but if there is a general asymmetry of the mask, an arbitrary placement of a defect, or typical lens aberrations, a three-dimensional dissolution simulation is needed. Similarly, in pattern transfer or etching, once an asymmetry is encountered it must be carried throughout the remainder of the etching and deposition process flow.

7.6 INFORMATION ON MODELS AND SIMULATORS

The availability of tools with extended physical models, rapid algorithms, design options, and convenient user interfaces is improving rapidly so that the technologist should not be without a good selection of tools. These tools are evolving rapidly, so this section will take a broader view of the general classes of capabilities to look for when selecting a simulation tool. Issues to consider are the computer environment, the user interface and design options, the physical models, and numerical algorithms. The level of sophistication of physical models and typical options for numerical algorithms go hand-in-hand at each of the key substeps in lithography of aerial image formation, exposure, post-exposure bake, and resist dissolution.

Modern simulation tools can offer convenient PC computer environments, task managers to make systematic assessments, and integration with process flow information. The commercial tools generally emphasize user friendly and task oriented interfaces suitable for turnkey operation. Universities offer source codes, making these tools suitable for incorporation of user-supplied physical models into the simulation. Special tools for difficult problems such as analysis of large mask areas, 3D dissolution effects, vector and nonplanar mask effects, interactions with nonplanar substrates, and simulation from empirical dissolution measurements are also emerging. Since the technologist is a good judge of what capabilities best meet current needs and programming resources, little more need be said other than to point out that the level of resources required rises sharply with the sophistication of the physical models.

In aerial image simulation the classical formal approaches are Abbe's method, which traces one source ray at a time, and Hopkins' method, which integrates the combined mask/lens effect before considering the mask. The latter is built on the assumption that the diffraction efficiency of the mask is independent of the illumination. Methods for including resolution enhancement techniques within this method are becoming important and include: (1) arbitrary illumination functions (annular, quadrupole, generalized quadrupole, nonuniformities, etc.), (2) phase-shifting masks, and (3) lens properties (exposure at two focus positions, in-lens filtering, and aberrations). The Abbe method, which numerically adds images calculated for each point in the source, is ideally suited for arbitrary source shapes and can easily accommodate phase-shifting masks and lens properties.

Speed is also important, particularly for making optical proximity corrections (OPC). Initial OPC studies have shown that these corrections can give significant lithographic improvement.[96] Simple rule based approaches show promise for first order optical effects and even process corrections.[97,98] Using Hopkins' method for image calculation requires rapid methods for integrating in wave space. With direct integration the simulation time goes up as the fourth power of the diameter of the area considered. This can be reduced significantly by using techniques similar to fast Fourier transform methods but which are more general and allow mask edges and sizes to have arbitrary location and sizes.[99] For speed in OPC a technique of reusing precalculated lens effects over a small window and then scanning that window over the sizable mask area of interest can be used.[100] A method for representing the partially coherent illumination with an ensemble of strictly monochromatic oscillations was introduced by Wolf,[101] has been used in OPC[102] and is now common to many approaches.[100] The number of monochromatic excitations that must be included for sufficient accuracy can be as low as five for low sigma and likely increases as the square of sigma.

Some second order physical effects in imaging are also of concern in aerial and latent image simulation.[103-111] One is accounting for propagation through the lenses with high numerical apertures of 0.5 and above.[106] Here it becomes important to (1) account for large ray angles (beyond the paraxial assumption) in formulating ray path effects (such as defocus), (2) consider obliquity factors to account for flux at various takeoff angles, and (3) account for the rotation of the electric field component vectors with propagation angle in the transverse magnetic (TM) polarization. This rotation is quite pronounced in air: for NA = 0.7, the ray angle can be 44° off axis and the large ray angles are only 71% as effective as the vertical rays. However, in a recording media such as photoresist the rays become more vertical and the relative effectiveness only drops to 92%. This drop is also partially offset by the lower transmission coefficient for transverse electric (TE) rays. So with appropriate high-NA tuning, scalar imaging can be used up to NAs of 0.7.[108] A second problem is properly accounting for the effects of propagation through the mask. Some simulators offer more rigorous analysis in that the distribution of electric fields emerging from the mask is not a square wave, polarization dependent, and varies with incident angle.

The physical models and algorithms used in post-exposure bake are another important consideration, particularly with chemically-amplified resists. It is well known that bake conditions affect resist contrast and even standing waves in conventional resists. In resists having a chemical reaction, the PEB becomes even more important. Most simulators simplify the PEB to a diffusion step or avoid the need for a PEB step by using a dissolution model derived under the process conditions. The concern today with t-BOC resists is that there is significant diffusion simultaneous with the reaction which affects linewidth bias. Currently, both Fickean and concentration dependent diffusion models are proposed, and the use of the latter will have major consequences on the algorithms and CPU time needed.

Finally, dissolution requires its own set of physical models and algorithms.[112-118] Most simulators assume that the correct physical model is that of anisotropic inhomogeneous material being removed by surface etching. Each point on a resist profile at time t_1 puts out a small sphere of influence into the remaining material, and the resist profile at time t_2 is the locus of these spheres. The simulators for this dissolution process in either two or three dimensions can be thought of as using cell removal, ray tracing (or its equivalent string algorithm), or a level set approach. Cell removal methods are generally easy to implement, have no loop problems, are often limited by the accuracy with which the local surface normal can be estimated, and often use 'spillover' of removal of one cell to the next to allow fewer and larger tie steps. In ray-tracing methods[113,116] either the nodes or faces (segments in 2D) are advanced and recursive subdivision of time steps can be used to locally adapt to sharp inhomogenities. Removal of loops and regularization of the node spacing on the resulting surface[117] require sophisticated methods, particularly in 3D. The level-set method over comes the difficulty of loops by generalizing the problem to computing a function over the entire resist region and then using a contour program to select the constant level of that function which represents the resist profile at the development time of interest.[118] This method is quite robust and can be sped up by making a physically reasonable guess for the initial function and by adapting most of the CPU effort to just that region near the resist profile. Today most simulators are based on ray tracing in 2D and cell removal in 3D. However, robust ray trace and level-set methods are emerging in 3D.

A considerable selection of simulators for optical lithography is available and the physical models, algorithms, design options and user interfaces are undergoing rapid growth. Aside from the references mentioned already, Refs. 119-138 describe lithography simulators and Refs. 139-147 discuss rigorous electromagnetic scattering relevant to lithography and metrology applications. A snapshot of simulators available in 1995 is given below in Table 1. References for further information are also included.

Table 1:   Summary of lithography simulators circa 1995.

Program	Organization	Developers	References
DEPICT	TMA	Pack, Bernard	105
FAIME	Vector Technologies	Barouch	128
METROPOLE	CMU	Strojwas & Students	133, 144
OPTOLITH	Silvaco		
PROLITH	FINLE Technologies	Mack	6, 110
SAMPLE	UC Berkeley	Neureuther & Students	5, 129
SOLID	Fraunhofer/Sigma-C	Henke	127

7.7 FUTURE PERSPECTIVE

Looking at optical lithography from the modeling and simulation point of view has helped create a new appreciation for the complexity of the exposure tool, mask, resist, and substrate interactions and the large number of parameters involved.   Modeling naturally provides an efficient framework for characterization, while simulation can identify the quantitative impact of various parameters. Together, they can make the assessment of practical performance and the investigation of technology innovation much more effective. A rapid evolution of models and growth of applications is expected as optical lithography is pushed to linewidths in the wavelength regime. It is important to grow the associated models and simulation tools along with the technology innovations being introduced to make these small feature sizes possible. The growth of modeling and simulation will likely see more integration with manufacturing information and modeling systems.

REFERENCES

1. F. H. Dill, "Optical Lithography," *IEEE Trans. Electron Devices*, ED-22, No. 7, pp. 440-444 (1975).

2. F. H. Dill, W. P. Hornberger, P. S. Hauge, and J. M. Shaw, "Characterization of Positive Photoresist," *IEEE Trans. Electron Devices*, ED-22, No. 7, pp. 445-452 (July, 1975).

3. K. L. Konnerth and F. H. Dill, "In-Situ Measurement of Dielectric Thickness During Etching or Developing Processes," *IEEE Trans. Electron Devices*, ED-22, No. 7, pp. 452-456 (1975).

4. F. H. Dill, A. R. Neureuther, J. A. Tuttle, and E. J. Walker "Modeling Projection Printing of Positive Photoresists," *IEEE Trans. Electron Devices*, ED-22, No. 7, pp. 456-464 (1975).

5. W. G. Oldham, S. N. Nandgaonkar, A. R. Neureuther and M. O'Toole, "A General Simulator for VLSI Lithography and Etching Processes: Part I - Application to Projection Lithography," *IEEE Trans. Electron Devices*, ED-26, No. 4, pp. 717-722 (April, 1979).

6. C. A. Mack, "PROLITH: A Comprehensive Optical Lithography Model," *Optical Microlithography IV, Proc.*, SPIE Vol. 538, pp. 207-220 (1985).

7. H. I. Smith, N. Efremow and P. L. Kelly, "Photolithographic Contact Printing of 4000 A Linewidth Patterns," , *J. Electrochem. Soc.*, Vol. 121, No. 11, pp. 1503-1506 (Nov., 1974).

8. B. J. Lin, "Deep-UV Conformable Contact Photolithography for Bubble Circuits," *IBM J. Res. and Dev.*, Vol. 20, pp. 213-221 (May, 1976).

9. J. H. Lee and A. R. Neureuther, "Proximity Effect in Contact Printing", *Proceedings 1983 International Symposium on VLSI Technology, Systems and Applications*, Taipei, Taiwan, ROC, pp. 260-264 (1983).

10. H. R. Rottman, "Advances in Contact and Proximity Printing," *Kodak Microelectronics Seminar Proceedings*, pp. 79-87 (1974).

11. L. White, "Proximity Printing of Contact-Hole Maskings," *Kodak Microelectronics Seminar Proceedings*, pp. 98-108 (1981).

12. J. H. Bruning, "Optical Imaging for Microfabrication," *Semiconductor International*, Vol. 4, No. 4, pp. 137-156, (April 1981).

13. D. Markle, "The Future and Potential of Optical Scanning Systems," *Solid State Technology*, Vol. 27, No. 9, p. 158 (1984).

14. B. J. Lin, "Partially Coherent Imaging in Two Dimensions and the Theoretical Limits of Projection Printing in Microfabrication," *IEEE Trans. Electron Devices*, Vol. ED-27, No. 5, pp. 931 (1980).

15. M. Born and E. Wolf, Principles of Optics, 6th edition, Pergamon Press (Oxford: 1980).

16. M. M. O'Toole and A. R. Neureuther, "The Influence of Partial Coherence on Projection Printing", *Semiconductor Microlithography IV*, SPIE Vol.174, pp. 22-27 (April 1979).

17. H. H. Hopkins, "On the Diffraction Theory of Optical Images," *Proc. Royal Soc. London*, Vol. A217, pp. 408-432 (1953).

18. K. H. Toh, C. Fu, K. Zolinger, A. R. Neureuther and R. F. Pease, "Understanding Voting Lithography Through Simulation *Optical/Laser Microlithography, Proc.*, SPIE Vol. 922, pp. 194-202 (1988).

19. M. G. Rosenfield, D. S. Goodman, A. R. Neureuther and M. D. Prouty, "A Comparison of Backscattered Electron and Optical Images for Submicron Defect Detection," *J. Vac. Sci. Technol. B.*, Vol. 3, No. 1, pp. 377-382 (Jan./Feb., 1985).

20. C. A. Mack, "Analytical Expression for the Standing Wave Intensity in Photoresist", *Applied Optics*, Vol. 25, No. 12, pp. 1958-1961 (June, 1986).

21. D. A. Skoog and D. M. West, Fundamentals of Analytical Chemistry, 3rd edition, Holt, Rinehart, and Winston (New York :1976), pp. 509-510.

22. J. M. Koyler, et al., "Thermal Properties of Positive Photoresist and their Relationship to VLSI Processing," *Kodak Microelectronics Seminar Interface '79*, pp. 150-165 (1979).

23. J. M. Shaw, M. A. Frisch, and F. H. Dill, "Thermal Analysis of Positive Photoresist Films by Mass Spectrometry," *IBM Jour. Res. Dev.*, Vol 21, pp. 219-226 (May, 1977).

24. C. A. Mack, "Absorption and Exposure in Positive Photoresist," *Applied Optics*, Vol. 27, No. 23, pp. 4913-4919 (Dec., 1988).

25. M. A. Narasimham and J. B. Lounsbury, "Dissolution Characterization of Some Positive Photoresist Systems," *Semiconductor Microlithography II, Proc.*, SPIE Vol. 100, pp. 57-64 (1977).

26. C. A. Mack, "Development of Positive Photoresist," *Jour. Electrochemical Society*, Vol. 134, No. 1, pp. 148-152 (Jan. 1987).

27. C. A. Mack, "New Kinetic Model for Resist Dissolution," *Jour. Electrochemical Society*, Vol. 139, No. 4, pp. L35-L37 (Apr., 1992).

28. P. Trefonas and B. K. Daniels, "New Principle for Image Enhancement in Single Layer Positive Photoresists," *Advances in Resist Technology and Processing IV, Proc.*, SPIE Vol. 771, pp. 194-210 (1987).

29. F. H. Dill and J. M. Shaw, "Thermal Effects on the Photoresist AZ1350J," *IBM Jour. Res. Dev.*, Vol. 21, No. 3, pp. 210-218 (May, 1977).

30. T. R. Pampalone, "Novolac Resins Used in Positive Resist Systems," *Solid State Tech.*, Vol. 27, No. 6, pp. 115-120 (June, 1984).

31. D. J. Kim, W. G. Oldham and A. R. Neureuther, "Development of Positive Photoresist," *IEEE Trans. Electron Dev.*, Vol. ED-31, No. 12, pp. 1730-1735 (Dec., 1984).

32. C. Zee, W.R. Bell II, and A. R. Neureuther, "Effect of Developer Type and Agitation on Dissolution of Positive Resist" *Advances in Resist Technology and Processing V, Proc.*, SPIE Vol. 920, pp. 154-161 (1988).

33. J. W. Bossung, "Projection Printing Characterization," *Developments in Semiconductor Microlithography II, Proc.*, SPIE Vol. 100, pp. 80-84 (1977).

34. C. A. Mack, "Understanding Focus Effects in Submicron Optical Lithography: a Review," *Optical Engineering*, Vol. 32, No. 10, pp. 2350-2362 (Oct., 1993).

35. E. W. Charrier and C. A. Mack, "Yield Modeling and Enhancement for Optical Lithography," *Optical/Laser Microlithography VIII, Proc.*, SPIE Vol. 2440, pp. 435-447 (1995).

36. C. A. Mack, "Photoresist Process Optimization," *KTI Microelectronics Seminar, Proc.*, pp. 153-167 (1987).

37. M. D. Levenson, D. S. Goodman, S. Lindsey, P. W. Bayer, and H. A. E. Santini, "The Phase-Shifting Mask II: Imaging Simulations and Submicrometer Resist Exposures," *IEEE Trans. Electron Devices*, Vol. ED-31, No. 6, pp. 753-763 (June 1984).

38. H. J. Levinson and W. H. Arnold, "Focus: the critical parameter for submicron lithography," *Jour. Vac. Sci. Tech.*, Vol. B5, No. 1, pp. 293-298 (1987).

39. W. H. Arnold and H. J. Levinson, "Focus: the critical parameter for submicron optical lithography: Part 2," *Optical Microlithography VI, Proc.*, SPIE Vol. 772, pp. 21-34 (1987).

40. C. A. Mack, "Understanding Focus Effects in Submicron Optical Lithography," *Optical/Laser Microlithography, Proc.*, SPIE Vol. 922, pp. 135-148 (1988), and *Optical Engineering*, Vol. 27, No. 12, pp. 1093-1100 (Dec., 1988).

41. C. A. Mack, "Comments on 'Understanding Focus Effects in Submicrometer Optical Lithography'," *Optical Engineering*, Vol. 29, No. 3, p. 252 (Mar. 1990).

42. C. A. Mack and P. M. Kaufman, "Understanding Focus Effects in Submicron Optical Lithography, part 2: Photoresist effects," *Optical/Laser Microlithography II, Proc.*, SPIE Vol. 1088, pp. 304-323 (1989).

43. C. A. Mack, "Understanding Focus Effects in Submicron Optical Lithography, part 3: Methods for Depth-of-Focus Improvement," *Optical/ Laser Microlithography V, Proc.*, SPIE Vol. 1674, pp. 272-284 (1992).

44. C. A. Mack, "Algorithm for Optimizing Stepper Performance Through Image Manipulation," *Optical/Laser Microlithography III, Proc.*, SPIE Vol. 1264, pp. 71-82 (1990).

45. C. A. Mack and J. E. Connors, "Fundamental Differences Between Positive and Negative Tone Imaging," *Optical/Laser Microlithography V, Proc.*, SPIE Vol. 1674, pp. 328-338 (1992), and *Microlithography World*, Vol. 1, No. 3, pp. 17-22 (Jul/Aug 1992).

46. C. A. Mack, "Optimization of the Spatial Properties of Illumination," *Optical/Laser Microlithography VI, Proc.*, SPIE Vol. 1927, pp. 125-136 (1993).

47. C. A. Mack, A. Stephanakis, R. Hershel, "Lumped Parameter Model of the Photolithographic Process," *Kodak Microelectronics Seminar, Proc.*, pp. 228-238 (1986).

48. R. Hershel and C. A. Mack, "Lumped Parameter Model for Optical Lithography," Chapter 2, <u>Lithography for VLSI, VLSI Electronics - Microstructure Science</u>, R. K. Watts and N. G. Einspruch, eds., Academic Press (New York: 1987) pp. 19-55.

49. C. A. Mack, "Enhanced Lumped Parameter Model for Photolithography," *Optical/Laser Microlithography VII, Proc.*, SPIE Vol. 2197, pp. 501-510 (1994).

50. C. A. Mack, "Lithographic Optimization Using Photoresist Contrast," *KTI Microlithography Seminar Interface '90, Proc.*, pp. 1-12 (1990), and *Microelectronics Manufacturing Technology*, Vol. 14, No. 1, pp. 36-42 (Jan. 1991).

51. M. P. C. Watts and M. R. Hannifan, "Optical Positive Resist Processing II, Experimental and Analytical Model Evaluation of Process Control," *Advances in Resist Technology and Processing II, Proc.*, SPIE Vol. 539, pp. 21-28 (1985).

52. C. A. Mack, "Optimum Stepper Performance Through Image Manipulation," *KTI Microelectronics Seminar Interface '89, Proc.*, pp. 209-215 (1989).

53. D. L. Fehrs, H. B. Lovering, and R. T. Scruton, "Illuminator Modification of an Optical Aligner," *KTI Microelectronics Seminar Interface '89, Proc.*, pp. 217-230 (1989).

54. M. Noguchi, M. Muraki, Y. Iwasaki and A. Suzuki, "Subhalf Micron Lithography System with Phase-Shifting Effect," *Optical/Laser Microlithography V, Proc.*, SPIE Vol. 1674, pp. 92-104 (1992).

55. N. Shiraishi, S. Hirukawa, Y. Takeuchi and N. Magome, "New Imaging Technique for 64M-DRAM," *Optical/Laser Microlithography V, Proc.*, SPIE Vol. 1674, pp. 741-752 (1992).

56. K. Tounai, H. Tanabe, H. Nozue and K. Kasama, "Resolution Improvement with Annular Illumination," *Optical/Laser Microlithography V, Proc.*, SPIE Vol. 1674, pp. 753-764 (1992).

57. D. A. Bernard, "Simulation of Focus Effects in Photolithography," *IEEE Trans. Semicon. Manufacturing*, Vol. 1, No. 3, pp. 85-97 (August, 1988).

58. D. A. Bernard and H. P. Urbach, "Thin-film Interference Effects in Photolithography for Finite Numerical Apertures," *Jour. Optical Society of America A*, Vol. 8, No. 1, pp. 123-133 (Jan., 1991).

59. M. Born and E. Wolf, Principles of Optics, 6th edition, Pergamon Press, (Oxford, 1980) pp. 113-117.

60. D. C. Cole, E. Barouch, U. Hollerbach, and S. A. Orszag, "Extending Scalar Aerial Image Calculations to Higher Numerical Apertures," *Jour. Vacuum Science Tech.*, Vol. B10, No. 6, pp. 3037-3041 (Nov/Dec, 1992).

61. D. G. Flagello, A. E. Rosenbluth, C. Progler, J. Armitage, "Understanding High Numerical Aperture Optical Lithography," *Microelectronic Engineering*, Vol 17, pp. 105-108 (1992).

62. M. Newmark, E. Tomacruz, S. Vaidya, and A. R. Neureuther, "Investigation of Proximity Effects for a Rim Phase-Shifting Mask Printed with Annular Illumination," *Optical/Laser Microlithography VII, Proc.*, SPIE Vol. 2197, pp. 337-347 (1994).

63. E. Walker, "Reduction of Photoresist Standing-Wave Effects by Post-Exposure Bake," *IEEE Transactions on Electron Devices*, Vol. ED-22, No. 7, p. 464, (July, 1975).

64. H. Moritz and G. Paal, U.S. Patent #4,104,070 (1978).

65. S. MacDonald, R. Miller, C. Willson, G. Feinberg, R. Gleason, R. Halverson, W. McIntire, M. Motsiff, "Image Reversal: The Production of

Negative Images in Positive Photoresist," *Kodak Microelectronics Seminar Interface '82*, pp. 114-117 (1982).

66. J. W. Thackeray, G. W. Orsula, E. K. Pavelchek, D. Canistro, L. E. Bogan, A. K. Berry, K. A. Graziano, "Deep UV ANR Photoresists for 248 nm Excimer Laser Photolithograhy," *Advances in Resist Technology and Processing VI, Proc.*, SPIE Vol. 1086, pp. 34-47 (1989).

67. D. Seligson, S. Das, H. Gaw, "Process Control with Chemical Amplification Resists Using Deep Utraviolet and X-Ray Radiation," *J. Vac. Sci. Technol.*, Vol. B6, No. 6, pp. 2303-2307 (Nov./Dec., 1988).

68. R. A. Ferguson, J. M. Hutchinson, C. A. Spence, and A. R. Neureuther, "Modeling and Simulation of an Acid Hardening Resist," *J. Vac. Sci. Tech.*, Vol B8, No. 6, pp. 1423-1427 (Nov./Dec., 1990).

69. C. G. Willson, H. Ito, J. M. J. Frechet, T. Tessier, and F. M. Houlihan, "Approaches to the Design of radiation Sensitive Polymeric Imaging Systems with Improved Sensitivity and Resolution," *J. Electrochem. Soc.*, Vol. 133, No. 1, p. 181 (1986).

70. R. Tarascon, E. Reichmanis, F. Houlihan, A. Shugard, and L. Thompson, "Poly(t-BOCstyrene sulfone)-Based Chemically Amplified Resists for Deep-UV Lithography," *Polymer Engineering and Science*, Vol. 29, No. 13, pp. 850-855 (July, 1989).

71. R. A. Ferguson, C. A. Spence, E. Reichmanis, and L. F. Thompson, "Investigation of the Exposure and Bake of a Positive-Acting Resist with Chemical Amplification," *Advances in Resist Technology and Processing VII, Proc.*, SPIE Vol. 1262, pp. 412-424 (1990).

72. W. D. Hinsberg, S. A. McDonald, N. J. Cleack, and C. D. Synder, and H. Ito, "Influence of Polymer Properties on Airborne Chmeical Contamination of Chemically Amplified Resists," *Advances in Resist Technology and Processing X, Proc.*, SPIE Vol. 1925, pp. 43-52 (1993).

73. J. Sturtevant, S. J. Holmes, and P. Rabidoux, "Post-exposure Bake Characteristics of a Chemically Amplified Deep-Ultraviolet Resist," *Proc. SPIE*, Vol. 1672, p. 114 (1992).

74. N. Eib, E. Barouch, U. Hollerbach, and S. Orszag, "Characterization and Simulation of Acid Catalyzed DUV Positive Photoresist," *Advances in Resist Technology and Processing X, Proc.*, SPIE Vol. 1925, pp. 186-196 (1993).

75. T. H. Fedynyshyn, C. R. Szmanda, R. F. Blacksmith, W. E. Houck, "The Relationship between Resist Performance and Acid Diffusion in Chemically Amplified Resist Systems," *Advances in Resist Technology and Processing X, Proc.*, SPIE Vol. 1925, pp. 2-13 (1993).

76. A. A. Kranosperova, M. T. Reilly, S. Turner, L. Ocola, and F Cerrina, "Prebake and Post-Exposure Bake Effects on the Dissolution of AZ-PF," *Advances in Resist Technology and Processing X, Proc.*, SPIE Vol. 1925, pp. 323-334 (1993).

77. M. Zuniga, G. Walraff, E. Tomacruz, B. Smith, C. Larsen, W. D. Hinsberg and A. R. Neureuther, "Simulation of Locally-Enhanced Three-Dimensional Diffusion in Chemically-Amplified Resists," *J. Vac. Sci. Technol.* Vol. B11, No. 6, pp. 2862-2866, (Nov/Dec, 1993).

78. M. Zuniga, G. Wallraff, and A. R. Neureuther, "Reaction Diffusion Kinetics in Deep-UV Positive Tone Resist Systems," *Advances in Resist Technology and Processing XII, Proc.*, SPIE Vol. 2438, pp. 113-124 (1995).

79. G. Wallraff, W. D. Hinsberg, F. Houle, J. Opitz, D. Hopper, J. Hutchinson, "Kinetics of Chemically Amplified Resists," *Advances in Resist Technology and Processing XII, Proc.*, SPIE Vol. 2438, pp. 182-190 (1995).

80. J. S. Petersen, C. A. Mack, J. W. Thackery, R. Sinta, T. H. Fedynyshyn, J. M. Mori, J. D. Byers and D. A. Miller, " Characterization and Modeling of Positive Acting Chemically Amplified Resist," *Advances in Resist Technology and Processing XII, Proc.*, SPIE Vol. 2438, pp. 153-166 (1995).

81. J. S. Petersen, C. A. Mack, J. Sturtevant, J. D. Byers and D. A. Miller, "Non-constant Diffusion Coefficients: Short Description of Modeling and Comparison to Experimental Results," *Advances in Resist Technology and Processing XII, Proc.*, SPIE Vol. 2438, pp. 167-180 (1995).

82. M. Newmark, E. Tomacruz, S. Vaidya, and A. R. Neureuther, "Investigation of Proximity Effects for a Rim Phase-Shifting Mask Printed with Annular Illumination," *Optical/Laser Microlithography VII, Proc.*, SPIE Vol. 2197, pp. 337-347 (1994).

83. A. R. Neureuther, P. Flanner III, and S. Shen, "Coherence of Defect Interactions with Features in Optical Imaging," *J. Vac. Sci. Technol.* Vol B5, No. 1, pp. 308-312, (Jan./Feb., 1987).

84. J. Wiley and J. Reynolds, "Device Yield and Reliability by Specification of Mask Defects," *Solid State Technology*, Vol. 36, No. 7, pp. 65-77 (July 1993).

85. H. Watanabe, E. Suguria, T. Imoriya, and M. Inoue, "Detection and Printability of Shifter Defects in Phase Shifting Masks, II. Defocus Characteristics," *Jpn. J. of Appl. Phy.*, Vol. 31, pp. 4155 (1992).

86. R. Socha, M. Yeung, A.R. Neureuther, and R. Singh, "Models for Characterizing Phase-Shift Defects in Optical Projection Printing," *IEEE Trans. on Semiconductor Manufacturing*, Vol. 8, No 2, pp. 1-11 (1995).

87. S. Y. Shaw, S. Palmer and S. J. Schuda, "Printability Study of Opaque and Transparent Defects using Standard and Modified Illumination," *Optical/ Laser Microlithography VIII, Proc.*, SPIE Vol. 2440, pp. 878-890 (1995).

88. H. Fukuda, N. Hasegawa, T. Tanaka, and T. Hayashida, *IEEE Electron Device Letters*, Vol. EDL-8, pp. 179-180 (1987).

89. H. Fukuda, N. Hasegawa, and S. Okazaki, *J. Vac. Sci. Technol. B*, Vol. 7, pp. 667-674 (1989).

90. H. Y. Liu and C. Yu, "Contributions of stepper lenses to systematic CD errors within exposure fields," *Optical/Laser Microlithography VIII, Proc.*, SPIE Vol. 2440, pp. 868-877 (1995).

91. A. K. Wong, and A. R. Neureuther, "Examination of Polarization Effects in Photolithographic Masks Using Three-Dimensional Simulation," *Optical/Laser Microlithography VII, Proc.*, SPIE Vol. 2197, pp. 521-528 (1994).

92. C. Pierrat, A. Wong, and S. Vaidya, "Phase-Shifting Mask Topography Effects on Lithographic Image Quality," *IEDM Tech. Digest*, pp. 53-56 (1992).

93. R. A. Ferguson, A. K. Wong, T. A. Brunner, and L. W. Liebmann, "Pattern-Dependent Correction of Mask Topography Effects for Alternating Phase-Shifting Masks," *Optical/Laser Microlithography VIII, Proc.*, SPIE Vol. 2440, pp. 349-360 (1995).

94. Bakeman et al., VLSI Symposium (1991).

95. M.P. Karnett and M.C. Sarnoff, "Optimizations and Characterizations of Single Layer Resist," *Optical/Laser Microlithography II, Proc.*, SPIE Vol. 1088, pp. 324-338 (1989).

96. Y. Liu and A. Zakhor, *IEEE Trans. Man. Sci.*, Vol. 6, No. 1, p. 1 (1993).

97. O. Otto, et. al., "Automated Optical Proximity Correction: a Rules-Based Approach," *Optical/Laser Microlithography VII, Proc.*, SPIE Vol. 2197, pp. 278-293 (1994).

98. R. Pforr, A. Wong, K. Ronse, L. Van den hove, A. Yen, S. Palmer, G. Fuller, "Feature biasing versus feature assisted lithography - a comparison of proximity correction methods for 0.5 (λ/NA) Lithography," *Optical/ Laser Microlithography VIII, Proc.*, SPIE Vol. 2440, pp. 150-170 (1995).

99. E. Barouch, U. Hollerbach, and R. Vallishayee, "Optimask: An OPC Alogrithm for Chrome and Phase-Shift Mask Design," *Optical/Laser Microlithography VIII, Proc.*, SPIE Vol. 2440, pp. 192-197 (1995).

100. N. Cobb and A. Zakhor, "Fast, Low-Complexity mask Design," *Optical/ Laser Microlithography VIII, Proc.*, SPIE Vol. 2440, pp. 313-327 (1995).

101. E. Wolf, "New Spectral Representation of Random Sources and of the Partially Coherent Fields That They Generate," *Optics Communications*, Vol. 38, No. 1, pp. 3-6 (July, 1981).

102. Y. C. Pati and T. Kailath, "Phase-Shifting Masks for Microlithography: Automated design and Mask Requirements," *J. Opt. Soc. Am. (A)*, Vol. 11. No. 9, pp. 2438-2452 (1994).

103. M. D. Prouty, "Optical Imaging with Phase Shift Masks," *Optical Microlithography III*, SPIE Vol. 470, pp. 228-232 (1984).

104. K. H. Toh and A. R. Neureuther, "Identifying and Monitoring Effects of Lens Aberrations in Projection Printing," *Optical Microlithography VI, Proc.*, SPIE Vol. 772, pp. 202-209 (1987).

105. D. A. Bernard, "Simulation of Focus Effects in Photolithography," *IEEE Trans. Semicon. Manufacturing*, Vol 1, No. 3, pp. 85-97 (Aug., 1988).

106. D. C. Cole, E. Barouch, U. Hollerhach and S. A. Orsag, "Extending Scalar Aerial Image Calculations to Higher Numerical Apertures," *J. Vac. Sci. Technol. B*, Vol. 10, No. 6, pp. 3037-3041, (Nov/Dec, 1992).

107. D. G. Flagello, A. E. Rosenbluth, C. Progler, and J. Armitage, "Understanding High Numerical Aperture Optical Lithography," *Microelectronic Engineering*, Vol 17, pp. 105-108 (1992).

108. M. S. Yeung, D. Lee, R. Lee, and A. R. Neureuther, "Extension of the Hopkins' Theory of partially Coherent Imaging to Include Thin-Film Interference Effects," *Optical/Laser Microlithography VI Proc.*, SPIE Vol. 1927, pp. 452-463 (1993).

109. D. G. Flagello and R. Rogoff, "The influence of Photoresist on the Optical performance of High NA Steppers," *Optical/Laser Microlithography VIII, Proc.*, SPIE Vol. 2440, pp. 340-348 (1995).

110. C. A. Mack and C. B. Juang, "Comparison of Scalar and Vector Modeling of Image Formation in Photoresist," *Optical/Laser Microlithography VIII, Proc.*, SPIE Vol. 2440, pp. 381-394 (1995).

111. Q. D. Quan and F. A. Leon, "Fast Algorithms for 3D High NA Lithography Simulation," *Optical/Laser Microlithography VIII, Proc.*, SPIE Vol. 2440, pp. 371-380 (1995).

112. R. Jewett, R Hagouel, A. Neureuther, and T. Van Duzer, "Line-Profile Resist Development Simulation Techniques," *Polymer Eng. Sci.*, Vol. 17, No. 6, pp. 381-384 (June, 1977).

113. P. I. Hagouel and A. R. Neureuther, "Modeling of X-ray Resists for High Resolution Lithography", *ACS Organic Coating and Plastics Preprints*, Vol. 35, No. 2, pp.258-265 (August 1975), and Ph.D. Thesis, UC Berkeley (1976).

114. F. Jones and J. Paraszczak, "RD3D (Computer Simulation of Resist Development in Three Dimensions)," *IEEE Trans. Electron Devices*, Vol. ED-28, No. 12, pp.1544-1552 (Dec., 1981).

115. A. Moniwa, T. Matsuzawa, T. Ito, and H Sunami, "A Three-Dimensional Photoresist imaging process Simulator for Strong Standing-Wave Effect Environment," *IEEE Trans. on CAD*, Vol. CAD-6, No. 3, pp. 431-437 (May 1987).

116. K. K. H. Toh, A. R. Neureuther, and E. W. Scheckler, "Three-Dimensional Simulation of Optical Lithography," *Optical/Laser Microlithography IV, Proc.*, SPIE Vol. 1463, pp. 356-367 (1991).

117. J. J. Helmsen, "A Comparison of Three Dimensional Photolithography Simulators," Ph.D. Dissertation, University of California, Berkeley, (1995), and Memorandum No. UCB/ERL M95/25.

118. J. Sethian, "Numerical Algorithms for Propagating Interfaces: Hamilton-Jacobi Equations and Conservation Laws," *J. Diff. Geometry*, pp. 131-161 (1990).

119. M. Narasimham and J. Carter, "Effects of Defocus on Photolithographic Images Obtained with Projection Printing Systems," *Developments in Semiconductor Microlithography III, Proc.*, SPIE Vol. 135, p. 2-9 (1978).

120. K. Matsumoto, K. Konno, K. Ushida, "Development and Application of Photolithography Simulation Program for Step-and-Repeat Projection System," *Kodak Microelectronics Seminar*, pp. 74-79 (1983).

121. T. Matsuzawa, T. Ito, and H. Sunami, "Three-dimensional Photoresist Image Simulation on Flat Surfaces," *IEEE Trans. Electron Devices*, Vol. ED-32, No. 9, pp. 1781-1783 (Sep., 1985).

122. L. Jia, W. Jian-kun and W. Shao-jun, "Three-Dimensional Development of Electron Beam Exposed Resist Patterns Simulated Using Ray Tracing Model," *Microelectronics Engineering*, Vol. 6, pp. 147-151 (1987).

123. Y. Hirai, S. Tomida, K. Ikeda, M. Sasago, M. Endo, S. Hayama, and N. Nomura, "Three Dimensional Process Simulation for Photo and Electron Beam Lithography and Estimation of Proximity Effects," *Symposium on VLSI Technology, Digest of Technical Papers*, p. 15 (1987).

124. E. Barouch, B. Bradie, H. Fowler and S. Babu, "Three-Dimensional Modeling of Optical lithography for Positive Photoresists," *KTI Microelectronics Seminar Interface '89*, pp. 123-136 (1989).

125. J. Bauer, W. Mehr, and U. Glaubitz, "Simulation and Experimental Results in 0.6 um Lithography using an i-Line Stepper," *Optical/Laser Microlithography III, Proc.*, SPIE Vol. 1264, pp. 431-445 (1990).

126. H. P. Urbach, and D. A. Bernard, "Modelling Latent Image Formation in Photolithography using the Helmholtz Equation," *Optical/Laser Microlithography III, Proc.*, SPIE Vol. 1264, pp. 278-293 (1990).

127. W. Henke, D. Mewes, M. Weiss, G. Czech, and R. Schiessl-Hoyler, "Simulation of Defects in 3-Dimensional Resist Profiles in Optical Lithography," *Microelectronic Engineering*, Vol. 13, pp. 497-501 (1991).

128. E. Barouch, J. Cahn, U. Hollerbach and S. Orszag, "Numerical Simulation of Photolithographic Processing," *J. Sci. Computing*, Vol. 6, No. 3, pp. 229-250 (1991).

129. E. W. Scheckler, K. K. H. Toh, D. M. Hofstetter, and A. R. Neureuther, "3D Lithography, Etching, and Deposition Simulation (SAMPLE-3D)," *1991 Symposium on VLSI Technology Technical Digest*, pp. 97-98 (1991).

130. K. Lee, Y. Kim and C. Hwang, "New Three-Dimensional Modeling of Optical Lithography for Positive Photoresists," *1991 International Workshop on VLSI Process and Device Modeling*, pp. 44-45, Oiso, Japan. (1991).

131. H. Tanabe, "Modeling of Optical Images in Resists Using Vector Potentials," *Optical/Laser Microlithography V, Proc.*, SPIE Vol. 1674, pp. 637-649 (1992).

132. M. Komatsu, "Three Dimensional Resist Profile Simulation," *Optical/ Laser Microlithography VI, Proc.*, SPIE Vol. 1927, pp. 413-426 (1993).

133. C. M. Yuan, "Calculation of One-Dimensional Lithographic Aerial Images Using the Vector Theory," *IEEE Trans. Electron Devices*, Vol. 40, No. 9, pp. 1604-1613 (Sep., 1993).

134. E. W. Scheckler and A. R. Neureuther "Models and Algorithms for Three-Dimensional Simulation with SAMPLE-3D," *IEEE Trans. CAD*, Vol. CAD-13, No. 2, pp. 219-230 (Feb., 1994).

135. K. K. H. Toh, A. R. Neureuther and E. W.Scheckler, "Algorithms for Simulation of Three-Dimensional Etching," *IEEE Trans. CAD*, Vol. CAD-13, No 5, pp. 616-624, (May, 1994).

136. J. J. Helmsen, M. Yeung, D. Lee and A. R. Neureuther, "SAMPLE-3D Benchmarks Including High-NA and Thin-Film Effects," *Optical/Laser Microlithography VII, Proc.*, SPIE Vol. 2197, pp. 478-488 (1994).

137. D. Adalsteinsson and J. A. Sethian, "A Level Set Approach to a Unified Model for Etching, Deposition, and Lithography I: Algorithms and Two-Dimensional Simulations," *J. Comput. Phys.*, Vol. 120, pp. 128-144 (1995).

138. D. Adalsteinsson and J. A. Sethian, "A Level Set Approach to a Unified Model for Etching, Deposition, and Lithography I: Algorithms and Three-Dimensional Simulations," *J. Comput. Phys.*, Vol. 120, to appear.

139. D. Nyyssonen, "Theory of Optical Edge Detection and Imaging of Thick Layers," *J. Opt. Soc. Am.*, Vol. 72, No. 10, pp. 1425-1436 (Oct. 1982).

140. D. Nyyssonen, and C. P. Kirk, "Optical Microscope Imaging of Lines Patterned in Thick Layers with Variable Edge Geometry: Theory," *J. Opt. Soc. Am.*, A, Vol. 5, No. 8, pp. 1270-1280 (Aug., 1988).

141. G. M. Gallatin, J. C. Wedster, E. C. Kintner, and F. Wu, "Modeling The Images Of Alignment Marks Under Photoresist," *Optical Microlithography VI, Proc.*, SPIE Vol. 772, pp. 193-201 (1987).

142. T. Matsuzawa, A. Moniwa, N. Hasegawa, and H. Sunami, "Two-Dimensional Simulation of Photolithography on Reflective Stepped Substrate," *IEEE Trans. CAD*, Vol. CAD-6, No. 3, pp. 446-451 (May 1987).

143. G. Wojcik, D. Vaughn, and L. Galbriath, "Calculation of Light Scatter from Structures on Silicon Surfaces," *Optical Microlithography VI, Proc.*, SPIE Vol. 772, pp. 21-31 (1987).

144. C. M. Yuan and A. J. Strojwas, "Modeling Of Optical Alignment And Metrology Schemes Used In Integrated Circuit Manufacture," *Optical/Laser Microlithography III, Proc.*, SPIE Vol. 1264, pp. 209 (1990).

145. E. Barouch, B. Bradie, G. Karniadakis, and S. Orszag, "Comprehensive 3D simulator with non-planar substrates," *Optical/Laser Microlithography III, Proc.*, SPIE Vol. 1264, pp. 334-342 (1990).

146. R. Guerrieri, J. Gamelin, K. Tadros, and A. Neureuther, "Massively Parallel Algorithms for Scattering in Optical Lithography," *IEEE Trans CAD.*, Vol. 10., No. 9, pp. 1091-1100 (Sep., 1991).

147. M. S. Yeung and A. R. Neureuther, "Three-Dimensional Refective-Notching Simulation using Multipole Accelerated Physical-Optics Approximation," *Optical/Laser Microlithography VIII, Proc.*, SPIE Vol. 2440, pp. 395-409 (1995).

CHAPTER 8

Issues in Nanolithography for Quantum Effect Device Manufacture

Martin C. Peckerar
F. Keith Perkins
Elizabeth A. Dobisz
Orest J. Glembocki
Naval Research Laboratory

CONTENTS

8.1 INTRODUCTION

In 1975 Moore published his now famous curve depicting the shrinkage of device critical dimensions over time.[1] A version of this curve is shown as Fig. 8.1. Over the past 20 years, this curve has provided a remarkably good means of projecting future processing requirements. As extrapolation of the curve indicates, in the first decade of the next century, the size of minimum features will be at or less than 100 nm. No matter what the governing principle of device operation is, "quantum effects" (i.e., tunneling and particle-wave interference effects) must be accounted for in predicting device performance.

MOORE CURVE:
PRESENTED AT IEDM 1975 INTERNATIONAL ELECTRON DEVICES MEETING

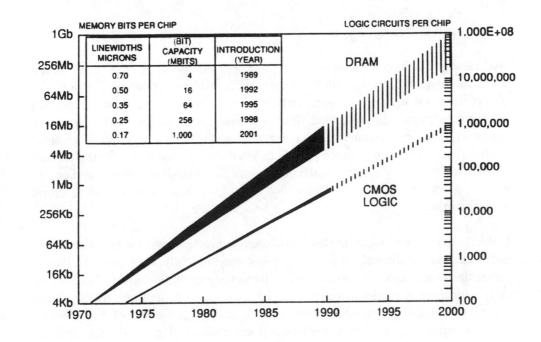

LINEWIDTHS MICRONS	(BIT) CAPACITY (MBITS)	INTRODUCTION (YEAR)
0.70	4	1989
0.50	16	1992
0.35	64	1995
0.25	256	1998
0.17	1,000	2001

Fig. 8.1 Moore curve illustrating IC evolutionary trends.

This is true even in the natural evolution familiar devices such as MOSFETs. As critical gatelengths go to 90 nm, the MOS oxide thickness will be 4.5 nm.[2] This is well within the quantum-tunnelable range. As is demonstrated in Ref. 2, even though the source-drain separation may be about 100 nm, the effective channel lengths will be far smaller than this, also approaching tunnelable distances. Quantized conduction effects are visible in quantum-wire devices with minimum feature sizes above 100 nm, if these devices are cooled to 80 K.[3] Quantum-wire lasers start becoming feasible with linewidths of 35 nm.[4] Based on these considerations, we consider the entry-point to the quantum-device arena to be at or below the 100 nm minimum feature size level.

Thus, it seems likely that we will be entering an era of "quantum effect" electronics over the next 10 to 15 years. Along with this era comes a whole host of new economic and physical challenges to manufacturing. In practice, there are three steps to manufacture for a given part generation: proof-of-principle, prototype development, and volume manufacture. Each of these phases is accomplished in roughly 2 to 3 years. It is appropriate for us to begin to address these challenges now if we are to proceed along our previously established development timelines. In this chapter, we separate manufacturing demands into three areas:

- •impossible feats
- •unlikely events
- •self-defeating purposes.

By "impossible feats," we mean those processes that press or attempt to exceed some fundamental physical limit for their successful performance. It is a trivial observation to say that quantum-effect device process technology presses fundamental physical limits! Such limits represent clear, impenetrable barriers to manufacture. From a manufacturing point of view, throughput - generally given as the number of wafers processed per hour - must also be considered in the context of ultimate limits. If we cannot make the devices fast enough in a given time period using a given technology, that technology will be precluded from use in future manufacturing.

Unlikely events are "yield limited" processes—those processes whose successful outcomes are so unlikely that they are not economically attractive. As device dimensions scale, more and more random flaws become "killer defects," destroying device function. A single dislocation can become such a killer flaw. As the size of critical volumes of device material reduce with scaling, the number of dopant atoms in these volumes goes down. Statistical fluctuations in dopant distributions can become a significant yield limiter in sub-100 nm circuit manufacture.

Frequently in manufacturing technology we encounter processes that are, themselves, feasible but that so compromise device performance that they lead to critical yield loss. This represents the self-defeating purposes category of manufacturing demand. The increasing reliance on processes that produce ionizing radiation, such as plasma etch and deposition processes, creates a more hostile environment for devices in fabrication. Damage mitigation measures must be invoked.

It is not possible to summarize all of the issues associated with these three concerns in a single chapter, but it is possible to outline their major features (at least for patterning technology) and to provide concrete examples of each, based on today's manufacturing experience.

8.2 IMPOSSIBLE FEATS: FUNDAMENTAL LIMITS IN TECHNOLOGY

The major "impossible event" in quantum-effect processing is pattern definition of the sub-100 nm minimum features required for device function. Of course, issues relating to minimum feature size also impact yield; these are discussed in the next section as "unlikely events." It is generally agreed that there are two parts to pattern definition: resolving the image in some photosensitive medium (lithography) and transferring these patterns into underlying material (etching). Both are key ingredients in nanolithography.

In this section, we primarily consider the physical limits issue for lithography in manufacturing. We also describe throughput from a physical limits viewpoint. Limits on throughput are as serious a manufacturing concern as physical limits on any process parameter.

We review five types of possible lithographies from a fundamental limits viewpoint: optical lithography, charged particle lithographies employing lens systems, x-ray lithography, lens-free e-beam lithography (based on proximal probes), and extreme ultraviolet (EUV) lithography. These technologies are presented in an order that is indicative of the magnitude of the paradigm shift each represents. Optics, of course is a current, evolutionary technology; proximal probe approaches are a radical shift in current methodology. On the surface EUV lithography appears to be a simple extension of optical lithography, careful consideration of the materials used, the physics behind the imaging systems employed, resists, etc. lead us to dispense with this notion.

Each approach is discussed in terms of ultimate resolution, pattern placement capability, and throughput. In most cases, we are forming images in etch-resistant layers, such as photoresist. Resists themselves can provide limits on the ultimate attainable resolution, as we will show. In addition, as dimensions are scaled, previously benign issues such as resist adhesion and aspect ratio become critical. The same can be said about our ability to "measure" success. In the past, our metrological capabilities were more than adequate. As we attempt to measure linewidths below 100 nm, major problems arise. Our notion of linewidth itself must be questioned and made more rigorous when we are dealing with measurements requiring the precision of a few atomic diameters. These issues are discussed in Sec. 8.2.6.

8.2.1 Optical Lithography

Most manufacturing lithography is done using flood photon exposure. Usually, an image of the desired pattern is projected through a lens system operating at or near diffraction limits \cite {PROC:Murarka}. For such systems, the minimum resolved feature size is

$$d_{min} = k_1 \frac{\lambda}{NA} \; . \tag{8.1}$$

Here, λ is the wavelength of the exposing light and k_1 is a constant pre-factor (generally equal to 0.7), that takes into account variables in resist exposure characteristics and development conditions. NA is the numerical aperture of the optical system. This is the cone half-angle subtended by the entrance (angle-limiting) aperture of the system when viewed from the point at which the optical centerline intersects the imaging plane.

Depth-of-focus (DOF) is of equal importance in defining whether an optical projection system is physically realizable. Depth-of-focus, here, is taken to be the minimum displacement of the image plane that will cause an ideally focused point to expand to the spot-size resolution limit defined in Eq. (8.1). DOF is given as

$$DOF = k_2 \frac{\lambda}{NA^2} \; . \tag{8.2}$$

Once again, λ is the wavelength of the exposing light and k_2 is a constant prefactor (taken, by experience, to be about 1), that takes into account variables in resist exposure characteristics and development conditions. These relationships are derived in the appendix to this chapter (Secs. 8.A.1 and 8.A.2). There is a significant statistical component relating to the impact of DOF on yield. This is discussed in Sec. 8.3.1 as it pertains to the concept of process latitude.

It must be emphasized that *both* equations lead to hard physical limits. Processing surfaces can have significant topography creating large image plane displacements (displacements on the order of or greater than 1 μm). Such height variations are local to a single chip and are referred to as "intra-field" topography. To tell a designer that he or she must limit out-of-plane displacements provides a significant limitation on design flexibility. Also, a natural amount of bowing and deformation takes place during wafer handling and processing. It is unlikely that we will achieve better than a few tenths of a micron control of image plane placement. Very aggressive development programs are now under way to "planarize" printing surfaces using chemical-mechanical polishing (CMP) techniques.[6] Such aggressive processing steps may achieve an intra-field flatness of a micron (or better), but they are costly and introduce new types of defects, reducing yield. And, they do little for "inter-field" flatness problems, such as wafer bowing.

An example of how these factors interrelate is shown in Fig. 8.2. Here we see numerical aperture plotted as a function of resolution for various wavelengths of exposing light (solid lines). The 436 nm and 365 nm curves are for the mercury arc g and i lines; the 248 nm and 193 nm curves are for KrF and ArF excimer laser emission wavelengths. The thick shaded bars refer to the depth-of-focus obtained for a given NA and resolution combination. Processing k-factors are 0.7 (k_1) and 1 (k_2). Of course, these factors can be improved with process changes (i.e., use of different resists, tighter control of surface reflectivities, etc.). In fact, (k_1) factors below 0.5 have been discussed, but these improvements usually occur at the expense of process latitude. That is, process parameters must be controlled better than what is reasonably expected. This creates the "unlikely event" discussed below.

Figure 8.2 offers some insight into the difficulties of entering the quantum effect world with light. An ArF exposure source used in conjunction with an optical system of NA = 0.52 can realize a DOF of about 0.7 μm with a resolution of 0.26

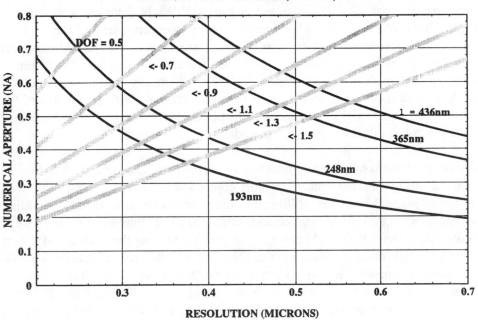

NA vs RESOLUTION FOR VARIOUS DEPTHS-OF-FOCUS

$R = k1 \ /NA \quad DOF = k2 \ /NA^2, \quad k1 = 0.7, k2 = 1$

FIG. 8.2 Numerical aperture vs resolution for various exposure wavelengths (436 nm Hg G-line; 365 nm Hg G-line; 248 and 193 nm are excimer sources). Also on the curve we see lines of constant depth of focus plotted. Regions to the left of each line have smaller depth of focus for a given resolution. Gray lines: depth of focus boundaries. Black lines: NA vs resolution curves for different wavelengths.

μm. A KrF system working to achieve the same design rule would need an NA of about 0.7. While such NAs are difficult to obtain in practice, their availability does not violate some basic physical principle. What we find, though, is a physical constraint. The KrF system has a DOF of 0.5 μm. The ArF system offers about 40% more DOF. This 40% change converts directly to a similar increase in process latitude. In a manufacturing environment, such improvements in process latitude are critical.

Current optical theory holds that it is possible to extend resolution limits without compromising process latitude by using a variety of contrast enhancements techniques. These include phase-shifting[7] and off-axis illumination.[8] Phase shifting has been of particular interest in this area, offering, for certain figure geometries, a doubling of obtainable resolution.[9] Based on Fig. 8.2, this would bring optical lithography close to satisfying the sub-100-nm quantum-effect device requirement discussed above.

In phase shift lithography, the optical path length of light transmitted through the mask is altered in various portions of the mask. The idea is to allow diffracted side-lobes to interfere destructively in-between adjacent features, thus improving exposure contrast. This is illustrated in Fig. 8.3. While there are a number of approaches to phase shifting, all add considerable process complexity. "Hard" phase shift approaches require additional deposition or etching processes to control optical path through the mask (such as the "phase shift material" in Fig. 8.3). "Soft" phase shift approaches are a bit less complex.[10] Here, the normally opaque portion of the mask is made semi-transparent. A phase reversed, attenuated negative image of the pattern is presented to the image plane. The phase reversal at the feature boundary greatly increases the contrast at the boundary.

Even in the "soft" phase shift approaches, novel absorbers must be used. Materials such as "leaky chrome" must transmit controlled amounts of light with well controlled phase. Thus, phase shifting introduces another process variable requiring precise control: the optical path length through the mask. Furthermore, extra steps in the "hard" approaches add process complexity, making the problem of defect control more difficult. In all cases, in addition to particulate related defects, a whole new class of "phase defects" works to lower yield. It is very difficult to detect and to repair such defects.[11] Even though this process was disclosed in 1992, this approach has not reached the manufacturing floor (at least for critical gate levels) at this time. Experimentation in new "defect-free" phase shift materials continues. Recent work on MoSi absorbers appears promising at this writing.[12]

In an effort to extend the usefulness of a given optical tool, pattern shape modification can be used to compensate for diffraction effects. This is referred to as optical proximity correction (OPC). An example of OPC is as follows: Consider the termination point of an exposed line. Light "leaks out" from around the

THE LEVENSON PHASE-SHIFTER

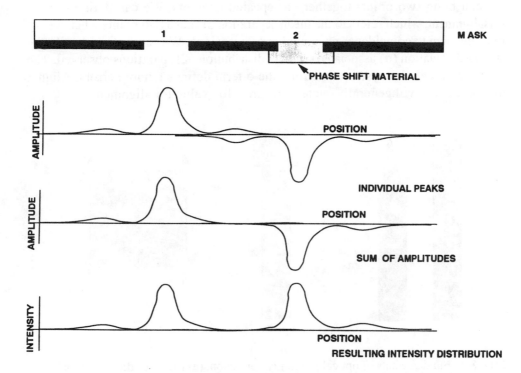

FIG. 8.3 The physics of phase shift contrast enhancement as illustrated by a Levenson phase shifter. The important point to note is the absence of summed amplitude and total intensity between the main feature peaks caused by the phase shifter.

boundary of the line in an amount proportional to the line perimeter present in a given amount of surface area. The effect of this on imaging in positive resist is shown in Fig. 8.4(a). Thus, by opening the clear portion of the line a bit more at the point of line termination, this light loss is compensated for. Such a terminator is called an anchor. Windows may be "squared" by adding serifs to the corners; these are shown in Fig. 8.4(b). This approach clearly improves pattern fidelity, but data-base management may be an issue (more "shapes" must be added to the database and more database sorting must be done to apply the corrector). In addition, feature correction must be based either on empirical studies or on accurate computer simulation of the diffracted exposure intensity profile. A recent review of the status of this field was provided by Kornblit et al.[13] That paper also addressed some of the issues associated with OPC pattern etch fidelity in achieving desired goals.

Of course, resolution and throughput are important considerations in determining the feasibility of a given lithographic approach, but alignment precision is important, even for quantum-effect devices. Let us briefly review the basic principles of optical alignment. What we say here applies equally well to all of the types of lithography described here. In general conversation "alignment" actually refers to a number of aspects of boundary registration. Strictly speaking, alignment refers to how closely we can bring two points together on repeated attempts. We can define a mean misalignment, which refers to the mean separation of these two points when we try to bring them to coincidence on a number of occasions. We can further specify the standard deviation (σ) associated with the distribution of separations observed. The mean term includes systematic errors; the σ term defines random errors. Aligner manufacturers will generally quote a "mean + 3σ" value for alignment.

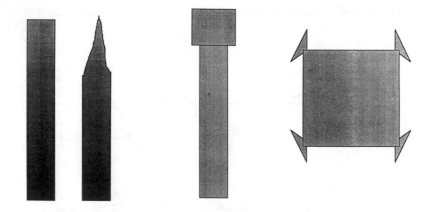

FIG. 8.4 Fundamentals of optical proximity correction. (a) Effect of diffraction on a line termination. (b) Implementations of OPC.

Overlay is important in systems in which we must bring large numbers of points (or boundaries) into registry all at the same time. Once again, we can define a "mean + 3σ" for overlay, but now we must cite the worst case mean and σ values for ensembles of points sampled across a pattern field. We can talk about three types of overlay measurement. The first (and easiest to obtain) is called "tool-to-itself" overlay. Here, we attempt to bring the same pattern, exposed in the same system, into coincidence with itself on two successive alignment operations. "Tool-to-tool" overlay refers to an attempt to bring the same pattern into registry on two different aligners. We can further define a "tool-to-grid" overlay as the separation of fiducial points on a pattern from some "ideal" reference grid. How such a grid is obtained is a matter of considerable debate.

In the past, most alignment systems worked using an optical pattern recognition approach. Here one pattern (on the mask) would be placed in another (on the wafer) by appropriate stage motion. The alignment percision would be set by the optical resolution of the microscopes used to sense the marks. A typical "cross-on-cross" alignment pattern is shown in Fig. 8.5(a). Using advanced computer-aided pattern

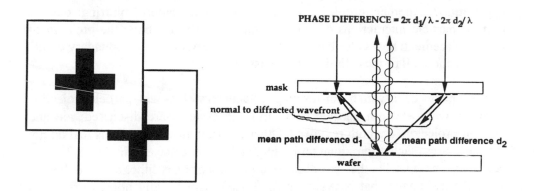

FIG. 8.5 Two methods of optical alignment suitable for use with UV projection steppers or x-ray steppers: (a) "Cross-in-cross" mark for use with pattern recognition; (b) A typical interferometric alignment scheme.

recognition techniques, alignment accuracies of better than 100 nm (3σ) could be obtained with this approach.[14]

Better alignment capability can be achieved using interferometric techniques. One such example is shown in Fig. 8.5(b). Here we see a laser beam diffracted onto a single on-the-wafer grating from two on-the-mask alignment marks. The angle of incidence on the wafer grating from both mask gratings is set to diffract in a direction normal to the wafer plane. The optical phase of the light from each mask grating depends on the path length from the mask grating to the wafer grating. When the wafer grating is just midway between the two mask gratings, we have constructive interference of the ± 1 diffracted orders from the left and right mask gratings. Movement left or right causes destructive interference. Our ability to sense the center position is excellent. Displacements of one-thousandth of a wavelength of incident laser light are detected with modest effort. For He:Ne laser illumination, this converts to about 6 nm. Of course, mechanical stability of the stage and the mechanical control system also determine the final alignment accuracy, but 10-nm 3σ alignment seems possible with such a system.[15]

8.2.2 Charged Particle Approaches to Lithography

For many years, particle beams of various types have been used in lithography. E-beams and ion beams have been used, or at least proposed, for both direct-write and mask making projects. E-beam technology has been used as the mask making tool of choice when resolution is the key factor defining quality. Both electron and ion sources have been used as area image projectors. All particle sources have the benefit of extremely high diffraction-limited resolution.

There are three "physical" limits one encounters. One pertains to throughput—how fast can one expose requisite patterns either in mask making or in direct-write application? Another problem refers to apparent pattern blurring due to particle-particle interactions of various types. Finally, we have the problem of pattern placement: how close can we place a boundary to a target boundary point? In this section, all of these limits are discussed.

The text below deals with beam-forming systems, as well as with particle projection systems. Only electron particle beam forming systems are discussed; ion and e-beam projectors are both reviewed. Other electron imaging systems that do not employ lenses (such as proximal probe approaches) are discussed in Sec. 8.2.4. Ion beam systems exist, but they have limited utility as a primary lithography tool. They have the advantage that backscatter components are usually not significant in degrading resolution. Forward scatter is an issue as one enters the quantum regime. But ions are more massive, more slowly moving, and generally require electrostatic lenses for focusing. "Stiffer" ion beams are less easily steered and throughput remains a considerable issue. Ion beams are used in lithographic technology, though. Ion beam repair tools exist that can remove mask (and wafer) defects. Ion beams are used to expose device cross sections for reliablity and process studies.[11, 16] In addition, these tools provide a local implant capability that may be useful in some device applications (such as asymmetric channel doping in MOSFET structures[17]). As write implanters, these tools may suffer from some of the statistically imposed limitations described in Sec. 8.3.2 for small total dose implants.

8.2.2.1 E-beams for direct-write and mask making applications: throughput considerations

It might be hoped that particle-beam direct-write tools could meet the demanding resolution requirements of quantum-device technology. E-beam technology, in particular, has the requisite resolution and has been used on the production floor in a direct-write capacity in the past.[18] The issue of throughput has been addressed in the open literature.[19, 20] Since most e-beam systems operate with a constant beam current density J (A/cm^2), an e-beam tool can "flash" expose a fundamental area in a time t:

$$t = \frac{D}{J} \quad ,$$

(8.3)

where D is the sensitivity of the resist (in Coulombs/cm^2). The exposure time is just t times the number of "flashes" used to expose the pattern. Of course, this is a lower limit to the actual exposure time. Machine overheads, such as electronic and mechanical settling times, must be accounted for. These can, in some circumstances, double the write time.

In addition, there are basic differences in e-beam writing strategy from tool to tool.

Early e-beam tools (and most mask makers in use today) are raster scan machines. That is, the beam is moved along some predetermined, uniform "address grid," in a manner similar to the way a television beam sweeps a picture on a cathode ray tube. As the beam sweeps across each exposure site (called an *exel*), it is either blanked or unblanked at that site. An alternative approach is to "vector" the beam to a feature, moving to the next feature once the first write is accomplished; this is called the vector scan approach. In both cases a Gaussian "round" beam of fixed diameter exposes each exel.

It might be thought that the speed of the vector scan approach scales inversely with the pattern density, and thus must be much faster than the raster scan approach (at least when resist sensitivity doesn't limit throughput). This isn't quite the case, though, since electronic settling times are longer for the long vector jumps. Thus, the write time becomes dependent on the "write strategy"—how the pattern is written.

An alternative type of e-beam tool makes use of a variable shaped beam (VSB).[21] Here, a large-area collimated beam is incident on a series of apertures. The shadowing effect of these apertures, placed at various points in the electron column, is to project a geometrical shape (a square, a rectangle, or even a triangle) onto the writing surface. Thus, rather than exposing a large shape exel by exel, a portion of this shape can be exposed in a single flash. The shapes can be vectored to the appropriate site and significant speed advantage can be obtained.

Let us assume for a moment that we can ignore machine overheads but that the pattern is exceptionally dense and full of features at or near minimum resolution levels for the tool employed, throughput can then be calculated from Eq. (8.3). In addition, we assume that ultimate resist speeds will be limited by statistical fluctuations in the number of electrons incident on a given exel. Assuming Poisson statistics (as elaboratated on in Sec. 8.3.2 and Sec. 8.A.4), 100 electrons incident on an exel will have a 10-electron (10%) uncertainty. In the following discussion, we compute throughputs for future e-beams based on this statistical requirement.

It is possible, even in a rasterscan context, to expose a single feature with one pass of the e-beam. Certainly, this is the case for narrow lines. Increasing beam dwell times will cause the exposure volumes to balloon outward, shrinking or expanding the feature (depending on the sign of the resist used). Even so, "single-pass" features are rarely used in e-beam lithography. Breaking the feature up into smaller subunits (exels) allows finer control of resist sidewall slopes and boundary placement. This is described in somewhat greater detail below. Typically, the beam diameter is four or five times smaller than the minimum feature size. If we take this factor to be four, a 50-nm minimum feature would correspond to a 12.5-nm exel size. For a square exel 12.5 nm on a side, these 100 electrons convert to about 10 $\mu C/cm^2$ sensitivity. (Note: this is about 30 times the sensitivity of standard high-resolution resists such as PMMA, which are currently used to obtain sub-100-nm resolution and is slightly more sensitive than PBS or chemically amplified resists like SAL-601.) Thermal field emitting electron sources just coming into use today provide current densities of 1000 A/cm². Equation (8.3) gives us the flash exposure time for each exel: 10 ns.

This corresponds to a system clock speed of 100 MHz—a clock speed consistent with today's raster scan technology but five or ten times faster than that obtained with VSB tools. Thus, considerable tool and resist improvement will be required if we are to reach this resist-limited throughput level. Let us now calculate the projected throughput for this hypothetical current-limited e-beam tool and ultra-sensitive (statistics-limited) resist system. This is a "best case" analysis for e-beam direct write.

The quantum-effect electronics of the future will be every bit as complex as current VLSI, and there will be significant demand for volume chip production. Thus, the 8"–12" wafer coverage requirement will prevail in the future, as it does now. An 8" wafer holds about 300 printable square centimeters, or about 2×10^{14} 12.5-nm exels. This would lead to an exposure time greater than a week, even allowing for a 50% "fill factor" (the fraction of the wafer actually written). This is considerably below current production standards of 20 to 30 wafers per hour. For mask-making applications, though, perhaps an order of magnitude smaller area must be written. Relaxation of the "4-pass line requirement" speeds the process more. Mask throughputs may be on the order of an hour. Thus, for wafer-writing applications, e-beam throughputs are too low and will remain too low for the foreseeable future. Mask-making requirements are somewhat different. Typical mask throughputs in today's maskshops are about a plate an hour for complex patterns. If we assume wafer level complexity is on the order of mask -level complexity, throughput would not be an issue for the mask maker.

8.2.2.2 Multibeam direct-write systems and stochastic effects

For many years, "multibeam" approaches have been proposed.[22] These could involve some kind of array of field-emitting sources, or locally addressed photocathodes. Much larger current densities are realizable from such sources. For example, in a recently proposed semiconductor-on-glass approach, individual beamlets with light-generated current densities greater than 300 A/cm² could be produced.[23] Individual beams could be driven by dense arrays of solid-state lasers or photodiodes. Here, the thin semiconductor photoemitter was a GaAs film with a cesiated surface to achieve negative electron affinity. Even a relatively small 100×100 array would give 10,000 times the total current provided by a typical single current source. Based on the above calculations, this would seem to solve the throughput problem for direct-write e-beam technology.

The full benefits of such parallelism may not be realized as a result of "stochastic" effects. Here, the discrete nature of charges in the particle beam leads to apparent blurring of the final image. This blurring can be visualized as two separate processes. The first is the so-called Boersch effect. As the beam propagates through the column, it appears to "thermalize" due to interparticle interaction. Some of the forward momentum of the beam is converted to transverse momentum and the effective "temperature" of the beam increases, increasing chromatic aberration.

In the second case, the random nature of particle interactions in the column leads to random trajectory displacements that further blur the beam in the image plane. In the limit of high-density beams, such displacements create a "global space charge" that can be well characterized by particle ensemble averages. This is a problem that can be eliminated by use of a stigmator. In beams of charge densities practical for lithography, though, the uncertainty in the trajectory of an individual particle is too large to deal with in terms of ensemble averages. Thus, an optical refocus cannot eliminate trajectory displacement blurring.

Models of the Boersch effect and trajectory displacement are usually based on Monte Carlo approaches.[24] More recent efforts attempt analytic solutions.[25] Current work indicates that stochastic effects significantly limit the number of beams in these multibeam systems. Schneider et al. recently studied these issues.[26] In the study cited, about 256 beams were possible (with a total current of about 2.5 μA at the wafer) before the stochastic blur rose above 10 nm (10% of a 100-nm line). This yields about 10-nA per beamlet. Let us, for a moment, relax the "multipass" requirement envoked above for improved control of resist sidewall slope and for dose-modulation proximity correction. This, in turn, leads to a relaxation of the "statistical limit" on resist sensitivity. For a high-sensitivity resist like SAL-601, exposed by the 50-KeV source described in Ref. 26, a sensitivity of about 10 μC/cm² may be workable. This corresponds to an exposure time of 20 minutes per 8" wafer. These results, though, were obtained from as yet untested calculations and must be verified.

In Schneider's work,[26] he showed that increasing the beamlet count from 64 to 256 beamlets increases the blurring by more than a factor of 3. Total blurring below 16 beamlets is always small (less than 1.5 nm). Thus, there is an apparent knee to the blur vs beamlet count curve. The exact position of this knee must be ascertained experimentally, allowing throughputs to raise (or fall) accordingly. Further difficulties arise in maintaining an adequately cesiated surface over time (over the area of the sources) and in preventing cathode poisoning. These factors may further limit the number of beamlets possible. These questions should be resolved over the next few years through work ongoing at Stanford University.

If all the beamlets were brought to a separate focus and were never brought close to one another, stochastic problems would not be an issue. Chang et al.[27, 28] proposed such a multibeam alternative using "microcolumns." These are miniature e-beam columns on the order of centimeters high. Such short columns have the advantage of exhibiting very small optical aberration coefficients. Also, there is less interior surface to keep clean, and column charging effects may be less significant. Perhaps a hundred such columns would be feasible, elevating system throughput to current stepper levels.

Mechanical assembly is a considerable problem for such systems, and recent proposals involve the use of silicon micromachining to fabricate the electrostatic lens components for this tool. A more significant problem turns on the issue of power generation in the electron gun. Gun cathodes run hot (temperatures greater than 1500 K are obtainable). This creates thermal distortion in these tiny columns. Considerable work is ongoing to reduce the operating temperatures of the electron.[29]

8.2.2.3 Beam heating, charging, and other miscellaneous effects

It should be noted that the discussion above did not include any allowance for machine overheads, and it was based on a write time for the finest features only. In addition, it must be noted that there are other physical limits to e-beam manufacture that production lines of the future must deal with. These include beam-heating effects and beam interation effects.

Even though the total energy deposited in each flash is small, this energy is absorbed in a very small volume. If we take the resist thickness to be 0.5 μm and assume that 0.15% of the incident energy is absorbed, 100 electrons, each with an energy of 100 eV, will deposit 34.5 J/cm^3 in each exel. For a resist specific heat of 1.17 J/gm-K and a resist density of 1 g/cm^3, this would lead to about a 29.5 K temperature rise.[30] In addition, heat spreading from the substrate could add to the final temperature. But temperature increases in the 10- to 20-K range have been shown to create marked effects in development characteristics. This is further elaborated on in Sec. 8.2.5. Even though the resist postbake temperatures are higher, this heating during irradiation causes marked effects on apparent resist sensitivity[31] and attendant linewidth variation. The latter reference reports as much as 36% sensitivity variation for a 20-K temperature rise.

During the writing process the resist can become inhomogeneously charged. This can lead to significant feature displacement,[32, 33] which may necessitate the development of conducting resists and/or discharge layers for use in nanolithography for manufacturing. Such discharge layers have not been successfully implemented yet.

8.2.2.4 Alignment

Beam displacement falls in the category of beam alignment problems. In the past, most researchers concentrated on issues relating to resolution in quantum-device manufacture, but alignment can be of equal importance. For example, consider submicron grating structures. These are in wide use today in distributed feedback laser systems, where they serve to better couple emitted light to the laser cavity. Small displacements of the grating at writing field boundaries destroy coupling efficiency. Many future applications, such as potential optical resonating devices, will require sub-10-nm feature placement accuracy,[34] as will transistor-type devices requiring 50-nm design rules.

Currently, e-beam systems reference beam position with respect to some aligment fiducial well away from the writing field. The large mechanical and/or electrically induced beam displacements needed to register these fiducials add considerably to machine overhead and reduce the accuracy of the calibration. These types of alignment are performed relatively infrequently; in the time between alignments, column charging or mechanical drifts compromise the reference alignment accuracy.

A number of approaches are currently being pursued to alleviate this problem. Smith et al. have proposed a technique called "spatial phase-lock" (SPL) alignment.[35] Here, grating structures are actually placed on the mask (as shown in Fig. 8.6). The beam can perform a local alignment with respect to these marks. Also, the phase of the alignment signal as it sweeps over the grating can be recorded and used to provide instantaneous beam positioning information. As originally envisioned, this process requires the placing of a fiducial grating over the writing field using a holographic technique, and working with beams of current too small to expose underlying resist during the alignment phase. More recent implementations work with segmented gratings placed in noncritical writing areas. Smith reports alignment accuracies better than 5 nm with this technique.[34] Other techniques, compatible with x-ray mask technologies, are described below.

8.2.2.5 Ultimate resolution

Let us turn our attention to the ultimate resolution achievable with e-beam technology. While beam diameters on the order of 1 nm are possible,[36] beam-matter interaction degrades this limit significantly. When the beam strikes the solid, it splays out, sending energy into adjacent exels. This is shown in Fig. 8.7, which plots energy deposition as a function of position from the point of electron incidence on the resist surface. Such point spread functions are usually obtained using Monte Carlo simulations.[37]

This curve appears to be composed of two parts: a relatively narrow (roughly Gaussian) peak and a broad background deposition caused by backscatter and secondary processes. The width of the narrow peak, referred to as the forward scattering peak, sets resolution limits for isolated features. These peaks are on the order of 10 nm for 100-eV electrons in 0.5 μm of resist for state-of-the-art e-beam columns.

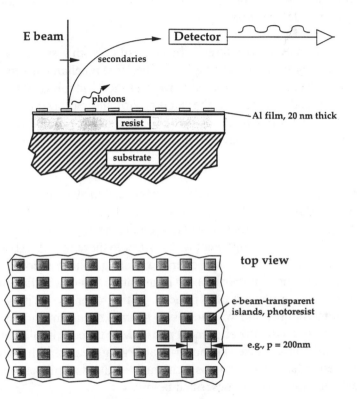

The Continuous Global Fiducial Grid

Fig 8.6 Schematic illustration of the MIT spatial phase-lock alignment approach: (a) Beam strikes secondary emitter alignment pattern placed on mask's photoresist depth; (b) Plain-view of alignment mark. (Courtesy H.I. Smith, MIT)

When we try to create a series of closely spaced features, other issues surface. The broad backscatter and secondary components start to sum to significant amounts of energy deposited in exels far from the regions of desired exposure. This is called proximity effect.[18] There are a variety of approaches for dealing with proximity effect. First, we may use a thin interlayer between the resist and the substrate to effectively "filter" the secondary and some of the backscatter component.[38]

In addition, features may be presorted in the pattern database and "scaled" to alleviate electron scattering problems.[39] However, scaling can take place only over

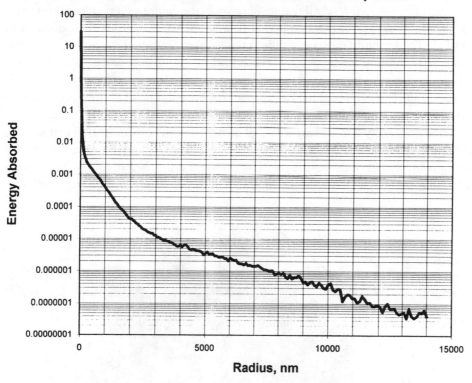

500 nm Resist on 500 nm W on 2000 nm Si, 50kV

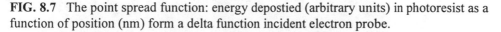

FIG. 8.7 The point spread function: energy depostied (arbitrary units) in photoresist as a function of position (nm) form a delta function incident electron probe.

exel dimensions and would be useless for single-pass features (one other reason for multipass writing). In addition, scaling represents an enormous database management issue.

One other "noncomputational" approach in widespread use is the so called GHOST technique.[40] Here, the normally unwritten part of the image field is written by a defocused, attenuated beam. The beam is adjusted so that it looks like the broad backscatter peak of the beam in the written portion of the field. Thus, the forward scattering peaks of the written field stand out sharply over a uniform, "dose-equalized" background. Since this background has been homogenized over the whole image plane, it can be accounted for and, in practice, eliminated in the development processes. Thus, the resolution limit approaches that set by the forward Gaussian beam width. This is shown in Fig. 8.8.

Recent work has indicated that for lithographic features below 0.25 μm, it becomes increasingly difficult to break up the problem into one of separation into forward and backscattering peaks.[41, 42] Thus, the ability to GHOST at sub-quarter-micron dimensions is in question. For background equalization it seems to work quite well for features embedded in homogeneously dense patterns, but consider the following "real world" situation. A small feature appears in the center of a pattern far removed from other features. This corresponds to a small written point embedded in an annulus. The second Gaussian in the standard two-Gaussian model drops off too rapidly to accurately model dose at this central feature. Thus, as far as the GHOST technique is concerned, this broad background does not exist. And yet, if the annulus is thick enough, all the backscatter background dose at the central feature will be the result of this unaccounted for component. This can create significant dose inhomogeneity and attendent loss of critical dimension control.[43] Furthermore, no one has yet succeeded in writing GHOST patterns with negative resists. Scumming problems become too severe. The elimination of negative resists from the manufacturing line would represent a practical throughput problem in some cases.

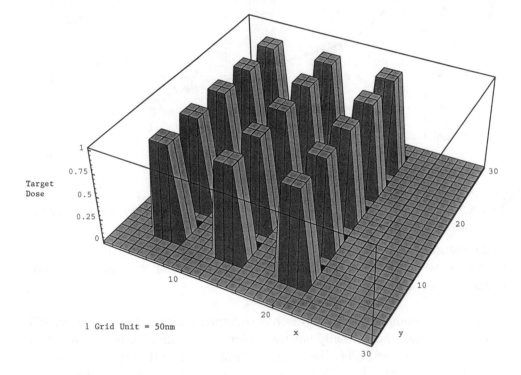

FIG. 8.8(a) GHOSTing flattens the background field and creates uniform feature peak heights for homogeneously dense patterns. Contrast is degraded as the field dose rises to meet the peak. (a) Target pattern.

Thus, we see that interfacial films deal only with a small part of the problem (forward scatter). Feature size scaling is impractical for small feature sizes, and GHOST correction does not work well for certain classes of features. The most likely path for universally applicable proximity correction is by "dose modulation."

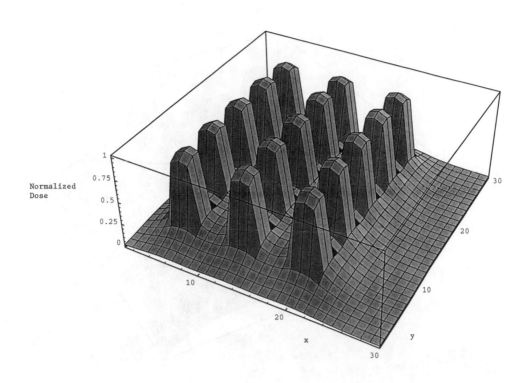

FIG. 8.8(b) UnGHOSTed dose distribution. Note intensity gradients at the edges of the pattern field.

This is a technique in which individual exel doses are varied in such a way as to compensate for forward and backscattering distortions.[44-46] Here, it is recognized that the total dose applied to any one exel is the weighted sum of contributions from all addressed exels. The weighting function, M_{ij}, depends on the separation of the exel under study from the other exels addressed. It tells us how much energy is deposited in exel i by the dose administered to exel j. Once M_{ij} is known, the absorbed dose in the ith exel, a_i can be calculated from the applied doses d:

$$\sum_{j=1}^{N} M_{ij} d_j = a_i \quad , \tag{8.7}$$

where the sum runs over all N exels.

It might be thought that it would be easy to invert this equation to obtain the necessary dose file (the d'_i to achieve an arbitrary absorbed dose pattern), but row and column dependencies in the M_{ij} matrix make the solution to the resulting equation set nontrivial,[47] as well as computationally daunting. Solutions derived from Eq. (8.4) always include negative dose requests at feature boundaries—an unphysical result. In order to achieve a self-consistent, physically realizable

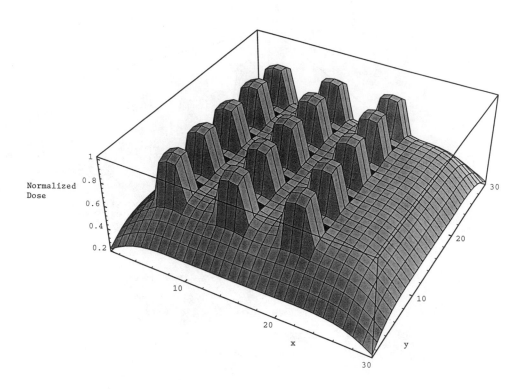

FIG. 8.8(c) GHOST levels background and maximum absorbed dose.

solution, additional constraints must be imposed. These constraints compromise the "quality" of the final image obtained.

As discussed in Ref. 47, the ultimate resolution achievable depends on the global distribution of features and the only rule of thumb is that pattern density degrades ultimate resolution. Even the definition of what one calls an "isolated" feature is open for debate. Arbitrarily, we can say that features spaced farther than a backscatter radius are truly isolated. The backscatter radius in silicon is roughly 100 nm per eV of the incident beam. Thus, at 50 eV, the interfeature separation should be 10 μm to be sure of eliminating interfeature proximity effect. For such situations, optimizing the dose and development time leads to minimum feature sizes within 20% of the beam diameter.[48]

8.2.2.6 Particle projection tools

Two classes of particle projection tools have emerged in recent years as possible high-throughput nanolithography tools. Electron projection has been pioneered at AT& T (Lucent) under the acronym SCALPEL (scattering with angular limitation for projection electron lithography).[49] Ion projection lithography (IPL) has been pioneered by Ion Microfabrication Systems GmbH.[50] In both cases broad beams of particles (either ions or electrons) are collimated and are incident upon a type of transmission mask. In SCALPEL the mask is a thin nitride film (about 100 nm thick) on which a thin metal pattern (50 nm W) is placed. The pattern does not absorb the beam but, rather, creates enough angular deflection to prevent these particles from being brought to focus in the image plane (see Fig. 8.9). IPL uses a stencil mask. Pattern slots are etched in silicon membranes that are close to 2 μm thick.

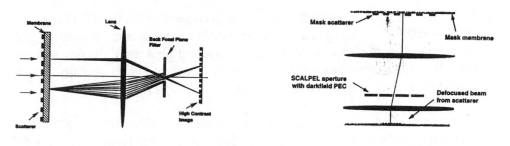

FIG. 8.9 (a) Basic principles of SCALPEL operation. (b) SCALPEL aperture modified for proximity effect control.

Both technologies allow for a reduction image: the mask features are *n* times the feature size on the wafer (*n* is typically 3 or 4). This makes imaging at the mask level and inspection easier. IPL exhibits a lack of particle backscatter, reducing proximity effect. An interesting "dark field" effect can be employed in SCALPEL to provide a GHOST image proximity corrector (described in greater detail below). Key issues associated with both types of projection system are very similar:

- mask stability and robustness
- stochastic effects.

In the case of IPL, features cut in the membranes cause in-plane distortion in the pattern field. This distortion will be far greater than the 10 nm required for 100-nm lithography. In addition, certain patterns cannot be fabricated in a single mask step. Consider the case of an annulus. The central opaque portion cannot be levitated in space and will drop out. It is hoped that a mathematical "inverse" technique will be developed that will correct pattern placement by intentionally displacing features in the pre-etched image. While it is hoped that such correction will be possible, no demonstration exists to confirm feasibility. In addition, the stencil mask will heat during exposure and some form of cooling will also be required.

The delicacy of the SCALPEL mask is also a problem. In order to ensure survival of the mask in processing and to prevent feature displacement, rib supports to the membrane must be provided. These rib supports form a kind of "grillage" that must be accounted for as the particle beam is swept over the mask.

As stated above, the SCALPEL approach does allow for a novel approach to proximity effect control.[51] This is illustrated in Fig. 9(b). A portion of the scattered beam is admitted to the image plane through lateral apertures cut into the final SCALPEL aperture. This, in effect, creates a dark field image that mimics (to some extent) the GHOST background dose. Some degree of contrast equalization is achieved. But, as discussed in Sec. 8.2.2.5, departures from the "two-Gaussian" model do give rise to significant issues in GHOST linewidth control.

It should also be pointed out that most optical masks are "pelliclized." That is, the mask is sealed in a transparent box whose interior contains a particle-free environment. Foreign matter falling on the transparent surfaces of the box will not be brought to focus in the image plane. This renders some particle immunity to such plates. Such protection is not possible for SCALPEL or IPL masks.

Recent computations for stochastic effects indicate that neither IPL or SCALPEL optical systems will meet 30-wafer-per-hour throughputs for 0.1-μm geometries. Of course, computations are debatable since experimental confirmation is lacking. New designs are continuously proposed to alleviate the problem.

8.2.3 X-Ray Lithography

X-ray lithography[52] is a proximity printing technique in which the mask image is created by a patterned heavy metal (such as gold) superimposed on a thin, x-ray transparent membrane (Fig. 8.10). The mask is held in proximity to the image plane, and the image appears as a shadow cast by a flood x-ray exposure. Proximity printing does not require focusing optics. Resolution is limited by two factors: diffraction and photoelectron-induced spreading of the exposure field.

First, consider the effect of diffraction on imaging a grating (see Fig. 8.11). The length of the clear portion of the grating equals the length of the opaque portion. The sum of the clear and opaque lengths is called the period of the grating. As we will show in the appendix, we can view the problem of the minimum resolved period in a manner consistent with the Rayleigh approach to calculating the minimum resolved spacing between two closely spaced apertures. We use $p_{\min}$ as a measure of resolution in a proximity printer. As shown in Fig. 8.11, the image of a single slit shadow is "interfered with" by the ± 1 order diffraction components. By using the definition of $p_{\min}$ described above, it can be shown that (when $\lambda \ll p_{\min}$),

$$p_{\min} = k_3 \sqrt{\lambda s} \quad , \tag{8.5}$$

where λ is the exposing wavelength, s is the wafer-to-mask separation, and k_3 is another process-related constant (generally taken as 1.6 in x-ray lithography[53]). Equation (8.5) indicates that as the gap increases, the minimum resolved period degrades. If the exposing wavelength is 8 Å (1.5 eV, typical of a synchrotron

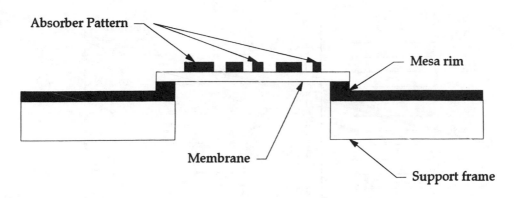

Absorber Pattern

Mesa rim

Membrane

Support frame

FIG. 8.10 Illustration of an x-ray membrane mask. (Courtesy H.I. Smith, MIT)

source) and the mask-to-wafer separation is 10 µm, the minimum resolved period is 0.14 µm. Sophisticated diffraction models confirm that the minimum resolution obtained under these conditions is consistent with this result.[54] A derivation of Eq. (8.5) is given in the appendix to this chapter (Sec. 8.A.3).

Of course, variation in the mask-to-wafer spacing also changes the minimum resolved period and, effectively, blurs the image. The following relationship can be derived from Eq. (8.3) (assuming a constant k_3):

$$\frac{\Delta p_{\min}}{p_{\min}} = \frac{\Delta s}{2s} \quad, \tag{8.6}$$

where Δs is the uncertainty in the mask-to-wafer separation. If the mask-to-wafer target separation is 25 µm, a 2-µm image plane positioning error causes a 4% change in the minimum resolved period. This is a significant improvement over the image degradation that a similar focal-plane displacement would create in optical lithography.

In the past it was felt that finite x-ray source size would lead to penumbral blurring of features. Recent work indicates that some penumbra has the beneficial effect of washing out diffraction peaks at exposure boundaries caused by coherence effects. Current point x-ray sources have sufficient intensity to provide optimum penumbral blur for a given design rule. Synchrotron sources can be designed around similar principles. This is reviewed in Ref. 52.

Next, consider the effect of secondary electron processes on image degradation. In the range of x-ray energies useful for lithography, inelastic (or Compton) interaction between the incident photon and the absorbing medium are unlikely. A single photon is absorbed by an atom, giving rise to a relatively high energy photoelectron. A 1-eV x-ray photon can, therefore, give rise to a photoelectron that is near a eV in

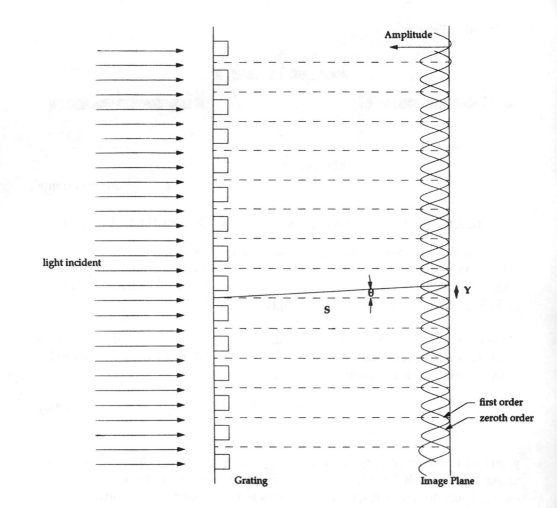

FIG. 8.11 Zeroth and first-order diffraction form a grating at the condition of minimum contrast.

energy. As a rule, for absorbing substrates with a density and composition similar to photoresist, the photoelectron will travel 0.1 μm from its point of origin for every kilovolt of kinetic energy. This "rule-of-thumb" was confirmed for soft x-rays in a classic experiment by Feder et al.[55] Thus, a 10-Å x-ray beam should produce photoelectron-induced exposure in the normally opaque region of a mask, under the feature by about 0.12 μm on a side.

The apparent inconsistency between experiments has recently been resolved by Early et al.[56, 57] Feder's experiment involved the use of erbium films deposited on photoresist and illuminated by a soft x-ray beam. The heavy metal produced many high-energy secondary electrons. The energy distribution of these photoelectrons was very narrow and close to the incident photon energy. Feder then measured the *maximum* extent of the resist damage as indicated by a change in resist dissolution rate. In actual resist exposures, the high-energy secondaries from the heavy metal absorber are absent. Also, early work in x-ray lithography was accomplished with tube sources containing a high-energy continuum component of radiation, which is absent in current-day plasma or synchrotron sources. Even with this high-energy component present, it was possible to control resist exposure profile by optimizing development cycles for the resist.[58]

In low Z materials such as carbon and oxygen, the energy distribution of x-ray-generated secondary electrons in a resist source volume is quite different from that of a heavy-metal source volume. The Auger processes tend to weight the mean of the energy distribution toward low energies.[59, 60] The effect of these low-energy Auger electrons on the energy deposited as a function of distance from a point probe incident on the resist is dramatic. The "cooler" electron ensemble cannot travel as far once it is released in the resist, preventing the spreading of the exposure volume.

Calculations by Ocala and Cerrina[61] indicate that secondary range considerations will not become important until sub-50-nm resolution is required. As is clearly seen in Fig. 8.12, 30-nm lines should be easily resolved using x-ray beams with photon energies as high as 1.49 eV (the Al K emission series). This demonstrates the feasibility of the approach well below 100 nm.

Based on the considerations presented above, it appears that x-ray proximity lithography is suitable (both from the point of resolution process latitude and from fundamental principles), for minimum feature sizes below 0.1 μm. This makes the approach applicable to quantum-effect device fabrication (from the point of view of resolution and process latitude). However, resolution and process latitude in resist exposure are not the only elements determining the desirability of a given process. Other critical issues evolve from consideration of the mask-making process.

Two outstanding issues remain for x-ray lithography to achieve production line compatibility. First, the problem of mask defects must be eliminated. Since x-ray lithography is a "1X" lithography, it is more difficult to locate and repair defects. It is possible to place particle protecting membranes over the mask surface during exposure, but such membranes will not act as a true pellicle. Depth-of-focus in x-ray lithography is enormous, and particles at pelliclelike standoff distances will leave some printable image. Thus, the masks must be kept in an environment in which they will not pick up particles after fabrication. Mask-cleaning techniques must also be developed. Furthermore, particles below a certain critical radius will be x-ray transparent and will not print. This may get around the small-radius particle "explosion" discussed in Sec. 8.3.2.

Image placement in the 1X mask must also be improved. Obviously, *n*X projects have some advantage here, but it should be pointed out that image placement in

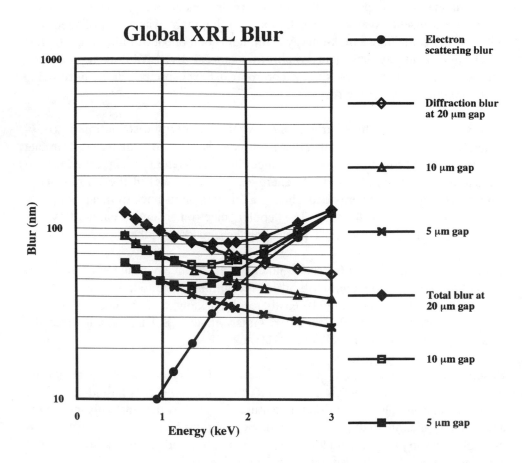

FIG. 8.12 Total image "blur" due to diffraction, penumbra, and photoelectron spreading for x-ray exposure of an isolated line with a synchrotron (x-ray energy, peak at about 1.2 nm). Total blur for a 5-μm gap (solid squares) is about 45 nm. (Courtesy F. Cerrina, CXrL)

mask making may be easier for membranes than for thick quartz optical masks.[63] Arrays of alignment fiducials can be placed behind the membrane during the mask-making process. These fiducials are just holes in metal absorbing material placed over a Schottky-contact detector element. The holes collect all of the incident energy from the beam and supply a nearly continuous reference for beam position during the writing process. This is shown in Fig. 8.13.

In order for industrial practitioners to feel that this approach is a natural extension of its existing optical systems, x-ray systems should, ideally, be "stand-alone." That is, each aligner should be capable of independent operation. In the current, most-developed approach to x-ray lithography, a number of aligners (usually more than 10) all share a common source—a synchrotron. In the past, it was thought that such

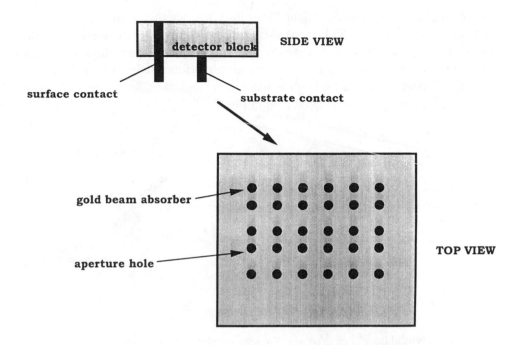

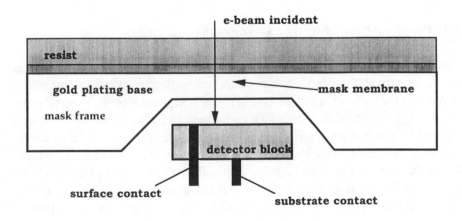

FIG. 8.13 Behind-the-wafer alignment block for x-ray mask manufacture: (a) Plain-view and cross-section; (b) Mask and block assembly.

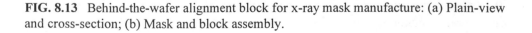

a source was overly expensive (about $25 million). Since this is but a small part of the total factory cost (over $1 billion), this is no longer an issue. But there is still the fear that if the single source were to fail, large production delays would be encountered. Thus, there is an active research effort under way to create a "granular," stand-alone source. High-power lasers incident on solids can yield large numbers of x-ray photons,[64, 65] as shown in Fig. 8.14.

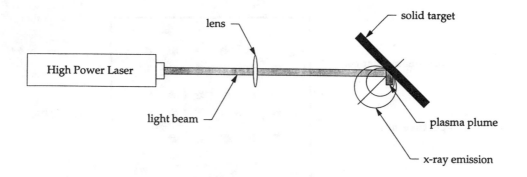

FIG. 8.14 A typical laser plasma source. Sources currently produce about a quarter of the energy needed for full production systems. Other plasma sources, such as dense-plasma focus tools, are also being developed. All of these sources produce debris that can damage masks.

8.2.4 Proximal Probe Electron Lithography

An emerging technology with potential applicability to quantum-device fabrication is based on proximal probes, such as the scanning tunneling microscope (STM)[62] and the atomic force microscope (AFM).[66] In both instruments, a small probe is brought to proximity with the surface under test, and some error signal is used to drive a servo mechanism controlling the probe-sample separation. In the case of the STM, the probe and surface electron wavefunctions overlap, and a bias voltage (<3 V) can generate a small tunnel current (typically 10 to 1000 pA) between the two. For the AFM, local repulsive forces deflect a probe mounted on a flexible microcantilever (spring force on the order of 1 Nm^{-1} away from the surface.

Again, in both cases, these processes are extremely sensitive to separation and can be resolved laterally on an atomic scale with sufficiently sharp probes, which are conducting for the STM and mechanically hard for the AFM. A proportional-integral-differential feedback network maintains the probe at a constant height over the surface by driving an electromechanical transducer, typically a piezoceramic material. This probe is rastered over the surface, and the signal driving the height transducer is recorded, thus mapping the surface.

There is no question that these devices have the appropriate resolution and stability for quantum-device patterning. Indeed, workers have long investigated atomic manipulation with the instruments.[67–70] An AFM using a metal-coated tip has been used to locally oxidize hydrogen-passivated Si, with the resulting patterns transferred into GaAs[71] and Si[72] through wet etching.[73] Amorphous Si has also been patterned through selective oxidation in this fashion, with pattern transfer into selected metals, Si_3N_4, and SiO_2 through dry etching in both positive and negative tone. Three-terminal devices have been fabricated.[74, 75] More-versatile application requires patterning of conventional resist materials. This allows pattern transfer into the substrate through etch processing or onto the substrate through metal lift-off. Scanned proximal probe instruments have a clear advantage over high-energy e-beams here in that proximity effects are absent, and so pattern density, minimum feature size,[76] and process latitude[78, 77] are all improved.

Another advantage is that the chemical processes involved in resist exposure all require an energetic threshold; operation below that threshold allows imaging without writing. Accurate alignment within 0.2 nm (or as limited by the stage!) is then possible, since fiducials and features from earlier "mask layers" may be identified. Polymeric resists such as PMMA,[79] polydiacetylene,[80] and Microposit SAL-601-ER7[81, 82] have been patterned successfully with the STM operating in field emission mode (bias voltage greater than 10 V, field determined by tip-sample separation and tip radius of curvature). However, once the films are made thin enough to be compatible with STM operating conditions, they are too thin to withstand sufficient reactive ion etching to be useful and are prone to pinholes. Recently, self-assembled monolayer films have shown great promise as a means of recording a pattern generated by an STM, which may be suitably transferred into a substrate.[83, 84] Trenches etched into Si as narrow as 12 nm have been demonstrated with these materials.[84] As a *very* rough approximation, linewidths in nanometer vary as twice the bias voltage in V, with an exposure threshold of 6 V in these materials. Linewidth is generally insensitive to current and writing speed, and so applied dose. It should also be noted that to date, a lower limit to the dose required for exposure with an STM has not been observed in these materials. The combination of high current density and extreme local electric field allows very efficient exposure.

Unfortunately, the compromises among writing speed, field size, and positioning accuracy of these instruments as typically configured require significant modifications in their implementation for manufacturing. A common configuration of a scanned proximal probe instrument uses a piezoceramic transducer in the shape of a long, thin-walled tube. The ends are bare, the inner surface is metal plated, and metal plating on the outer surface is divided into four equal sections along the tube axis. A positive or negative bias across the wall induces a proportionate strain. In operation, lateral displacement is due to extending and contracting opposite quadrants of the tube. Vertical displacement, i.e., normal to the surface, is due to uniform strain on all four quadrants. Either the probe or the sample may be mounted on the transducer; both are in wide use. A typical STM designed for vacuum operation displaces the probe.

Thus, this instrument is essentially a probe at the end of a rigid rod swept at high speeds a few tenths of a nanometer over a surface. A strain is electrically induced in the rod to move the probe in response to variations in surface topography; failure to contract can cause a catastrophic event known as a "tip crash." Mechanical resonances of the transducer thus limit the scanning speed of a typical STM to be on the order of a few $\mu m\ s^{-1}$ when scanning over a typically rough surface. These resonance modes v_n are given by

$$v_n = \frac{n - \frac{1}{2}}{2L} \sqrt{\frac{Y}{\rho}}, \tag{8.7}$$

where L is the length, Y is Young's modulus, and ρ is the density. The flexible cantilever in a commercially available AFM can accommodate considerable surface roughness and allows speeds up to 100 $\mu m\ s^{-1}$ at the cost of resolution and variations in contact force. Faster designs have been demonstrated, but at the cost of scan field size and/or positional accuracy. However, lateral displacement of a piezoceramic actuator as a function of voltage contains several high-order terms; furthermore, effects such as hysteresis and creep become significant at higher absolute displacements and scanning speeds.

However, an advantage of proximal probe lithography that may be exploited is the inherent small size and simplicity of the writing instrument. It should be possible to take advantage of existing manufacturing technology to fabricate a parallel scanned probe system optimized for lithography. Clearly, there are several advantages in separating lateral and vertical displacement actuators. Each may be designed for the appropriate displacement range and frequency response. While some sort of feedback is needed to maintain a constant probe-sample separation at each probe, presumably all probes would be moved laterally as a unit. While linearity, hysterisis, and creep are important considerations for lateral motion, due to the presence of feedback, they are less important for vertical motion. The ideal system would then be a combination of a large array of extremely small probes mounted on individual vertical motion actuators with an integrated sensing and feedback network, and a stage capable of high-speed, high-precision lateral motion. Quate and coworkers have made considerable progress in this area, developing a technology based on the AFM. By measuring probe displacement using a piezoresistive material for the cantilever and inducing vertical probe displacement with an integrated piezoelectric thin film, they have been able to scan surfaces with 2-μm corrugation at speeds up to 3 mm/s.[85] They have fabricated parallel arrays of five cantilevers of a similar design, with a 100-μm spacing.[73, 86]

What are the inherent physical and realistic limits to this sort of technology, corresponding to diffraction limits in optical lithography? Consider for example an array of field emitters fabricated on a single die.[87] A single axis of motion at each tip, that is, control normal to the writing plane, is needed to respond to local variations in surface roughness on the atomic scale. This has also been shown to be necessary for parallel AFMs, since oxidation linewidths are very sensitive to the applied force.[73] Thus, each probe would need to be grown on a column of piezoelectric material and have its own servo-control electronics fabricated on the die. This entire die could then be rastered over the product wafer. While the field size of the probe is essentially infinitesimal, stitching errors still come about in the one-time manufacture of the instrument and can be minimized there. The true field size is determined by the stage moving the writing die over the wafer. Reducing the size of the transducer raises the mechanical resonance frequency, allowing the raster speed to be increased. Independent "on/off" control allows an arbitrary pattern to be written. For dense lithography patterns, tip utilization can approach 100% and effective speed increase is limited by the number of probes. At this point it is worth considering whether such a device is feasible.

Let us consider, in detail, the fabrication of one possible scheme, based on column transducers driving an STM. The tips could be based on the field emitters described earlier. Manufacturing variations between one probe and the next can be corrected with local deposition of material by means of selective operation in an organometallic atmosphere, e.g., WF_6. The relevant expression for piezoelectric motion Δl of an element of height l and width w due to applied bias V across dimension w is

$$\frac{\Delta l}{l} = d_{31} \frac{V}{w} \ , \tag{8.8}$$

where d_{31} is the appropriate piezoelectric tensor element for strain in response to an applied field.

One commonly used piezoelectric material in scanned probe instruments is lead zirconate titanate,[88] with parameter $d_{31}=2.62\times10^{-10}\,mV^{-1}$. The achievable aspect ratio lw^{-1} is difficult to determine at this time. Since this is a crystalline, ordered oxide and the absolute width is likely to be substantial, for purposes of discussion we will use a value of 20. If we limit the applied voltage range to what is achievable with existing off-the-shelf linear electronic components, 40 V, we find that Δl can be 200 nm. This is the maximum of combined peak-to-peak roughness of the sample, topography, bowing error, and *dynamic* skew error between the writing die and the sample wafer. An upper limit to the field that may be applied to this material without depolarization is approximately 0.5 V μm^{-1}. This establishes a lower limit for the width of 80 μm, and thus a height of 1600 μm.

The lowest mechanical resonance of this element, using a value of $(Y/\rho)^{1/2} = 1105$ Hz m[88] is at 170 kHz. However, this is in a lateral mode, orthogonal to the direction of the tip, and can be strongly damped by ρ, e.g., embedding the transducer array in polyimide. The resonances that are a problem are along the long axis of the structure, excited by the feedback network moving the tip up and down in response to surface topography. We then use a higher frequency constant of 1981 Hz m,[88] which gives a resonance of 310 kHz. Allowing a 10% error due to phase lag δ near resonance, given by

$$\tan \delta = \frac{\dfrac{1}{Q}}{\dfrac{\nu_0}{\nu} - \dfrac{\nu}{\nu_0}} , \tag{8.9}$$

where the mechanical Q of this material is 65,[88] we can see that we need to limit driving frequencies to 285 kHz. The minimum spatial period of concern is that associated with amplitudes on the order of the tip-sample gap. For the biases used in lithography, the gap will be roughly 1 nm. Assuming a clean surface of a thin-ordered organic film (as from self-assembled monolayer films), we can estimate a spatial period of 2 nm. This forces the writing speed to be less than 570 μm s^{-1}. This design offers the minimum footprint of any, without regard to manufacturability of either the towerlike transducers or the stage. Other designs may be significantly easier to fabricate or place less stringent requirements on the stage. A further consideration is that when a number of such transducer elements are closely packed, capacitive coupling becomes a significant problem and must be shielded against. However, this scheme is based on existing materials. Improvements in piezoelectric or other transducer materials can likely lead to enhanced writing speeds.

The electronics needed to control such an element (in one dimension) can be fabricated using two op-amps and an instrumentation amp, one large integrating capacitor, six power supply decoupling capacitors, and six large resistors, as shown in Fig. 8.15. The overall closed loop gain of the system is determined by the need to control the tunnel gap from the tunnel current, which through most thin organic films is not significantly above 1 nA. Closed loop gain calculations for an STM include not only the electronic feedback but also the transducer mechanical gain, the displacement per unit volt, and the gain of the tunnel-gap, current-per-unit displacement. The gap gain is an intrinsic quality of the tip-sample system and is largely determined by the operating conditions. We have found in our laboratory that for a tungsten tip at -10 V relative to a Si sample coated with one of the self-assembled monolayer films, and a current of 1 nA, there is a gap gain of 0.044 nA nm^{-1}. Multiplying this by the piezo gain of 5 nm V^{-1} and inverting, we find that for unity gain of the closed loop we must have a gain of 4.5×10^9 Ω. A gain at IC1 of 2 and IC2 of 10^2 requires a 22.5 MΩ resistor at R_{pi}.

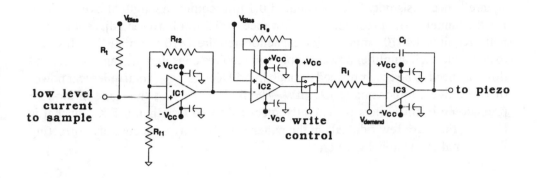

FIG. 8.15 Schematic of proportional-integral-differential feedback circuit needed to drive a piezoelectric transducer in response to a low-level current, i.e., a tunnel current.

The addition of six transistors allows any particular tip to be blanked off, if so addressed. This circuit would not be useful for topographic characterization,[89] but it doesn't need to be: The tunnel current is unipolar; except for a few token test points, the error signal does not need to be recorded; the tunnel current does not have to be linearized; and all gain and sensitivity adjustments may be done at the point of manufacture by resistor trimming. This circuit is closed loop and on-chip: except for the input binary write control lines and any output tunnel signal monitoring lines, only five signals are needed for the entire chip. For illustrative purposes, we can use commercial, off-the-shelf components. Appropriate selections might be an OPA111BM for IC1, an INA111 for IC2, and an AD820 for IC3. These devices satisfy requirements of low noise, high bandwidth, and moderate power. The 3-dB point of the first two stages is at 400 kHz. The time constant of the integrator is generally selected to roll off the feedback at the resonance, 1.6×10^{-5} s, and if we use a 100 pF capacitor for C_{pi} then R_{pi} need be only 157 kΩ.

A potential problem here is that the die sizes for these devices are 4.3, 7.5, and 2.7 mm^2,[90] respectively. Since a substantial portion of the (commercial) chip area is taken up by bonding pads, it may be assumed that some shrinkage will occur upon integration. Furthermore, chip size is generally not a consideration in linear device manufacturing, except as it affects thermal transfer. There is very little industrial impetus driving analog VLSI. Nevertheless, the same principles apply to integration here as with digital. Design rules do not set absolute limits on dimensions, only relative limits.[91] As fabrication technology improves, devices may be made smaller, whether or not they operate in saturation. It is therefore likely that substantial reductions in size could be achieved through application of the state of the art of fabrication technology. Further reduction could come from redesign toward a more application specific circuit.

Other potential integration limits are posed by the finite real estate needed for the passive components and power dissipation of the circuit. Thin metal oxide film resistors, needed for an improved temperature coefficient, are limited to only 2 kΩ square^{-1} sheet resistivity.[92] A linewidth of 0.1 μm requires a length of over 1.1 mm for the tunnel current resistor! An allowance of 0.3 μm between adjacent current paths requires 0.4×10^{-3} mm^2. Oxide capacitors require 2×10^{-3} mm^2 pF^{-1}.[92] The need to decouple each amplifier power supply still requires considerable capacitance, and this may need to be accomplished through hybrid technology. Total quiescent power of the active components is 195 mW. Thermal drift ΔT of the stage in modern precision e-beam writers is generally specified to be less than 0.1 K, but that is because there are few fiducials in the pattern and no way of accurately correcting for thermal expansion Δx, where

$$\Delta Tx = \Delta \frac{x}{\alpha} \qquad (8.10)$$

and α, the thermal expansion coefficient, is 3×10^{-6} K^{-1} for Si. The consequences of the rather substantial but steady and distributed thermal load can be minimized if the wafer and the writing die are allowed to reach thermal equilibrium. Some sort of active cooling may also be needed, although the mechanical vibrations induced by such a system have to be weighed against the electrical noise of the electronics.

We have determined the writing speed s of an STM tip configured as described here to be 570 μm s^{-1}. Let us suppose that optimal circuit design and advances in passive component fabrication technology will allow us to integrate these tips to 100 mm^{-2}. Let us further suppose that this instrument will be used to write features of width $\lambda = 30$ nm, with a fill factor $\phi = 0.1\%$, as described in Sec. 8.2.2. Larger features will be generated using some other technology. Define some parameter ϵ, which represents the utilization effeciency of any particular tip, where

$$\epsilon = \frac{\rho \text{ time "on"}}{\text{time "on" + time "off"}}. \qquad (8.11)$$

This parameter is not quite independent of ϕ but sets considerable constraints on the pattern layout if we wish we optimize throughput in such a massively parallel lithographic tool. If we grant that it is possible to optimize both the pattern layout and the raster pattern to achieve 10% efficiency, the time τ needed to write a single "unit cell" subject to these constraints is

$$\tau = \rho \frac{\text{area}}{\text{tip}} \frac{\phi}{s\epsilon\lambda}. \qquad (8.12)$$

Using the values we have generated here, we find that $\tau = 5.8$ s. Provided that the stage could be fabricated, there is no reason why an entire wafer could not be filled up with such tips, as shown in Fig. 8.16, although this is likely to affect ϵ. Such a system as described has the general advantage of low unit cost, low overhead, and

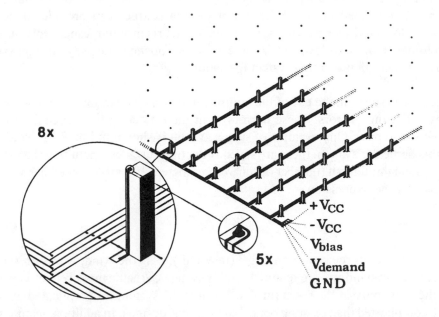

FIG. 8.16 Example of integrated parallel procimal probe lithography system, here designed for response to tunnel current. Lateral distribution is primarily determined by integration level of the control electronics.

at least partial self-replication. The number of such units engaged in simultaneously writing the same pattern is largely limited only by consumer demand of the finished product.

To conclude, we have seen that there is no physical limitation preventing implementation of a massively parallel probe system for high-resolution lithography on an industrially competitive scale. A practical limitation may be the stage, which is required to move the wafer over a "unit cell" of 100 μm × 100 μm in an arbitrary pattern at speeds of up to 570 μm s^{-1} while keeping vibrations below 265 kHz and a total position error below 200 nm. On the other hand, since this is inherently a "closed-loop" tool, engineering of the system should be in the realm of the possible.

8.2.5 Extreme Ultraviolet Lithography

All lithographic systems today make use of transmissive lenses or combinations of transmissive lenses and mirrors (catadioptric systems). These systems function at wavelengths as short as 193 nm. Catadioptric systems using fluoride ion lasers emitting at 157 nm may be possible, but to move to even shorter wavelengths, all-reflecting systems must be used. Absorption on transmission and lens damage (already evident at 193 nm) precludes these more conventional systems. In fact, good sources of extreme ultraviolet radiation already exist for wavelengths as short as 7 nm. Synchrotrons are potentially good sources, but they are not "granular" in the sense described in Sec. 8.2.3. Laser plasma sources can provide sufficient radiation for stand-alone systems, provided the debris problem can be solved. The development of multilayer EUV reflectors has opened the way to a possible short-wavelength projection system for nanolithography.

Projection systems make use of the multilayer approach described above to create x-ray reflecting mirrors. These mirrors form the reflecting lenses required for a projection stepper. One example of such a system is shown in Fig. 8.17. Note that for this particular system, there are seven reflecting surfaces (including the mask). The system throughput (number of square centimeters of resist exposed per second) is given by the expression

$$T = P_s W^3 R^7 / S \quad , \tag{8.13}$$

where P_s is the source power on target (mW/cm^2), W is the window transmissivity (three vacuum windows are required for this system), T is the mask reflectivity, and S is the sensitivity of the resist (mJ/cm^2). While the system is complicated, it is no more complicated than existing optical stepper lens designs. In addition, feature size reduction factors of 20 are possible with this system operating as an 0.15-μm printer.

Note that the throughput goes as the seventh power (seven reflecting surfaces) of reflectivity. This, of course, assumes that the mirror reflectivities are all equal. For reflectivities as high as 70%, the intensity at the wafer plane is reduced to about 8% of that available from the source. Thus, high mirror reflectivities over the exposure field are required. In addition, in order to create an aberration-free system, aspheric lenses with surface figure control of better than 20 Å are required. These are daunting feats by the standards of curent technology. The hope is to have some preliminary success with small-field systems by the middle of the decade and to be ready for production of large-field systems by the end of the decade.

Some further description of the performance of such systems is necessary in order to understand their appeal and their limitations. Resolution and depth-of-focus are still given by Eqs. (8.1) and (8.2). Thus, we would like to use exposure wavelengths that are as short as possible. However, shorter wavelengths are more penetrating, and the thickness of the multilayer stack must increase. Even so, absorption in the stack reduces reflectivity.

Surface roughness is also a key performance limiter. Interface roughness acts to create a d-layer spacing uncertainty. This is similar to the effect of thermally induced interplanar vibrations in a crystal, the so-called Debye-Waller effect.[94] The net effect is to broaden the spectral width of the reflection and to lower the peak reflectivity. The reflectivity is lowered by a factor f, given by

$$f = \rho \, \exp[\frac{-16\pi^2\sigma^2}{\lambda^2}] \ , \tag{8.14}$$

where 2σ is the interplanar layer thickness variation and λ is the incident wavelength. A 2σ value of 13 Å and an incident wavelength of 100 Å will halve reflectivity.[95]

Existing mirrors fall just short of the nominal 70% minimally acceptable value discussed above. Current thinking holds that a 130 Å incident wavelength gives the best compromise between resolution and reflectivity. Silicon/molybdenum composites deposited by magnetron sputtering are most common as interlayer materials. The 2σ value for these interlayers is usually better than 14 Å.

Under the restrictions described above, optical systems of relatively low numerical aperture can be constructed. Numerical apertures of less than 0.2 are typical of today's designs. Since the wavelength is so short, the resolution of these systems easily extends to 0.1 μm for modest k_1 values of 0.7. Depth of focus is also large

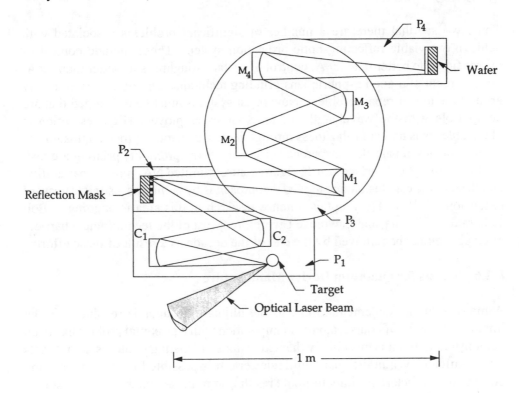

FIG. 8.17 An EUV optical system.

(greater than a micron) for a system of such high resolution. In addition, multilayer mirror systems have limited ranges of reflecting angles. Numerical aperture is related to the range or reflecting angles avalable to the system. Thus, it is very hard to design aligned lens systems with NAs greater than 0.25. This would yield a resolution close to the 30 nm obtainable using proximity x-ray lithography.

Despite these excellent resolution and depth-of-focus properties, mask repair remains a significant issue. The mask in these systems is a planar reflector with a heavy metal absorber patterned on its surface. Pinholes and pin-dots can be repaired using ion beam etches or ion-beam-induced deposition processes.[96] There is currently no technique available to repair a damaged region of the underlying multilayer reflector.

Furthermore, existing resists do not have sufficient transparency in the projection x-ray wavelengths to sustain useful exposures. Top-layer imaging systems will be required. Such systems are currently under development for x-ray and deep-UV lithography.[97]

Thus, we see that there are a number of significant problems associated with achieving a viable reflecting-optic projection system. These include control of mirror figure to better than 20 Å, control of surface roughness to better than 10 Å, development of a good aspheric lens grinding technique, and discovery of a way around the mirror repair problem. New resist systems must be developed that are compatible with the wavelengths of the sources employed. High resolution is obtainable, as is process robustness through increased depth of focus, but some of the advantages of particle transparency offered by x-ray proximity printing are lost. Problems in optical surface finish, figure control, limited NA, mask repairability, etc., lead to the conclusion that the EUV approach is by no means a straightforward extension of UV or DUV optics. Intense research in this area is ongoing at Bell Laboratories, Sandia, and Lawrence Livermore. Most of the technological barriers cited above may be removed by the end of the decade as a result of these efforts.

8.2.6 Resists for Quantum Device Manufacture

A major issue in fundamental limits is the ultimate sustainable resolution in the imaging medium. Of course, atomic manipulation with a proximal probe[68] has been demonstrated. But it is most likely that extensions of existing photoresist materials will dominate even in the quantum-device era. It is possible for us to make some comments on factors limiting ultimate resolution in these materials as a result of advanced research performed to date. The highest resolution lithography reported was achieved by e-beam lithography.[98–102] Therefore, the bulk of the discussion to follow centers around e-beam lithography, but parallels to other lithographic formats is evident. It is almost certain that e-beam lithography will continue to play a major role in nanofabrication, whether it is used to form mask patterns for x-ray or other parallel lithographies or to directly pattern a device.

Whatever the resist of the future might be, it not only must be high contrast and high resolution but must also be compatible with nanofabrication processing. In the extension to commercial production, the development of a viable resist and lithographic technology becomes even more challenging. In addition to the requirements for laboratory nanofabrication, the nanofabrication process must have high yield with few defects and high throughput. Small dimensions also exacerbate mechanical strains and adhesion problems in resists. A certain minimum thickness of resist is required to safely protect underlying material from the pattern-defining etch process. If this minimum thickness is fixed, reducing the size of the patterns etched means that the resist aspect ratio (ratio of resist height to the smallest feature dimension) increases. Such structures become mechanically unstable.

A large-moment arm is presented to external forces that naturally are present in processing. Development and drying processes provide forces that bend the resist until it breaks or delaminates from the underlying substrate (adhesion loss). As we learn in the course of practical application, the maximum useful aspect ratio is 3.5:1. Stronger, better adhering resists might improve this ratio in the future. Shown in Fig. 8.18 are examples of failures in nanolithography with conventional resist processes.[103] In the top photo is an array of dots defined in 0.75-μm-thick resist by e-beam lithography. On the edge of the array the dot diameters are <100 nm and the dots are not freestanding. In the bottom photo are adhesion problems with sub-50-nm lines in a 70-nm-thick negative resist.

Whether making a mask or a device directly, nanolithography must be closely integrated with nanofabrication. This means that the lithographic process must be incorporated into the sequence of selective material deposition and removal processes to make the desired end product. One direct consequence of going from submicron lithography to nanolithography is that one must use thinner resists. One reason is illustrated in Fig. 8.18(a). In the "hard core" nanofabrication area, all cases in which linewidths ≤20 nm have been demonstrated, resist thicknesses of ≤ 70 nm were used.[98–102] In high-voltage beam lithography, the resolution is degraded by both elastic and inelastic scattered electrons. Thin resists minimize the beam broadening in the resist due to elastic scattering. At somewhat larger critical dimensions (≈ 50 to 100 nm) resist thicknesses of up to 300 nm can be used. (These numbers are for the feature size of the remaining resist on the substrate. For an isolated positive resist trench in a large area of resist, much thicker resists can be used.) Thin resists present challenges to both nanofabrication and defect control. Usually manufacturers do not want resist less than 0.75 μm due to these issues.

The application of thin resists to nanofabrication of an electronic device or structure becomes very challenging when conventional processes are employed. This is

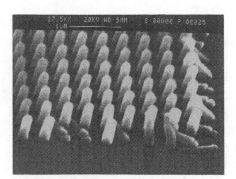

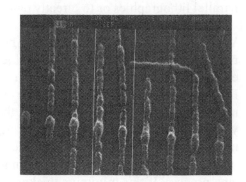

Mechanical stresses
become severe

Adhesion becomes severe

SAL-601, a negative chemically amplified resist from Shipley
patterned with a 50 kV, 15 nm diameter e-beam probe

FIG. 8.18 Illustration of resist "tip-over" due to (a) excessive aspect ration and (b) adhesion loss.

illustrated schematically in Fig. 8.19. For a given fabrication level, the resist pattern is most often used in one of two approaches: (1) as a template for subsequent deposition of a material or (2) as an etch mask for etching the underlying material.[77, 82, 104] In the first case,[105] the resist and the material deposited on the resist are "lifted off" in a solvent. The lift-off technique played an important role in the early development of high-speed integrated circuits (VHSIC) in silicon. Top-layer metallization in the IBM CMAC chip was accomplished through lift-off, since etch techniques for aluminum-copper were not perfected at the time. Lift-off may serve such a bridge in quantum-device manufacture.

The deposited material that remains after lift-off can serve as an electronically active component or as an etch mask. The disadvantage of the approach is nonselectivity; the material is deposited on the resist and the substrate. Success with this technique requires a discontinuity between the material on top of the resist and the material in the pattern channels on the substrate. In practice, with few notable exceptions[79] this translates to a deposited layer thickness of one-third to one-half the resist thickness. In addition, the deposition technique is limited to one that does not coat the sidewalls or heat the resist, causing it to flow. If a plated metal is desired and a plating base on the substrate is tolerable, higher metal:resist thickness ratios can be achieved. This is because the metal plates selectively in developed resist pattern in regions of exposed plating base. In the second approach, the required etch resistance in the thin resist case is frequently higher than found that in most polymeric resists.[106, 107]

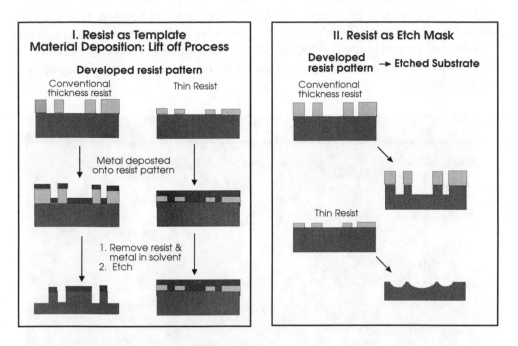

FIG. 8.19 Schematic diagram of nanofabrication with commercial resists. In both approaches the use of a manufacturing thickness (500 nm to 1 μm) resist is compared to a thin (<50 nm) resist.

For e-beam lithography, the highest resolution has been reported with the inorganic resists, with minimum features ≈1 to 5 nm.[98] However, these resists have sensitivities of ≈1 cm². Under current manufacturing requirements of desired sensitivities of about 10 μC/cm², these resists are deemed impractical. PMMA has been the "happy medium" resist for years. In laboratories it has been used to make 10- to 20-nm features for over 10 years.[98] Although PMMA has enjoyed the position of being *the* high-resolution resist for years, there are both fundamental and practical considerations with its use. With the exception of one report by Chen and Ahmed,[102] the smallest lines reported for PMMA have been ≈10 nm and the smallest grating period ≈50 nm.[99–101] These dimensions can be produced on a variety of substrates including membranes, Si, and high-atomic mass substrates such as tungsten and with tools with probe sizes from 1 nm up to 10 nm and voltages from 50 to 300 kV. For example, shown in Fig. 8.20 are 12-nm lines in a 60-nm period grating written in PMMA on a high-density W substrate.[101]

The practical implication is that these results are not described by our current models for electron scattering, which form the basis for our proximity effect corrections. Other technological considerations include nanofabrication with PMMA and sensitivity. Nanofabrication processes with PMMA are limited and offer little flexibility. Since PMMA is a positive resist and does not offer very good etch resistance, it is used almost exclusively with a lift-off process. On some occasions, if the desired etch depth is not very deep, a wet etch can be used. It is not very robust to dry etching. In addition, if used under high-contrast, high-resolution conditions, its sensitivity is ≈ 300 to $400 \ \mu C/cm^2$, which is generally too slow for manufacturing applications. This is particularly critical since e-beam lithography is a serial process.

FIG. 8.20 12-nm lines in PMMA on W, exposed with a 50 kV e-beam.

Over the past few years, chemically amplified resists have become commercially available. The chemically amplified resists have become technologically important because they offer high resolution, very high sensitivity and much better etch resistance than PMMA. The resist is doped with an acid generator that catalyzes a cross-linking reaction during postexposure processing. The cross linked resist offers more resistance to chemical processing than the non-cross-linked resists. In negative resists, the exposure itself liberates the acid catalyst. In a positive resist, the exposure liberates an acid inhibitor that "inhibits" cross linking in the exposed region during postexpoure processing. The acid generator (or inhibitor) has much higher sensitivity than non-acid-catalyzed resists. In fact, the radiation-sensitive resist dopant is chosen to be selective or to have a sharp absorption peak for the incident radiation. This can be photons for UV, DUV, EUV, or x-ray lithographies or electrons for e-beam lithography. Upon exposure the acid generator (inhibitor) releases an acid (base). Following exposure the resist is processed and the second part of the "chemical amplification" occurs, namely the acid catalyzes a cross-linking reaction in the resist. For reference at 50 kV, the dose to expose a large area of SAL-601 is 5 to 20 $\mu C/cm^2$, which is ≈ 15 to 100 times faster than PMMA (depending on process conditions). Other chemically amplified resists are even faster, but one runs into problems with the clock speed of an e-beam tool. Furthermore, there are serious concerns about shot noise when applying faster resists to nanolithography.

A chemically amplified e-beam resist that has been commercially available for a few years is the Microposit SAL-601/603 series from Shipley.[81] It is a negative novolac-based resist. First, the resist is spun onto a wafer and the wafer is baked prior to exposure, usually $\leq$ 90°C. This step removes solvents and water and allows the resist to flow through viscous creep to increase the uniformity of coverage. This is standard for most polymeric resists. The resist is then exposed. Following exposure the resist is baked at 100°C to 115°C for 1 to 10 min (depending on desired results). Finally, the resist is developed. A schematic diagram of the processsing of this resist is shown in Fig. 8.21.

The major disadvantage of chemically amplified resists is that they are very sensitive to the process conditions.[108, 109] Environmental factors such as airborne amines, time in the vacuum system, and time to hydrate as well as extremely tight bake cycle tolerances can strongly affect the resist performance. In some cases the acid can diffuse out of the resist and substantially lower its sensitivity over times on the order of one hour. This is the time frame to load a wafer (mask), expose the wafer (mask) and unload the wafer (mask) in an e-beam mask-writing process.[110] If a batch of workpieces were loaded for sequential e-beam exposure overnight or over the weekend this would be a severe problem. Several negative chemically amplified resists have been used in manufacturing prototype laboratories. Although there are some exciting new positive resists in development,[111, 112] they have been slow in commercialization because they have exhibited greater sensitivity to processing variables.

Microposit SAL-601 is one of the resists that are less process criticaland has been widely used.[81] It shows some sensitivity to prebake, preexposure, and postexposure (prior to postexposure bake) times and is somewhat sensitive to time spent in the vacuum of the beamwriter.[113] However, the most critical step is the postexposure bake, which the determines the resist sensitivity, contrast, and resolution of this material. According to the manufacturer's guidelines, the tolerance of the postexposure bake temperature is ±0.5°C over a time of 1 min. The manufacturer further recommends baking wafers on a vacuum chuck hotplate. This type of control is still difficult to achieve, even with the vacuum hotplate. Area temperature uniformity becomes most critical as the size of the workpiece is made larger. Any loss of contact between the workpiece and the plate on which it rests can cause a significant temperature change. This is a particular problem for x-ray membranes, and special fixturing is being designed to overcome it.[114]

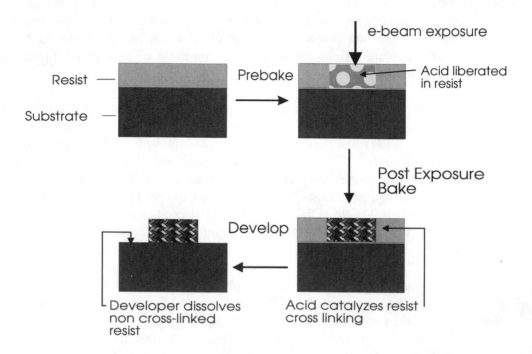

FIG. 8.21 Schematic diagram of processing SAL-601, a chemically amplified resist.

As the dimensions shrink to the nanofabrication area, the process latititude of currently available resist material also shrinks. This is illustrated in Fig. 8.22. Shown in Fig. 8.22(a) is the variation in resist sensitivity with PEB time on a hot plate at 105°C. Plotted are curves for resist thickness vs dose for a large pad in SAL-601 exposed at 50 kV with three different PEB times. The data at the 70% resist insolubilty point show that the difference between the 10-min PEB and the 3-min PEB is less than the difference between the 1-min and 3-min PEBs.The implications of approaching the resolution limit of this particular resist are shown in Fig. 8.22(b). Here linewidth is plotted as a function of dose. One can see that as the linewidths become ≤140 nm, there is a dramatic change in slope and the linewidth varies rapidly with dose at smaller dimensions. That is because at these dimensions one is reaching the lateral escape depths of fast secondary and photoelectrons, which are important exposure mechanisms for the resist. In the resolution limit, the critical dimension strongly depends on the dose, and the sensitivity in turn depends on the baking conditions. It is clear that this resist is not suitable for any viable manufacturing process at<140 nm.

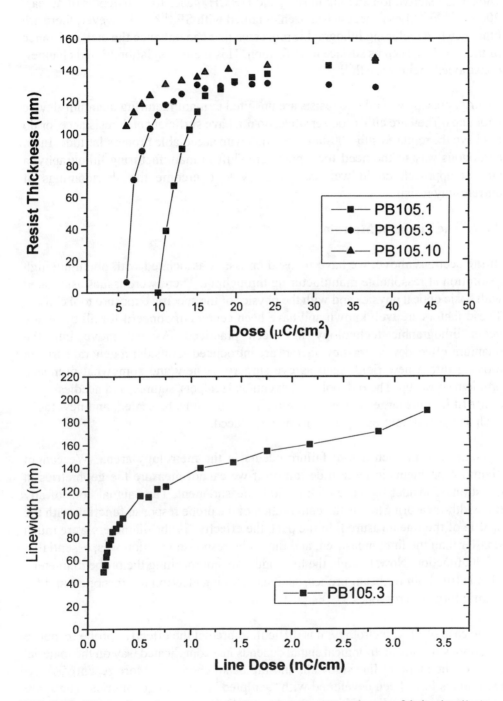

FIG. 8.22 Illustration of process sensitivity with SAL-601 at the nanofabrication limit: (a) Contrast curve: resist thickness vs exposure dose; (b) Linewidth vs exposure dose.

For quantum-device processes, thin (50 to 70 nm thick) layers of SAL-601 have withstood reactive ion etching in BCl_3 and H_2+CH_4 gases to etch depths of at least 200 nm.[77, 82, 104] Less success has been obtained with SF_6.[106, 107] However, there are other newly developing lithographic processes that can enhance the etch resistance of these resists, such as surface modification. This include sylation,[115] and channel-constrained metal growth.[116, 117]

Clearly, presently available resists are unsuited to manufacturing quantum-device structures. They are either too sensitive, do not have sufficient etch resistance, or do not have the required sub-100-nm resolution with acceptable process latitude. In an analogous way to the need for a paradigm shift in manufacturing lithography, a similar approach could well be necessary to overcome the shortcomings of current-day resists.

8.2.7 The Limits of Metrology

In the sections above, we have focused on issues associated with obtaining high resolution at reasonable manufacturing throughput. There were issues associated with basic optical physics and with the physics of the specific exposure tool chosen. These factors are well known and have been centers of concern for all the many years lithographic technology has been practiced. As we move into the quantum-effect device era, new factors are introduced. Consider a general surgeon moving into a new field, such as eye surgery. A new and somewhat mundane problem shows up: The old tools—conventional scalpels, sutures, and needles—just don't fit into the target spaces anymore! The tools must be scaled, and new tools, such as surgical microscopes, must be introduced.

A clear case of such a tool failure occurs in the metrology arena. We cannot diagnose problems in pattern definition if we cannot measure the geometries in question. Consider the case of linewidth measurement. The "signal" out of any linewidth-measuring tool is the convolution of the probe response function with the profile of the line measured. In the past, the effective probe diameters were much smaller than the lines measured, and the probe response function was, essentially, a delta function. Now, though, the linewidths are approaching the probe diameters. This is true both for beam probes, such as scanning electron microscopes, and for atomic force microscopes.[118, 119]

In the case of SEM metrology, it is difficult to predict the effective probe response function, since the metrological signal depends in a complicated way on the material probed, the slope of line sidewall, and the beam energy.[120] More recently, AFM techniques have been developed with "sculpted" probe tip geometries. These are being optimized for linewidth measurement.[121] Still, surface roughness shows up as "noise" in the metrology system. Recent data indicate the linewidth measurement reproducibilities better than 6 nm are possible with the AFM approach.[119] Still, this technique may not be able to probe accurately in the interstices between closely spaced lines—the probes themselves may be too fat!

8.2.8 Summary

From the above discussion, it appears that new lithographic techniques will have to be developed for the quantum-effect device manufacturing facility. Standard UV and decp-UV optical approaches will not allow sub-0.1-μm production. E-beam lithography has the potential for sub-0.1-μm resolution, but conventional, serial e-beam tools do not have the requisite throughput for the manufacturing floor. Conventional e-beam lithography, though, will always be a critical element in the mask-making process.

More advanced techniques look promising. X-ray lithography has the requisite resolution and the potential for satisfying pattern placement goals. Defects (discussed in greater detail below) remain a problem. The massively parallel e-beam writers described in Sec. 8.2.4 offer the possibility for speed and resolution. Fabrication issues remain outstanding for this emerging technology. Three approaches to parallel beam e-beam lithography were discussed in the text. These were multibeam e-beam (either using the photocathode or the Chang microcolumn approach), the SCALPEL projection approach, and the proximal probe approach. Ion projection approaches were also discussed. These approaches suffered from a host of problems ranging from delicate masks that were hard to fabricate and maintain to problems in basic physics of operation (stochastic blurring). A radically new approach, EUV lithography, was summarized. The differences between this technique and conventional DUV optics were clearly demonstrated. All of these issues are summarized in Table 8.1. Resolution records for these approaches (as well as the status of device demonstrations) are shown in Table 8.2.

In addition to the fundamental limits associated with the physics of the exposure system, new manufacturing problems arise on device scaling. These include mechanical failures in the resist systems in current widespread use. Also, our ability to measure the results of our work is limited. New metrological techniques must be developed to allow the visibility necessary for good process control.

In order to achieve quantum-effect device dimensions in a manufacturing environment, one of the advanced lithographies outlined above must be employed. Each of these new techniques has a number of as yet unsolved problems associated with it. Costs for develpment in this area will be large (past experience has indicated that generational improvements in *conventional* lithography have cost in excess of $100 million). It is the authors' viewpoint that all of these approaches can be made to work, providing requisite resolution, pattern placement, and alignment goals. What is necessary is a firm commitment to bear the costs and time required.

8.3 UNLIKELY EVENTS: QUANTITATIVE TECHNIQUES FOR YIELD ASSESSMENT

In any manufacturing operation, yield converts directly to profit. High yields are essential in real-time cost recovery for the billion-dollar fab lines of today. It is reasonable to assume that the same will hold true for tomorrow's fabs. Yield is essentially a statistical issue, some details of which are summarized in this section. In the first part of this section, some fundamental concepts and tools relating to the quantitative discussion of how process control impacts yield are described. In particular, the concept of process latitude is developed. In the second part of the section, some basic yield statistical calculations are presented.

8.3.1 Quantifying Process Latitude

There are two sources of yield loss. The first relates to process latitude. Suppose we run a given process close to its theoretical performance limit. Take for example one of the optical systems studied in Sec. 8.2. If we run any one of these systems close to its resolution or DOF limit, critical dimension (CD) control starts to suffer. Our ability to create lines of a fixed dimension with good control suffers well before the optical image "washes out."

At present, no literature exists modeling the impact of process latitude on yield. The parameter space of such a model would be extremely large and experimental verifications would be costly. The basic mechanisms whereby DOF creates boundary placement uncertainty have not been elucidated since they depend critically on development times, pre- and postbake temperature variabilities, and resist contrast. The ability to compute the impact of boundary placement error on yield requires a knowledge of how this error impacts device performance.

However, one thing is certain: it is possible to define a latitude parameter for any process variable. Again, consider the CD issue. Let us take all relevant parameters in the system to be fixed, and then vary one. The change in CD for a small change in the varied parameter gives us this latitude parameter. For example, we can define a focus latitude as the change in CD accompanying a small displacement in image plane position. In general, we can define a latitude parameter L_p as

$$L_p = \frac{\partial P}{\partial v_i}\Big|_{...v_k...\neq v_i} , \tag{8.15}$$

where P is the parameter studied (CD, feature centroid placement, etc.), and the v are the various parameters that affect P. A large L_p means that the parameter under consideration is strongly influenced by small loss of process control.

Closely related to the concept of a latitude parameter is the exposure-defocus (ED) plot.[122] For purposes of illustration, let us again take CD as the control parameter of interest. Let us further define a process variable (one of the v discussed above). We could take image plane displacement as one of the v for optical lithography, or we could take mask-to-wafer gap as the v for x-ray lithography. The change in CD created by changing v can be compensated by overexposing (increasing incident dose) or by underexposing (decreasing incident dose). For a given CD target and a given v, it is possible to compute (or to measure) the dose required for a +10% or a −10% CD variation.

Making this calculation (or measurement) for a series of values of v results in two rather wavy lines in a v versus log(exposure dose) plot. This is illustrated in Fig. 8.23(a). The region between the ±10% curves represents a "safe space." A large safe space represents a process with wide latitude. Usually, a number of ED ±10% curves for various types of features (lines of various width, contact windows, etc.) are superimposed. A "safe window" emerges in the plot, as illustrated in Fig. 8.23(b).

8.3.2 Particle-Limited Defect Yield Statistics

In the past, the concept of the "fatal flaw" played a dominant role in chip-manufacturing yield studies. The idea behind the concept is easily grasped. Somewhere in the manufacturing process, a defect is introduced. This defect could be extrinsic, originating from the environment outside the chip. Particles, chemical stains, and heavy metal impurities all fall in this category. Flaws could also be intrinsic, such as mechanical defects introduced within the chip volume during processing. In any event, a single defect destroys the functioning of the entire chip.

In this section we focus on a single type of extrinsic defect, the particle-induced defect. Such defects are highly significant in chip manufacture, since particles are everywhere: dust in the air, human dander, and residue from broken wafers are just a few particle sources. The processing tools themselves are notorious sources of particles. Any time metal rubs on metal during a machine operation, particles can be introduced. Particles form as a result of the reaction chemistry taking place in CVD systems and in plasma etchers.[123] As device dimensions scale, these flaws will become more significant. Lithographic process selection plays an important role in determining the net effect of particles on yield. In addition, much about what is said here has bearing on the other types of fatal flaws mentioned.

We begin by examining the density of particles in various manufacturing environments. Relevant information can be summarized in a cumulative density distribution (CDD) plot. The y-axis of the CDD plot represents the number of particles per cubic foot of air with a diameter greater than a certain diameter recorded on the x-axis. The CDD usually appears as a log-log plot, since most CDD curves are power-law relationships.

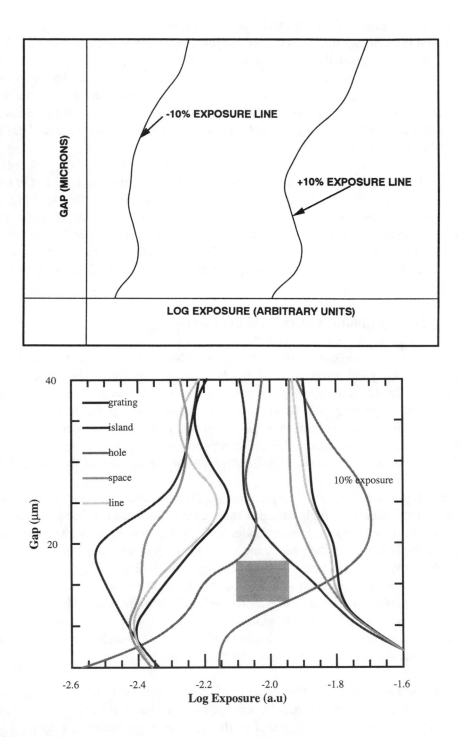

FIG. 8.23 The exposure-defocus approach to assess process latitude: (a) ±10% curves for a single feature; (b) ED plots for a number of different shapes. Note that a "safe-window" (shaded square) emerges.

Chip manufacturing operations take place in a clean room, in which air is filtered continuously to remove particles. The "class" of the clean room is mainly determined by a landmark feature on the CDD plot: the log of the number of particles per cubic foot with diameter greater than half a micron is the main indicator of cleanroom class. Of course, other factors are required for cleanroom certification, but these are not as relevant for the current discussion. For a discussion of all these factors, the reader is referred to Ref. 124.

We take the "model" CDDs of Ref. 124, extrapolated to class 1 and class 0.1, as the basis of the quantitative discussion to follow. These extrapolations are shown in Fig. 8.24.

The first thing to note is that the cumulative density varies as a high power of particle diameter. In fact, the model curves shown can be expressed as a formula:

$$C_d = AD^{3.32} \quad , \tag{8.16}$$

where A is a constant reflecting the clean room class, and D is the particle diameter, and C_d is the number of particles with diameters greater than D.

Two things stand out when examining the CDD curves. Clearly, changing the cleanroom class by an order of magnitude changes the cumulative density at any diameter by almost an order of magnitude. Also, changing the particle diameter has a huge effect on the cumulative density. Both of these factors have profound effects on yield.

To calculate the effect of the CDD curve on yield, we postulate a simple test structure. This structure is composed of equally sized metal lines and spaces: a grating structure. The linewidth is designated as w. There are two ways to envision a "fatal flaw" in such a configuration: a bridging flaw may span the space between the metal lines. Or a "break" may occur within a single line. We refer to these two types of flaws as cases A and B, respectively. Both are illustrated in Fig. 8.25.

The essential assumption that we make for subsequent analysis is that the number of particle-induced fatal flaws is proportional to the density of particles in the environment. The probability p that a single particle will create a fatal flaw is equal to a constant multiplied by the probability of drawing a particle of some diameter D or greater from the ensemble of particles. D differs, depending on the nature of the fatal flaw. For example, we may consider a 50% reduction in linewidth to be a fatal flaw. Thus, w_1 would equal $0.5w$ for case B. For simplicity, though, we consider $w_1 = w$ here. Thus, p will be proportional to the cumulative density at $D = w$.

Actual values of p are obtainable from the CDD plots if they are suitable "normalized." That is, the y-axis would be the number of particles of diameter greater than D *divided by the total number of particles in the ensemble*. These

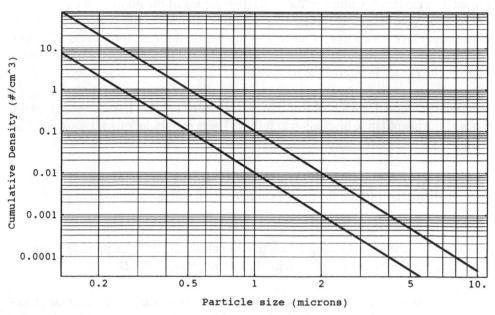

FIG. 8.24 Model cumulative particle size distribution curve for class 1 and 0.1 clean rooms.

concepts are summarized in Ref. 125. One may then determine the number proportionality factor by experiment. For the case at hand, though, it is merely enough to know that the probability of failure is proportional to the cumulative density for some minimum diameter.

We begin by considering the case of a fixed number n of fatal flaws distributed over a wafer. The flaw density will then be n/N, where N is the number of square centimeters on the wafer. Let us take the die size to be 1 cm². An 8" wafer contains 300 such die. We are interested in calculating the fraction of good die (the so-called "fractional yield") obtained under these conditions. We make use of the formula for yield Y (based on Poisson statistics) derived in Sec. 8.A.4:

$$Y = e^{-np} \quad .$$

(8.17)

Here, p is the probability that a fatal flaw (a particle incident somewhere on the wafer) will land on a given die. Thus, p is the ratio of the die area to the wafer area (1/300 for the case at hand). The results of this calculation are shown in Fig. 8.26. Here, we see yield plotted as a function of fatal flaw density n/N for the 1 cm² die. As expected, yield drops off steeply as the number of fatal flaws exceeds an average of one per die.

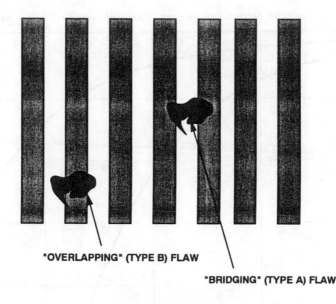

"OVERLAPPING" (TYPE B) FLAW

"BRIDGING" (TYPE A) FLAW

FIG. 8.25   Fatal flaws in a grating array.

We can now relate fractional yield to the cumulative density distributions for particles discussed above. We continue along with our assumption that the fatal flaw density is proportional to the cumulative density of particles above some critical diameter. By making this assumption, the impact of changes in the CDD on fractional yield become apparent. For example, changing the class of the clean room by an order of magnitude lowers the CDD curves by a factor of five. This, in turn, reduces the fatal flaw density by an order of magnitude and pushes the fractional yield curves to the right by an order of magnitude.

From a lithographic point of view, it is interesting to observe the effect of increasing the minimum fatal flaw diameter w by some amount. For example, let us take the minimum flaw diameter to be 0.2 μm, as opposed to 0.1 μm. Doubling the critical diameter decreases the cumulative densities by almost an order of magnitude. The effect of this decrease is shown in Fig. 8.26 as the curve on the right.

Lithographically, it is possible to change w without relaxing the design tolerances. Suppose, for example, the lithographic exposure source provides high-energy, penetrating radiation. Particles below some critical diameter would be transparent to such radiation. The $1/e$ attenuation diameter for 1.2-nm x-ray radiation incident on a silicate particle is about 1 μm; for a heavy metal, such as gold, this diameter is 0.1 μm. Thus, one might conclude that x-ray and, perhaps, high-energy e-beam lithography would skirt the "particle explosion" that occurs as minimum fatal flaw dimensions decrease.

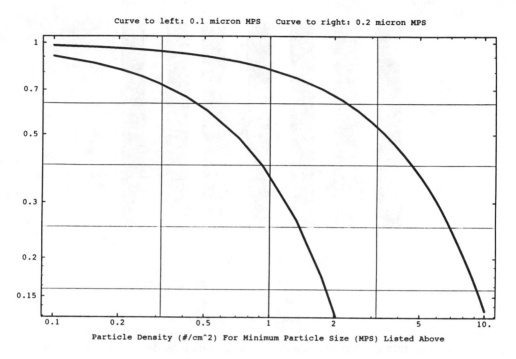

FIG. 8.26 Yield vs flaw sensitivity for a 1 cm² die for 0.1- and 0.2-µm minimum printable particle diameters.

In addition to lithographically induced damage, mechanical defects in the bulk crystal may occur with some density. The basic statistical formalism developed above for surface defects can be extended to such volume defects, once we decide on what constitutes a fatal mechanical flaw. It is perhaps less well appreciated that statistical issues associated with the doping process itself will become critical in quantum-device manufacture. This has particular bearing on the ion beam direct-write implantation lithography mentioned above. The mathematical basis for our discussion is presented in Sec. 8.A.4. In the paragraphs below, we see how these results are applied.

Specifically, consider some significant doping volume in a device. If we take this volume to be a cube 100 nm on a side, the resulting volume is 10^{-15} cm³. Typical background dopings in silicon starting material are 10^{15} cm³. Thus, this significant volume contains one doping atom! This is too small a number to do meaningful statistics on. Let us say that the background doping rises to 10^{17} cm³. This yields 100 atoms in the critical volume. Let us further assume that the statistics associated with doping obey Poisson statistics. Here, we are concerned with the "statistics of success." That is, on average, we'd like to get 100 atoms into our dopant box.

To demand, on average, 100 doping atoms in each critical volume is to say that the target expectation value for the doping is 100. Based on the above analysis, this is just the λ referred to in Sec. 8.A.4. We then calculate the standard deviation of the doping to be $\sqrt{\lambda}$, as shown in Eq. (8.47). The anticipated standard deviation of repeated samplings of the "dopant boxes" would be $\sqrt{100}$, or 10 atoms. This implies that we would see doping fluctuations that are 10% of the mean doping a marginal result at best!

All of the above results assume random processes described in terms of the Poisson formalism. It is necessary to discuss the validity of this basic formalism. The underlying assumption is that the events studied (i.e., the introduction of a fatal flaw or a dopant atom) are statistically uncorrelated. The introduction a single flaw somewhere on the wafer doesn't cause a pileup of similar flaws at that point or close by.

Furthermore, we understand that the essential physics that "activates" the flaw doesn't change through the process. For example, in the case of oxide film growth, as one thins the oxide, the number of oxide asperities (or "weak spots") seems to increase. Thus, one is tempted to create a model for the asperity that distributes a constant number of defects through the volume of the film. The volume density of defects goes up as the film thickness declines, giving rise to more oxide weak spots per unit area. But when oxides get very thin, the basic film-forming process seems to change; the total number of volume defects is not constant, and yields appear better than one would extrapolate from purely random distributions.

Perhaps more to the point is the issue of general perceptions in the field. One "just knows" that thinner films have more pinholes. Thus, if we go to ultrathin resists for high resolution, we will see enormous amounts of defects introduced. This defect explosion isn't as evident as one would predict from simple extrapolation for the "self-organizing" systems described in Sec. 8.5. Here the film-forming mechanisms are different from those encountered in conventional resist work. Simple extrapolation represents an "apples and oranges" comparison. It might be postulated that a similar mechanism would be encountered in x-ray lithography. Below some critical radius, particles are rendered ineffective for creating defects, since they become transparent. No such transparency mechanism holds for UV or EUV lithography. Here, the essential physics of the defect generation process changes below a given particle size.

8.3.3 Summary

In this section, we explored the issue of yield, looking at yield from two points of view. First, we examined the impact of how our desire to stress critical limits affected the probability of defining a successful outcome in a process. While such information is highly process specific, we described tools useful in quantifying yield loss near the process limit. Specifically, we described the concept of the latitude parameter and the use of the ED curve to obtain reasonable processing windows.

Finally, we discussed some basis issues in yield statistics. We showed how yield was a sensitive function of the cumulative particle distribution curve. We also showed how the choice of lithographic tools influenced the effective critical flaw dimension without a change in design rules. The statistical principles developed in this section could be extended to volume mechanical defects and to doping fluctuations in junctions and in contacts. We continue our study by demonstrating how conflicting goals might arise in nanoelectronic processing.

8.4 SELF-DEFEATING PURPOSES: PROCESS-INDUCED DAMAGE IN DEVICE MANUFACTURE

Suppose we would like to obtain a given level of performance in a device. This level depends on our ability to achieve or control a given process step. Frequently, in the course of such aggressive processes, defects are introduced that compromise or destroy device function. Key examples of this occur in both the pattern definition and pattern-transfer processes. High-speed device requirements convert into demands for small transport lengths. As discussed above, novel lithographic and etch techniques are required to realize process goals.

X-ray and e-beam lithographies involve exposure of sensitive device areas to ionizing radiation.[126] These radiations are known to cause lasting effects on the electrical characteristics of component devices. In silicon devices, damage appears in sensitive insulating layers that are part of the active device or that constitute surface passivation layers. In the quantum-effect device arena, compound semiconductors are frequently used. The dry-etch techniques used to define these compound semiconductors create surface charging. Surface charging tends to deplete these materials of mobile charge, prohibiting device function.

In the text below, both of these types of damage are described, and their effects on device performance are outlined. Techniques for controlling process-induced damage are also illustrated.

8.4.1 Ionizing Radiation Effects

The basic mechanism of ionizing radiation damage involves the formation of electron-hole pairs in materials that can retain a fixed charge. This is illustrated in Fig. 8.27. Here we see a beam of ionizing radiation incident on a radiation sensitive insulator. In silicon technology, this insulator is usually SiO_2. Electrons are mobile in this medium, but the holes are almost immediately self-trapped. Electrons leave the medium by diffusion or as a result of local electric fields. The oxide is then positively charged. Enough positive charge will create surface leakages in bipolar devices or offset turn-on voltages in MOSFET devices.

X-ray, e-beam, and ion beam lithographies all produce ionizing radiation that can

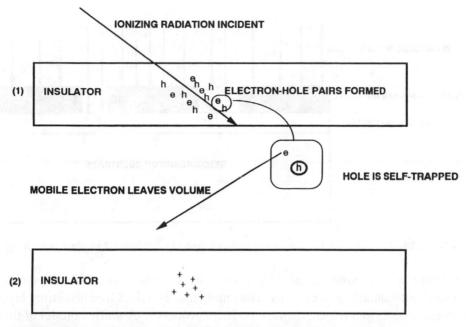

FIG. 8.27 Schematic of the ionizing radiation damage mechanism for insulators used in microelectronics.

damage sensitive parts.[127] As minimum feature sizes reduce, device-scaling principles demand that gate thicknesses reduce. This provides less-sensitive insulator volume for damage, and bulk charging becomes less of a factor. But interface state generation and the generation of "neutral" traps can degrade device performance and reliability.[128] As Hsu points out, special forming-gas anneals may be necessary to fully eliminate these insulator defects, but it is generally agreed that such damage can be controlled within acceptable limits.

Another source of difficulty occurs as a result of the lateral inhomogeneity of insulator charging. Even when the ambient doesn't contain ionizing radiation, plasma processes can charge insulators. This is illustrated in Fig. 8.28. Here we see the result of such charging introduced in a plasma etcher. If the differential charging is sufficient across the wafer, local breakdown can occur, permanently damaging the device. Also, in the case of MOS gate etching, the gate itself can charge during the process, creating an electric field sufficient to break down the gate oxide.[129–133]

At this point in technology development, etch process optimization can control most etch damage problems. In addition to high fields, an ability to source high currents is also a requisite for damage to occur.[129] A parallel plate reactive ion etcher, for example, can develop relatively large fields, but low plasma currents limit damage.

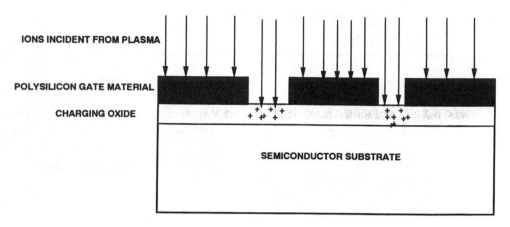

FIG. 8.28 Damage due to area-charge inhomogeneity introduced in plasma etching.

Similarly, a magnetron etcher with a lower bias voltage can offset the damage caused by potentially high currents. The importance of defect-free insulating layers also contributes to a robust, damage resistant process.[132] A unified model of these damage processes remains to be developed. Thus, it is uncertain how far our ability to do damage-free etching will extend in the future.

8.4.2 Dry Etch Damage in Compound Semiconductors

In the fabrication of recessed gates and nanostructures that will utilize quantum effects, the definition of a three-dimensional structure is critical. This requires not only epitaxial growth of the component materials but also the ability to selectively and anisotropically remove material. Except for special circumstances, liquid chemical etching is isotropic and thus is limited to large lateral dimensions. As a result of these limitations, dry-etching techniques have become popular. In dry etching, anisotropy is attained through the use of energetic and directional ions, while high etch rates are obtained from the reactive gas. This combination results in an enhancement of etch rates beyond those of the chemical or sputtering processes. Because of this control, etching techniques such as reactive ion etching (RIE), chemically assisted ion beam etching (CAIBE), and plasma etching with an electron cyclotron resonance (ECR) source have replaced liquid etching in many areas of device fabrication.

Since most dry-etching processes involve energetic ions, they naturally produce an undesirable side effect in the form of etch-induced electronic damage. The ions not only remove surface atoms but also penetrate below the surface, where they can create point defects that can act as traps in electronic devices. On the surfaces of semiconductors, the etch process can change the stoichiometry of the surface, which controls the chemical and electrical behavior of the surface during subsequent process steps such as oxidation and metalization. In particular, the Fermi-level pinning position and the density of interface states will be functions of the stoichiometry of the surface. This translates to modifications in the Schottky barrier heights and in both the ac and dc responses of the gate.

In semiconductors, bulk damage levels as low as 1 part in 106 can render a device inoperable. Because of this, it is desirable to use low-energy ions. In CAIBE and RIE, ion energies are typically well above 100 eV. Lowering ion energies to reduce damage, however, also reduces the etch rate, which is undesirable. With an ECR source, ion energies below 50 eV are possible. In addition, ion fluxes and energies can be controlled separately, and it is possible to have high etch rates with low ion energies. Therefore, an ECR source offers possibilities for control of etch damage without significant loss of throughput.

All of the effects noted above become particularly important when we are dealing with nanostructures such as quantum wires and dots. These are characterized by a high surface-to-volume ratio, and thus surfaces can dominate their electrical properties. In this regard, a wonderful demonstration of the effects of dry etching can be found in the work of Ko et al., who studied the use of ECR in forming conducting wires in GaAs.[134] The conducting wires, with thicknesses in the range of 40 to 1000 nm, were fabricated using e-beam lithography to create the etch mask and then transferring the pattern into n^+GaAs through ECR with an Ar/Cl$_2$ plasma. Fig. 8.29 shows the conductance of the wires as a function of linewidth. The solid line is the normal etch, while the dashed line represents the postetch removal of etch damage with low-energy Cl ions. Ws is the effective wire width for zero conductance. The solid line is for the case of ECR using 200 W rf power and no postetch treatment, while the dashed line corresponds to a similar etch but treated with a low-energy Cl$_2$ plasma after the etch. It is clear from the data that the minimum wire width attainable is not zero, but either 4 nm or 13.1 nm, depending on the postetch treatment. This occurs because of the Fermi-level pinning at the walls of the wire. Electrons from the wire are transferred into the surface to satisfy surface states, and this leads to a depletion layer and a reduction in the effective thickness of the wire. The Cl$_2$ postetch treatment removes the damage and in turn reduces the density of surface states. This decreases the depletion from the surface and leads to conductance at lower thicknesses of the wire.

Using contactless optical characterization of surface electric fields, Glembocki and coworkers have directly shown that the dry-etch process has detrimental effects on density of surface states and the Fermi-level pinning position at the surface.[135] Dry etches that utilize energetic ions, such as CAIBE and RIE, were shown to cause not only surface but also subsurface damage. In the case of the more gentle ECR etching, both types of damage can occur at high rf powers, with the significance of subsurface damage greatly reduced at low rf biases. Surface effects in ECR are evident even at very low ion energies ($E < 150$ eV).

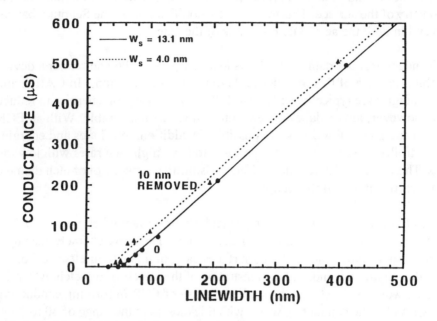

FIG. 8.29 The conductance of n+GaAs wires etched to a depth of 0.5 μm using an electron cyclotron resonance source as a function of wire width.

Glembocki et al. showed that in ECR etching, the source of the ion damage is the formation of excess As at the GaAs/oxide interface.[135] It was demonstrated that the Fermi-level pinning is controlled by the Ga/As ratio of the GaAs/oxide interface: the more As at the interface, the greater the pinning. Using the work of Ref. 136, it is possible to control the nature of the oxide through P_2S_5 chemical passivation. Studies of this phenomenon indicate that the Ga/As ratio increased and that this procedure reduced the pinning, thereby reversing the surface damage caused by ECR etching. Interestingly, the treatment was much more effective in p^+ GaAs, virtually unpinning the Fermi level. Figure 8.30 shows the Fermi-level pinning position of p-GaAs as a function of the ion energies used in the etch.[137–139] It is quite clear that the ions cause defects and change the pinning position to near midgap. This represents a significant increase in the density of surface states. These surface states have been associated with excess As at the GaAs/oxide interface. The stoichiometry of this interface can be modified through P_2S_5 chemical passivation as shown in Fig. 8.30. The effect is to partially unpin the Fermi level. Note that the etch damage has been removed.

The above results show how sensitive the surfaces of compound semiconductors are to the stoichiometry and chemical composition of the native oxide. In order to reproducibly manufacture nanostructures, we must be able to control the semiconductor/oxide interface. The studies cited above show two different

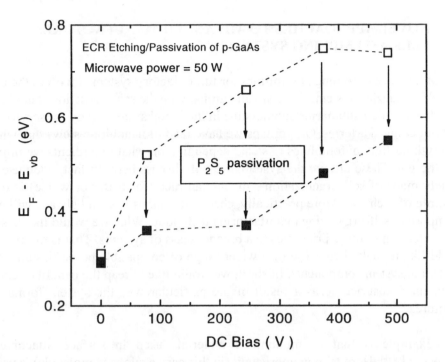

FIG. 8.30 The Fermi-level position of p-GaAs (relative to the top of the valence band) as a function of ion energy. The open squares represent the as-etched data, while the solid ones are the data after P_2S_5 chemical passivation.

techniques for controlling that interface. The work of Ko et al. gives us an in-situ postetch treatment, while the work of Glembocki et al. provides an alternative chemical means of controlling the semiconductor/oxide interface. Clearly the nature of the treatments will depend on the material being processed.

8.4.3 Summary

Increased reliance on "aggressive" processes leads to a host of new manufacturing issues. Going to higher-resolution lithographies invariably means exposure to ionizing radiation. High-resolution etch processes lead to surface charge inhomogeneities that can create local breakdowns. As we move into the quantum-device arena, we will be using broader classes of materials in processing. There will be greater reliance on compound semiconductors. In this section, we see how much a more detailed understanding of the process and the way the process influences underlying material is required as device technology evolves. New anneals may be necessary to quench neutral traps induced by x-ray or beam lithographies. New guidelines for the construction of plasma etch chambers may be required to limit currents and lower electric fields during anisotropic etching. New surface passivation techniques based on a detailed understanding of the electronic structure of the compound semiconductor surface will have to be developed.

8.5 NOVEL APPROACHES TO ADVANCED MANUFACTURE: SELF-ASSEMBLING SYSTEMS

One of the most intriguing possibilities for future factory systems involves the use of "self-organizing systems." We are all familiar with the science-fiction "nanobot" —perhaps a microsubmarine submersible in the blood system ready for service to rout out clogged arteries. Or, perhaps we have read of nanobiotic stews that whip up anything from automobiles to steak, depending on what ingredients we throw into the pot. These are *not* the systems we will discuss here. In fact, wide use is already made of self-organization—the human race (for better or worse) is one example of such use. More practically, chemical synthesis is, and always will be, an important self-organizing manufacturing technique. What we would like to see is an extension of these principles to a broader class of systems. That is to say, we would like to make better use of a wider range of compatible properties already resident in system components. In short, we would like to reap the benefit of extra "dividends" that accrue as a result of the particular way the system forms its structure.

One example is that of monomeric materials used in surface attachment processes.[140–142] Here, a monomeric unit (in this case a siloxane molecule) is used as a base to which other functional groups are attached. Head group and tail group additions are possible. In one configuration,[142] the tail group is engineered to "grab onto" and adhere to a variety of substrates. The head group is capable of initiating electroless metal deposition (plating) from solution. The head group can be cleaved from the siloxane by deep-UV irradiation. Thus, the material has applications as a photoresist.

The resist application of the monomer is illustrated in Fig. 8.31. Here, we see the patterning accomplished by head-group cleavage. The resulting "pattern" of active groups takes up an electroless metal plate. This plate is reactive-ion etch hard, allowing for transfer of the pattern into underlying material. As shown in Ref. 42, this approach can be used in conjunction with DUV imaging at 193 nm.

In studying this example, we must answer the question: *what's new here?* Clearly the basic process is a novel approach to providing a reactive etch hard mask. By separating the adhesion from the photosensitivity functions (through two separate molecular modifications) we have the possibility of independently optimizing both, but this still doesn't give us any extra "dividends" as discussed above. But, in fact, there is a dividend that comes about as a result of the way the system organizes itself. The issue is lateral association. When the system "self-organizes," after the surface attachments are made, there is some degree of lateral bonding. Thus, the monomers would extend, through side-by-side attachments, over surfaces to which they would not adhere in their monomeric form. Also, in the plating process, lateral plating could "heal" pinholes. Thus, in addition to the normal film-forming properties of the material, we derive added gains in surface passivation and in film perfection.

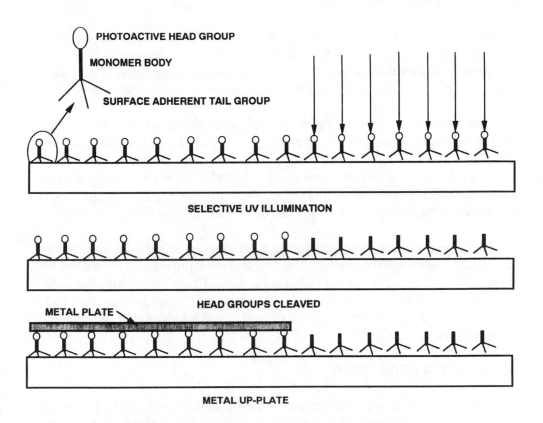

PHOTOACTIVE HEAD GROUP

MONOMER BODY

SURFACE ADHERENT TAIL GROUP

SELECTIVE UV ILLUMINATION

METAL PLATE

HEAD GROUPS CLEAVED

METAL UP-PLATE

FIG. 8.31 Application of self-organization to metal patterning.

8.6 CONCLUSIONS

Current electronic device manufacture is accomplished using interlocking arrays of diverse technologies. Overall product goals are set, and these ripple through to the individual process stations as error limits, precision goals, and tolerances. The fact that a given processing tool cannot perform to a certain specification means that other interrelated tools must perform better. For example, if the alignment "error budget" in a stepper is x, and the feature placement in the completed product is y, all of the other error terms associated with related processes (such as feature placement on the stepper mask) must be adjusted to accomplish y in the presence of x. Thus, relatively poor control in a single step may not lead to ultimate failure of manufacture.

Also, as we approach hard physical limits in a given process, proponents of a given technology will exercise enormous amounts of creativity in extending its capability. This is usually accompanied by enormous cost expenditure, and more of a demand would be placed on other parts of the technology array assembled to meet a given manufacturing goal. We see this in attempts to extend optical technologies through phase shifting and off-axis illumination.

As we scale into the quantum-effect regime, the hard physical barriers are more obvious, and new, previously unencountered difficulties become evident. But demands for high-volume production with high yield will not abate. These challenges will not be overcome by continued refinement of the existing process base.

Thus, the main conclusion to be drawn from this discussion is that we must be more willing to adopt new manufacturing paradigms. We must develop techniques for bringing new technologies on-line in a rapid, cost efficient manner. Some of these new manufacturing approaches (massively parallel e-beam lithography such as self-organizing systems) are exciting and intellectually challenging. Bringing these techniques to the manufacturing floor will be neither cheap nor easy.

The lead time to develop a new manufacturing technology is almost always underestimated. Even though the quantum-effect regime is not projected to be reached for at least a decade, now is the time to address the scientific and engineering issues for viable and robust quantum-device manufacturing. The challenges may be significant, but the potential rewards in terms of intellectual achievement and economic prosperity are enormous.

8.7 ACKNOWLEDGMENTS

The authors wish to express their thanks to Drs. C.R.K. Marrian, A.K. Rajagopal (NRL), Prof. H.I. Smith (MIT), and Prof. R. Fabian Pease (Stanford) for many useful discussions relating to this work.

8.8 APPENDIX: DERIVATION OF KEY EQUATIONS

In this appendix, we give derivations of the resolution and depth-of-focus formulae presented in the text. It is important to understand the physical bases and the critical assumptions made in arriving at these widely used relationships.

8.8.1 Resolution of a Projection Optical System: $d_{min} = k_1 \lambda / NA$

A number of resolution criteria have appeared in the literature over the years. The so-called Rayleigh or Sparrow criteria are frequently cited in elementary optics texts.[143] These derivations are usually based on the physical optics associated with optical point-spread functions. In the discussion below, we take Fourier optics as our starting point. Here, we decompose a two-dimensional image in terms of its spatial frequencies.[144] We assume that the minimum resolved feature will be equivalent to the minimum period spatial frequency resolvable by the system. In order to resolve such a period grating, the optical system must transmit *at least* the +1, 0, and -1 orders of diffracted light emanating from the grating. First, we consider the case of coherent illumination. This is followed by a discussion of how this result is modified by incoherent illumination.

The discussion proceeds with reference to Fig. 8.32. We assume that the minimum of resolution is set by the entrance aperture of the optical system (defined as the aperture that limits the angular range of acceptance for the system). As seen in the figure, if the aperture was smaller, the -1 and +1 orders of diffraction would be clipped. Only the zeroth order would pass through the system, giving a uniform, unmodulated illumination to the imaging plane. Higher pitch gratings, composed of narrower lines and spaces, would have larger diffraction angles Θ. These would clearly be filtered from the image plane.

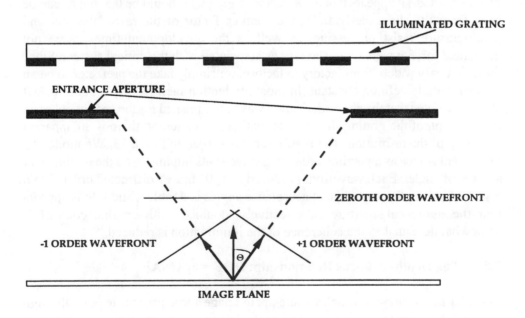

FIG. 8.32 Exit aperture limits resolution by spatial-frequency filtration. Here we see the +1, 0, and -1 orders "just barely" admitted. Closing the exit aperture would clip the +1 and -1 orders, destroying resolution. Tightening the grating pitch increases Θ, preventing the +1 and -1 orders from reaching the image plane.

We know from basic diffraction physics[145] that the following (Bragg) relationship prevails:

$$n\lambda = p_{min}\sin(\Theta) \quad , \tag{8.18}$$

where n is the order of the diffraction (which we take as 1), λ is the wavelength of the illuminating light, and Θ is the diffraction angle (of the first-order diffraction for the case at hand). Furthermore, from our definition of numerical aperture, we have (again referring to the figure)

$$\rho NA = k\sin(\Theta) \approx \sin(\Theta) \quad , \tag{8.19}$$

where k is the optical index of the medium surrounding the optics (usually air, for which k is close to 1). Combining Eqs. (8.18) and (8.19), and realizing that $p_{min} = 2d_{min}$, we have

$$p_{min} = 2d_{min} = \frac{\lambda}{NA} \quad . \tag{8.20}$$

From Eq. (8.20) it appears that the k_1 factor of Eq. (8.1) should be 0.5, but as can be seen from the above derivations, the contrast factor of the resist, the pre- and postexposure resist processing, as well as the development times, were not accounted for, nor were other aberrations associated with the optical system. Since these can vary widely from factory to factory, we simply take the prefactor to be an experimentally defined constant. In most production facilities, k_1 is about 0.7. It should be noted that the above derivation was accomplished for the case of coherent illumination of the grating. If the illumination is incoherent, there is an apparent doubling of the resolution. The reason for this is seen in Fig. 8.33. We model the incoherent source as an emitter producing wavefronts impinging on the grating over a range of angles. Each wavefront produced a -1, 0, and +1 diffracted order. From Fig. 8.34 we see that the range of acceptance angles is doubled, and it thus appears that the numerical aperture is, effectively, doubled. Lithographic contrast is somewhat degraded as the coherence of the illumination is reduced.[146]

8.8.2 The Depth-of-Focus Relationship: $DOF = k_2 \lambda/NA^2$

Consider the situation in which we attempt to image a true geometric point through an optical system by integrating light in the image plane over the whole entrance aperture of this system. Let us further assume that the point is placed on the geometric centerline of the system, as shown in Fig. 8.34. Thus, light will be brought to focus over a range of angles Θ. As is shown in the diagram, and as can be derived from the discussion in Sec. 8.A.1, Θ is the NA of the system. As Y increases, the spot will blur, as indicated in the figure.

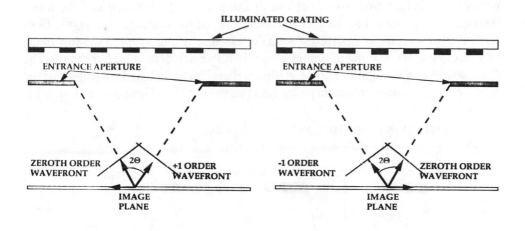

FIG. 8.33   Incoherent illumination doubles acceptance angle.

With these definitions and assumptions in hand we proceed, referring to Fig. 8.33. Throughout, we make the small angle approximation

$$Y \approx Y'$$

(8.21)

$$X \approx 2\Theta Y = 2\mathrm{NA}Y \quad .$$

(8.22)

When X "blurs" to just equal the $d_{\min}$ derived above, Y becomes the DOF. Thus,

$$k_1 \frac{\lambda}{\mathrm{NA}} = 2(\mathrm{NA})(\mathrm{DOF}) \quad .$$

(8.23)

Solving for DOF, we have

$$\mathrm{DOF} = \left(\frac{k_1}{2}\right) \frac{\lambda}{\mathrm{NA}^2} = k_2 \frac{\lambda}{\mathrm{NA}^2} \quad .$$

(8.24)

Equation (8.24) would lead you to believe that k_2 is 0.35. This number provides DOFs that are far too small, as shown by practical experience. The reason for this is the same as that given to show that k_1 is not 0.5, as the simple theory of Sec. 8.A.1 would predict. In fact, k_2 is closer to 1 in most practical systems.

8.8.3   Minimum Resolved Period in Proximity Printing: $p_{\min} = k_3\sqrt{\lambda s}$

Again, we consider imaging a grating of minimum line and space widths. We proceed along the lines of Lord Rayleigh in his definition of the minimum separation for resolution of two closely spaced spots. The essential idea of the calculation is to observe how the image of a single slit in the grating is interfered with by diffraction from other regions of the grating. The strength of the interference is proportional to the diffractive "throw" of the grating. That is, the first-order diffracted image "interferes" with the essentially uncorrupted shadow inage of a single slit. We assume minimum power in diffracted orders higher than 1. The first-order diffraction pattern of the grating is identical to the shadow pattern of the near-field image, only shifted in phase with respect to it. This is shown in Fig. 8.11.

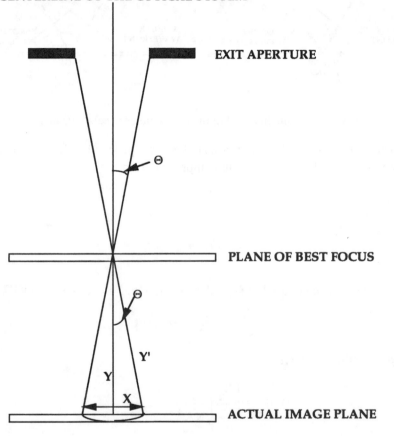

FIG. 8.34 "Blurring" a minimum-sized spot by image plane displacement.

We see that Fig. 8.11 defines a condition of minimum accepted contrast: the first-order diffracted peaks align with the undiffracted, near-field intensity minima. Other criteria are possible, and these would lead to other k-factors. So the situation is as follows: When the image plane and the grating are coincident, we are clearly in the near-field regime, and no diffraction occurs—we obtain a perfect image of the grating. As we move the image plane farther from the grating, the first-order diffraction pattern slides in phase with respect to the near-field image to destroy its contrast.

An alternative way of looking at this keeps the gap setting fixed, shortens the period of the grating, and looks for the minimum contrast condition. This defines the shortest pitch grating observable. Again, we assume that the Bragg relationship [Eq. (8.18)] holds. Making the small-angle approximation, we find

$$\sin(\Theta) \approx \tan(\Theta) \approx \frac{y}{s} \quad . \tag{8.25}$$

The minimum acceptable overlap occurs when

$$y = \left(\frac{1}{2}\right) P_{min} \quad . \tag{8.26}$$

In that case:

$$\sin(\theta) = \left(\frac{1}{2}\right)\frac{P_{min}}{s} \quad . \tag{8.27}$$

Substituting Eq. (8.27) into the Bragg relationship [Eq. (8.18)] yields

$$P_{min} = 1.4\sqrt{\lambda s} \quad . \tag{8.28}$$

Once again, the 1.4 prefactor does not include subsequent processing effects, and we generally write

$$P_{min} = k_3\sqrt{\lambda s} \quad . \tag{8.29}$$

In practice, we find k_3 to be about 1.6.

8.8.4 Derivation of Key Statistical Formulas

8.8.4.1 Basic yield formula

Define p as the probability that a fatal flaw incident on the wafer (a particle, for the study done in the body of this text) will land on a given die. Thus, the probability that a given die is flaw free is $f = (1-p)$. The "fractional yield," which is the fraction of good die, is thus

$$Y = \prod_{i=1}^{n} (1-p) = (1-p)^n \tag{8.30}$$

for the case of n particles incident on the wafer. This is one example of the binomial distribution for the probability P of k successes in n trials

$$P_n(k) = \binom{n}{k} p^k (1-p)^{n-k} \tag{8.31}$$

for the case at hand, $k = 0$. Since we are evaluating the number of surviving die, any "successful" particle detracts from yield. Thus, the pure binomial form with $k = 0$ gives us a tool to study the statistics of failure. We frequently require the study of the statistics of success. When we evaluate dopand density fluctuations below, we will have to resort to a nonzero k value.

For nonzero k values it is useful to resort to the Poisson distribution approximation to the binomial distribution. To see this, we define the parameter λ as the product of n and p. Expanding the binomial coefficient [the bracket function in Eq. (8.31)], we find

$$\binom{n}{k} p^{k}(1-p)^{n-k} = \frac{n(n-1) \cdots (n-k+1)}{n^{k}} \frac{\lambda^{k}}{k!} \frac{\left(1-\dfrac{\lambda}{n}\right)^{n}}{\left(1-\dfrac{\lambda}{n}\right)^{k}} . \tag{8.32}$$

We see that the right-hand side of Eq. (8.32) breaks into three terms. As n becomes large, the first of these terms becomes unity. Since λ/n is small, the denominator of the third term also becomes 1. Furthermore, since we know that

$$\lim_{n \to \infty}\left(1-\frac{\lambda}{n}\right)^{n} = e^{-\lambda} , \tag{8.33}$$

the right-hand side of Eq. (8.32) simplifies to

$$P(k) \approx \frac{e^{-\lambda}\lambda^{k}}{k!} . \tag{8.34}$$

For our simple $k = 0$ example, we thus derive the basic yield expression as

$$Y = e^{-np} . \tag{8.35}$$

8.8.4.2 The standard deviation of a Poisson distribution

Our goal here is to get a feel for the "width" of a Poisson distribution. Issues in statistical fluctuation in number density can be studied in light of the Poisson formalism when k is not zero. We begin by calculating the expectation values and variances for k based on Eq. (8.34). The expectation value of k, denoted as $<k>$, is gotten by multiplying the probability [as given by Eq. (8.34)] of achieving a given k by k and summing over all k:

$$<k> = \sum_{k=1}^{\infty} k \frac{e^{-\lambda}\lambda^{k}}{k!} .$$

$$(8.36)$$

Next, we observe that the exponential can be expanded,

$$e^{\lambda} = \sum_{k=1}^{\infty} \frac{\lambda^k}{k!}$$

$$(8.37)$$

and the expansion can be differentiated with respect to λ,

$$e^{\lambda} = \frac{d}{d\lambda} \sum_{k=1}^{\infty} \frac{\lambda^k}{k!} = \sum_{k=1}^{\infty} k \frac{\lambda^{k-1}}{k!} \quad .$$

$$(8.38)$$

Multiplying both sides through by $\lambda e^{-\lambda}$, we find:

$$\sum_{k=1}^{\infty} k \frac{e^{-\lambda}\lambda^k}{k!} = \lambda \quad ,$$

$$(8.39)$$

but the left-hand side of Eq. (8.39) is the same as the right-hand side of Eq. (8.36). Thus, we have shown that

$$<k> = \lambda \quad .$$

$$(8.40)$$

In order to find the variance of the Poisson distribution, we must evaluate $<k^2>$. To do this, we take a second derivative of Eq. (8.38):

$$e^{\lambda} = \frac{d^2}{d\lambda^2} \sum_{k=1}^{\infty} \frac{\lambda^k}{k!} = \sum_{k=1}^{\infty} k(k-1) \frac{\lambda^{k-2}}{k!} \quad .$$

$$(8.41)$$

Multiplying both sides of Eq. (8.41) by $\lambda^2 e^{-\lambda}$ and separating terms yields

$$\lambda^2 = e^{-\lambda} \sum_{k=1}^{\infty} k^2 \frac{\lambda^k}{k!} - e^{-\lambda} \sum_{k=1}^{\infty} k \frac{\lambda^k}{k!} \quad .$$

$$(8.42)$$

Combining Eq. (8.42) with the definition of $<k^2>$,

$$<k^2> = \sum_{k=1}^{\infty} k^2 \frac{\lambda^k}{k!} e^{-\lambda} \quad ,$$

$$(8.43)$$

we have:

$$\lambda^2 = <k^2> - <k> \quad .$$

$$(8.44)$$

From the definition of variance (the square of the standard deviation),

$$\sigma^2 = <k^2> - <k>^2 \quad ,$$

$$(8.45)$$

and from Eqs. (8.40) and (8.44), we have our final result:

$$\sigma^2 = \lambda \quad .$$

(8.46)

Thus, the standard deviation σ of the Poisson distribution is

$$\sigma = \sqrt{\lambda} \quad .$$

(8.47)

REFERENCES

1.   M. Lundstrom, S. Datta, "Physical device simulation in a shrinking world," *IEEE Circ. and Dev.*, **6**(1), p. 32 (1990).

2.   C. Hu, "ULSI device scaling and reliability," *J. Vac. Sci. Technol.*, **12**(6), p. 3237 (1994).

3.   K. Yoh, A. Nishida, and M. Inoue, "Quantized conductance and its effects on non-linear current-voltage characteristics at 80 K in mesa-etched InAs/AlGaSb quantum wires with split gates," *Solid State Electron.*, **37**(4–6), p. 555 (1994).

4.   Y. Arakawa, "Fabrication of quantum wires and dots by MOCVD selective growth," *Solid State Electron.*, **37**(4–6), p. 523 (1994).

5.   S. P. Murarka and M. C. Peckerar, *Electronic Materials: Science and Technology,* Academic Press, Cambridge (1989).

6.   M. A. Fury, "Emerging developments in CMP for semiconductor planarization," *Solid State Technol.*, **38**(4), p. 47 (1995).

7.   M. D. Levenson, "A novel technique for contrast enhancement iiin optical lithography," *IEEE Trans. Electron Devices*, **ED-29**(12), pp. 1828–1836 (1982).

8.   S. Asai, I. Hanyu, and K. Hikosaka, "Improving projection lithography image illumination by using sources far from the optical axis," *J. Vac. Sci. Technol.* B, **9**, p. 2788 (1991).

9.   S. Okazaki, "Resolution limits of optical lithography," *J. Vac. Sci. Technol.* B, **9**, p. 2829 (1991).

10.   K. Ronse, R. Pforr, K.-H. Baik, R. Jonckheere, and L. Van den Hove, "Extending the limits of optical lithography for arbitrary mask layouts using attenuated phase-shifting masks with optimized illumination," *J. Vac. Sci. Technol.* B, **12**(6), p. 3783 (1994).

11.   L. R. Harriott, J. G. Garofalo, and R. L. Kostelak, "Focused ion beam repair of phase shift photomasks," in *Electron-Beam, X-Ray, and Ion-Beam submicrometer Lithographies for Manufacturing II*, M. Peckerar, ed., Proc. SPIE **1671**, p. 224 (1992).

12.   R. Jonckheere, K. Ronse, O. Popa, and L. Van Den Hove, "Molybdenum silicide based attenuated phase shift masks," *J. Vac. Sci. Technol.* B, **12**(6), p. 3765 (1994).

13. A. Kornblit, J. J. De Marco, J. Garofalo, D. A. Mixon, A. E. Novembre, S. Vaidya, and T. Kook, "The role of etch pattern fidelity in the printing of optical proximity corrected photomasks," *J. Vac. Sci. Technol.* B, **13**(6), p. 2944 (1995).

14. F. Gabeli, H.-L. Huber, A. Kucinski, H.-U. Scheunemann, K. Simon, and E. Cullmann, "Alignment and overlay accuracy of an advanced x-ray stepper using an improved alignment system," in *Electron-Beam, X-Ray, and Ion-Beam Submicrometer Lithographies for Manufacturing II*, M. Peckerar, ed., Proc. SPIE **1671**, p. 401 (1991).

15. M. Nelson, J. L. Kreuzer, and G. Gallatin, "Design and test of a through-the-mask alignment sensor for a verticle stage x-ray aligner," *J. Vac. Sci. Technol.*, **12**(6), p. 3251 (1994).

16. D. Stewart, T. Olson, and B. Ward, "0.25 µm x-ray mask repair with focused ion beams," in *Electron-beam, X-Ray, and Ion-Beam Submicrometer Lithographies for Manufacturing III*, D.O. Patterson, ed., Proc. SPIE **1924**, p. 98 (1993).

17. A. Hiroki, S. Odanaka, and A. Hori, "A high-performance 0.1 µm MOSFET with asymmetric channel profile," Proc. IEEE IEDM, p. 439 (1995).

18. T. H. P. Chang, "Proximity effect in electron-beam lithography," *J. Vac. Sci. Technol.*, **12**, p. 1271 (1975).

19. L. H. Veneklasen, "A high-speed EBL column designed to minimize beam interactions," *J. Vac. Sci. Technol.* B, **3**, p. 185 (1985).

20. L. H. Veneklasen, "An optimizing electron beam writing strategy subject to electron optical, pattern and resist constraints," *J. Vac. Sci. Technol.* B, **9**, p. 3063 (1991).

21. T. R. Groves, H. C. Pfeiffer, T. H. Newman, and F. J. Hohn, "EL3 system for quarter-micron electron beam lithography," *J. Vac. Sci. Technol.* B, **6**(6), p. 2028 (1988).

22. T. H. Newman, W. DeVore, and R. F. W. Pease, "Dot matrix electron beam lithography," *J. Vac. Sci. Technol.* B, **1**, p. 999 (1983).

23. A. W. Baum, J. E. Schneider, and R. F. W. Pease, "High-performance negative electron affinity photocathodes for high-resolution electron beam lithography and metrology," Proc. IEEE IEDM, p. 409 (1995).

24. G. H. Jensen, *Coulomb Interactions in Particle Beams*, Academic Press, Boston (1990).

25. M. M. Mkrtchyan, J. A. Liddle, S. D. Berger, L. R. Harriott, A. M. Schwartz, and J. M. Gibson, "An analytical model of stochastic interaction effects in projection systems using a nearest-neighbor approach," *J. Vac. Sci. Technol.* B, **12**(6), p. 3508 (1994).

26. J. E. Schneider, A. W. Baum, G. I. Winograd, R. F. W. Pease, M. McCord, W. E. Spicer, K. Costello, and V. Aebi, "Semiconductor on glass photocathodes as high-performance sources for parallel electron beam lithography," *J. Vac. Sci Technol.* B, in press (Dec. 1996).

27. T. H. P. Chang, D. P. Kern, and L. P. Murray, "Microminiaturization of electron optical systems," *J. Vac. Sci. Technol.* B, **8**(6), p. 1698 (1990).

28. E. Kratschmer, H. S. Kim, M. R. G. Thompson, K. Y. Lee, S. A. Rishton, M. L. Yu, and T. H .P. Chang, "Sub-40 nm resolution 1 KeV scanning tunneling microscope field-emission microcolumn," J. Vac. Sci. Technol. B, **12**(6), p. 3503 (1994).

29. H. S. Kim, M. L. Yu, E. Kratschmer, B. W. Hussey, M. G. R. Thompson, and T. H. P. Chang, "Miniature schottky electron source," *J. Vac. Sci. Technol.* B, **13**(6), p. 2468 (1995).

30. N. K. Eib and R. J. Kvitek, "Thermal distribution and the effect on resist sensitivity in e-beam direct write," *J. Vac. Sci. Technol.,* **7**, p. 1502 (1989).

31. E. Kratschmer and T. R. Groves, "E-beam lithography on glass masks," *J. Vac. Sci. Technol.,* **8**, p. 1898 (1990).

32. K.D. Cummings and M. Kiersh, "Charging effects from e-beam lithography," *J. Vac. Sci. Technol.,* **7**, p. 1539 (1989).

33. H. Itoh and K. Nakamura, "Charging effects on trilevel resist with an e-beam lithography system," *J. Vac. Sci. Technol.,* **8**, p. 185 (1990).

34. V. V. Wong, J. Ferrara, J. N. Damask, J. M. Carter, E. E. Moon, H. A. Haus, and H. I. Smith, "Spatial phase locked electron beam lithography and x-ray lithography for fabricating first order gratings on rib waveguides," *J. Vac. Sci. Technol.,* **12**, p. 3741 (1994).

35. H. I. Smith, S. D. Hecter, M. L. Schattenburg and E. H. Anderson, "A new approach to high-fidelity e-beam and ion beam lithography based on an in-situ global fiducial grid," *J. Vac. Sci. Technol.,* **9**, p. 2992 (1991).

36. O. C. Wells, *Scanning Electron Microscopy*, McGraw-Hill, New York (1974).

37. R. J. Hawryluk, A. M. Hawryluk, and H. I. Smith, "Energy dissipation in a thin polymer film by electron beam scattering," *J. Appl. Phys.,* **45**, p.2551 (1974).

38. K. W. Rhee and M. C. Peckerar, "Proximity effect reduction using thin insulating layers," *Appl. Phys. Lett.,* **62**(5), p. 533 (1993).

39. B. D. Cook and S.-Y. Lee, "Fast proximity effect correction: an extension of PYRAMID for thicker resists," *J. Vac. Sci. Technol.,* **11**(6), p. 2762 (1993).

40. G. Owen and P. Rissman, "Proximity effect correction for e-beam lithography by equilization of background dose," *J. Appl. Phys.,* **54**(6), p. 3573 (1983).

41. M. A. McCord, R. Viswanathan, F. J. Hohn, A. D. Wilson, R. Naumann, and T. H. Newman, "100 KeV thermal field emission electron beam lithography tool for high-resolution x-ray mask patterning," *J. Vac. Sci. Technol.* B, **10**, p. 2764 (1992).

42. S. J. Wind, M. G. Rosenfeld, G. Pepper, W. W. Molzen, and P. D. Gerber, "Proximity correction for electron beam lithography using a three-gaussian model of the electron energy distribution," *J. Vac. Sci. Technol.* B, **7**, p. 1507 (1989).

43. M. C. Peckerar and C. R. K. Marrian, "Pattern density dependent contrast in commonly used dose-equalization schemes," in *Electron-Beam, X-Ray, EUV, and Ion-Beam Submicrometer Lithographies for Manufacturing VI*, D.E. Seeger, ed., Proc. SPIE **2723**, p. 134 (1996).

44. M. Parikh, "Corrections to proximity effect in electron beam lithography," *J. Appl. Phys.*, **50**(6), p. 4371 (1979).

45. T. R. Groves, "Efficiency of e-beam proximity effect correction," *J. Vac. Sci. Technol.* B, **11**(6), p. 2746 (1993).

46. W. T. Lynch, T. E. Smith, and W. Fichner, "An algorithm for proximity effect correction with e-beam exposure," in *Proc. Int'l. Conf. on Microlith. and Microcircuit Engin.*, Grenoble, France, p. 309 (1982).

47. M. C. Peckerar, S. Chang, and C. R. K. Marrian, "Proximity correction algorithms and a co-processor based on regularized optimization. I. Description of the algorithm," *J. Vac. Sci. Technol.*, **13**(6), p. 2518 (1995).

48. E. A. Dobisz, F. K. Perkins, S. L. Brandow, C. R. K. Marrian, W. J. Dressick, J. Kosakowski, J. M. Calvert, and T. Koloski, "Self assembled monolayer films for nanofabrication," *Proc. Mat'l Res. Soc.*, S, to appear.

49. S. D. Berger, J. M. Gibson, R. M. Camarada, R. C. Farrow, H. A. Huggins, J. S. Kraus, and J. A. Liddle, "Projection electron-beam lithography: a new approach," *J. Vac. Sci. Technol.* B, **9**(6), p. 2996 (1991).

50. H. Loschner, G. Stengl, I. L. Berry, J. N. Randall, J. C. Wolfe, W. Finkelstein, R. W. Hill, J. Melngailis, L. R. Harriott, W. Brunger, and L. M. Buchmann, "Ion projection: the successor to optical lithography," in *Electron-Beam, X-Ray, and Ion-Beam Submicrometer Lithographies for Manufacturing IV*, D.O. Patterson, ed., Proc. SPIE **2194**, p. 384 (1994).

51. G. P. Watson, S. D. Berger, J. A. Liddle, and W. K. Waskiewicz, "A background dose proximity effect correction technique for scattering with angular limitation projection lithography implemented in hardware," *J. Vac. Sci. Technol.* B, **13**(6), p. 2504 (1995).

52. M. C. Peckerar and J. R. Maldonado, "X-ray lithography: an overview," Proc. IEEE **81**(9), p. 1271 (1993).

53. S. D. Hector, M. L.Schattenburg, E. H. Anderson, W. Chu, V. V. Wong, and H. I. Smith, "Modeling and experimental verification of illumination and diffraction effects on image quality in x-ray lithography," *J. Vac. Sci. Technol.* B, **10**(6), p. 3164 (1992).

54. H. K. Oertel, M. Weisz, H. L. Huber, Y. Vladimirski, J. Maldonado, "Modeling illumination effects in resist profiles in x-ray lithography," Proc. SPIE **1465**, p. 244 (1991).

55. R. Feder, E. Spiller, and J. Topalian, "Electron ranges in photoresist," *J. Vac. Sci. Technol.*, **12**, p. 1332 (1975).

56. K. Early, M. L. Schattenburg, and H. I. Smith, "Absence of resolution degradation in x-ray lithography for $\lambda = 4.5$nm to 0.83 nm," *Microelectronics Engineering*, **11**, pp. 317–321 (1990).

57. K. Early, M. L. Schattenberg, D. B. Ulster, M. I. Shepard, and H. I Smith, "Diffraction in proximity x-ray lithography: comparing theory and experiment for gratings, lines and spaces," *Microelectronics Engineering*, **17**, pp. 149–152 (1992).

58. J. R. Maldonado, G. A. Coquin, D. Maydan, and S. Somekh, "Spurious effects caused by the continuum radiation and ejected electrons in x-ray lithography," *J. Vac. Sci. Technol.*, **12**, p. 1329 (1975).

59. K. Murata, D. F. Keyser, and C. H. Ting, "Monte Carlo simulation of fast secondary electron production in electron beam resists," *J. Appl. Phys.*, **52**, p. 5985 (1982).

60. K. Murata, "Theoretical studies of the electron scattering effect on developed pattern profiles in x-ray lithography," *J. Appl. Phys.*, **57**, p. 575 (1985).

61. L. E. Ocola and F. Cerrina, "Parametric modeling at resist-substrate interfaces," *J. Vac. Sci. Technol.*, **12**(6), p. 3986 (1994).

62. G. Binnig, H. Rohrer, Ch. Gerber, and E. Weibel, "Surface studies by scanning tunneling microscopy," *Phys. Rev. Lett.*, **49**, p. 57 (1982).

63. F. K. Perkins, D. McCarthy, C. R. K. Marrian, and M. C. Peckerar, "Position measurement of high-energy e-beams for pattern placement improvement," in *Electron-Beam, X-Ray, EUV, and Ion-Beam Submicrometer Lithographies for Manufacturing VI*, D.E. Seeger, ed., Proc. SPIE **2723**, p. 91 (1996).

64. M. C. Peckerar, J. R. Greig, D. J. Nagel, Pechacek, and R. R. Whitlock, "Plasma sources for x-ray lithography," in *Eighth International Conference on Ion, Electron, and Photon Beam Technology*, R. Bakish, ed., Electrochemical Society Press, Pennington, N.J., p. 432 (1978).

65. P. Alaterre, H. Pepin, R. Fabbro, and B. Faral, "Modeling of x-ray emission created by short wavelength laser target interaction," in *Laser Interaction and Related Phenomena*, vol. 7, H. Hora and G. H. Miley, eds., Plenum Publishing Corp., New York (1986).

66. G. Binnig, C. F. Quate, and Ch. Gerber, "Atomic force microscope," *Phys. Rev. Lett.*, **56**, p. 930 (1986).

67. R. S. Becker, J. S. Golovchenko, and B. S. Swartzentruber, "Atomic-scale surface modifications using a tunneling microscope," *Nature*, **325**, p. 419 (1987).

68. D. M. Eigler and E. K.Schweizer, "Positioning single atoms with a scanning tunneling microscope," *Nature*, **344**, p.524 (1990).

69. J. W. Lyding, T.-C. Shen, J. S. Hubacek, J. R. Tucker, and G. C. Abeln, "Nanoscale patterning and oxidation of H-passivated Si(100)-2×1 surfaces with an ultrahigh vacuum scanning tunneling microscope," *Appl. Phys. Lett.*, **64**, p. 2010 (1994).

70. C. T. Salling and M. G. Lagally, "Fabrication of atomic-scale structures on Si(001) surfaces," *Science*, **265**, p. 502 (1994).

71. J. S. Dagata, J. Schneir, H. H. Harary, J. Bennett, and W. Tseng, "Pattern generation on semiconductor surfaces by a scanning tunneling microscope operating in air," *Appl. Phys. Lett.*, **56**, p. 2001 (1990).

72. E. S. Snow and P. M. Campbell, "Fabrication of nanostructures with an atomic force microscope," *Appl. Phys. Lett.*, **64**, p. 1932 (1994).

73. S. C. Minne, Ph. Flueckiger, H. T. Soh, and C. F. Quate, "What are the inherent physical and realistic limits to this sort of technology, corresponding to diffraction limits in optical lithography? Consider for example an array of field emitters," *J. Vac. Sci. Technol.* B, **13**, 1380 (1995).

74. S. C. Minne, H. T. Soh, Ph. Flueckiger, and C. F. Quate, "Fabrication of 0.1 μm metal oxide semiconductor field effect transistor with the atomic force microscope," *Appl. Phys. Lett.*, **66**, p. 703 (1995).

75. P. M. Campbell, E. S. Snow, and P. J. McMarr, "Fabrication of nanometer-scale side-gated silicon field effect transistors with an atomic force microscope," *Appl. Phys. Lett.*, **66**, p. 1388 (1995).

76. E. A. Dobisz and C. R. K. Marrian, "Sub-30-nm lithography in a negative electron beam resist with a vacuum scanning tunneling microscope," *Appl. Phys. Lett.*, **58**, p. 2526 (1991).

77. E. A. Dobisz and C. R. K. Marrian,"STM lithography: a solution to electron scattering," *J. Vac. Sci. Technol.* B, **9**, p. 3024 (1991).

78. F. K. Perkins, E. A. Dobisz, and C. R. K.Marrian, "Determination of acid diffusion rate in a chemically amplified resist with scanning tunneling microscope lithography," *J. Vac. Sci. and Technol.* B, **11**, p. 2597 (1993).

79. M. A. McCord and R. F. W. Pease, "Lift-off metallization using polymethylmethacrylate exposed with a scanning tunneling microcope," J. Vac. Sci. and Technol. B, **6**, p. 293 (1987).

80. C. R. K. Marrian, E. A. Dobisz, and R. J. Colton, "Lithographic studies of an e-beam resist in a vacuum scanning tunneling microscope," *J. Vac. Sci. and Technol.* A, **8**, p. 3563 (1990).

81. Shipley Corporation, Marlborough, MA.

82. C. R. K. Marrian, E. A. Dobisz, and J. A. Dagata, "Electron-beam lithography with the scanning tunneling microscope," *J. Vac. Sci. and Technol.* B, **10**, p. 2877 (1992).

83. F. K. Perkins, E. A. Dobisz, S. L. Brandow, T. S. Koloski, J. M. Calvert, K. W. Rhee, J. E. Kosakowski, and C. R. K. Marrian, "A proximal probe study of self-assembled monolayer resist materials," *J. Vac. Sci. and Technol.* B, **12**, p. 3725 (1994).

84. F. K. Perkins, E. A. Dobisz, S. L. Brandow, J. M. Calvert, J. E. Kosakowski, and C. R. K. Marrian, "12-nm linewidths observed in self-assembled monolayer resists and etched into Si," submitted (1995).

85. S. R. Manalis, S. C. Minne, and C. F. Quate, "Atomic force microscopy for high-speed imaging using cantilevers with an integrated actuator and sensor," *Appl. Phys. Lett.*, **68**, p. 871 (1996).

86. S.R. Manalis, S.C. Minne, C.F. Quate "Atomic force microscopy for high-speed imaging using cantilevers with an integrated actuator and sensor," *Appl. Phys. Lett.*, **68**, p. 871 (1996).

87. K. Derbyshire, "Beyond AMLCDs: field emission displays?" *Solid State Technology*, p. 55 (Nov. 1994).
88. EBL No. 3, Staveley Sensors Inc., East Hartford, CT 06108.
89. E. I. Altman, D. P. DiLella, J. Ibe, K. Lee, R. J. Colton, "Data acquisition and control system for molecule and atom-resolved tunneling spectroscopy," *Rev. Sci. Inst.*, **64**, p. 1239 (1993).
90. This is one-half the dual device die size.
91. C. Mead, *Analog VLSI and Neural Systems*, Addison-Wesley Publishing Co., Reading MA, p. 305 (1989).
92. P. R. Gray and R.G. Meyer, *Analysis and Design of Analog Integrated Circuits*, 2nd ed., John Wiley and Sons, New York (1984).
93. S. Akamine, T. R. Albrecht, M. J. Zdeblick, and C. F. Quate, "Micro-fabricated scanning tunneling microscope," *IEEE Elect. Dev. Lett.*, **10**, p. 490 (1989).
94. B. E. Warren, *X-Ray Diffraction*, Addison-Wesley, Reading, MA (1969).
95. D. G. Stearns, N. M. Ceglio, A. M. Hawryluk, R. S. Rosen, and S. P. Vernon, "Multilayer optics for soft x-ray projection lithography: problems and prospects," in *Electron-Beam, X-Ray, and Ion-Beam Submicrometer Lithographies for Manufacturing*, M. Peckerar, ed., Proc. SPIE **1465**, p. (1991).
96. D. M. Tennant, K. E. Early, L. A. Fetter, L. R. Harriott, A. A. MacDowell, P. P. Mulgrew, and W. K. Waskiewicz, "Defect repair for soft x-ray projection lithography masks," *J. Vac. Sci. Technol.*, in press (Dec.1992).
97. J. M. Calvert, C. S. Dulcey, M. C. Peckerar, J. M. Schnur, J. H. Georger, G. S. Calabrese, and P. Sricharoenchaikit, "New surface imaging techniques for sub-0.5-micrometer optical lithography," *Solid State Technol.*, 34(11), p. 77 (1991).
98. M. Isaacson and A. Murray, "In-situ vaporization of very low molecular weight resists using 1/2-nm diameter electron beams," *J. Vac. Sci. Technol.*, **19**, p. 1117 (1981).
99. A. Broers, "Electron beam lithography at 350 KeV," presented at the International Symposium on Nanostructure Physics and Fabrication, March 13–15, College Station, TX (1988).
100. H. G. Craighead, R. E. Howard, L. D. Jackel, and P. M. Mankiewich,"10-nm linewidth electron beam lithography on GaAs," *Appl. Phys. Lett.*, **42**(1), p. 38 (1983).
101. E. A. Dobisz, C. R. K. Marrian, R. E. Salvino, M. A. Ancona, F. K. Perkins, and N. H. Turner, "Reduction and elimination of proximity effects," *J. Vac. Sci. Technol.* B, **11**, p. 2733 (1993).
102. W. Chen and H.Ahmed, "Fabrication of 50-7 nm wide etched lines in silicon using 100 KeV electron-beam lithography and polymethlymethacrylate resist," *Appl. Phys. Lett.*, 62, p. 1499 (1993).

103. E. A. Dobisz, F. K. Perkins, S. L. Brandow, J. M.Calvert, and C. R. K. Marrian, "Self assembled monolayer resists for nanofabrication," in Proc. of the 1995 Spring Materials Reseach Society (1995).

104. C. R. Eddy, E. A. Dobisz, C. A. Hoffman, J. R. Meyer, "ECR-RIE of fine features in $Hg_xCd_{1-x}Te$ using Ch_4/H_2 plasmas," *Appl. Phys. Lett.*, 62, p. 2362 (1993).

105. For example, E. A. Dobisz, H. G. Craighead, E. D. Beebe, and J. Levkoff, "Lithographic fabrication of transmission electron microscopy cross sections in III-V materials," *J. Vac. Sci. Technol.* B, **4**, p. 850 (1986).

106. L. M. Shirey, K. W. Foster, W. Chu, J. Kosakowski, K. W. Rhee, E. A. Dobisz, C. R. Eddy, D. Park, I. P. Iaacson, D. McCarthy, C. R. K. Marrian, and M. C. Peckerar," Reactive ion etching of tungsten for high-resolution x-ray masks," in *Electron-Beam X-Ray, and Ion-Beam submicrometer Lithographies for Manufacturing IV*, D.O. Patterson, ed., Proc. SPIE **2194**, p. 169 (1994).

107. E. A. Dobisz, C. R. Eddy, J. Kosakowski, O. J. Gelmbocki, L. M. Shirey, K. W. Foster, W. Chu, K. W. Rhee, D. Park, C. R. K. Marrian, and M. C. Peckerar, "Comparison of dry etch approaches for tungsten patterning," in *Electron-Beam, X-Ray, and Ion-Beam submicrometer Lithographies for Manufacturing IV*, D.O. Patterson, ed., Proc. SPIE **2194**, p. 178 (1994).

108. See entire section on Chemically Amplified Resists from the 35th International Symposium on Electron, Ion, and Photon Beams, *J. Vac. Sci. Technol.* B, **9**, pp. 3338–3398 (1991).

109. D. Seeger, R. Viswanathan, C. Blair, J. Gelorme, and W. Conley, "Single layer chemically amplified resist processes for device fabrication by x-ray lithography," *J. Vac. Sci. Technol.* B, **10**, p. 2620 (1992).

110. J. Grimm, J. Chlebek, T. Schulz, and H.-L. Huber, "The influence of post exposure bake on linewidth control for the resist system RAY-PN in x-ray mask fabrication," *J. Vac. Sci. Technol.* B, **9**, p. 3392 (1991) and references therein.

111. K. Y. Lee and W. S. Huang, "Evaluation and application of a very high performance chemically amplified resist for electron beam lithography," *J. Vac. Sci. Technol.* B, **11**, p. 2807 (1993).

112. T. Sakamizu, H. Yamaguchi, H. Shiraishi, F. Murai, and T. Ueno, "Development of positive tone electron-beam resist for 50 kV electron-beam direct-writing lithography," *J. Vac. Sci. Technol.* B, **11**, p. 2812 (1993).

113. T. H. Fedynyshyn, M. F. Cronin, and C. R. Szmanda, "The relationship between critical dimension shift and diffusion in negative chemically amplified resist systems," *J. Vac. Sci. Technol.* B, **9**, p. 3380 (1991).

114. D. J. Resnick, K. D. Cummings, W. A. Johnson, H. T. H.Chen, B. Choi, and R. L. Engelstad, "Temperature uniformity across an x-ray mask membrane during resist baking," *J. Vac. Sci. Technol.* B, **12**(6), p. 4033 (1994).

115. C. W. Jurgenson, R. S. Hutton, G. N. Taylor, "Resist etching kinetics and pattern transfer in a helicon plasma," *J. Vac. Sci. Technol.* B, **10**, 2542 (1992) and references therein.

116. J. M. Calvert, G. S. Calabrese, J. F. Bohland, W. J. Dressick, C. S. Dulcey, J. H. Georger, J. Kosakowski, E. K. Pavelcheck, K. W. Rhee, and L. M. Shirey, "Photoresist channel-constrained deposition of electroless metallization on ligating self-assembled films," *J. Vac. Sci. Technol.* B, **12**, p. 3884 (1994).

117. E. A. Dobisz, to be published.

118. H. M. Marchman, J. E. Griffith, J. Z. Y. Guo, J. Frackoviak, and G. K.Celler, "Nanometer-scale dimension metrology," *J. Vac. Sci. Technol.* B, **12**(6), p. 3585 (1994).

119. J. E.Griffith, H. M. Marchman, L. C. Hopkins, "Edge position measurement," *J. Vac. Sci. Technol.* B, **12**(6), p. 3567 (1994).

120. H.Yamashita, K. Nakajima, H. Nozue, "Highly accurate critical dimension measurement for sub-0.5µm devices," *J. Vac. Sci. Technol.* B, **12**(6), p. 3591 (1994).

121. J. E. Griffith, H. M. Marchman, G. L. Miller, L. C. Hopkins, M. J. Vasile, and S. A. Schwalm, "Line profile measurements with a scanning probe microscope," *J. Vac. Sci. Technol.* B, **11**(6), p. 2473 (1993).

122. B. J. Lin, "The paths to sub-half micrometer optical lithography," in *Optical/Laser Microlithography*, Proc. SPIE **922**, p. 257 (1988).

123. G. S. Selwyn, K. L. Haller, E. F. Patterson, "Trapping and behavior of particulates in a radio-frequency magnetron plasma etching tool," *J. Vac. Sci. Technol.* A, **11**(4), p. 1132 (1993).

124. P. R. Austin and S. W. Timmerman, *Design and Operation of Clean Rooms*, Business News Publishing Co., Detroit (1965).

125. C. Ash, *The Probability Tutoring Book: Revised Edition*, IEEE Press, New York (1994).

126. G. C.Messenger and M. S. Ash, *The Effects of Radiation on Electronic Systems*, Van Nostrand Reinhold Co., New York (1986).

127. J. R. Maldonado, A. Reisman, H. Lezec, C. K. Williams, and S. S. Iyer, "X-ray damage considerations in MOSFET devices," *J. Electrochem. Soc.*, **133**(3), p. 628 (1986).

128. C. H. Hsu, K. L. Wang, J. Y.-C. Sun, M. R. Wordeman, and T. H. Ning, "Hot electron induced instability in 0.5-µm CMOS patterning using synchrotron x-ray lithography," Proc. International Reliability Physics Symposium (1989).

129. W. M. Greene, J. B. Kruger, and G. Kooi, "Magnetron etching of polysilicon: electrical damage," *J. Vac. Sci. Technol.* B, **9**(2), p. 336 (1991).

130. C. T. Gabriel, "Gate oxide damage from polysilicon etching," *J. Vac. Sci. Technol.* B, **9**(2), 370 (1991).

131. S. Fang and J. P. McVittie, "A model and experiments for thin oxide plasma damage from plasma induced wafer charging in magnetron plasmas," *IEEE Electron Device Lett.*, **13**, p. 347 (1992).

132. S. Fang and J. P. McVittie, "Oxide damage from plasma charging: breakdown mechanisms and oxide quality," *IEEE Trans. Electron Devices*, **ED-41**(6), p. 1034 (1994).

133. T. Gu, R. A. Ditizio, O. O. Awadelkarim, and S. J. Fonash, "Reactive ion etching induced damage to SiO_2 and SiO_2-Si interfaces in polycrystalline Si overetch," *J. Vac. Sci. Technol.* A, **11**(4), p. 1323 (1993).

134. K. K. Ko, S. W. Pang, T. Brock, M. W. Cole, and L. M. Casas, "Evaluation of surface damage on GaAs etched with an electron cyclotron resonance source," *J. Vac. Sci. Technol.* B, **12**, p. 3382 (1994).

135. O. J. Glembocki, J. A. Tuchman, K. K. Ko, S. W. Pang, A. Giordana, R. Kaplan, and C. E. Stutz, "The effects of electron cyclotron resonance etching on the ambient (100) GaAs surface," *Appl. Phys. Lett.*, May 22 (1995).

136. J. A. Dagata, W. Tseng, J. Bennett, J. Schneir, and H. H. Haray, "P_2S_5 passivation of GaAs-surfaces for scanning tunneling microscopy in air," *Appl. Phys. Lett.*, **59**, p. 3288 (1991).

137. O. J. Glembocki, J. A. Dagata, E. S. Snow, and D. S. Katzer, "Optical characterization of the electrical properties of processed GaAs," *Appl. Surf. Sci.*, **63**, p. 143 (1993).

138. O. J. Glembocki, J. A. Dagata, K. K. Ko, S. W. Pang, J. A. Tuchman, A. Giordana, R. Kaplan, and C. E. Stutz, "Photoreflectance study of the chemically modified (100) GaAs surface," in *Spectroscopic Characterization Techniques for Semiconductor Technology V*, O.J. Glembocki, ed., Proc. SPIE **2141**, p. 96 (1994).

139. O. J. Glembocki, J. A. Dagata, K. K. Ko, S. W. Pang, J. A. Tuchman, A. Giordana, R. Kaplan, and C. E. Stutz, "The effect of P_2S_5 passivation on the electrical properties of (100) GaAs surfaces," to be submitted to *Appl. Phys. Lett.*

140. J. M. Calvert, "Self-assembled films for semiconductor technology," in *Organic Thin Films and Surfaces*, edited by A. Ulman, Academic Press, New York (1994).

141. A. Ulman, *Introduction to Ultrathin Organic Films: From Langmuir-Blodgett to Self Assembly*, Academic Press, San Diego (1991).

142. J. M. Calvert, "Lithographic patterning of self assembled films," *J. Vac. Sci. Technol.* B, **11**(6), p. 2155 (1993).

143. G. O. Reynolds, J. B. Devilbis, G. B. Parent, and B. J. Thompson, *The New Physical Optics Notebook: Tutorials in Fourier Optics*, SPIE Optical Engineering Press, Bellingham, WA (1989).

144. J. W. Goodman, *Introduction to Fourier Optics*, McGraw-Hill, New York (1968).

145. K. W. Ford, *Classical and Modern Physics*, vols. 1 & 2, Xerox Press, Lexington, MA (1973).

146. L. F. Thompson, M. J. Bowden, and C. G. Willson, *Introduction to Microlithography*, Americal Chemical Society Symposium Series No. 219, ACS Press, Washington, D.C., 16 (1983).

INDEX